CAAD futures 1997

CAAD futures 1997

Proceedings of the 7th International Conference on
Computer Aided Architectural Design Futures
held in Munich, Germany, 4-6 August 1997

Edited by

Richard Junge
Faculty for Architecture,
Technical University Munich,
Munich, Germany

SPRINGER-SCIENCE+BUSINESS MEDIA, B.V.

A C.I.P. Catalogue record for this book is available from the Library of Congress.

ISBN 978-94-010-6350-0   ISBN 978-94-011-5576-2 (eBook)
DOI 10.1007/978-94-011-5576-2

Printed on acid-free paper

Table of Contents

viii

Virtual Reality

Preface

Since the establishment of the *CAAD futures* Foundation in 1985 CAAD experts from all over the world meet every two years to present and at the same time document the state of art of research in Computer Aided Architectural Design. The history of *CAAD futures* started in the Netherlands at the Technical Universities of Eindhoven and Delft, where the *CAAD futures* Foundation came into being. Then *CAAD futures* crossed the oceans for the first time, the third *CAAD futures* in '89 was held at Harvard University. Next stations in the evolution where in '91 Swiss Federal Institute of Technology, the ETH, Zürich. In '93 the conference was organized by Carnegie Mellon University, Pittsburgh and in '95 by National University Singapore. *CAAD futures* '95 marked the world wide nature by organizing it for the first time in Asia.

The seventh *CAAD futures* is the first being organized by a German University. For the as small as newly and only provisional established CAAD group at the Faculty for Architecture at Technical University München it is honor and challenge at the same time to be the organizer of *CAAD futures* '97.

Keynote Papers
At *CAAD futures* '97 four keynotes papers are delivered. All four are dealing with quite diverse aspects of the CAAD universe.
Gerhard Schmitt in his paper 'DESIGN MEDIUM- DESIGN OBJECT" is reflecting on the impact that computer applications of design, if not only regarded to be 'a tool only', that partially replaces the traditional pencil, but regarded and used as a medium for design that is offering new and before unknown possibilities, will have on the designed object. This is an issue that is discussed by far too little and not many facts are known.
Thomas Bock in 'CAD- SO WHAT?' points at a weak point of the CAAD community: the prefix 'CA-' in connection with architecture should not only be thought of as referring to design. Design should be seen as a step in creating the built environment, it leads to the construction process. In other industries 'CA-' in design and 'CA-' production are a closely inter-woven system. What if not a human bricklayer is interpreting a drawing but a robot - a type of computer machine- has to do the bricklaying?
Chuck Eastman in his paper is dealing with 'INTEGRATION OF DESIGN APPLICATIONS WITH BUILDING MODELS'. In technical detail he is discussing issues of conversions and of translations between heterogeneous applications using different data and naturally different semantics as well. Proposed is to use a common building data model

John Gero's paper 'WHAT ARE WE LEARNING FROM DESIGNERS AND ITS ROLE IN FUTURE CAAD TOOLS' is focusing on how knowledge about the activities and behaviors of human designers could find a way into applications where computational processes are modeled after the human design processes. He is providing examples from research on design and some early implementations.

For reasons of structuring the conference program papers had to be divided into ten major thematic categories, although these assignments sometimes may seem to be somewhat arbitrary.

Shapes, Grammars and Types
This part contains four papers with topics ranging from: the discussion of the existence of grammars and art as a base for computation, an prototype implementation based on shape grammars, the use of features for architectural design, to the design of typical buildings without the use of 'types'.

Design Support
This section is made of six papers on issues of: a prototype of a generic CAD tool supporting cooperative design, an integrated design support systems based on new design methodologies, using lines and geometric relations as a framework for design, the computability of design diagrams as in early sketching, the coordination of hand and eye in VR systems, and information survey as base for design applications.

Exploration and Evaluation
This chapter contains ten papers dealing with a wide range of issues starting from: software tools for the analysis of spatial architectural design, a prototype using the designers own hypothesis to evaluate his design, a simulation- assisted building control strategy, a discussion of potential of aggregated space-time performance indicators for operative information. Very demanding issues are: the investigation into the quantification and computability of architectural aesthetics in a framework of a descriptive approach to architectural analysis and design, the application of neural networks as a means of human- computer interaction in architectural design, and a technique to generate flexible building simulators using proposed 'patterns' as basis. The remaining three papers are more strictly on evaluation issues: the use of 'virtual acoustic' techniques for navigating and escape of unfamiliar smoke filled buildings, a prototype implementation exploring designed pedestrian walkways, and a study of simulating the optimal location of fire egress signs.

Environment and Simulation
This section contains four papers dealing with issues of: applying a KB CAAD system for pre-conceptual design of bio- climatic and low energy buildings, to a discussion on the reliability of computational daylight modeling, to a computer evaluation tool for visual aspect of architectural design for high- density and high- rise buildings, and

using a mobile agent oriented method for simulation of interaction between a built environment and occupants actions.

Collaboration in Design

In this section nine papers are dealing with a topic that is still of growing interest for the research community. A system that is exploring participants understanding and negotiation in the process of urban design is among one of the few papers that is not using the web as a base. A group of papers are concentrating on models for collaboration: an integrated computing environment for collaborative, multi-disciplinary building design, a model for web- based design interaction based on a human- machine- design matrix, research on a multi- user workspace for inter university collaboration, the implementation of shared virtual reality on the web for a large- scale project, and a conceptual network for web representation of design knowledge. Others papers are dealing with: an advanced groupware approach for an integrated planning process in building construction, research on an intra- and interorganizational design information system, and an experiment on data exchange in the design and realization process.

Product Model and Virtual Enterprises

This chapter contains six papers reflecting on topics of more abstract to reasonably pragmatic nature. Themes are: the existence of a generic building object, questions of life cycle models of buildings, a model for a computer integrated construction process. A second part is presenting three interrelated papers based on work carried out in the ESPRIT project VEGA. Included are architectures of product models for an environment of distributed applications and models, to the necessary IT base for enabling 'virtual enterprises' to implementation of a dynamic product model and it's application. Two more papers are dealing with the issue of a distributed architectural model for co-operative design, and with a concept for multiple, flexible representations of design.

Virtual Reality

This section has nine papers dealing with the buzzword 'virtual reality' which, as the nature of buzzwords, is used for a very diverse set of issues. The presentations start with things like: Images and arguments in an anthropology web site which is a VR of an archaeological finding in EL Salvador, and with a web museum of architecture and history. Part two is dealing with 'virtual cities': with a web based model of the City of Bath an d London's West End, and with issues of abstraction, accuracy and realism especially in large scale computer urban models. Another two papers are dealing with explorations of design systems based on the web: a network based kit-of-parts virtual building system, and the role of language in design based on virtual communities. Other papers are discussing the use of precedents from architecture, urban design and film for improvements of navigation in virtual environments, the development of a three dimensional semantic codes as a proposal in the discussion of a missing language of cyberspace, and a examination of the legacy of surrealism in the electronic design studio.

Learning and Reasoning

This section contains four papers dealing with issues from: dynamic decision making based on conjoint measurement, a system for managing information using fuzzy reasoning methods, the discussion of human analogical reasoning and case adaptation and it's applicability to computers, and formalized situated learning applied to an agent based framework.

Design Media

This section has six papers dealing with: the extension of a modeling tool to deal with architectural space, and with feature based qualitative modeling of shapes for use in early stages of design, and with the level of detail in interactive walkthroughs and the elimination of useless details in large scale models. With topics ranging from an experiment in hybrid architectural form- making, to an universal structuring model of technical documentation in architectural design using hypermedia structures, and a proposal to use hypersketching in early design phases creating a graphical hyperdocument are dealing three more papers.

Generative Systems and Emergence

This section has five papers. A first group is dealing with issues ranging from: an evolutionary approach to generating constraint- based space layout topologies, to emergent shape generation in design using the boundary contour system, and to a genetic programming approach to the space layout planning problem. The other two papers are about the use of genetic programming in exploring 3D design worlds, and the application of genetic algorithms to intuitive designs as Gehry's Bilbao Museum.

Thanks to all members of the team that helped to make *CAAD futures 97* become reality, especially to: Christof Lintl, Dimitris Economides, Fabian Scheurer, Brigitte Rampp, Martin Muschkiet; Alexander Pilz, Manfred Lang, Norbert Ruf, Melanie Meyer, Tonja Schütz,Kathrin Leuze and Martha Hipp.

Richard Junge
Mai, 1997
München

International Review Committee

Akin, Omer.	Carnegy Mellon University, USA
Augenbroe, Godfried	Technical University Delft, Netherlands
Bridges, Alan	University of Strathclyde, UK
Caneparo, Luca	Politecnico di Torino, Italy
Christiansson, Per	Lund Institute of Technology, Sweeden
Coyne, Richard	University of Edinburgh, UK
Eastmann, Charles	Georgia Tech, USA
Flemming, Ulrich	Carnegy Mellon University, USA
Gero, John	University of Sydney, Australia
Junge, Richard	Technical University München, Germany
Kalay, Yehuda E.	University of Berkeley, USA
Kohler, Niklaus	Universität Karlsruhe, Germany
Koutamanis, Alexander	Delft University of Technology, Netherlands
Maher, Mary Lou	University of Sydney, Australia
Maver, Thomas	University of Strathclyde, UK
McCullough, Malcolm	Harvard University, USA
Tan, Milton	National University of Singapore, Singapore
Oxman, Rivka	Technion, Israel Institute of Technology, Israel
Oxman, Robert	Technical University Eindhoven, Netherlands
Sasada, Tsuyoshi	Osaka University. Japan
Schmitt, Gerhard.	Swiss Federal Institute of Technology, Switzerland

1

Shariv, Edna	Technion, Israel Institute of Technology, Israel
Wiesmann, Stefan	Fachhochschule Darmstadt, Germany
Woodbury, Robert.	University of Adelaide, Australia
Zreik, Khaldoun.	University Nantes, France

DESIGN MEDIUM - DESIGN OBJECT

GERHARD SCHMITT
Architecture and CAAD, Swiss Federal Institute of Technology Zurich

The impact of computing on architecture receives little reflective judgment, its role either being negated or over-emphasized. To one group of architects, it is not desirable that the machine should influence the object. Another, mostly younger group, takes the impact for granted, without much reflection on its underlying reasons. The smallest group - mostly in academia - is interested in actively defining the impact of computing on design and in defining a new kind of architecture. The paper will explore the relation between computer and architecture on three levels, in which the machine has the role of an instrument, a medium, and a partner. It will demonstrate a serious deficit in education regarding the new roles of computing in design.

1. Status Quo

At the end of the first phase of computerization - employing computers as tools - we find office automation and CAD software in each and every architecture office. The second phase of computerization and networking - employing the computer as a medium - is about to begin. One is surprised to find so little hard data showing the effects on architecture caused by the most capital intensive investment since design offices exist. Contradictory statements such as „The machine has a tremendous impact", „Computers have no impact on my thinking or design", „We tried computers, it did not work, so we went back", are well known. In spite of architects working with computers at various stages of the design, it seems that they have - with few exceptions - little or not changed the design object to date.

To measure the impact of computing on architecture, at least two criteria are of interest. The first criterion is the visible influence on the built environment. It could be measured by comparing projects and built structures before and after the use of computers, based on buildings of similar size and purpose by the same design office.

The second criterion is the impact of the medium on design education. This can be observed - albeit incompletely - by studying yearbooks and studio reports of architecture schools. Most of the schools produce the documents digitally, but in spite of massive increases in the use of the computer in the studio, they contain only little final work produced with CAD (Rowe, 1996). It might be that the computer contributes during the

3

R. Junge (ed.), CAAD Futures 1997, 3-13.
© 1997 *Kluwer Academic Publishers.*

design process and that CAD output is not used in the final presentations due to time pressure or graphic presentation standards, but it is certainly astonishing, how little inroads computer presentations have made in education.

2. Computer Evolution: Tool, Medium, Partner

One reason for the lack of observed impact of computers on architecture is that, opposite to the intentions of many CAAD pioneers, computerization has taken a much more brutal course than originally anticipated. Certain terms make this clear: office automation instead of office work support, and drafting automation instead of intelligent drafting support (see the software name AutoCAD, for example). Computerization followed the lines of automation earlier this century. Computers today are used as tools, replacing mental work as machines replaced physical work. This era of replacement is coming to an end because of the high degree of automation achieved already in post-industrial society. The era of the computer as a medium is beginning and will be followed by that of computers as partners.

For simplification, the term computer will be used for the entire working environment, consisting of hardware, software, and network. In most languages, the computer is treated as a living being rather than a lifeless object: a work horse, an enemy, a friend, and many meanings in between. If common language is an indicator, the computer is already considered as a medium and will be accepted as a partner soon. This is an expression of hope, given the quality of today's design software and hardware. Whereas the next evolutionary steps of computer environments seem to be clear for research and parts of practice, education has recently concentrated on teaching mainly the tool aspects of computers. Widely abandoned or forgotten are the previous attempts to use the machine as a medium by supplying it with design strategies, ranging from shape grammars to case-based reasoning (Schmitt, 1993). This is where architectural design education should place more emphasis.

3. The computer as a tool or instrument

The computer as a tool or instrument has been most successful in terms of efficiency. One reason is that a clear paradigm or *Leitbild* exists for this purpose. Yet the definition of the computer as a mere tool does not take into account that it not only emulates office instruments, but that it also simulates new design instruments unthinkable without the computer. This instrument can quickly turn into a self-generating and self-referential system.

Werner Oechslin in his essay *Computus et Historia* describes the close relation between instrument and architecture, in particular between computing devices and design thinking (Oechslin, 1993). One can observe that today's technically highly developed physical and intellectual instruments have gradually replaced the philosophical instruments and measuring devices in classical antiquity. But similar to antiquity, control over the instrumentarium is restricted to a small percentage of the population. A few understand and develop, the others apply and execute.

Many architects, made nervous by the emergence of the new computing environment, found solace in the hope that the computer was *just* an instrument. In the early days of transition from the industrial to the information society, this may have been correct. But this view of a computer can correspond to a regressive *Leitbild,* associated with a large number of problems: A computer tool under a regressive *Leitbild* must prove itself in eliminating previously human activities with less cost and higher quality. While this might be a goal for short term gains, it is a dangerous development in the long term. If one considers the computer as an instrument only, it must pay for itself in that it perfectly replaces activities that were previously difficult or boring or expensive to achieve. Examples are word processors, when seen as replacing secretaries; spread sheets, when seen as replacing calculators; CAD programs, when seen as electronic pencils; office automation programs, when seen as a collection of desktop activities; rendering programs, when seen as a way to impress clients. Those instruments pose the problem that they directly aim at eliminating expensive human labor. Even worse from an intellectual standpoint is that human skills are directly transported to the computer which is then „personalized" or „humanized". The computer becomes the repository of activities that were once human. Because machines cannot yet reason about, improve or question their own activities, it also implies, that these activities are then frozen in their present state. With this, the computer could become a retarding, regressive instrument that hinders progress.

4. Example: Useful in the Office, Useless for Design?

There are positive examples of using the computer as an instrument, supporting and strengthening a particular design method: A design method being the overall problem solving strategy, and the instruments, in a subordinate role, supporting that method (Schmitt, 1993).

Jim Glymph, principal with Frank O. Gehry & Associates, reported on the past, present and future use of computers in the office (Glymph, 1997). Frank Gehry's office is a positive example of using the computer as an instrument to strengthen a particular, already existing design philosophy. As Glymph pointed out, the earlier, small scale work of Frank Gehry showed many of the characteristics of his later buildings, but it was the intelligent use of the computer that enabled the actual physical construction of these designs on a large scale. The machine contributed to the constructability by acting as a communication and translation instrument between architect and contractors, and as a CAD/CAM instrument for the material producers. Glymph countered the repeated questions concerning the dependency of the office on a particular solid modeler with the statement that the software was only selected *after* the fields in which it could be helpful were decided on, and that the software could be changed in the future. He also used the expression of computers as electronic pencils and remarked that they were useless for design at present - a statement that is logical and correct if computers are seen as an instrument only.

Hans Kollhoff has employed the computer as an instrument for visualization and communication for several years. The machine has influenced the presentation and delivery of his work. He used telephone lines early on to transfer files between Berlin and Zürich and employed artists from other fields to produce computer generated videos of his designs. But, as in Frank Gehry's case, he uses the machine to strengthen his particular design philosophy and position, rather than being influenced by it. For him, the machine is just another tool.

5. The computer as a medium

A medium is more than a tool or a method. It is an interactive counterpart, not necessarily an intelligent being, but something that has to offer knowledge and capabilities in the area we are interested in. A computer aided architectural design environment, equipped with the necessary components and in cooperation with a competent designer, can achieve the status of a design medium already today. In this case, we can enter a discourse on the level of a medium that will not take away work from us, but that will allow us to deal with the important questions of future architecture more competently.

Architectural design must be more than playful interaction with geometric forms, supported by increasingly more attractive computer tools. Architecture must use all available technology on the best possible level during its creation. Architecture today does not suffer from too much technology, but from the inappropriate application of it. It cannot be the *Leitbild* for the computer to design the most complex artifacts of our civilization with drafting programs under the electronic pencil paradigm. The missing competence in realizing and controlling certain actions and relations in the building process must not serve as an excuse for retreating to a position of decorating. Should architects keep and improve their role in the building process in the future, the computer must achieve the role of a medium that supports designers in areas were they do not have sufficient knowledge or competence themselves. The most obvious application for the computer as a medium is simulation. More advanced applications are computer supported methods (Schmitt, 1993) and agents (Schmitt, 1996). This is where university education in architecture should place more emphasis.

6. Example: Phase X

Phase X is the first design course at ETH Zürich using the computer as a medium. It is the newest in a series of network-based teaching experiments which involved more than 600 students since 1993. Its purpose it to pose and explore fundamental questions concerning design, modeling and authorship. Phase X expands the idea of the paperless studio by building more-dimensional computer models, by networking the designs and by focusing on abstract concepts such as Types & Instances (Madrazo, 1996). Adding new modeling instruments such as Sculptor (Kurmann, 1996) offered students additional opportunities to explore different design approaches, based on playful interaction with design objects. The complete description of the results can be found under http://caad.arch.ethz.ch/teaching.

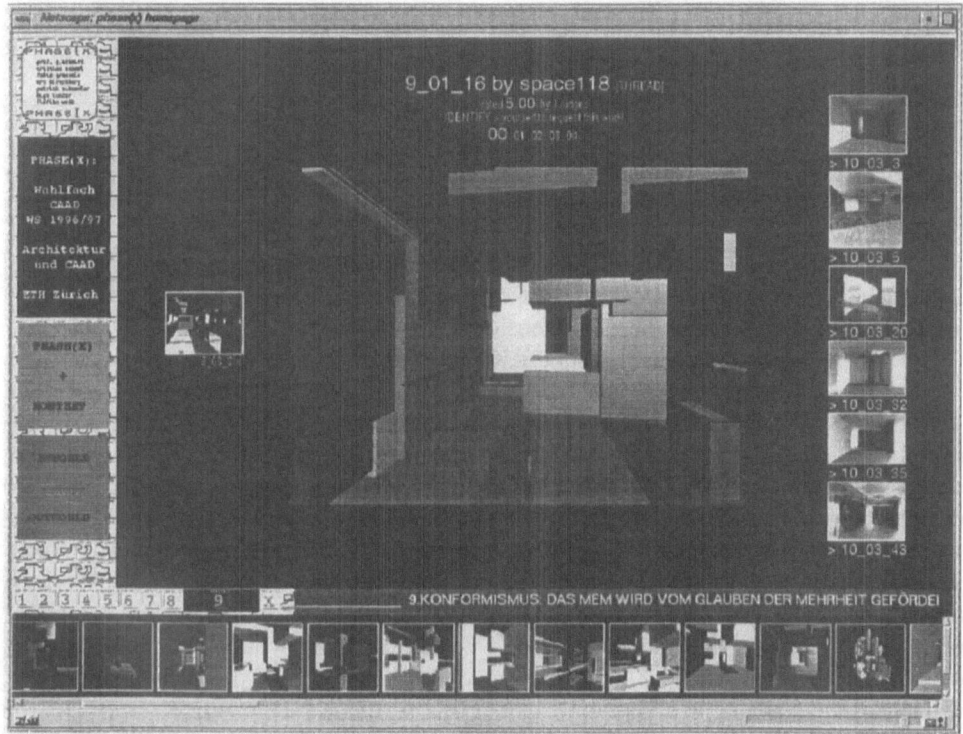

Figure 1. Snapshot of the Phase X InWorld interface. Parent design from the previous phase on the left and children designs on the right of the object. F. Wenz, F. Gramazio, U. Hirschberg, C. Besomi, B. Tunçer, 1996.

Phase X treats authorship in a way that is only possible in a networked, cooperative design environment. After each phase, students do not proceed with their own design but continue with one produced by their colleagues. In Phase 1, they choose from three different designs as a starting point. After saving the results they designed based on those examples into a data base, they are made public to all other students. In Phase 2 and in all following phases, students can choose freely from the examples in the data base. They check out a design and continue to work on it. Students progressively refine the objects in the following design stages. The end result are complex objects with shared authorship that can be traced back to the contributing authors and co-authors. As a result, two important views of the process and the products developed: InWorld and OutWorld.

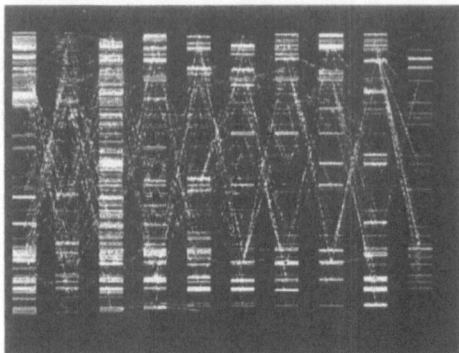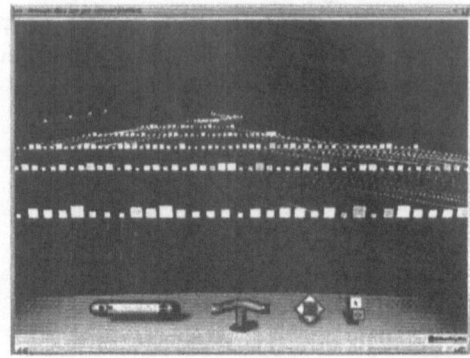

Figure 2. Left: View on a two-dimensional presentation of OutWorld. Each horizontal bar stands for an individual design. Diagonal lines depict the path that students chose from phase to phase. Right: Three-dimensional view of the same data base. F. Wenz, F. Gramazio, U. Hirschberg, C. Besomi, B. Tunçer, 1996.

InWorld describes the perspective of the participant from within the structure of the experiment. The Interface supporting the creation of memes placed the observer and designer into an introverted position. He can only see what is directly before or after his design: its parent and its children. He is caught in the system in a genetic tree structure, without horizon or perspective. Only navigation from one branch to the next sheds some light on the system structure. The relations between objects, although rigidly maintained by the data base, remain subjectively connected only through the memories of individual images and models. InWorld is the plane on which design ideas were developed and stored. Phase X objectively keeps the memory of the individual designs and makes them available in real time. Out of this develops the OutWorld.

OutWorld is the name for the presentations that develop from the entire data set. Based on these overviews, cross comparisons emerge, assumptions and theses can be studied. The OutWorld replaces the sequential view of the InWorld with a parallel view. The interface produces the presentation - consisting of lines and surfaces in three-dimensional space - in real time. The different overview presentations give partial objectivity to the OutWorld. The observer influences the views by choosing parameters (see figure 2). More than 700 student works from one semester form a large quarry of design data. Together, InWorld and OutWorld form an environment that could not be created or exist without the computer. They are therefore an example for the use of the computer as a medium.

7. The computer as a partner

Architecture in the information territory is a radical alternative to existing, physical architecture. In this territory, information is the raw material and the only reality is virtual reality. The computer is at the same time instrument, infrastructure and design environment. It has received a role as a partner that is able to accept responsibility and is able to execute certain tasks independently.

Design in the information territory will rely on similar methods and instruments as traditional computer aided architectural design. Methods are in particular abstraction, generation of models, and simulation (Schmitt, 1996). The instruments made for the information territory include more abstract capabilities than those for traditional CAAD. But contrary to using a computer in traditional design, the creation of physical structures is not the highest goal. Rather, the goal is to overcome the typical shortcomings of physical architecture. Results could be the elimination of the energy consumed to build and maintain physical buildings, and the reduction of transportation energy. At the same time, the advantages of personal meetings between people and of physical architectural environments must not be lost.

Design with the computer as partner is a mental activity in which computing offers methods and instruments absolutely necessary in the information territory. In this mental modeling territory - a human invention - the designer meets computer based modeling instruments that are also human inventions. In this abstract world of ideas, man and machine can achieve a high degree of compatibility. This can lead to drastic improvements of the design process and ultimately to better design. Working in the information territory must not involve a humanization of the machine or a computerization of humans. Rather, it is a neutral, abstract territory to which both man and machine have access. In this environment the computer is a natural, intellectual instrument. Without it, certain tasks are as impossible to achieve as an eye surgeon could work without a laser instrument. Although the eye surgeon has neither built the machine nor can service it, he has to rely on it.

8. Example: TRACE City

The design and construction of a new city in the information territory is certainly an occasion where humans and machine must interact as partners. The opportunity for this experiment came with the exhibition *The Archeology of the Future City* that opened in July 1996 in the Museum of Contemporary Art in Tokyo (Tokyo, 1996). Florian Wenz and Fabio Gramazio developed TRACE, an interactive computer installation. There was

10

purposely no attempt to reconstruct visionary city models of the past. TRACE was built to present and test ideas concerning the virtual aspects of cities. The location is the Internet, the material is information, the builders and inhabitants are the visitors of the exhibition.

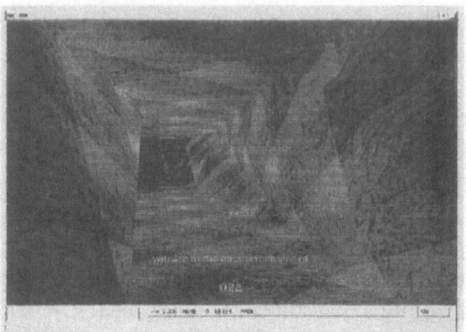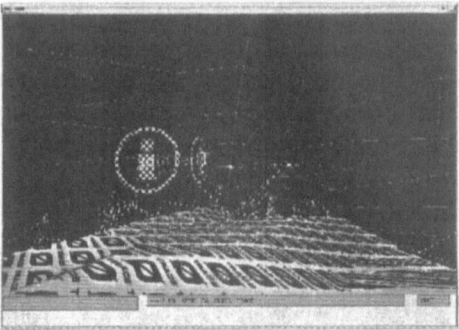

Figure 3. A view of the TRACE interface. Left: *private_in.world*, the navigation space connecting and containing nodes. Right: *public_out.world*, an overview that shows the growing city of information. F. Wenz, F. Gramazio, 1996.

The interaction between natural systems - such as a city - and virtual systems - such as the Internet - is not well understood, as they do not share a common language which could describe phenomena in both systems. Virtual environments must mediate between those two worlds. TRACE generates spaces by registering activities of local and networked visitors, by interpreting and representing them (Wenz and Gramazio 96). In analogy to the real city the structural substance - the code of the space - is generated by the activity of the users. The energy from which TRACE originates is the motivation of the visitors to present themselves in the environment, to leave traces and to read and interpret traces of previous visitors. The space develops further by a constant information exchange between a data base and a geometry generator. The data base saves the traces with the help of an event agent, a program that acts on the behalf of the visitor. The geometry generator translates the traces into perceivable, three-dimensional spaces. The visitor experiences space in two different time axes: in synchronous real time as the condition of the system at the time of login, and as an evolving system time in the asynchronous superposition of the visitor's activities.

TRACE uses an abstracting architectural syntax in the form of icons to demonstrate specific aspects of the whole system. The spatial icons represent the actions of the visitors and the resulting matrix of impressions. The two fundamental forms of interaction in TRACE are abstract in exterior, public spaces and immersive in interior, private spaces. Following the metaphor of the city, TRACE replaces the urban fabric of a private and public space by a dialectic system of two complimentary spatial experiences, *public_out.world* and *private_in.world*. Those are at the same time separated and networked, both represent interaction, navigation, geometry and aesthetics in different ways. TRACE is not a simulation, although it uses methods and instruments of simulation. It is, in the sense of Jean Baudrillard, a substitution of the real (Baudrillard 85).

One enters TRACE through one of many possible variations of *public_out.world* (see figure 3, right). The navigation space consists of a deformable, closed volume (blob), generated by a single nurbs surface, that in its ideal state has the shape of a perfect sphere. The actual form of the blob is defined by a fluid balance between pulling and pushing forces, which press as symbolized Internet sides on the control points of the surface. The visitor moves on the surface similar to the movement on an artificial landscape, without having to understand the actual form of the blob. This illustrates two basic properties of information space: as three-dimensional coordinate space, the system is scaleable, but in contrast to this general feature, it is formulated differently in certain areas. Because it always leads back to itself, it has no noticeable external limitations such as a beginning or an end.

In contrary to *public_out.world* as complex superpositions, *private_in.world* is specific and seemingly simple. Here, the geometry generator translates traces of previous visitors in a network of containers with connecting corridors. Containers hold one unit of media (image, sound, model or text), while the connections contain a specific pattern of movement (straight, zigzag, up and down, curved). The user is caught in this labyrinth and navigates through it by moving continuously forward and by choosing in each container one of four options: straight ahead, back, right or left. If he moves back, the system will stop the generation cycle and will produce a new *public_out.world* which now contains its previously created traces. A more complete description of the TRACE environment can be found under http://caad.arch.ethz.ch/trace.

9. Tool - Craft, Medium - Abstraction?

Analogies exist between the products originating from the computer used as a tool and the products of craft, and so do analogies between the products of the computer used as a medium and the products of abstraction. The interdependencies are complex and not causal. Malcolm McCullough in his book *Abstracting Craft* (McCullough, 1996) treats in-depth the fascinating relations between hands, eyes, and tools in the human context; between symbols, interfaces and constructions in the technological context; and between medium, play and practice in the personal context. What clearly emerges is the role of digital craft as an entirely new field which could elevate the architectural discourse to a higher level of abstraction than ever before.

With the computer increasingly assuming a role as a medium, several terms important to architecture must be expanded or re-defined, in particular *material*, *structure* and *firmitas*. The conclusion of a discussion on *firmitas* in November 1996 between Mario Botta, Roger Diener, Frank Gehry, Jacques Herzog and Jean Nouvel at ETH Zürich was

- not unexpected - that *firmitas* is not a function of the physical building matcrial, its thickness or weight, but rather a conceptual and structural property. As the essence and quality of architecture depends on its structure and material, an expansion of the definition of both terms is needed. The definition of material must include information, the definition of structure must include organizations possible only in computer networks, and the definition of firmitas must include stability and character in the emerging information space.

10. Conclusions

Computers used as instruments act as amplifiers: They help to perfect known tasks, they enable to expand on existing theories, and they support strong personal architectural and design philosophies. On the negative side, modern software overpowers weak philosophies with default assumptions not generated by architects but by developers with other agendas. Computers used as tools tend to eventually replace all activities and tasks that can be automated.

Computers used as a medium create the unexpected in cooperation with human designers. Genuinely new findings may develop from this, as the Phase X experiment has demonstrated. Here, the entire design object has become much more than the sum of its parts. Employed as a medium, the machine can influence the architectural product to a degree that the question of ownership must be re-defined. Because of the combined competence of program and designer, it can lead to better architecture.

Computers as partners are the least explored area but it is safe to assume that this constellation will make most sense in the information space. Here, machines and humans can interact meaningfully, and the design objects will have a higher information content than physical structures. In the information space, the influence of design medium on the product will be strongest.

As long as computers do not influence the architectural design process, their impact on built architecture will be minimal. As soon as a partnership between architects and computers develops, the impact will become visible and result in an improvement of the built environment. Increasingly, a digital architecture emerges that can only exist in the information territory. Whether we like it or not, this architecture will not be designed by architects, unless we rapidly begin to educate architecture students for this task. This implies that architectural education should less concentrate on teaching computers as tools, but rather research, teach and thus influence the way the computer will be used as an instrument and partner.

Acknowledgments

The author wants to thank his junior faculty for the development and implementation of the Phase X teaching environment: Florian Wenz, Fabio Gramazio, Urs Hirschberg, Cristina Besomi and Bige Tunçer. For building TRACE city, special thanks to Florian Wenz and Fabio Gramazio.

References

Baudrillard, J. (1985) Simulacres et simulation, Editions Galilee, Paris.

Glymph, J. (1997) Discussion with Jim Glymph at Harvard University, GSD, April 15, 1997.

Kurmann, D. and Engeli, M. (1996) Modeling Virtual Space in Architecture, VRST '96. Virtual Reality Software and Technology, M. Green, K. Fairchild and M. Zyda (eds.), ACM, Hong Kong University, pp. 77-82.

Madrazo, L. (1996) Typen & Variationen., Types & Instances, in Schmitt, G., Architektur mit dem Computer, Vieweg, Wiesbaden, pp. 126.

McCullough, M. (1996) Abstracting Craft - The Practiced Digital Hand, The MIT Press, Cambridge, Ma.

Oechslin, W. (1993) Computus et Historia, in: Schmitt, G., ed. Architectura et Machina, Vieweg, Wiesbaden, p. 14-23.

Rowe, P. (1996) Studio Works 4, Brooke Hodge and Linda Pollak, eds., Harvard Graduate School of Design, Cambridge, Ma.

Schmitt, G. (1993) Architectura et Machina, Vieweg, Wiesbaden.

Schmitt, G. (1996) Architektur mit dem Computer, Vieweg, Wiesbaden.
http://caad.arch.ethz.ch/projects/acm/.

Tokyo (1996) The Archaeology of the Future City, Exhibition Catalogue, Museum of Contemporary Art, Tokyo, pp. 219-236.

Wenz, F. and Gramazio, F. (1996) The Archaeology of the Future City: TRACE, in Schmitt, G., Architektur mit dem Computer, Vieweg, Wiesbaden, p. 178.

CAD-So What?

Thomas Bock, Professor for Construction Automation,

Seeweg 4, D 74259 Widdern, Germany

Abstract

Computers were applied in construction towards the end of the 50s. In the meantime CA-X technologies rapidly evolved in areas such as integration of application software, 3D modelling and simulation, multimedia systems, artificial intelligence, CADCAM, robotics, and computer-based integration of design, construction and facility management. The structural changes under way in the construction industry ask for a transition from mere CAD, where „D" stands for design and drafting, towards CAC, where the second „C" represents construction , thus further processing the previously generated CAD data .

Introduction

Computer applications began evolving in larger construction companies in the late 1950s primarily in the USA. Breakthrough computer-based project management

R. Junge (ed.), CAAD Futures 1997, 15-43.
© 1997 Kluwer Academic Publishers.

techniques, such as the critical path method (CPM) for scheduling, were developed at about the same time, and also moved quickly into the construction industry

The 60s and 70s saw steady progress mostly in the USA, particularly in home-office applications such as accounting and finance, though the impact did not match the revolutions that took place in other industries. Project-oriented applications such as CPM scheduling, estimating, simulation and plotting evolved, but till the 80s their overall success remained far below early expectations. Initial problems related to costs of and access to computer hardware, complex software, user-unfriendly human-machine interfaces and unpractical interpretations of the results. But the most important omittance was the negligence of qualifying on-site crews for this computer technology.

In the 1980s microcomputers became widespread and rapidly decreased in cost; finally easy-to-use application-development software, such as spreadsheets and databases, was introduced. We now more often see CAD and microcomputers at work on construction sites, and the people using them effectively know how to apply software tools for estimating, scheduling and cost control. Due to the imminent structural changes in the construction industry the further processing of CAD data will be a strategic key for survival of many small offices and SME's during increasingly chaotic market conditions to come.

1. Present CAX Trends

While it is more difficult to predict the dissemination considering human and organizational barriers the integration of existing applications, 3D electronic modelling and simulation, and multimedia-based systems have overcome the technical obstacles for widespread implementation.

2. Integrated Applications

In the 1970s and 1980s many problems resulted from the development of mainframe and minicomputer databases which frequently tried in vain to integrate construction applications. CAX-construction applications have been implemented in relatively self-contained packages, and there are some major compatibility barriers to exchanging even file-level data among many of them. Finally now developers of system and application-development software are making it easier for designers of application software to interface to other applications. Furthermore commercial developers of related packages, such as estimating, scheduling, cost control and CAD, have been working together to provide better interfaces for moving data between their packages. It became possible to take a set of standard package applications and link them together in a manner that resembled an integrated construction information system.

The planning, scheduling and performance-measurement functions can be handled by network-based project management packages. The interface from CAD to estimating facilitates the calculation of lengths, areas and volumes (e.g., earthwork and concrete), and transfers a bill of materials for items that can be directly priced (e.g., plumbing and light fixtures) and entered into the procurement system. Dynamic CAD, if it is used as an „as-built" model during construction, can potentially be used by owners and operators for their facility management. As far as estimates are concerned we need to move from mere statistical estimates, which can only react to the estimate overrun, towards an active realtime estimate control. This type of estimate has several useful outputs to other systems. Its crew production estimates send resource requirements and time durations to the scheduling system; its categorized estimate of costs forms the basis of a budget for the job-cost system; and its quantities and materials feed into the procurement system. Procurement information flows into both the job-cost system and into the accounts-payable component of the accounting system; both of these are well established links that are available in most high-quality accounting packages. Job cost provides input data to the accounting system, particularly for the payroll and accounts-receivable billing modules.

But these applications all fell short, since the CAD data could not efficiently be used for real time control of the actual construction process. Basically what has happened to CAD was that it has been used in the conventional way of thinking and designing with

„paper and pencil" and existing design habits prevailed without restructuring the design and construction process according to the potential of CAX technologies. What is really needed is the redesign of existing organizational, technical and qualifying habits in such a way that CAX technologies can increase the efficiency of the integrated construction process.

3. CAD-VR (Virtual Reality)

Computer-aided design software is finding its way onto construction project sites, but already designers in the AEC offices are making use of three-dimensional "walk-through" simulations. These graphical simulations provide realistic views into designed space long before it is constructed. Problems ranging from component interferences to subtle problems with HVAC can be caught well before their real-world costs and resources are determined.

By adding a time base to the simulation, architects and managers can see how the components are installed over time, experiment with different sequences, and play the result like a video animation to clearly communicate the plan to field supervisors, crew members and even Do-it yourself builders. Some systems also add scaled 3D images of major pieces of construction equipment, such as cranes and trucks, robots, ABCS, in order to help users explore with the computer actual materials-handling methods. Byproducts of these simulations can be short-interval schedules, resource requirements, and materials logistics details, which can be sent to scheduling and database software.

As workstations become faster and "virtual reality" devices become practical, more complex, continuous and realistic images will look and behave almost like the construction process itself. It will have important applications in operations planning, worker training, and facility management for future construction projects. Even though VR looks promising, it will not be effective as long as it is just used as a nice animation toy. What we need here is the further processing of VR data for programming construction machinery and controlling the actual construction process.

4. CAD-MM (Multimedia)

Multimedia computing systems have been moving rapidly into training functions in many industries. Their economics and teaching effectiveness are complementory and maybe even superior to traditional training methods, even though they are still in a fairly early stage of development. As the technology becomes an increasingly standard part of future workstations, this training function can be extended to teach people how to use other types of computer applications in remote, inaccessible, rural or third world regions.

In prototypic industries like construction, where the logistics have to cover considerable dimensions asking for unprecedented management decisions, it is difficult to make the methods and techniques developed on one project available to people who could use them on other projects; the industry is therefore criticized for its prototyping attitude whenever a contract has been aquired. Just imagine what would happen, if the car industry would sell prototypic cars through their dealers! Multimedia technology could provide a means to record innovative and productive methods to build up a corporate memory that can be accessed by the planners and managers of future projects.

Another strategic benefit of multimedia technology could be the temporary establishment of virtual offices or companies in order to aquire design and build contracts. This would be a strategic tool for SME's and small architectural and engineering offices to compete in future construction markets even on a global scale by the internet.

5. CAD-AI (Artificial Intelligence)

Examples to be considered here include artificial intelligence, CADCAM, robotics, and fully integrated construction project planning, design, management and production control systems.

Already there have been many efforts to apply the evolving computer science software technologies loosely called "artificial intelligence" (AI) to construction. So far, most construction investigators have focused on techniques called "expert systems" and "knowledge-based systems," although some have recently been moving into a new area called "neural networks."

The main reason for trying to apply such methods to construction is to deal with the qualitative and ad-hoc based types of problems that are so prevalent in those parts of the industry where the construction is mostly executed on-site. The most valuable career asset for a construction professional is not mathematical or scientific skill of the type taught in engineering schools, but rather it is a methodology to use the not yet existing human experience to solve new problems. Another objective of construction is to capture the knowledge of experienced architects and engineers in computer programs so that other construction engineers and managers can access it and apply it, perhaps even after the experts who provided the knowledge are no longer available due to personnel fluctuations. Such programs also provide a means to integrate and validate the knowledge and experience of many experts, and thus provide a means for accumulating and improving a body of knowledge over time. A good example of this sort of knowledge are the 800 precut systems currently in action in Japan, where the craftmanship of the famous „daikusan" or carpenter, who could carf all the traditional japanese joinery, has been programmed and generated CADCAM production of traditional joinery at affordable cost. Furthermore this type of knowledge could be very helpful considering the long product life cycle of the built environment and retrieving it for future recycling.

Implicit in this type of computer application is the need to deal with uncertainty in the information needed to design, build and operate a building. For example, design starts with only general conceptual knowledge of what a project will look like when it is completed. Yet as early design decisions evolve into commitments for configurations, materials and systems, they can adversely affect constructions costs and schedules, and compromise the efficiency and effectiveness of facility operation during the whole life cycle of the built environment.

AI techniques can capture knowledge of construction methods and management making this available at the design stage during which about 70% of the construction costs are determined. For example, if a designer has a choice of configurations for a wooden structure, an expert system could provide advice as to which would be most economical to build.

A promising area for future applications of AI technology is in planning, monitoring and controlling the construction process and costs in real time. Up to date we controll construction costs statistically instead of adjusting the process in real time. Already there have been some good attempts to build construction planners of various types, and other applications have been made to analyzing construction contracts, preparing construction cost estimates, and select construction methods. Other important applications for intelligent agents will be in helping people to coordinate the vast amount of documentation that is generated in a large construction project, and to assist in negotiating the long and complex permitting procedures that are now required for most projects. Probably the most interesting applications will occur when AI techniques supplement or replace the procedural programming that is now used for one of a kind prefabrication, automated machinery and construction robots.

6. CADCAM (manufacturing)

Development in the field of construction being predominantly characterized by increasing shortage of skilled labour, this shortage will have to be compensated for by an increase in the level of prefabrication to be achieved in the manufacture of pre-cast concrete, wooden, steel frame and brick wall building elements. As an example I describe here the increasing market demand for pre-cast concrete ceiling elements and pre-cast concrete wall elements to pre-cast concrete columns and beams, but due to shortage of paper length I can not explain the same for the other materials, but I will show advanced precut systems for wooden elements , steel element factories and automated brick wall facilties using my slides .

On the basis of the development of the European market for that product, the intention to invest in that product can be regarded as farsighted and promising of future success. The application is available for the most advanced CAD and CAM technologies currently available to the manufacture of double wall elements, massive walls, ceiling and roof elements, column and beam elements. As there has never been any plant automated to this degree, development tasks will continue to account for a large percentage of the work to be done.

In traditional manufacturing in high wage countries the share of labour cost increases faster due to low mechanization rate. Applying mechanized manufacturing methods allows labour cost share reduction up to 30% by increasing mechanization rate. According to the mechanization ratio a minimum lot size of about 30 elements is required. Most gains can be achieved by automated manufacturing using robots, CNC machines and FMS in order to further reduce the significance of labour costs. Traditionally automated factories required a minimum lot size of 1000 or even 10.000 pieces to guaranty ROI. Through the use of FMS (Flexible manufacturing systems, robots, off line programming methods, hybrid control systems etc.) it became possible to run a one of a kind production efficiently. Most present day CADCAM factories reach their ROI point after 3-5 years. They can run 1-3 shifts, producing 1500 to 2000 m^2 ($16,145.88$ ft^2 to $21,527.84$ $ft^{2)}$ of floor-wall panels per shift.

Another substantial advantage in favour of the pre-cast con-crete elements consists in the job efficieny of the workforce. As the building site personnel is to a far lesser extent concerned with somewhat more complicated tasks, such as, for example, moulding, insertion of reinforcement steel, etc. than is normally the case with regular construction workers, job efficiency at the plant level reaches an optimum that cannot possibly be arrived at on a building site. Costs of transportation are approximately the same both for the prefabricated elements and for corresponding quantities of site-mixed concrete.

An increasingly competitve construction market asks for new flexible production as well. Due to the notion of robotic handling devices any shutters can be freely positioned. Recent trends in flexible manufacturing technologies offer solutions for PC production placed on a platform using magnetos. This development allows to cast free shaped and designed panels in concrete and produce one of a kind elements very efficiently as required. What we have to do next with the CAD data is to use them for rapid precast concrete production by robotics as strategic advantage for improuving quality and staying competitive during chaotic market conditions. Similar technologies are available for wooden structures using precut-CADCAM-production systems or for autoclaved lightweight concrete which can becarved by threedimensional milling center.

7. CAD-C R (Robotics).

Considerable benefits of CAD will be realized, as soon as we link it up to any kind of machine or robot in order to implement computer integrated construction. Here I describe the progress and obstacles faced by the first generation of construction automation and robotics which have been developed and tested during the 80's mostly in Japan and the greatly increased potential that will be evident in automated building systems, which are tested in the 90's and will be furthermore implemented in the first half of the 21st century.

The challenges of developing robots for construction jobsite are much greater than those of most factories. First the products of construction are much more complex and ill structured.Second, in contrast to the repetitive products that flow down production lines, the design of the construction product and the process to build it are individually adapted in each case. While the manufacturing process is highly repetitive once production starts, that in construction is always changing. The physical environment of construction is often much more hostile to machines as well as people, so machine design must be stirdy and robust accounting for extremes of weather, dust and unexpected forces.

Given the difficult and complex environment of construction, it is remarkable that robots and automated machines are already performing routine tasks on some jobsites. The first construction robots have either been derived by adding sensors and computer-based controls to existing construction equipment (e.g., to control the cutting edges or screeds on various types of earthmoving and paving equipment, robotic tower crane etc.), by adapting the comparatively rigid factory-type robots to construction (e.g., for spraying fireproofing material or painting), or by developing hybrids of the two (e.g., robot arms mounted on tunnel machines). While the sophistication of their mechanisms and sensors has often been quite high, these robots have had only the most basic forms of on board "intelligence."

Most of the construction robots developed to date are stand-alone devices designed to perform narrowly defined tasks without the need to communicate or cooperate with other machines. However, coordinated teams of robots quite commonly perform sequential operations on factory assembly lines, and there are some formal communication mechanisms linking them together and similar technology also moves

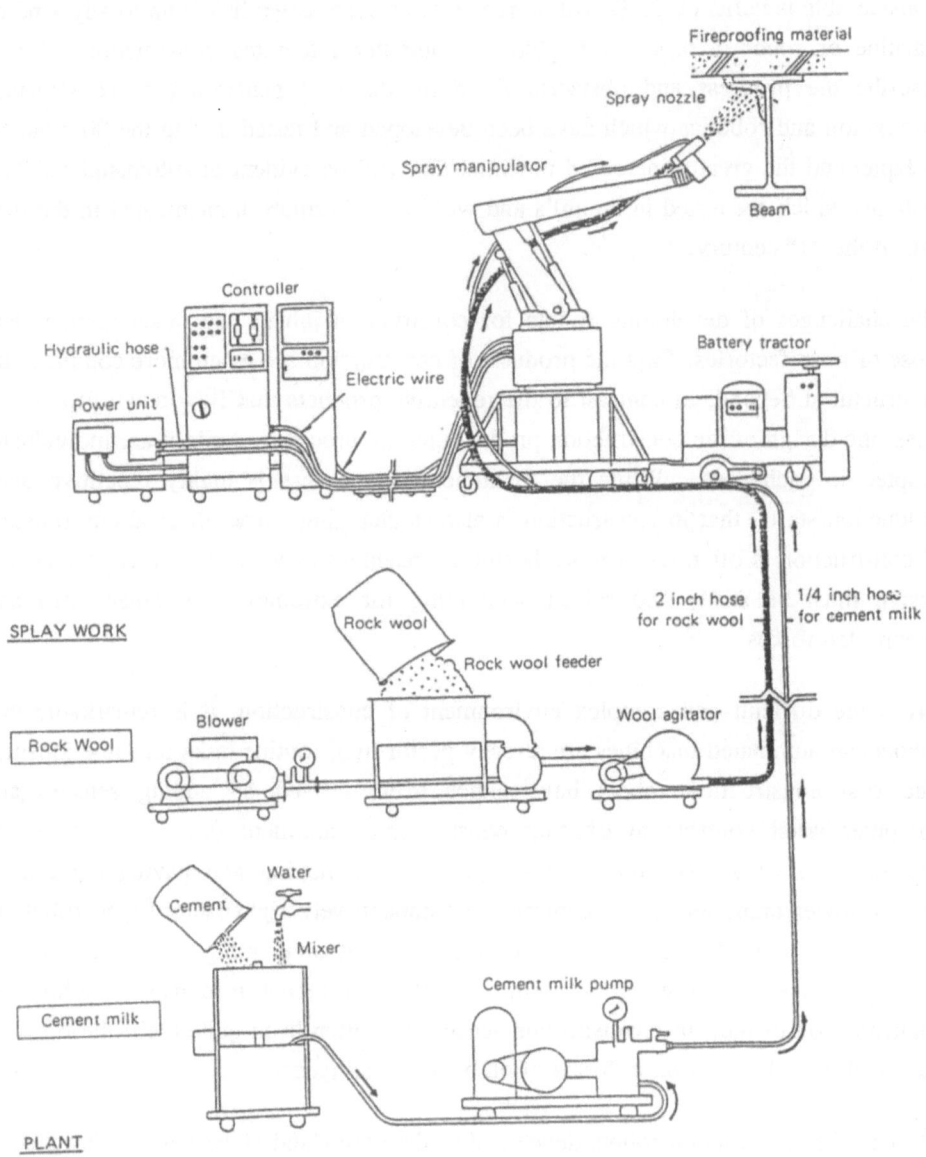

Fireproofing material

Spray nozzle

Spray manipulator

Beam

Controller

Hydraulic hose

Power unit

Electric wire

Battery tractor

2 inch hose
for rock wool

1/4 inch hose
for cement milk

SPLAY WORK

Rock wool

Rock wool feeder

Blower

Wool agitator

Rock Wool

Water

Cement

Mixer

Cement milk pump

Cement milk

PLANT

Rock wool spray system with SSR-1

to construction in the dozen or more automated building construction systems or the EU project „ROCCO" which stands for robotic assembly system for computer integrated construction and has been scientically guided by the author:

7.1. ROCCO: Robotic Assembly System for Computer Integrated Construction

Here I describe only the IT part of the ROCCO project: The system deals mainly with the construction oriented modification of existing technologies and with closing the gaps between them through intelligent interfaces and IT-based tools, in order to provide the necessary flexibility for one of a kind building production, robust design and user friendly programming. To achieve the requirements, the above mentioned basic strategies for automated construction are applied: the information integration , the transfer to the pre-fabrication (only where applicable) and the redesign of the used construction materials. The main emphasis lies on the creation of automated system, which enables the complete and continuous automation and the integration of computer based construction systems, without restricting the freedom of the design of the architects. With that system one can build in a shorter time with fewer personnel more and better buildings.

Within the ROCCO project a mobile robot system will be developed for the assembly of masonry on-site. Therefore a suitable robot under development as well as the integration in a computer based system for the working preparation and programming. The robot system with a reach of 5.50 m and a load capacity of 350 kg consist of a vehicle , the actual robot or manipulator and a gripping and assembly tool. In the framework of the working preparation, the necessary data for the pre-fabrication of the costumized blocks and for the robot programming are generated.

Based on a CAD representation of the building, first the walls are divided into the single blocks automatically by a software tool. The next step contains the planning of the construction site layout, i.e. calculation of the optimal working points of the mobile robot systems, the space for the pallets, the configuration of the blocks on the pallets and the sequences of the block's assembly. With the then available information the customized blocks for realizing individual wall dimensions can be produced, cutted and palletized on stationary plants. The last step of the working preparation is the generation of the robot programs.

26

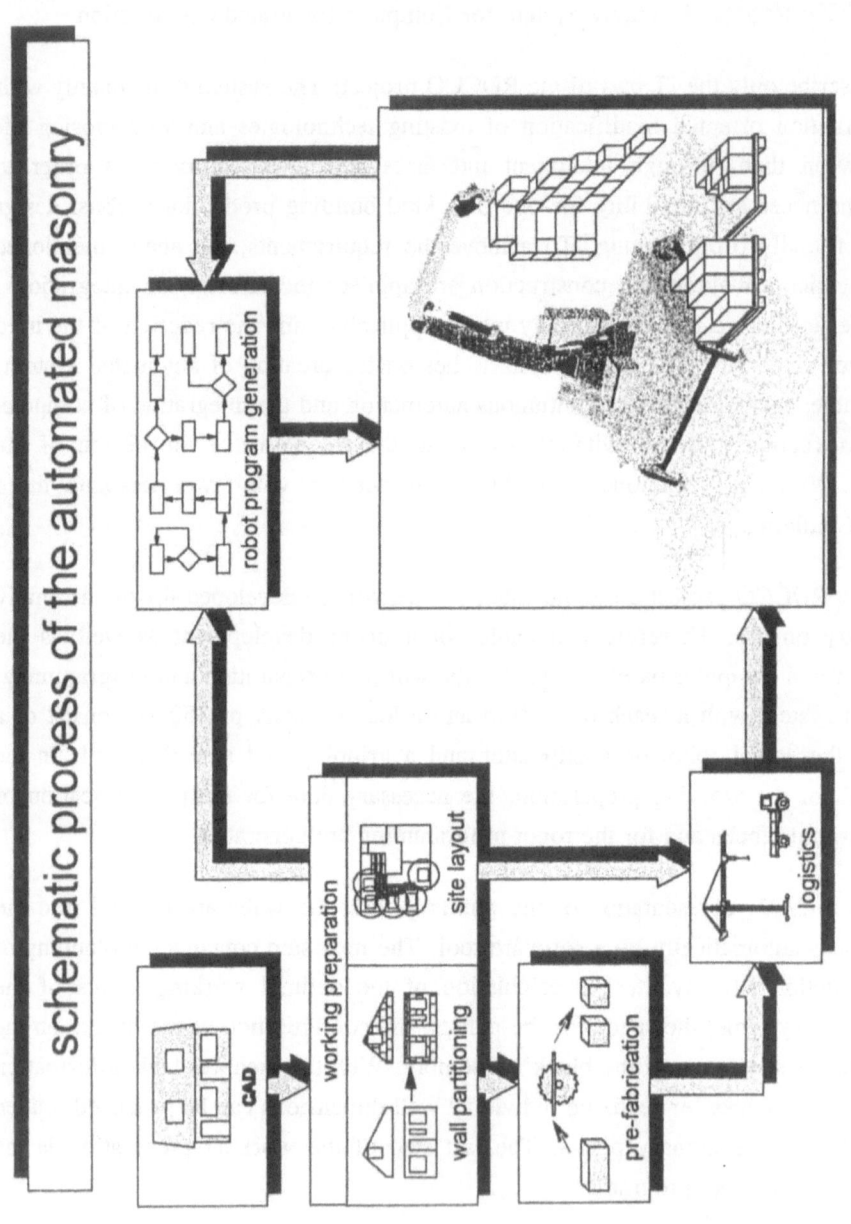

schematic process of the automated masonry

7.1.1. ROCCO's integrated information management

The chosen approach bases on the idea of Computer Integrated Manufacturing (CIM), which is already successfully implemented in other industries and which shows there its efficiency. The idea is a continuous information flow from the architectural design to the automated execution of construction process on-site and in consideration of the construction elements. This procedure, called Computer Integrated Construction (CIC), makes it possible to automatically process all once collected data without loosing the data consistency. This enables all participants to stay as flexible as necessary during short-term changes with as low error rates as possible.

The different development stages of the information flow between the parties participating in a construction process will be shown exemplary by means of the chosen application: masonry. The concept of CIC leads into the components and tools described as follows.

7.1.2. ROCCO's information flow

Conventional information flow. Until the introduction of CAD-systems for designing buildings, all information necessary for masonry construction was included in the manually drawn architectural design plans. These plans where send to the executing construction company, who ordered with them the necessary building materials as non-costumized prefabricated standard products, adapted it according the plan sin a handicraft manner on construction site and assembled the walls manually based on the information of the architectural plans. This procedure implied in each of its steps big sources for errors, which resulted in delays, subsequent work and bad quality of the construction.

Advanced flow information. In the last years, computer aided design gains more and more importance. The use of CAD in architectural offices is increasing rapidly and with this the availability of electronically processed data, which causes more exact and consistent plans. An additional trend can be observed in the building material industry. They are increasingly able to produce custumized non-standard blocks through computer aided working preparation and production, which can easily assembled on the site according to automatically generated assembly plans without the necessity of a manual re-shaping. Hereby the information flow is reversed by the building material industry. They get the information in pre-fabrication directly from the architect and pass the assembly plans and parts lists to the construction companies.

7.1.3. The ROCCO CIC information flow

To integrate the complete construction process in an IT-framework, as many as possible process steps should be based on electronical data processing. For the masonry we describe following a complete integrated information chain from the architect's design to the robotized execution of the tasks on-site, where all tasks are based on electronical data processing. This concept is IT backbone of the ROCCO project and represents the state of the art in European computer integrated construction systems.

The plans in the architectural offices are created with CAD-systems. On these base the production of standard and non-standard blocks in the pre-fabrication with the computer based production scheduling and numerically controlled production. This enables again the use of programmable assembly tools and systems on the construction site. The corresponding information flow is shown on the picture below.

To use the architectural design data for production planning and programming in the prefabrication and on the construction site, it is necessary to process and to extend the data. The necessary software systems will be presented in the next chapter. The development of the software tools is realized within an interdisciplinary group on the base of already available knowledge.

7.1.4. ROCCO's systems components

A basic difference between conventional robot applications, e.g. in mechanical engineering, and the robotized masonry is the number of repetitive robot motions. In conventional applications the same set of motions is repeated thousands of times, whereas every house has a unique design and every block has to be placed with an individually set of motions. However, at the same time we can reduce the structure of the motion process to some basic pattern like gripping, positioning, placing. This helps to reduce the complexity of the task.

A crucial importance has the programming of the robot. Normal robots can be taught or programmed with off-line programming-systems, where much more time is necessary to program the sequences than to execute it (but for the efforts are distributed to the number of cycles). This is not possible with a one of a kind production. Here an automatic program generation based on the available geometry data is compelling.

For this the start point on the pallet and the end point in the wall of each block must be known. Additional the position of the robot in relation to the floor must be known. This leads to another difference between industrial robots and on-site robots. The on-site robots are exposed to permanently changing environmental conditions. It is necessary to re-calibrate the robot at each working point. In addition unknown obstacles and events may occur, which make it necessary to adapt the robot easily to the environmental conditions.

During the whole working sequence, it is necessary, that the operator knows exactly, what statement is executed by the robot and how this can be manually modified, if environmental changes or material breaking occurs. The main difficulty lies in the re-entrance of the generated program after a manual modification, since the decision is necessary, at what point and with what parameters the program will re-start after unpredictable manual operations. To solve the problem, advance programming methods must be applied, which allow the simple and fast modification of the generated program.

We choose a distributed database approach to be able to run the different parts of the software system on different platforms (adapted to the respective requirements) and at different places, i.e. at the working preparation office, at the pre-fabrication plant and directly at the robot's working points on-site.

7.1.5. ROCCO's off-line components

The used computer platform for the working preparation at the building material producer is a high level PC running the OS/2 operation system with the integrated OS/2 database manager. The relational database serves as the central storage and information distributor for following applications:

- Different CAD-format converters, which convert the architectural design information into a process-able format

- The graphical user interface, which enables the user to collect data from manually designed houses and to show the result s of the following information processing systems.

- The wall partitioning software, which divides the architectural walls into the necessary blocks under the consideration of windows , doors , lintels, etc. The outputs are the dimensions and positions of each block in the respective walls. During the segmentation procedure, optimization criteria have to be considered under hard boundary conditions. The number of non-standard blocks should be minimized to keep the costs low and the dimensions should be well balanced to keep the waste during cutting low. Simultaneously official and technical prescriptions should be kept concerning the bearing capacities, the joints' positions, the walls' connection, etc.

- The sequence and task planning software, which is responsible for different calculation and optimization procedures. In the first step, the software has to determine the possible assembly sequences of the blocks, that is to generate an assembly precedence graph. In the second step, the optimal sequence has to be determined concerning the optimization criterion of minimizing the number of vehicle movements respectively of maximaizing the number of blocks built from one working position. This is mainly dependent on the reach of the manipulator. During the calculation of the optimal sequence, one can simultaneously determine the number and positions of the necessary working points. In the next step, this information is used to generate a collision -free path of the vehicle from the first working point to the last one along the sequence of working points.

- The palletizing software, which determines the position of the blocks on the pallets and the sequence and positions of the pallets on the construction site. Necessary information are the assemble sequences related to the perspective working points, the dimensions of the blocks and pallets, the position and dimension of the free storage space around the robot and the specific properties of the gripper, i.e. the gripping direction and for that the free reachable block surfaces. The main intention of the tool is to minimize the number of pallets under the condition of guaranteeing the necessary free reachable surfaces of the blocks for gripping. To achieve this goal advanced operation research algorithms as heuristics and genetic algorithms are applied.

After processing the incoming CAD-data with the above described tools, all necessary geometrical information as the positions and the dimensions of the blocks on the pallets and in the wall as well as the positions of the pallets and the robot system are available. Together with the assembly sequence, all data is onhand to be able to generate the control programs for the pre-fabrication production and cutting machines as well as for the on-site assembly robot system.

7.1.6. CAD controlled pre-fabrication of building components

The operation in the pre-fabrication plant can be divided into two basic procedures. First the production of standard blocks in different formats. The production is not specific for a certain order. No custumization is necessary. Therefore the only necessary information for the production scheduling is the number of blocks and their main format. So non specific software tool is necessary to process the data before using it for the production of standard blocks.

Another situation is in the production of the non-standard blocks as the second main procedure in the pre-fabrication. To get an efficient production of these blocks, it is necessary to have a continuos information flow from the above mentioned off-line tools to the production machine controls. If one wants to assemble the tools automatically, not only the dimensions should be known for correct cutting but also the positions on the pallets for correct palletizing. To get an efficient production procedure, both, the cutting and the palletizing, should be automated. For both numerical controls with well-defined interfaces are necessary. In both cases, the geometrical information must be converted into motion information of the respective axises. Additional the sequence of cutting must be optimized to minimize the cuts and waste under the boundary conditions of the features of the used cutting equipment and of the succeeding palletizing station. So, three different software tools are necessary described below:

- The cutting sequence optimization, which minimizes the waste number of cuts considering different boundary conditions: The features of the used cutting equipment as the number of saw-blades or the thickness of the saw-blades, the features of the palletizing equipment as the ability to palletize randomly or the maximum number of the simultaneously available pallets. The software uses similar algorithms as the palletizing software in the one-line field.

- The cutting program generation, which processes the geometrical information of the blocks into the motion information of the respective NC-controlled cutting axises.

- The palletizing program generation, which generates the motion programs for the palletizing devices. The generator has to consider not only the necessary final positions on the pallet. Depending on the design of the gripper, it is also necessary to consider the already palletized blocks in order to determinate the approach direction to avoid collisions.

The fully automated and computer-integrated pre-fabrication of the blocks represent an important prerequisite for a smooth assembly process on the construction-site.

7.1.7. Construction Site IT Components

After a conversion, the geometrical and process data are available in a format, which is suited to serve as the base for the robot program generation and the user interface for the robot control. The figure below shows the complete structure of the software components. Here we chose an object oriented approach to implement a hierarchical implicit programming system to meet the following requirements with simultaneous ease of use, extendibility and programming power.:

Requirements: Integration of simulation packages, input and administration of world, sequence and program data, integration of path planning algorithms, providing of programming features, real time capability, interfaces to robot controls and measurement systems. The system consists of an implicit layer, were the working sequence is described in an abstract, robot independent manner on three hierarchical levels: Mission, task, action. Each of the levels offers different commands to be used for the description of robot operation. With the representation of the different operations as icons in a precedence graph, it is possible to provide an easy to use programming feature without the necessity of the knowledge of a programming language. Also the automatic program generation is possible, since at the point only the assembly sequence information is necessary. The picture below shows the operation hierarchy.

The second layer converts the implicit instructions into explicit elementary operations (EEOs), still robot language independent, but now provided with the explicit geometrical parameters for robotic motions. It is possible to integrate in this layer the automatic off-line planning, where the respective geometrical information can be used to parametrize each elementary move of the robot under the boundary conditions of a collision free path and consideration of the gripper features.

Through a succeeding interpreter layer, where the universal motion commands are transferred to the robot control dependent commands, it is possible to integrate different on-line control, sensor and simulation systems. Through the real-time process monitoring, it is possible to include sensor information in order to follow the execution and to update the world database even after switching from automatic to manual mode and vice-versa. One can integrate alternative manual operations either in the feed forward and nominal feedback line, if the assembly of difficult parts is necessary, or in the non-nominal feedback line, if the assembly of difficult parts is necessary, or in the non-nominal feedback line, if the sensor detect an irregularity or problem during execution. This strategy enables a flexible reaction to all unforeseeable situations on the construction site, which can be handled either through manual re-programming of the automatic operation.

By describing this ROCCO project I wanted to show how we have to make efficient use of CAD data, because then we can increase the productivity of the whole design and build process.

8. CAD-ABCS: Automated Building Construction Systems

A major step toward an integrated system of robots is now being undertaken by some of the mid-size to large contractors who promote the development of systems that will substantially automate the construction of mid-to-high-rise buildings, and about a dozen of these systems are being deployed. Basically, they consist of a jack-up frame or push up jacks on which or below which a variety of robots for materials handling (e.g., cranes, hoists); fabrication (welding, cutting, finishing); and inspection are installed. The frame will have an all-weather enclosure to enable work to continue around the clock, at any season of the year. This framework will initially be positioned at the first of a series of repetitive floors to be built, and the robots will do about 30- 70% of the work to construct that floor. Next, the whole frame jacks itself

Construction   Maintenance

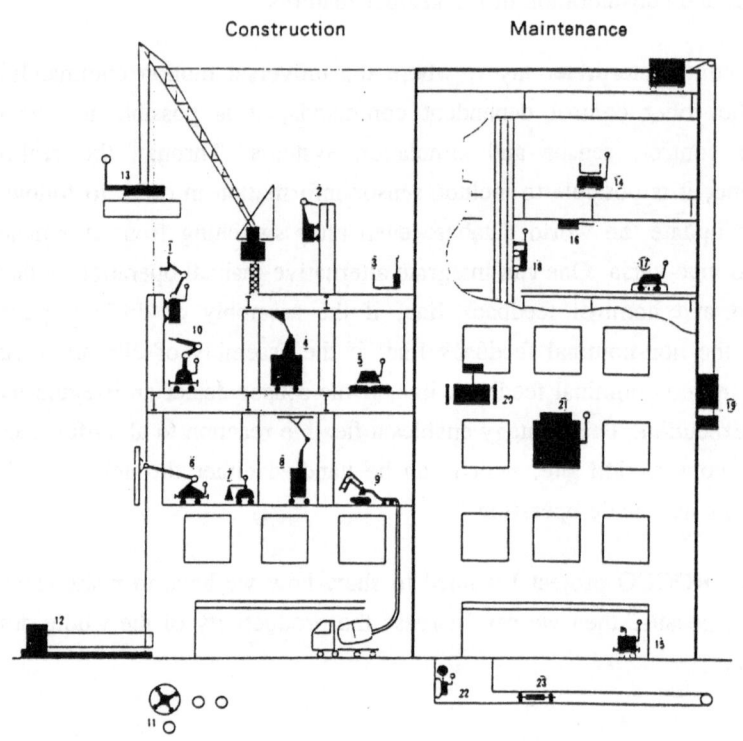

1   Steel frame erecting robot
2   Steel frame welding/bolt tightening robot
3   Material transport robot
4   Fireproof spraying robot
5   Floor finishing robot
6   Robot to manipulate exterior wall board
7   Floor cleaning robot
8   Robot to manipulate wall board
9   Concrete placing robot
10 Interior finishing robot
    (spraying, painting)
11 Pipe embedding robot
12 Reinforcement bar erecting robot

13 Position controlling robot
14 Package transport robot
15 Security patrol robot
16 Duct cleaning robot
17 Floor cleaning robot
18 Piping system diagnosing robot
19 Exterior wall maintenance robot
    (painting, cleaning)
20 Exterior wall diagnosing robot
21 Window cleaning robot
22 Tank cleaning robot
23 Pipe inspection robot (gas, sewage)

up to the next level, and builds another floor. The idea is somewhat like a slip-form for constructing a concrete structure, except that a whole building, not just a concrete structure, is "extruded" from the system. This process continues until the building is done, then the automated components are removed, leaving the frame in place to become the structure for the top floor of the building.

The systems are in part motivated by an expected shortage of skilled labor in Japan, but over time will have economic and quality advantages similar to those of an automated factory. About 90% of present labor requirements will be replaced by automation. Those workers who remain will probably be highly skilled technicians who can program and maintain the robots. The systems provide for substantial integration of structural, mechanical, electrical and finishing operations that are used in the construction of a building. There are also obvious interfaces to and interactions with design. In this way they represent the computer-integrated-construction (CIC) systems of non manufacturing industries.

Impressive as such automated building systems will be, there remain many challenges facing the advancement of construction automation and the development of more capable construction robots. Perhaps the most difficult is that of developing the intelligent software to integrate future machines into the complex environment where they will work.

Before considering what should go into the core of construction robot software, it is important to think about some bounds on this software. Relative to the intelligence to support the successful execution of construction tasks and to the intelligence and human dimensions of a typical construction worker, we are still looking at a most rudimentary kind of "intelligence" to form the core of a construction robot's software.

In general, what is needed is some way of modelling within robot agents some feeling of their environment, such as key characteristics of objects and other agents, in ways useful for reasoning. We have to reduce the knowledge that needs to be encoded in machine systems a priori by enabling them to tap the vast knowledge sources in their environment when needed. This is extensibility, which some might call a simple form of machine learning. Robot societies or groups should be able to assemble knowledge and enlist other agents needed to perform a task and respond dynamically to change. Robot reasoning and control software should deal with unexpected obstacles, road

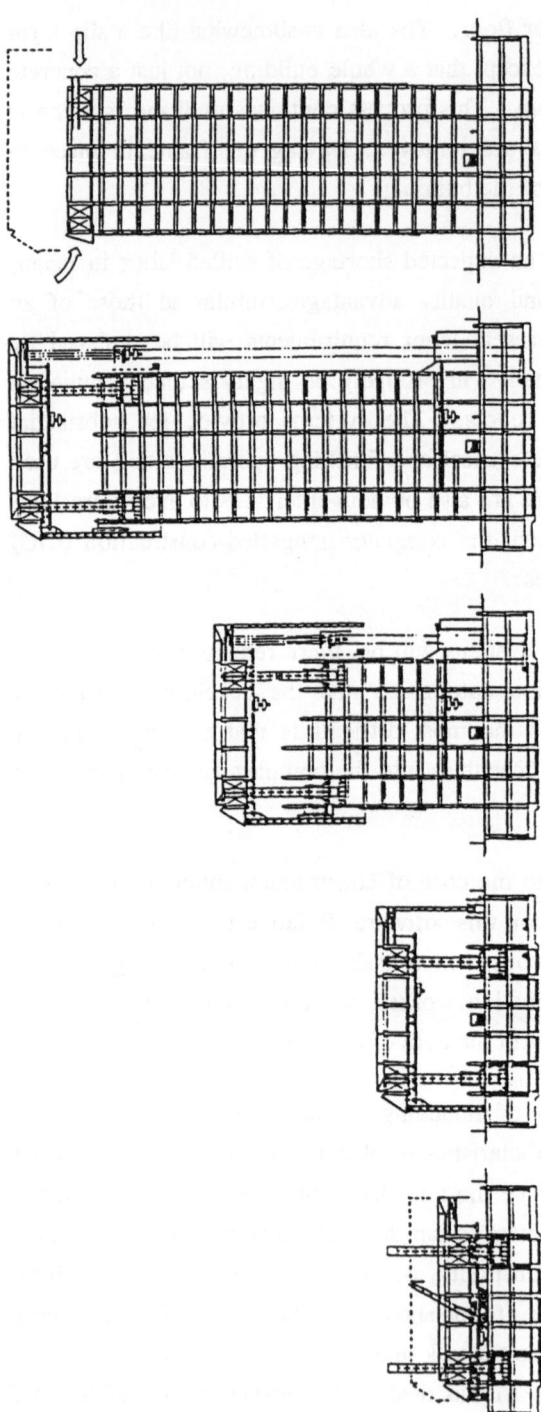

conditions, failure of a machine-positioning system, damaged material, improper tools, or imprecise instructions.

9. CAD-SR/ Service Robotics

9.1. CAD/AGV (Automated Guided Vehicles)

In order to use CAD information for future service systems and service robots we collected experience in the technical of AGV (automatic guided vehicles) concerning kinematics, navigation guidance control and design of a control of conditions. The kinematics of vehicles known for automatic movement on construction sites. Such a kinematics was developed by my center for technology transfer, but further alternatives should be worked out.

9.2. CAD/ SAFEMAID (Semiautonomous facade maintenance device)

Another promising application of CAD data into service systems and robotics is the field of half -respectively full-automatical maintenance of façades under difrent surface-conditions:

Fields of examination are:

- Height reaching devices
- Ergonomic investigation of facade maintenance
- Cleaning and maintenance methods of different facade surfaces
- Prerequisites for the use of automatic or telemanipulated facade maintenance devices
- Economics , serviceability
- Testing and control by simulation

Aim of the researching-efforts is the development of a semi-automatical façade-service implement for the maintenance and diagnostics of skycraper- façades and other areas of building which are accessible from outside. The implement shall be programmable as well as navigateable by remote-control, and it shall be flexible concerning the tasks (cleaning, diagnostics by camera, maintenance, ...) and the use (suitability for buildings already existing).

Beside first studies concerning the pracability several computersimulations about possible kinematics and working-processes were worked out at the department.

10. CAD-TR /CAD-CSCW: TeleRobotics/Computer Supported Cooperative Work

Global AEC projects require a multi-designer and multi-construction-machine system which has a realizeability capacity. It also provides a cooperative creation environment from design activity to prototype construction, which is indispensable for product development and the predictive simulation of complex construction projects. The requirements, the necessary functions and the implementation of a "cooperative tele-designing and tele-construction system" which is distributed on a computer network . The necessary technologies which have been implemented for a cooperative tele- designing and tele-construction system using the Internet are as follows: (1) visual information display, such as predictive display of geometrical information to compensate for time delay and real-time construction assembly display using multi-axis force information, (2) predictive auditory information presentation using a physical model of block assembly and an information transformation technique, (3) predictive force information presentation, (4) tactile presentation of the state joining as high frequency vibration, and so on. The software system was implemented as a multi-construction -machine system. Necessary agents and their functions are discussed based on the system, as implemented and tested.

11. CAD/Integration of Intelligent Agents as the Missing Link.

The factory automation already reached a sufficient level of automation, but as for the on site environment, which is ill defined, unstructured and shape-changing, there may not be accurate correspondence of an agent's knowledge about the environment to its real state at any time. So we have to prefabricate as much as possible in the structured environment of factories and only realize the final assembly on-site. In future construction field environments, intelligent machines, like their human counterparts, will thus need to gather knowledge to plan and control autonomous tasks. Not only the robots, but most of the intelligent agents will need a unifying core of intelligent software and a framework for defining and communicating knowledge about designs and field operations in a way that can effectively be utilized for their production tasks.

IMS Intelligent Manufacturing Systems International Conference on Rapid Product Development	Prof. Thomas Bock Universität Karlsruhe (TH) 

Inter World Intelligent Manufacturing System in Building Construction

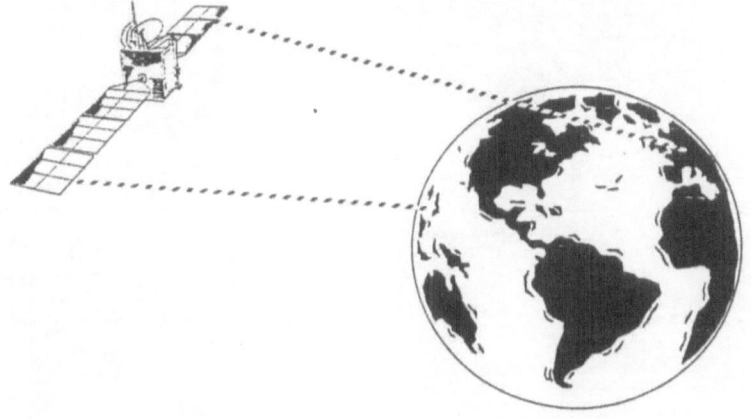

- high precision and quality, three shifts per day
  Präzision und hohe Qualität, 3-Schicht-Betrieb

- favours the realization of an CIM-concept
  begünstigt die Realisierung eines CIM-Konzeptes

- efficient use of expert knowhow / simultaneous monitoring of more than one construction site
  Effektive Nutzung von Expertenwissen / gleichzeitige Überwachung von mehreren Baustellen

- further application: maintenance of supply installations (e.g. pipelines) in inaccesible areas
  Weitere Anwendung: Wartung von Versorgungs-einrichtungen (z.B. Pipelines) in unzugänglichen Gebieten

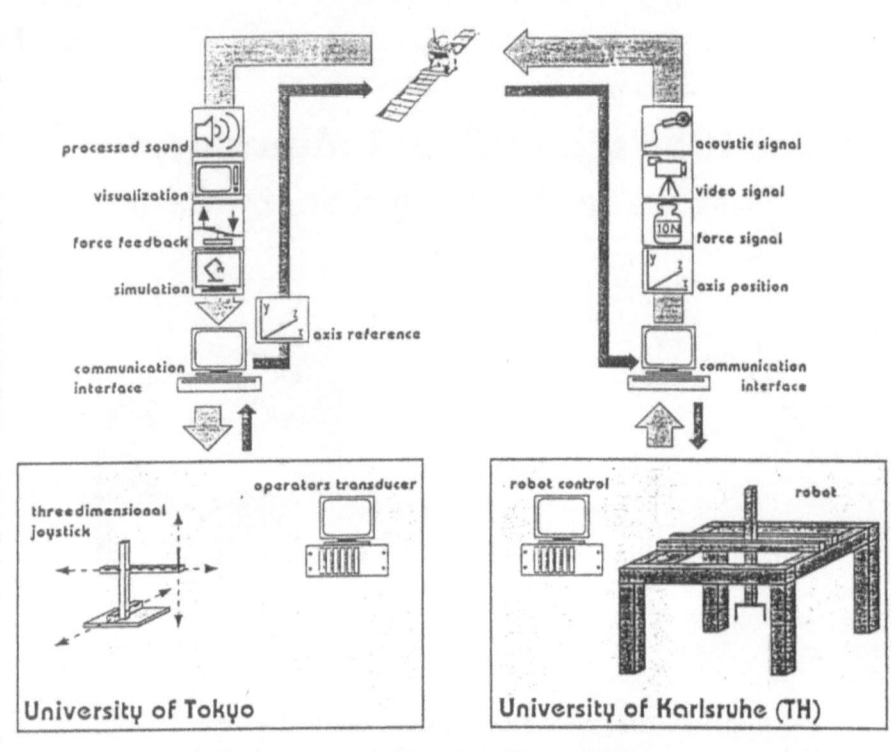

University of Tokyo

University of Karlsruhe (TH)

- reality sensation based on a predictive
  information display method
  Zustandsdarstellung durch prädiktive Simulation

- six-axis force / tork sensor
  Sechs Achsen Kraftmomentensensor

- fail safe system / force monitoring
  Störfallbehandlung / Kraftüberwachung

The future could be the construction knowledge environment, and some ways in which the core software of an intelligent robot might interact with the environment. The organizational context in which the robots might be working, the interfaces to computer-aided design (CAD) databases and reasoning, interactions with other field agents—both human and machine—and interfaces to knowledge sources in the world beyond the field.

In an environment of this type, all of the agents—both human and machine—could be working in the context of an integrated model—possibly one that may evolve from today's research on distributed databases and knowledge bases, object-oriented systems, constraint-based systems, neural networks and other advanced telerobotic computer science and engineering. In this context, from project conception through design and construction, and on into facility management and recycling over the life cycle of the project, the virtual model would evolve and change to accurately reflect the history, present state, and future plans for the facility.

The scope of research needed to build theories and core software to support integrated design, construction and facility management is huge. Each step in this research should lead toward a general architecture handling the knowledge agents need to function productively in a knowledge environment. The resulting software could then be extended by developers of applications-oriented robots to handle particular areas of expertise, whether in performing design, managing other machines, in doing specific physical tasks, or monitoring and controlling a facility's operation. The ultimate objective should be to design and develop the general theory and software core for machine agents which can then be embedded in agents specialized for particular tasks. Thus may evolve parallel virtual and physical models of our built environment, and intelligent agents who could work cooperatively to sustain our ecosystem.

Conclusion

It is clear that CAD technologies will continue to advance quickly in other industries. But what we need is how effectively the construction industry will adapt to and exploit this technology for its own advancement. If we do so successfully then we can cope with the structural changes that are now taking place in the construction industry. If CAD data will not only be used for designing and drafting but also for controlling the whole construction process, then we can stay competitive during the chaotic market conditions that lie ahead of us. Due to the fact that about 2/3 of the construction costs are decided during the design stage it becomes a strategic necessity to further process the CAD data for the efficient realization of the construction process.

References:

CIMA Personal Notes from the Centre Informatique Methodologies en Architecture 1989, Paris

Company brochure of CADCAM prefabrication of low cost housing components (SÜBA Corporation), 1996, Mannheim , Germany

ROCCO-ESPRIT III 6450 (Robotic Assembly for CIC), 1997

Various informations on ABCS-Automated Building Construction Systems, 1988-1997

Personal conversation with Mr. Nishigaki of Hazama Construction on Intelligents agents in CM, 1996

T. Bock and M. Mitsuishi, „Development and Trial of a Teleconstruction System", in: Proceedings 14th ISARC, International Symposium on Automation and Robotics in Construction, Carnegie Mellon University, Pittsburgh, June 1997

Sheridan, T.B., „"Space Teleoperation Through Time Delay: Review and Prognosis," IEEE Trans. Robotics and Automation, Vol.9, No. 5, pp.592-606, 1993

Sato, T., et al., „"MEISTER: A Model Enhanced Intelligent and Skillful Teleoperational Robot System," Robotics Research: The Fourth Intern. Symp., The MIT Press, pp.155-162, 1988.

Mitsuishi,M.,et al., „"Development of Teleoperated Micro-Handling/Machining System Based on Information Transformation," Proc. IEEE/RSJ Intern. Conf. on Intelligent Robots and Systems (IROS'93), pp.1473-1478,Japan, 1993.

Mitsuishi, M., et al., „"Predictive Information Display for Tele-Handling/Machining System,"Proc. IEEE/RSJ Intern. Conf. on Intelligent Robots and Systems(IROS'94), pp.260-267, Munich, 1994.

Hatamura, Y., et al., „"Actual Conceptual Design Process for an Intelligent Machining Center," Annual of the CIRP, Vol.44/1, pp.123-128, Enschede, The Netherlands, 1995.

Bock,Thomas

Long Distance Telemanipulating, in: Proceedings of the IMS (Intelligent Manufacturing Systems International Conference on Rapid Product Development), Stuttgart, 31.1.-2.2.94

Bock,Thomas;Müller,Christian

DISYC-Distributed Intelligent System in Construction, in: Proceedings of the sixth international conference on computing in civil engineering, Berlin, 12-15.07.95, S.435-437

Bock, Thomas; Weingartner, Harald:

Roboter und Telemanipulatoren für Reinigungs-, Inspektions-, und Wartungsaufgaben an Fassaden und auf Dachbereichen; in: Deutsches Architektenblatt; 11/93

Innnovative Fassadentechnologie - Fassadenroboter für Montage, Diagnostik, Instandhaltung; Kongreß: Internationales Forum Innovative Fassadentechnologie; 2. und 4. November 1994

Frederic Vester, „Ecopolicy", 1997, Studiengruppe Biologie und Umwelt, München

INTEGRATION OF DESIGN APPLICATIONS WITH BUILDING MODELS

CHUCK EASTMAN, TAY SHENG JENG, ROY CHOWDBURY
Design Computing, College of Architecture, Georgia Institute of Technology, Atlanta, GA. USA
KIM JACOBSEN,
Department of Planning, Technical University of Denmark, Denmark

Abstract:
This paper reviews various issues in the integration of applications with a building model. First, we present three different architectures for interfacing applications to a building model, with three different structures for applying maps between datasets. The limitations and advantages of these alternatives are reviewed. Then we review the mechanisms for interfacing an application to a building data model, allowing iteration execution and the recognition of instance additions, modifications and deletions.

1. Introduction

As the various professionals in the building industry become more experienced in the use of Information Technology (IT), they will utilize an increasing number of intelligent design applications. They are likely to include computer applications for designing special facilities, for designing with and detailing a range of construction technologies, and applying a wide range of performance analyses and/or simulations. Numerous other applications will emerge for design and detailing of proprietary products. Overall, this is the anticipated transition to knowledge-based design, where each of these applications encapsulate the knowledge in some specific area.

Given this growing diversity of applications, design will become increasingly characterized by the use of different specialized representations. We are skeptical of efforts to define a single representation or data structure able to support all important applications. Each varied representation has some shared data that is organized differently for each use. Some applications also require unique data not used in the others. Each organization of data is called a *class*. Because of the need to use many of these applications in an integrated way, we assume that integration facilities will eventually become available to support the use and dynamic extension of a suite of applications. We assume that these integration technologies will be centered around a building product model. Here, we consider a *building product model* to be a set of technologies supporting the integrated representation of a building.

R. Junge (ed.), CAAD Futures 1997, 45-59.
© *1997 Kluwer Academic Publishers.*

Because of the need for applications to use different classes, an important aspect of a building model are the routines that convert data from one class to another, called *maps*. Maps have received major attention recently in research [Verhoef, Liebich and Armor, 1995], [Khedro, Eastman, Junge and Liebich, 1996] and within the ISO STEP community [Bailey, 1996], [RPI, 1996].

In this paper, our interest is in the effective integration of applications with a building model during design. We are especially concerned with the process coordination involved in integrating intelligent applications, and to support the necessary information flows found in design processes, such as iteration, coordination and collaboration. In the next section, we examine and compare different methods of integrating applications with a building product model and their different capabilities and limitations. Then we turn to a finer grained examination of the structuring of application interfaces needed to support typical design processes. The work is presented as part of the ongoing work of the EDM research group, now at Georgia Institute of Technology, Atlanta.

2. Architectures for Integrating Applications

There are a range of methods for interfacing applications to a building model. Here, we review three different architectures. Each is described, then their capabilities and limitations are reviewed.

Because a building product model may include more than just a description of a building, but also maps and other data exchange technology, we distinguish between the *building data model*, consisting only of the structure describing the building, from the *model environment*, which includes the building data model and other associated exchange technologies. We also distinguish between two types of maps. Some maps are between an application and the building data model. They are partially outside of the model environment and thus are called *external maps*. Other maps are between entities that are within the model environment and are called *internal maps*. All maps are of one of these two types.

2.1 INTERFACE ARCHITECTURE CRITERIA

There are a wide variety of issues that can affect the usefulness of how applications are interfaced with a building product model. This review is based on five kinds of criteria:

Visibility of Conversion Logic: A shortcoming of most data exchange processes is the lack of visibility regarding what data will be produced. For example, in a system that uses only NURBS surfaces, when mapping to another system using a variety of surface types do all surfaces remain NURBS or converted to their simplest equivalent? Often, more than one conversion is possible for an entity. One conversion may be preferable during early design stages (the simplest) and another (the most accurate) at later design stages. In practice, hand tuning is often required and often some custom programming when such exchanges are expected to be repeated. Also, the instances to be exchanged may not be visible for review. By making all class conversions internal to the model environment, the user potentially has access to controls for all aspects of the exchange process.

Allowing Updates From Multiple Sources: Data exchange processes have focused almost exclusively on file-to-file exchange among pairwise applications. In design, a repository allowing multiple applications to build up a building model, from which others can read and then write, has a long history as a desired alternative[Carrera and Kalay, 1994]. However, the updating of such a model is complex, and supported differently by each approach.

Allowing Incremental Updates: In addition to the general mode of update, part of a practical working design environment is to control the entity instances examined within an application, by selecting or filtering the desired set. This reduces the size of exchanged datasets and allows focus on specific issues, for example in dealing with coordination of changes.

Extensibility of the Building Data Model: Extensions to the building data model allow changes to the building's enclosed activities, to the construction technologies used to build it, or to the kinds of analysis applied to it. Several researchers, including the co-authors, believe this is a fundamental requirement for a building model supporting design [Eastman,1991],[Galle,1995]. Yet some particular difficulties of making extensions to the building data model will be described, for which easy solutions do not exist.

Support for Collaboration: Collaboration methods include identifying changes made to the design by a user, or the differences among multiple proposed changes by different users. Other collaboration actions include methods to flag or propagate changes, automatic routing of changes for review and "whiteboard" tools allowing multiple people to share and update design data. Each of these collaboration capabilities requires certain facilities which may not be provided by the integration architecture. For example some types of collaboration require explicit representation of multiple versions of the data. Others require change propagation.

Each of these criteria are affected by the integration architecture. We now turn to the three architectures for linking applications with a building data model.

2.2 DIRECT MAPS TO/FROM A BUILDING DATA MODEL

What appears to be the simplest and most direct way of interfacing applications with a building data model is to implement maps directly to and from it. The approach is characterized in Figure One. Direct mapping means that maps both read and write the data structures of the application program and also must convert data to and from the structure of the building model. Such an approach can be used for pairwise data exchange as well as a repository for general integration. Direct mapping is the most common means to exchange data between two or more applications through a neutral representation. It is the kind of mapping most strongly supported by the ISO-STEP technologies, using application developed maps to an EXPRESS-defined building model [Bloor and Owens, 1995, Part 4].

48

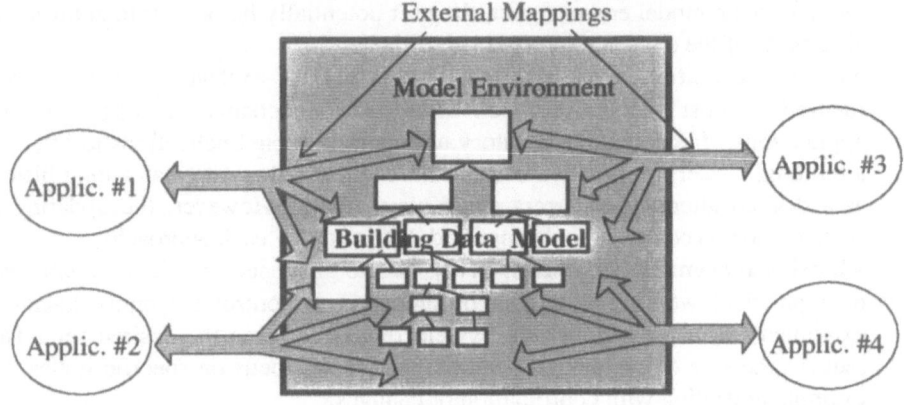

External Mappings

Figure One: A central building data model, with multiple applications mapping directly to/from it.

In direct mapping, the building data model is structured to support a specific set of applications. Typically, it will carry any information that is shared by two or more applications. The maps must do all class conversions needed between the model classes and the application classes. Because application data structures are only accessible in limited ways, the maps will be custom coded typically in a low level language such as C or C++ and the conversion of instances from one class to another will not be visible to users. The maps require significant coding effort and are not easily modifiable. To the degree that the new applications can access the data they need in the building model, new applications can be added. However, extending the central model to support new applications is usually not attempted, because the changes are likely to make incompatible the already existing maps.

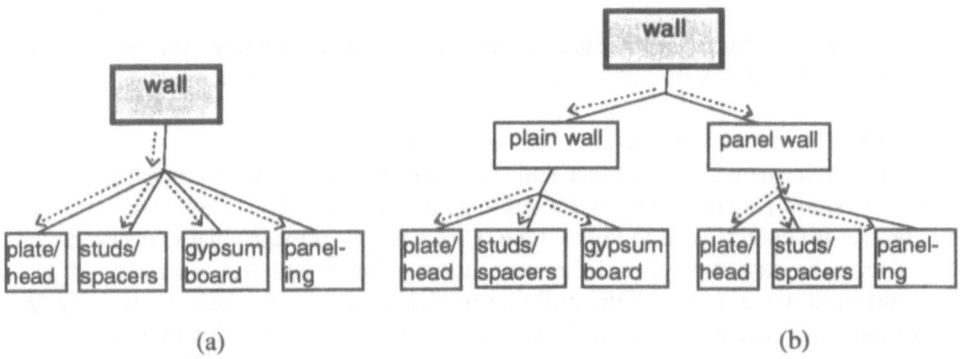

(a)   (b)

Figure Two: A wall model which is modified to accept an additional application. The addition may make maps for existing applications invalid.

To explain this limitation, we offer a small example. Suppose a wall is initially defined with a composition of its construction components, as shown in Figure Two(a). Access to the part attributes of a wall are through pathnames that access the wall then its part attributes, following the dotted arrows shown in Figure Two(a). Later, zones within the wall have to be added, representing, for example, areas with different energy transmission properties. Adding these areas changes the access paths from the wall to its components, probably invalidating maps using them. Such changes would invalidate all mapping schemes known to the authors, including those proposed as extensions to EXPRESS. Thus while this approach allows limited extensions to be made, others are not supported.

Since the data stored in the building data model may be used by multiple applications, updates by applications should be cognizant of all the relationships in the building model that are needed by the applications. Because of the potential impact of such updates, in practice they need to be carefully reviewed before they are applied. For example management procedures might be developed to review all updates before they are accepted [Eastman and Shirley, 1994]. Such building models support strong vertical organizational structures, but not wide shallow networks emphasizing collaboration.

Maps typically extract all instances of the entity types used in an application. They may have conditions, for example in deriving loads, that only consider loads larger than some figure; thus they may have capabilities to filter selections of data instances. However, these filters are coded into the map and are not user tailorable. There is talk of developing facilities for user selection of entities to which maps apply [Hartwick,1997]. Currently, if an update is made by one application, then other applications must extract a full dataset and check manually what has been changed or learn through external messages.

2.3 MAPS TO DERIVED VIEWS FOR APPLICATION INTERFACES

A standard feature of relational database management systems is derivable views, a computed subschema of the database. The data in a view may be any subset of the database or derivable from it. A view is custom-built to provide the data needed for an application. In conventional use, the database manages the consistency of all view data, in that views are not stored, but computed from the building model as needed. Thus the view data is not redundant and cannot be inconsistent. This approach is different from the direct mapping method, described above, in that data extraction is a two-step process: (1) the first derives the needed data into a view, then (2) the data is mapped to the application. Both maps may do class conversions.

Application updates to the building model can be approached in either of two ways, with different impacts. One way is to allow an application to make an update back to its view. This is only possible in certain cases, in that many complex derivations cannot be reversed. For example, if an application uses the areas of spaces and these are derived from a floorplan and the application changes the area of some spaces, how should the floorplan be updated? Similar problems are encountered in view updates in object-oriented databases [Kim and Kelly, 1995]. The alternative way is for all updates to be limited to one or a few special applications, which can write to the building data model. This can work when a CAD system is used to create all data and all other applications read the data in order to analyze it. It is assumed that the special application can invert derivations and check that other relations are consistent. This is the approach characterized in Figure Three.

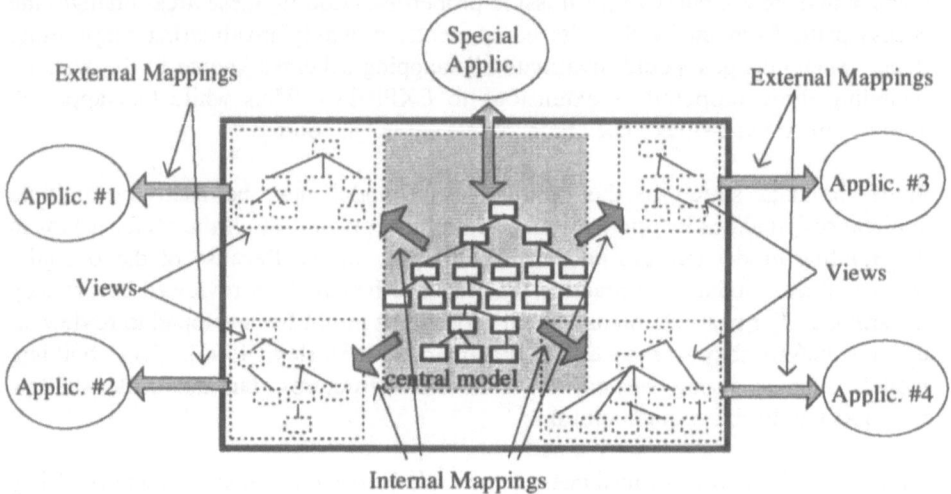

Figure Three: A Building model with derived views. The views can extract data from the building model. However, if updates are made, they must be done centrally, so the building model can manage the consistency

The map deriving a view, which is built into most database management systems, greatly simplifies the data extraction from the building model and can make some of the class conversions. Using the query language of the building data model, the model environment can extract and format the information making it easy to convert to the application through the external map. The external map is much smaller than in direct mapping, but still needs to be written in a low level language. Adding an application can be supported by writing new view generators, to the degree that the data they need can be derived from the building data model. Extension to the building data model requires massive changes, for the same reasons as the direct mapping approach.

The query language also supports user selection of instances, based on criteria such as timestamps that identify just-changed data. This can significantly facilitate coordination among a team of designers, because an update can carry just the data of significance, not the complete dataset. Additional, stronger forms of collaboration are not supported however; application data is not duplicated in the model; there is no notion of versions. Like the first approach, all design conflicts should be resolved before an update is made to the building model, forcing issues of coordination and collaboration to be made outside of the model.

It should be noted that even though SQL supports views, EXPRESS, used in ISO-STEP [Schenk and Wilson, 1993], does not. Bailey has defined an extension to EXPRESS, called EXPRESS-M, that supports views [Bailey, 1996]. Recently, EXPRESS-M has been merged into a larger mapping extension called EXPRESS-X [RPI, 1996].

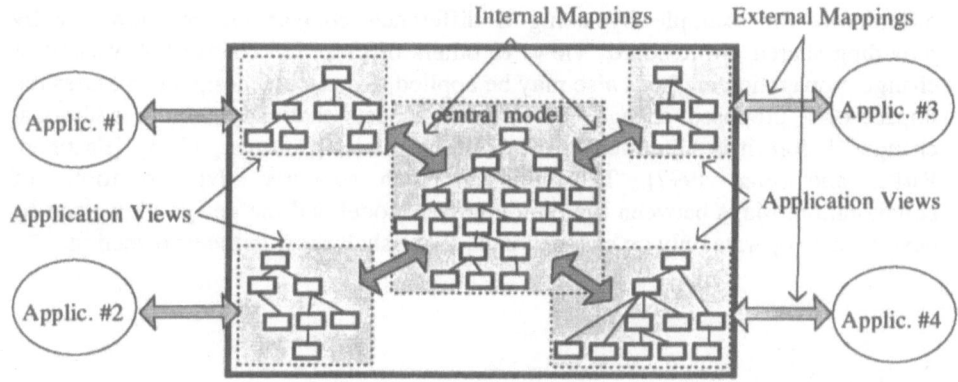

FIGURE Four: A schema architecture relying on design views which have their own models, and a central building model that carries equivalent DE classes.

2.4 SEPARATE DESIGN VIEWS WITH A CENTRAL MODEL

A problem with direct mapping is that the external map, which is probably written by the application developer in a low level language and thus is not public, makes all class conversions. What if, instead, the external map is simply a direct transfer of the application model into a format readable by the building model, and the building model makes all type conversions in internal maps. In this approach, external maps involve defining an isomorphic map from the application to/from an equivalent structure in a shared building model language. No class changes are required here. All the class conversions are made in the internal maps to/from the central model. The internal maps are part of the building model environment, and thus are visible and possibly adjustable. In such an approach, we call the application data model's view in the building model the *design view*.

There are several ways that the data within an design view may be organized. One is to define separate design views and a central building data model. To extract information into an application, it is first derived from the building model, with any class conversion needed, to the application's view, similar to standard derived views. However, the design view is stored (sometimes called in the database literature a "materialized view"). The application then accesses the design view data, does its design operations and eventually updates back to the design view. At this point, this view can be compared with the building data model or with other views -- as an issue of coordination. After review, all or some of its data may be mapped to the building model for later mapping to other design views. The central model carries the currently authorized building definition. This schema architecture is shown in Figure Four.

Here, all class conversions are made in internal mappings, which are part of the model environment. This allows them to be made visible and adjustable. Because redundant and potentially inconsistent design views may be carried in the model environment, collaboration tools can be developed to support resolving the differences, for example by doing a difference comparison on views or by providing shared "whiteboard" views to others of proposed changes. A variety of change propagation methods also may be applied that allow a designer to assess the impact of a proposed change, by flagging aspects that may require additional changes, before it is submitted to the building model [Eastman, 1996], [Eastman, Parker and Jeng, 1997]. Thus this approach supports advanced forms of collaboration. Maps between the building data model and the design views can be developed to operate only on changed data, as we shall show in the next section.

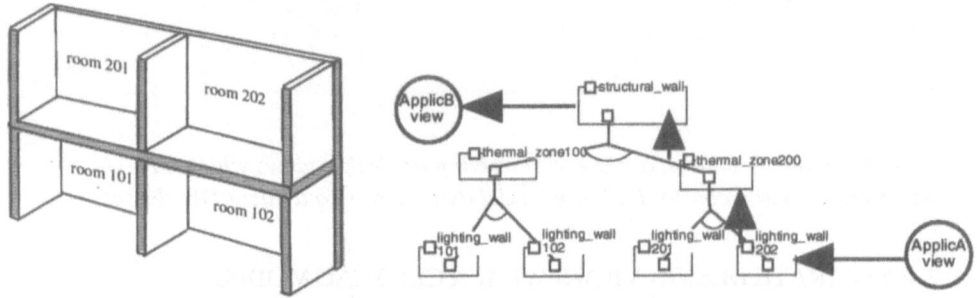

Figure 5: A wall is defined at several level of aggregation. Updates are made to the walls at all levels, by different applications. Maps are needed within the building data model to maintain the consistency of the different descriptions.

A problem not addressed in the first two approaches involves some issues of consistency maintenance within the building model. When some part of the design is created or modified, some other parts must also change. When the derivation is fixed and clearly defined, then there is not a problem. All data models are able to support derived data. However, in other cases, there is no single derivation path. For example, a wall may be part of several aggregated larger walls and any of these walls may be updated. This implies that in some cases, maps must exist within the building model, going in both directions between various entities in an aggregation lattice. A change in one level in the lattice may require updates to both the aggregations of the changed entity and its parts, and also derivable data. An example is shown in Figure 5. In the figure, a large wall is defined for dealing with structural shear, which is decomposed into the wall bounding each floor, representing energy zones. These are further decomposed into walls bounding singles spaces for lighting. Then if a single wall model, Application A, receives a window change, the translation to the aggregate wall in Application B, requires two maps between levels of aggregation and two class conversions, as shown in Figure 5b. In this arrangement, the translation of an entity from one design view to another may be a sequence of both changes of class within the same generic entity and a change in aggregation level. Translation becomes the concatenation of a series of individual class-to-class maps.

These maps allow changes to be managed up and down the aggregation lattice within the building model. For further examples, see [Eastman and Jeng, 1997]. Such maps are especially important if the building model is considered a repository for multiple applications, which incrementally update the model over time. For a detailed presentation of the schema architecture based on design views, see Jeng and Eastman [1997].

Most often entity classes close to those in each design view will exist in the central model, particularly with concern to level of aggregation. Maps between design view and building data model will only deal with class conversions. If a new aggregation level is introduced by an added application, it must be added to the building data model when using the other two approaches. However, here the building model classes need not be revised if a new application requires some special classes, not supported in the central model. The classes at different levels of aggregation can be optionally defined within the design view and the maps placed there rather than the central model. For such cases, the central model structure does not change and the access paths of other application maps need not change. However, if multiple design views need the class data, it should probably be moved to the central model, allowing only one set of aggregation maps to serve all applications, rather than a separate set within multiple design views.

2.5 SUMMARY ANALYSIS OF INTERFACE ARCHITECTURES

The three architectures for interfacing applications to a building data model are summarized in the columns of Figure 6. The desired capabilities are shown in each row, along with the capabilities and/or limitations of each approach.

As seen in Figure Six, direct mapping is primarily a file-to-file exchange that does not support well iterative translation between applications or collaboration. It seems most appropriate during transitions between design stages. It requires strong review processes and policies if multiple updates are to be made to a central model. A derived view approach supports updates by one or a limited number of privileged applications. Some maps are visible, while others are not. In other ways, it is similar to direct mapping. The approach based on design views provides strong visibility of all aspects of the conversion process. It also strongly supports all forms of updates and collaboration. It does not strongly support model extensibility, though it is better in this regard than the others. EDM-2 relies on a building model architecture using design views, the third alternative [Eastman and Jeng, 1997].

	Direct Mapping	Derived Views	Design views w/ Central Model
Visibility of Conversion Logic	not visible; all in external maps;	limited visibility of those maps in view generators;	completely visible in internal maps;
Allowing Updates From Mutiple Sources	should reflect all aspects of the design; requires careful preview;	must be made by special application; updates by multiple applications not supported	updates made to isomorphic view; any application can make updates
Allows Incremental Updates	not supported	supports incremental reading, but all updates are limited	supported

Extensibility of Building Data Model	not practical	not practical	some extensions supported
Support for Collaboration	no support	limited support	strong support for collaboration

Figure 6. Table depicting the three application architectures reviewed and their response to the five criteria introduced in Section 2.1.

3. INTERFACING OF APPLICATIONS

Once an application architecture has been selected, many functional aspects of data exchange are determined. Other details of the application interface, however, determine the fine grained flow pertaining to the functionality of an individual update. Our interest is in the development of application interfaces that particularly support collaboration among a group of designers. We assume that there are a number of applications that both read data from the building data model and write to it, possibly multiple times. These interfaces must support iteration and design revision. Here, we review the issues and identify criteria for such an interface. Then we describe alternative strategies for achieving the criteria. We assume that the building data model is based on the multiple design views approach presented in the last section, and shown diagrammatically in Figure 4. That is, data exchange consists of isomorphic external maps and class varying internal maps.

General criteria for an application interface should include:
a)   an application should be able to generate an initial part of a design and submit it to the building model; if not changed by another application, the initiating application should be able to retrieve the submitted work and continue, as if it was stored locally. That is, *the mappings between the building model and application should be isomorphic and complete*;
b)   the application must be able to update part of a design, in a manner allowing modification of specific design entities that are identifiable within the building model, and that can be distributed back to other applications. To do this, *the external mappings must allow maintenance of the identity of the entities it receives and updates*;
c)   updates by internal maps to the building data model and then to other applications must maintain the identity of entities, so that changes made by other applciations can be propagated back to other applications that use the data; that is, *entity identity musy be maintained throughout the building model;*
d)   *the application interface should support making incremental design changes to the product*, which includes modifying previously defined entities, adding new entities, deleting entities.

A higher level and more general requirement is that the model environment be able to provide significant aid in managing the integrity of the design, as relations are defined, assessed or analyzed and satisfied. Some aspects of this capability is provided by EDM-2's constraint management capabilities [Eastman, 1996], [Eastman, Parker and Jeng, 1997]. This capability can apply to any data carried within the building model.

3.1 SEMANTIC DEFINITION OF A DESIGN VIEW

An important top level issue is the semantic characterization of a design view. Is it simply an interface to an application, supporting all uses of that application? What about multiple tasks undertaken by different team members, using the same tool? What about using the same application on different parts of the design?

We consider an design view to be a unit of integrity, a unit of process and possibly a unit within the organization. A design view is a unit of integrity because an update made to a design view is assumed to be consistent within the view but possibly not across the whole building model environment. It is a unit of process, corresponding to a design task, in that two different design tasks may be undertaken with the same tool, but arrive at inconsistent designs, which must be resolved. Thus each task defined for a project, even though using the same tool as other tasks, should have its own design view. A design view is a unit of organization, in that it is notified when the integrity state of its data changes. It also has responsibilities in maintaining the consistency of the description within its view. Reporting and responsibilities suggest that it corresponds to a unit of organization.

In summary:
- a separate design view should be defined for each assignable design task;
- the same design view should be used when revisions are made based on that task
- sometimes tasks get revised, when the task scope significantly changes, it should be reflected in a new design view;
- multiple tasks using the same application should each have their own design views, for collaboration purposes.

3.2 INSERTS FROM AN APPLICATION AND LATER UPDATES

Applications can update a building data model in at least two different ways. One is that each time an application executes, including updates, it writes out the complete dataset, as if it were a new one. Another way is to make incremental updates to just the modified entities. Regardless of how the application update works to its own saved files, we consider it important that within the building model, support is given to incremental modifications to an existing set of entities. This has two implications: (1) partial updates designate only the design data that has changed, and (2) the identity of modified entities can be maintained across design views and the building data model.

Identity identification in incremental updates can be accomplished in two ways:
(1) use the entity IDs that are carried within the application itself. These will exist in all applications that themselves do incremental updates. The application IDs

may change from session to session if incremental updates are not supported. In this case, a table of ID conversions is required when entities are read from a design view to the application[1].

(2) Alternatively, a new ID may be associated with entities when they are read from a design view. This requires that someplace within each entity data structure there exists a slot that can carry the ID.

These two methods can be mixed, in that different application interfaces operating on the same product model can use different strategies to maintain object identity. In EDM-2 the initial interfaces have been defined using an additional ID added to each entity. The ID is added by the building data model whenever it receives an instance without an ID.

3.3 TWO-WAY MAPPINGS

Once an application has created an entity instance, all applications that read that data and possibly update it, must be able to report back to the building data model what instance has been modified, which ones deleted and also identify new entities that are added to the design view.

Since the entity IDs are assigned by the building model, not by the application, this provides a means to identify newly added entities. All entities written by the application to its design view that have IDs were earlier read from the building model; these are not new. All those without entity IDs are new and require the insertion of a new entity instance.

In reading from a design view, only some of the entity instances in the view may be read. When an update is made, all those changed can identified by doing a comparison of the old values with the new ones. If no change was made by the application, no update is made. All updated values also are flagged. In addition, however, some entity instances may be deleted and these cannot be distinguished from those not Read into the application. In neither case are they written back. In order to make this distinction, each entity instance within a design view has a flag field. If it is read for a session, the flag is set. All updates involving a change are flagged one way, while no change turns off the flag. Any entity instances whose flag remains set after an update have been deleted by the application.

---

[1] It is important that no operations change the entity ID numbers during an application session. For example, the Bentley Microstation© COMPACT operation changes IDs when it deletes flagged entities. This would corrupt the ability to read and write back entities while maintaining their identity.

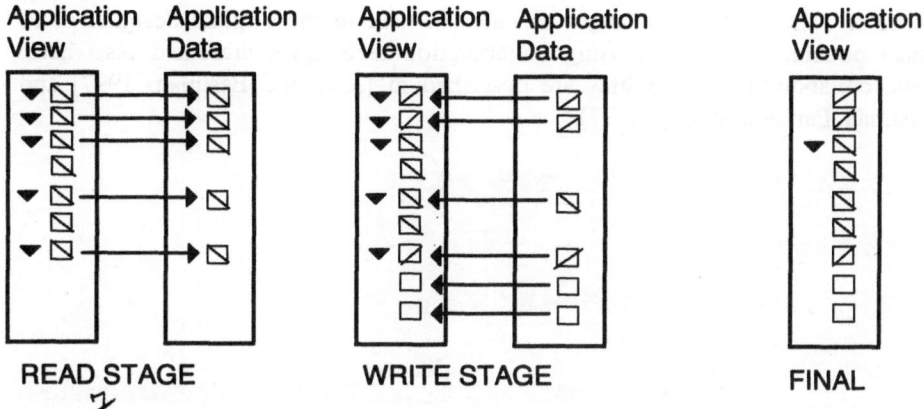

Figure 7. The management of updates from an application back to a design view.

Figure 7 diagrams an example of a design view and its transfer to the external application, its update and final condition. Within each stage, the design view and application dataset are shown as enclosing boxes. Within them, each entity is shown as a small box. In the Read Stage, each entity instance read is flagged, shown to the left of each entity. A slash in a box indicates that it has an ID. Because entities are assigned IDs by the building model, all entities read from the design view have them. After the application runs, it makes an update back to the design view. If the entity instance written back has been modified, its slash has been reversed. Three entities were modified. All entities with IDs are written back and the flags (triangular on left) in the design view are eliminated. Some entities are written back without IDs and these must be new. One entity is not written back and after the Write Stage, it is left with its flag on. This indicates deletion.

These bookkeeping methods allow unambiguous updates from an application to a building model, supporting addition, modification and deletion. They cover all cases of incremental updates, except one. If an incremental update is to be made, the subset of design needs to be determined at the time of reading from the application's design view. When data is written back to the design view, all the data must be written, so as to allow deletions to be distinguished. Partial write-backs from the application are not supported.

3.4 EXAMPLE
The initial capabilities of the incremental update operations and their support for collaboration can be demonstrated in their implementation in EDM-2. In the example, an initial designer has laid out the service core of an office building. This is written to the application's design view, then passed onto the building model. This base design is then passed to several other designers who will review and detail various parts. One of these is a person who will review and detail the stairway. The stairway portion of the design is mapped into this person's application design view, then externally mapped into the application. The stairway designer determines that one landing must be widened and another stair tread added. The person updates the design view, where the changes and additions are identified. At this point, the person making the changes is notified, as shown in

58

Figure 8.   The record of changes can also be sent to the original designer or to others potentially affected, using collaboration techniques presented elsewhere. Some collaboration capabilities are described in [Jeng and Eastman, 1997] and [Eastman, Parker and Jeng, 1997].

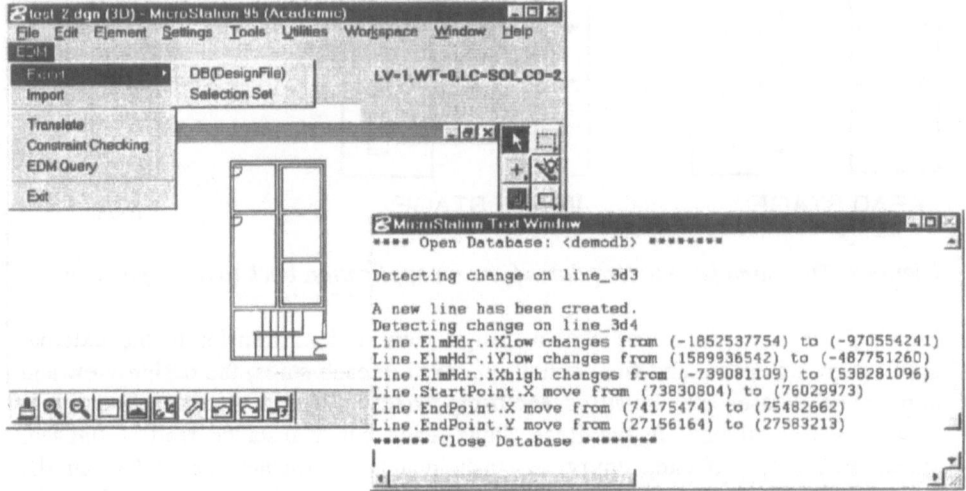

Figure 8. Example screen after a designer has updated a portion of the design of a building's service core, with a building model window in the application environment showing the effects of the update.

4. Summary

As technology advances to support multiple designers using heterogeneous applications, many issues arise in the realization of an effective multi-user collaborative design environment. These issues include: supporting visibility of the conversion logic going to an application; allowing updates from multiple sources; supporting incremental updates; facilitating extensions of the building model to support new applications, and support for collaboration. We have shown that the architecture how applications are integrated with a building model greatly affect these issues. We believe that an architecture based on design views, our appraoch number three, is best able to support the needs of collaborative design. We have also presented details showing how to support incremental updates and flagging low level design changes.   This capabilitiy appears to provide a framework potentially supporting high-level forms of colaboration.

Our overall goals are to develop the needed capabilities for creative team design both at a distance and temporally separated, and also for enhancing the support for designprocesses that respond to much higher levels of complexity.

NOTE: This work has been supported by the National Science Foundation, grant No. IRI-9319982.

REFERENCES

Assal, H and C. Eastman, [1995] "Engineering Database as a Medium for translation", 1995 CIB W-78 Symposium, Stanford, Calif.

Bailey, Ian, [1996] EXPRESS-M Reference Manual, ISO TC184/sc4/wg5 N243, CIMIO Ltd, Brunel Science Park, Surry, England, 12 August, 1996.

Bloor, Susan, and J. Owens [1995] Product Data Exchange, UCL Press, 1995.

Carrera, G. and Y. Kalay (eds.), [1994], Knowledge-Based Computer-Aided Architectural Design, Butterworths-Heinemann Press, N.Y..

Eastman, C.M., [1992], "A data model analysis of modularity and extensibility in building databases", Building and Environ., 27:2, pp. 135-148.

Eastman, C.M. and G. Shirley, [1994], "The management of design information flows", in S. Dasu and C. Eastman (eds.) Management of Design: Engineering and Management Processes , Kluver Press, N.Y.

Eastman, C.M., M.S. Cho, T.S. Jeng and H.H. Assal, [1995] "A Data Model And Database Supporting Integrity Management",1995 ASCE Intern. Computing Congress, Atlanta, GA.

Eastman, C.M., H. Assal, and T. Jeng [1995] "Structure Of A Product Database Supporting Model Evolution", 1995 CIB W-78 Symposium, Stanford, Calif.

Eastman, C.M. [1996] "Managing Integrity in Design Information Flows" Computer Aided Design (May, 1996), 28:6/7, pp.551-565.

Eastman, C.M., D. S. Parker, T.S. Jeng, [1997] "Managing the Integrity of Design Data Generated by Multiple Applications: The Principle of Patching", Research in Engr. Design (in process).

Eastman, C.M. and and T.S. Jeng, [1997] "A Database Supporting Evolutionary Product Model Development for Design" Automation and Construction (in process).

Galle, Per, [1995] "Towards integrated 'intelligent' and compliant computer modeling of buildings", Automation in Construction, 4: 3, (October), pp. 189-211,

Hartwick, M. D.Spooner, T. Rando and K. Morris [1997], "Data Protocols for the Industrial Virtul Enterprise", IEEE Internet Journal. 1:1 available on
http://www.computer.org/nternet/9701/harwick970.html

T.S. Jeng and Eastman, C.M. [1997], " A Database Architecture for Design Collaboration" TeamCAD Conference Proceedings, Georgia Institute of Technology, Atalanta May 12-13, 1997.

Khedro, T. C. Eastman, R. Junge, and T. Liebich, [1996] "Translation Methods for Integrated Building Engineering, ASCE Conference on Computing, Anaheim, CA, Ju, 1996.

Kim, Won, and Wm. Kelly, [1995] "On View Support in Object-Oriented Database Systems", in Modern Database Systems: the Object Model, Interoperability and Beyond, W. Kim (ed.) Addison Wesley, 1995, pp. 108-129.

Rensselear Polytechnic Institute[1996] ISO TC184/SC4/WG5, EXPRESS-X Reference Manual, Working Draft, May 28, 1996, Lab for Industrial Infrastructure, Rensselear Polytechnic Institute, Troy, NY.

Schenk, D.A. and P:R, Wilson [1994], Information Modeling the EXPRESS Way, Oxford U. Press, N.Y.

Verhoef, M., T. Liebich, R. Armor, [1995], "A Multi-Paradigm Mapping Method Survey", Proc. CIB Workshop on Computers and Information in Construction, M Fisher, K. Law, B. Luiten (eds.), Stanford, U, pp. 233-247.

WHAT ARE WE LEARNING FROM DESIGNERS AND ITS ROLE IN FUTURE CAAD TOOLS

JOHN S GERO
Key Centre of Design Computing
University of Sydney

Recent research into the activity and behaviour of human designers as they design has provide an impetus to carry out research which underpins the development of new CAAD support tools. However, there are computational processes of interest in designing which are not modeled on human design processes. This paper outlines some of the design processes which are being researched based on our understanding of human designers and provides examples from some early implementations.

1. Introduction

Given the large body of design research it is surprising how little we know about designing: the activity carried out by designers. There has been an upsurge in interest in studying human designers formally. Much of this interest has been driven by the efforts of the artificial intelligence in design research community, which has developed a range of computer-implementable designing processes (Gero and Sudweeks, 1994; 1996). However, more recently, studying human designers has been founded on the development and later formalisation of experimental methods based on protocol analysis (Ericsson and Simon, 1993; van Someren et al., 1994; Cross et al., 1996).

Protocol analysis provides a rich source of information on designing as a time-based activity. This then allows design researchers to develop richer models of designing based on the behaviour of human designers. These models in turn provide the basis for a better understanding of designing. Such an understanding then feeds into the development of computer-based support tools for computer-aided architectural design.

Section 2 briefly introduces some results from studying human designers. Section 3 proceeds to describe a computer implementation of a model of designing based on developing cross-domain analogies drawn from the putative behaviour of human designers. Section 4 introduces the notion ambiguity and emergence in images and describes a computer implementation of a model of designing based on utilising such notions.

R. Junge (ed.), CAAD Futures 1997, 61-70.
© *1997 Kluwer Academic Publishers.*

2. Results from Protocol and Related Studies

Protocol studies are a means of obtaining data from verbal utterances. Designers are asked to "think aloud" while they are designing. While they are designing they are video and audio taped. The designer's verbal utterances are transcribed. The transcription is then used, along with design theory, to develop a coding scheme. The transcription in then coded and finally analysed. The step are listed below:

- taping
- transcription
- code development
- coding
- analysis

Protocol studies are providing detailed evidence of how designers spend their time as they are designing. At a gross level a designer's time can be spent either on postulating solutions, called structure, or in reasoning about the function and behaviour of possible or postulated designs. Figure 1 shows a typical distribution of the time spent between these two large classes of activities by experienced designers. It is interesting to note that it is almost twenty minutes into the session, for this design, before any structure is proposed.

Such behaviour is in contrast to that of inexperienced designers who appear to need to "put pencil to paper" very early in order to have something to work with. This is exemplified in Figure 2.

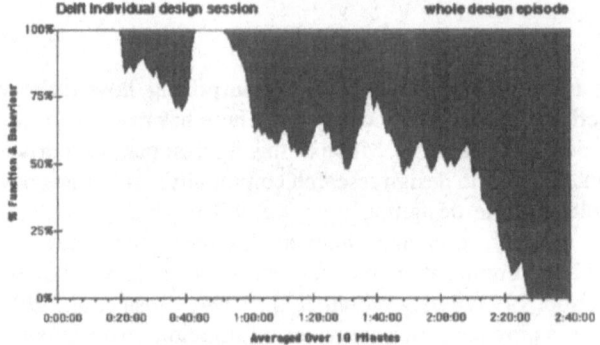

The percentage of time spent in reasoning about Function and Behaviour as compared to Structure, calculated at one minute intervals and averaged over periods of ten minutes.

Figure 1. Typical plot of distribution of time spent on function and behaviour (light), as against structure (dark), for experienced designer (Gero and McNeill, 1997).

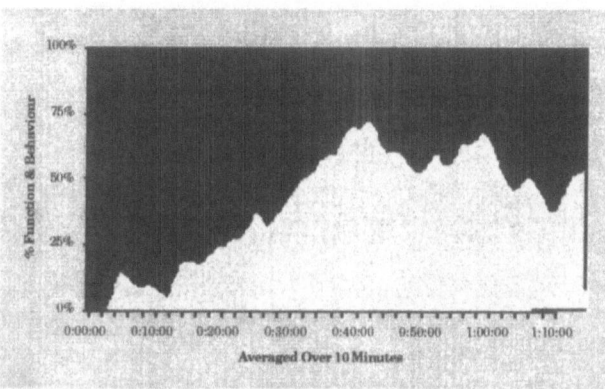

Figure 2. Typical plot of distribution of time spent on function and behaviour (light), as against structure (dark), for inexperienced designers (Gero and McNeill, 1997).

Such studies of designers support previous cognitive studies of human problem solving in areas such as analogy, fixation, emergence and visual ambiguity. They provide detailed information on how designers use such concepts in the development of their design proposals.

3. Using Cross-Domain Analogies

Analogy has long been recognised as an approach in the elicitation of design ideas. In architecture, the most obvious analogies are visual analogies. Here, the surface similarities of the structure are used and transferred to a new design, as shown in Figure 3. However, deep rather than surface similarities require a different approach to their recognition. Similarities associated with function and behaviour provide the locus for deep similarities. While function specifies what a design does, behaviour specifies how it achieves it functions. We may extend the definition of analogous designs to:

"Two designs are analogous if they have a similar function or behaviour; they may or may not have a similar structure" (Qian and Gero, 1996).

Thus, the aim of analogy-based design is to obtain new ideas of possible structures from an existing design with similar functions and/or behaviours. The use of deep similarities has the potential to allow analogies to be drawn not just from designs in the same domain but from designs in quite different and apparently unrelated domains. It appears that humans may reason about function and behaviour in order to draw cross-domain analogies if we subscribe to the function-behaviour-structure (FBS) characterisation of designing (Gero, 1990).

64

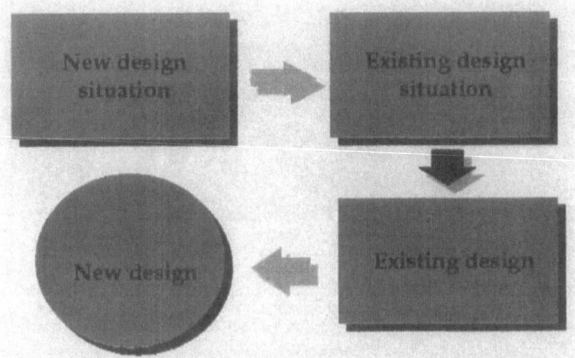

Figure 3. Model of design by analogy based on concepts derived from human designers.

There are at least four modes of reasoning about function (Qian and Gero, 1996):

FBS Type I: achieved by static behaviour, Figure 4
FBS Type II: achieved by a state of dynamic behaviour, Figure 4
FBS Type III: achieved by a set of behaviours occurring contemporaneously, Figure 5
FBS Type IV: achieved by a set of behaviours occurring sequentially, Figure 5.

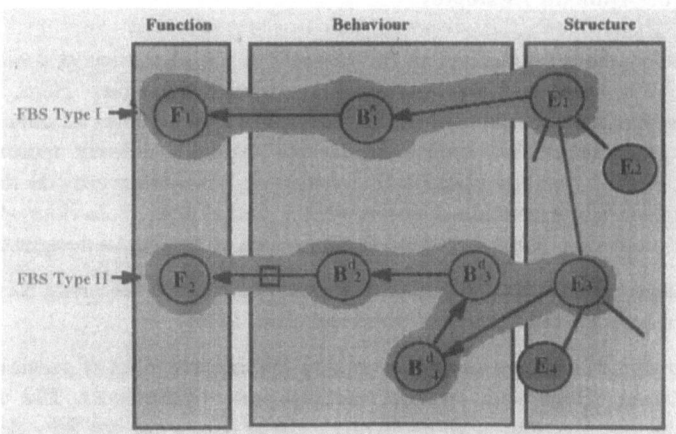

Figure 4. Achieving functions by behaviour paths Types I and II.

The structure is composed of elements, E, from which behaviours, B, are derived. In turn, the functions, F, are derived through the behaviours. The role of the different modes of reasoning about function is to develop a basis for looking for similarities amongst designs which are dissimilar on the surface or structure level.

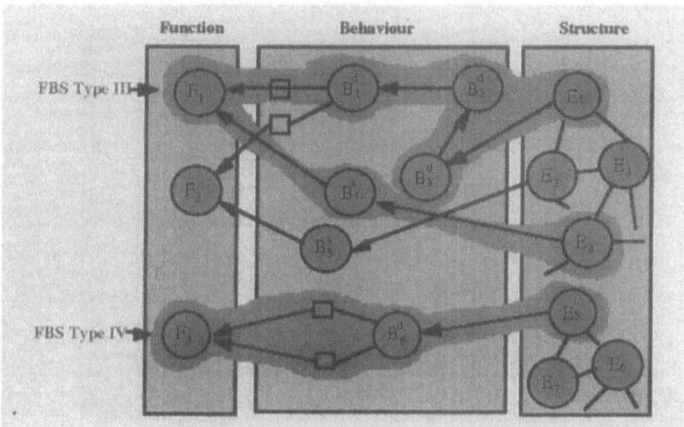

Figure 5. Achieving functions by behaviour paths Types III and IV.

A computer-aided design system built using this paradigm demonstrates in Figures 6 and 7 that it produces cross-domain analogy conjectures. Figure 6 shows that in redesigning doors the system has made a conjectured analogy with the design of an existing domestic water tap. From this figure the proposal is unclear, however, in Figure 7 the proposed analogy is clarified at the structure level. The conjecture is that the hinges of the door could be replaced by the screw mechanism of the water tap.

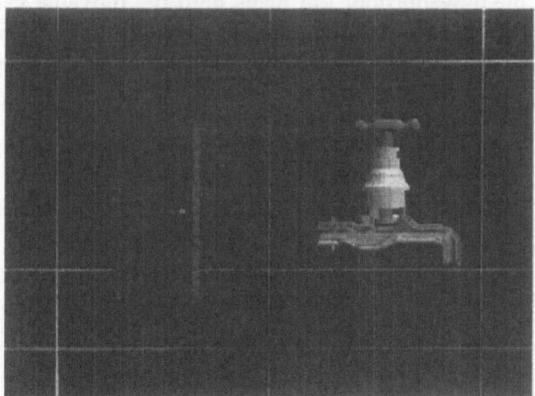

Figure 6. Suggestion by computer system when designing a door that a design analogy could be drawn with a water tap.

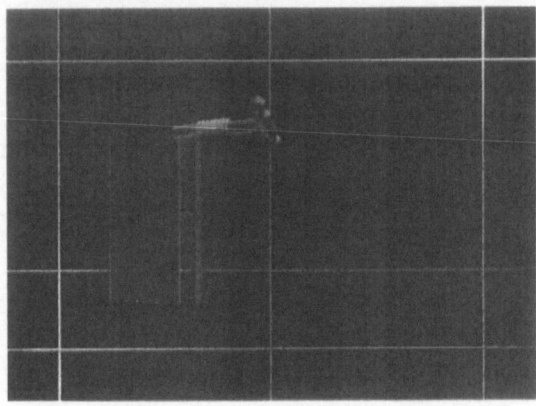

Figure 7. Proposal by computer system to replace door hinges by water tap screw mechanism to produce a new door design.

Figure 8 shows in graphical form the system's reasoning behind the conjecture. This graphical form is derived from a reasoning process concerning the ways both designs achieve their functions and then examining their respective behaviour paths. From this isomorphism the structure mapping is generated. It is only the structure mapping that is shown in Figure 8, ie the door's door frame maps onto the tap's pipe as fixed elements; the movable door leaf maps onto the movable valve; and the hinges that produce rotation which controls the amount of door opening map onto the screw that produces linear motion which controls the amount of tap opening.

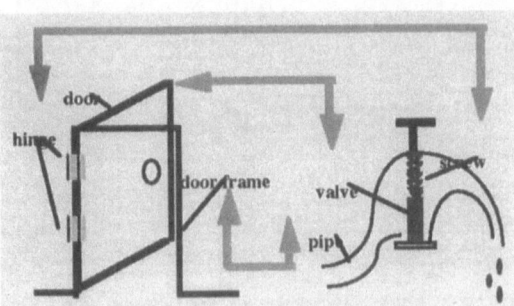

Figure 8. Explanation by computer system, in graphical form, of how the door hinges can be replaced by the water tap screw (after Qian and Gero, 1995).

4. Using Ambiguities and Emergence

Architects when they are designing use shapes to outline the spaces of interest. At the conceptual stage of designing whilst the designer is still exploring possibilities, it is important that the shapes not be fixed and unique but rather that they be open to a variety of interpretations (Schön and Wiggins, 1992). Some artists have used the concept of multiple interpretations as the basis of their work. One of the most effective artists of this genre is M. C. Escher, Figure 9. There are many interpretations possible of this image: white angels on a black background; black 'devils' on a white background; rotationally symmetric shapes; scaled shapes; and so on. Each of these interpretations may be viewed as *emergent*. The interpretations are a result of ambiguities in the representation.

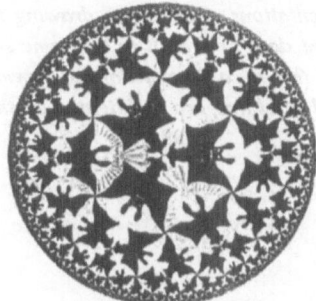

Figure 9. M. C. Escher's Circle Limit IV.

Figure 10 demonstrates this issue. Both the drawings in Figures 10(a) and (b), produced using AutoCAD, appear to be the same, however, selecting the same point in each drawing does not result in the same lines be selected. The actual lines that are consequentially selected depend entirely on the underlying representation and not on the image of the drawing. In paper and pen 'representations' of drawings this issue does not arise since human designers construct in their heads the representation necessary to support the activity they wish to carry out.

The need to support a variety of interpretations creates new requirements for CAAD systems. These new requirements change the expected behaviour of any computational system which is used to support designing. Such computational systems must have the capacity to provide multiple representations of the objects of interest and have the capacity to allow for ambiguous representations. The difference between multiple and ambiguous representations is that with multiple representations there is a set of unique representations for the image, each representation may be used directly. With ambiguous representations there is an interpretation phase required before the representation can be used. Figure 11 shows an example of multiple representations of a square. Each of the representations can be used differently.

68

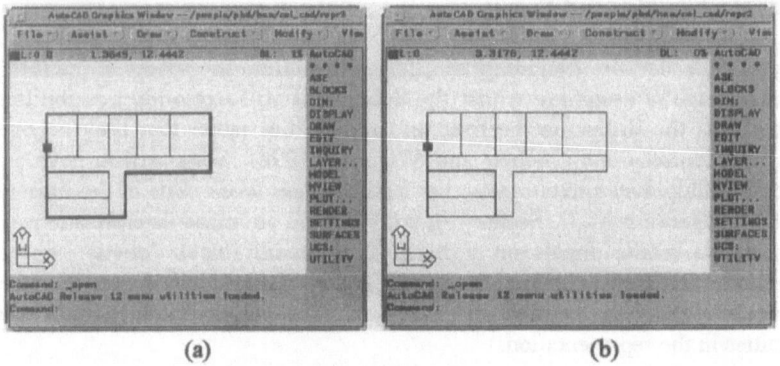

Figure 10. Different representations of the same drawing in a computer-aided design system. Selecting the same point does not result in the same set of lines being selected. The resulting selected lines(a) and (b) are a function of the representation used and not of the image as it appears. The selected line are highlighted by thicker lines (Jun and Gero, 1997).

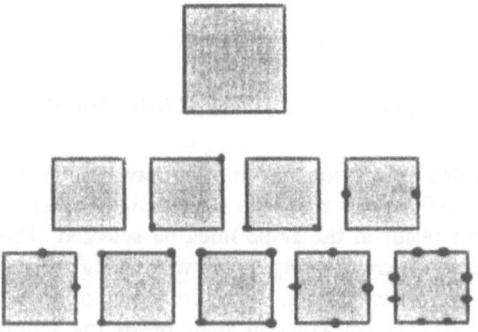

Figure 11. Multiple representations of the image of a square.

Figure 12 shows an example of an ambiguous representation of a triangle, where the triangle is represented by numerous competing 'edgelets'. With such a representation there is a class of possible triangles represented rather than just the single instance which was drawn initially. The effect of representing classes instead of instances that multiple instances can be derived from the class. This plays an important role at the early stages of designing when the designer does not want to be specific about the details and wants to work at the general level. Class based representations provide an opportunity for this to occur. Previously, it was considered that ambiguous representations should be avoided but it appears that they have a useful role to play in designing.

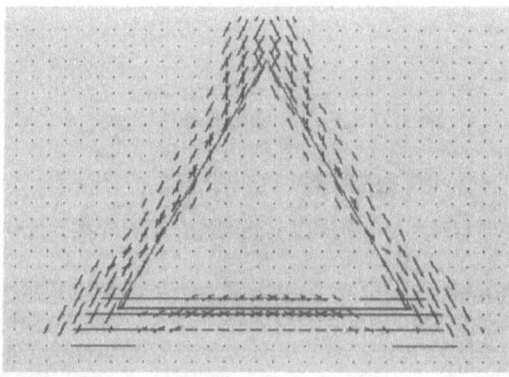

Figure 12. Ambiguous representation (Tomlinson and Gero, 1997).

There are a number of class-based representations being developed. The most common is based on the object-oriented paradigm, although this has some difficulties. Other approaches are based on qualitative representations.

5. Discussion

Recent research into how humans design is providing a rich source of both well-supported as well as anecdotal evidence on which to base computational design paradigms. The effect of utilising such as an approach is not to model human designing processes but rather to use such processes as the bases of analogies for the development of computational processes. Humans are not computers so we should not expect that computational design processes would be models of human designing processes. As we attempt to develop such computational tools we are also developing a better understanding of designing as an activity. We are not limited, however, to basing our tools on analogies with human designing processes. We can construct computationally-based tools founded on conjectures with processes unrelated to human designing: evolutionary systems represent one such approach. We can look to other concepts unrelated to such conjectures and postulate processes founded of axioms and derivations from those axioms.

Acknowledgments

The research described here has been supported by grants from the Australian Research Council, the Australian Postgraduate Awards, the Overseas Postgraduate Awards and the University Research Grant. Computing resources have been provided by the Key Centre of Design Computing.

70

References

Cross, N., Christiaans, H. and Dorst, K. (eds) (1996) Analysing Design Activity, Wiley, Chichester.

Ericsson, K. and Simon, H. A. (1993) Protocol Analysis: Verbal Reports as Data, MIT Press, Cambridge, Massachusetts.

Gero, J. S. (1990) Design prototypes: a knowledge representation schema for design, AI Magazine 11(4): 26-36.

Gero, J. S. and McNeill, T. (1997) An approach to the analysis of design protocols, Design Studies (to appear).

Gero, J. S. and Sudweeks, F. (eds) (1994) Artificial Intelligence in Design'94, Kluwer, Dordrecht.

Gero, J. S. and Sudweeks, F. (eds) (1996) Artificial Intelligence in Design'96, Kluwer, Dordrecht.

Jun, H. and Gero, J. S. (1997) Representation, re-representation and emergence in collaborative computer-aided design, in Maher, M. L., Gero, J. S. and Sudweeks, F. (eds), Preprints Formal Aspects of Collaborative Computer-Aided Design, Key Centre of Design Computing, University of Sydney, Sydney, pp. 303-319.

Qian, L. and Gero J. S. (1995) An approach to design exploration using analogy, in J. S. Gero, M. L. Maher and F. Sudweeks (eds), Preprints Computational Models of Creative Design, Key Centre of Design Computing, Department of Architectural and Design Science, University of Sydney, Australia, pp. 3-36.

Qian, L. and Gero, J. S. (1996) Function-behavior-structure paths and their role in analogy-based design, AIEDAM 10: 289-312.

Schön, D. and Wiggins, G. (1992) Kinds of seeing and their functions in designing, Design Studies 13(2): 135-156.

Tomlinson, P. and Gero, J. S. (1997) Emergent shape generation in design using the boundary contour method, CAAD Futures'97 (this proceedings).

van Someren, M., Barnard, Y. and Sandberth, J. (1994) The Think Aloud Method: A Practical Guide to Modelling Cognitive Processes, Academic Press, London.

GRAMMARS AND ART

A Contingent Sense of Rules

DEAN BRUTON
Department of Architecture, Landscape and Urban Design
University of Adelaide
North Tce Adelaide South Australia 5000

This paper contributes to the debate about the utility of the grammatical paradigm in art and design. It reports an investigation of the contingent sense in which grammars and grammatical design apply in the practice of form making in art using two complementary research strategies: the examination through a perspective of grammatical design of some selected bodies of art work, including interviews with artists, theorists and designers; and the reflective practice of image making with computer media in my own work as an artist. The major hypothesis is that a contingent sense of grammar can facilitate the creation, understanding, and discussion of form-making in art. The sub-hypotheses are that (1) An understanding of grammatical design can enhance a reflective design activity, and that (2) Revealing the contingency of grammars can expose moments of inspiration and redirection in a reflective design activity.

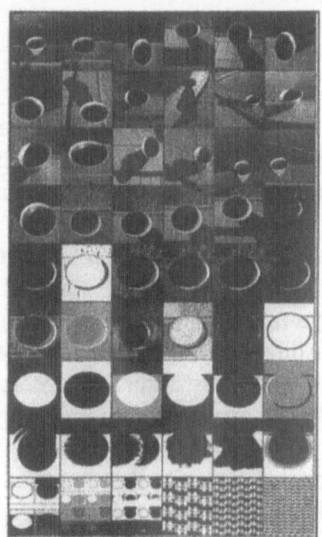

Figure 1. *Bowl Grammar Series*

71

R. Junge (ed.), CAAD Futures 1997, 71-82.
© 1997 Kluwer Academic Publishers.

1. Introduction

Formal computational grammatical programs are few while many critics of grammars are often without personal hands on experience. This paper explores the first hand use of computational grammatical programs to generate two- and three-dimensional forms. It investigates the idea of grammars through examining bodies of work from a grammatical perspective, accepting the idea of grammar and rules. It concentrates on the views of some of those who use the idea of grammars: how these ideas relate to art. The work of some selected artists that have been the subject of grammatical analysis are discussed to illustrate the application of grammatical perspective in a formal way.

Special consideration is given to the field of art, particularly painting to enhance links between design fields, especially art and architecture. Ideas of *grammar* and *gram - matical design* and the associated idea of *rule* are used and located within discourses of artists and designers. Ideas of *contingency*, are highlighted, first as a concept and then within the specific discourses of art, architecture, and grammatical design. The idea of rules in art is then discussed. It then reviews some of the positions taken about the nature of grammars as computational systems; as encapsulating knowledge of style and form; as a frame of reference in the composition of shape and form in designing and in discourse about art and design. The contingency of grammars is then discussed to show how rules are used to develop awareness of hermeneutical moments and metaphorical truths.

Contingency is understood as both "in addition to" and "depend upon" the moment of grammatical design decisions and judgements. A plurality of ideas of grammars in art and design is emphasised: when two people refer to grammars, they may be referring to very different concepts. It concludes by supporting a view of the utility of grammars contingently in art and design.

2. Grammar

The idea of grammar in art and design draws on analogies between "visual languages" and "natural languages". The role of grammar in natural language is to independently facilitate the communication of meaning by providing and organisational structure. Universal principles of natural language were put forward by Noam Chomsky (Chomsky 1957). His theory of transformational grammar was based on a system of internalised rules capable of generating an infinite number of grammatical sentences. For generative grammarians in linguistics, a grammar refers to the entire system of structural relationships in a language, viewed as a set of rules for the generation of sentences. This idea of transformational generative grammars emerged with the use of computers for the analysis of natural language, concentrating on formalist approaches.

The use of *grammar* as an analogy or metaphor in art and design arose with the mathematical work on production systems of Post (Post 1943), and was developed for spatial design by George Stiny and James Gips (Gips and Stiny, 1980). Pioneering work on picture languages (Kaneff 1970) was initiated by Russell Kirsch (Kirsch 1966). The formalisation of grammars of art and design as a computational system following Chomsky's ideas was developed in the early work of Stiny (Stiny 1975; Stiny 1976; Stiny and Gips 1972; Stiny and Gips 1973), Downing and Flemming 1981), Flemming (Flemming 1978; Flemming 1981; Flemming 1986; Flemming 1987a; Flemming 1987b; Flemming et al. 1988), (Oksala 1979) and Gips (Gips and Stiny 1975; Gips and Stiny 1980). In 1966, Lionel March used grids to restrict the set of dimensional values used in his art work (March 1981). By 1980, Stiny's proposal "develops the idea that a language of designs can be defined from scratch by rules which apply to a vocabulary of building elements" (Stiny and Mitchell 1980, 416). Using Froebel's building gifts, George Stiny proposed a *constructive* approach to languages of design by means of shape grammars declaring it "will ultimately replace the kindergarten method in the studio and in practice" (Stiny and Mitchell 1980, 416). The formal theoretical machinery for the definition of languages of two- and three dimensional spatial design was established. A revised history of grammars in art and design is reported in (March and Stiny 1985). Raymond Lauzzana documents a general history of colour and visual formalism (Lauzzana 1993; Lauzzana 1994).

Although these ideas were subsequently explored and applied in architecture and design by the CAAD research community (Chase 1989; Eastman 1973; Fawcett nd, Gero 1984; Hanson and Radford 1986; Tapia 1996; Valkaló and Liou 1996; Welsh 1989; Wojtowicz and Fawcett 1986), artists and art discourse has been slow to develop the metaphor of shape grammar as a basis for studio practice. Examples in art are often early explorations of proof of concept (Cohen and Cohen 1984; Colan 1995; Edmonds 1992; Grabska 1995; Kirsch and Kirsch 1988; Knight 1993; Knight 1994; Lansdown and Earnshaw 1989; Lauzzana and Pocock Williams 1988; Makkuni 1986; Makkuni 1988; Petrovich and Tanaka 1994; Stiny 1981; Stiny and Gips 1972; Tarransky 1995).

3. Contingency

A contingent sense of rules is proposed as a vital component for the development of art practice. A *contingent sense* refers to an ability to appreciate insights into a specified state of affairs in designing. Both self-judgment of design through reflection on designing, and self-reflection on one's values as one judges the design as it develops, maintain a distinct sense or acquired meaning of the moment of insight.

In adopting the term *contingent*, I am following the work of Barbara Herrnstein Smith in literature who emphasises that contingency is a starting point , not a reason for abandoning a paradigm: "If we recognise that literary value is "relative" in the sense of *contingent* (that is, a changing function of multiple variable) rather than *sub - jective* (that is personally whimsical, locked into the consciousness of individual subjects and/or without interest or value for other people), then we may begin to investigate the dynamics of that relativity. Such an investigation would, I believe, reveal that the variables in question are limited and regular—that is, that they occur within ranges and that they exhibit patterns and principles—and that in that sense, but only in that sense, we may speak of "constancies" of literary value" (Herrnstein Smith 1988, 12). Arguing that literary value is "radically relative and therefore 'constantly variable'", Herrnstein Smith reconsiders notions of relativism as not a conviction but a conceptualisation: "Relativism, in the sense of a contingent conceptualisation that sees itself and all others as such, cannot be found, ground, or prove itself, cannot deduce or demonstrate its own rightness, cannot even lead or point the way to itself" (Herrnstein Smith 1988, 183).

For Paul-Alan Johnson contingency is 'the anticipation of uncertainty, the chance that something might or might not happen, the possibility of an accident or unforeseen event or circumstance; we say for example, " we must be prepared for any contingency." In mediaeval Latin it meant circumstance, the context which surrounds an event and, like context, makes contact with it, hence its etymology from *tangere*, to touch, such that to be *contingent* is to depend upon. Its opposition with uncertainty seems to be reconciled by the tangent, the fleeting touch of one upon another, accidental or incidental yet consequential"(Johnson 1994, 343).

Using this conceptualisation of relativism contingency may be applied to metaphors of grammar to instigate new vistas of exploration. It is in the nature of grammars in design that they are continually modified, adapted and invented depending on the situation, and this contingency is what makes the idea of grammars rich and productive. Thus the context of the design situation determines the extent that grammars may be found and the grammar or grammars that appear to be operating. Similarly, when seen from a contingent perspective, grammars are never a complete representation of a design; they are contingent to design.

The first sub-hypothesis focuses on the process of form making:

(1) An understanding of grammatical design can enhance a reflective design activity. Jürgen Habermas describes how rational reconstructions may contribute to self knowledge: "Self reflection leads to insight due to the fact that what has previously been unconscious is made conscious in a manner rich in practical consequences: analytic insights intervene in life. . . . A successful reconstruction also raises an "unconsciously" functioning rule system to consciousness in a certain manner; it renders explicit the intuitive knowledge that is given with competence with respect to the rules in the form of 'know how'" (Plant 1984, 23). This contention that interpretation of each moment of the design continuum is enhanced by recognition and understanding of rules for form making is particularly relevant in education. The novice is encouraged to be aware of patterns and codes of behaviour in her/his own and other's work, and hence to be in a better position to reflect on (and perhaps change) these patterns and codes.

A recognition that the contingency of grammars means that these may be turning or change points in the progress of design leads to the second sub-hypothesis: (2) Revealing the contingency of grammars can expose moments of inspiration and redirection in a reflective design activity. While grammatical design understanding can contribute through the identification of codes and patterns of behaviour, the points where these patterns change is also significant. These are points where the design is being re-framed, where different possibilities are seen.

4. Rules

In natural language grammatical "rule" construct a sentence. Grammatical metaphors may use rule in a variety of formal guises. Visual language analogies such as shape grammars use rules that describe a vocabulary, operations, a start state for transformations that produce derivations in a design space. March and Stiny recognised that rules were a basis for languages of designs: "designs derived from a properly specified rule system, a grammar, constitute a language. Transformations of rules provide a translation from one language to another" (March and Stiny 1985) . A discourse emerged that developed an articulate system of rules that describe design behaviour, a behaviour previously tacit and opaque. Using that system to enhance artistic practice requires an acceptance of the metaphor of rule. Writers on grammars typically use a language based on the metaphor of rules that may choose to incorporate intervention and displacement. Schön's work links design behaviour and reflection uniting the concepts of art, understanding, hermeneutical circle and historical relativism (Schön 1988). Despite criticism that the foundations of shape grammars are logical and atomistic, shape grammarians utilise many discourses by adhering to a pragmatist view of design that contributes to all kinds of picture language discourses. Stiny urges a reconsideration of grammars and computation: "That is one of the things that in many ways is distressing about some of the negative commentary that we get about grammars and computation, because people think too narrowly about it. They tend to think that it is a combinatorial activity, where you have got a bunch of your little pieces that you are moving around a board, in a mechanistic way, or that it is something like playing chess. But it is not like that at all, computation is a much more general kind of enterprise where as much as the process is to do with picking out what you move and refining it and constantly reconfiguring dynamically what you are playing with. It is an entirely different kind of situation" (Stiny 1996).

4.1 ARTISTS THAT USE GRAMMARS IN THEIR WORK

Three types of artists use grammars. First there are those that use grammars explicitly. Few artists fit this category because a knowledge of computer systems such as LISP has been required. Exponents such as Lionel March, Harold Cohen (Fifield 1995), Raymond Lauzzana and Alvy Ray Smith, use grammatical systems to generate conventional experimental patterns and illusions. Russell and Joan Kirsch described a grammar for the work of Richard Diebenkorn and Joan Miro. The idea of a visual pattern language was established but not developed by artists except informally in mainstream image production.
The second group of artists use grammars informally. When they discuss their work

they can identify elements, rules, operations and derivations of an image. Often ideas of themes, sequence transformation and documentation guides their work, especially in painting. For example, in the figurative painting, internationally renowned New York artists, Jennifer Bartlett (Johnson 1995) and Philip Pearlstein (Bowman (1983) work in these terms. In the digital image field, Los Angeles artist, Bill Barminski describes his work in traditional painting and internet media as corresponding to rules and visual relationships (Bar. An informal use of grammar as a production metaphor seems to guide artists of this type.

Thirdly, a group of artists that use grammars for the grammatical production of forms is developing. These artists seek a refinement of their understanding of self-creation. Form making is a means of self reflection and discovery. This group merge both the logic of grammatical formalism with philosophical hermeneutics. Holistic notions of an event are reflected upon in conjunction with the metaphor of grammatical design. This approach offers significant gains in terms of formalisation of major concerns and development of key form making ideas and personal interests. Grammars are understood as a catalyst as well as a logical system for form analysis and generation. For example the *Bowl Grammar Series* (Figure 1) represents an early experiment that reveals moments of insight through reflection on the narrative and arrangement of the forms, using a grammatical frame of reference. Elements of the image became identifiable for further exploration. A vocabulary of images refined future possibilities. Operations were recorded and available for evaluation and repetition. Meaning and understanding develop through play. Elements of the *Bowl Grammar Series* were selected and used in Tartan Worlds, a grammatical computer program that generates forms based on defined rule sets. The results are then used for further experiments outside of Tartan Worlds.

By documenting changes in personal ideas and practice through reflection on grammars and grammatical formalisms in art, moments of insight are found such as the formal possibilities of black and white; the colour filter variations; fragmentation and distortion operations; and repetition as abstraction. Further computer operations could grammatically develop many of these early images and also provide scope for refinement and addition to holistic understandings. Images that are generated from rule sets without intervention may be contrasted or combined within a manipulated grammar adding to the complexity and unpredictability of outcomes.

In reference to Donald Schön's (1983) suggestion that designers know more than than can say, Johnson suggests intuition is a key factor in design decisions: "This knowledge in action is the one ability architects and designers have that artificial intelligence will probably never achieve: the capacity to both change minds and to accommodate the mind changes of others, or changes obliged by shifting circumstances, and to then intuit the right path in the midst of wrong clues" (Johnson 1994, 345). My experiments use Schön's approach, one that reflects on action as it happens and symbiotically experiments with the situation adjusting received theory and counteraction. Rather than argue against the use of algorithmic interpretations of the world, a metaphor that incorporates a reflective hermeneutical dimension is assumed.

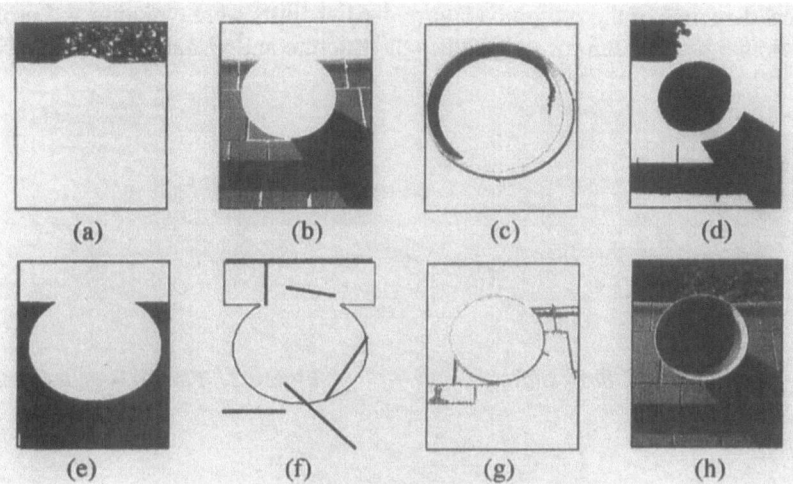

Figure 2. *Finding elements of a vocabulary*

4.2 EXPERIMENTS IN FORM MAKING

Using a grammatical frame of reference, the *Bowl Grammar Series* demonstrates the influence of the shape grammar mechanism. Finding elements in the vocabulary requires reflection on the key thematic directions for the communication. Decisions depend upon a holistic understanding of the moment as much as upon the mechanism in play. Interpretation was necessary for selection of fundamental qualities of the image (h) from a range of possibilities (e-g). The large range of possibilities requires a selection of a vocabulary to refine the qualities and overall direction of the work. Early Tartan World experiments established the possibility of using the entire image as a symbol in conjunction with red and blue shapes (Figure 3).

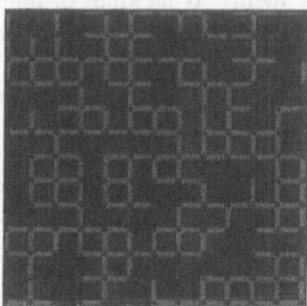

Figure 3. *Bowl Grammar Series used as a single vocabulary element*

Four elements of a vocabulary were chosen and used as symbols in the exploration of the computer program Tartan Worlds. Four simple rules were decided upon in rela-tion to the circumstances surrounding the initial form making experiment and the technical demands of the program (Figure 4–6). Results were dynamic and prolific. Derivation#8 was chosen for its qualities of structure and visual interest (Figure 6).

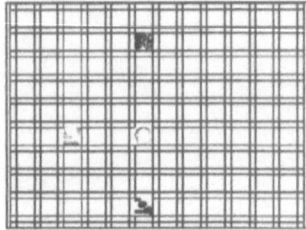

Figure 4. *Start State*

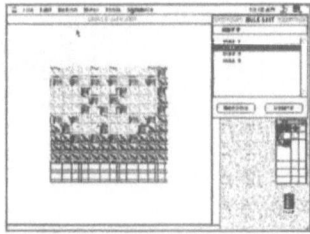

Figure 5. *Tartan World Screen*

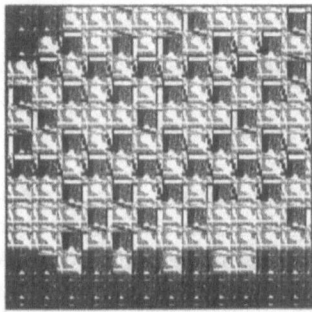

Figure 6. *Derivation #8 from BG Series*

Following the work of Terry Knight, (Knight 1980; Knight 1981; Knight 1981; Knight 1983; Knight 1983; Knight 1983; Knight 1988; Knight 1992; Knight 1993; Knight 1994; Knight 1996), key concerns are: How do artists go about formulating rules and how do they change them? What kinds of rules are kept and what sorts of rules are dropped out? The intention is not to prove or disprove that art or design is grammatical. The intention is to look at these sites from a perspective of grammars, without denying that they can be looked at from other perspectives. In looking at work from a grammatical perspective, there are certain qualities that stand out and are sought by the viewer. Clearly the interest centres on form and formal qualities such as continuity in the flow of ideas, colour and spatial relations. Because grammars imply a language there is an interest in membership of a set of similar works, and some kind of formal consistency amongst the works in the set.

Associated with grammars is the idea of rules. So a set of work is viewed with an interest in consistent compositional rules that might be inferred from the work. Derivations, particularly in serial art suggest a sense of progression of ideas from one member of the series to another. For Stiny ". . . the serious issue is what you have to do to add to the grammar or change the grammar to get another derivation that tells you something else that gives you more insight to carry on or expand what you have already. It is just like criticism. . ." (Stiny 1996). An acknowledgment of rule sets is often retrospective since artists tend to make up production rules as they produce (whereas designers tend to be more constrained by the rules of a client's brief). Rule sets are usually not consciously featured in an artist's production repertoire due to the view that invention occurs despite rather than because of, previous rules and traditions.

5. Discourse

Discourse is speech or language. It comes from the Latin *discurrere* and suggests movement "back and forth" or "to and fro". In linguistics, discourse analysis is often applied to the study of those linguistic effects — semantic, stylistic, syntactic — whose description needs to take into account sentence sequences as well as sentence structure . Discourse through speech and language develops through shared concerns, values and communication . The shared use of codes enables a linguistic community to evaluate through comparison and generate democratic values through notions of free information exchange. Thus discourse fosters shared perception and engenders reflection on aspects of self-creation.
Engagement in diverse discourses between communities is an integral part of the learning process. Communities of scholars debate the value and validity of discourses in relation to rational, privileged and hermeneutic qualities. As Richard Rorty and others have noted , discourse is impossible without some rationality and depends upon the nature of contingency within language(Rorty 1989). Discourses often overlap, interweave and shift ground. Thus the idea that there is a scientific art and an artful science is a mix of discourses that may enable new understandings.
Tacit knowledge provides rules about meaning: "Wittgenstein's well-known definition, "the meaning of a word is its use in the language," should be understood neither as a denial of meaning nor as an insistence on meaning as pragmatics. Instead, it suggests that we know the use or rule of language practically but not theoretically. 'And hence also "obeying a rule" is a practice. A nd to think one is obeying a rule is not to obey a rule. Hence it is not possible to obey a rule "privately": otherwise thinking one was obeying a rule would be the same thing as obeying it.' Even if I believe that I know the rules of a foreign language, I cannot prove that I really know them unless the other acknowledges it" (Wittgenstein 1958, 81e.) Quoted in (Karatani 1995, 137). Traversing from practical to theoretical realms through reflection encourages acknowledgement of language interpretation. Studies that search for grammars of creativity and principles of visual language follow a Cartesian discourse although some claim to maintain a completely "neutral" objectivity. Karatani submits there is no universal, neutral position between languages: "In the case of Saussure, however, he was aware that in the comparison or translation of two languages, the translator, even if bilingual, necessarily places himself or herself within one or the other of these languages at a time. There is no universal, neutral position between languages" (Karatani 1995, 138).

This view suggests there is no universal, neutral position between grammars. A position is contingently assumed for each utterance.

Discourse on art is often concerned with authenticity and interrogative practice that questions a status quo. Nevertheless, artists often seek acknowledgement of art institutions for some validation of their practice. Twentieth century distinctions between art and art so-called contribute to essential theoretical discourse. Sociological questions about the role of art establish a vital discourse based on notions of plurality and complexity which displace many mainstream aesthetic activities. Grammars provide a means for further recognition of these changes in the understandings of art practice.

6. Conclusion

This paper suggests that for artists, finding methods to discuss and use complex visual information can be assisted by formal grammars framed by the metaphor of a contingent sense of grammar. Through reflective action some strategies for identifying and defining moments of inspiration when rule-transformation or invention are demonstrated. By documenting changes in personal ideas and practice through reflection on grammars and grammatical formalisms in art and design, moments of insight are found to be more transparent and educationally germane. Through contingently incorporating ideas of vocabulary, rules, frames of reference and metaphors of grammars, a significant prototype is demonstrated for a more transparent art practice and vital visual education. As Steven Holtzman suggests, "Visual art can be thought of in terms of grammars just as can other languages and systems of communication. Though to many it is less intuitive to think of the visual arts as languages, to the extent that they are forms of communication they must have sets of rules that permit interpretation. . . Experiments to date have barely begun to define the rich visual languages that address colour, texture, the rules of composition, dimensionality, and visual structures in time" (Holtzman 1994, 190-1).

Chomsky's lead in the pursuit of a universal natural language grammar has been superseded by notions of many languages that are culturally dependent for their form making. The academic division between making forms and discerning their significance has extended since the early days of Chomsky's universalist structuralism. A contingent sense of grammar may contribute to self-judgement grounding art and design behaviour in legitimate societal concerns.

All of this concentrates on form. I end by quoting from Paul Klee: "We are artists, practical craftsman, and it is only natural that in this discussion we should give priority to matters of form. But we should not forget that before the formal beginning, or to put it more simply, before the first line is drawn, there lies a whole prehistory: not only man's longing, his desire to express himself, his outward need, but also a general state of mind (whose direction we call philosophy), which drives him from inside to manifest his spirit in one place or another. I emphasise this point to avoid the misconception that a work consists only of form" (Spiller 1970, 99).

References

Bowman, R. (1983). Philip Pearlstein : the complete paintings, Alpine Fine Arts Collection, New York.

Chase, S. C. (1989). Shapes and shape grammars: from mathematical model to computer implementation. Planning and Design, 162, 215-241.

Cohen, H., and Cohen, B. (1984). The First Artificial Intelligence Colouring Book, William Kaufman, Inc, Los Altos, CA.

Colan, C. (1995). Paul Klee Grammar, Masters Thesis, Univeristy of Massachusetts, Amherst.

Eastman, C. M. (1973). Automated Space Planning. Artificial Intelligence, 41, 41-64.

Edmonds, E. A. (1992). Knowledge-based systems and new paradigms for creativity. Modelling Creativity and Knowledge-based Creative Design, J. S. Gero and M. L. Maher, eds., Lawrence Erlbaum, New Jersey.

Fawcett, W. (nd) The Design Engine. , Working paper, Cambridge, UK.

Fifield, G. (1995). AARON (computer programmed to generate images). Art New England, 16, 5.

Flemming, U. (1978). Wall representations of rectangular dissections and their use in automated space allocation. Environment and Planning B, 5, 215-232.

Flemming, U. (1981). The Secret of the Casa Giuliani Frigerio. Environment and Planning B8, 87-96.

Flemming, U. (1986). On the representation and generation of loosely packed arrangements of rectangles. Planning and Design, 13, 189-205.

Flemming, U. (1987a). More than the sum of parts: the grammar of Queen Anne houses. Planning and Design, 14, 323-350.

Flemming, U. (1987b). The Role of Shape Grammars in the Analysis and Creation of Designs. Computability of Design, Y. Kalay, ed., Wiley Interscience, London, 245-272.

Flemming, U., Coyne, R., Glavin, T., and Rychener, M. A Generative Expert System for the Design of Building Layouts - Version 2. Artificial Intelligence in Engineering; Design (Proc. Third International Conference, Palo Alto, CA), New York, 445-464.

Gero, J. S. (1984) Knowledge Engineering in Computer-Aided Design. , North-Holland, Amsterdam.

Gips, J., and Stiny, G. (1975). Shape Grammars and their Uses: Artificial Perception, Shape generation, and Computer Aesthetics, Birkhauser, Basel and Stuttgart.

Gips, J., and Stiny, G. (1980). Production Systems and Grammars: a Uniform Characterisation. Environment and Planning B7, 399-408.

Grabska, E. J. (1995). Visual Evaluation in Design Space. (forthcoming).

Hanson, N. R., and Radford, A. D. (1986). On Modelling the Work of the Architect Glenn Murcutt. Design Computing, 1, 189-203.

Herrnstein Smith, B. (1988). Contingencies of Value, Harvard University Press, London.

Holtzman, S. R. (1994). Digital Mantras, The MIT Press, Cambridge Massachusetts.

Johnson, K. (1995). Jennifer Bartlett at Paula Cooper. Art in America, 83, 130.

Johnson, P.-A. (1994). The Theory of Architecture, Van Nostrand Reinhold, New York.

Kaneff, S. Picture Language Machines. Proceedings of Conference, Australian National University.

Karatani, K. (1995). Architecture as Metaphor: Language Number, Money, The MIT Press, Cambridge, Massachusetts.

Kirsch, J. L., and Kirsch, R. A. (1988). The Anatomy of Painting Style: Description with Computer Rules. Leonardo, 214, 437-444.

Kirsch, R. A. Picture syntax. Pattern Recognition, Proceedings of IEEE Workshop, Las Croabas and Dorado, Puerto Rico, 183-5.

Knight, T. W. (1980). The generation of Heppelwhite-style chair-back designs. Environment and Planning B, 7, 227-238.

Knight, T. W. (1981a). The forty-one steps. Environment and Planning B, 8, 97-114.

Knight, T. W. (1981b). Languages of designs: from known to new. Environment and Planning B, 8, 213-238.

Knight, T. W. (1983a). Transformations of languages of designs: part 1. Environment and Planning B, 10, 125-128.

Knight, T. W. (1983b). Transformations of languages of designs: part 2. Environment and Planning B, 10, 129-154.

Knight, T. W. (1983c). Transformations of languages of designs: part 3. Environment and Planning B, 10, 155-177.

Knight, T. W. (1988). Comparing Designs. Planning and Design, 151, 73-110.

Knight, T. W. (1989). Colour grammars: designing with lines and colours. Environment and Planning B: Planning and Design, 16, 417-449.

Knight, T. W. (1992). Designing with Grammars. CAAD Futures, G. N. Schmitt, ed., Vieweg, Wiesbaden, 33-48.

Knight, T. W. (1993). Colour Grammars: The Representation of Form and Colour in Design. Leonardo, 262, 117-124.

Knight, T. W. (1994). Transformations in Design, Cambridge University Press, Cambridge.

Koen, B. V. (1987). Definition of the Engineering Method, American Society for Engineering Education, Washington D C.

Lansdown, J., and Earnshaw, R. A. (1989). Computers in Art, Design and Animation, Springer-Verlag, New York.

Lauzzana, R. (1993). Chronological Bibliography of Visual Formalism. Languages of Design1, 279-283.

Lauzzana, R. (1994). Chronological Bibliography of Color Formalism. Languages of Design2, 353-364.

Lauzzana, R. G., and Pocock-Williams, L. (1988). A Rule System for Analysis in the Visual Arts. Leonardo, 214, 445-452.

Makkuni, R. (1986). A Representing the Process of Composing Chinese Temples, Working Paper, Palo Alto Research Centre.

Makkuni, R. (1988). Diagrammatic Interface to a Database of Thangka Imagery, Working Paper, Xerox Palo Alto Research Centre.

March, L. (1981). A class of grids. Environment and Planning B, 8, 325-332.

March, L., and Stiny, G. (1985). Spatial Systems in Architecture and Design: Some History and Logic. Environment and Planning B, 121, 31-53.

Oksala, T. (1979). The Language of Formal Architecture. Environment and Planning B, 6, 269-278.

Petrovich, L., and Tanaka, K. (1994) Visual Proceedings: The Art and Interdisciplinary Programs of SIG-GRAPH 1994. , Association for Computng Machinery, New York.

Post, E. (1943). Formal reductions of the general combinatorial decision problems. American Journal of Mathematics, 65, 197-268.

Rorty, R. (1979). Philosophy and The Mirror of Nature, Princeton University Press, Princeton, N J.

Rorty, R. (1982). Consequences of Pragmatism, University of Chicago Press, Chicago.

Rorty, R. (1989). Contingency, irony, and solidarity, Cambridge University Press, New York.

Schön, D. A. (1988). Designing: Rules, types and worlds. Design Studies, 93, 181-190.

Spiller, J. (1970) Paul Klee Notebooks Volume 1 The Thinking Eye. , Lund Humphries, London, 99-100.

Stiny, G. (1975). Pictorial and Formal Aspects of Shape Grammars on Computer Generation of Aesthetic Objects, Birkhauser Verlag, Basel, Switzerland,.

Stiny, G. (1976). Two Exercises in Formal Composition. Environment and Planning B, 3, 187-210.

Stiny, G. (1981). Design Machines. Environment and Planning B, 8, 245-255.

Stiny, G. (1996) George Stiny: Interview by Dean Bruton. .

Stiny, G., and Gips, J. (1972). Shape grammars and the generative specification of painting and sculpture. Information Processing, C. V. Freiman, ed., North Holland, Amsterdam, 1460-1465.

Stiny, G., and Gips, J. (1973). Formalization of Analysis and Design in the Arts. Basic Questions of Design Theory, W. R. Spillers, ed., North Holland Publishing Co, Amsterdam, 507-530.

Stiny, G., and Mitchell, W. J. (1980). Kindergarten grammars: designing with Froebel's building gifts. Environment and Planning B, 7, 409-462.

Tapia, M. (1996). From Shape to Style, Shape Grammars: Issues in Representation and Computation, Presentation and Selection, Ph.D dissertation, University of Toronto, Toronto.

Tarransky, A. (1995). Stella, Newman and Fra Angelico, Masters Thesis, MIT, Cambridge, Boston.

Vakaló, E.-G., and Liou, S.-R. (1996) Speculations on the Morphology of the Plans of Seven Ando Houses. , College of Architecture and Urban Planning, University of Michigan.

Welsh, M. A. (1989). Computer-aided Conceptual Ship Design System Incorporating Expert Knowledge, PhD Thesis, University of Newcastle Upon Tyne (United Kingdom).

Wojtowicz, J., and Fawcett, W. (1986). Architecture: Formal Approach, Academy Editions, London.

Woodbury, R., Radford, A. D., Taplin, P., and Coppins, S. Tartan Worlds: A generative symbol grammar system. ACADIA, 211-220.

SPATIUM

a system for the definition and design of shape grammars

MICHAEL HELLGARDT* and SOURAV KUNDU**
* *Architect, Amsterdam, The Netherlands.*
** *Bionic Design Laboratory, Kanazawa University, Japan.*

Abstract. It is shown how Augmented Transition Networks (ATN) can be gradually programmed with shape grammar structures. This work is inspired by natural language parsing. Another major reference is the space-between or spatium assumption. An application is given with a simulation of Palladio villas. Then is shown that ATN frames can be encoded in a way that allows their use without specific knowledge of computer modeling. Connections between human and machine learning are touched on.

1 Introduction

Research in natural language processing draws on our knowledge of grammatical inference, which is used for computational cognitive modeling, but acquired independantly from computational modeling. Though problems of grammar are an issue of the debate in architecture since long, a comparable body of knowledge, *shape grammar knowledge*, is neither given nor really studied in the realm of architecture. Here *shape grammar* is rather imported from foreign fields - namely computational modeling, AI and its application in design. This is presumably the main reason that the impact of shape grammar research on architectural practice and theory is modest, if existing at all. This situation, which we are not the first to raise (e.g. Coyne, McLaughlin and Newton, 1996), is challenging and impeding at the same time. Challenging it is because the possibility to mechanise knowledge is a challenge to knowledge itself. Here we refer to the knowledge about grammatical and syntactic principles underlying architecture, which is our specific case. Impeding the situation is to the extent that the new section of AI in design, shape grammar, is dominated, or even undermined, by the requirements of formal and computer modeling, rather than by its proper subject matter: *shape grammar*.

Some architects work with design formalisms (Eisenman for instance). This is still the exception, but perhaps the future architect will design essentially rather by computer progr-ammation than with a pencil etc. Computer modeling will increasingly play a role in architectural education, but we cannot expect any architect, or architectural firm, to be equipped with the formal knowledge required for computer programming and techniques of computational modeling. Shall these architects (presumably the majority) be excluded

83

R. Junge (ed.), CAAD Futures 1997, 83-96.
© 1997 *Kluwer Academic Publishers.*

from the exploration and investigation of shape grammar in architecture, irreversibly associated with computer modeling? The literature on shape grammar is rich but either too concrete, or too abstract. There exist grammars of particular architectural styles, or building types. And there are more fundamental investigations which have no answer, however, to the question what such singular grammars have in common. Both ways do not grasp the matter *architecture* seriously, and do not really meet the situation sketched above. Obviously a tool is required which neither already is a ready-made grammar nor too abstract, but which allows, or rather challenges the architectural practitioner, student and theorist to explore and enrich his own design experience and intentions by computer modeling and to transcribe that into computational devices. Such a tool encodes and defines grammars, and it supports the design of them. It is based on and enables computer modeling and programming, but it should not require special qualification in formal computational modeling.

Augmented Transition Networks (ATNs) are nondeterministic graph-searching programs, widely used for parsing natural language expressions (Winston, 1984; Graham, 1994). This formalism is an appropriate tool for the representation of architectural shape grammars too (Kundu and Hellgardt, 1996). The present paper proceeds with his approach and shows its transcription into the device proposed: a Shape Grammar Design and Definition System.

2 A shape grammar ATN

2.1 THE EMPTY ATN FRAME

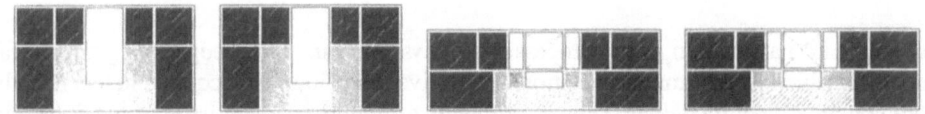

Figure 1

Figure 1 shows paraphrases of Palladio villa floorplans, generated by rules, called *patch-rules*, that substitute color values in pixel-maps (for reasons of screen printing represented as hatched patches). The leftmost figure, a paraphrase of the villa Badoero, underlies a rule sequence loggia-staircase: fill a bottom zone of a specified debth entirely with value loggia, then fill an angle region, if given, entirely with value staircase. The same sequence reversed leads to the obviously unacceptable result to the right. In the angle context of the rightmost figure, a paraphrase of a planimetric study by Palladio from the RIBA archive, the same reversed sequence results in the probably acceptable appearence to the left. Such reversals are not possible in deterministic, or top-down models.

The nodes of nondeterministic networks, such as ATNs, are surrounded by *clouds* of possible choises. The notation of a node reflects that (figure 2). A node has one or more outgoing arcs which invoke functions to be evaluated. When a node is called in a network its first arc is evaluated and its remaining arcs are stored. These stored options are activated by a *continuation-function* as long as no result from other arcs is given, or in case that more than one solution is desired, for instance for exhaustive search. In our application all arcs are of the form f(g(x)), where x is the register containing all information available at a given stage of network evaluation, g is the current and f the next function to be evaluated with the result of g(x). In terminal arcs the f, the next node is empty, or of the class of the network where a network is embedded in (subsection 2.4.3). To this arc notation a tree notation of the network corresponds. The branches, or arcs of this tree are bidirectional, which means that we climb up and down in it from category to instance, and vice versa.

```
(defnode node                (defnode category_index
   (arc_1)                       (continue category_1 <category-rules_index>)
   ...                           ...
   (arc_n))                      (continue category_n <category-rules_index>))
```

Figure 2, Basic ATN structure (empty ATN frame)

Nodes and arcs are defined as macro functions. Obviously all next functions in the arcs must be encoded as nodes. For the current functions that is less obvious. Corresponding to grammatical structures to be discussed below, rules are addressed as rule categories by the next-nodes (f), and as instances of these categories by the current functions (g). These rule instances are clustered to rule-sets (marked by <...>; brackets, 'defnode' and 'continue' are omitted in figure 5 and 7). They can be encoded as ordinary functions. This appears to have technical advantages. Its main advantage, however, in terms of codification is that rules, more exactly: calls of rules, defined independantly from some ATN, can be inserted into that ATN. With that condition, to be discussed separately in section 3, an ATN is user-friendly. All the user has to do is to enter names of rule-categories and -sets in a user dialog.

2.2 AN EMPTY SPATIUM GRAMMAR ATN

2.2.1 *A remark on grammar*
In some influential papers of the 'first generation' (cf Stiny, 1980a) is assumed that grammars can design. This may meet a colloquial use of the word *grammar*, and probably it is an adequate view in the context of technical languages and the design of machines for given problems (Schmitt and Cagan, 1994). In the light of *living languages*, producing *discursive*, or cultural expressions - regardless if verbal or non-verbal - it is not correct. Here a grammar is a device not to solve problems, but to control the *acceptability* of expressions, the *reading* or *speaking* of which is another matter. Many of the objections against premature claims of AI focus on the assumption that the latter can be done by machines. An ATN can be confused with such machines too. In fact, it is a machine, not a grammar, a machine that can produce shapes - in our case- with the aid of grammatical knowledge. A machine that can be programmed with grammars.
Obviously some assumption about shape grammar structures must be available which is more than just a collection of rules. The distiction between Verb Phrase and Noun Phrase applies in many, perhaps even all languages. This is a *linguistic universal,* not a technical

assumption. Following the well established argument in language-theory and semiology (Barthes, 1964/83) that such universals apply not only in all languages but also in non-verbal fields of expression, we can assume that this is the case in architecture too. However, we do not know if a single comparable universal law is given in this field. Whatever the answer may be, we will present here how an empty ATN frame can be filled, or programmed with a shape-grammar structure, the *space-between* structure, which can be associated with linguistic principles and which allows application in various directions.

2.2.2 *The basic spatium mechanism*

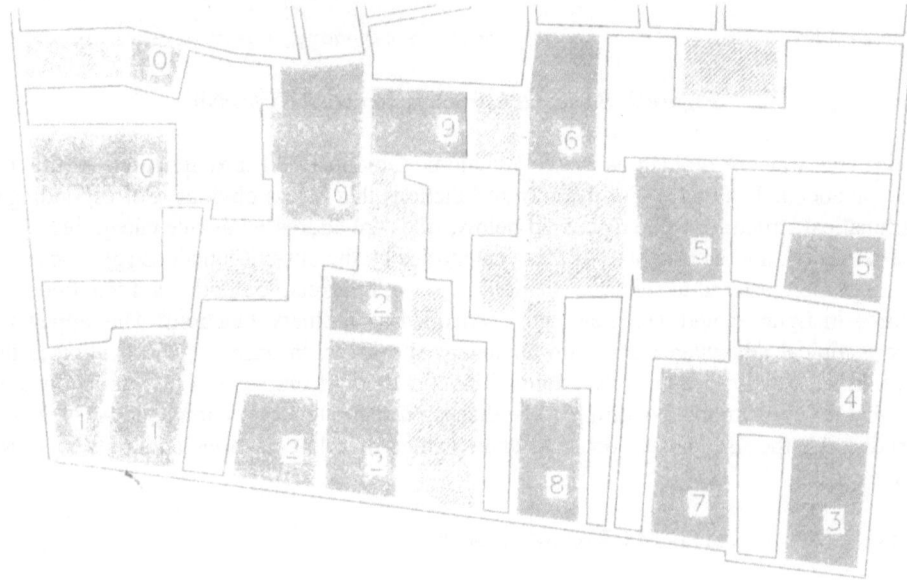

Figure 3

The partially numbered hatching represents an interpretation of a settlement in Greece (Traulou, 1960) in terms of environmental growth expanding in time and space. The image can be read as a sequence of the allocation of solids, here 1-3 storey houses or parts of them (narrow hatched) which we call *extensio*'s, in a way that voids (blank), space-be-tweens between solids which we call *spatium*'s, result. To configurations resulting new extensio's can be added creating new spatium's. The image shows that not always a recog-nisable, in terms of computer modeling: *evaluable*, spatium results from a single extensio. Sometimes more extensio's must be added in order to make an only virtually given spatium real. We can resume this mechanism as a simple network evaluation tree (figure 4).

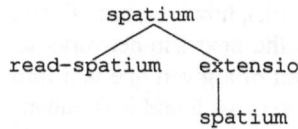

Figure 4, spatium top network

Following the 'read-spatium' option first is tested, for instance by means of a delimitation of some potential, but still endless spatium, if an evaluable spatium is given. If no result is returned the search goes back and enters the 'extensio' option by which a new spatial context is created and recursively passed to another 'spatium' evaluation. Whenever a 'read-spatium' test has been successful the spatium network proceeds with the evaluation of the spatium identified in order to see if it is in some way acceptable. If this returns 'nil' again, the search goes back again and another extensio is entered again. Otherwise possibly *extensio-in-spatium's*, for the ease called *fillings*, of the kind of the wider hatched patches in figure 3 are added. The remaining blank fields resulting from this operation, regardless if some fillings were added or not, obviously have passed some acceptability test in real life in the case of figure 3. Computer modeling and simulation *clones* such human performances. The final remaining patches are called *rest* in our model. If we map these observations onto the frame of section 2, we get the following ATN frame:

```
spatium                                 read-spatium_index
    read-spatium_1                          read-spatium_1 <read-rules_index>
    ...                                     ...
    read-spatium_n                          read-spatium_n <read-rules_index>
    extensio_1                              fill-cat_1 <read-rules_index>
    ...                                     ...
    extensio_n                              fill-cat_n <read-rules_index>
extensio_index                              rest-cat_1 <read-rules_index>
    extensio_1 <extensio-rules_index>       ...
    ...                                     rest-cat_n <read-rules_index>
    extensio_n <extensio-rules_index>   fill-cat_index
    spatium <extensio-rules>_index          fill-cat_1 <fill-rules_index>
rest-cat_index                              ...
    rest-cat_1 <rest-rules_index>           fill-cat_n <fill-rules_index>
    ...                                     rest-cat_1 <fill-rules_index>
    rest-cat_n <rest-rules_index>           ...
    extensio_1 <rest-rules_index>           rest-cat_n <fill-rules_index>
    ...
    extensio_n <rest-rules_index>
```

Figure 5, SPATIUM ATN frame

With the exception of the spatium node, each node of this network provides options for continuation in the same or in the next subnetwork. The tree structure resulting from an actual evaluation is the one described in the previous section, with the addition that at its bottom the first category of a sub-network to be addressed next can be specified. Sub-networks can also be addressed by all other nodes, for example the staircase network referred to in subsection 2.4.3. Such networks, however, are of another order. We would rather call them *spatial* networks, whereas the next sub-networks are of a *grammatical order*. To enter a new subnetwork at the end of a given one is a form of notation which is different from Natural Language ATN Parsers we found in (Graham, 1994) and (Winston, 1984). It results from our distinction between rule-categories and rule-sets.

If a design starts with a not empty spatial context, given as *initial shape*, we can speak of *environmental design*, otherwise we can speak of *tabula rasa* design. Strictly spoken, this kind of design generates no environmental but a design context. In both options a new extensio can be addressed, controlled by the continuation-function mentioned in subsection 2.1, as an arc-option of a rest-category. This continuation with a new extension at the end of a spatium evaluation reflects environmental growth, which in real life is endless. In design, or its simulation through computers it stops at the moment that a given design program or *brief* of desired extensio's is consumed, or if such a brief is empty. The extensio part of the brief can be entered as a start-up argument, or as a global variable. The spatium part of the brief is expressed by the actual categories and rule-sets of the spatium network, which can be selected from a vocabulary. This part can be entered by the ATN programming user (see section 3) directly, or by means of global variables too.

The order of sequence of categories can be more or less fixed. Grammars are more or less deterministic. The network structure discussed here is both. The sequence read-spatium_fill/rest is fixed, the rest not. Such structures represent grammatical properties. We come back to that with an example in the next subsection.

Some more examples from non professional architecture would show that structures as given with figure 3 are wide-spread, if not omnipresent in the built environment. Some experimentation with the work of the architect Scharoun has demonstrated that the space-bewteen assumption is relevant in professional architecture too. The space-between assumption also appears in architectural theory and comments by architects, though rather only occasionally and not really elaborated explicitely. In philosophy the spatium-extensio distinction is a well established matter. This reference can be associated with the distinction between Verb- and Noun-Phrase mentioned above. In W. von Humboldt's work (1795/6, 1973) we found explained that the verb has a unique syntactic power which enables the constitution of meaningful units. The space-bewteen has a similar syntactic function. To expand on that is fascinating and necessary, but exceeds the limits of this paper.

All design and shape-grammar formalisms we know are rather extensio than spatium approaches. Occasionally the space-between appears in this literature too, but rather as by-products such as the non-trivial holes in (Flemming ed al., 1992), or a space remaining in a boundary in (Cagan and Reddy, 1992)

2.3 AN APPLICATION: PALLADIO VILLAS

2.3.1 *The spatium grammar ATN applied*

Background. With the previous section not a yet shape grammar, but an ATN frame of a basic grammatical assumption, the space-between assumption, is outlined. This frame can be filled with various variants of spatium grammars. An extension into layers of increasingly refined grammars is imaginable. At the bottom of such a branching we may find individual grammars. Our modeling of spatium structures began with the work of the architect Scharoun (Hellgardt, 1993; 1994). Figure 6 shows in schematic representation a detail of Scharouns Salute block in Stuttgart and computer-generated paraphrases of dwelling units in this block. The white wings of these paraphrases are extensio's resulting from an *extensio-generator* for the allocation of accessible furniture mats. The spatium between them is identified and generated by means of delimitation rules drawing internal and external contours. The resulting spatium sections are tested by tentatively inserting furniture-pieces or -mats in them, and/or by the surface values of these sections. A readjustment of the angle between the wings and deformations of wing components and furniture-pieces can be involved with these operations.

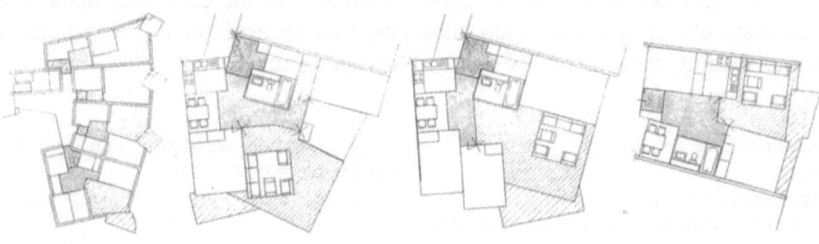

Figure 6

This mechanism was modeled by a series of algorithms written in AUTOLISP. To some extent this worked, but we did not manage to establish a consistant and reliable model. We discovered that the formalisms we were defining increasingly produced their own laws, generally not completely in line with what we intended to achieve and to model. Obviously some more general framework in terms of grammar theory and formal modeling was required and presumably also another programming tool. Above that, in its complexity the architecture of Scharoun is a challenge, but perhaps not the most appropriate material in the stage of fundamental experimentation that we were about to skip over. We switched to a phenomenon which is simpler in terms of computer modeling: Palladio villas, and started programming in Common Lisp (discovering later that Allegro Lisp with its graphic module and its object system seems to be the most appropriate tool).

The Palladio ATN. Palladio's comments on his villas testimony his commitment for environmental design. Other comments and some planimetric studies not contained in his 'The Four Books...' (1570, 1983) speak another language: *tabula rasa* design, experimentation with a kind of design as such. Our report here follows this view. It is linked to the literature on two levels: the impact of the local, in particular Venetian building type (Ackerman, 1977) and Palladio's harmonic theory (Wittkower, 1973).
A main feature of the Venetian palace, umambigously recognisable in Pallido's villas, is a central hall between mirrored wings. Palladio's almost explicitely declared intention was to

cultivate that by means of firstly the elimination of irregularities in the symmetry, and secondly by tuning floor-plan and other proportions. More or less explicitly articulated we find also a kind of villa-*brief* in the 'Four books ...': a small number of chambers, most three, with a central hall and loggia between them. All of these rooms should be shaped and combined harmonically. Following that we assume that the wings are *chords* of chambers. Such *wing-chords* can be relatively easy generated by means of a multiple Cartesian product. This is a top-down approach which we do not enter, resulting in some hundreds of wing instances reflecting Palladio's way of harmonic dimensioning. Staircases are less explicitly mentioned in the 'Four books ...', but they can be assumed to be part of the 'brief' too. Hardly mentioned are auxillary elements which some of the villa's contain, such as niches, passages and mass fill-blocks (figure 8). The shape and positioning of the not-wing elements, as the mere occurrence of auxillary elements, seems to be resulting in some way *automatically* and more or less un- or sub-conscious from the conscious manipulation of the main design objectives. If we map these observations again onto the spatium ATN of the previous section, we get an ATN of the following form:

```
p-spatium                                    wing
   read-spatium                                 p-spatium <select-wing-pitch₁>
   wing                                       vault-articulation
read-spatium                                     fill-main/aux₁ <tunnel/croos-vt>
   vault-articul <delimit-by-mirror>             ...
fill-main/aux_index                              fill-main/auxₙ <tunnel/croos-vt>
   fill-main/aux₁ <fill_main/aux_index>      hall
   ...                                           final-test <hall-test-tules>
   fill-main/auxₙ <fill_main/aux_index>      lggia-to-hall
   fill-aux/main₁ <fill_aux/main_index>         hall <lggia-to-hall-rules>
   ...                                       final-test
   fill-aux/mainₙ <fill_aux/main_index>         () <test-rules>
   hall <fill_main/aux-rules_index>
   lggia-to-hall <fill_main/aux_index>
staircase                                     read-spatium
   read-spatium                                  landing₁ <sub-spatium-rules>
landing₁                                       landing₂
   landing₂ <landing-rules₁>                     flight <landing-rules₂>
flight
   fill-aux/main₁ <flight-rules>
   ...
   fill-aux/mainₙ <flight-rules>
```

Figure 7, Palladio ATN

We have two types of fill nodes: main fill categories (loggia and staircase, 'fill-main$_{index}$') and auxillary fill categories (see above, 'fill-aux$_{index}$'). This is resumed here in one notation by appending '-main/aux$_{index}$'. The 'staircase' node opens a subnetwork . The aux-category is added to the general frame of subsection 2.2.2, as well as a 'vault-articulation' node which envokes the articulation of a spatium for tunnel- or cross-vault elaboration. All other nodes are the same as those of the general frame. We only used other names for a better illustration of the functioning of the ATN in applied form. The application of this general network frame is largely a question of inserting appropriate rule-sets, in other words. For example, we have only one type of a read-spatium category, invoking 'delimitation-by-mirror' rules. With that is assumed that by mirrowing a wing a spatium is identified and delimited. Our Scharoun modeling was more elaborated in that regard.

To some extent the functioning of the network may require no further comment. A more detailed description of the network in action is given with the register the network produces with the evaluation of each path. We call such a register a shape-marker because it is inspired by, or rather an application of Chomsky's (1965) concept of a phrase-marker. We proceed with an example. Some main aspects involved with that are resumed in subsection 2.4.

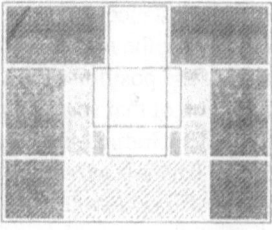

```
(P-SPATIUM
    (EXTENSIO
        (WING_LR  ((16 16)(16 24)(27 16))(VI V I)))))
    (P-SPATIUM
        (READ-SPATIUM
            (DELIM-BY-MIRROR_Q1  (3.7))))
        (VAULT-ARTICUL
            (VAULT_V2  (CROOSV)))
        (FILLER
            (MAIN
                (LOGGIA_L2  ((18 43)  (36 58))  (1 1 2)))
                (STAIRC
                    (READ-SPATIUM
                        (SUB-SPATIUM_T15  ((18 18)  (23 33))  (2 2 2))))
                    (FILLER
                        (HLANDG_D1  ((18 18)  (23 20))  (2 2 2)))
                        (HLANDG_D2  ((18 18)  (23 33))  (2 2 2))))
                    (REST
                        ((FLIGHT_F0  ((0 0)  (0 0))  (1 2 2)))
                (AUX
                    (NICHE_N1  ((18 34)  (23 42))  (1 2 2)))
                    (PASSAGE_P1  ((0 0)  (0 0))  (2 2 2)))
                    (BLOK_B1  ((24 35)  (27 42))  (1 2 2)))))
            (REST
                (HALLN_H0  ((0 0)  (0 0))  (1 0 1))))))))
```

Figure 8, Shape-marker of a Pisani paraphrase

2.3.2 *A Shape-Marker*

Figure 8 shows the register of a path in the Palladio ATN resulting in a Pisani paraphrase. The bracketed representation is equivalent with a tree representation. As it contains all required data, it is also equivalent with its graphic representation.

First a remark on Chomsky's understanding of a phrase-marker: A phrase- or shape-marker reports a Structural Description. It is the result of the application of grammatical rules which are based on a vocabulary of symbols including *formatives* and *category symbols* which can be subdivided in *lexical* and *grammatical items*. As syntactic (as distinct from phonetic) units sentences are strings of formatives, rather than strings of *phones*. We can directly apply this view: 'Sentences' of buildings are *spatial configurations*, not *strings*, of 'formatives' which we call *space-partitions* not yet transformed, or *materialised* into building materials, surfaces and similar.

Our rule-instances produce such 'formatives'. They appear in the shape-marker as indexed expressions, the names of rules. After these names the result of rule-evaluation is entered in terms of the position of the space-partition generated, and a list of scores resulting from an evaluation of that space-partition. The not indexed symbols in the Shape Marker are lexical categories. Grammatical items do not appear in the shape marker directly.

On this background we can resume the Palladio network in action. As we follow only the *tabula rasa option* the result of the first arc, 'read-spatium' is nil, there is no context to be interpreted as possible spatium. The second p-spatium arc is addressed then, a wing-extension. The $_{LR}$-index specifies a wing-type in terms of its shape (L-shape). '(VI V I)' marks a 'chord' consisting of a fifth, a sixth and a prime. 'Read-spatium' is entered then, this time resulting in a successful recognition and delimitation. This spatium is articulated then as a cross-vault disposition and filled with 'aux' and 'main' components, one of which is a spatial subnetwork.

Our example ends with a rest rule. Rest rules are not productive, they only evaluate a remaining patch, a hall of a villa, or a flight of a staircase. We can also think of rest rules testing two or more remaining patches to see if they are in some way exchangible. A hall can become a loggia and vica-versa, for example. This way paths can be re-interpreted. The evaluation functions of rules that produce patches have to be correspondingly flexible. All test results contained in the register returned by a terminal arc can be submitted to a final test (not reported in the shape marker).

2.4 MAIN ASPECTS OF MODELING SHAPE GRAMMARS

2.4.1 *Grammatical relations*

Grammatical items appear in the shape marker only in terms of their order of sequence. The sequence wing-extension_read-spatium_vault-articulation_fill-category_rest-category is fixed, or deterministic. Non-deterministic are the fill-subnetworks, addressed as next-expressions by the 'vault-articulation' node in figure 7. These subnetworks simulate the presumably mainly un- or sub-conscious filling of a spatium described above. The whole is a grammatical structure, the fixed parts of which represent grammatical relations, comparable with relations as subject-verb in sentences.

2.4.2 *Evaluation*
The scores resulting from the 'HALLN$_{H0}$' rule in figure 8 contain a "0" whereby the result might be rejected. Other scores of this example are also only modest. Other rule combinations come up with a better results. These details illustrate that evaluation functions are a precarious matter. Until now we work with three test criteria: surface, shape and topology. Such properties can be seen as either use-values, or as aesthetic or perceptional values. A shape-value, for instance, can be rejected on aesthetic grounds, or on reasons of use-value. Architectural floor-plans must satisfy use-value standards to some minimum extent. This is neccessary, but trivial. It does not meet the nature of architecture which lies - to put it hyperbolically - in the reconciliation of such trivialities with standards of what we call, in line with assumptions of lingustics and semiology raised in subsection 2.2.1, *perceptional acceptability*.

2.4.3 *Spatial subnetworks*
Natural languages are recursive. Sets of sentences defined by such languages are infinite consequently. Sentences can contain sentences, and so on. This applies to all kinds of discursive expressions. Networks generating such expressions are recursive to the extent that networks are addressed within them. The staircase network embedded in the main-fill section of the Shape-Marker of figure 8 is an example. It starts with a 'read-spatium' node addressing a 'sub-spatium' rule which generates a subfield for a staircase filling. It is planned to extend that to the recognition of subfields. The terminal node, flight, continues with a node of the network on top of it, here a 'fill-main/aux' node.

2.4.4 *Learning search*
The Shape-Marker model is inspired by natural language parsing. In this light it is a *reading*, not a *speaking* device. Design, can be objected, is rather 'speaking' than 'reading', you should better look at language generation systems. Our answer at the moment is that a design can be seen as the 'reading ' of a design-brief of the kind raised in the subsection. 2.2.2. The categories of such a brief are translated into lexical items and possibly space-partitions which are possibly entered into a Shape-Marker. This is based on and enables another kind of reading, evolving with the design itself: at any stage of network evaluation a given rule-instance inquires the register of already evaluated instances together with the shape-configuration given at that stage in order to decide if the rule and operators within that rule can be applied. This can be compared with *thinking speaking*. Before speaking the next word in a sentence, provided it is no stereotype, we look back in order to understand what we really have been saying before, in order to decide then how we can proceed with that preliminarily uttered meaning.

Such decisions do not only address the register of an evolving expression, which we might call a short-term memory. In a speaking performance other medium- or long-term memories of knowledge, grammatical knowledge and experience are inquired too. In this light we can think of three prospectives of how a design formalism might clone learning.

If the reasoning of the register of a path is unrelated to the mechanism of search organised by a network there is no learning search. Exhaustive search is such a kind of search. Though partially only in similar appearance, evaluated exhaustively the Palladio ATN produces all instances, actually contained in Palladio's legacy, of non-bisymmetric villas for one *pitch* of dimensioning. The state space is large (depending on the length of rule-sets, thousands of legal paths for one of some hundreds of wings). This method is ineffective, but reliable compared with insufficiently elaborated formalisms of machine learning.

The second prospective is stochastic search, which ranges from plain stochastic optimisation to Multicriteria GA (Genetic Algorithm) directed search (Kundu, 1996). It is difficult to define adequate objective functions in architecture. Multicriteria methods make that easier. Application of GA's in our field is possible when cross-over and mutation formalisms can be reconciled with partially deterministic, partially non-deterministic structures of the kind discussed (we are not the first to encounter such a problem, cf Davidor, 1991). Experimentation has shown that this seems not to be a fundamental problem and that GA directed search for acceptable populations of Palladio villas is possible. To what extent such populations can compete with exhaustive search, perhaps even in a more effective way, we have not yet shown.

The third prospective is learning search directed by experience and related logical inference. We started experimentation with analogic pattern recognition. A medium-term memory of stored accepted and and rejected populations can be matched against evolving candidates in order to accept, reject or convert them.

3 Rule definition and connection to existing software

The value of a shape grammar ATN is restricted if no associated rule definition system is available. As raised in subsection 2.1, in our approach calls of rules defined independantly from ATNs can be inserted into the arcs of such ATNs. The combination of program modules of different provenance is the most important aspect with respect to applicability and user-friendliness. It opens the connection to commercial CAD software, an obviously urgent matter and current argument (cf Fawcett, 1990).

A remark on rules is useful first. We have to distinguish between rule types in terms of grammatical properties, as discussed above, and rule types in terms of how a human design act is combined with an act of representation which can be a drawing act, but also one of sculptural modeling etc. Our experimentation draws on only one type of such rules, preliminarily called *representation rules*: a kind of patches-distribution or *brushstroke* technique (see subsection 2.1) which reflects an assumed way of spatial orientation. Another representation rule-type defines line-drawing rules. This type is addressed as delimitation-rules in our model. Another type of representation rules, to some extent contained in commercial software, are array rules. Rules that combine labelled shapes (Lego or Froebel building blocks for example, cf Stiny, 1980b) can be can be subsumed under this type too.

If the functioning of rule-types of the second kind can be generalised, rule definition macros can be defined which encode and write rules on the basis of user dialogs. Generally rules, not necessarily all, are of the form IF ... THEN ... To some extent the applicability of a rule in a path can be tested by the 'continue' macros in the arcs (see subsection 2.1 and last par. below). This makes the rules ATN-dependant which, however, is also a virtue because it forces rule-design into an overall grammar structure. The rule defining user is challenged to design rules and grammar structures connectedly. But not all features of rules can be generalised this way. These features are tested on the IF side of the instances of rules (1). On the THEN side, in our approach, two modules are addressed: the execution of a rule (2) and optionally the evaluation of the result of that (3). On all three levels rule modules can be addressed which are defined independantly from the 'top'-rule itself in other programming languages or, last not least, in commercial CAD packages. Rules on these three levels are not neccessarily of the same class of grammars.

With the IF ... THEN structure described a body is given which can be encoded and written away by a rule-definition macro. This macro can be addressed in a user-dialog asking for the type of a rule to be defined and for all specifications that are required to define a rule of that type. For certain rule-types, such as our *patch*-rules, graphic intervention on the monitor can be part of this dialog. All rules of our experiment are defined with this rule-definition macro, with the arguments: rule-name, rule-type, test-type, accessability-arguments, executability-arguments, test-arguments, destination-file. The actual rule is called with only one argument, the register of the evolving network which does not appear in this list but in the code of the rule defined by it. This rule-definition system can be addressed independantly from some CAD package in case the code it generates is under-standable by a language associated with that package, for example AUTOLISP. There are other possibilities, but this way seems to be the easiest and most feasible one, at least for us in the present situation.

The communication between ATNs programmed with such rules and existing software is similar but different. The mechanism described above and in subsection 2.1 performs two tasks: the selection of legal sequences of rule-instances, excluding undesired duplicates for example, and the actual evaluation of these rules. These tasks can be separated. The compilation of legal rule-instance sequences can be elaborated by a separate program, written in Lisp for instance. If the result can be imported the actual evaluation can be done by a commerical CAD program.

4 Conclusion: human and machine learning

Heidegger has been explaining at the München Politechnics (1952) that any technological innovation is a challenge to intellectual and artistic curiosity. The ATN frame is presented predominantly as a tool for human learning. With that the threshold to Artificial Intelligence is not yet passed. But the possibility to mechanise human knowledge, given with such a tool, is a challenge of the kind raised by Heidegger. The tool presented enables the user in practice and education to approach the old, but unsettled account of *grammar in architecture*, raised initially. As discussed in particular in section 3, no specific computer programming knowledge is required that for. Some grammar knowledge, however, is required. To the same extent grammar knowledge results from experimentation with the tool.

96

Mechanisms of machine learning as raised in subsection 2.4 can be addressed without specific computer modeling knowledge too. Their development, however, is a matter of computer modeling and programmation. This work can be related to design experience and research in architectural history and theory. If adequate forms of co-operation cannot be established the danger of machine learning independantly from human learning is given.

References

Ackerman, J.S. (1977) Palladio, Penguin Books, Harmandsworth, Great Britain.

Barthes, R. (1964/83) Elemente der Semiologie, edition suhrkamp 1171.

Cagan, J. and Reddy, G. (1992) An improved shape annealing algorithm for optimally directed shape generation, in: J.S. Gero and F. Sudweeks (eds), Artificial Intelligence in Design '92, Kluwer Academic Publishers, Dordrecht, pp. 307-324.

Chomsky, N. (1965) Aspects of the theory of syntax, the MIT Press, Cambridge, Massachusetts.

Coyne, R., McLaughlin, S, Newton, S. (1996) Information technology and praxis: a survey of computers in design practice. Environment and Planning B, 23, pp. 515-551, in part. 533.

Davidor, Y. (1991). A genetic algorithm applied to robot trajectory generation, in: Davis, L., Handbook of genetic algorithms, Van Nostrand Reinhold, New York, pp. 144-165.

Fawcett, W.H. (1990). Shape rules for architecture, paper presented at EuropIA conference, Liège, Belgium.

Flemming, U. Baykan, C.A., Coyne, R.F. and Fox, M.S. (1992) Hierarchical generate-and-test vs constraint-directed search: a comparison in the context of layout synthesis, in: J.S. Gero and F. Sudweeks (eds), Artificial Intelligence in Design '92, Kluwer Academic Publishers, Dordrecht, pp. 817-838

Graham, P. (1994) On LISP, chapter 23, Parsing with ATNs, in: Prentice Hall Inc., Englewood Cliffs, New Jersey.

Kundu, S. (1996). A multicriteria genetic algorithm to solve optimisation problems in structural engineering design, Proceeding of International Conference on Information Technology in Civil and Structural Engeneering Design , 14th - 16th August 1996, Glasgow, Scotland. Civil-Comp Press Ltd. Edinburgh, Scotland, pp. 225-233.

Kundu, S. and Hellgardt, M. (1996) A networks approach for representation and evolution of shape grammars, in: J.S. Gero and F. Sudweeks (eds), Artificial Intelligence in Design '96, Kluwer Academic Publishers, Dordrecht, pp. 291-310.

Heidegger, M. (1954/85) Die Frage nach der Technik, in: Vorträge und Aufsätze, Neske Verlag Pfulligen, pp. 9-40.

Hellgardt, M. (1993) Syntaktische Aspekte der Arbeit von Hans Scharoun, Begründungen und Erläuterungen zu den im Begleitprogramm der Ausstellung 'Werkschau Hans Scharoun' vorgeführten Fragmenten einer Scharoun-Syntax, Akademie der Künste, Abteilung Baukunst, Berlin.

Hellgardt, M. (1994) Dentro l'architettura di Scharoun, Housing 6, Etaslibri, Milano.

Humboldt, W. von (1830, 1973) Akt des selbsttätigen Setzens in den Sprachen, Verbum, Schriften zur Sprache, Reclam Verlag, Stuttgart, pp. 169-188.

Palladio, A. (1570, 1983) Die vier Bücher zur Architektur, Artemis Verlag, Zürich und München.

Schmitt, L.C. and Cagan, J. (1996) Grammars for machine design, in: J.S. Gero and F. Sudweeks (eds), Artificial Intelligence in Design '96, Kluwer Academic Publishers, Dordrecht, pp. 325-344.

Stiny, G. and Mitchell, W.J. (1978) The Palladian grammar, and: Counting Palladian Plans, Environment and Planning B 5, pp. 5-15 and 189-98.

Stiny, G. (1980a) Introduction to shape and shape grammars, Environment and Planning B 7, pp. 343-351.

Stiny, G. (1980b) Kindergarten grammars: designing with Froebel's building gifts, Environment and Planning B 7, pp. 409-62.

Traulou, v.I. (1960) Poleidomiki exelixis ton Athinon, in: Caniggia, G. (1976) Strutture dello spazio antropico, page 41, Uniedit, Firenze.

Winston, P.H. (1984) Artificial Intelligence, chapter 9, Language Understanding, Addison-Wesley Publishing Company, Reading Massachusetts.

Wittkower, R. (1973) Architectural principles in the age of humanism, Academy Editions , London.

ARCHITECTURAL DESIGN-BY-FEATURES

JOS P. VAN LEEUWEN [a]
HARRY WAGTER [a+b]

[a] *Eindhoven University of Technology, The Netherlands*
Faculty of Architecture, Building, and Planning
Building Information Technology
http://www.calibre.bwk.tue.nl
email: j.p.v.leeuwen@bwk.tue.nl

[b] *Origin International Consultancy*
Eindhoven, The Netherlands.
email: harry.wagter@nlehvips.origin.nl

Abstract

Design tasks, in particular architectural design tasks, have been found hard to support by means of computers. The main reason for this is that design is a problem solving process, which requires a dynamic way of handling information involved in the design process. The research presented in this paper focuses on this aspect of CAAD: the support of design tasks with dynamic, flexible information modelling techniques. The basic concepts for the developed approach is taken from the field of Feature-based modelling. We briefly review these concepts and then interpret and transport them to the context of architectural design. In defining types of Features, a distinction is made between domain-specific Features and generic Features for which we propose a classification. A framework for the definition and modelling of Features is discussed as well as a prototype Feature-based Modelling Shell based on this framework.

1. IT Support for Design

Computer support for processes of design has been subject of research for nearly as long as computers have been around. The great amount of today's approaches to supporting architectural design ranges from the development of computer-aided drafting to the application of artificial intelligence in shape-grammars, rule-based design, case-based design, and many more. Common to these approaches is the need for representing and manipulating a multitude of information that is somehow involved in the tasks of design. Modelling information for support of design tasks is subject of research presented in this paper.

Product Modelling (PM) is an approach that concentrates on information modelling for the support of communication between the many different participants

97

R. Junge (ed.), CAAD Futures 1997, 97-115.

in design processes. It addresses problems such as the distinct views of participants on the product of design and the definition and structure of data involved in these views. This results in the definition of data models that represent the information in various domains of design and engineering. Much of the research in the field of PM focuses on the standardisation of these information models in processes of communication [ISO 1994], also in the Building & Construction (B&C) industry [Tolman and Wix 1995]. However, information models do not merely serve the communication at certain stages in the design and building process. The representation of information during tasks of design, e.g. for support of the analysis, generation, and evaluation of information, is an equally important role for information models. Yet this role of information models imposes fundamentally more complex requirements on the definition and structure of models, mainly due to the dynamic nature of the process of design. Successful research in relation to this subject is found in the work of Eastman et al. [1991, 1995] and Ekholm and Fridqvist [1995, 1996].

1.1 DYNAMIC NATURE OF DESIGN

One of the main problems that need to be addressed when defining information models for architectural design is caused by the dynamic nature of design. During the process of design, information is not treated as static data, but is invariably subject to change. This is due to the problem-solving character of design, which involves the search for information, analysis, structuring, interpretation, and evaluation of information in repeating cycles. Most important, it involves the generation of information during this cyclic process and combining this new information with known data, defining new information-structures that lead to design-solutions. We conclude that the definition and structuring of information during the cyclic process of design is a key issue in supporting design tasks with information models and design information management systems. The dynamic nature of design is a precondition for creative design; its preservation is perhaps the most challenging requirement for the development of design information models and design support systems.

1.2 REQUIREMENTS FOR MODELLING DESIGN-INFORMATION

One of the aspects that form the dynamic nature of design is the flexible manner of handling information. In the problem-solving process of design, the search for solutions, alternatives, and optimisations require that information is constantly analysed, interpreted, generated, and restructured. Information models that are meant to support design tasks will have to show sufficient flexibility, allowing for definition and redefinition, structuring and restructuring of information. Information models will need to evolve in order to reflect the evolution of the design.

Evolution in design does not only occur during the course of a single design project. The fact that designers learn makes them change their approach to solving design problems, finding new techniques, new rules, new concepts. Mitchell [1990] recognises this stylistic evolution as an essential component of creative design which must be addressed by future CAD systems. For information models, this means that the definition and structure of information should enable adaptation to changes in requirements, insights, methods, and conceptual basis, etc.. On a larger scale in the building industry, there is a similar form of evolution that is manifested in the development of new products, materials, or construction-techniques. This evolution too requires information models to adapt to changing requirements.

1.3 EXTENSIBILITY AND FLEXIBILITY OF INFORMATION MODELS

The requirements for information models in terms of evolution and adaptation along the design process and the development of the design discipline, can be translated into required extensibility and flexibility of information models.

Extensibility   Extensibility of information models allows designers to extend the set of definitions that constitute a conceptual information model with definitions that represent specific concepts and notions. In this manner, the domain of an information model can be extended into areas that are specified by the user of the model. This extension may concern, for instance, a specific style of design, style-rules or conventions for a particular building project. Also, new information-definitions will represent new technologies of construction, new products and materials, and their individual characteristics. For construction-systems that are based on industrial production of components, the definition of information will accurately represent the characteristics, requirements, and assembly-procedures, etc. of the components and these systems as a whole.

Flexibility   Enabling the definition of entities of information that accommodate specific requirements in design-tasks is one aspect of supporting the dynamic nature of design. A second aspect addresses the flexibility of information models. Flexibility is required when new definitions of information are added to a conceptual information model: extension of the model requires flexibility. Newly defined information entities will define relationships to existing entities, and reversely, existing entities will need to define relationships to new entities. This requires a significant level of flexibility in the definition of information entities, allowing properties or attributes defining relationships to be modified or added.

Flexibility is also necessary to accommodate the design process as a problem solving process, which, apart from generating information, also continually involves interpreting and restructuring available information. Therefore models that are to represent this information, not only at a final stage but also during the course of design, will have to follow this process of restructuring. Again it requires properties or attributes representing relationships to exhibit a sufficient level of flexibility.

Feature-based approach   The research presented in this paper addresses the requirements of extensibility and flexibility of information models by applying concepts and techniques from Feature-based Modelling (FBM) in the context of architectural design. Feature-based Modelling is developed in areas of mechanical engineering and industrial design, concentrating on the description of part-geometry using high-level elements that correspond closely to the terminology of the specific domain of design. The structure of the models that are used in this approach, and the openness offered to modelling systems that employ this approach meet the requirements of flexibility and extensibility, which we have recognised to be important for architectural design systems as well. In order to investigate the applicability of the Feature-based approach in the context of modelling architectural design information, we first analyse the concepts of Feature-based modelling in the area of mechanical engineering. These concepts are then recapitulated in the context of architectural design, leading to the formulation of a Feature-based approach to modelling architectural design information.

2. Concepts of Feature-Based Modelling

2.1 ORIGINS OF FEATURE-BASED MODELLING

In disciplines of mechanical engineering and industrial design, the practice of using computers in the design and production processes has given rise, in the early eighties, to the need for richer information models. Geometric models of a product did not suffice for advanced evaluation of designs, such as manufacturability evaluation, or for integration in for instance process planning tasks. For these purposes it was necessary to develop models with a higher level of information than just geometry. This resulted in the development of high-level information entities which are called Features. Since form is the major aspect of interest in the disciplines mentioned, a strong focus is still today on the development of Form Features.

Many definitions of the term Feature have been given in literature, depending on the context and the purpose for which Features are applied. A definition for Form-Features given by Shah [1991a] seems to cover their common notion: '[Form] Features are generic shapes with which engineers associate certain properties or attributes and knowledge useful in reasoning about a product.' In research and practice of Feature modelling Features generally describe characteristics or parts of a product, mostly concerning the manufacturing of the part, by describing its surface or shape and the technological attributes that are associated to for instance tools and operations.

One quality of Feature modelling that appears to be important for our research is that parts are not represented by a complete, rigidly defined model. The definition of Features is independent of their future context. Therefore the structure of a part in a Feature model is not prescribed, but is composed by creating Features and structuring the Feature-model during design.

Feature modelling technology focuses on the description of the shape of parts, using Form Features. However, there are many other kinds of Features involved in the description of shape, as well as in other aspects of interest in modelling a product. Thus, also more generic definitions of the term Feature are found in literature, such as 'a set of information related to a part's description'. In an assessment of Feature technology, Shah [1991b] gives the generic definition: 'Features are elements used in generating, analysing, or evaluating designs', and in [Shah 1991a]: 'A Feature is any entity used in reasoning about the design, engineering, or manufacturing of a product.'

There are probably as many classifications of Features as there are definitions of the term Feature, again depending on the context of using Feature technology. Some of the classes of Features generally found are: Form Features, Precision Features, Material Features, Assembly Features, Performance Features, Pattern Features, Connection or Constraint Features, Application Features, etc. An example of some typical Form Features is given in figure 1. Although there are efforts to standardise the classification of Features, it is generally accepted that no collection of Feature definitions can be complete. An important requirement for Feature modelling systems is the level of extensibility of these systems with the definition of new types of Features.

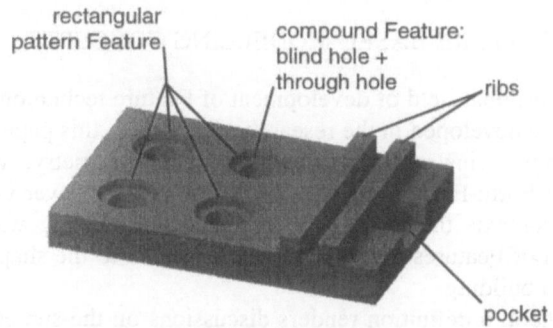

Figure 1   Some typical Form Features.

2.2 FEATURE RECOGNITION AND DESIGN-BY-FEATURES

Feature technology has developed into two main approaches towards high-level modelling of products. One approach is Feature recognition [Henderson and Chang 1988, Laakko and Mäntylä 1993, Meeran and Pratt 1993]. Using geometric models of product-parts, a model of Features is built up after analysis of the geometry. At first this was a human-assisted, interactive process, later Feature recognition was automated. The Feature model could then be used for evaluation and creation of for instance process plans or NC programming.

Among the main problems with the Feature recognition approach are the limited type of information that can be recognised from geometry, and the very complicated procedures that are necessary to extract meaningful Features. Another important drawback is that the geometric model needs to be completed before the recognition process can start. After changes in the geometry, this process needs to be repeated. Moreover, valuable information that is already available during geometric modelling cannot be included in the model and will be lost. An important advantage of Feature recognition is that geometric models created without specific context can be evaluated and interpreted using knowledge from a particular context, e.g. a specific manufacturing process.

The logical successor of the recognition approach is the design-by-Features approach [Shah and Rogers 1988, van Emmerik 1990, De Martino 1994]. Geometry is no longer the basis for the model that is now built up by creating Features. In Feature models, geometry now forms one aspect of the product-information: also non-geometric information is included from the beginning of the modelling process. Some of the main advantages of the design-by-Features approach are that high-level entities are used to model a design, that these entities correspond (better than geometry) to the terminology of the domain of design, and therefore that the resulting model will more accurately represent the actual design. The error-prone procedures of recognising high-level information from geometry is no longer necessary.

Recent research has resulted in systems that combine the two approach of Feature recognition and design-by-Features [De Martino 1994].

2.3 ABSTRACTION OF THE FEATURE-BASED MODELLING CONCEPTS

The definition of Features in the original field of development of Feature technology is an important basis for the theory developed in the research presented in this paper. Since the focus in mechanical engineering is on describing shape and geometry, we have to consider the definition of Form-Features as was quoted above. However we anticipate that in architectural contexts the technology of Feature modelling will require a more general definition of Features than one that is restricted to the shape or geometry of physical parts of a building.

Shah warns that too general a definition renders discussions on the subject meaningless [Shah 1991a], he then concentrates solely on the development of Form-Features. We have concluded that architectural information modelling cannot permit itself this luxury and will have to deal with a more general concept of Features. Architectural products cannot be described by an apparent decomposition of physical parts. Architectural design involves many concepts and notions that are not directly, not at all times, or even not at all related to physical parts in a building. Some indicative examples are concepts of space, function, costs, safety, comfort, etc.. Moreover, these concepts apply to both early and later stages of design, and are relevant to different levels of abstraction in looking at a building-design.

In this research, we have abstracted an understanding of Features from the original Feature technology, which allows us to involve both physical and abstract concepts in the definition and modelling of Features. This notion of Features will be applicable to multiple levels of abstraction in architectural design information.

In the next section we introduce a general definition of the term Feature in the context of architecture and propose a generic classification of architectural Features.

3. A Feature-based approach to modelling architectural design information

3.1 DEFINITION OF ARCHITECTURAL FEATURES

In earlier papers [van Leeuwen et al. 1995, 1996] on this research we have defined a Feature as follows:

A Feature is a collection of high-level information defining a set of characteristics or concepts with a semantic meaning to a particular view in the life-cycle of a building.

Breaking down this definition into three aspects, it can be noted that:

* Features represent semantics of a building (or its design);
  As in other Product Modelling approaches, we attempt to use definitions of information entities that closely relate to the terminology, and therefore semantics, of the domain of application. Important however is how these entities are chosen and how their inter-relationships are defined. This is framed by the second aspect:

* Features are the formal definition of characteristics or concepts;
  In this aspect, the Feature-modelling approach is fundamentally different from the 'traditional' product modelling approaches. In the latter approaches, physical components often form the basic entities in the structure of information models, their properties being the attributes of these entities. In the Feature-based approach, properties or, more general, characteristics and concepts, are entities of information themselves. As a result, the relationships between components and properties of components are much more flexible. This means, for example, that a property that has not been previously assigned to a certain type of component can, at any given moment, be added to the information structure representing the component and its properties. Also, in this manner properties can be shared by different components. It should be noted that characteristics and concepts are not restricted to static data, but may define behaviour or procedural knowledge as well.

* Features are related to particular views in the life-cycle of a building.
  The concept of views is one very common to product modelling approaches. In the Feature-based approach, this leads to the definition of libraries of Feature-types that represent a particular domain of information modelling. In this context we define Generic Feature Types as those types that are part of the domain of B&C as a whole, and Specific Feature Types defining those Features that are specific to a particular view or sub-domain in the B&C industry. This can be the view of one participant, a view related to a particular project, or for instance a domain deriving from a particular industrial construction technology.

104

3.2 LEVELS OF ABSTRACTION

The definition of a Feature given above does not limit information to a particular level of abstraction or detail. Although not explicitly so stated in Feature-definitions in mechanical engineering, the Feature-technology in that area follows a strict hierarchy of the description of a product, namely the decomposition of a product (or assembly) into parts, and of parts into Features (see figure 2).

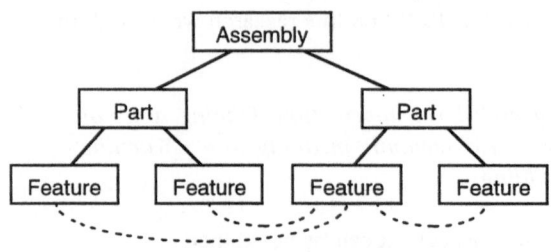

Figure 2   Product hierarchy in mechanical engineering.

Levels of abstraction are often related to different stages in design. In architectural design information is generally involved in several levels of abstraction at the same time, or shifted between these levels. This research does not assume a predefined hierarchy, thus allowing Features to appear at any level of semantic abstraction, and allowing Features to be migrated from one level to another. The levels of abstraction are not intrinsically predefined in our theory, therefore abstraction layers from other approaches, design-methods, or standards can be adopted and incorporated in the structure of information models.

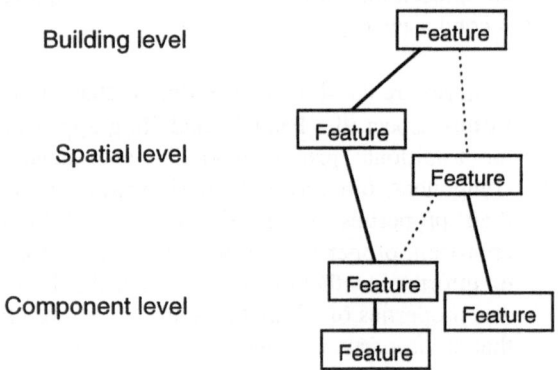

Figure 3   Features at different levels of abstraction.

Figure 3 shows how a network of Features covers multiple levels of abstraction. This network incorporates information concerning a building as a whole, as well as information at the more detailed levels. Different types of relationships between Features establish the structure of information. This approach has also been chosen to allow information modelled at early stages, often not very detailed, to evolve as more details become known during later stages of design.

3.3 GENERIC VERSUS SPECIFIC FEATURES

Although the approach advocated in our research aims at supporting dynamic design-tasks, this does not mean that we oppose any kind of standardisation of data-definition. An important role in standardisation is appointed to so-called core models which have the function of intermediate model between communicating or data-sharing participants. However, standardisation should not conflict with the requirements of extensibility and flexibility. Therefore, the structure of core models need to meet these same requirements.

Generic Feature Types are those types that form a core model that maintains sufficient flexibility. They are formalised, common concepts and represent terminology accepted and used by the B&C industry. Although typical relationships between Generic Feature Types do occur, they do not form a rigidly defined structure but retain a great deal of independence. The exact and complete context of Features is defined only during the modelling process itself. The next section addresses the classification of Generic Feature Types and includes a proposal for the main categories. However, the task of classifying and defining the Generic Feature Types clearly needs to be left to international standardisation institutes, such as the ISO.

Feature definitions that are not part of the common domain in B&C are called Specific Feature Types. They represent particular views or disciplines, or are related to particular styles, projects, applications, or any other specific context. Standardisation may also play an important role in the definition of Specific Feature Types, e.g. for the standardisation of branch-models. But Specific Feature Types also enable the extension of standardised models with definitions of information-entities that serve certain specific needs.

3.4 CLASSIFICATION OF ARCHITECTURAL FEATURES

We present a proposal for the main categories in a standardisation of Generic Feature Types. Table 1 lists 5 main categories with some examples of sub-categories and a brief indication of the contents of each of the categories. The configuration of the categories is based on an analysis of building- and design-related information and on a survey of existing classification tables, including a proposal by [Woestenenk 1995] for international conversion tables for parts and functions.

This classification is not intended to be complete. Even when standardised, the set of Generic Feature Types cannot be presumed to be complete and must retain a structure that supports the required flexibility for adaptation to future developments. However, the proposed categories are believed to be a valid starting point for standardisation and for the definition of Specific Feature Types.

3.5 A FRAMEWORK FOR FEATURE MODELLING

The notion of Generic Feature Types and Specific Feature Types, and the proposed classification have been the basis for the development of a framework for Feature modelling. From the original technologies of Feature modelling, the design-by-Features approach has been found most appropriate for the problems that challenge us in architectural design. Feature recognition, as in mechanical engineering would soon appear to be too limited, since it excludes valuable information in early stages from the modelling process.

106

Table 1   Proposed classification of Generic Feature Types [van Leeuwen et al. 1996].

GENERIC FEATURE TYPES	BRIEF DESCRIPTION

Form Features
   Morphological Features
   Topological Features
   Geometrical Features
	Form Features describe the form, shape, or topology of other entities in the building model, which can be physical entities, but abstract as well.

Physical Features
   Compositional Features
   Material Features
   Composition performance Features
	Physical Features form the group of Features that describe the physical qualities, performances, and requirements of entities in the building model.

Context Features
   Design conceptual relation Features
   Interface Features
   Performance dependency Features
   Constraint Features
	Context Features define characteristics and concepts that form relationships between entities, such as dependencies, adjacencies, and for instance constraints, such as tolerances.

Procedural Features
   Planning Features
   Preparation Features
   Staging Features
   Integration Features
	Procedural Features include the type of information that somehow describes procedures related to the construction process, from the preparation of the work to the activities on site.

Life-cycle Features
   Functionality Features
   Operation Features
   Quantitative Features
   Maintenance Features
   Re-usability Features
   Security Features
	Life-cycle Features are the ones that describe concepts and characteristics that are especially relevant during the complete life-cycle of the building, particularly when it is being used, maintained, revised, renovated, or is given new functions. Also quantitative information such as costs falls within this category.

The design-by-Features approach, employing Features as the fundamentals for representing design information, seems to offer much more potentials in the development of a design-system that corresponds to the terminology of the design-domain. However, it is expected that techniques of Feature recognition, when combined with the design-by-Features approach, will be of significant importance in the support of design-tasks, especially when integrated in design-systems' interfaces to the underlying Feature-models. Our research currently focuses on the design-by-Features approach, recognition techniques will be subject of study at later stages.

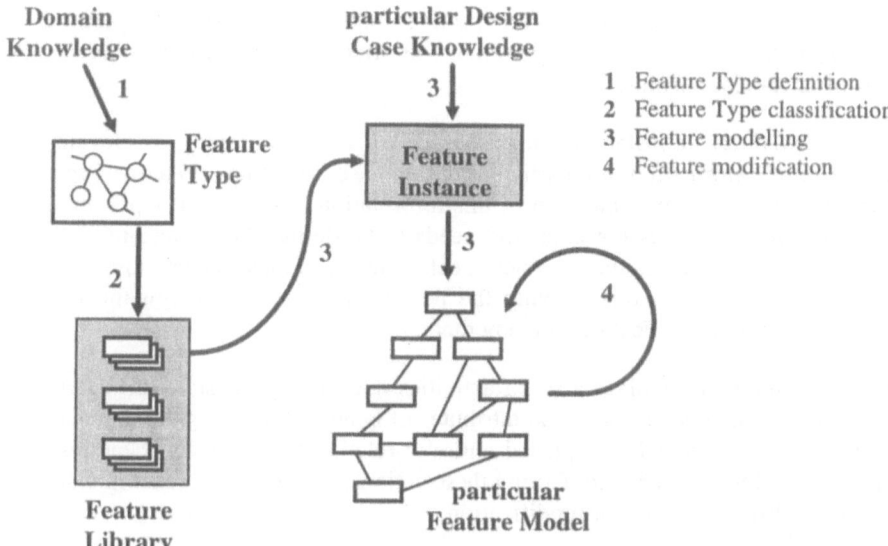

Domain
Knowledge

particular Design
Case Knowledge

1 Feature Type definition
2 Feature Type classification
3 Feature modelling
4 Feature modification

Feature
Type

Feature
Instance

Feature
Library

particular
Feature Model

Figure 4   Activities in Feature-based modelling [van Leeuwen et al. 1996]

The framework for Feature modelling describes the activities involved in defining types of Features and creating models of Features. To accommodate these activities, a schema for the definition of Feature Types and their instances has been developed. Figure 4 shows the activities of respectively Feature Type definition (1), Feature Type classification (2), Feature modelling (3), and Feature modification (4).

Feature Type definition is the activity of formalising domain knowledge into the definition of information entities. When a collection of information with relevant coherence has been identified, it can be formally described as a new Feature Type, either from scratch or based on previously defined Feature Types. Feature Types define the structure of their instances, by their state and behaviour. The state of Features is contained in a set of variables or attributes that obtain their values during the modelling activities. Behaviour of a Feature is a set of procedures that implement how a Feature reacts, either to events from outside the model, e.g. user-interactions, or to events from within the model, e.g. modifications to related Features. Relationships to other Feature Types can be expressed in attributes or be included in the behaviour of a Feature. Relationships are to be included in the type definition inasmuch as they are generic relationships, i.e. relevant for every Feature instance that will be based on this new type. Relationships that may occur for certain instances of the new type but that do not have generic relevance, should not be defined as part of the Feature Type, but will be defined during the modelling activities. This distinction between typical and occasional relationships is crucial for the flexibility of the Feature model that will result from Feature Type definitions.

Feature Type classification is the activity of classifying Feature Types into Feature Libraries which have the function of representing a particular domain. Feature

Libraries may be further divided into sections and mainly serve the purpose of organising collections of Feature Types that will often display a certain coherence by means of inter-relationships.

Feature modelling, or instantiation, is the sequence of identifying a relevant and coherent set of information in a particular design case; selecting a Feature Type that is appropriate for the representation of this information (however, it is possible that such a Feature Type does not exist and needs to be defined first); instantiating the Feature Type and its attributes (the new Feature instance may display some kind of behaviour in reaction to this); giving the new Feature its position in the Feature model, in relation to the Features already there.

Feature modification involves a set of activities that modify the state of Features or the structure of the Feature-network: attributes of Features may be given new values; Features may be removed or replaced entirely; relationships between Features may be modified, added or removed. During these actions, Features may react in response to the many different events of modification.

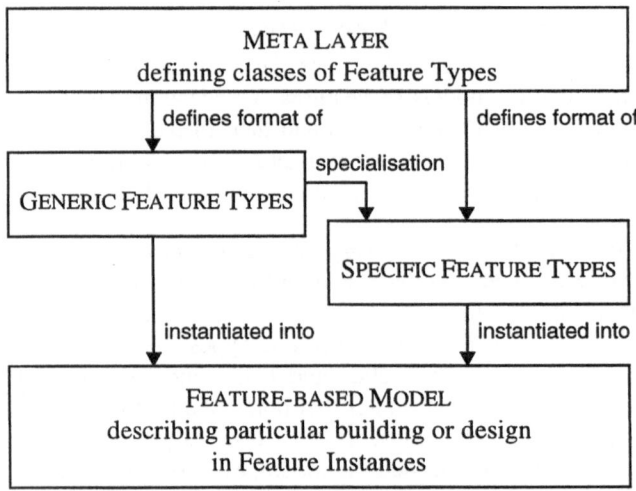

Figure 5   Infrastructure of Feature-based modelling

Schema for the definition of Feature Types and Feature instances   The   above activities require an infrastructure that provides the formal definition of Features as well as Feature Types. In a Feature modelling system, Features Types define the format for their instances, the actual Features in a particular model. Feature Types themselves are defined by either standardisation organisations or users, e.g. designers. However, the format of the definition of Feature Types needs to be defined independently of their contents. This format establishes the possibilities (and limitations) for formalising domain knowledge into Feature Types.

Figure 5 shows three layers of information-definition. The middle layer defines the Feature Types, both Generic and Specific. This is where the domain knowledge is formalised. The format of these type-definitions is laid out in the upper layer: the Meta Layer. This layer includes a formal description of what may be contained in a Feature Type: it defines the kind of attributes and procedures that make up a Feature Type's state and behaviour respectively. An EXPRESS-G representation of the main components of the Meta Layer is shown in figure 6.

The lower layer in figure 5 forms the level of Feature-based models that represent a particular building or design-case. Feature-based models consist of Features: instances of Feature Types. The domain knowledge which is generically formalised into Feature Types is here particularised with information from a specific case.

The following section discusses a prototype implementation of a so-called Feature-based Modelling Shell. This shell implements the infrastructure for Feature-based modelling and demonstrates the principles of extensibility and flexibility of information models.

4. Feature-based Modelling Shell: a prototype

The purpose of the prototype is to demonstrate the concepts of Feature-based modelling in the context of architectural design. The prototype will implement a selection of the contents of the meta layer and the infrastructure that is required for the related activities: definition and classification of Feature Types, and Feature instantiation and modification. We will use the prototype as a test-environment for the definition of Feature Types, especially we will start building a generic Feature Library based on the proposed classification of Feature Types. Flexibility of Feature models will be a major aspect in the demonstration of the Feature modelling processes.

4.1 DESIGN OF THE PROTOTYPE

Within the context of the above objectives for the prototype, the main requirements of the system are the following. The system supports the definition of the classes of Feature Types that are defined in the meta layer (see figure 6): simple Feature types, enumeration types, complex types, specialisation types, geometric types, and constraint types. The system organises Feature Types in Feature libraries. Feature models are created by instantiation of Feature types and their relationships, forming a network of Features. These models can be modified in terms of Feature attributes and by modification of the relationships between Features.

110

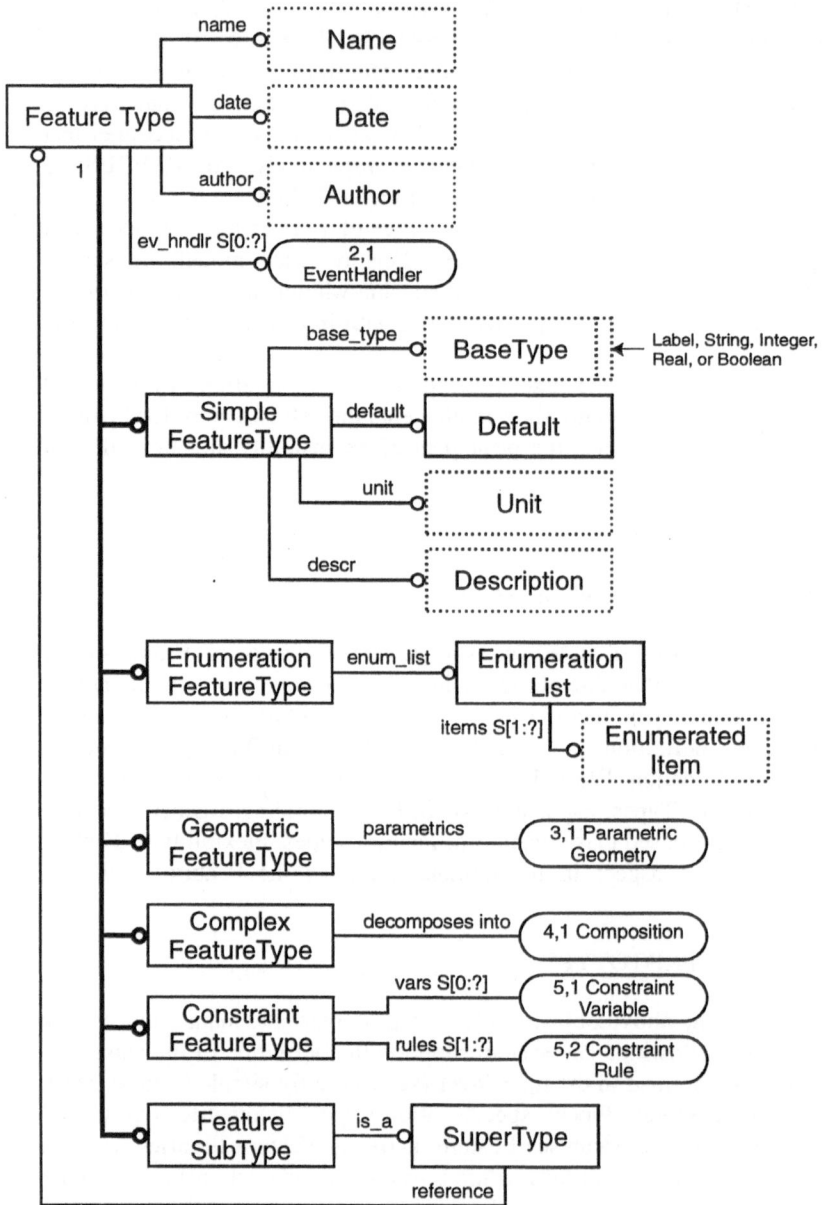

Figure 6 Schema of the Meta Layer representing the format for Feature Type definition (incomplete EXPRESS-G schema)

The system's design is divided into 4 modules which implement the activities of definition, classification, and instantiation, plus some processes required by the incorporation of knowledge into Feature definitions and the Feature modelling process. Some aspects of these modules are described below.

Feature Type definition module   The definition of Feature types involves the selection of a class from the meta layer, possibly the selection of a parent Feature Type, and the specification of the attributes and behaviour of the Feature Type. The attributes include some administrative data of the type such as *name, author, time-stamp, documentation*, etc., and the definition of the instances that the type allows. In the case of a simple Feature Type, the attributes include *data-type, unit*, and *default*. In the case of complex Feature Type, attributes involve the selection of Feature Types that form the components of the complex Feature Type. Geometric Feature Types contain a parametric description of the geometry.

Interface is important aspect in the definition of all Feature Types. A considerable part of the interface that will be used in the instantiation of new Feature Types, needs to be designed during Feature Type definition. This aspect ranges from the design of dialogue-based interfaces to complex graphic user-interaction for instance with existing geometry.

The behaviour incorporated in Feature Types is yet another aspect of Feature Type definition. Behaviour of Features is defined as their reaction to particular events occurring in the modelling system. In order to allow users to define this behaviour, some kind of programming facility will need to be supplied by the system, for instance an interpreted language such as basic or lisp. This facility needs to have access to the attributes of the Feature Type being defined as well as other existing Feature Types. The complexity of this behaviour-programming will increase with the requested possibilities it should offer, such as access to the user-interface, etc..

Feature Type classification module   After the definition process, a new Feature Type will need to be stored in an organised library of Feature Types. The organisation and management of Feature Types is very important throughout the complete system: for Feature Type definition, because some relationships between Feature Types may be included in their definition, which makes them inter-dependent; for Feature modelling, obviously, because the selection and searching for Feature Types is a crucial aspect of the modelling activity.

Feature modelling module   Initially, modelling Features involves the selection of an appropriate Feature Type, creating an instance thereof, positioning it in the network of Features that forms the Feature model, and possibly processing its immediate behaviour. Modelling Features and modifying Features or their relationships may invoke more reactive behaviour of Features, for example: a Feature that defines the structural dependency of a beam would react when one of the bearing components, say a column, would be removed from the model.

Constraint handling module   A constraint defines a relation that should hold, defined on one or more entities or constraint variables [Dohmen 1995]. We include constraint modelling by implementing a class Constraint Feature Types. In the Feature model, a Constraint Feature defines a relation between other Features which is a condition that needs to be satisfied for the model to be valid. The process of constraint satisfaction is subject of many research efforts, in architectural context [Gross 1990, Yoon 1992], in mathematical context [Saraswat and Van Hentenryck 1995], and in the context of Features technology, as reviewed by Dohmen [1995].

For the satisfaction of constraints, a so-called Constraint Handling Engine (CHE) needs to be integrated into the modelling system. A CHE interprets sets of constraints, using several mechanisms. These mechanisms are aimed at attempting to either satisfy the set of constraints, simplifying constraints or combinations of constraints, or adding new constraints that may help the process of satisfaction (constraint propagation).

The constraint handling module will be a separate module in the Feature-based Modelling Shell, however the concepts of constraints need to be addressed in the definition module as well.

4.2 ASSUMPTIONS AND BASIS FOR IMPLEMENTATION

We will briefly address some of the assumptions and points of departure for the implementation of the prototype. From the requirements for the system we concluded that the system best be based on an object oriented database management system (OODBMS). Geometry, although not addressed in early phases of prototyping, will eventually play an important role in the resulting Feature modelling system, the prototype should allow strong relations to a geometric modelling system, or even be integrated in one. The design-by-Feature approach of our research requires that, in the latter case, the Feature modelling system controls the geometric modelling system. Furthermore, for the scope of the prototype it appeared important to base any form of implementation as much as possible on available experience.

These assumptions have led to the following choices for the implementation environment. The prototype will be integrated in Autocad R13 using ARX technology. This choice implies the choice for Windows NT and MS Visual C++ for development environment. The ARX technology (Autocad Runtime eXtension) allows us:

- to have full control over the geometric modelling functionality of Autocad;
- to have direct access to the OO database that makes an Autocad drawing;
- to define new object classes in the OO database, geometric and non-geometric classes, which are fully integrated in the modeller;
- to use the OODBMS facilities in Autocad (serialisation);
- the full usage of C++ and Windows facilities.

4.3 SOME IMPLEMENTATION ISSUES

Adding new classes to an object oriented system generally requires compilation of the code defining these classes. Since the prototype Feature-based Modelling Shell has the purpose of allowing users of the system to add new Feature Types, this means that the system either offers very user-friendly capabilities for compiling code and linking the new software to the system, or avoids compilation of code and offers an alternative way of defining Feature Types.

The first approach, compiling new classes and linking them into the system (for instance using parameterised classes or templates), is the most elegant and robust one, however, it requires considerable technical knowledge and effort from the user. The second approach, avoiding compilation of new classes, has led to an approach which implements Feature Types as objects instantiated from classes that represent the meta-layer described in section 3.5. Feature Instances are implemented as objects instantiated from classes that are defined in correspondence to the classes for Feature Types. The instantiation of a Feature from a Feature Type, the activity of Feature modelling, is implemented as instructing a Feature Type to create a new Feature object of the corresponding class. In this approach, the functionality of Features in the model depends on the relationship and joint performance of the two objects 'Feature' and 'Feature Type'.

5. Conclusions and discussion

The concept of Feature modelling is experienced as being refreshing in discussions with experts, when attempting to break the discussion on product models which seems to have stranded on the issue of rigidity of standardised core models. The notion of definable information entities being applied in (early) design phases is attractive to designers who do not find adequate tools for innovative computer-aided design. The customisation of modelling systems for particular purposes, e.g. a specific industrial building system, is known to have great attention. Feature modelling can be expected to give this field of software-development a new stimulus.

The problem of data-explosion will not be solved by applying the Feature-based approach. However, this problem, often incorrectly indicated as a disadvantage of computer support for design, is inherent to the design discipline itself, not a result of using computers. The problem is being given attention since computers slowly start to provide ways of handling the vast amount of data involved in design.

Working with different levels of abstraction requires more detailed research. Many design theories or methodologies employ levels of abstraction or levels of detail in a flexible manner. By means of case-studies the implication and potentials of Feature-based modelling in working with abstraction levels will need to be investigated. First attempts will include an integration with the work of Achten [1996, 1997].

The representation and manipulation of networks of Features, the structure of Feature relationships, is an important subject for further research. Recently we have started research on this subject, investigating the possibilities of using a Virtual Reality (VR) interface to Feature models [Coomans 1997]. In the VR environment, both the abstract entities in the Feature model and the geometric components in the model will be visualised in an integrated approach. This approach will allow us to represent and manipulate physical and non-physical aspects of design in both modes.

An important issue in the definition of Feature Types is that Feature Types will act as knowledge bases in the modelling system. Designers will formalise generic domain knowledge into the behaviour or constraints defined in Feature Types. The principles and methods in knowledge engineering need to be studied and applied in the design and implementation of Feature-based modelling concepts.

Classification and standardisation of the Generic Feature Library will be the necessary basis for wide-spread application of the concept of Features technology in architecture. Although outside the scope of this paper, the relation with standardisation efforts in ISO 10303 (STEP) [ISO 1994] is evident, and the authors are eager to open the discussion on this subject and collaborate with experts in this field.

Acknowledgements

The research presented in this paper is a PhD. research under the supervision of Prof. Harry Wagter and Prof. Robert M. Oxman who's input to the project has been an invaluable support. Thanks also go to our colleagues Bauke de Vries, Henri Achten, and Marc Coomans for the many useful comments and discussions, and to the members of staff at Building Information Technology for their substantive technical and otherwise support.

References

Achten, H.H., Bax, M.F.T., and Oxman, R.M. (1996) Generic representations and the generic grid: knowledge interfaces, organisation and support of the (early) design process. Proceedings of the 3rd Conference on Design and Decision Support Systems in Architecture and Urban Planning, pp. 1-19. Spa, Belgium, August 18-21, 1996.

Achten, H.H. (1997) Generic representations - typical design without the use of types. These proceedings of the CAAD Futures 1997 Conference, München, Germany, August 3-7, 1997.

Coomans, M.K.D. (1997) Visualisation and manipulation of dynamic data structures in Virtual Reality. PhD-research proposal in preparation, Eindhoven University of Technology.

DeMartino, T., Falcidieno, B., Giannini, F., Hassinger, S., and Ovtcharova, J. (1994) Feature-based modelling by integrating design and recognition approaches. Computer-Aided Design vol. 26(8).

Dohmen, M. (1995) A survey of constraint satisfaction techniques for geometric modelling. Computers & Graphics vol. 19(6): pp. 831-845.

Eastman, C.M., Bond, A.H., and Chase, S.C. (1991) A data model for design databases, in J.S. Gero (ed.), Artificial intelligence in design '91, pp.339-365. Sydney, Butterworth Heinemann.

Eastman, C.M., Assal, H.H., and Jeng, T.S. (1995) Structure of a product database supporting model evolution. Modeling of buildings through their life-cycle (proceedings workshop CIB W78 '95), pp.327-338. Stanford University, CIB W78.

Ekholm, A., and Fridqvist, S. (1995) Object-oriented CAAD, Design object structure, and models for building, user organisation and site. Modeling of buildings through their life-cycle (proceedings workshop CIB W78 '95), pp.553-564. Stanford University, CIB W78.

Ekholm, A. (1996) A conceptual framework for classification of construction works. ITcon, electronic journal at http://www.fagg.uni-lj.si/~itcon/ vol. 1.

van Emmerik, M.J.G.M. (1990) Interactive design of parameterized 3D models by direct manipulation. Ph.D. thesis. Delft, Delft University Press.

Gross, M.D. (1990) Relational modelling: a basis for computer-assisted design, in McCullough, Mitchell, and Purcell (ed.), The electronic design studio, pp.123-136. Cambridge MA, The MIT Press.

Henderson, M.R., and Chang, G.J. (1988) FRAPP: Automated Feature Recognition and Process Planning from solid model data. Computers in Engineering vol. 1: pp.529-536. ASME.

ISO. (1994) ISO TC184/SC4 10303. Industrial automation systems and integration - Product data representation and exchange. ISO TC184/SC4.

Laakko, T., and Mäntylä, M. (1991) Feature modelling by incremental feature recognition. Computer-Aided Design vol. 25(8): pp.479-492.

van Leeuwen, J.P., Wagter, H., and Oxman, R.M. (1995) A Feature based approach to modelling architectural information. Modeling of buildings through their life-cycle (proceedings workshop CIB W78 '95), pp.260-269. Stanford University, CIB W78.

van Leeuwen, J.P., Wagter, H., and Oxman, R.M. (1996) Information modelling for design support - a Feature-based approach. Proceedings of the 3rd Conference on Design and Decision Support Systems in Architecture and Urban Planning, pp. 304-325. Spa, Belgium, August 18-21, 1996.

Meeran, S., and Pratt, M.J. (1993) Automated feature recognition from 2D drawings. Computer-Aided Design vol. 25(1): pp.7-17.

Mitchell, W.J. (1990) A new agenda for computer-aided design, in McCullough, Mitchell, and Purcell (ed.), The electronic design studio, pp.7. Cambridge MA, The MIT Press.

Saraswat, V., and Van Hentenryck, P. (eds.) (1995) Principles and practice of constraint programming - the newport papers. Cambridge, Massachussetts: The MIT Press.

Shah, J.J., and Rogers, M.T. (1988) Functional requirements and conceptual design of the Feature-Based Modelling System. Computer-Aided Engineering Journal vol. 5(1): pp.9-15.

Shah, J.J. (1991a) Conceptual development of form features and feature modelers. Research in Engineering Design vol. 1991(2): pp.93-108. New York, Springer-Verlag.

Shah, J.J. (1991b) Assessment of features technology. Computer-Aided Design vol. 23(5): pp.331-343.
Tolman, F.P., and Wix, J. (1995) Building Construction Core Model, ISO/WD 10303-106. Industrial automation systems and integration - Product data representation and exchange.

Woestenenk, K. (1995) Proposal for international conversion tables for parts and functions. The Netherlands, STABU.

Yoon, K.B. (1992) A constraint model of space planning. Southampton, UK: Computational Mechanics Publications.

GENERIC REPRESENTATIONS

Typical design without the use of types

HENRI ACHTEN
Eindhoven University of Technology. Faculty of Architecture, Building and Planning. Design Methods Group. P.O. Box 513. 5600 MB Eindhoven, The Netherlands

Abstract
The building type is a (knowledge) structure that is both recognised as a constitutive cognitive element of human thought and as a constitutive computational element in CAAD systems. Questions that seem unresolved up to now about computational approaches to building types are the relationship between the various instances that are generally recognised as belonging to a particular building type, the way a type can deal with varying briefs (or with mixed functional use), and how a type can accommodate different sites. Approaches that aim to model building types as data structures of interrelated variables (so-called 'prototypes') face problems clarifying these questions.
It is proposed in this research not to focus on a definition of 'type,' but rather to investigate the role of knowledge connected to building types in the design process. The basic proposition is that the graphic representations used to represent the state of the design object throughout the design process can be used as a medium to encode knowledge of the building type. This proposition claims that graphic representations consistently encode the things they represent, that it is possible to derive the knowledge content of graphic representations, and that there is enough diversity within graphic representations to support a design process of a building belonging to a type.
In order to substantiate these claims, it is necessary to analyse graphic representations. In the research work, an approach based on the notion of 'graphic units' is developed. The graphic unit is defined and the analysis of graphic representations on the basis of the graphic unit is demonstrated. This analysis brings forward the knowledge content of single graphic representations. Such knowledge content is declarative knowledge. The graphic unit also provides the means to articulate the transition from one graphic representation to another graphic representation. Such transitions encode procedural knowledge. The principles of a sequence of generic representations are discussed and it is demonstrated how a particular type - the office building type - is implemented in the theoretical work. Computational work on implementation part of a sequence of generic representations of the office building type is discussed. The paper ends with a summary and future work.

R. Junge (ed.), CAAD Futures 1997, 117-133.
© *1997 Kluwer Academic Publishers.*

1. Introduction

Both architectural theory (Argan 1963; Colquhoun 1981; Rossi 1982, p. 40-41; Westfall and van Pelt 1991, p. 140-144) and design methodology (Heath 1984, p. 121, p. 133; Habraken 1985, p. 23-36; Rowe 1987, p. 85-88, p. 190-194; Schön 1988) pose the building type as a constitutive element of architectural thought. Formulated generally, a building type constitutes knowledge of classes of buildings. It plays an important role in architectural design as it aids architects in both generating designs that belong to a specific class of buildings and to recognise buildings as belonging to a specific type. Types constitute a major source of architectural knowledge. CAAD systems therefore, can profit considerably if such kind of knowledge is implemented in design aid systems. Work on computational approaches towards type-like structures in architectural design (Gero 1990, Coyne *et al.* 1990, Oxman 1990, Rosenman and Gero 1993) faces problems about the relationship between the various instances that are generally recognised as belonging to a particular building type, the way a type can deal with varying briefs (or with mixed functional use), and how a type can accommodate different sites. This is due to the emphasis on the data structure and downplaying the importance of the design process. A more balanced relation between knowledge structure and design process seems required.

1.1. GRAPHIC REPRESENTATIONS

In the design process, graphic representations are a predominant element. They are a generally acknowledged medium through which the architect develops the design. The relationship between graphic representations, design decisions, and knowledge of building types is outlined in Figure 1. A single graphic representation depicts the state of the design in a particular stage of the design process. Through a sequence of graphic representations the design is worked out. Establishing a graphic representation requires taking design decisions (*e.g.* a contour of a building envelope requires decisions upon the shape, relative dimensions, and major building parts).

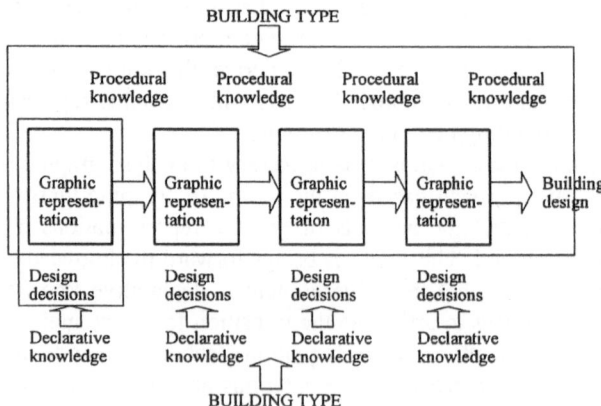

Figure 1. Relations between graphic representations, design decisions, and procedural and declarative knowledge of building types.

Making design decisions requires knowledge of the design task. This knowledge encompasses factual knowledge of the kind of building that is being designed. Such knowledge generally is indicated by the term declarative knowledge. Since in each graphic representation design decisions are involved, declarative knowledge is required in each graphic representation. During the design process, the design evolves through a number of graphic representations. Such a sequence of graphic representations reflects a sequence of design decisions. Knowledge of the order of design decisions is generally termed procedural knowledge.

1.2. GRAPHIC UNITS AND GENERIC REPRESENTATIONS

If it is possible to consistently describe design decisions in graphic representations and transitions between graphic representations (arrows in Figure 1), graphic representations can be the basis for modelling design processes. In this paper, an approach is presented that describes knowledge content and transitions by means of graphic entities present in the graphic representation. Graphic entities such as lines, shapes, and symbols conform to conventions of depiction and conventions of encoding. This means that a certain set of graphic entities (*e.g.* a set of lines) has a particular meaning (*e.g.* a system of axes, a grid, zoning). Such a set is connected to different design decisions (*e.g.* a basic ordering of spaces along axes, a field defining place and measure, and areas with specific properties). Therefore, it is necessary to distinguish between sets of graphic entities that have different meanings. Such a set is termed a "graphic unit": a set of graphic entities with a specific meaning. Description by means of graphic units results in generalised graphic representations. Graphic representations that share the same graphic units deal with the same design decisions. Such generalised graphic representations are termed "generic representations."

2. Identifying Graphic Units in Graphic Representations

By means of analysing graphic representations it is possible to identify graphic units. For this purpose, it is necessary to compare graphic representations and identify similarities and differences. In the research, 220 graphic representations selected from books are analysed. This is demonstrated by the examples in Figure 2.

- Circle, triangle, and square are instances of regular n-sided polygons (n=3, 4, 5,...), including the circle. They are characterised by the term "simple contour." Under the assumption that the graphic representation depicts a building, the "simple contour" represents the building envelope.
- The forms that are part of the layout are closed polygonal shapes. Under the assumption that the graphic representation depicts a building, this drawing represents a differentiated building layout, where the lines of the shapes indicate borders between major spaces.

120

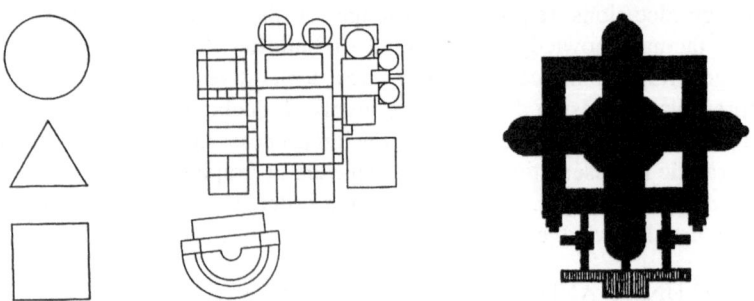

Figure 2. Three graphic representations. Left: three contours (Ching 1979, p. 54). Middle: layout (Mitchell and McCullough 1991, p. 136). Right: figure-ground drawing (Zevi 1974, p. 51).

- The shapes of the layout are not always "simple contours," but in all cases they are "contours." The act of drawing a *simple contour* indicates the decision to limit the possible shape to a specific class of shapes (regular n-sided polygons). Drawing a *combination of contours*, as in the layout, indicates the decision to use particular shapes and to establish their relations concerning place and relative scale.
- Therefore, two graphic units can be distinguished: *simple contour* and *contour*, which are instances of two generic representations: simple contour (which has one graphic unit, the *simple contour*), and combination of contours (which has one graphic unit, the *contour*).

Although both the layout (Figure 2; middle) and the figure-ground drawing (Figure 2; right) depict combinations of contours, they are different from each other:

- In the layout, all shapes are represented by lines only. Under the assumption that the graphic representation depicts a building, the drawing represents a differentiated building layout, where the lines of the shapes indicate borders between major spaces.
- In the figure-ground drawing, graphic distinctions are made by colour (black or hatching pattern) in complex shapes. Under the assumption that the graphic representation depicts a building, it represents the mass-space distribution of the building, where the lines and edges indicate borders between space and mass, and the colours (black, white, hatched) identify either mass or space.
- In the layout, there is no distinction between mass and space. In the figure-ground drawing there is a distinction between mass and space. Both drawings imply different design decisions. In the figure-ground drawing, the filled-in black and hatched drawing imply the decision how to articulate mass and space and their edges.
- Therefore, a new graphic unit can be identified: *complementary contours*, and a new generic representation: complementary contours (which has one graphic unit: *complementary contours*).

Describing the form aspects of graphic units results in a vocabulary which uses terms such as regular n-sided (n=*3, 4, 5, ...*) polygonal shapes, closed polygonal shapes, filled-in (black, white, hatched, etc.) polygonal shapes, interlocked surfaces, etc. for shapes; single line, double line, line weight, linetype, etc. for lines; and direction, parallel, module, irregular distance, colour, and hatching, etc. for describing sets of graphic entities. The degree of specification of the vocabulary demonstrates that it is important to carefully distinguish between graphic entities that occur in a drawing.

3. Generic representations

As the examples above show, graphic representations are very diverse in appearance. However, it appears that if a number of graphic representations have the same graphic units, then no matter how different they may seem, it is possible to state that they deal with the same design decisions. Such groups of graphic representations with the same graphic units are generic representations. In the manner outlined above, graphic representations are described in terms of graphic units. In the analysis 220 graphic representations taken from architectural sources are analysed, resulting in 24 graphic units and 50 generic representations.

A generic representation can be described by the following features: (1) *name*, (2) *source*, (3) *graphic representation*, (4) *textual description*, (5) *graphic units*, and (6) *iconic representation* (see Figure 3). The following pages show two cases.

Name of generic representation	
Graphic representations: - picture of case - *Image from the source list on right side of table* - source of picture - *Text identifying the image*	**Sources:** - list of sources and pictures - *Sources of the images, place in the source, and brief description of the image included in the graphic representation section* **Description:** - description of graphic representations - *The use in the design process, related design decisions, and graphic units found. Graphic units are named and numbered.*
Graphic units: - drawings of graphic units - *The graphic units as they occur in the graphic representations section above. It is possible that graphic representations have more than one graphic unit. The numbers correspond to the text in the description section.*	**Icon:** - schematic representation - *The salient features of the graphic representations are shown by a drawing consisting of graphic units. In this way, the properties of the diverse drawings in the graphic representations section are made clear. The icon therefore, shows the generic representation that can be derived from the graphic representations of the case study.*

Figure 3: The format of presentation of generic representations.

122

Contour in grid	
Graphic representations: 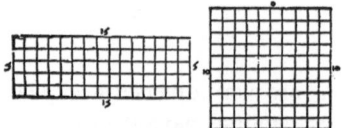 Serlio (1611), First Book First Chapter, Fol. 7 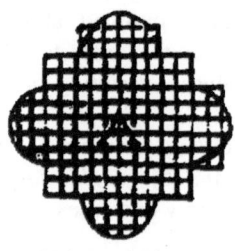 Cesariano (1521), p. 239 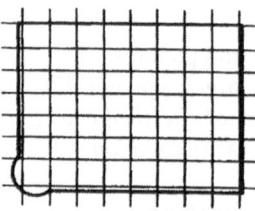 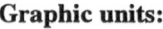 Sullivan in Clark and Pause (1985), p. 117	**Sources:** Serlio (1611), First Book, First Chapter Fol. 7. Grid argument for demonstrating different surface areas with the same perimeter. Cesariano (1521) in Tzonis (1986), p. 21 figure 10 top drawing. 'Grid pattern.' Ching (1979), p. 239 figure B, D, H Sullivan in Clark and Pause (1985), p. 117 figure F. Geometry in Carson Pirie and Scott Store, Chicago, Illinois. **Description:** The grid structures the place of elements, such as the perimeter, or columns. Not every part of the contour has to conform to the grid. When the perimeter follows the grid, this establishes a surface area unit that can be used in the building to co-ordinate rooms and spaces. If the grid is used for a structural system it is sometimes kept distant from the facade in order to resolve conflicts between columns and walls. Decisions concern the relationship between contour and grid, in particular the dimension of the module with which the contour is measured. Graphic units are the *grid*: orthogonal set of lines (16) and *contour*: closed polygonal shapes (2).
Graphic units: 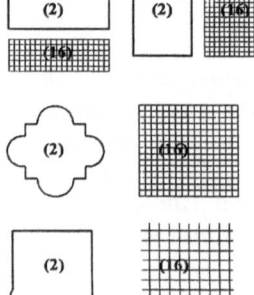	**Icon:** 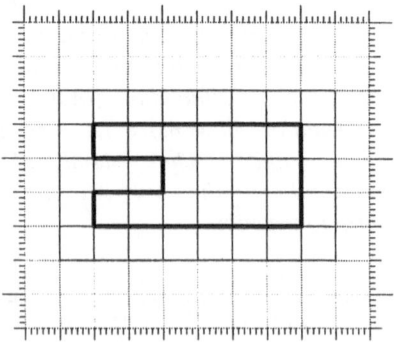

Schematic subdivision with function symbols

Graphic representations:	Sources:

Graphic representations:

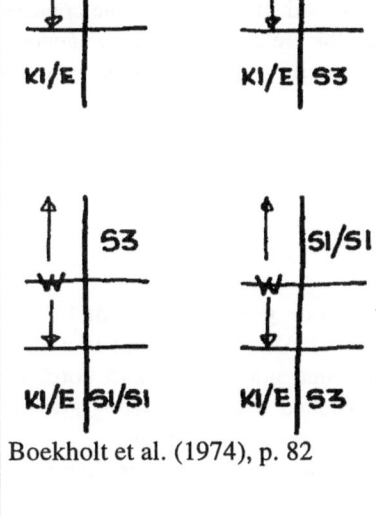

Boekholt et al. (1974), p. 82

Sources:
Boekholt et al. (1974), p. 82 figure 1 bottom figure.
Sector analysis of a basic variant in housing.

Description:
Schematic subdivision with function symbols demonstrates a general principle of subdivision in combination with the assignment of functions. The figure shows a SAR-representation of a so-called basic variant: different functional layouts within a particular subdivision. The symbols W, K1/E, S1/S2, and S3 denote functions: W (living), K1/E (cooking, eating), S1/S2 (single person sleeping), and S3 (master bedroom sleeping). The arrow denotes in which area of the subdivision the function is allocated.

Decisions in this graphic representation concern the principle relations of place between functions and their position in a general subdivision.

Graphic units are *schematic subdivision*: lines (10) and *function symbols*: letters and numbers indicating functions (7).

Graphic units:

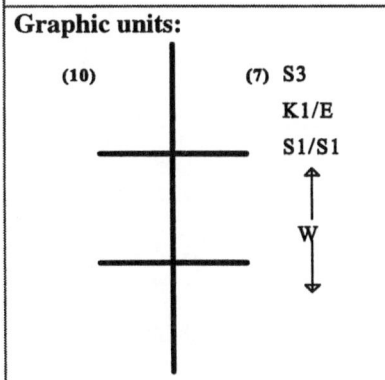

(10)   (7) S3
           K1/E
           S1/S1
           W

Icon:

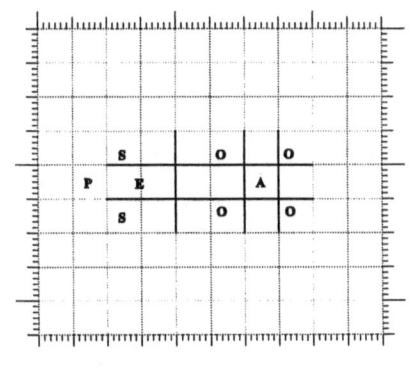

4. Relations Between Generic Representations

The list of graphic units and generic representations is presented in Appendices I and II. At this point it is necessary to identify how in a design process transitions from one generic representation to the next are established. The claim of the hypothesis of the research work is that graphic representations are the medium for supporting design processes. Therefore, the properties of graphic representations identified in this work must be used in order to show how sequences of generic representations are established. It means that the concept of the graphic unit must be used for elaborating the relations between generic representations. Three kinds of relations between generic representations can be distinguished:
1. Addition.
2. Themes.
3. Succession.

4.1. ADDITION

Addition of graphic units identifies all transitions from one generic representation to another by means of addition of graphic units. In each 'step' a graphic unit is added. Therefore, the transition is by definition from simple (one graphic unit) to complex (more graphic units). The new graphic unit has to be matched with the existing graphic units of the previous generic representation. Figure 4 illustrates such relations.
Applying the notion of addition of addition to generic representations identifies sequences of generic representations that become more complex.

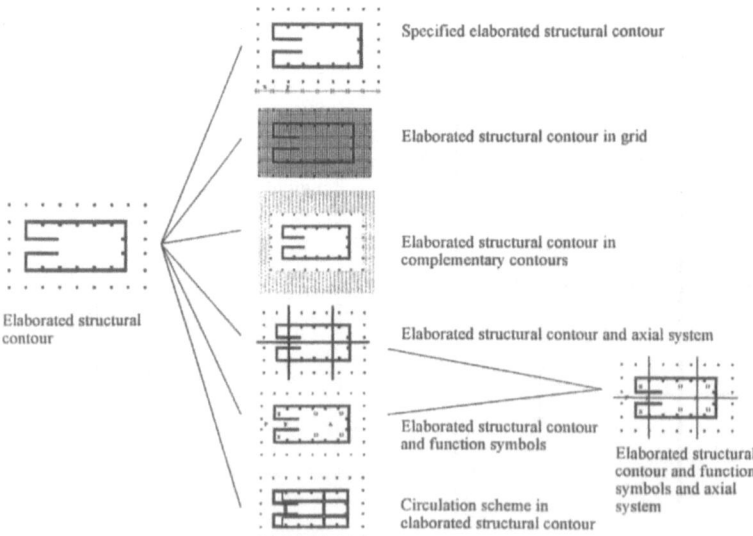

Figure 4. Addition of graphic units to a generic representation.

4.2. THEMES

Similarity between generic representations indicates if they deal with similar design decisions. For example, the generic representations <u>simple contour</u>, <u>combination of contours</u>, and <u>complementary contours</u> are the only generic representations that deal with the shape and place of the building edge exclusively. They constitute a theme, which is called "shape." In the same manner, other themes can be established. The generic representations in a theme develop independent from generic representations in other themes. By combining generic representations from themes (addition of graphic units), it is possible to establish more complex generic representations that deal with more sophisticated design decisions. Themes that are found on the basis of generic representations are: "shape," "system," "structure," and combinations of themes "shape and system," "shape and structure," "structure and system," and "shape and system and structure." Applying the notion of themes to generic representations results in groups of generic representations that deal with the same design issues.

4.3. SUCCESSION

A generic representation provides preconditions for more elaborate generic representations if one or more of its constituent graphic units provides such preconditions. A generic representation implies more schematic or less specific generic representations if one or more of its graphic units implies more schematic or less specific graphic units. For example, before the particular length of a wing is decided upon, the decision has been taken that the shape of the building actually consists of a number of wings. In terms of graphic units this means that the *contour* (a shape with no particular dimensions) is established before the *specified form* (a contour with particular dimensions). By analysing graphic units, it is possible to identify such sequences. These are called "successive graphic units":

1. *Contour → specified form → combinatorial element vocabulary → elaborated structural contour*
2. *Simple contour → specified form*
3. *Contour → complementary contours*
4. *Function symbols → zone → functional space → element vocabulary*
5. *Modular field → grid → refinement grid → tartan grid → structural tartan grid*
6. *Structural tartan grid → structural element vocabulary*
7. *Measurement device → proportion system*
8. *Schematic subdivision → partitioning system*
9. *Schematic axial system → axial system*
10. *Circulation scheme → circulation*

Applying successive graphic units to generic representations of a theme results in sequences of generic representations that develop a particular issue (shape, system, structure, etc.) from general to specific (less defined to more strict defined).

126

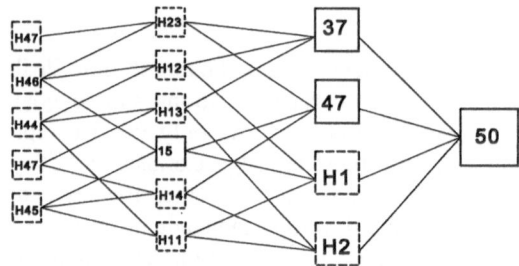

Figure 5. Identifying hypothetical generic representations (H1, H2, etc.) by addition of graphic units ending in found generic representations.

5. Missing Generic Representations

The relations of addition, theme, and succession can be applied to the set of generic representations found in the analysis. A number of generic representations is not related to each other by these relations. For example, the generic representation <u>schematic subdivision</u> in <u>zone</u> in <u>contour</u> with <u>function symbols</u> (number 49 in Appendix II) is not established by addition of either *function symbols, contour, zone,* or *schematic subdivision* to any generic representation 35 - 47 (see list of Appendix II). Given the number of possible generic representations[1] this leads to the conclusion that there are generic representations missing in the survey.

Given the aim to use generic representations as a medium for encoding knowledge of building types, it does not seem very productive to add all possible generic representations to the set of found generic representations. It is possible to limit this number by adding the constraint that any sequence of addition of graphic units must terminate in a generic representation that is part of the set found in the survey or in a hypothetical generic representation that is required for reaching a generic representation found in the survey (see Figure 5). This strategy yields a total of *106* generic representations of which *50* generic representations are found in the survey and *56* are hypothetical generic representations (see Appendix III). These hypothetical generic representations are embedded in the set of generic representations found in the survey. It is possible to state their properties on the basis of the constituent graphic units. In the remainder of this paper hypothetical generic representations are included. Figure 6 shows how found and hypothetical generic representations are related to each other in the theme "structure." Five groups of structures are identified. From left to right: grid, proportion, axial system, subdivision, and zone. Each group of generic representations within the theme deals with an increasing level of specification with its structures.

---

[1] The analysis results in the identification of 24 graphic units. Generic representations identified in the analysis have no more than four different graphic units. Therefore, all possible generic representations are generated by combining one up to four graphic units. This gives a total of $24! / (4! \cdot 20!) + 24! / (3! \cdot 21!) + 24! / (2! \cdot 22!) + 24 = 12950$ possible generic representations.

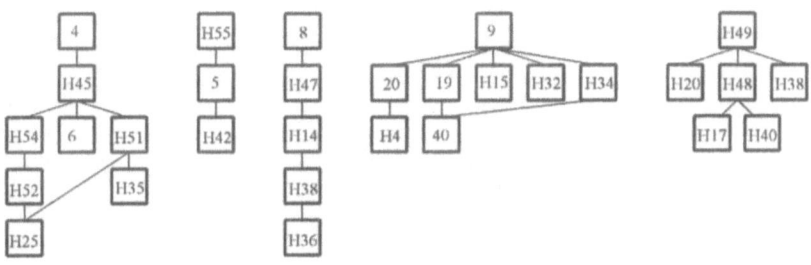

Figure 6. Found and hypothetical generic representations in theme "structure."

6. General Sequences of Generic Representations

With the set of found and hypothetical generic representations it is possible to establish sequences of generic representations. By means of the relations of additional and successive graphic units generic representations have been linked to each other within themes. Therefore, a sequence of generic representations can be based on a sequence of themes. This is illustrated in Figure 7.

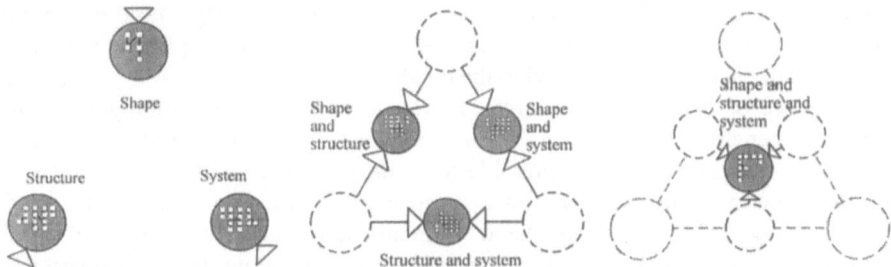

Figure 7. Three major steps in a sequence of generic representations.

In the first step (Figure 7; left) for each of the themes "structure," "shape," and "system," the relations between generic representations are established. These provide the starting points for a sequence. It means that any sequence must start either with generic representations dealing with "structure" (grid, zone, axes, subdivision), "shape" (contour), or "system" (circulation, functional, structural). In the second step (Figure 7; middle) for each of the combinations of themes, the relations between generic representations are established. The sequence continues with generic representations that combine "shape and structure" (matching the building layout with its internal organisation), "shape and system" (matches the building layout with the systems), or "structure and system" (matches the systems with the internal organisation). In the third step (Figure 7; right) for the combination of all themes, the relations between generic representations are established. The building layout is matched with its internal organisation and systems. All sequences terminate in this theme. By means of this strategy, it is possible to define six different sequences of themes that establish a sequence of generic representations (see Figure 8).

128

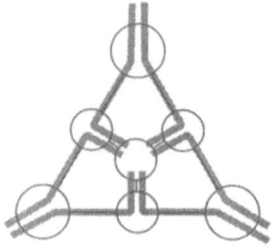

Figure 8. Six possible sequences of successive themes (bold gray lines).

7. A Particular Sequence of Generic Representations

The general sequence of generic representations becomes particular when a design task is given. This makes it possible to test to which extent generic representations can support a design process. As stated in Figure 1, the building type provides declarative knowledge required for making design decisions in each generic representation of the sequence. Procedural knowledge of the building type is encoded in the sequence of generic representations. The building type chosen for the design task is the office building. It is a type which is well covered in publications which simplifies the process to acquire declarative knowledge related to the building type.

The procedure of mapping declarative knowledge of the office building type on the sequence of generic representations is discussed in detail in Achten et al. (1995). For each generic representation, it is possible to identify the pieces of knowledge required from the office building. When put in a sequence, this models a design process of the office building. The work by hand demonstrates that the theory of generic representations aids in knowledge acquisition of a building type. Furthermore, it demonstrates that application of generic representations to the office building type leads to a sequence of different graphic representations (see Figure 9; Achten et al. 1996 shows the complete list).

Implementation issues are presented and discussed in Achten et al. (1995). It shows the computational work and contains also contains source code of the program implemented in a CAD system. To summarise, the computational approach underlying the structure of the system has been the use of frames for generic knowledge. Each slot of the frame refers to one generic representation. The sequence of slots encodes the sequence of generic representations. The particular sequence of generic representations uses seven generic representations of the general sequence. The system is programmed in AutoLISP in an AutoCAD environment. At the end of the sequence the program aids the architect in developing orientation, size, basic layout, and functional organisation via zoning of the office building. Although the implementation uses only a limited set of generic representations, it is large enough to provide additional weight to the conclusion that the theory of generic representations can model procedural and declarative knowledge.

Generic representation	Name and some characteristics
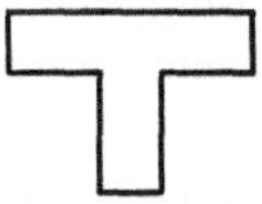	*Contour* (H46; see Appendix III) Defining the outward form of the building. Establishing the T-shape; triple-winged building. Surface area. Parametrise wing-length.
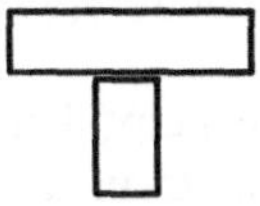	Combination of contours (2; see Appendix II) Composing ensemble of simple contours to establish overall shape. Define internal proportions and place of simple contours. Explore emergent forms.
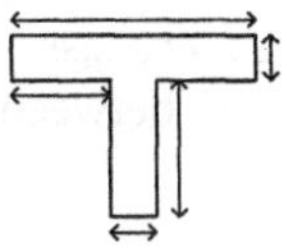	*Specified form* (H50) Establish tentative dimensions for wing length and depth, and orientation of the building.

Figure 9. Three generic representations of the office building.

8. Discussion

At this point, the following is established in the research work:
- The graphic unit is defined and identified in an analysis of graphic representations.
- The generic representation is defined and identified on the basis of graphic units.
- The knowledge content of generic representation is identified.
- Relations between generic representations are defined and applied.
- Missing generic representations are identified on the basis of graphic units.
- Six general sequences for design processes are identified.
- A particular sequence for the office building is established.

The work demonstrates that it is possible to establish knowledge content of graphic representations by means of the graphic unit. It identifies such issues as building envelope, structuring devices such as grid, zone, subdivision, axial systems, systems such as circulation, function, structural systems, and combinations of these issues. It is demonstrated that generic representations develop these issues from schematic to more specific. These issues are relevant especially in the early stages of design which leads to the concept design of the building.

The fact that generic representations are derived from architectural sources such as books, indicates that the notions of knowledge encoding and knowledge transferring via graphic representations is well-established within the architectural community. The analysis shows that such architectural ways of communicating knowledge can be consistently described and used in computational environments.

Both procedural and declarative knowledge of the building type are encoded in the sequence of generic representations. It is not clear at this point however, how far in the design process generic representations can be supportive. In the work on generic representations, attention is given to the process and the role of knowledge in the process. It is demonstrated that combination of a particular process laid down in the sequence of generic representations and declarative knowledge of a building type leads to designs belonging to that type. This is done without the use of an explicit type-like structure such as a prototype.

The work demonstrates that the strategy of combining process and knowledge has potential for developing typical design with the use of overt types, that is, abstract data structures which have to be sequentially particularised. It may prompt another approach to the role of building type in computational environments.

By showing how generic representations encode design issues that are helpful in the early stage of design, the work points to directions to support architects in a more architectural fashion in design aid systems. A CAAD system that operates via generic representations or that could identify generic representations in a drawing is able to identify the knowledge required at that particular stage in the design process. The implementation of generic representations would result in an extra 'layer' between the architect, the graphic user interface, and the computer (Achten 1996a).

Future work has to address the following questions:

- Can generic representations be applied to other building types.
- Is it possible to consistently interpret sketches by means of graphic units.
- How can generic representations can be implemented in a responsive and interactive manner.
- To which extent do different knowledge bases generate different designs
- Finding cases of graphic representations that belong to hypothetical generic representations proposed in the research work.

Acknowledgements

The research work is a Ph.D. project undertaken in the Design Methods Group of the Faculty of Architecture, Building, and Planning under supervision of Prof.dr.ir. M.F.Th. Bax and Prof.dr. R.M. Oxman.

Appendix I: Graphic Units Found in the Analysis.

1. *Simple contour*
2. *Contour*
3. *Specified form*
4. *Elaborated structural contour*
5. *Complementary contours*
6. *Functional space*
7. *Function symbols*
8. *Modular field*
9. *Grid*
10. *Refinement grid*
11. *Tartan grid*
12. *Structural tartan grid*
13. *Proportion system*
14. *Measurement device*
15. *Zone*
16. *Schematic subdivision*
17. *Partitioning system*
18. *Schematic axial system*
19. *Axial system*
20. *Element vocabulary*
21. *Structural element vocabulary*
22. *Combinatorial element vocabulary*
23. *Circulation scheme*
24. *Circulation*

Appendix II. Generic Representations Found in the Analysis.

Generic representations with one graphic unit:

1. Simple contour
2. Combination of contours
3. Complementary contours
4. Modular field
5. Proportion system
6. Multiple grids
7. Functional spaces

8. Schematic axial system
9. Schematic subdivision
10. Elaborated structural contour
11. Element vocabulary
12. Combinatorial element vocabulary
13. Circulation scheme

Generic representations with two graphic units:

14. Proportion system in contour
15. Contour in grid
16. Zone in specified form
17. Function symbols in combination of contours
18. Axial system in specified form
19. Schematic subdivision in grid
20. Schematic subdivision with function symbols
21. Schematic subdivision in contour
22. Partitioning system in contour
23. Specified elaborated structural contour
24. Elaborated structural contour in grid
25. Elaborated structural contour in complementary contours

26. Elaborated structural contour and axial system
27. Elaborated structural contour and function symbols
28. Element vocabulary in grid
29. Element vocabulary in multiple grids
30. Combinatorial element vocabulary in grid
31. Combinatorial element vocabulary in specified form
32. Circulation in contour
33. Circulation scheme in elaborated structural contour
34. Structural element vocabulary in structural tartan grid

Generic representations with three graphic units:

35. Proportion system in elaborated structural contour in tartan grid
36. Zone in contour in grid
37. Axial system in contour in grid
38. Axial system in contour in tartan grid
39. Axial system in specified form in structural tartan grid
40. Schematic subdivision in grid and refinement grid
41. Schematic subdivision and schematic axial system in contour

42. Elaborated structural contour and function symbols and axial system
43. Element vocabulary in zone and contour
44. Circulation in contour in grid
45. Structural element vocabulary in contour in modular field
46. Structural element vocabulary in structural tartan grid and refinement grid
47. Structural element vocabulary in axial system in contour

Generic representations with four graphic units:

48. Element vocabulary and function symbols and grid in specified form
49. Schematic subdivision in zone in contour with function symbols
50. Structural element vocabulary in axial system in contour in grid

132

Appendix III. Hypothetical generic representations

Hypothetical generic representations with one graphic unit:

H44. *Structural element vocabulary*
H45. *Grid*
H46. *Contour*
H47. *Axial system*
H48. *Zone*
H49. *Function symbols*
H50. *Specified form*

H51. *Refinement grid*
H52. *Structural tartan grid*
H53. *Circulation*
H54. *Tartan grid*
H55. *Measurement device*
H56. *Partitioning system*

Hypothetical generic representations with two graphic units:

H11. *Structural element vocabulary* in *grid*
H12. *Structural element vocabulary* in *contour*
H13. *Structural element vocabulary* in *axial system*
H14. *Axial system* in *grid*
H15. *Schematic subdivision* in *zone*
H16. *Zone* in *contour*
H17. *Function symbols* in *zone*
H18. *Function symbols* in *contour*
H19. *Element vocabulary* and *function symbols*
H20. *Function symbols* in *grid*
H21. *Function symbols* in *specified form*
H22. *Specified form* in *grid*
H23. *Axial system* in *contour*
H24. *Structural element vocabulary* in *refinement grid*
H25. *Structural tartan grid* in *refinement grid*
H26. *Structural element vocabulary* in *modular field*

H27. *Contour* in *modular field*
H28. *Circulation* in *grid*
H29. *Element vocabulary* in *zone*
H30. *Element vocabulary* in *contour*
H31. *Function symbols* and *axial system*
H32. *Schematic subdivision* and *schematic axial system*
H33. *Schematic axial system* in *contour*
H34. *Schematic subdivision* in *refinement grid*
H35. *Grid* in *refinement grid*
H36. *Axial system* in *structural tartan grid*
H37. *Specified form* in *structural tartan grid*
H38. *Axial system* in *tartan grid*
H39. *Contour* in *tartan grid*
H40. *Zone* in *grid*
H41. *Proportion system* in *elaborated structural contour*
H42. *Proportion system* in *tartan grid*
H43. *Elaborated structural contour* in *tartan grid*

Hypothetical generic representations with three graphic units:

H1. *Structural element vocabulary* in *contour* in *grid*
H2. *Structural element vocabulary* in *axial system* in *grid*
H3. *Schematic subdivision* in *zone* in *contour*
H4. *Schematic subdivision* in *zone* with *function functions*
H5. *Schematic subdivision* in *contour* with *function symbols*

H6. *Zone* in *contour* with *function symbols*
H7. *Element vocabulary* and *function symbols* in *grid*
H8. *Element vocabulary* and *function symbols* in *specified form*
H9. *Element vocabulary* in *specified form* in *grid*
H10. *Function symbols* in *specified form* in *grid*

References

Achten, H.H. and Dijkstra, J. and Oxman, R. and Bax, M.F.Th. (1995) *Knowledge-Based Systems Programming for Knowledge Intensive Teaching*. BIT-Note Publication 1995/3. Department of Architecture, Eindhoven University of Technology

Achten, H.H. (1996a). Generic Representations: Intermediate Structures in Computer Aided Architectural Composition. Asanowicz, A. and Jakimowicz, A. (eds.), 1996 *Approaches to Computer Aided Architectural Composition*, Technical University of Bialystok, Bialystok

Achten, H.H. and Bax, M.F.Th. and Oxman, R.M. (1996). Generic Representations and the Generic Grid: Knowledge Interface, Organisation and Support of the (early) Design Process. Timmermans, H. (ed.), 1996 *Proceedings of the 3rd Design & Decision Support Systems in Architecture & Urban Planning Conference*, Spa, Belgium

Akin, Ö. (1986) Psychology of Architectural Design, Pion Limited, London

Argan, G.C. (1963) On the typology of architecture. Papadakis, A. and Watson H. (eds) 1990 New Classicism, SDU Publishers, The Hague, 117-118

Boekholt et al. (1974). Denken in Varianten. Samson Uitgeverij, Alphen aan den Rijn

Clark, R.H. and Pause, M. (1985) Precedents in Architecture. Van Nostrand Reinhold, New York

Ching, F.D.K. (1979) Architecture: Form, Space and Order. Van Nostrand Reinhold, New York

Colquhoun, A. (1967). Typology and design Method. Essays in Architectural Criticism: Modern Architecture and Historical Change, The MIT Press, Cambridge, Massachusetts, 43-50.

Coyne, R.D. and Rosenman, M.A. and Radford, A.D. (1990) Knowledge-Based Design systems, Addison-Wesley Publishing Company, Reading, Massachusetts

Mitchell, W.J. and McCullough, M., (1991) Digital Design Media. Van Nostrand Reinhold, New York

Gero, J.S. (1990) Design prototypes: a knowledge representation schema for design, AI Magazine, Winter, 26-36

Heath, T. (1984) Method in Architecture, John Wiley & Sons Ltd., Chichester

Habraken, N.J. (1985) The Appearance of the Form, Awater Press, Cambridge, Massachusetts

Oxman, R. (1990) Architectural knowledge structures as "design shells": a knowledge-based view of design and CAAD education. McCullough, M. and W.J. Mitchell and P. Purcell (eds) (1990) The Electronic Design Studio, The MIT Press, Cambridge, Massachusetts, 187-200

Rosenman, M.A. and Gero, J.S. (1993) Creativity in design using a design prototype approach. Gero, J.S. and Maher, M.L. (eds.) Modelling Creativity and Knowledge-Based Creative Design, Lawrence Erlbaum Associates, Hillsdale, 111-138

Rossi, A. (1982) The Architecture of the City. The MIT Press, Cambridge, Massachusetts

Rowe, P.G. (1987) Design Thinking, The MIT Press, Cambridge, Massachusetts, London

Schön, D.A. (1988) Designing: rules, types and worlds, Design Studies, Vol 9 No 3, July 1988, 181-190

Serlio, S., (1611) The Five Books of Architecture. Reprint 1982 Dover, New York

Tzonis, A. and Lefaivre, L., (1986) Classical Architecture. MIT Press, Cambridge, Massachusetts

Westfall, C.W. & Pelt, R.J. van (1991) Architectural Principles in the Age of Historicism, Yale University Press, New Haven/London

Zevi, B., (1948) Architecture as Space - How to look at architecture. Revised edition, Da Capo Press, New York

ARMILLA5 - Supporting Design, Construction and Management of Complex Buildings

LUDGER HOVESTADT
University of Kaiserslautern - ARUBI:CPE - Germany - lhov@rhrk.uni-kl.de
VOLKMAR HOVESTADT
University of Karlsruhe - IFIB - Germany - volkmar@ifib.uni-karlsruhe.de

Abstract: ARMILLA5 is a generic computer aided design system, which supports the cooperative design of complex buildings (such as labs, offices or schools) over multiple levels of abstraction. It follows the metaphor of a virtual building site. The designers and engineers meet at a spatial location on the InterNET and prepare the building construction by simulating the building site. This article describes the three essential components of the ARMILLA5-model: the geometric model which describes the spatial and physical aspects of the building site, the semantic model which implements passive building components as objects and active building components as applets or applications, and the planning model, which organizes the work steps of the individual engineers and their cooperation. The model is described using different software prototypes written in Objective_C, CAD systems and HTML/JAVA.

Keywords: Dynamic Buildings, CAAD, CSCW, VRML, Casebased Reasoning, Facility Management, Augmented Reality

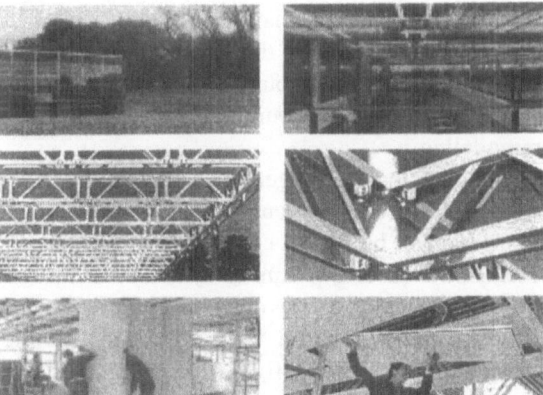

Figure 1: The Swiss railroad's instruction building in Murten, CH in the MIDI construction system. Exterior and interior views, MIDI construction, detail of support column, assembly of interior wall system and suspended ceilings. All construction components used can be dismantled, and their configuration can easily be changed. The building was completed in 1982 and has been renovated four times since.

R. Junge (ed.), CAAD Futures 1997, 135-150.
© 1997 *Kluwer Academic Publishers.*

136

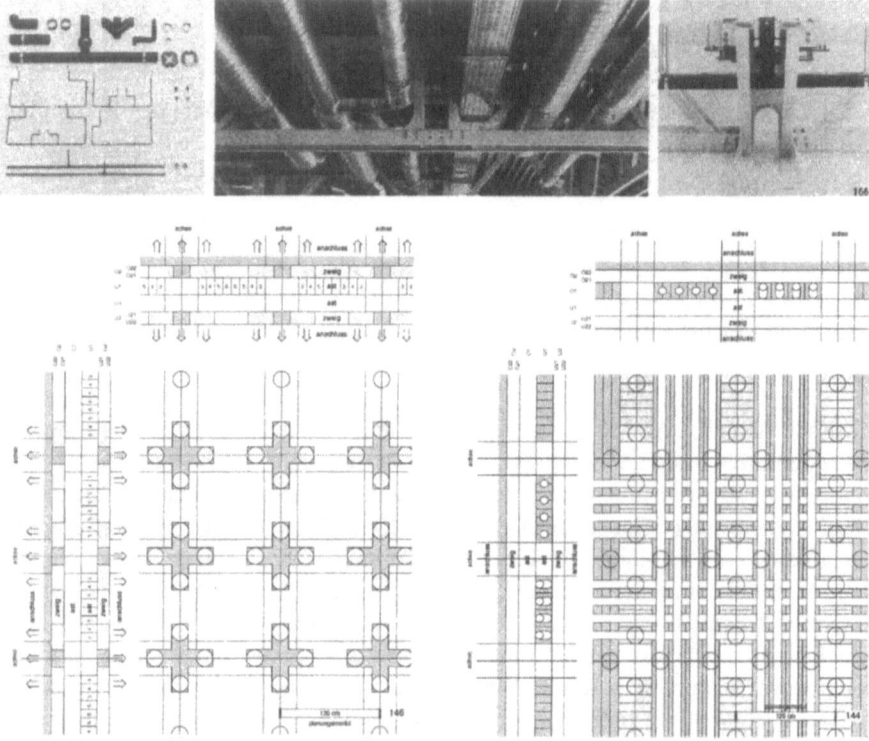

Figure 2: *Views of the general installation model of ARMILLA. The goal of ARMILLA is to modularize the technical systems in dynamic buildings. The views in the first row show the integration of the constructive system (MIDI) and the technical system (ARMILLA) using the example of plumbing. The lower views show charts with planning rules that make it possible the designing of reconciled modules of the technical systems needed for a building [Haller 97].*

1   The Context of ARMILLA5

ARMILLA5 is based on the construction modules MINI, MIDI and MAXI, which were developed by Prof. F. Haller in the 1960s and have widely been used in Switzerland [Haller 74, 88]. The main characteristic of this building is that it can be completely disassembled and reconfigured. The construction and use of the building is based upon a permanent design of a dynamic building. New construction is seen as a special category of remodeling.

These construction modules are supplemented by the generic installation system ARMILLA [Haller 85]. The goal of ARMILLA is the control of technical systems in complex, dynamic buildings (such as office buildings, schools and laboratories). ARMILLA is not a closed system, but rather a planning system that allows technicians to confer regarding numerous planning phases of their technical systems in a given building as building blocks. They attain a high degree of planning reliability, faster assembly and they can simply reconfigure their systems for a change in use without necessitating the destruction of building components.

The MIDI and ARMILLA systems are currently in use for new construction in the architectural firm of Prof. F. Haller, however without computer support.

ARMILLA5 is the fifth software prototype to result from the planning and building system of MIDI/ARMILLLA. In the concepts presented in ARMILLA5, examples of MIDI/ARMILLA are used, but the systems presented are easily transferred to other architectures. In the following, three different partial models of ARMILLA5 are presented: the geometric model, the semantic model and the planning model.

The geometric model (GM5) describes the geometric physical context of the building planned with ARMILLA 5 . As the lowest common denominator of all modelings in ARMILLA5, it also has the role of the integration model. It is supplemented as needed by the so-called subject models in the semantic model of ARMILLA5 (SM5). For previously modeled buildings, GM5 can be kept very simple: in a worldwide coordinate system, orthogonal spaces (containers) are assigned at specific locations - virtual construction sites. The semantic model of ARMILLA5 (SM5) describes in subject models how these containers are used with information regarding construction components, planning tools or representations of people. The planning model (PM5), as a third partial model of ARMILLA5, describes how these containers are configured and arranged on a "virtual construction site", so that they simulate real construction sights in preparation and can assist in control of the real building. Descriptions of the various levels of abstraction are primarily required in order to do so.

2   The Geometric Model

Here the geometric model of ARMILLA5 is described with the editors of the prototype implementation of ARMILLA5 system: figure 3 shows the graphic editor of the ARMILLA5 system with a so-called personal planning section (PPS) in figure 4 by hyperlinks, for example from a mail system on a particular planning issue. By means of hyperlinks to PPSs, can not only planning issues be formulated, as in the example, but also new procedures for work processes or interactive tutorials can be created.

Figure 5 shows the graphic editor, produced by the hyperlinks mentioned above, tailored to a particular virtual construction site. Each hyperlink first creates an A5 navigator, represented at the lower left margin by the rectangle surrounding a small interface. Two A5 navigators are pictured in figure 5. In a second step, they create a link to an A5 InterNET server. A5 servers are installed by specialized planners and manage the various component systems of the ARMILLA5 building as an classification of subject models (see SM5). In a third step, A5 navigators load certain planning elements/containers into the graphic editor, where they can be

138

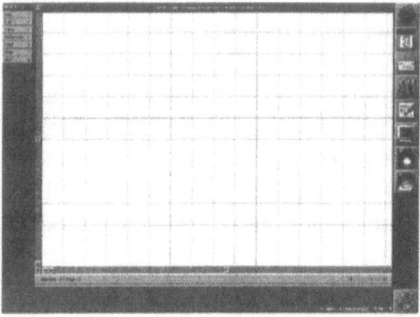

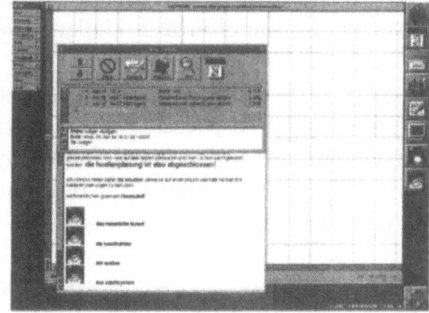

Figure 3: The graphics editor of ARMILLA5 in the worldwide coordinate system.

Figure 4: The positioning of the personal excerpt of the plan, e.g., through the hyperlinks of a mail system.

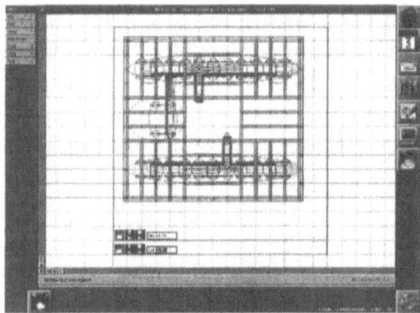

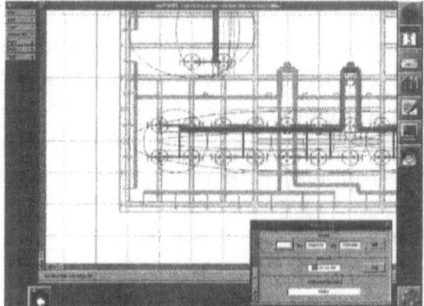

Figure 5: A5 navigators build links to A5 Inter-NET servers so that construction components of the „virtual construction site" can be edited. Here, two A5 navigators with links to a construction server and a return are server are installed.

Figure 6: The graphic editor is used here to enlarge an excerpt from the plan and to bring the construction components of the return air system to the foreground for editing.

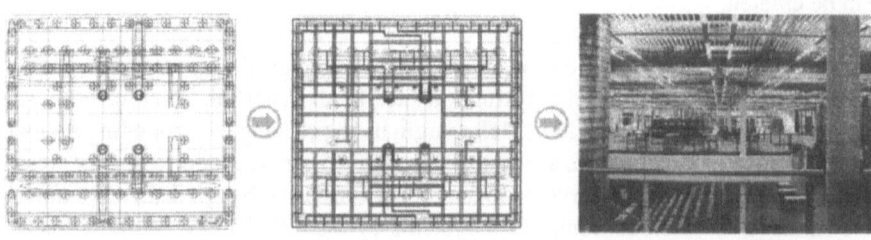

Figure 7: ARMILLA5's geometric model describes buildings at various levels of abstraction. Shown here are: a strategic planning stage at the level of sketches in the virtual construction site, a more detailed planning stage of the virtual construction site, and, at the last level of detail, the real construction site/building.

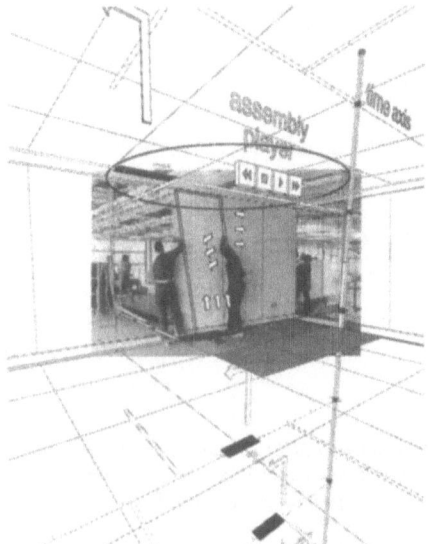 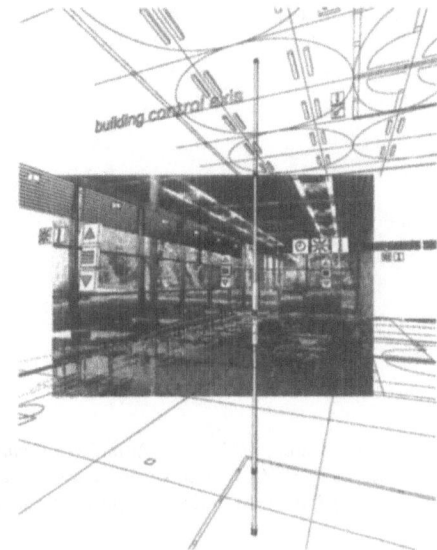

Figure 8: The components of the virtual construction site can expand the real construction site through appropriate projection processes (e.g., semitransparent glasses with stereoscopic projections), here, for example, instructions for erecting walls.

Figure 9: With the same projection technologies, the virtual construction components can be used for, e.g., facility management. Here, for example, a janitor's view with various control and indicator elements.

edited. In figure 5, links to the A5 servers of the return air and the load-bearing elements have been installed and planning elements have loaded an intermediate level of abstraction. During the work in the graphic editor, A5 navigators serve as „parent" to their planning elements, make read/write access possible, and manage A5 server addresses.

ARMILLA5 is based on the concept that the virtual construction components are a preliminary simulation of the actual building process, and that the last concrete stage of the model represents the real building site (figure 7). Figure 8 shows how ARMILLA5's preliminary virtual structures can support ARMILLA5's real structures. Shown here is an abstract virtual construction component, which projects installation instructions into the real construction site. Figure 9 uses a Facility Management Application to show that ARMILLA5's virtual structures can also be meaningfully expanded even during actual construction. This conception is particularly useful for the idea of a „permanent construction site" for „dynamic buildings". ARMILLA5 formulates an expanded concept of architecture, in which real and virtual structures complement each other and can be treated in essentially the same way.

The ARMILLA5 model is not limited to proprietary software. With reduced functionality, it can also be implemented with a conventional CAD system (figure 10 and 11) or in the WWW (figure 12).

140

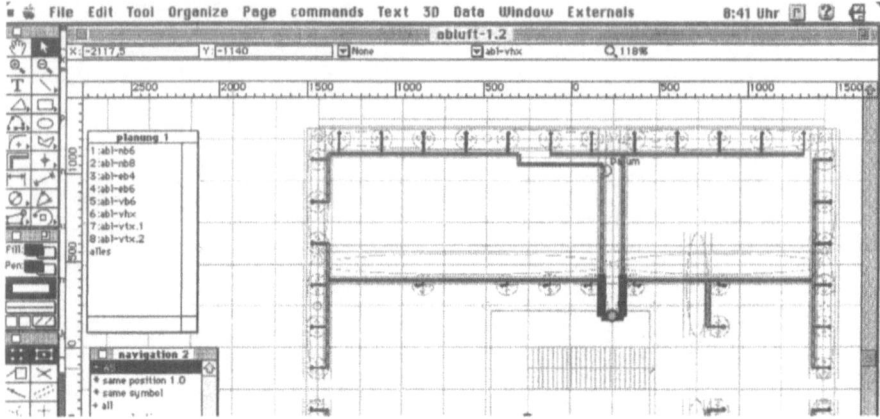

Figure 10: *View of a CAD system (MiniCAD, Graphsoft) on the ARMILLA5 model. The outcome of the planning model (PM5) is symbol, class and layer structures (upper panel at left), and the A5 navigators functionality is made available through a collection of smaller programs (here MiniPascal) (lower left panel).*

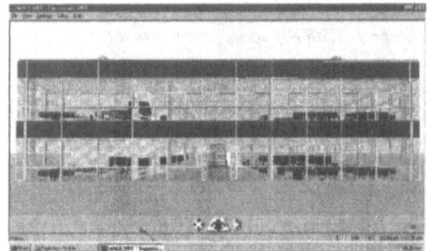

Figure 11: *A VRML view of an ARMILLA5 building.*

Figure 12: *This figure is an example for the same Facility Management System which in this case is usable in the form of maps in the InterNET.*

3   The Semantic Model

The semantic model of ARMILLA5 (SM5) describes the information filling the containers of the geometric model (GM5). SM5 employs an object-oriented modeling method using taxonomies, inheritance, polymorphism, relationships, etc.. Each of ARMILLA5's planning elements is anchored as a container in the geometric model and as a determinant in the semantic model. In contrast to the geometric model, the semantic model works from partial/subject models, since it does not seem realistic to develop a consistent model for all buildings. The examples in this article use the MIDI and ARMILLA subject models, which follow the systematology of the corresponding construction system. However, other subject models can easily be envisioned.

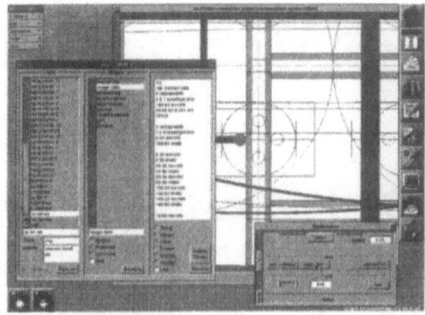
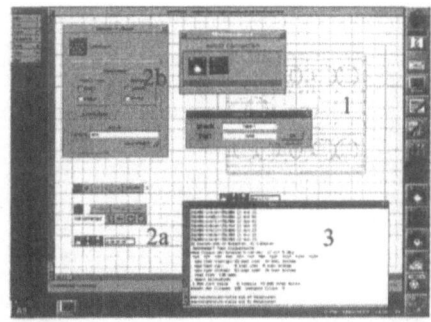

Figure 13: *The semantic editor A5Browser, like the graphic editor, has access to the A5 servers. However, it shows a semantic view of the A5 server's contents, rather than a graphic view. In this example, the A5Browser is launched by the selection of a construction component in the graphic editor and the command „edit instance, " and causes these construction components to appear and be displayed on the A5 server. In the first column, the semantic editor shows a list of all construction components of the A5 server; in the second column it shows the attributes of the construction component selected in the first column; and in the third column it shows the contents of the attribute selected in the second column. Selected here is the „object-instance" attribute that contains a reference to the class information of the selected construction component in another A5 server.*

Figure 14: *This illustration shows the three levels of semantic modeling in ARMILLA5. 1: Passive construction components are created as objects with attribute-value pairs and are treated as graphic objects. 2: Simple active construction components behave differently and offer the user several levels of interfaces inside (a) and outside (b) of the drawing surface. 3: Complex functions are offered on the InterNET server and are controlled by active construction components.*

Most of the semantic model's planning elements are passive objects such as supports or girders. A few components of the ARMILLA5 system are implemented as active objects: in addition to a graphical representation (which passive objects also have), they offer the planner interfaces with which other ARMILLA5 system planning elements can be edited. This modular concept is based on the idea that the planners go to the „virtual construction site" with their „tool boxes" in order to be able to carry out their planning tasks. Simple active components correspond to JAVA applets and are installed and implemented on the user's host. One example is the A5 navigator described above. In ARMILLA5, the A5 navigator is implemented with Objective_C under NextStep. Complex active objects are implemented as stand-alone applications (e.g., in CommonLisp or Objective_C) and installed on an Inter-NET host: As a server, they offer their services on the network and can be used as simple active objects. The A5 servers described above are one example.

142

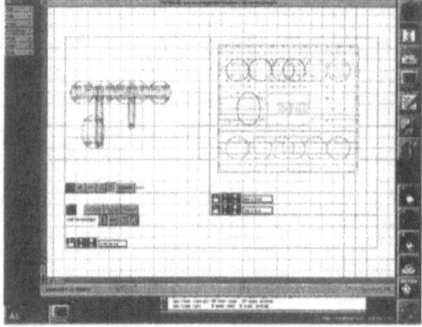 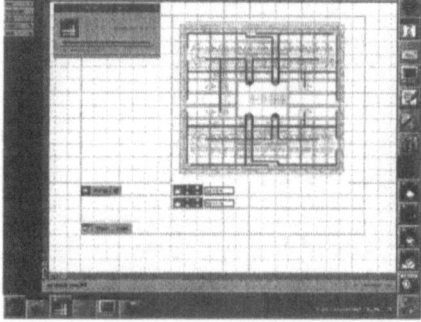

Figure 15: *Case-retrieval and adaptation by ASM and TOPO*

Figure 16: *Arrangement of ducts and pipes in a given construction grid by ANOPLA.*

In the FABEL project [FABEL 93 - 96], a large number of complex active construction components was developed [FABEL 93-96, Report 40]:

Retrieval Tools

- RABIT: A versatile retrieval system using similarity measures based on case attributes.
- ASM: A fast retrieval tool using an associative memory; the similarity is also based on attributes.
- ODM: Geometric arrangements of objects are retrieved by comparing object density maps.
- TOPO: Cases are transformed into graphs representing topological relations.
- ASPECT: Combines pre-computed similarities of stored cases with respect to different similarity measures (aspects) using the „fish and shrink" heuristics to effectuate retrieval.

Adaptation tools

Retrieved cases will generally not fit as they are; some adaptation has to take place. This adaptation uses other types of knowledge in addition to the source case. Depending on the knowledge used, an adaptation tool may be more or less powerful, highly specialized or widely applicable.

- TOPO: After retrieving a useful case, Topo is able to transfer objects (of a given type) from the source case into the query case maintaining the topological relations between matching objects.
- AAAO: Arrangement of columns in a steel frame construction given a floor plan obeying static and aesthetic restrictions.
- SYN: Transfers branch lines (e.g. for return air) from clustered cases into the incomplete query case in order to connect the trunk pipe with the outlets.

Construction and assessment tools

Some tools work, or can work, without using source cases to design certain parts of a building. AAAO can also be classified here. Other tools find failures or inconsistencies in a complete or partial design. Topo can also be used for this purpose because it finds unusual relations in a given query case.

• ANOPLA: Arranges provisionally placed pipes so as to obey a given construction grid (template) and to avoid collisions.
• CHECK: This tool checks whether a set of topological predicates holds for objects of certain types; this knowledge was derived (learned) from cases.

4   The Planning Model

The ARMILLA5 (PM5) planning model describes the course of planning. It orders the planner's work through the various stages of elaboration of a plan and orders way in which the various planners can work together. PM5 describes the planning in the form of a tree structure: the roots of this structure indicate the total planning process (GP), and the leaves describe the basic planning steps.

Planning Scale

Overall planning consists of planning at various planning scales (PM). It begins with rough overviews and ends with fine details. Planning scales are the general milestones of planning and can be counterparts of planning phases, of the German HOAI, for example. Seven planning scales are currently formulated in ARMIL-LA5:

• In *Preplanning 1000* (v10), planning elements the size of main buildings and streets are brought together. This corresponds roughly to a plan on a scale of 1:1000.
• In *Preplanning 500* (v5), planning elements the size of secondary buildings and smaller roads are brought together. This corresponds roughly to a plan on a scale of 1:500.
• In *Preplanning 200* (v2), planning elements the size of stories and main exploitation are brought together. This corresponds roughly to a plan on a scale of 1:200.
• In *Preplanning 100* (v1), planning elements the size of use zones and corridors are brought together. This corresponds roughly to a plan on a scale of 1:100.
• In the *Line Plan* (lp), all construction components present are described in their approximate spatial location and are fitted together. This corresponds roughly to a plan on a scale of 1:50.

144

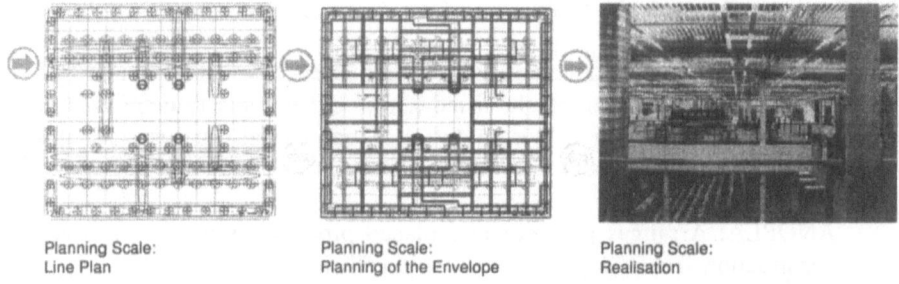

| Planning Scale: | Planning Scale: | Planning Scale: |
| Line Plan | Planning of the Envelope | Realisation |

Figure 17: An ARMILLA5 building is described in numerous abstraction levels.

- Durin the *planing of the envelope* (pe), all construction components present are described in their exact spatial location and are fitted together. This corresponds roughly to a plan on a scale of 1:20.
- During the *planing of the components* (pc), the real construction components are chosen and arranged. This corresponds roughly to a plan on a scale of 1:10.
- With the actual building, the *realization plan* (rp) is the last stage of making the ARMILLA5 model concrete.

Planning Area

For planning in ARMILLA5, subject models must be used. They describe specific conceptions of architecture and construction in their components and development. In ARMILLA5, the following subject models have previously been used:

- *SPACE*: This subject model formulates the transformation of the space catalogue into a spatial layout.
- *MIDI*: This planning model is taking over the construction planning area.
- *ARMILLA*: The ARMILLA planning model describes the technical building features planning area.

The names of the planning areas arise from both the abbreviations of the planning scales and the names of the drafting models: e.g., „lp-midi" planning area indicates the Line Plan (lp) planning scale of the MIDI planning model.

Planning Aspect

A planning area is worked on by several planners. They take on various component tasks called „planning aspects." (The term „planner" is used here in a general sense. Naturally, it is conceivable and even possibly sensible for a single planner to work on planning aspects such as, for example, return air and exhaust or façade and lower slab. Cooperation of the various planners involved is described at the planning aspects level.

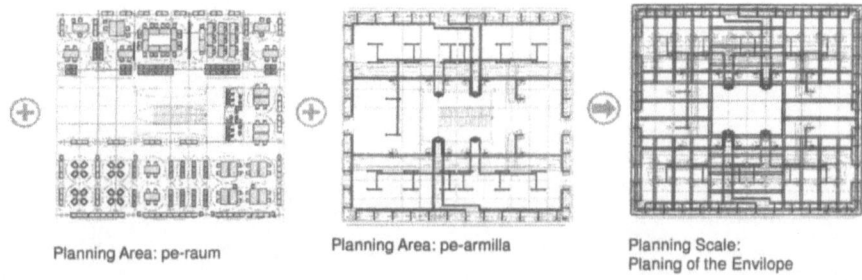

Planning Area: pe-raum   Planning Area: pe-armilla   Planning Scale:
                                                                   Planing of the Envilope

Figure 18: *The plans of a scale follow various subject models, here for spatial ground plan, construction, and building technology.*

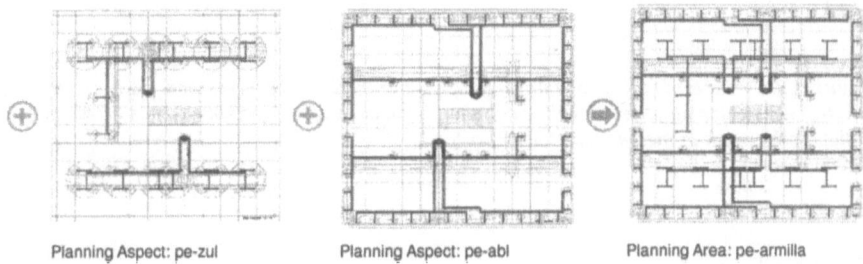

Planning Aspect: pe-zul   Planning Aspect: pe-abl   Planning Area: pe-armilla

Figure 19: *A subject model consists of planning aspects, which define the work of the various planners needed.*

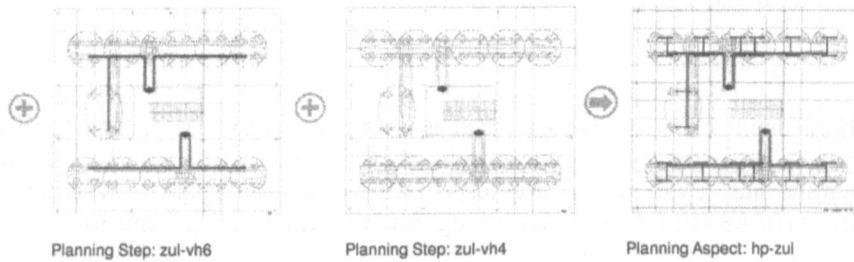

Planning Step: zul-vh6   Planning Step: zul-vh4   Planning Aspect: hp-zul

Figure 20: *The work of a planner consists of various planning steps. The basic planning elements from catalogs are classified and configured in a planning step.*

For the RAUM subject model, planning aspects are currently formulated as rau (spatial layout) and lct (lighting). The MIDI subject model formulates the planning aspects as bdn (floor), dch (roof), fas (façade), trl (stairs, elevator), trw (construction), udk (lower slab) and wnd (interior walls). The ARMILLA planning model consists of the planning aspects abl (exhaust), abw (sewage), zul (return air) and approx. 20 other technical systems. The names of the planning aspects are formed by combining the planning scale descriptor and the responsible party: lp-abl indicates planning for exhaust air (abl) at the Line Plan (lp) level of detail.

146

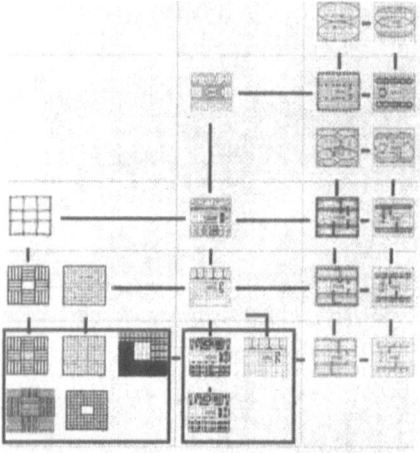

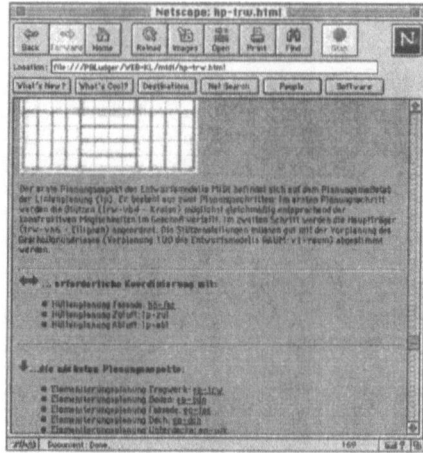

Figure 21: *Excerpt from a chart with ARMILLA5's planning aspects, with which the collaborative work of the various planners can be coordinated.*

Figure 22: *Resources and instructions for a planning aspect, presented on an HTML page.*

Planning Step

Each planning aspect consists of various planning steps. Planning steps are altered exclusively within a planning aspect. Planning steps include the tasks that can be performed with one operation and by one person.

Planning Element

By „planning elements" is meant the container/construction components familiar from the geometric and semantic models. They can be assigned to a single planning step and thus lie within the area of responsibility of a specific planner. The current version of ARMILLA5 includes approx. 600 different planning elements.

5   Tools of the Planning Model

Planning support tools can be derived from the described system for a planning model.

Figure 21 shows an excerpt from a chart on which the planning aspects of an AR-MILLA5 plan and the tasks of the individual planners are noted. Horizontal links on this chart indicate that a planning aspect must be reconciled with other planning aspects. Vertical connections vertical links lead a planner through the various phases of his planning process. This chart can also give an overview of the actual state of development of a plan, since completed planning aspects are highlighted in comparison with the planning aspects which have not yet been completed.

Figure 22 shows a tool that complements the planning chart, produced in HTML. For each planning aspect, an HTML page is created making all necessary information, instructions, previous experiences, and resources available to the planner in one place.

6   Working Methods in the ARMILLA5 Model

Working With Catalogs

As shown, a single planning step can be assigned to several planning elements. Within a planning step, it is a planner's job to classify planning elements on the virtual construction sites, arrange them, and reconcile them horizontally and vertically with the planning elements of the neighboring planning steps. For this purpose, the possible planning elements are offered to the planner in a catalog corresponding to the planning step. This process is very general, technically easy to carry out, and theoretically possible for any plan. However, it has the disadvantage that, primarily in the case of complex planning steps, the use of the planning elements does not follow intuitively from the planning elements themselves. The planner must therefore be trained outside of the system and before actual planning takes place. Furthermore, the working method is reduced to many small actions and becomes nothing more than a search for the appropriate component in extensive catalogs.

Working With Cases

Another working method presents itself in the use of cases, which means that successful parts of planning processes or planning episodes are stored in a case database and, thanks to appropriate indexing, can be proposed as a solution for new but similar planning situations. Case-based inference was the research emphasis of the FABEL project. It was divided into „retrieval" (how the appropriate case for a planning problem is found) and „adaptation" (how a case can be adapted to the new planning problem). It can also not be decided whether a case must be divided into „problem" and „solution." In the majority of cases, in FABEL, the path selected was that a case simply describes a large number of planning elements that can be approached as the solutions for a number of problem formulations, (arising from the planning interrelations of the current plan).

148

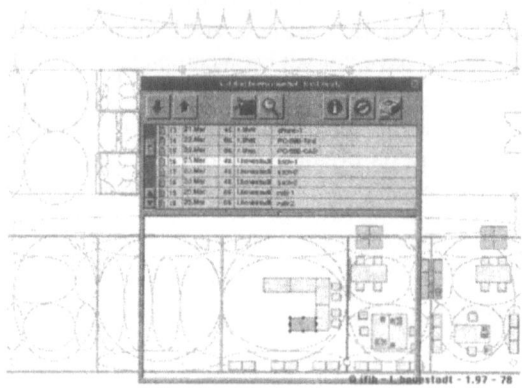

Figure 23: A simple editor for working with catalogs in ARMILLA5. An active construction component offers the planning elements of a planning step on a list for classification and organisation on the virtual construction site.

Cases have many advantages in comparison to catalogs:

- Planning can be carried out in larger, interrelated units and a planner need not carry out so many elementary actions.
- The planning elements of a case are already interrelated in a meaningful way (or, to give a negative example, in a way known to be detrimental). This interrelationship no longer needs to be explicitly formulated by a planner, but can be implicitly assumed.
- Cases are not offered as isolated cases by a retrieval process, but rather in a cluster of comparable cases for similar question formulations. A planner is thus informed about his possible actions and can make an informed decision about a solution.
- The user also has the possibility to learn by doing while working on customary planning structures. As long as he has the basic knowledge and works with positive examples, he can assume that he has made no funda-mental errors. In this case, unlike that of catalogs, it is possible to work within defined planning structures, without learning them thoroughly before planning takes place.
- Because this technology allows work in outside models, case „marketplaces" for planning services can in principle be established from the models of various planners combined relatively freely. The CaseBank (figure 25) is an implementation of this concept on the InterNET, using HTML and JAVA.
- Cases are first and foremost a construct by which plans are handed down and by which planning processes can be established without models formulated before each planning process. In combination with conventional modeling technologies, their exceptional capability unfolds: to economically describe the basic framework of a plan through models in the form of catalogs and prototypes (see next page). With cases, the many exceptions and special cases that come up in general use are formulated and made available for later plans.

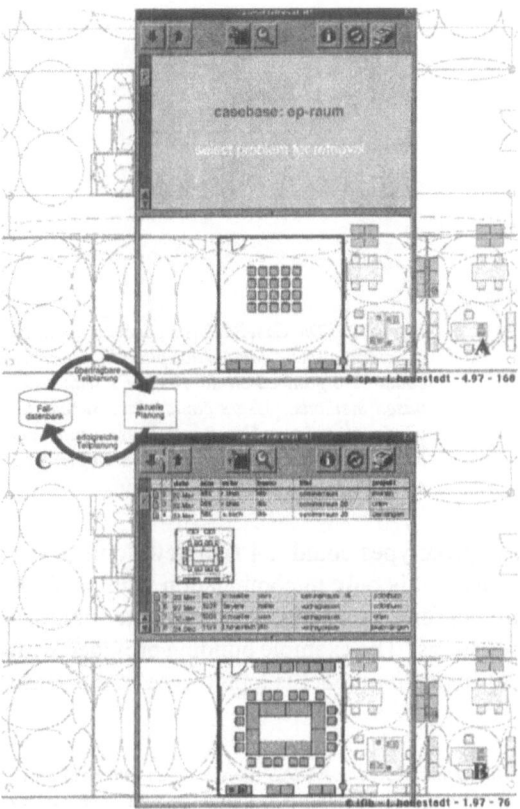

Figure 24: A simple editor for working with cases in ARMILLA5. A: On the basis of a selection of planning elements from the virtual construction site (corresponds to the problem formulation) from a case data bank (here, the ep-raum planning aspect data bank), an active construction component offers a selection of similar cases, determined on the basis of qualitative, quantitative, topological and other characteristics. B: A user can transfer one of these cases or parts of different cases to the virtual construction site and - possibly supported by a planning assistant - use them after adaptation as a solution to the current problem. C: The successful solution of the planning problem can then be put in the case data bank with the same tool, and will be available as a case for future plans. In this illustration idealized representations of the AR-MILLA5 prototype are depicted.

Working with Prototypes

A third form of working with complex models is the use of prototypes: completely planned-out buildings serving as models for a current plan. In this case, the actual planning does not consist of the creation and organization of new planning elements, but of a revision planning of this prototype building to make it match the requirements of the current plan. The prototype buildings are thus designed with the didactic goal of overcoming the greatest possible number of planning situations with a typical solution and to structure the revision planning as simply as possible. Even more than working with cases, this manner of proceeding makes it possible to use an „outside" model for the prototype, without the latter having to be familiar with all of the details of this model from the outset. Seminars with students have shown that the effort needed to get used to complex models is substantially reduced through the use of prototypes in planning. Through this process, the use of complex models such as MIDI or ARMILLA by „outsiders" is, for the first time, possible without excessive frustration.

150

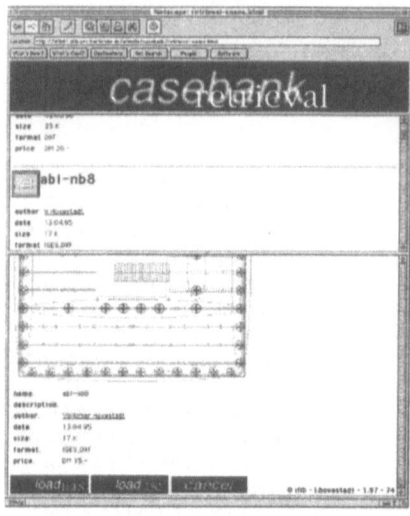

Figure 25: The CaseBank is a concept, by which a „marketplace" for partial planning services could be organized on the InterNET with case-based methods. Users could place successful partial plans in the CaseBank. These would then be distributed to other users with similar planning problems via the CaseBank.

As an isolated process, working with prototypes could not find new solutions to structurally new queries. For this reason, it is only in combination with catalogs and cases that it becomes a powerful tool for the elaboration of adequate solutions to practically relevant problem formulations. The example building provided with the planning steps of the ARMILLA5'planning model is a prototype in the sense described here.

In the last four years ARMILLA5 had been the domain of the FABEL project, titled ‚Integration of model-based and case-based approaches to the development of knowledge-based Systems' (contract no 01IW104 - BMBF) [Fabel 93 - 96] [Gebhardt et. al. 97].

FABEL (1993-96) BMBF- Verbundvorhaben 01W 104- Das FABEL-Konsortium: FABEL-Reports 1-40, GMD, St.Augustin

Gebhardt F., Voß A., Gräther W., Schmidt-Belz B. (1997) Reasoning with Complex Cases, Kluwer Academic Publishers

Haller F. (1974) MIDI - ein offenes system für mehrgeschossige bauten mit integrierter medieninstallation. USM baussystme haller, Münsingen, CH

Haller F. (1985) ARMILLA - ein Installationsmodell. Institut für Industrielle Bauproduktion der Universität Karlsruhe, Germany

Haller F. (1988) bauen und forschen. Dokumentation der Ausstellung. Solothurn, CH, 1988, or: Wichmann, Hans (Ed.) (1989) System-Design: Fritz Haller Bauen - Möbel - Forschung. Birkhäuser Berlin
Haller F. (1997) MIDI-ARMILLA, gesamtbaukasten und installationsmodell. Solothurn CH 1997.

Hovestadt L. (1994) A4- Digitales Bauen - Ein Modell für die weitgehende Computerunterstützung von Entwurf, Konstruktion und Betrieb von Gebäuden. Dissertation, Universität Karlsruhe (TU), Institut für Industrielle Bauproduktion, Forschrittberichte VDI, Reihe 20 Rechnerunterstützte Verfahren Bd. 120, ISBN 3-18-31 2020-8, Düsseldorf Germany

Hovestadt V. (1997) Informationsgebäude. Ein Integrationsmodell für Architektur und Informationstechnologien. Dissertation, Institut für Industrielle Bauproduktion Universität Karlsruhe, to appear 1997.

IMPROVING CAAD BY APPLYING INTEGRATED DESIGN SUPPORT SYSTEMS AND NEW DESIGN METHODOLOGIES

SEVIL SARIYILDIZ, HARRY VÖLKER, MATHIAS SCHWENCK
Delft University of Technology (The Netherlands)

ABSTRACT:
This paper deals with the improvement of the current design practice by means of ICT (Information and Communication Technology) tools in the field of architectural design.

In the first part we make suggestions which can contribute significantly to improvements in the mentioned field. This includes:
- the development of Integrated Design Support Systems (IDSS)
- the application of new design methodologies in relation to IDSS.

In the second part we will discuss the topic more generally. Which other aspects have to be considered regarding the development of support software for architectural design? Which improvements can be reached by introducing advanced information and communication technology? Which changes are necessary in the promising relationship between architecture and computer science?

KEY WORDS:
Computer Aided Architectural Design, Support Systems, Integrated Design Environments, Design Methodology, Architecture and Computer Science

1 Introduction

The building process has been changed significantly in the last years. Recent developments include the following trends, that are analysed in [StBR96]:

- principals work more professional
- increasing industrialisation of the building process
- demand for sustainable buildings
- increased use of information and communication technology.

Some other issues like the increasing importance of financial aspects or new strategies with respect to process management can be added. We can conclude that, at the moment, the building process is changing with respect to many different aspects.

Considering the architectural design process, we can make a similar statement. There are changes within the design process itself. Furthermore, some of the changes in other phases of the building process have direct effects on the design stage. Consequently, architectural design has to be adapted to the new situation.

R. Junge (ed.), CAAD Futures 1997, 151-162.
© 1997 *Kluwer Academic Publishers.*

With the conclusion that current developments in the building sector require new concepts for architectural design, the question arises in which fields solutions can be found. In this paper we will take into account two important aspects - the use of Integrated Design Support Systems (IDSS) and the application of new design methodologies in relation to these IDSS.

Furthermore, we will more generally analyse the relationship between architecture and computer science to indicate the reasons for its description as "difficult but promising".

When we look at the historical developments of computer science applications for the building sector, we can state that the use of computers in the field of architecture came relatively late compared with other scientific disciplines. The first CAD tools were developed and used by engineering disciplines such as mechanical engineering or aerospace engineering. Later at the beginning of the 60's some of the architects have also seen the importance and the necessity of these tools for their work. However, these tools were not directed to the needs of designers and architects. They were far more isolated from the architectural practice. They were developed for other purposes than architects needed. Another fact was that these tools were developed by computer scientists who naturally have no significant knowledge concerning architectural design. These tools could replace the drawing tables so far. Gradually the need and the use have grown. In the 70's we talked about "building informatics" and later in the 80's it became information technology for the building sector.

In the 90's the use of these tools was more spread out and the applications of ICT were implemented in the whole building sector. The range of applications differs a lot. Some use it still only for drawing purposes, others for representations. Furthermore, there even are attempts to imitate human intelligence to let the machine design as humans do. Another positive development is the fact that the designers came closer to the computer science subjects or to the computer scientists. Finally the knowledge about each other's domain increased and therefore the communication became easier between these two scientific fields. This resulted in many software and hardware products which are more convenient to use by architects and building engineers.

Generally, it is a matter of fact that the technological developments in every field of science have an influence on the society and therefore on the design and the design process itself. We are forced to think fundamentally about the influence of the rapid developments of ICT in architectural design. [VSSD96] [Sari96]

What will be the way and the method to integrate the new tools in a design process to increase its efficiency and to reach better design results? What will be the most necessary, relevant developments for our sector? Where do we have to go with these ongoing developments? What are most essential needs for the building sector, especially for the designing architect?

We believe that there is no unanimous answer for these questions. In this paper we will introduce our approach which we consider an important contribution to solve existing problems within the concerning field of ICT for architectural design including conceptual (spatial form findings) and materialisation (building techniques related) phases of the process.

2 Integrated Support Systems for Architectural Design

We consider a complete automation of architectural design as an unlikely proposition and undesirable for the architect. Therefore, the general objective is to give support to the architect to improve the quality and to increase the efficiency of the design process.

So far there are different tools providing such functionality. Many software systems are in common use in the field of architectural design. The following statement is also valid for the field of architectural design:

> *"In the past four decades, civil engineering in general and structural engineering in particular have achieved remarkable successes in adapting successive generations of computational support technologies and developing computational tools for specific process steps." [Fenv95]*

Nevertheless, there are no appropriate tools for many of the sub-processes. [Fenv95] describes this phenomenon as "readily identifiable individual process steps, that have been heavily computerised, to a point where these processes are, in effect, isolated "islands of automation" in a vast sea of essentially manual processes."

Furthermore, we can state that there is a significant lack of integrated systems providing a general support for the designer during the whole design process. So, the two key features are:
- development of tools for sub-processes where no appropriate software support is available so far, so that for every sub-process of architectural design where the use of suited software can lead to improvements, these tools has to be available.
- the integration of these tools into a framework in order to realise an open, modular, distributed, user friendly and efficient operating environment.

We have already stated that the main characteristic of the system is to give support during the whole design process. "Support" means, that the tools should provide functions to free the architect of routine tasks, to avoid faulty actions and to detect errors as early as possible, to support the architect by increasing the amount of available information, to support the exchange of information between different partners participating in the building process, etc. [SaSc96]

In [ScBr93] the term "integration" is described by distinguishing integration with respect to the following three dimensions:
- Data (Information)
- Control (Communication)
- User Interface (Presentation).

The *data integration* aspect of tools determines the degree to which data generated by one tool is made accessible and is understood by other tools.

The *control integration* aspect of a tool determines its communicational ability, i.e. the degree to which it communicates its findings and actions to other tools and the degree to which it provides means to other tools to communicate with it.

The *user interface integration* aspect is the degree to which different tools present a similar external look-and-feel and behave in a similar way in similar situations.

Integration has to be realised in all three dimensions. This avoids situations where limitations occur because of incompatible file formats, incompatible communication protocols or because of user interfaces that are not suited for the people working in the field of architectural design.

Instead of only developing design tools integrated system are also addressing the problem of the operating environment of these tools. The development concept is described in greater detail in [ScSa97].

Figure 1 gives a schematic view of an integrated environment as used in [Wolf93].

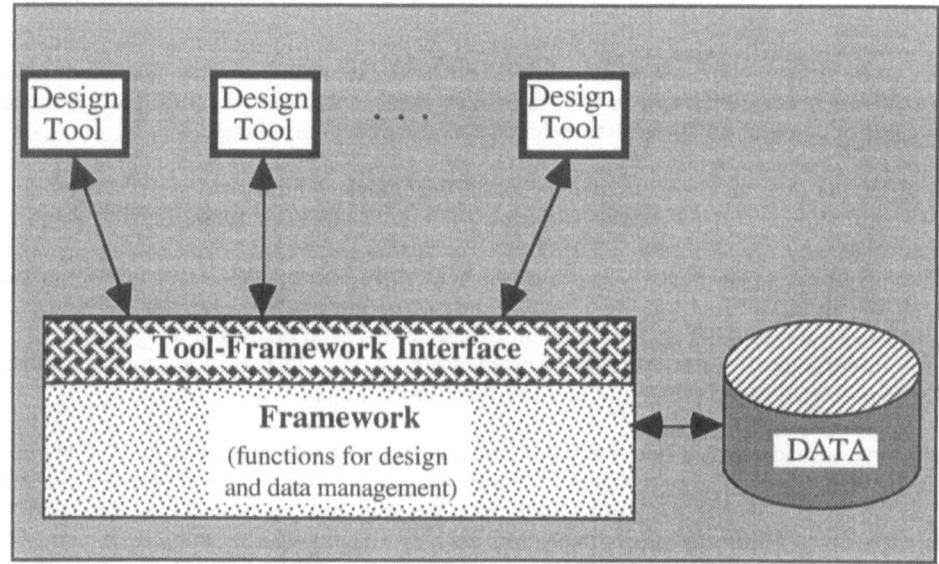

Figure 1: An integrated software environment

The framework provides general services for the tools. It realises functions for data management in order to organise design descriptions and to provide access as well as design management functionality guiding the designer through the design process.

The interactions between tools and framework take place according to functions of the Tool-Framework Interface. The definition of the interface is a key issue in tool integration because the effectiveness of tool integration depends on it.

Within the integrated environment different tools can be used. Tools that are newly developed can be implemented according to the facilities of the framework, whereas existing software may be integrated using known tool coupling methods.

3 New Methodologies for Architectural Design

One may argue that the development of design methodologies and design support systems are completely different fields which only have very few aspects in common. Besides the "trivial" reason, that every design tool is applied in a process following a certain methodology, others arguments can be given for our point of view, that it is essential to address both fields simultaneously. The automation of the traditional way of architectural design may lead to some improvements, but fundamental problems are not addressed e.g. communication, co-operation and management issues. Therefore, a lot of potential is not used. In order to come to fundamental improvements the application of advanced computer technology in the field of architecture has to be co-ordinated with improvements in the area of design methodologies.

Many architectural design processes have been successfully performed. Different methodologies have been applied, that are characterised by working without computer tools or by applying them only in a very limited role. In general these methodologies, that we will refer to as "traditional methodologies", can be considered a suited way of designing. On the other hand there are some inherent problems like for example:
- problems to combine different drawings in order to check the possibilities and to tune the various technological solutions
- impossibility to take into account the distinguished alternatives immediately during the discussions between architect, principal, consultants or authorities
- fast and simple exchange of information between architect and the other partners is almost impossible (at least very limited)
- architect has to take over the tasks of a co-ordinator and therefore less time available for his creative tasks because of the effort needed for the management of the design process
- calculations during the modelling phase require much effort to be taken, but the results remain relatively rough and insecure.

A more detailed description is given in [ScVö96]. These problems and shortcomings may lead to significant quality and efficiency problems. Therefore we have to look for improvements to overcome these limitations. The fast development of science and technology offers some solutions.

In order to use information technology for architectural design effectively in an integrated way we will approach a new design methodologies is approached in [VöSc97]. "Integrated Architectural Design on the Basis of 3D Computer Models" is characterised by the following basic concepts:

- The process is based on an integrated manner of designing. Decisions are made as a result of discussions in a design team, where possible alternatives have been carefully evaluated.

- The whole design process is executed on the basis of a 3D model which is handled by means of computers.

- The availability of support software corresponding to the needs of the architect is one of the key features determining the success of the idea.

- In the design process there are different types of models. These models contain all relevant information generated in the design process. This includes the possibility to deal with several alternatives. Because of their availability at later stages of the design process, definitive decisions can be made "better" and highly qualified.

This methodology provides good possibilities for the integration of design and construction process. The co-operation between designers and consultants from different fields can lead to synergetic effects because of the different points of view and the different areas of knowledge. An extension of this methodology where the contractor participates in the architectural design process is also covered in [ScVö96].

So far we have discussed three different methodologies for architectural design. They will be referred to by using the following symbols:

A - Traditional Methodology of Architectural Design

B - Integrated Architectural Design on the Basis of a 3D Computer Model

C - Integrated Architectural Design on the Basis of a 3D Computer Model with Participation of a Contractor

The most significant differences occur with respect to the level of integration of the different participants. Consequently, the question arises if higher levels of integration can be reached. Such an approach is discussed in [VöSc97a] where we deal with architectural design within a "realisation team". In contrast to the building team as assumed in methodology C the principal participates in the design process as a team member. Therefore, his role is significantly more active. We will refer to this methodology as

D - Integrated Architectural Design on the Basis of a 3D Computer Model with Participation of Principal and Contractor.

All new approaches for design methodologies covered so far, i.e. B, C and D are applied with the objective to increase the degree of co-operation and integration between the participants. Issues of process control are not especially taken into account. Recent developments in the building practice show that the field of design management becomes more and more essential. This is illustrated by the fact, that managers are involved in many design processes. In general they take over tasks from the architect related to the process management, i.e. they act at a very important position in the decision making process during architectural design. Consequently, the whole design process changes significantly.

How can this be related to the methodologies covered so far? The first idea of extending a building team with a manager fails because incompatibility of the dominant position of a manager with the decision making processes on the basis of equal co-operation. It would more or less automatically destroy the team.

Another idea is the separation of the manager from the team, i.e. a manager who "manages" the building team of architect, consultants and contractor. The relative strong position of the team, as a group consisting of several people, compared to the manager as single person makes it impossible to guarantee the manageability of the process. Additionally, the character of the team changes in any case as for example the manager takes over the communication with the principal.

As consequence of the failure of these two ideas we have to discuss the participation of a manager in the design process at a low level of integration of the different participants, i.e. similar to the traditional way of designing. Such a methodology is described in [VöSc97a]. It will be referred to as

E - Management-Oriented Methodology of Architectural Design.

As a consequence, many advantages of the integrated way of designing are lost. Obviously, we would like to combine the primary advantages of both types of methodologies discussed so far, i.e. to realise a design process where the decisions are made on the basis of interdisciplinary co-operation in an integrated way and to guarantee the "manageability" of the process. As stated before, this requires to solve the problem of combining the key position of a manager with an integrated co-operation of the partners in the design process.

In [VöSc97a] we approach a methodology that fits this demand. It is based on the application of integrated support software, i.e. information and communication technology is used to combine the two dimensions of "manageability" and "integration between the partners involved". We will refer to this methodology as

F - Computer Supported Management-Oriented Methodology for Architectural Design.

The relation to the other methodologies is illustrated in figure 2.

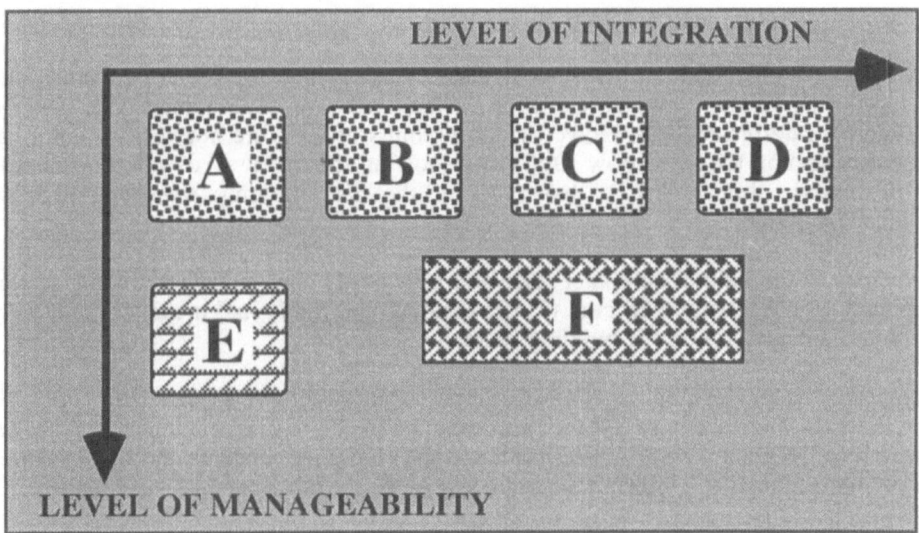

Figure 2: The relation between the covered design methodologies

Very generally, this methodology combines an "open" (separation of partners because of management-orientation) and "closed" (co-operation within a realisation team) approaches for architectural design. It only becomes possible by using integrated support software. [VöSc97a] describes this methodology in general and gives an overview about the sub-processes and data involved. Future research will extend and evaluate these results.

4 Future Developments in the Difficult but Promising Relationship between Computer Science and Architecture

The developments in the field of computer science have an inevitable influence on architecture. Therefore we have to deal with chances and problems of computer application in the building process. How did the computer science influence the architecture till now, and what is going to be the future of the architecture with this ongoing of computer science developments? In which way will these developments affect the position and the responsibility of an architect? We will discuss these questions and present our "vision of the future".

> *"Architecture is a science which is a mixture of an exact science and the art. The combination of these two important items makes architecture a difficult task. ... An architect has to combine these both primary elements in the design and in the same time while expressing the feeling of art, must take very good care of many other factors which play an important role in the building and design environment. The technical aspects on one hand, the social aspects on the other hand." [Sari91]*

Comparing this statement with common descriptions of the field of computer science where exact sciences like mathematics and electronics are considered to be the most significant elements, the contrast to the creative nature of architecture seems to be obvious.

One may argue that computer science is also related to interdisciplinary demands like perceptive psychology and ergonomics or that software development can also be considered a design process. However, these points of view are not very common, at least this is not the way how architects look at the field of computer science. Unfortunately, this "conflict" has effected the use of computer technology in architectural design.

Innovation in the field of computer science is only partly applied in the building sector. Applications are mostly restricted to conventional and traditional areas, consequently limiting the innovation in the building sector.

Regarding recent developments in computer-aided structural engineering [Fenv95] identified the interface between the problems and needs of civil engineering practice, research and education, on the one hand, and the emerging concepts and opportunities offered by computer and information technologies, on the other hand, as a key feature.

What are the factors creating a gap between these aspects? Some issues that we consider essential will be discussed in the following section:

- Architectural design is a mixture of various disciplines and very different tasks. Therefore, very different kinds of support software are necessary. Software has to be adapted to the specific needs of these tasks. In order to achieve this, these specific needs have to be compiled completely. So far they are at least partly "hidden" and consequently not considered in current software.

- One of the important advantages computers have to offer - the possibility to transfer information digitally over long distances in a very short time without copying drawings etc. is at least not commonly used so far. "Digital communication" could be a driving force for using computers in architectural design, i.e. a way, where computers become a communication medium for the partners in the building process. Recent research on areas like product data models and data exchange formats will contribute to this development.

- [Kler97] describes a difference in problem solving between engineers and architects. "Typical" engineers want to have a fixed framework and rules within in order to solve the problem, whereas architects usually start when everything is uncertain and try to stretch the limits to be able to use the solution in their imagination. The method of problem solving by architects is summarised as "sketching". Current computers do not have this ability , i.e. no computer support is available.

- In our opinion, architects have to develop a better "feeling" for the possibilities and impossibilities of computers. This must include the acceptance of computer technology as a factor in architectural design, i.e. to overcome positions of ignoring this technology. On the other hand, "euphoric" positions must be changed. A computer is only able to solve problems which are clearly defined. One may not expect the computer and the computer science professional ,respectively, to solve a problem in architectural design, if the architect cannot describe problem solving process. Modelling the problem is a fundamental requirement for its computational solution.

- Development and application of new ICT-techniques, methods and tools in the building sector is a very complex problem. In [SaSc97] three categories of tools are identified - design related tools, building techniques related tools and building process related tools. Contributions can be made by developing tools in the three categories as well as by dealing with their integration. As a consequence of the complexity of the process and the various interdependencies, many factors determine, whether a new development will be successful which are not always directly related to a specific development.

What will happen in the future? We expect a very positive development because of various reasons.

First, the necessity the use appropriate computer support in architectural practice will increase significantly. Buildings are becoming more and more complex, the influence of financial aspects on decisions in the design process will be stronger, the competition will force attempts to make architectural design as effective as possible. All these problems are very well-suited for computational support.

Secondly, apart from the application of information and communication technology new possibilities based on developments in fields like building technology and material sciences are applied in current architectural practice. Some architects go even further and participate in the development of new technologies. This trend will continue and also influence future software development projects. In this sense, we can consider architecture as a significant stimuli for future developments in these fields with a significant feedback to the degree of innovation in the building sector.

Finally, another factor will contribute in closing the gap between architecture and computer science. [Mave97] stated referring to the use of computers in architectural design:

"It is clear that as our young graduates enter the profession we can anticipate a massive increase in the use of the technologies in architectural practice."

So, the role of computers in architectural practice will increase in every case. It is the task of architectural education to prepare the students to these changes. There is actually no space to decide whether to support this process or to defend a design philosophy that avoids the use of computers, it is an irreversible process, but its progression will depend on current research as addressed in this paper.

5 Summary and Conclusions

In this paper we have discussed two possible contributions for improvements in architectural design - the developments of integrated design support systems and the application of new design methodologies that are directed to make the use of computers as effective as possible.

To overcome the current situation, where individual process steps have been heavily computerised, to a point where these processes are, in effect, isolated "islands of automation" in a vast sea of essentially manual processes [Fenv95], two aspects have to be covered. First, research has to be directed to evaluate the possibilities of support software for processes, where no software tools are existing yet or where only inappropriate tools are available. Secondly, these tools should be integrated in an open, modular, distributed, user friendly and efficient environment.

Integrated design support environments are a possibility to improve the quality and to increase the efficiency of architectural design. Compared to single tools their development is a complex process. On the other hand, the effort will be compensated by significant advantages as discussed in the paper.

If discussing the development of software for architectural design, the relationship between software support and developments of design methodologies has to be taken into account. In the first place, this is necessary to take into account current developments in design practice. Secondly, an identical transformation of the current design practice into software tools will limit their efficiency and consequently decrease the improvement that can be reached.

Architects have to participate actively in innovative development processes in the field of support software for CAAD. A more active position of architects will contribute to the solution of the problems in the relationship between computer science and architecture. The use of computers in architectural design will increase in the future. The challenging task is to have an effect on this development with own contributions and to take care of possible obstacles.

These ICT developments will have significant impact on the building sector. In addition to general quality improvements and efficiency increases other advantages are:

- better communication between the partners involved in the building process
- participation of the designing engineer in the whole design process becomes possible
- contributions in the industrialisation of the building sector
- development of assembly techniques
- robotising of construction processes
- contribution in the solving of environmental problems by sustainable buildings
- increased flexibility in the architectural design
- more variety in the architectural design

The incredibly rapid developments in the field of computer science and the emergence of this technology in all fields will have their effect on our subject area, architecture, our way of living, our habits and our cities; this will create fresh challenges, fresh concepts and finally new buildings and urban designs in the 21st century.

References

[Fenv95]   Fenves S.J., "Successes and further challenges in computer-aided structural engineering"
             In: Pahl P.J., Werner H. (eds.), "Computing in Civil and Building Engineering" (Volume 1)
             Balkema, Rotterdam, 1995

[Kler97]   Klercker J. af, "Implementation of IT and CAD - what can architects schools do?"
             Proceedings of the First International AVOCAAD-Conference on the "Added Value of
             Computer Aided Architectural Design", Brussels, Belgium, April 1997

[Mave97]   Maver T.W., "Added Value for CAAD in Eduaction, Highlights of an Exciting 25 Years"
             Proceedings of the First International AVOCAAD-Conference on the "Added Value of
             Computer Aided Architectural Design", Brussels, Belgium, April 1997

[Sari91]   Sariyildiz S., "Conceptual Design By Means Of Islamic-Geometric-Patterns Within A
             CAAD-Environment"
             PhD-thesis, TU Delft, Faculty of Architecture, 1991

[Sari96]   Sariyildiz S., "Influence of Computer Science in the Next Generation of Architecture"
             Yearbook, TU Delft, Faculty of Architecture, 1996

[SaSc96]   Sariyildiz S., Schwenck M., "Integrated Support Systems for Architectural Design"
             Proceedings of the 3rd Conference on "Design and Decision Support Systems in
             Architecture and Urban Planning", Spa, Belgium, August 1996

[SaSc97]   Sariyildiz S., Schwenck M., "Tools of an Integrated Software Environment for
             Architectural Design"
             Proceedings of the First International AVOCAAD-Conference on the "Added Value of
             Computer Aided Architectural Design", Brussels, Belgium, April 1997

[ScBr93]   Schefström D., Broek G. van den, "Tool Integration - Environments and Frameworks"
             John Wiley & Sons, Chichester, 1993

[ScSa97]   Schwenck M., Sariyildiz S., "An Integrated Software Environment for the Architectural
             Design Process"
             Proceedings of the International Conference on Applications of Computer Science and
             Mathematics in Architecture and Building Science (IKM 1997), Weimar, Germany,
             February 1997

[ScVö96]   Schwenck M., Völker H., "Modelling the Architectural Design Process - Analysis of
             Traditional Methodology and Approaches to the Future"
             Report, TU Delft, Faculty of Architecture, 1996

162

[StBR96]   "Geïntegreerde bouwconcepten - Van partijen tot partners: voorbeelden van kansrijke
             samenwerkingsvormen in de bouw"
             Stichting Bouwresearch, Rotterdam, The Netherlands, 1996

[VSSD96]   Völker H., Sariyildiz S., Schwenck M., Durmisevic S., "The Next Generation of
             Architecture within Computer Science"
             Proceedings of the 6th Conference of the European Full-scale Modelling
             Association (EFA)", Vienna, Austria, September 1996

[VöSc97]   Völker H., Schwenck M., "Approaching a New Methodology: Integrated Architectural
             Design on the Basis of 3D Computer Models"
             Proceedings of the First International AVOCAAD-Conference on the "Added Value of
             Computer Aided Architectural Design", Brussels, Belgium, April 1997

[VöSc97a]   Völker H., Schwenck M., "Modelling the Architectural Design Process - Part II - Effects
             of Management Issues and Formal Descriptions"
             Report, TU Delft, Faculty of Architecture, 1997

[Wolf93]   Wolf P. van der, "Architecture of an Open and Efficient CAD Framework"
             PhD-thesis, TU Delft, Faculty of Electrical Engineering, 1993

REGULATING LINES AND GEOMETRIC RELATIONS AS A FRAMEWORK FOR EXPLORING SHAPE, DIMENSION AND GEOMETRIC ORGANIZATION IN DESIGN

BRANKO KOLAREVIC
University of Hong Kong
Department of Architecture
Knowles Building, Room 234
Pokfulam Road, Hong Kong

Abstract. The paper introduces regulating lines and geometric relations as a framework for shape delineation and dynamic drawing manipulation. It describes a relations-based graphic environment that can provide a qualitatively different way to explore shape, dimension, and geometric organization in design. It also presents ReDRAW, a limited prototype of the relations-based graphic system, and discusses some implications of its use in conceptual architectural design.

1. Introduction

In architectural design much of the creative discovery takes place in the two-dimensional realm of study drawings. An apparent contradiction, however, emerges in closer examination of the intertwined acts of drawing and designing. The act of drawing is inherently static—it produces drawings, snapshots of an evolving design concept. The act of designing, however, is intrinsically dynamic; shapes depicting an evolving design concept change constantly, i.e., they are seldom static.

The recognition of this disparity between drawing as a static and design as a dynamic activity, and the inability to adequately represent and manipulate design relationships using traditional means, provided an impetus to search for a computer-based drawing and design medium that can provide a qualitatively different way to explore shape, dimension, and geometric organization. The result of that search, a computer-based graphic context for shape delineation based on regulating "pencil" lines and their geometric relations, is described in this paper, its prototype implementation presented, and application implications discussed.

2. Regulating Lines, Geometric Relations, and Shapes

Shapes are fundamental to the act of drawing. Through shapes designers express and examine ideas and represent elements of design. Shapes denote edges and boundaries, spaces, building elements, or abstract concepts such as diagrams. Their role in design is significant—they represent and inform.

163

R. Junge (ed.), CAAD Futures 1997, 163-170.

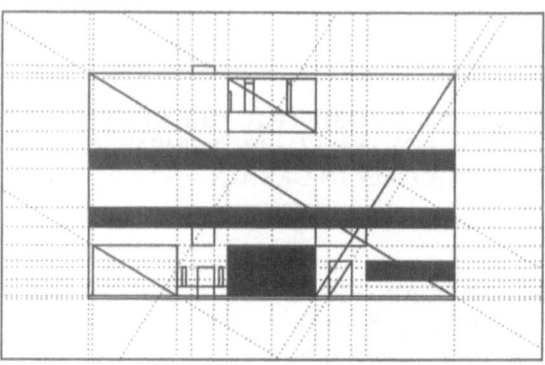

Figure 1. Le Corbusier's "les tracés régulateurs."

In architectural design, as in other design disciplines, shapes are frequently constructed within some graphic context, which is at a basic compositional level set by some abstract organizational devices, such as grids, axes, and regulating (or construction) lines. For example, Durand and Sullivan relied heavily on grids (patterns of regulating lines) and axes (regulating lines of specific importance). Le Corbusier's work from the purist period, both in architecture and painting, was guided by the application of regulating lines—"les tracés régulateurs" (Figure 1).

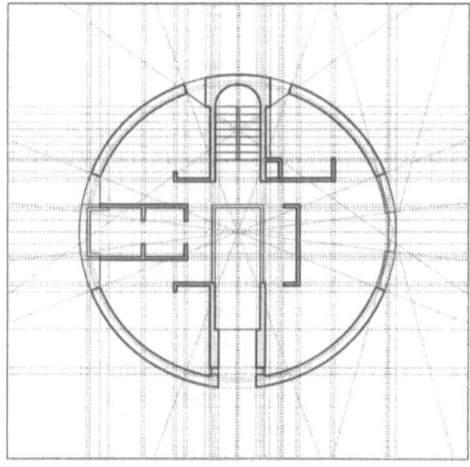

Figure 2. "Pencil" (regulating) lines and "inked" line segments. An interpretation of Mario Botta's Casa Rotunda based on regulating lines and their geometric relationships. Geometric shapes and relations are abstracted and translated into a relational drawing. New designs can be created by applying the transformations of translation and rotation (figures 4 and 5).

Regulating "pencil" lines therefore often provide, at a basic compositional level, an organizing framework for establishing positions and relations of "inked" line segments within and between shapes. Those "pencil" lines, however, can become much more useful and interesting when they are used not just as a rigid skeleton for the delineation of shapes, but to regulate the behavior of a drawing and to maintain its essential structure as its parts are manipulated. In other words, by allowing some "pencil" lines to control positions and orientations of other lines through their geometric relations and dependencies, we can structure the behavior of the object being designed under transformations. A computer based design "assistant" can record and maintain once established relationships, recognize the emergent ones, and compute the consequences of design transformations while preserving the semantic integrity of the drawing.

In this scenario, regulating "pencil" lines[1] define a compositional framework for establishing positions and relations of shapes. Shapes are constructed as combinations of shape primitives—"inked" line segments—delimited by intersecting regulating lines (Figure 2). Each "inked" line segment has an underlying regulating line as a baseline, and two regulating lines that intersect the baseline. This process of delineation is very similar to traditional manual drafting practice, whereby "pencil" regulating lines are laid out first, followed by "inking" of the selected portions between intersections.

A rather small repertoire of geometric relations[2], which are present or recognizable in any architectural composition, can be used to establish dependencies between "pencil" lines and "inked" line segments:

> CONNECTED AT a point
> INTERSECTED AT a point
> ALIGNED ALONG a curve
> PARALLEL TO a curve
> PERPENDICULAR TO a curve
> ANGLED TO a curve
> SYMMETRICAL (bilaterally) TO a curve

The architectural composition then essentially becomes a process of forming geometric relations between "pencil" and "inked" lines. Shapes are constructed by delineating underlying and intersecting "pencil" lines. Design begins by first laying out inter-related "pencil" lines—its organizing framework. It proceeds with the designer adding new regulating "pencil" lines, relations and shapes or changing the existing ones. In the process, many different options can be explored. As design evolves, shapes depicting an evolving design concept are manipulated and changed dynamically.

3.   ReDRAW – A Relations-Based Drawing System

ReDRAW (RElational DRAWing), a working, but very limited prototype of a relations-based drawing system (Figure 3), was developed to explore some of the computational and application issues associated with the relational description of shapes (Kolarevic

---

[1]   The regulating lines are not necessarily linear. We can classify regulating lines as straight (linear) or curved. Curved lines can be broken into subclasses: circular, elliptical, parabolic, sinusoidal, etc.

[2]   It is important to note that the number of geometric relations is indeed quite large and cannot be determined in advance. The hypothesis is that a fairly small set of carefully selected relations could provide an appropriate compositional repertoire. New relations could be defined as combinations of already defined relations. For a detailed discussion of geometric relations and their properties see (Kolarevic 1993).

166

1993, 1994). It is partly modeled on traditional drawing practice, as previously described. A user lays out infinite "pencil" regulating lines and simultaneously specifies positional relations (none, parallel, perpendicular, or angled) and dependencies (none, uni- or bi-directional) between them.[3]

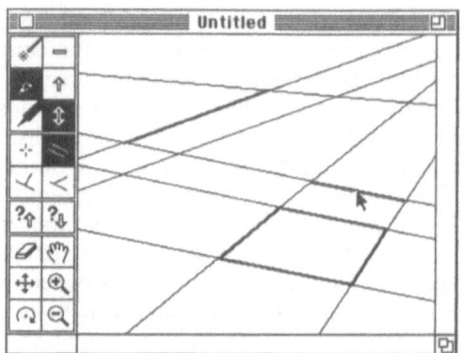

Figure 3. ReDRAW's drawing window with icon menu.

To construct shapes, user "inks" selected portions of "pencil" lines that are bound by intersections with other regulating lines.[4] The user manipulates created compositions by applying editing operations (erase, move, rotate) to selected regulating lines. ReDRAW automatically propagates changes while maintaining previously established relations. If some of the relations cannot be maintained during transformation, it can automatically establish new relations (in the "Smart Mode") or delete them. The user can also change once established relations, either by changing the type of the relationship or dependency.

ReDRAW supports only hierarchical, uni- or bi-directional dependencies. Its maintenance mechanism is based on simple, direct propagation through recursive traversal up and down the tree database structure (because of the bi-directional dependencies).[5] The conflicts in propagation are resolved in two ways, i.e., two modes: inactive and active. In the inactive mode, ReDRAW simply eliminates invalidated relations. In active ("smart") mode, it establishes new uni-directional relationships based on an angle between the two lines. In short, invalidated relations are either eliminated or new relations are established. This simple strategy eliminates extensive user intervention in solving potentially numerous low-level conflicts, which may be too distracting and unimportant in the design process. (After all, if the results of

---

[3] Currently, ReDRAW supports straight (linear) regulating lines only. Its repertoire of positional relations is also purposely limited to only three binary relations—parallel, perpendicular and angled. Ternary relations, such as symmetry and intersection, are not currently supported, since they can introduce cycles into ReDRAW's database representation. For more information about ReDRAW's data structures, important algorithms, interface, and usage rules see (Kolarevic 1993, 1994).

[4] Connectivity and alignment relations between shape segments are implicitly supported through the database structure. See (Kolarevic 1993, 1994) for a detailed description of the drawing database structure.

[5] ReDRAW employs an incremental propagation technique, i.e., relations are satisfied sequentially. In its current capacity, ReDRAW does not involve any equations to satisfy geometric relations—relations are simply satisfied by only two actions: translation and rotation.

propagation are unacceptable, user can always use the "undo" command.) ReDRAW also provides for substitution of once established relationships. Both the relationship and dependency can be changed by using the "magic wand" tool.[6]

Since hundreds or thousands of geometric relations can be established in a typical architectural parti, a designer will need some ability to anticipate the consequences of propagating changes through the composition after some transformation. The problem is that the compositional complexity, or a number of relations alone, will make the "mental" tracking of dependencies almost impossible. A computer-based graphic context, such as ReDRAW, should therefore aid designers in visualizing dependencies within the drawing. ReDRAW supports four types of queries of dependencies and relationships established in the composition. First, a user can query the database for a parent relationship of a selected "pencil" line—the type, dependency, and reference (i.e., parent) construction line will be graphically displayed. Second, a user can request that direct "dependents" of a selected "pencil" line be displayed. Third, users can query the drawing database to display all regulating lines to be affected by a certain transformation. Lastly, users can request a display of all regulating lines whose transformation will affect a selected line.[7]

The existing version of ReDRAW is limited in its features. The next version should add two very important ternary relationships: bilateral symmetry and intersection.[8] The next version should also provide circular "pencil" lines and parametric definition of relations. By incorporating shape recognition capabilities of Tan's ECART (Tan 1991), it could also support "search and replace" function of shape grammars.[9]

Like most prototype developments, ReDRAW evolved from assumptions and expectations which would require some change in order for ReDRAW to develop into a more fully-implemented design tool. The introduced concept of shape delineation based on regulating lines and their geometric relations can be extended into three-dimensional modeling. Regulating planes can become primary constructs— their intersections can define regulating lines.

4. Drawing and Designing Using Relations

> *"After all, nothing is more fundamental in design than formation and discovery of relationships among parts of a composition."*
> — *William Mitchell and Malcolm McCullough (1991)*

As a design "tool," ReDRAW is seen as an active agent in a design process rather than a passive record of the design development. It is envisioned as a tool that can efficiently and effectively generate new information within the design task through graphic

---

[6] Changing, or substituting an existing relationship can introduce cyclical dependencies. If ReDRAW recognizes a dependency cycle, it cancels the substitution and informs the user of its action.

[7] See (Kolarevic 1993) for more information about database queries in ReDRAW.

[8] Implementing ternary relations will require a slightly different database structure and probably a very different database maintenance mechanism, which will become increasingly more complex. It will probably rely on relaxation to resolve potential conflicts.

[9] Tan's ECART prototype for shape recognition (Tan 1991) and ReDRAW share a similar database representation. By incorporating the results of Tan's study, ReDRAW's value as a design tool can be considerably expanded.

processes, i.e., dynamic manipulation of architectural compositions. Its capability to generate new information, however, is highly dependent on designer's perceptual and cognitive abilities. Its generative role is accomplished through the designer's simultaneous interpretation and manipulation of a graphic image in a complex discourse that is continuously reconstituting itself—a 'self-reflexive' discourse in which graphics actively shape the designer's thinking process.

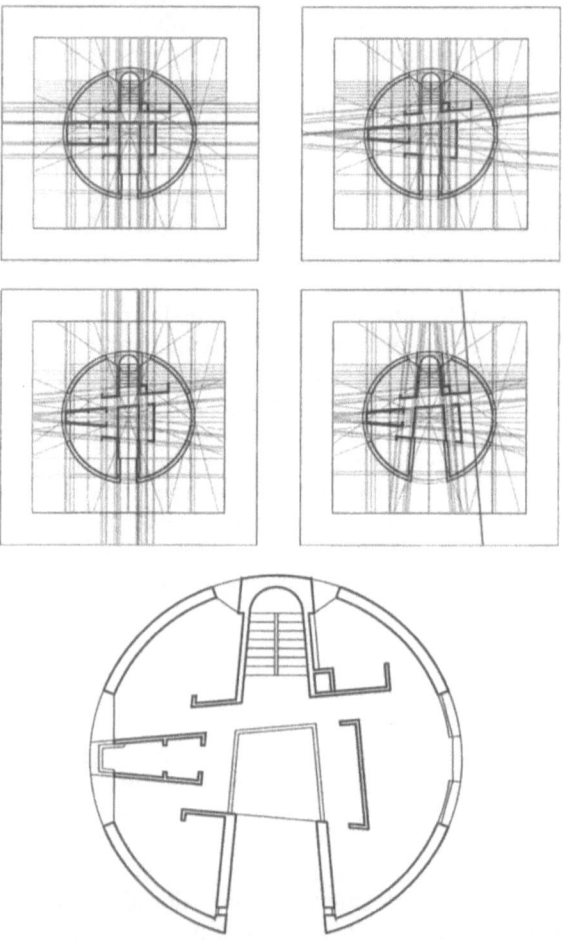

Figure 4. A possible transformation of Mario Botta's Casa Rotunda, based on an interpretation illustrated in figure 2.

Using geometric relations, a designer can enforce desired spatial configurations of building components and spaces (Figure 2). The established relations constrain the design possibilities—they structure possible manipulations. The choice of relationships applied in a composition (parti) may result in a dramatically different designs even though a small set of possible relations and a few transformations are available. How the composition is assembled, structured, or re-structured, determines its developmental potential. As William Mitchell (1989) observes:

"[T]he choice of modeling conventions and organizational devices that will structure the internal symbolic model [...] will determine how the model can be manipulated, and what can be done with it."

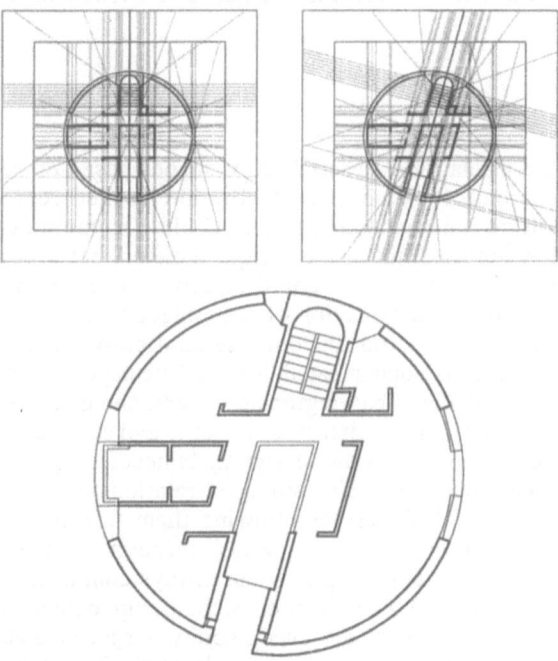

Figure 5. Another possible transformation of Mario Botta's Casa Rotunda, based on an interpretation illustrated in figure 2.

The relations, however, do not prescribe a particular form—they bound a space of alternatives without specifying a solution to the design task. "Composition often becomes a game of translating and rotating shapes to vary their spatial relations," writes William Mitchell (1990b). By applying different transformations, such as translation or rotation, to the parts of the composition, designers explore various alternatives (Figures 4 and 5).

Relationships and dependencies determine the behavior of the model. A designer must understand them to operate successfully upon them. This understanding is required on a basic, pragmatic level—if an object is moved, what other objects will move too. However, if a composition is too complex, applying a transformation to it might be difficult to control and envision. In other words, the consequences of propagating changes to the composition after applying a transformation can be very surprising. Resulting configurations can be genuinely new, and, in some instances, might trigger innovation and creativity. If the results of the operations are absolutely predictable, there would be little room left for creative discovery. "Imagination needs something to play with," asserts Mitchell (1990a). A drawing can become a vehicle on a path from known to unknown, from predictable to unpredictable. One formal universe might collapse into another, order can turn into chaos.

One of the major features of creativity "is the way in which it pioneers new contents—less in magically 'creating' something out of nothing, than a re-creation or re-framing" (Tan 1991). It is precisely this re-framing or re-structuring that is in the focal point of this work, which foresees geometric relations and transformations as a vehicle to support it.

5. Conclusions

This paper presented a relational description of shapes based on regulating lines and their geometric relations. It demonstrated how interrelated regulating lines, as an organizing device in design conceptualization, could become much more useful and interesting when they are used not just as a rigid skeleton, but to regulate the behavior of a drawing and to maintain its essential structure as its parts are manipulated. Designers could structure the behavior of the object being designed under future transformations; drawings could become semantically charged and could be manipulated in a semantically sophisticated fashion. The paper also presented ReDRAW, a limited prototype of a relations-based graphic system, and discussed its application in conceptual architectural design as a dynamic, versatile and stimulating medium.

The principal conclusion is not that designing is necessarily done as proposed, but that it might and beneficially be. The proposed relations-based approach to design conceptualization benefits designers by allowing them to efficiently and effectively generate new information within the design task through graphic processes, i.e., by providing graphic means of generating new but always contingent information within the design task through dynamic manipulation of the design object's relational structure. The proposed approach expands the designer's ability to speculate about possibilities. It places value on explicit formulation—its use requires "discipline" and an understanding of the relation-based approach to design as a method. Once the approach is understood, it can be used effectively to "program" the "behavior" of a design object.

References

Kolarevic, B. (1993) *Geometric Relations as a Framework for Design Conceptualization*, Doctoral thesis, Harvard University Graduate School of Design.

Kolarevic, B. (1994) Lines, Relations, Drawing and Design, *in* Anton Harfmann and Michael Fraser (eds.), *Reconnecting, Proceedings of the Association for Computer Aided Design in Architecture (ACADIA) 1994 Conference, Washington University, St. Louis.*

Mitchell, W. J. (1989) Architecture and the Second Industrial Revolution, *Harvard Architectural Review* 7, Rizzoli, New York.

Mitchell, W. J. (1990a), Introduction: A New Agenda for Computer-Aided Design, *in* M. McCullough et al. (eds.), *The Electronic Design Studio*, MIT Press, Cambridge, MA.

Mitchell, W. J. (1990b), *The Logic of Architecture*, MIT Press, Cambridge, MA.

Mitchell, W., and McCullough M. (1991) *Digital Design Media*, Van Nostrand Reinhold, New York.

Tan, M. (1991) *Themes for Schemes: Design Creativity as the Conceptualization, Transformation, and Representation of Emergent Forms*, Doctoral thesis, Harvard University Graduate School of Design, Cambridge, MA.

COMPUTABILITY OF DESIGN DIAGRAMS

an empirical study of diagram conventions in design

ELLEN YI-LUEN DO

College of Architecture, Georgia Institute of Technology,
Atlanta, GA 30332 - 0155, U. S. A. ellendo@cc.gatech.edu
&
Sundance Laboratory for Computing in Design and Planning,
College of Architecture and Planning, University of Colorado,
Boulder, CO 80309 - 0314. U. S. A. ellendo@colorado.edu

Abstract. Designers draw diagrams to think about architectural concepts and design concerns. We are interested in programming a computer to recognize and interpret design diagrams to deliver appropriate tools for the design task at hand. We conducted empirical studies to find out if designers share drawing conventions when designing. In this paper we first discuss reasons to investigate design diagrams. Then we describe our experiment on diagramming for designing an architect's office. The experiment results show that designers use different diagramming conventions when thinking about different design concerns. We discuss and report our efforts to implement a freehand drawing program.

1. Why do computers need to understand design diagrams?

Diagrams play an important role in design practice. Designers draw diagrams to explore ideas and solutions in the early, conceptual phases of design. They use diagrams as objects to think about design concerns (Laseau, 1980) and to record their ideas (Graves, 1977). Designers find it hard to "think without a pencil" (Lawson, 1994) and "must interact with the drawing" (Herbert, 1993).

Several design studies discussed the connection between design drawing and design thinking. They argued that design drawing and verbal protocols are related (Eastman, 1968) and complementary (Akin & Lin, 1995). Designers "see information" from drawing to refine their ideas (Suwa & Tversky, 1996); they "move" after seeing (Schön & Wiggins, 1992), and they operate between "seeing as" and "seeing that" modalities (Goldschmidt, 1991).

R. Junge (ed.), CAAD Futures 1997, 171-176.

Many computer aided design systems have been built to support design by giving designers advice such as cases, suggestions and simulation. A problem with these systems is that in order to provide appropriate advice they must identify the design context. We argue that the diagrams designers use in the early conceptual design process is a good indication of design concerns and can be computed.

In the following sections, following a brief description of our previous diagram study, we report a new empirical study of design to discuss the implications for a computer system. Section two focuses on the experiment about extracting design intentions from freehand design diagrams. Section three describes the experiment results. Section four concludes with a brief discussion of computational approaches and future work.

2. Empirical Studies of Design Diagrams

In our earlier experiment to identify the association of drawing marks with design thinking, we conducted an empirical study with sixty-two designers (Do, 1995) using diagrams and stories from a case based design aid Archie (Domeshek & Kolodner, 1992; Kolodner, 1991; Zimring, et.al, 1995). We asked participating designers to 1) make diagrams from given stories, 2) write stories from given diagrams, 3) pair diagrams and stories together, and 4) comment on the diagram story pairs from Archie case base.

The experiment identified that 1) designers used drawing conventions when diagramming different design concerns (figure 1 shows a sample of symbols and configurations designers used for sun, person and lighting concerns), 2) designers preferred to use certain views for different design concepts, 3) keywords from the stories were often used as labels in diagrams, and vice versa, and 4) designers understand each others' diagrams. However, this study only involved diagramming from descriptions instead of real design tasks. Therefore we developed a new design experiment to find out if these findings also apply in the design process.

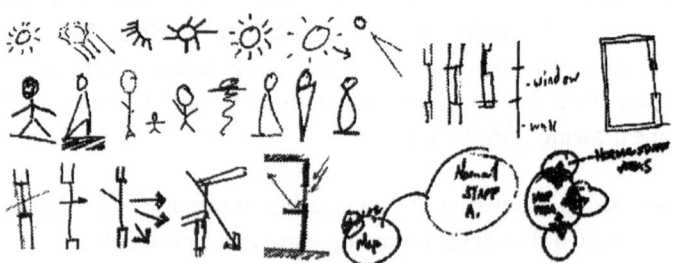

Figure 1. The lexicon of diagram symbols and configurations designers used for architectural concepts (Do, 1995). Top: symbols for architectural objects: sun, figure person and building elements such as walls and windows. Bottom: view preference: sectional views for lighting concerns and plan views for spatial layouts.

Three designers (one instructor and two students) participated in the diagramming design experiment at different dates. We gave participating designers a design program of an architect's office and video taped their design process while an observer took notes. The design program described the dimension of the site (70 ft. by 25 ft. one story warehouse) and required functions (work space for designing, CAD operations, drafting, meeting room, kitchenette, bathroom, etc.). The experiment has four tasks; each tasks called for a different focus. After reading the design program, designers were asked to do conceptual design starting with a new sheet of tracing paper and to focus on four different concerns for each task: (1) spatial layout for zoning of different work spaces, (2) lighting concerns, (3) visual access and privacy issues, and (4) fitting a large meeting table (4 ft by 10 ft) into a conference room and a minimal area requirement (800 square feet) for the designers' work space.

3. Mapping Drawing Conventions and Design Contexts

From the design experiment we verified that designers share drawing conventions not only when diagramming architectural concepts but also in their design process. Below we briefly describe three major findings that have influenced our current computational implementation.

3.1. DESIGNERS USE GRAPHIC SYMBOLS AND TEXT LABELING WHEN DESIGNING

We found from the experiment that designers arranged graphic shapes and symbols to compose architectural objects. Designers also used text labeling to identify architectural concepts (figure 2). They used simple geometric shapes such as circles and lines in their drawings, and composed them in conventional ways (e.g. parallel lines for walls and windows, an arrow and a letter N for North). We also found that key words from design concepts are included their drawings. For example, labels of functional spaces were written inside a containing shape (ovals, rectangles); an entrance was represented with a label and an pointing arrow.

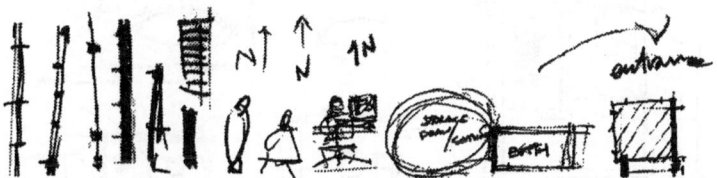

Figure 2. Designers share drawing conventions: They used lines, arrows, hatches and simple geometric shapes to compose symbols for architectural objects such as walls, windows and stairs. Words are used to indicate directions and functional spaces.

174

3.2. VIEW PREFERENCES FOR DIFFERENT ARCHITECTURAL CONCEPTS

Designers showed a preference for using different orthographic views (plan or section) when diagramming different design concepts for the experiment tasks: layout planning, lighting, visual access, and dimensioning. Participating designers drew plan views to illustrate spatial relations between spaces (figure 3a) and sectional views to illustrate lighting concerns (figure 3b, c , d). Another phenomenon worth noting is that designers used circling or overtracing to draw attention to an element, to refine a shape, or to explain to the observer (figure 3a, d).

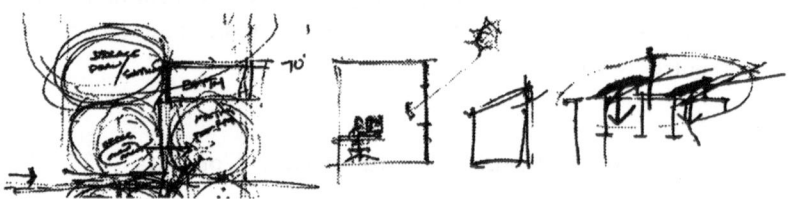

Figure 3. Designers revealed view preference for illustrating different architectural concepts. [a] plan view for spatial layout, [b, c, d] sectional view showing lighting by lines that penetrate the building envelope from windows and roof. Overtracing and circling showed a designer's attention to space [a] and concerns of reflecting light [d].

3.3. DESIGN CONTEXTS ILLUSTRATED BY DIMENSIONAL REASONING & FURNITURE

We found from the experiment that designers used symbols of architectural objects such as furniture and dimensioning figures to put themselves in mind of design concerns. When thinking about allocating objects or spaces with a required dimension, designers wrote down numbers beside the drawing to reason about size and calculate dimensions. For example, task 4 called for a work space of at least 800 square feet, and a conference room to accommodate a 4' by 10' table. One participating designer annotated his design drawing with numbers for calculation (figure 4a).

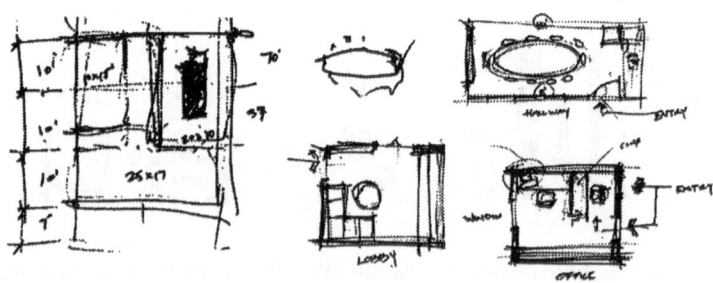

Figure 4. Designers wrote numbers for dimensions and drew furniture in their drawings to put themselves in the design context. [a] drawing documented a design process of calculating space requirements and dimensioning. [b, c, d, e] Symbols give clues for context. [b, c] conference space. [d, e] lobby and office space

When thinking about different functional spaces, designers drew simple shapes into the spaces to represent furniture, fixing in their minds the context to think about design. For example, when thinking about the placement of a conference table, designers drew chairs (small rectangles or dots) surrounding the table to see and test if the space is big enough (figure 4a, b). One designer also drew a door and windows along the wall, service counters, and white board (figure 4c). Similarly, designers drew symbols for furniture such as tables, chairs, and sofa in a lobby and office space (figure 3d, e).

4. DISCUSSION AND FUTURE WORK

Recognition and interpretation of design context is an interesting and complex problem. Our empirical studies showed that designers share a universe of conventional symbols and configurations when thinking about different design concerns. Therefore we can program a computer to recognize these conventions and use them to infer design context and intentions. We are conducting further empirical studies to understand more about design drawing conventions, their associated tasks and relevant design tools.

We are currently working on a freehand sketching program called the Electronic Cocktail Napkin (http://wallstreet.colorado.edu/Napkin) that recognizes and interprets hand drawn diagrams into relational propositions and communicates these to other knowledge based systems (Gross, 1996; Gross & Do, 1996). We are using this program to build a "Right Tool at the Right Time" manager (Do, 1996) that will activate different design tools based on the task at hand. The Napkin program uses a simple low level recognizer based on a 3x3 grid to identify low level glyphs, and employs rules built interactively by end users to parse configurations of low level glyphs to higher level configurations. Designers can train their personal language of diagram symbols and configurations by showing examples to the Napkin program. The right-tool-right-time manager uses Napkin's recognition abilities and sends requests to different programs. For example, when the designer is drawing bubble diagrams, the right-tool-right-time manager may infer that the designer is working on 'functional configuration' and call up a floor plan with similar layout from a case base or a slide library.

In sum, we have conducted empirical studies that show that designers share drawing conventions when designing. We have suggested that an "intention recognizer" in an integrated freehand sketching environment could support delivering the right tools at the right time. We have already connected various design tools with the drawing environment and are incorporating additional design tools into the system, automating tool activation, and improving the context detection mechanism.

Acknowledgments

Discussions with Mark D. Gross and Craig Zimring provided help in developing the test set-up and analysis of the experiment. Thanks also goes to the Archie Group at Georgia Tech (Craig Zimring, Janet Kolodner and Eric Domeshek) for the use of Archie program, and National Science Foundation grant DMII 93–13186 and IRI–96–19856. A more detailed version of this paper appeared in CAADRIA 97 conference proceedings.

References

Akin, O., & Lin, C. (1995). Design Protocol data and novel design decisions. Design Studies, 16(#2, April), 211-236.

Do, E. Y.-L. (1995). What's in a diagram that a computer should understand. In M. Tan & R. Teh (Eds.), *CAAD Futures '95: The Global Design Studio, Sixth International Conference on Computer Aided Architectural Design Futures* (pp. 469-482). Singapore: National University of Singapore.

Do, E. Y.-L. (1996). The Right Tool at the Right Time -- drawing as an interface to knowledge based design aids. In P. McIntosh & F. Ozel (Eds.), *ACADIA 96, Design Computation: Collaboration, Reasoning, Pedagogy* (pp. 191-199). Tucson, AZ: Association of Computer Aided Design in Architecture.

Domeshek, E. A., & Kolodner, J. L. (1992). A case-based design aid for architecture. In J. Gero (Eds.), *Artificial Intelligence in Design '92* Dordrecht: Kluwer Academic Publishers.

Eastman, C. M. (1968). On the Analysis of Intuitive Design. In G. T. Moore (Eds.), *Emerging Methods in Environmental Design and Planning* (pp. 21-37). Cambridge: MIT Press.

Goldschmidt, G. (1991). The Dialectics of Sketching. *Creativity Research Journal*, v.4(# 2), 123-143.

Graves, M. (1977). The necessity for drawing: tangible speculation. Architectural Design, 6(77), 384-394.

Gross, M. D. (1996). The Electronic Cocktail Napkin - working with diagrams. Design Studies, 17(1), 53-69.

Gross, M. D., & Do, E. Y.-L. (1996). Ambiguous Intentions. In *Proceedings, ACM Symposium on User Interface Software and Technology (UIST '96)* (pp. 183-192). Seattle, WA: ACM SIGGRAPH and SIGCHI.

Herbert, D. M. (1993). *Architectural Study Drawings*. New York: Van Nostrand Reinhold.

Kolodner, J. L. (1991). Improving human decision-making through case-based decision aiding. *AI Magazine*, 12(2), 52-68.

Laseau, P. (1980). *Graphic Thinking for Architects and Designers*. New York: Van Nostrand Reinhold.

Lawson, B. (1994). *Design in Mind*. Butterworth. Oxford.

Schön, D. A., & Wiggins, G. (1992). Kinds of Seeing and their functions in designing. *Design Studies*, 13(#2), 135-156.

Suwa, M., & Tversky, B. (1996). What Architects See in their Sketches: Implications for Design Tools. In *ACM Human Factors in Computing* (pp. 191-192). Vancouver, BC: ACM.

Zimring, C., Do, E. Y.-L., Domeshek, E., & Kolodner, J. (1995). Supporting Case-Study Use in Design Education: A Computational Case-Based Design Aid for Architecture. In J. P. Mohsen (Eds.), *Computing in Civil Engineering, A/E/C Systems '95* (pp. 1635-1642). Atlanta, GA: American Society of Civil Engineers.

HAND-EYE COORDINATION IN DESKTOP VIRTUAL REALITY

THEODORE W. HALL
Chinese University of Hong Kong

Abstract. For hand-eye coordination and intuitive interaction with virtual-reality displays, the projected image of a 3-D cursor in virtual space should correspond to the real position of the 3-D input device that controls it. This paper summarizes some of the issues and algorithms for coordinating the physical and virtual worlds.

1. Introduction

Contrary to Hollywood fantasies, virtual reality is not an out-of-body experience. Humans experience inertia and gravity and have a proprioceptive sense of body posture that is independent of vision. Moreover, many virtual-reality displays augment the user's view of the real world but do not completely mask it out or replace it. Thus, intuitive control and realistic interaction with virtual reality depend on accurate hand-eye coordination: the projected image of a 3-D cursor in virtual space should align visually with the real position of the 3-D input device that controls it. Though this may seem obvious, it is not automatic. Position input and graphic output rely on separate devices with independent coordinate references. Some systems provide neither the hardware nor the software necessary to map coordinates from one reference to the other, leaving the application programmer or end user to fend for themselves.

In the remainder of this paper, "desktop display" is short-hand for any non-head-mounted display. The discussion applies as well to wall-mounted displays. "Stereo glasses" are eye wear for filtering a desktop display into separate left and right views. These must also be equipped for position tracking. "3-D mouse" refers to a hand-held pointing device in a three-dimensional position tracking system. This may be a simple point-and-click device similar to the common 2-D mouse, or it may be incorporated into a more sophisticated system such as a data glove. "Tracker" refers to the reference device in the position-tracking system. The system reports the positions of the stereo glasses and 3-D mouse relative to this device.

2. Calibrating the Tracking System

The first step toward achieving any visible correspondence between the physical pointer and the virtual cursor is to calibrate the position tracking system relative to the display. For desktop displays, this presumes that the tracker is mounted in a stable position relative to the display, and that the display is within the tracker's operating range. The precise position of the tracker is arbitrary, and generally involves rotation and translation along all three axes. Moreover, the display hardware itself may be maladjusted, resulting in slightly different scale factors (pixels per millimeter) on the x

R. Junge (ed.), CAAD Futures 1997, 177-182.
© *1997 Kluwer Academic Publishers.*

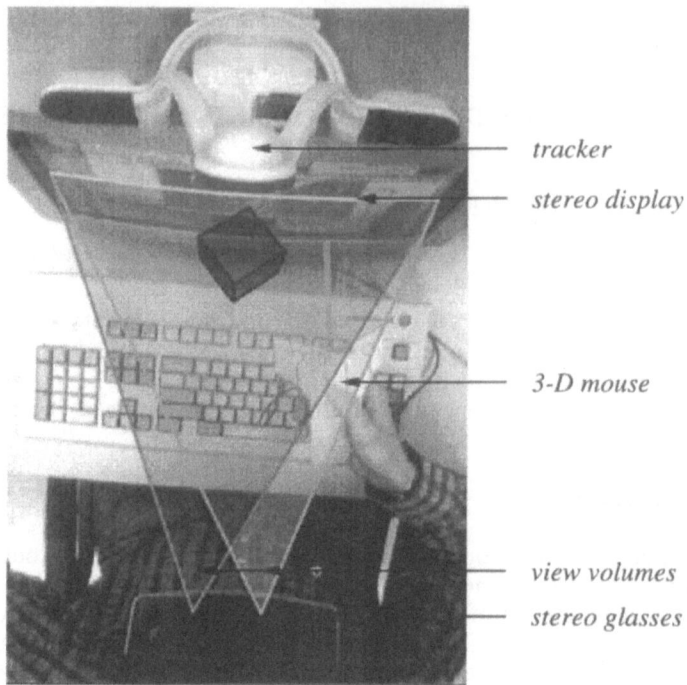

Figure 1: Correspondence of physical and virtual view volumes.

and y axes.

Figure 1 shows a plan view of a typical desktop virtual-reality system. The tracker is fastened to the top of the display (in this case, with adhesive tape). It monitors the positions of the 3-D mouse and stereo glasses relative to its own coordinate axes, and reports them to the application via the computer's serial ports. The application must derive and apply a transformation from tracker coordinates to display coordinates and thence to virtual world coordinates. These transformations are necessary to align the virtual 3-D cursor with the physical 3-D mouse. Similarly, the virtual projection plane and view points must correspond with the physical display plane and the user's eyes.

2.1. CALIBRATION PROCEDURE

In theory, three non-collinear points define a plane. In practice, most computer displays are not planar, but rather cylindrical or spherical. It's difficult to position three points on such a display without one of them bulging. Four points, located symmetrically at the center of each edge, avoid this problem. These points define horizontal and vertical vectors, parallel to the x and y axes. The cross product of these vectors defines the z axis and the orientation of the ideal plane. The location of the plane is such that the average z translation of the four points is zero.

For each of the four points, the calibration procedure draws a target and prompts the user to touch and click with the 3-D mouse. From the tracker coordinates reported for the targets, the procedure computes roll, pitch, and yaw rotations to bring the tracker axes into alignment with the display axes.

After aligning the axes, the procedure computes display scale factors. The display image size is easily altered by anyone who adjusts the hardware controls, and may differ significantly from the nominal values reported in the documentation or obtained from standard system functions. The calibration procedure uses the position tracking system to measure the actual display dimensions (in millimeters), and uses those dimensions to compute accurate scale factors.

Finally, the procedure subtracts the rotated-and-scaled tracker coordinates from the display coordinates to determine translations.

The transformation from tracker coordinates to display coordinates (roll, pitch, yaw, scale, translate) is constant as long as the tracker remains in a stable position relative to the display and the display image size doesn't change. The application may save the transformation in a file and retrieve it in subsequent runs.

2.2. TRANSFORMATION FROM TRACKER TO DISPLAY

In the following, the subscript mt denotes coordinates of the mouse relative to the tracker, td denotes coordinates of the tracker relative to the display, and md denotes coordinates of the mouse relative to the display. Matrices are defined to post-multiply row vertices. (To pre-multiply column vertices, transpose the matrices and reverse the order of multiplication.)

Let $\mathbf{T}_{td}$ represent the transformation matrix formed from the roll, pitch, yaw, scale, and translation of the tracker relative to the display (determined above). This matrix transforms the translation of the 3-D mouse from tracker to display coordinates:

$$\begin{bmatrix} xt_{md} & yt_{md} & zt_{md} & 1 \end{bmatrix} = \begin{bmatrix} xt_{mt} & yt_{mt} & zt_{mt} & 1 \end{bmatrix} \times \mathbf{T}_{td}$$

It's often useful to "transform" the rotation as well – to factor it into roll, pitch, and yaw components on the display's z, x, and y axes. These derive from the mouse's rotated basis vectors. In its own coordinate system, its basis vectors are the rows of an identity matrix. In display coordinates, they are the rows of a rotation matrix.

Let $\mathbf{R}_{mt}$ represent the rotation of the mouse relative to the tracker. Let $\mathbf{R}_{td}$ represent the rotation of the tracker relative to the display. Then:

$$\begin{bmatrix} \mathbf{i}_{md} \\ \mathbf{j}_{md} \\ \mathbf{k}_{md} \end{bmatrix} = \begin{bmatrix} i_x & i_y & i_z \\ j_x & j_y & j_z \\ k_x & k_y & k_z \end{bmatrix} = \mathbf{R}_{md} = \begin{bmatrix} 1 & 0 & 0 \\ 0 & 1 & 0 \\ 0 & 0 & 1 \end{bmatrix} \times \mathbf{R}_{mt} \times \mathbf{R}_{td}$$

In general, assuming the order of rotation is *roll* about z, *pitch* about x, *yaw* about y:

$$roll_{md} = \tan^{-1}\left(i_y / j_y\right)$$

$$pitch_{md} = \tan^{-1}\left(-k_y / \sqrt{k_x^2 + k_z^2}\right)$$

$$yaw_{md} = \tan^{-1}\left(k_x / k_z\right)$$

In the special case that i_y and j_y are both zero, or (equivalently) k_x and k_z are both zero, then the pitch is $\pm\pi/2$ (opposite to the sign of k_y). The roll can be set to zero, and the yaw can be computed from i_z and i_x:

$$roll_{md} = 0$$

$$pitch_{md} = \begin{cases} \pi/2 & \text{if } k_y = -1 \\ -\pi/2 & \text{if } k_y = 1 \end{cases}$$

$$yaw_{md} = \tan^{-1}\left(-i_z/i_x\right)$$

3. Stereoscopic Perspective Projection

In essence, stereoscopic perspective is simply a double application of the well-known procedure for monoscopic perspective. Extend rays from a viewing point through points in a scene. The intersections of these rays with a projection plane define the perspective projection of the scene on that plane. The ray from the viewing point perpendicular to the plane defines the center of projection. Computed perspective derives algebraically from the proportions of similar triangles: it scales width and height from the center of projection, according to the ratio of distances from the

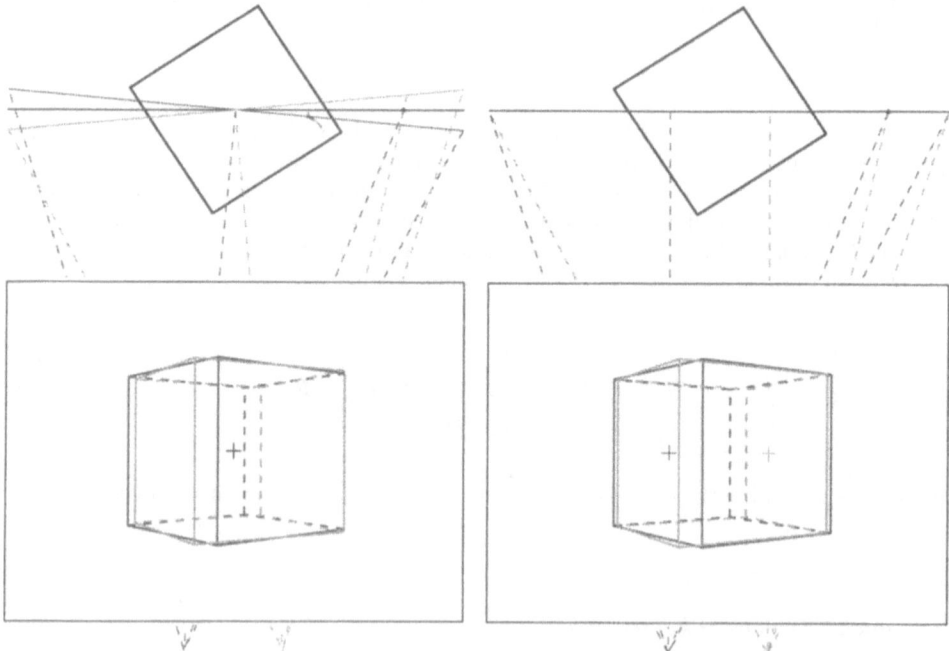

Figure 2: Incorrect construction of stereoscopic perspective (by rotation).   *Figure 3: Correct construction of stereoscopic perspective (by translation).*

viewing point to the projection plane and the point in the scene.

The center of vision can pan across the scene, from one ray to another, without any effect on the intersections of the rays with the plane – provided that the viewing point, the object, and the projection plane remain fixed in place.

Because the eyes rotate to converge their centers of vision on a point of interest, there is a temptation to compute stereoscopic perspective on the basis of a rotation between two views, as shown in Figure 2. This assumes that the eyes' centers of vision are also their centers of projection, resulting in two distinct, incongruous projection planes. It leads to vertical parallax between the two projections that should never occur as long as the eyes are level and equidistant from the display. Points in the plane of the display, which should project to the same pixel in both views, project to two different pixels. The error is small near the center, but becomes increasingly severe near the edges. These discrepancies confound depth perception and hinder hand-eye coordination in the virtual world.

Figure 3 shows the correct algorithm. Both views project onto a common plane, with only a horizontal offset between them. If one assumes a symmetric view frustum, then one must enlarge the horizontal angle and clip one side. Alternatively, one can

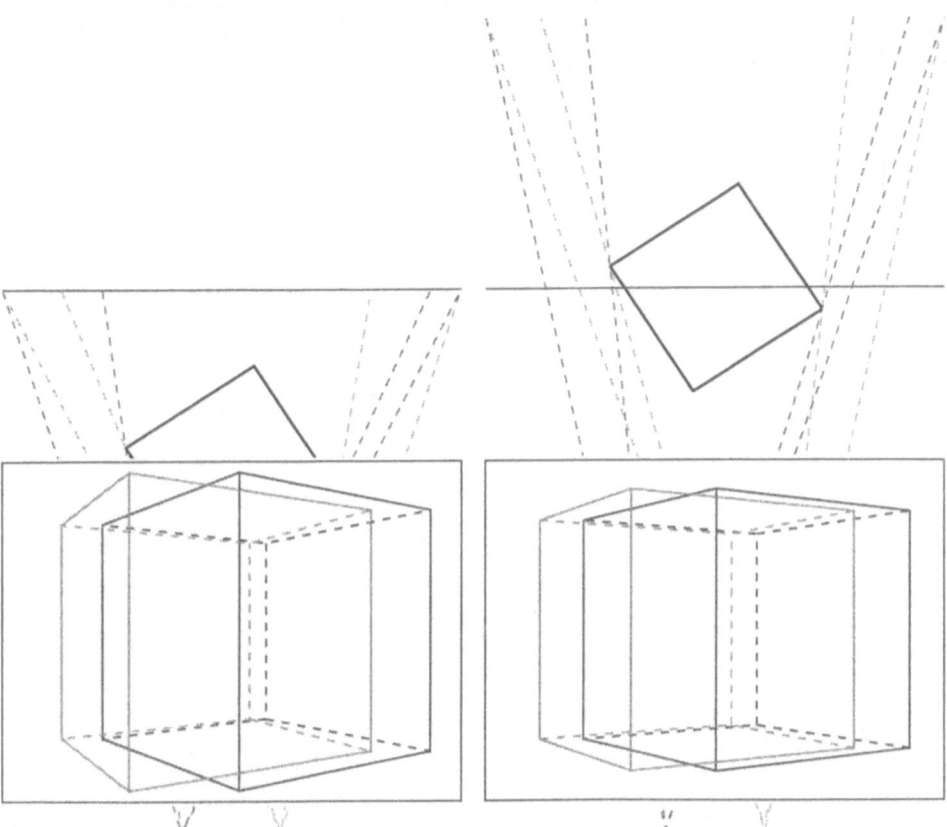

Figure 4: Moving the object closer.   *Figure 5: Zooming-in on the object.*

modify the equations for perspective projection to account for the asymmetric frustum.

It's desirable to control the apparent depth of the stereo image, to avoid large discrepancies between optical focus and binocular convergence, as well as to insure that interactive elements in the virtual world appear within reach of the 3-D mouse.

Problems may arise in trying to interact with a small image of a large object: the interactive elements may be too small to resolve, let alone control. There are several strategies for enlarging the image.

Figure 4 shows the effect of moving the object closer to the view points. Though it enlarges the image, it also increases the parallax. The result still appears as a small three-dimensional object, but now uncomfortably close to the user's face.

Figure 5 shows the effect of zooming-in on the object, by reducing the view angle and scaling the image to fill the display. This scales the width and height of the stereo image, but not its depth. Depth cues from both perspective and parallax make the object appear flattened.

Figure 6 shows the effect of enlarging the object. Alternatively, if the distance between the view points is reduced in proportion to their distance from the projection plane, the effect appears *as if* the object has been enlarged. Thus, the visual scale of the virtual world derives from the placement of the view points and projection plane. These should conform to the view angles and proportions of the real view volume, defined by the user's eyes and display plane. The application may apply any scale to the view volume, but should apply it uniformly.

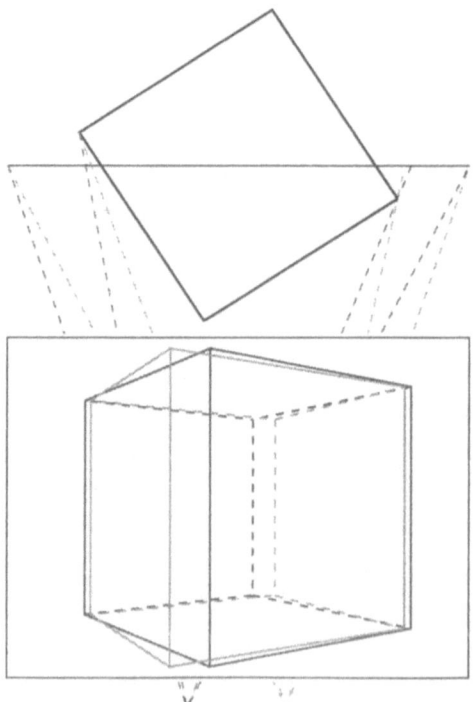

Figure 6: Scaling the object.

A digital way of planning based on information surveying

DONATH, DIRK
PETZOLD, FRANK

Bauhaus Universität Weimar
Informatik in Architektur und Raumplanung
D-99421 Weimar
Germany

The aim of this project is to develop a software system for generating complex digital models of existing buildings and structures, i.e. in the broadest sense a computer-supported surveying and management system for existing buildings.
The built environment is registered by surveying a series of geometrical and building-relevant information broken down into different levels of abstraction. The recorded data consists of a variety of geometric, multimedia and verbal - less structured - pieces of information.
The starting point for developing such a system is both an analysis and reworking of the methods used in architectural surveying, and the evaluation and use of current techniques and tools in the field of computer applications.

1 Starting point

A large part of current architectural and building practice is concerned with the field of restoration, modernisation and reconstruction of existing built structures, in particular the conservation, renovation and documentation of listed buildings and monuments.
Around 70% of current planning activity in the former East German states, and over 50% of building work in West Germany are work on existing buildings. Over 200 billion DM are spent yearly on the conservation of existing buildings.
A detailed building analysis and survey is an essential part of these planning activities. The statistics demonstrate the importance and relevance that the computer-supported surveying and management of existing buildings has.
The starting-point with which the architect or engineer is confronted is often the same.
Before the actual planning task can begin, a comprehensive consideration of the existing built situation has to be undertaken. This requires an intensive and detailed exploration and explanation of the existing situation, a survey of all building-relevant information necessary to describe and understand the task at hand.
The geometry of a structure is documented as a series of abstracted single 'views' - plan, section, elevation, detail. Further documentation, in the form of analyses, log books, statistical information, project descriptions, photographs or expert reports, follow or are

R. Junge (ed.), CAAD Futures 1997, 183-188.
© 1997 *Kluwer Academic Publishers.*

produced alongside the graphical survey. These are invariably detached from the graphical views, connected at most through cross references.

Our aim, therefore, is the conception of an integrated building information system, combined with a digitally-supported survey.

2 Building information system - GEBIS (GEBäudeInformationsSystem)

The aim of this research project is twofold: to design a practice-relevant software concept (GEBIS); and to develop a prototype system (GEBISexp), for a structured way of capturing and organising building-related information about existing buildings in digital form. The research is principally oriented towards existing buildings, in particular residential, office and commercial buildings.

2.1 THE REQUIREMENTS FOR A BUILDING INFORMATION SYSTEM BASED ON AN INTEGRATED SURVEYING SYSTEM

Practical experience in the field of digital architectural surveying has revealed clear deficiencies, in particular the absence of a systematic approach and its respective support in the software systems.

The deficiencies were identified, sorted and evaluated according to their importance. Ways of reducing the effort involved and factors influencing exactness were identified.

The following requirements for a computer-based building and surveying system were formulated as a result of our investigations:

- the development of a 'universal solution' based upon 'more intelligent' building models & interfaces, in order to enable a better interchange of data between the different planning phases.
- the integration of different technical and technological surveying procedures
- the capability to combine different recording techniques - hand measurements (the inch rule), exact measurements, photogrammetry, tachymetry, ...
- universal application in all fields, as well as in the field of conservation
- an efficient and cost-effective survey

2.2 RESEARCH - PRINCIPAL AREAS OF FOCUS

The research and development progressed as follows:
a) **Basic principles in architectural surveying and of information technology.**
b) **Specification and development of a prototype system (GEBISexp)**
   elementary functionality for computer-based capture and organisation of data in a practice-relevant environment.
c) **Evaluation of responses**
d) **The results will form the basis of a detailed specification.**

3 Selected research aims

A prototype system (GEBISexp) was designed based upon the multitude of requirements.

'GEBISexp' only considers specific objects of room and element structure. It attempts only to demonstrate the principle approach of a computer-supported surveying and information system.

In the following section selected aspects of the prototype system will be described in more detail.

3.1 DERIVING A ROOM AND BUILDING ELEMENT STRUCTURE FOR OBJECT-ORIENTED BUILDING-SURVEYING

Principal emphasis is laid on the systematisation of the built structure of a building, and the establishment of relevant planning and use-related information. Taken into account are the planning methods of architects working in the field of building restoration and reconstruction.

A typical problem when surveying existing buildings is that a large amount of information is recorded without an overview of the overall situation. The tendency is to concentrate on details, whereby simpler structural connections within the building go overlooked. These problems can be countered through the use of an ordering system that is used right from the start.

There are two primary ordering principles:
(A)   Room structure - the spatial subdivision of the building
           building complexes can be arranged both as entire buildings or individual
           rooms (i.e. building, floor level , room),
and
(B)   Element structure - the hierarchy of built elements in the building
           that which defines space and from which the geometry of the building is
           measured (i.e. wall, opening, window).

The structuring principles for both room and element structure have been developed. They are not independent from one another.

Rooms are defined, at least in an architectural sense, by spatial events. The room is therefore described only by the shape of the rooms boundaries, its perceived surfaces (Figure 1.). Thickness is not necessarily identifiable. Building elements can be interpreted as 'material rooms' and described by their actual surfaces.

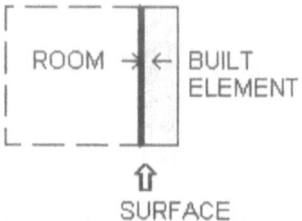

Figure 1. The relation between both structures

A geometric survey measures the building surfaces - wall, ceiling, floor etc. The connection between room structure (A) and element structure (B) is the building surfaces.

3.2 SPECIFIC PROPERTIES OF OBJECTS

The computer-supported building survey is not simply a geometric description of a building. It should also provide a multitude of features and characteristics relevant both to the buildings future use.
A series of attributes and their range of possibilities were identified, based upon practical experience in the field of architectural surveying.

The following are examples of object properties and characteristics relevant in working with existing buildings:
- method of construction, material, building damage
- constructional qualities
- thermal and technical details

The specific properties are descriptive qualities quantified as alphanumeric attributes with a range of possibilities (Figure 2). Defining these provides the architect or engineer with a consistent basis upon which to qualify his or her decisions.

When trying to define a 'static' attribute, it soon becomes clear that it is impossible to provide a comprehensive solution for all possibilities.
A 'variable attribute', however, allows the user to define custom attributes for the situation at hand. These can be defined and extended as required, both in terms of class and their evaluation methods.

At present the following types of 'variable attribute' are catered for:
- notional attributes
- numeric attributes
- textual attributes

attribute group	attribute	range
geometry		
	type	foundation / wall / floor / roof / ...
	...	
construction		
	material	tiling / cork / wallpaper / textile / ... plaster / stucco / ... boarding / plasterboard /... panelling / wood (pine, oak ...)/ ... smooth / profiled / rough / ... Metal plate (lead, zinc ...) / Metal sandwich panel / ...
	finish	none / opaque / transparent / glossy / matt / satin / ... oil-based / lime-based / wax / stain / ...
	colour	white / black / green / ...
	...	
history		
	condition	good / smooth / cracked / dry / damaged / repaired / ...
	extent of damage	total / internal / framework / nogging / ...
	...	

Figure 2. Relevant properties of the object "surface"

The integration of multimedia data into the building survey is of fundamental importance (Figure 3). It is an essential part of a comprehensive survey of any existing building.

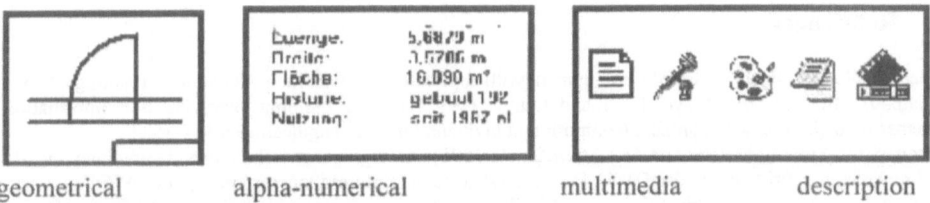

geometrical   alpha-numerical   multimedia   description

Figure 3. Multimedia functionality

3.3 ABSTRACTION LEVELS

Nevertheless, the architectural survey tends to be communicated in its graphical form. We have developed three different levels of abstraction corresponding to the phases in architectural practice:

- sketch orientated   size and orientation
- 2D-plan orientated   2D-representation
- 3D-model orientated   3D-representation

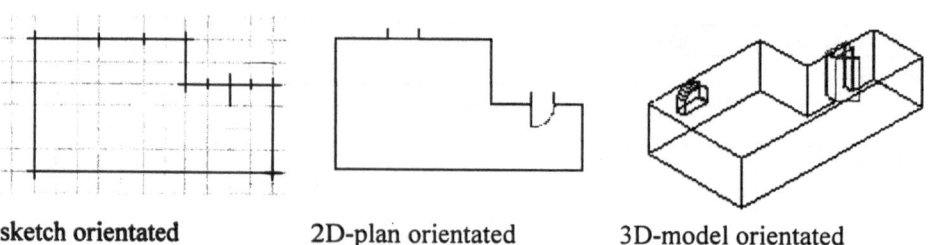

sketch orientated   2D-plan orientated   3D-model orientated
Figure 4 Abstraction levels of geometrical views

The transfer between different levels of abstraction - sketch orientated, 2D-planing orientated, or 3D-model orientated is always possible as a result of the higher density of information carried within each object. The object remains the same, the level of scrutiny changes.

4   Acknowledgements

Our current research work would be impossible without the great efforts of all the members of the chair of InfAR, especially Marion Schmitt, Holger Regenbrecht and Birgit Felsch. Jens Knüpfer turned our ideas into a real and working computer equipment, thank you. Thanks also to the students of applied computer science for programming different parts of the project.

5   References

Booch, G. (1994), Object-Oriented Development with applications -2nd Edition, Benjamin-Cummings, MA
Donath,D., Albrecht, W., Maye, H.-G.,and Ott, C (1995) Digital building surveying and information systemsAn objekt oriented approach, Computing in civil and building engineering,S.851-858
Mason, S.O.; Sinning-Meister,and M.; Streilein, A.(1995) Photogrammetrische Gebäudeerfassung für 3D Stadtmodelle in Schriftenreihe des DVW, Bd. 19/1995 ,Gebäudeinformationssysteme, S. 145-155
Rumbaugh, J., et al (1991) Object-Oriented Modelling and Design , Prentice Hall, Englewood Cliffs, NJ
Rüppel,U. (1997) Multimediale Gestaltung von Bestandsverwaltungsprozessen im Bauwesen, IKM'97, Weimar - Weimar , http://www.uni.weimar.de/~ ikm
Schmidt, W. (1989) Das Raumbuch als Instrument denkmalpflegerischer Bestandsaufnahme und Sanierungsplanung. Arbeitshefte des Bayerischen Landesamtes für Denkmalpflege Nr. 44/1989
Steinhage,V, (1997),Wissensbasierte Mustererkennung zur Erfassung von Bauplänen, IKM'97 - Weimar, http://www.uni.weimar.de/~ ikm
Streilen, A., Hirschberg, U.(1995) Integration of Digital Photogrammetry and CAAD: Constraint-Based Modelling and Semi-Automatic Measurement, CAAD futures 95 - Singapore, S.35 - 45

TOOLS FOR VISUAL AND SPATIAL ANALYSIS OF CAD MODELS

implementing computer tools as a means to thinking about architecture

ELLEN YI–LUEN DO
College of Architecture, Georgia Institute of Technology,
Atlanta, GA 30332-0155, U.S.A.   ellendo@cc.gatech.edu
& Sundance Lab for Computing in Design and Planning,
University of Colorado, Boulder, CO 80309-0314, U. S. A.

AND

MARK D. GROSS
Sundance Laboratory for Computing in Design and Planning,
College of Architecture and Planning, University of Colorado,
Boulder, CO 80309-0314, U. S. A.   mdg@cs.colorado.edu

Abstract. The paper describes a suite of spatial analysis programs to support architectural design. Building these computational tools not only supports the task of spatial analysis for designers but it also helps us think about the spatial perception. We argue that building design software is an important vehicle for understanding architecture, using our efforts to build various visual and spatial analysis tools as examples.

1. Tool building as a way to understand design concepts

The past twenty-five years have seen the development and refinement of tools for simulating, analyzing, and predicting the performance of building designs with respect to lighting, energy, acoustic, and structural behavior. Surprisingly, little work has been done on modeling the spatial characteristics of designs, although they strongly influence users' experience of buildings, and determining the spatial structure of a building is the central focus of architectural design. Tools for analyzing spatial characteristics have for the most part been neglected in CAAD research.

Plenty of work has been done outside the arena of CAAD on spatial and visual analysis of built environments. Architectural researchers have looked at the relation between physical features of the built environment and experiential qualities that users perceive. For example, Benedikt (Benedikt 1979), Thiel (Thiel 1961; Thiel 1981), Appleyard, Lynch, & Myer (Appleyard and others 1964) among others have proposed frameworks, theories, and specific analyses of perceptual space that can—and we argue ought—be made computable. Computations of these analyses would be immediately usable both to CAAD users directly as well as to intelligent CAAD programs of the type described by Koile (Koile 1997). We suggest however that implementing such analyses in a

R. Junge (ed.), CAAD Futures 1997, 189-202.
© *1997 Kluwer Academic Publishers.*

computer program is often nontrivial and raises issues may have been glossed over in the original, non-computational, formulation.

In this paper we aim to do two things. First, we describe a class of tools for spatial analysis of building designs based on Benedikt's isovist research and other models of the perception of architectural space. Second, we point out that the detailed implementation strategy chosen to build a tool for analysis can have a tremendous effect on the outcome, not only making the tool more or less efficient, but actually affecting the framing of architectural concepts. Thus we see the building of CAAD tools not only as a means to perform useful calculations, but also as a way to think about underlying and central concepts of space, place, and architecture.

In the following sections we discuss computational tools for spatial analysis in architecture. In particular, we consider how different implementation approaches influence the results of the computation, and more fundamentally, how the tools lead us to think differently about underlying architectural concepts. We focus here on models of the perception of spaces and places as exemplified by Benedikt's work on isovist (Benedikt 1979). Accordingly, section two reviews briefly the underlying work of Benedikt and others on spatial and visual analysis. Section three discusses our experience implementing visual and spatial analysis tools for CAD models. We briefly describe the algorithms, and presentations of these tools. Finally, in section four, we compare the strengths and weaknesses of these different methods and propose an agenda for future work.

2. Concepts of visual and spatial analysis

Good designers manipulate the placement and spatial relations of walls, screens, and partially defined boundaries to provide feelings of containment and privacy, opportunities for survey and outlook, and the sense of flow and direction in a building. A well designed building is an appropriate and varied arrangement of spaces whose spatial characteristics match intended uses and offer a range of physical definitions of habitable space. Analyses of the spatial and visual character are often done for the design of museums, office work group layouts, prisons, and religious and ceremonial buildings. The level of analysis ranges from informal inspection to a more careful checking of sight lines and boundaries. Often, though, architects do not perform spatial analysis, but simply work on the basis of their own experience when designing the spatial characteristics of a building.

Rooted in the principles of human perception (e.g., Ittelson's *Visual Space Perception* (Ittelson 1960)) this relationship between form and perception, environment and behavior is widely discussed throughout the discipline of architecture. Porter's *How Architects Visualize* (Porter 1979)) offers a good overview of the perceptual issues. Ashihara's *Aesthetic Townscape* used 'degree of enclosure' to describe and discuss the composition of townscapes, observing for example, that people pause and linger in the 'inside corners' of spaces (Asihara 1983). Lynch discussed 'imageability' (Lynch 1960), and changes of views in sequence when moving in an environment (Lynch 1972), and the use of 'viewshed' -- terrain maps for analyzing visual effects from a major viewpoint (Lynch 1976). Appleyard, Lynch, and Myer discussed spatial and visual perceptions associated with movement along a highway (Appleyard and others 1964).

Various theoretical and empirical research efforts have been made to account for the relation between physical form and human perception of spaces, which we review below briefly. These include Benedikt's work on isovists (Benedikt 1984; Davis and Benedikt 1979) and the work of Hillier and Hanson (Hillier and Hanson 1984) on space syntax and Peponis on various space partitioning schemes (Peponis and others in press). However these efforts to understand the relation between built form and perceived space have not been widely applied, in part because they involve tedious calculations from floorplans. These calculations can be easily made using computers, adding spatial analysis tools to the growing collection of tools for analyzing building performance.

Benedikt proposed that a space, or an environment, is perceived as a collection of visible surfaces not occluded by physical boundaries such as walls and partitions (Benedikt 1979). He defined an 'isovist' at a given point in the floorplan as the space visible from that point, looking around in 360 degrees. Quantifying the isovist areas, perimeters, and solid boundaries can be used to compare the quality of different spaces. He further described making physical (say, clear plexiglass) models of isovists along a defined path and stacking these models to visualize a moving user's changing perception of a built environment. In his papers (Benedikt 1979; Davis and Benedikt 1979) he described computational implementations that were done at the University of Texas at Austin in the 1970's.

In *The Social Logic of Space*, Hillier and Hanson presented 'space syntax' as a way to describe and analyze the character of spaces and predict human behavior patterns and cultural activities in these spaces (Hillier and Hanson 1984). In space syntax the connectivity between spaces—the integration value—captures the depths of topological values. Hillier and Hanson analyzed two dimensional planar views to draw and quantify connectivity and integration values. First, a convex space is indicated by drawing maximal 'space bubbles' between wall partitions. Then an 'axial line' is drawn to connect convex spaces. The number of intersections for each axial line in relation to overall area produces an 'integration value'. Empirical observations found that the frequency of activities coincides with integration values of an axial map, that is, in real life more highly integrated places tend to have more activities.

Researchers at University College, London have developed the Axman program to support axial map calculation. Users import a picture of the floor plan or bubble diagrams as an underlay and then draw axial lines on top of the plan. The Axman program finds intersections among all axial lines, calculates the integration values for each line, and displays axial lines in different (rainbow) colors: Red lines indicate the most integrated paths and blue or violet lines indicate the least integrated ones.

Peponis, WIneman, and others. (Peponis and others in press) recently proposed various spatial analyses to study the shape and spatial configuration of building plan. These analyses are based on the identification of the boundaries of spaces in a floorplan with certain properties, in particular with respect to a viewpoint moving through the floorplan. They recognized that a wall end or a corner can be an important index of spatial qualities because the 'endpoints' appear and disappear or remain invisible as the viewer moves. Their endpoint partition', for example, identifies convex areas that are "informationally stable" with respect to shape, i.e. all their points are visually connected to the same edges of the plan. They are supervising a computer implementation of spatial partition analysis in the Microstation environment (Peponis and Wineman 1997).

3. Implementation methods of spatial analysis tools

We describe our work on automated spatial analysis of CAD models, which includes two main styles of analysis. The first, a discrete approach, involves calculations of perceived space, by computing the projections and extensions of physical boundaries, and the computation of viewsheds and visual fields. The second, a continuous approach, involves the calculation of spatial gradients or fields. We describe several programs, built on a variety of platforms, including PL/I, Topdown Pascal on Macintosh, Tk/Tcl on Sun, AutoLISP with AutoCAD, and Macintosh Common Lisp, that perform these computations on floor plans and sections entered by a designer.

3.1. ENCLOSURE CALCULATION (ENCLOSURE)

As an undergraduate project in 1977 Gross built an enclosure measuring program in PL/I, (subsequently reproduced in 1982 by Gross and Fred Wu in Apple II Logo). The Enclosure program calculated numeric values for subdivided spaces in a floor plan, intended to model a user's feeling of enclosure or protectedness at each point in the plan.

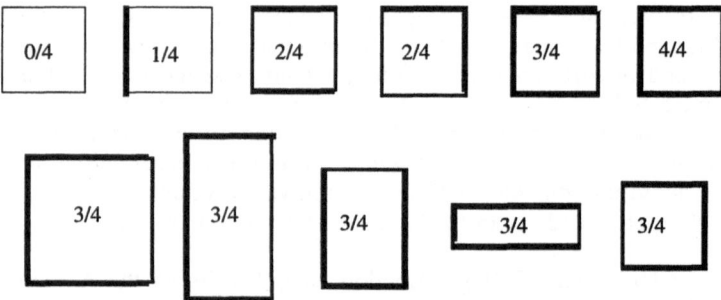

Figure 1. Top: Enclosure value counts number of adjacent walls. Bottom: Spaces bounded by three walls have same enclosure value 3/4 regardless of size.

As shown in Figure 1, enclosure values ranged from 0 (fully open) to 1 (fully enclosed). First, the program divided the floor plan into subspaces by extending the lines of walls and constructing perpendicular lines at wall endpoints (figure 2). This creates many subspaces with different sizes. Then it counted the number of walls in the boundaries of each subspace and divided by the number of boundaries to normalize the values from 0 to 1. For example, a four sided subspace has the value 1/4 if it has one wall boundary, 2/4 if there are two walls, and 4/4 if the space is surrounded, totally enclosed by four walls. The numeric enclosure value is displayed as gray scale intensities on the floor plan.

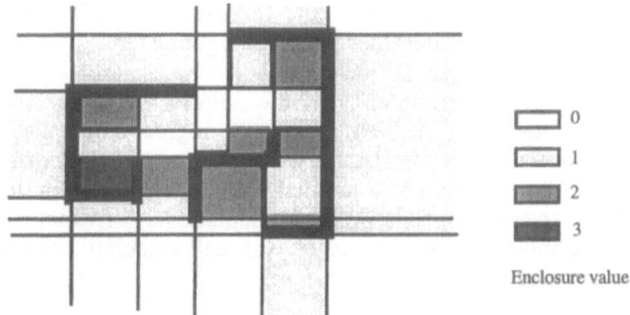

Figure 2. Values of subdivided spaces computed by Enclosure.

3.2. GRADIENT ANALYSIS - TOPDOWN ISOVIEW

A second approach to visual and spatial analysis is a discrete simulation program called Isoview that computes and displays "degree of openness" of cells in a space (Do 1993). Isoview was implemented in Topdown Pascal (Mitchell and others 1990), a parametric programming environment that provides interface widgets such as slider bars and buttons and a library of graphics routines. Users of the Isoview program used Topdown's parametric interface tools to position walls and openings and adjust the sizes of the rooms.

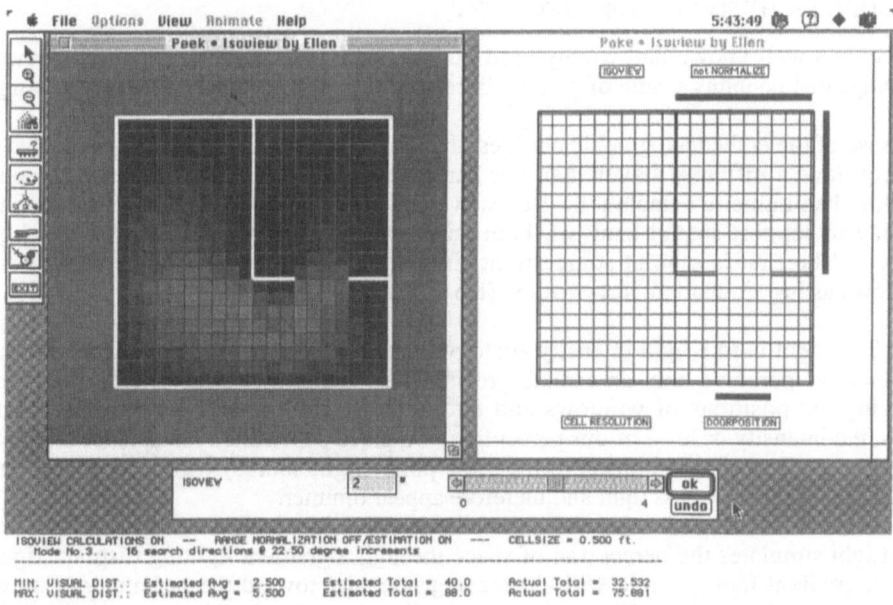

Figure 3. Gradient implementation of Topdown Isoview shows openness intensity in a floorplan. Left: graphic display. Right: user interface. Bottom: message to user.

A floorplan in Isoview is overlaid on a square grid whose resolution the user can set. Isoview allowed designers four degrees of grid resolution from a 10x10 to a 40x40 grid. Isoview's definition of "openness" is the average distance to the nearest surrounding walls. Isoview computes the openness value for each grid cell and displays them simultaneously as color values in the floor plan (see figure 3).

The designer can change the granularity of the angular calculations. The crudest approximation uses a four directional $\pi/2$ radian (90 degree) calculation, in which the cell value is the average distance horizontally and vertically in both directions to the nearest wall. In addition to the simple 4 cardinal directions calculation, Isoview can use 8, 16, or 32 directions to perform finer grained calculations, corresponding to $\pi/4$, $\pi/8$. and $\pi/16$ radian angle intervals for distance-to-wall cell calculations. That is, for each cell x,y, Isoview calculates

$$V_{x,y} = (\sum_{0}^{k} d_i)/k \qquad k = 4, 8, 16, 32$$

where $V_{x,y}$ represents the value in cell (x,y); d_i represents the distances in the k directions to the nearest walls; and k represents the resolution of the calculation (4, 8, 16, or 32 directions).

Color intensity is used to display the cell values. Higher openness values are represented by light red, lower values are displayed as dark red. A normalized version shows the relative value of each cell with the most open cell displayed as bright red, and the least open cells as black.

3.3. POINT LIGHT SIMULATION - TCL-LIGHT

Tk/Tcl is a well known and widely used human computer interaction tool kit in the C language that contains a suite of graphic library calls and widgets for rapid prototyping.

Because of the embedded graphic abilities, the spatial analysis programs implemented in Tk/Tcl have a different flavor than our earlier efforts. They focus more on graphic display than numeric calculation. Below we describe two programs that used graphic display to support spatial analysis, both based on the analogy of view point as light source. The first is a point source light simulation called Tcl-Light; the second is a shadow casting simulation, Tcl-Shadow (Do 1994a).

Do's Tcl-Light used a light intensity analogy to model the perception of space. From a distance we perceive only an outline profile of a facade; as we move closer we start noticing the positions of windows and doors, then detailing and building materials. Thus, the intensity or level of our perception of a place depends on its distance from our view point. The closest objects to the view point (light source) will be brightest, and farther objects receive less light and therefore appear dimmer.

Tcl-Light simulates the perception of space through a gradient lighting display (figure 4). A gradient fading circle with decreasing intensity toward the perimeter is drawn around each view point. The designer can adjust how far the circle reaches and the intensity gradient to represent different individuals' spatial awareness. Similar to illuminating a room by adding more light sources, a user's understanding of a room can be strengthened by perceiving it from more view points.

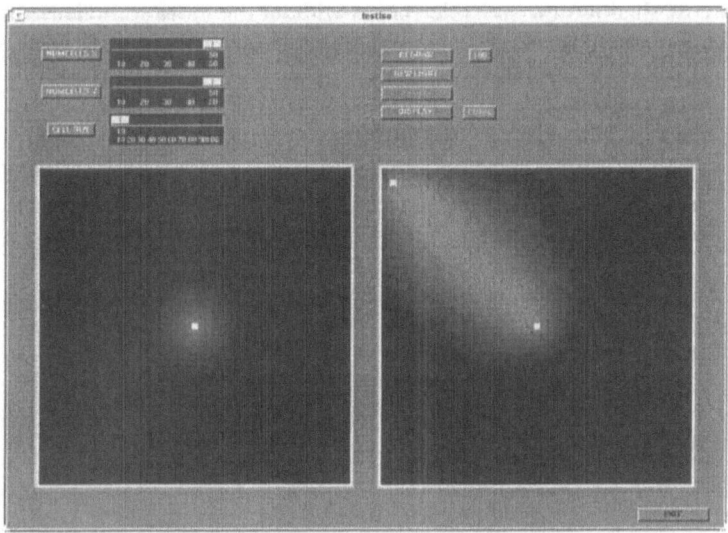

Figure 4. Gradient implementation of Tcl-Light models space perception as light sources.

3.4. SHADOW CASTING - TCL-SHADOW

Do implemented a second lighting analogy approach called Tcl-Shadow (Do 1994a) using a shadow casting technique: anything behind a wall receives a dark shading. Instead of painting the lighting effects from a light source like Tcl-Light, we simply plot shadows behind the walls.

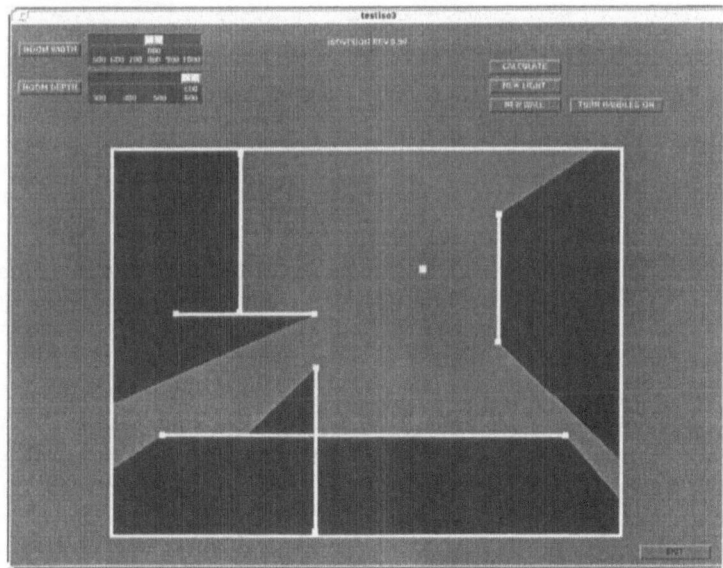

Figure 5. Tcl-Shadow displays an isovist by casting shadows.

Tcl-Shadow cast shadows on invisible spaces without the more complicated calculation of occlusions and visible field. By painting black shadows behind the walls, we are left with a field of visible surface, in effect displaying Benedikt's isovist (figure 5).

3.5. REMOVE OCCLUDED WALLS - AUTO-ISOVIST

Do implemented an isovist module (Do 1994b) in the AutoCAD drafting environment (using AutoLisp) to automate viewshed calculations.

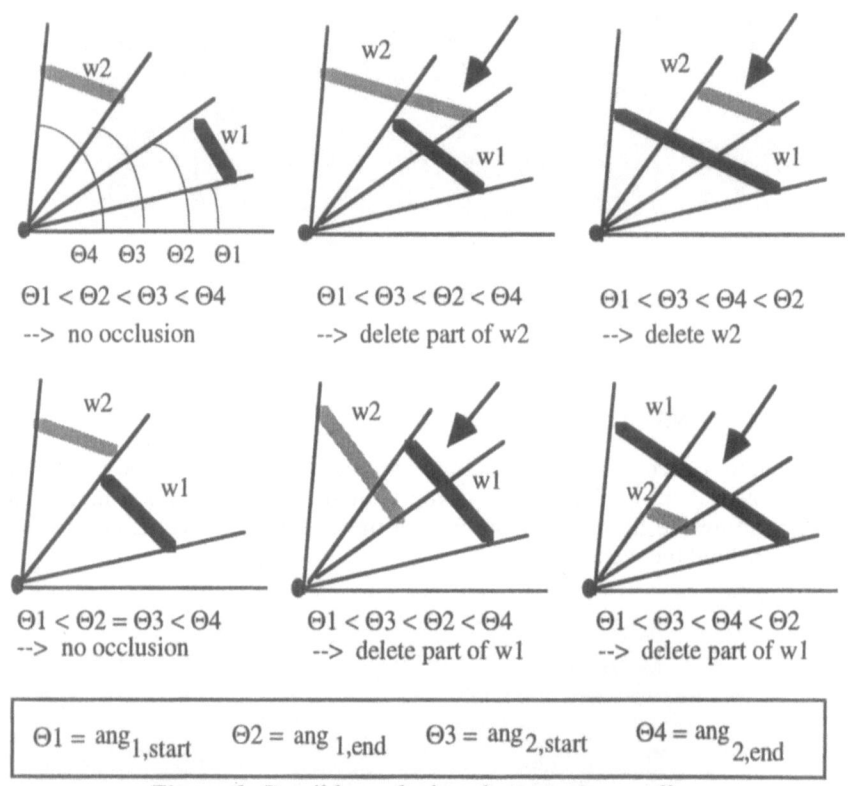

Figure 6. Possible occlusions between two walls.

The Auto-Isovist program maintains a database of the walls designers draw. A chosen view point becomes the origin for sorting the walls. First the program determines and tags 'start' and 'end' points for each wall according to their angles to the x-axis. Then it sorts the walls by their startpoint angles in a sequence: w_1, w_2,...w_n, such that the angles from the view point to their startpoints ($ang_{1,start}$, $ang_{2,start}$, ... $ang_{n,start}$) increase monotonically. Each wall w_i and the next wall in sequence w_{i+1} are then compared to find occlusions (figure 6). If the angle of w_i's endpoint is smaller than angle of w_{i+1}'s startpoint, then there is no occlusion, and the program moves on to compare the next walls w_{i+1}, w_{i+2}. If the angle ($ang_{i,end}$) of the end point of w_i is greater than the angle ($ang_{i+1,start}$) of the starting point of w_{i+1} then the wall segments between these angles are in occlusion. The next step checks the distance from the view point to the two occluding wall segments to determine which wall is occluded, and deletes the occluded portion from its original wall. The program continues, comparing w_i and w_{i+2}, and so on. If the angle of the end point of a wall w_j is larger than the angle of the end point of a second wall w_{j+k}, and the distance from view point to w_{j+k} is larger than w_j, then the farther wall w_{j+k} is completely blocked by the nearer and hence entirely deleted from the database. If w_j is further, then the occluded part will be deleted from w_j, deriving two visible wall segments.

After this round of checking, we are left with only wall segments that are visible from the view point. The isovist field is then the polygon constructed by connecting the wall segments in order. The sum of the wall segments is calculated and labeled as the real-surface perimeter (figure 7).

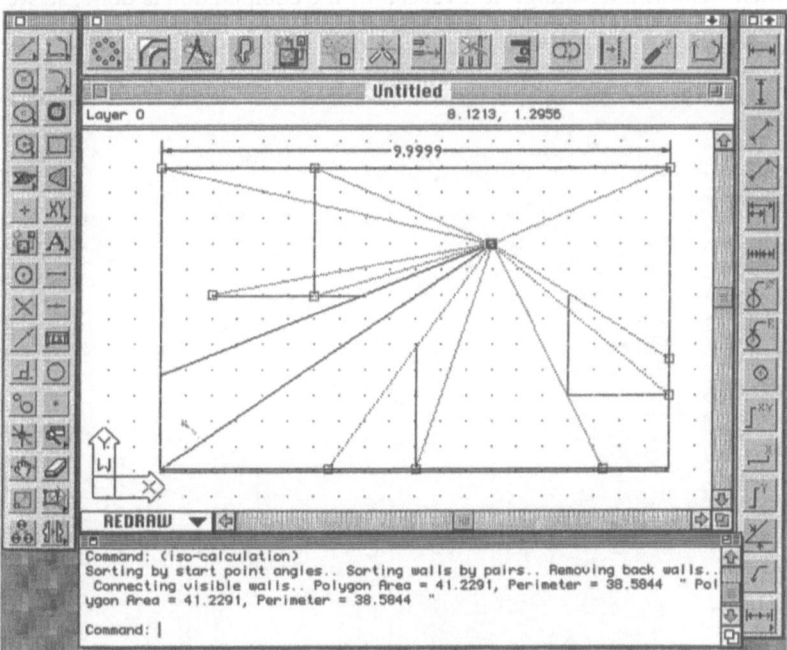

Figure 7. Isovist implementation in AutoCAD using startpoint angle sorting and occlusion checking .

3.6. FINDING INTERSECTIONS AND VISIBLE WALLS - MCL- ISOVIST

The last program we called IsoVist (Do 1995), named following Benedikt. It is implemented in Macintosh Common Lisp.

An internal database of walls is maintained and updated as the designer adds, deletes, and moves walls. The program constructs rays from the view point to the end points of all walls. The *find-intersections* routine computes all intersections of each ray with all other walls. It returns for each ray a data structure containing the intersection points and their distances to the view point, and with each intersection point, the intersected wall. Next it sorts the rays by their angles. Finally it constructs a representation for the isovist by making triangular polygons bounded by the rays and the walls they intersect (figure 8).

It constructs triangles clockwise from the first ray. It starts a triangle from the view point, finding the first wall intersection point on the first ray. It checks if that wall intersects the second ray. If so, it finds the wall's intersection with the second ray, and follows the second ray back to the view point to complete a triangle. If not, it finds the next wall intersection of the first ray. As above, it checks this wall to see if it intersects the second ray, and so on.

Initially for debugging, we displayed the construction of light rays to all the intersections, and displayed the intersections as dots. Designers found this display of rays and intersections interesting. It reveals how the computation is done, so we left the debugging display on.

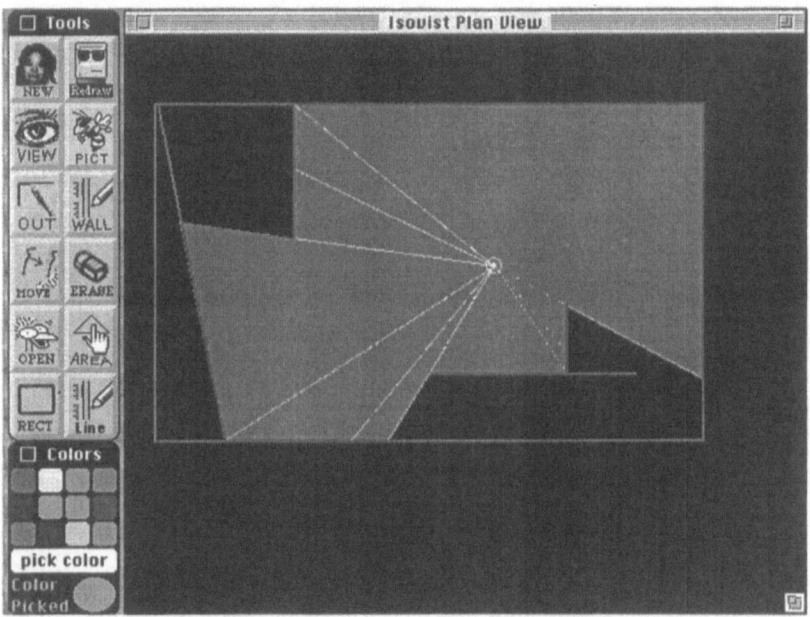

Figure 8. IsoVist displays the area visible from a given vantage point.

4. Discussion and Future Work

4.1. COMPARISON OF THE DIFFERENT APPROACHES

We have described six different computational implementations that support similar spatial analyses. Because we used different programming languages, platforms, data structures, algorithms and presentation methods, each tool addresses a different aspect of spatial analysis and encourages different ways of thinking about the problem.

Roughly, the six tools we described above can be classified into two categories: (1) accurate portrayal of the Isovist concept—this includes Tcl-Shadow, Auto-Isovist and MCL-IsoVist, and (2) a gradient and intensity approach—this includes Enclosure, Topdown Isoview, and Tcl-Light.

In the first category, the Tcl-Shadow, Auto-Isovist and MCL-IsoVist programs all portray the Isovist concept as defined by Benedikt, and they all consider and display a single view point depiction of the field of vision. Among them, the Tcl-Shadow version is a purely visual approach; it displays the isovist but does not maintain a computational representation of it. It simply displays the isovist field by shadowing out invisible places behind all walls. On the contrary, both Auto-Isovist and MCL-IsoVist all compute the visible field with sorting, intersection finding, and polygon constructions. Auto-Isovist takes a "behind" approach, finding occluded walls and eliminating them from the database to leave only visible walls. MCL-IsoVist takes a "front" approach, finding the nearest walls to construct visible triangles bounded by rays.

Second, the gradient and intensity approach programs all deal with more than one view point. Enclosure and Topdown Isoview both compute enclosure or openness values and they use a field display method to show the relative values of many different view points at the same time. This is quite different from the Isovist single view point approach. In other words, Isovist portraying programs plot a single isovist field at a time, while gradient programs like Enclosure and Topdown Isoview plot a multiple view point isovist field simultaneously (see figure 9).

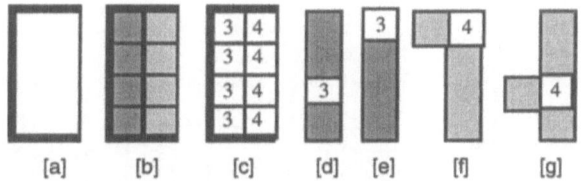

[a]   [b]   [c]   [d] [e]   [f]   [g]

Figure 9. Topdown Isoview depiction of a space:
[a] space bounded on three sides,
[b] Isoview gradient display, $\pi/2$, 90 degree angular resolution
[c] cell distance to walls is 3 (2 up + 1 down) before normalization,
[d]-[g] eamples of cell value calculations: sum of distance to wall.

Enclosure and Isoview both measure the space quality in response to the bounding walls. Enclosure divides space into uneven subspaces, and computes a value based on the number of adjacent walls in each subspace (see figure 1). Subspaces of varied dimensions all have same enclosure value if they are bounded by same number of walls. Topdown Isoview takes a grid approach to divide space into even size subspace cells. A finer gradient of values occurs when grid or angular resolution is increased (figure 10). The crudest resolution of Isoview that only calculates cardinal directions in a four sided subspace, however, obtains similar results as the Enclosure method: both compute the same value for spaces with the same wall bounding conditions (figure 11).

.

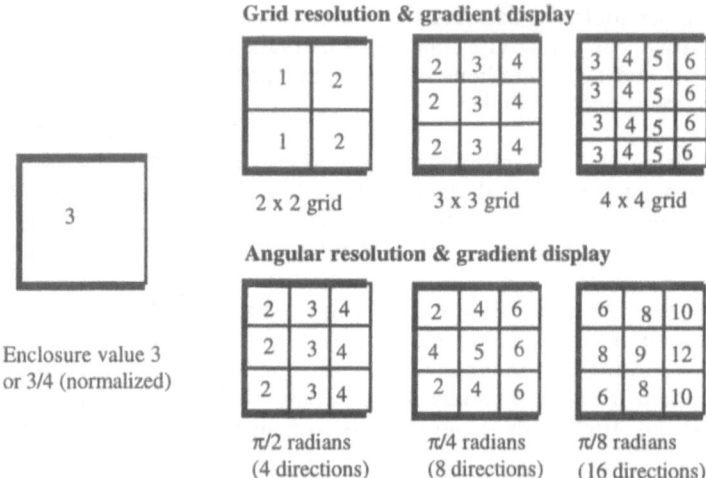

Figure 10. [a] Only one value derives from the number of walls of a bounded space in Enclosure, [b] The same space modeled in Topdown Isoview, gradient values with different granularity resulting from increased resolution: Top, grid resolutions from 2 x 2 to 4 x 4 with same angular resolution (4 directions); Bottom, increase of angular resolution (from 4 to 16 directions) with same grid resolution (3 x 3).

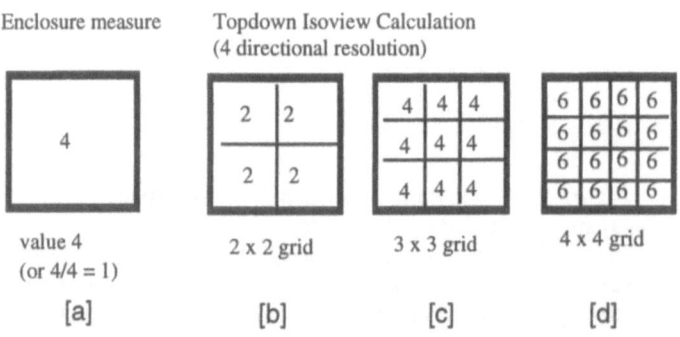

Figure 11. Enclosure and/or openness value for a space bounded by four walls. After normalization, all values in the space appear the same. [a] Enclosure computation derives value 4, [b - d] Topdown Isoview calculation with different grid resolutions.

The Tcl-Light approach is an outlying case; it supports multiple view points and also uses gradients like Enclosure and Topdown Isoview. However, it differs from these other implementations in that it uses gradient to display the fading intensity of perceptual space. Although Tcl-Light also supports single view point depiction like Isovist portraying programs it differs from these approaches in that it only plots a gradient perception circle instead of the shape of the visible field.

By using different programming environments and computation approaches, we have explored spatial analysis tools that address similar but varied concepts. The earlier efforts by Gross aimed to understand spatial character through building computational tools (Enclosure). Later Do's series of programming experiments began with an attempt to portray multiple view point isovist fields on the same floor plan and resulted in a discrete gradient openness display program (Topdown Isoview). Seeing that gradient displays can help a designer to visualize relative openness values and that the isovist idea focused on single viewpoint, the point light analogy with intensity gradient was explored (Tcl-Light), which supported both single and multiple view points. A shadow casting approach to display isovist fields (Tcl-Shadow) then emerged from the lighting analogy. The discovery that the visible field from a single view point can be displayed by darkening spaces behind all the walls led to the eliminating walls "behind" implementation (Auto-Isovist). Finally, a "front" wall finding program (MCL-IsoVist) resulted from efforts to simplify computation.

4.2. FUTURE WORK

The suite of spatial and visual analysis tools described here were developed over a span of years, hardly a short term research project. It is interesting to see how the issues of spatial analysis were implemented differently. The approaches differ not only in the platform and programming language used, but also in their data structure, algorithm and presentations. We have reexamined the approaches of these programs to revisit the issues of computational support for spatial analysis.

We are interested in implementing these concepts in a three dimensional space, instead of the plan and sectional views supported by the various prototypes we have described. We are currently investigating the use of a head mounted display to provide a simulated walk through of a space in connection with traditional plan and section isovist analyses as outlined here. We would like to provide an analysis tool for designers to see and explore a simulated virtual environment. The project will combine not only visual presentation of a space walk through but also analysis and feed back of spatial perception.

Acknowledgements

Chenning Hsi, Ali Malkawi provided valuable programming help on random occasions, and Jean Wineman, John Peponis, Craig Zimring, and Aaron Fleisher provided insightful discussions.

We sadly dedicate this work to the memory of Wade Hokoda (1957-1997), artist and master craftsman of digital media, teacher, and friend, who died unexpectedly shortly before this paper was submitted for publication. Wade worked with Ellen Do on the projects described in this paper; many of the best insights herein are due to his sharp intellect and programming talent. His untimely death is a personal loss as well as an impoverishment of our field.

References

Appleyard, D., K. Lynch, and J. Myer. 1964. *View from the Road*. Cambridge, MA: MIT Press.

Asihara, Y. 1983. *The Aesthetic Townscape*. Cambridge, MA: MIT Press.

Benedikt, Michael L. 1979. To take hold of space: isovist and isovist fields. *Environment and Planning B* 6:47-65.

Benedikt, Michael L. 1984. Perceiving Architectural Space: From Optic Arrays to Isovists. In *Persistence and Change*, edited by W. H. Warren and R. E. Shaw. Hillsdale, N.J.: Lawrence Erlbaum.

Davis, Larry S., and Michael L. Benedikt. 1979. Computational Models of Space: Isovists and Isovist Fields. *Computer Graphics and Image Processing* 11:49-72.

Do, Ellen Yi-Luen. 1993. *Imaging the concepts of isovist field in design process -- Topdown Isoview -- a tool for spatial analysis*. Georgia Institute of Technology. Special Topics, independent study ARCH8183E.

Do, Ellen Yi-Luen. 1994a. *Design and description of form -- using tool command language Tk/Tcl to visualize isovist by lighting and shadow casting analogy*. Georgia Institute of Technology. Computer program ARCH8193A4.

Do, Ellen Yi-Luen. 1994b. *Isovist calculation in AutoCAD*. Georgia Institute of Technology. Computer program and independent study.

Do, Ellen Yi-Luen. 1995. *Visual Analysis through Isovist -- building a computation tool*. Georgia Institute of Technology & Sundance Lab for Computing in Design and Planning, University of Colorado, Boulder. Working paper and computer program

Hillier, W., and J. Hanson. 1984. *The Social Logic of Space*. Cambridge, UK: Cambridge University Press.

Ittelson, W.H. 1960. *Visual Space Perception*. New York: Springer.

Koile, K. 1997. Design Conversations With Your Computer. In *CAAD Futures '97*, edited by R. Junge.

Lynch, Kevin. 1960. *The Image of the City*. Cambridge: MIT Press.

Lynch, Kevin. 1972. *What Time is This Place?* Cambridge, MA: MIT Press.

Lynch, Kevin. 1976. *Managing a Sense of Region*. Cambridge, MA: MIT Press.

Mitchell, William J., R. Liggett, and M. Tan. 1990. Top-Down Architectural Design. In *The Electronic Design Studio*, edited by M. McCullough, W.J. Mitchell, and P. Purcell. Cambridge, MA: MIT Press.

Peponis, John, J. Wineman, M. Rashid, S-H. Kim, and S. Bafna. in press (Environment and Planning B). On the description of shape and spatial configuration inside buildings: three convex partitions and their local properties.

Peponis, John and J. Wineman, 1997. SPATIALIST-PARTITIONS, computer program for Microstation software, Copyright Georgia Tech Research Corporation.

Porter, Tim. 1979. *How Architects Visualize*. New York: Van Nostrand Reinhold.

Thiel, Philip. 1961. A Squence Experience Notation for Architectural and Urban Space. *Town Planning Review* 32 (April):33-52.

Thiel, Philip. 1981. *Visual Awareness and Visual Perception*. Seattle: University of Washington Press.

DESIGN CONVERSATIONS WITH YOUR COMPUTER: EVALUATING EXPERIENTIAL QUALITIES OF PHYSICAL FORM

KIMBERLE KOILE
MIT Artificial Intelligence Laboratory
Massachusetts Institute of Technology
Cambridge, MA 02139 USA

Abstract. This paper describes a prototype system that evaluates an architectural design using the designer's theories about how to manifest experiential qualities in physical form. The system uses AI methods in conjunction with geometric and non-geometric knowledge to represent experiential qualities, e.g. privacy, in terms of concrete details of a design, e.g. wall dimensions and locations. This paper describes the organization and implementation of the system, and reports the results of an experiment in which the system was used to evaluate Frank Lloyd Wright Prairie houses.

1. Introduction

Imagine that you're sketching a design, pen in hand. You tell the computer near you that the design is for a family of four who want a house that feels both spacious and cozy; that invites the community to visit, but protects their privacy; that is just large enough for their needs, but not larger. You have ideas about circulation patterns and about physical forms that manifest feelings of privacy, and you're translating those ideas into lines and annotations on a page. You stop to assess your latest sketch and ask the computer for its comments. It graphically shows you access from exterior to interior and within the interior. It shows you regions defined by your proposed physical forms. It tells you that the main living space is not very private with respect to the exterior, but that it is spacious. It suggests moving the front door to increase the privacy. It suggests adding an entry territory between the front door and the main living space, and points out that if the area of the main living space decreases as a result, then the spaciousness may decrease. You like the first suggestion and accept it, but not the second. You continue sketching and evaluating.

This paper describes a prototype system capable of an important part of the above scenario: evaluation of an architectural design using the designer's theories and preferences about how to manifest experiential qualities in physical form.

2. Overview

The evaluation system described here is the first component of a prototype design support system for architectural design. The design support system adopts the prevalent view that architectural design is an iterative process of design generation, evaluation, and

203

R. Junge (ed.), CAAD Futures 1997, 203-218.
© 1997 *Kluwer Academic Publishers.*

modification. The system assumes that a designer will generate a design, e.g. via sketching, then will evaluate and modify that design iteratively, asking the system: Does this design adhere to a set of specified design goals? If not, how can it be modified so that it does? Such a design support system will engage a designer in a conversation of the sort in the opening scenario.

The design support system under development differs from other work in computer-aided architectural design in two ways:

1. The system focuses on architectural knowledge not often found in design systems: what architects know about manifesting experiential qualities in physical form. While some architects may design with only form in mind, most create spaces that people inhabit. They and their clients describe spaces as private, sunny, open, spacious, etc. Architects use their knowledge from past experiences, from environment behavior research, and from their own theories to create spaces with experiential qualities such as these. This knowledge can be articulated and structured as general design principles (e.g. Alexander, et al., 1977; Zeisel and Welch, 1981; Hertzberger, 1993), which can serve as a basis for a design support system that reasons about experiential qualities and physical form as in the opening scenario.

2. The system bridges the gap between knowledge-based systems and CAD systems by using AI methods in conjunction with both geometric and non-geometric knowledge. AI methods are ideal in domains rich with complex, subjective, heuristic knowledge, and design has been an important focus of AI research for many years (e.g. Coyne, et al., 1990; Pham, 1991; Gero and Sudweeks, 1996). Much of that research, especially in the domain of architecture, has focused on knowledge-based or case-based systems, with emphasis on non-geometric knowledge, such as topological relationships. Geometric knowledge has generally been left to CAD drawing systems. The use of AI methods in conjunction with both geometric and non-geometric knowledge will enable a tool to be built such as the one described in the opening scenario, which reasons about both experiential qualities and concrete details of physical form.

The design support system under development will be a step toward a design tool that acts as an intelligent design assistant. In particular, it will aid a designer by easily and quickly generating and keeping track of alternatives, by managing the complexity of many conflicting design goals, by providing more systematic exploration of a design space, and by serving as a repository for reusable design knowledge. The system also will contribute to the clarification of terms used in architectural discourse.

3. Evaluation System

The evaluation component of the design support system described in the previous section assesses architectural designs for the presence or absence of experiential qualities. For example: Is the main living space private? The evaluation system is organized around two ideas: that experiential qualities are manifested in physical form by means of the concrete details of a design; and that the knowledge relating qualities to details can be explicitly represented and manipulated in a reasoning system. The system, thus, represents and reasons about design details, experiential qualities, and the relationships between them.

3.1 DESIGN MODELS

A design is represented by several kinds of design models—a design-element model, a circulation model, a territory model, a use-space model, and a connectivity model. The design-element model contains information about design elements, which represent physical objects used to create form, e.g. walls, windows, doorways. The circulation model is a geometric abstraction of the design-element model, represented as a graph with nodes for doorways and links between doorways. The nodes and links retain dimension and location information. The territory model is another geometric abstraction of the design-element model; it contains lists of points, lines, and territories—regions of space derived from the design elements. The use-space model contains a list of use-spaces, which are activity areas defined by the user by assigning activity labels to particular territories. Representing use-spaces and territories separately enables the system to evaluate the physical form itself without necessarily tying it to intended use, and vice versa. Finally, the connectivity model is a representation of topological relationships between use-spaces; it is a graph with nodes for use-spaces and links between use-spaces directly connected to each other (e.g. by means of a shared doorway). See Figures 1 and 2 for diagrams of design models for the Mrs. Thomas Gale house.

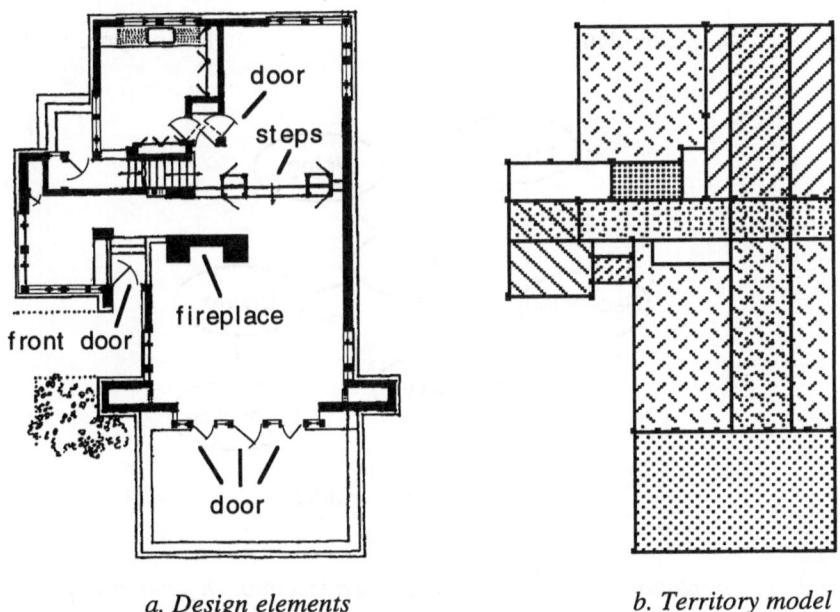

a. Design elements   *b. Territory model*

*Figure 1. a. Floorplan[1] showing several design elements in the design-element model.
b. Territory model: territories formed by design elements, including two
overlapping territories.*

---

[1](Storrer, 1993)

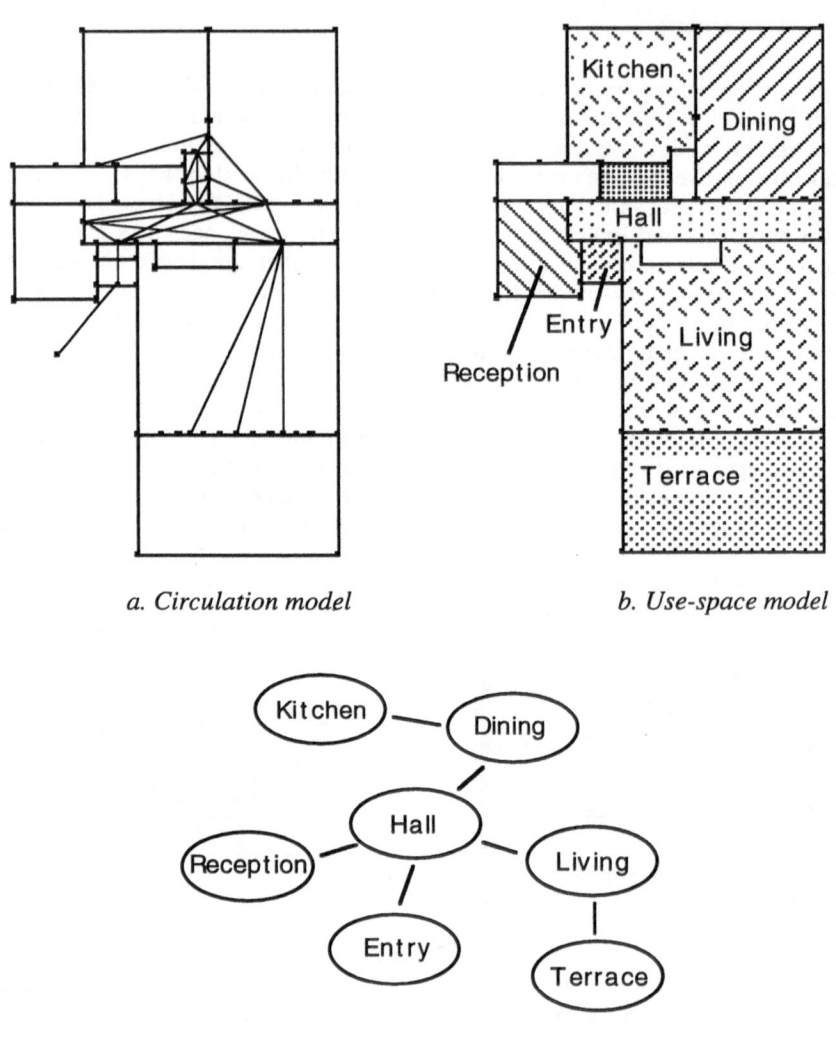

a. Circulation model

b. Use-space model

c. Connectivity model

Figure 2. a. Circulation model:[2] paths from experior approach point through interior.
b. Use-space model: use-spaces, which are territories plus activity labels.
c. Connectivity model: topological relationships between use-spaces.

[2]The circulation model is derived automatically from the design-element and territory models; the connectivity model from the use-space model. Design-element, territory, and use-space models currently are entered by hand. In future versions, design-element and use-space models will be derived automatically from a sketch (Gross, 1996); territory models will be derived automatically from the design-element model (Kincaid, 1997).

3.2 DESIGN CHARACTERISTICS

At the core of the system is a hierarchy[3] of design characteristics that provides mappings between experiential qualities and concrete architectural details that realize those qualities. The hierarchy represents knowledge such as: A place may feel private if it feels hidden. A place may feel hidden if it is reached via a circuitous path. It may also feel hidden if it is on a different level from other places, or if it is not entered directly from a place likely to be frequented by many people. In this example, feeling private is an experiential quality which can be achieved in a building by means of concrete architectural details, such as particular materials, locations and dimensions of walls, windows, etc. Both experiential qualities and concrete architectural details are represented as design characteristics.

Concrete details are referred to as *primary design characteristics*; they can be objectively measured. Experiential qualities are referred to as *secondary design characteristics*; they cannot be objectively measured; their values are based on primary design characteristics.

Both kinds of design characteristics are defined in terms of attributes of design objects[4] and relations between design objects, or as combinations of other design characteristics. Design characteristics can be combined via boolean connectives, arithmetic relations, arithmetic operators, or function composition.

The hierarchy described above provides mappings between secondary design characteristics and primary characteristics that realize them. A mapping between a secondary characteristic and primary characteristics need not be direct, but may pass through intermediate, more specific secondary characteristics. The most general secondary characteristics are near the top of the hierarchy, more specific ones closer to the bottom, with the bottommost level of secondary characteristics mapped directly to primary characteristics, which form the bottommost level of the hierarchy. See Figure 3 for an example showing a mapping from the secondary characteristic is-private to primary characteristics.

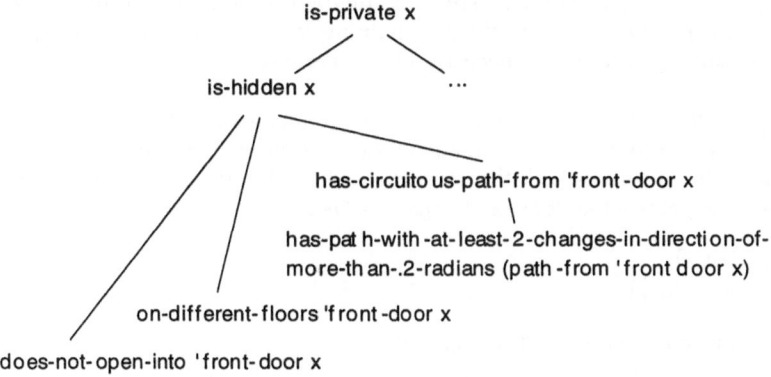

Figure 3. Portion of design characteristics hierarchy.

[3]It has multiple roots so is not a strict hierarchy. It was defined by the author with help from architect Richard Krauss, of Arrowstreet Inc.; future versions will be defined by system users.

[4]A design object is any design element, territory, use-space, or entire design.

3.3 EVALUATION

To evaluate a design, the user of the system specifies a design characteristic or list of design characteristics and any of the design's use-spaces, territories, or design elements. The system evaluates the design by traversing the design characteristics hierarchy to find and evaluate the appropriate primary design characteristics.

For example, given the design characteristic is-private and a territory, the system evaluates is-private for the territory by referring to the design characteristics hierarchy. It determines that is-private for a territory can be evaluated by looking for a circuitous path from the front door to the territory. A circuitous path is defined in terms of the circulation model: as a path containing at least two changes in direction of greater than .2 radians. So to evaluate is-private for a territory, the system evaluates the territory's associated circulation-model for the presence of such a path from the front door.[5] The system also checks whether the front door opens into the territory and whether the territory and front door are on different levels.

3.4. IMPLEMENTATION

Design models and their associated components (e.g. design elements, territories, points) are implemented as Common Lisp Object System (CLOS) objects (Bobrow, et al., 1988). Object classes are arranged in a hierarchy in which each class can inherit from multiple parents; methods, specialized by class on any number of arguments, are used to dispatch generic functions on particular objects.

Design characteristics are represented as predicates in a rule-based, backward-chaining system.[6] Predicates for primary characteristics are each associated with an evaluator function, which operates on a design model or models. The value of a primary characteristic is the value returned by its evaluator function. All secondary design characteristics are boolean-valued. Their evaluator functions are implicit: Secondary characteristics evaluate to true if backward chaining from the secondary characteristic to true-valued primary characteristics succeeds, false otherwise.

The mapping between secondary and primary characteristics is achieved by means of If-Then rules. Each link in the design characteristics hierarchy corresponds to a rule. For example, the following rules link the secondary characteristic is-private to the primary characteristic has-path-with-at-least-2-changes-in-direction-of-more-than-.2-radians:

    IF [is-hidden x]
    THEN [is-private x]

    IF [has-circuitous-path-from 'front-door x]
    THEN [is-hidden x]

    IF [has-path-with-at-least-2-changes-in-direction-of-more-than-.2-radians x y]
    THEN [has-circuitous-path-from x y]

---

[5]When computing a path to a territory, the center of the territory is used as an endpoint.
[6]The Joshua software system, by Symbolics Inc., was used.

The user invokes the evaluation mechanism by supplying the name of a design characteristic and applicable design elements, territories, use-spaces, or the name of a design as arguments to a top-level evaluation function.   To evaluate the Gale design for the presence of a private main living space, for example, the user types:

(has-characteristic 'private-main-living-space 'GALE)

The result is a textual explanation derived from the backtrace of the rule firings, the value (true or false), and the original backtrace. See Figure 4 for example output taken from the Prairie house experiment described in the next section.

The following Prairie secondary design characteristics are present in GALE:

1. The main living space is private:

The design has a main living space <LIVING>.
and
The space <LIVING> is private:
The space <LIVING> is hidden:
The front door does not open into <LIVING>.

The design has a main living space <LIVING>.
and
The space <LIVING> is private:
The space <LIVING> is hidden:
There is a circuitous path from the front door to <LIVING>.
...

Figure 4. Portion of evaluation output for the Mrs. Thomas Gale house;
<LIVING> denotes a particular use-space[7] named LIVING.
A main living space for a design may be derived or user-supplied.

A final note about territories and use-spaces in the current implementation: Territories may overlap, use-spaces may not.   This restriction simplifies the semantics and computation of topological relationships such as adjacency and geometric relationships such as distance or paths between use-spaces. Future versions of the system will remove the restriction.

4. Experiment

As a test of the evaluation system, six secondary characteristics, representing experiential qualities, for Frank Lloyd Wright Prairie houses were encoded, along with fifteen primary characteristics representing geometric details of two-dimensional floorplans. Six Frank Lloyd Wright Prairie houses were evaluated, along with six non-Prairie houses and three Frank Lloyd Wright houses considered to be transitions between his pre-Prairie and Prairie periods. As an exercise, a measure of Prairieness—the number of characteristics present—

---

[7]The terms "use-space" and "space" will be used interchangeably.

was derived for each design. (See Appendix for the list of design characteristics, designs, and results.)

Frank Lloyd Wright was chosen because he was prolific, has been well-studied, and is regarded as a master at manifesting experiential qualities in his buildings. His Prairie houses were chosen because they share many common features while also being quite varied, and because they have been extensively studied by architectural critics and historians (e.g. Manson, 1958; Brooks, 1972; Twombly, 1979; Hildebrand, 1991), and by researchers interested in computational systems for design (e.g. Koning and Eizenberg, 1981; Chan, 1992). Six representative Prairie houses were chosen, one from each of six Prairie house categories (Pinnell, 1990). The non-Prairie examples were chosen in order to minimize the differences that might be attributed to non-architectural issues. The examples were limited to single-family stand-alone houses, to minimize differences due to type of design; to approximately the same time period, to minimize differences due to societal changes, e.g. addition of a garage; to those about the same size, to minimize differences due to mismatch in number or sizes of spaces; to American designs, to minimize cultural influences. Examples considered transitions between Wright's Prairie and pre-Prairie periods were chosen to see whether the transition nature of the designs would be reflected in the evaluation.

The Prairie houses were evaluated as expected, with all but one exhibiting all six secondary characteristics; half exhibited all fifteen primary characteristics, with the rest exhibiting at least twelve. The transition houses were not distinguishable from either the Prairie examples or the non-Prairie examples; their differences were not captured by the design characteristics used in the experiment. The non-Prairie houses were evaluated as expected with respect to the primary characteristics: Only one exhibited ten of the fifteen; the rest exhibited nine or fewer. The non-Prairie houses and transition houses, however, exhibited many of the secondary characteristics. This result is unsurprising: Focusing on primary characteristics, which represent concrete details, yields a better measure of Prairieness, or perhaps any architectural type, since secondary design characteristics may be manifested through a variety of concrete means, with particular means favored by individual designers. In other words, different architects and their clients may want many of the same experiential qualities in their designs, but may prefer different means for achieving those qualities.

The experiment showed that a system can be built that evaluates an architectural design using knowledge about manifesting experiential qualities in physical form. It also showed that such a system can be used to represent and experiment with definitions of architectural type. It showed that systems can be built that bridge the gap between knowledge-based tools and CAD tools by representing and reasoning with both geometric and non-geometric knowledge.

The experiment also illustrated an obvious limitation in using boolean-valued characteristics for evaluation: How well a design exhibited a particular characteristic cannot be measured. Main living spaces of the non-Prairie examples were deemed private by the system, for example, but in reality are less private than those of the Prairie examples. With boolean-valued characteristics the system cannot make this comparison, nor can it rank designs or make suggestions about how to increase or decrease the values of characteristics. The next system will extend the representation of design characteristics to alleviate these shortcomings.

5. Next

The first version of the system evaluates a design, answering the question: Does this design exhibit a given set of design goals? A design goal is represented as a design characteristic; the desired value of the characteristic is implicitly represented as true. The next version of the system will represent a design goal as a design characteristic plus an explicit value, which may be boolean, qualitative, or quantitative. Quantitative characteristics such as visual openness will be added, providing alternative, and in some cases more precise, methods for evaluating such experiential qualities as privacy. Quantitative characteristics also will facilitate the ranking of design alternatives.

The second version of the system is to include modification, answering the question: How could a design be modified so that it adheres to a set of design goals? To add this capability new functionality must be implemented.

Non-opaque evaluator functions for design characteristics are needed so that derivation of values can be explained in detail. If the system is to suggest modifications to a design in order to change the value of a design characteristic, it must first know why the value failed to be the desired value. For example, if a territory is deemed not private because there is no circuitous path leading to it from the front door, the system must know something more than that the characteristic has-path-with-at-least-2-changes-in-direction-of-more-than-.2-radians was false. It has to know why it was false: Were there not enough changes in direction? Were the changes in direction not great enough? For the system to know this, the evaluator function for this characteristic can no longer be an opaque black box that just returns true or false; it must explicitly represent the number and magnitude of changes in direction in such a way that the system can infer corrections. The reasoning will be similar for non-boolean-valued characteristics as well. In reasoning about visual openness, for example, if the value is not as desired, the system will need to know how to change the value. The evaluator in this case must return a quantitative value and give clues about changes in a design that might change the value in the desired direction. The system, therefore, needs an explicit representation of *influences* on design characteristics, along with knowledge about the directions of influence. Such a representation will be similar to representations used in the qualitative reasoning literature (e.g. Faltings and Struss, 1992).

Reasoning about trade-offs between conflicting goals needs to be added. In trying to satisfy more than a single design goal, the system will have to reason about trade-offs between conflicting goals. The difficulty here is in determining the independence of the various goals. Does increasing the continuity between two territories affect the privacy of one with respect to the other? Does increasing the size of a territory affect the circuitousness of a path from the front door to that territory? Each design goal will need to be translated into a common vocabulary that refers to the design objects to be manipulated so that the effects of simultaneously satisfying particular goals can be judged. The constraint-based reasoning and planning literature contain examples of dealing with these issues (e.g. Allen, et al., 1990; Freuder and Mackworth, 1994).

6. Related Work

Many computer-based systems have been developed for evaluating various aspects of architectural designs. Most of these systems differ from the present work by not evaluating a design with respect to experiential qualities and/or by not operating on geometric representations of physical form. Many of the systems focus on engineering rather than experiential aspects, some employing well-known algorithmic metrics (e.g. Radford and Gero, 1988; Wiezel and Becker, 1992), others employing knowledge-based methods or simulation techniques (e.g. Shaviv and Peleg, 1991; Flemming and Mahdavi, 1993). Some of these systems represent a design's physical form explicitly as physical design elements having locations and dimensions; others instead represent a design only as a set of spaces. Evaluation systems that focus on non-engineering aspects often evaluate spatial organization, but not experiential aspects of the organization. Some of these systems represent a design's physical form explicitly (e.g. Carrara, et al., 1994); others only represent a design as a set of statements derived from a graphical representation of design elements (e.g. McCall, et al., 1990; Oxman, 1992).

A few systems have evaluated designs with respect to experiential qualities, but have not represented or reasoned about physical form, relying instead, for example, on human evaluators and statistical scoring techniques (e.g. Mortola and Giangrande, 1991). One such system (Cao and Protzen, 1994) goes a step further and explicitly represents mappings between experiential qualities and physically measurable properties, but relies on previously collected data rather than measuring properties dynamically from a representation of a design. Several case-based reasoning systems have represented experiential qualities, often derived from postoccupancy evaluations, as annotations, but have not related them to physical form or evaluation per se (e.g. Domeshek and Kolodner, 1992). One case-based reasoning system represents physical form and dynamically calculates values of design characteristics, hinting at the possibility of evaluating experiential characteristics of a design (Dave, et al., 1994). A topological evaluation technique that relates spatial organization to social behavior has been successfully paired with a geometric analysis of visibility (Hanson, 1994). Finally, a recent paper (van der Voordt, et al., 1997) suggests relating physical form to experiential qualities, as is done in work described in this paper.

Several systems are close in spirit to the current work. McLaughlin's system (McLaughlin, 1991) targets the same design task as the current work: evaluation of a developing design with respect to experiential qualities represented as design goals such as open, sunny, and private. The system is a rule-based backward-chaining system that relies on design recommendation literature for knowledge about how to achieve particular design goals in physical form. The system takes as input a design and outputs lists of goals achieved and goals not achieved, along with explanations. The major difference between this system and the current work is that McLaughlin's system does not represent or reason about physical form. The system instead represents a design as a topological arrangement of design elements—walls, windows, doors, openings, spaces—and reasons about the design using a fact base of assertions, as do most rule-based systems. Addition of a geometric representation of a design would simplify some computational tasks (e.g. computation of circulation paths), enable others (e.g. computation of visual barriers), and facilitate interaction with a designer (e.g. by enabling integration of a sketching system).

Two other systems are similar to the current work in their attempts to represent and reason about experiential qualities and physical form. They differ from the current work in ways that result from adopting different views of the design process.

Galle's system (Galle, 1994) aims to be a general design support tool, facilitating sketching and development of evolving designs. Its goals are similar to those of the design support system envisioned in the opening scenario of this paper. The design knowledge in Galle's system is that embodied in Alexander's pattern language (Alexander, et al., 1977), which represents general design principles as patterns, arrangements of physical design elements in service of particular design goals. As Galle points out, the patterns "can be of a technical, aesthetic, or social nature." The patterns of a social nature describe methods for achieving experiential qualities. One pattern, for example, suggests achieving what Alexander calls an intimacy gradient by creating a sequence of spaces arranged according to degrees of privacy, with least private near the entrance, followed by slightly more private spaces, leading eventually to the most private. Galle's system represents patterns and parts, which correspond to physical design elements such as walls, and distinguishes between types and instances. (Pattern types correspond to Alexander's patterns; pattern instances correspond to part instances arranged in the configuration specified by a particular pattern.) Patterns are represented as constraints on user-specified configurations of design parts. Sketching a design amounts to creation of pattern instances: Each part instance added to a design must be associated with at least one pattern instance.[8] If a part is moved, changes propagate to other parts so that all pattern instances associated with the moved part remain satisfied. The current work takes a different view of design: It aims eventually to support sketching, parsing of the sketch into design elements, followed by evaluation of the design with respect to design goals. Design elements will not have to be associated with a particular design goal; they will only be implicitly related to a design goal if they contribute to satisfaction of that goal. Further, the design goals may be stated in terms of secondary design characteristics, which correspond to Galle's (and Alexander's) patterns, or in terms of primary design characteristics such as lengths of walls or number of turns in a particular path. Primary design characteristics have no user-accessible counterpart in Galle's current system.

Gullichsen and Chang (Gullichsen and Chang, 1985) built a design generation system based on Alexander's pattern language. It is a rule-based backward-chaining system that relies on the interconnections between patterns, which are ordered in "decreasing morphological importance to ensure that a whole, imprecisely-specified form is successfully differentiated during the process of design." They adhere to Alexander's position that because of this ordering, conflicts do not arise and backtracking during the design process is unnecessary. The system, thus, generates designs in a top-down fashion, progressing from general patterns to more specific patterns which implement general ones. The user initiates the generation by specifying the list of patterns to be satisfied. Gullichsen and Chang's system is concerned with generation rather than evaluation, but parts of it can still be compared with the current work. Their system's view of design as a top-down process differs from the current work's view of design as an iterative process of evaluation and modification. Their system's representation of patterns as predicates in a rule-based system is similar to the current system's representation of design characteristics. Finally, their system's geometric representation of a design may be similar to the current

---

[8]Galle suggests in a footnote that later versions of his system will relax this restriction.

work's, since mention is made of "lower-level procedures [that] typically employ geometric methods." An example is given of computing positive space between buildings by calculating the "sum of areas enclosed by walls of buildings or segments which constitute the convex polygonal hull of the building's wings, weighted by the ratio of its enclosing perimeter of the hull," but no mention is made of the underlying representation.

7. Conclusion

The system described in this paper evaluates an architectural design using the designer's theories about how to manifest experiential qualities in physical form. It engages a designer in a conversation of the sort: Does a design adhere to a set of specified design goals? The next system, under development now, will add to this conversation: If not, how can it be modified so that it does? This conversation is an integral part of design; it occurs as a designer searches for a design solution by exploring the relationships between changes in a design and changes in the values of design characteristics. Current work focuses on extending the present system to include more experiential design characteristics, more non-boolean-valued design characteristics, a representation for influences on design characteristics, and a more general representation for design goals. Work also focuses on representations for architectural knowledge about how to "fix" a design, and on reasoning mechanisms that employ the knowledge to explore trade-offs between conflicting design goals. The knowledge is being acquired through discussion with architects.[9] The resulting system will make suggestions for redesign which will produce a design exhibiting as many specified design goals as possible. Such a system will be a step toward an intelligent computer assistant as design conversation partner.

Acknowledgments

The author thanks the following people for helpful conversations, suggestions, and arguments: John Aspinall, Randall Davis, Aaron Fleisher, Mark Gross, Duncan Kincaid, Richard Krauss, Tomás Lozano-Pérez. The author also thanks Bill Mitchell and Patrick Winston for discussion and recommendations that ultimately led to the National Science Foundation Graduate Fellowship funding this work.

References

Alexander, C., Ishikawa, S., Silverstein, M., Jacobsen, M., Fiksdahl-King, I., and Angel, S. (1977) *A Pattern Language,* Oxford University Press, New York.
Allen, J., Hendler, J., and Tate, A., eds. (1990) *Readings in Planning,* Morgan Kaufmann, San Mateo, CA.
Bobrow, D. B., DeMichiel, L. G., Gabriel, R. P., Keene, S. E., Kiczales, G., and Moon, D. A. (1988) "Common lisp object system specification," X3J13 document 88-002R.
Brooks, H. A. (1972) *The Prairie School,* Norton, New York.
Cao, Q. and Protzen, J.-P. (1994) "Deliberation and aggregation in computer-aided performance evaluation," in *Automation Based Creative Design,* A. Tzonis and I. White, eds., Elsevier, New York, 251-264.

---

[9]Machine learning programs may be of assistance here, though they are outside the scope of the present inquiry.

Carrara, G., Kalay, Y. E., and Novembri, G. (1994) "Knowledge-based computational support for architectural design," in *Knowledge-Based Computer-Aided Architectural Design*, G. Carrara and Y. E. Kalay, eds., Elsevier, New York, 147-201.

Chan, C.-S. (1992) "Exploring individual style through Wright's designs," *Journal of Architectural Planning and Research*, vol. 9, 207-238.

Coyne, R. D., Rosenman, M. A., Radford, A. D., Balachandran, M., and Gero, J. S. (1990) *Knowledge-Based Design Systems*, Addison-Wesley, Reading, MA.

Dave, B., Schmitt, B., Faltings, B., and Smith, I. (1994) "Case-based design in architecture," in *Artificial Intelligence in Design '94*, Proceedings of the Third International Conference on Artificial Intelligence in Design, J. S. Gero and F. Sudweeks, eds., Kluwer, Norwell, MA, 145-162.

Domeshek, E. A. and Kolodner, J. L. (1992) "A case-based design aid for architecture," in *Artificial Intelligence in Design '92*, Proceedings of the Second International Conference on Artificial Intelligence in Design, J. S. Gero, ed., Kluwer, Norwell, MA, 497-516.

Faltings, B. and Struss, P., eds. (1992) *Recent Advances in Qualitative Physics*, MIT Press, Cambridge, MA.

Flemming, U. and Mahdavi, A. (1993) "Simultaneous form generation and performance evaluation: a `two-way' inference approach," in *CAAD Futures '93*, Proceedings of the Fifth International Conference on Computer-Aided Architectural Design Futures, U. Flemming and S. Van Wyk, eds., North-Holland, New York, 161-174.

Freuder, E. C. and Mackworth, A. K., eds. (1994) *Constraint-Based Reasoning*, reprint of Artificial Intelligence 58 (1992) 1-3, MIT Press, Cambridge, MA.

Galle, P. (1994) "Computer support of architectural sketch design: a matter of simplicity?," *Environment and Planning B*, vol. 21, 353-372.

Gero, J. S. and Sudweeks, F., eds. (1996) *Artificial Intelligence in Design '96*, Proceedings of the Fourth International Conference on Artificial Intelligence in Design, Kluwer, Norwell, MA.

Gross, M. D. (1996) "The Electronic Cocktail Napkin--a computational environment for working with design diagrams," *Design Studies*, vol. 17, 53-69.

Gullichsen, E. and Chang, E. (1985) "Generative design architecture using an expert system," *The Visual Computer*, vol. 1, 161-168.

Hanson, J. (1994) "'Deconstructing' architects' houses," *Environmental and Planning B*, vol. 21, 675-704.

Hertzberger, H. (1993) *Lessons for Students in Architecture*, 2nd ed, Uitgeverij 010 Publishers, Rotterdam.

Hildebrand, G. (1991) *The Wright Space: Pattern and Meaning in Frank Lloyd Wright's Houses*, University of Washington Press, Seattle.

Kincaid, D. S. (1997) (in progress), MArch Thesis, Department of Architecture, MIT.

Koning, H. and Eizenberg, J. (1981) "The language of the prairie: Frank Lloyd Wright's prairie houses," *Environment and Planning B*, vol. 8, 295-323.

Manson, G. C. (1958) *Frank Lloyd Wright to 1910: The First Golden Age*, Van Nostrand Reinhold, New York.

McCall, R., Fischer, G., and Morch, A. (1990) "Supporting reflection-in-action in the Janus design environment," in *The Electronic Design Studio*, M. McCullough, W. J. Mitchell, and P. Purcell, eds., MIT Press, Cambridge, MA, 247-259.

McLaughlin, S. (1991) "Reading architectural plans: a computable model," in *CAAD Futures '91*, Proceedings of the Fourth International Conference on Computer-Aided Architectural Design Futures, G. N. Schmitt, ed., Vieweg, Wiesbaden, 347-364.

Mortola, E. and Giangrande, A. (1991) "An evaluation module for 'An Interface for Designing' (AID): a procedure based on trichotomic segmentation," in *CAAD Futures '91*, Proceedings of the Fourth International Conference on Computer-Aided Architectural Design Futures, G. N. Schmitt, ed., Vieweg, Wiesbaden, 139-154.

Oxman, R. (1992) "Multiple operative and interactive modes in knowledge-based design systems," in *Evaluating and Predicting Design Performance*, Y. E. Kalay, ed., Wiley, New York, 125-143.

Pham, D. T., ed. (1991) *Artificial Intelligence in Design*, Springer-Verlag, New York.

Pinnell, P. (1990) "Academic tradition and the individual talent: similarity and difference in the formation of Frank Lloyd Wright," in *Frank Lloyd Wright: A Primer on Architectural Principles*, R. McCarter, ed., Princeton Architectural Press, New York, 19-58.

Radford, A. D. and Gero, J. S. (1988) *Design By Optimization in Architecture, Building, and Construction*, Van Nostrand Reinhold Company, New York.

Shaviv, E. and Peleg, U. J. (1991) "An integrated KB-CAAD system for the design of solar and low energy buildings," in *CAAD Futures '91*, Proceedings of the Fourth International Conference on Computer-Aided Architectural Design Futures, G. N. Schmitt, ed., Vieweg, Wiesbaden, 465-484.

Storrer, W. A. (1993) *The Frank Lloyd Wright Companion*, University of Chicago Press, Chicago. Also available as *The Frank Lloyd Wright Companion CD-ROM*, Prairie Multimedia, Inc., West Chicago.

Twombly, R. C. (1979) *Frank Lloyd Wright: His Life and His Architecture*, Wiley, New York.

van der Voordt, T. J. M., Vrielink, D., and van Wegen, H. B. R. (1997) "Comparative floorplan-analysis in programming and architectural design," *Design Studies*, vol. 18, 67-88.

Wiezel, A. and Becker, R. (1992) "Integration of performance evaluation in computer-aided design," in *Evaluating and Predicting Design Performance*, Y. E. Kalay, ed., Wiley, New York, 171-181.

Zeisel, J. and Welch, P. (1981) *Housing Designed for Families: A Summary of Research*, Joint Center for Urban Studies of MIT and Harvard University, Cambridge, MA.

Appendix

DESIGN CHARACTERISTICS

The following secondary characteristics, representing experiential qualities, were used in the Prairie house experiment.

1. The design exhibits Wrightian group togetherness.
2. The design exhibits home/hearth symbolism.
3. The main living space is private.
4. The main living space is a place of refuge.
5. The main living space is a place of prospect.
6. A private exterior space is contiguous with the main living space.

The following primary characteristics, representing geometric properties, were used in the Prairie house experiment. The secondary characteristics to which each primary characteristic contributes are shown in parentheses. (The numbers correspond to those in the above list.)

1. The design has a main living space that is the largest living space. (1)
2. The design has a main living space containing a region from which all other living spaces are visible. (1)
3. The design has a main living space that is connected to all other living spaces. Two spaces are connected if they are no more than one space apart or if they have axially aligned doorways. (1)
4. The design has one fireplace location. (2)
5. The design has a fireplace on an interior wall.
6. The design has a fireplace in the main living space. (2)
7. The path from the front door to the private area does not pass within five feet of the center of the main living space. (3)
8. The front door does not open into the main living space. (3,4)
9. The front door and the main living space are on different levels. (3,4)
10. The path from the front door to the main living space contains at least two changes in direction of greater than .2 radians. (3,4)
11. The path from the exterior approach point to the main living space contains at least two changes in direction of greater than .2 radians. (3,4)
12. The main living space is elevated above the terrain. (5)
13. An exterior space at least 40% of the size of the main living space is contiguous with the main living space. (5)
14. The front door does not open into the exterior space contiguous with the main living space. (6)
15. The path from the exterior approach point to the front door does not pass through the exterior space contiguous with the main living space. (6)

The design examples used in the Prairie house experiment are shown on the next pages. Each floorplan is accompanied by counts of design characteristics exhibited. The first number is the count of secondary characteristics; the second is the count of primary characteristics.

DESIGN EXAMPLES

Prairie Houses: Cheney, Gale, Horner, Tomek, Willits, Roberts (Storrer, 1993)

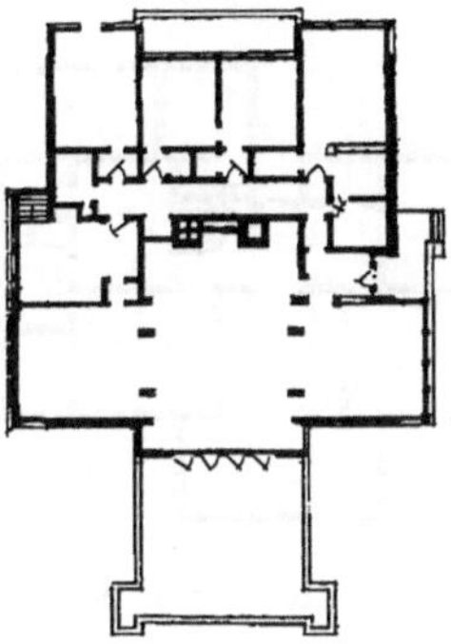

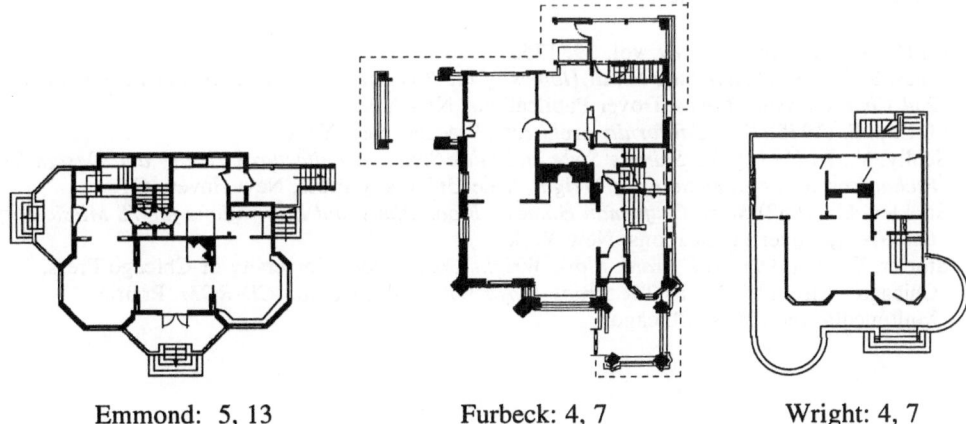

Cheney: 6, 13 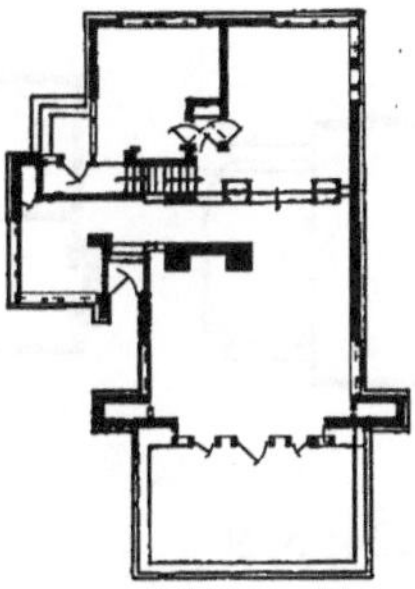   Gale: 6, 15 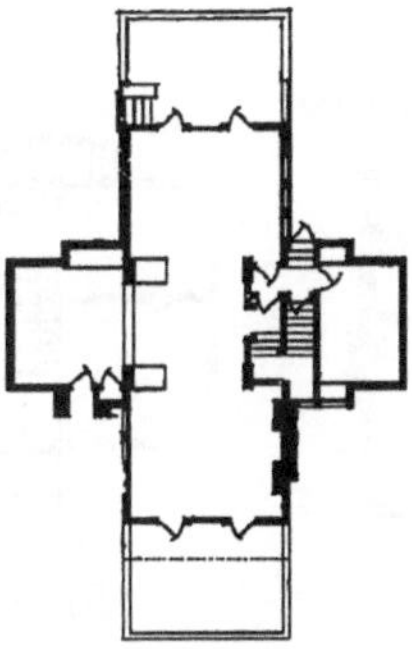   Horner: 6, 14 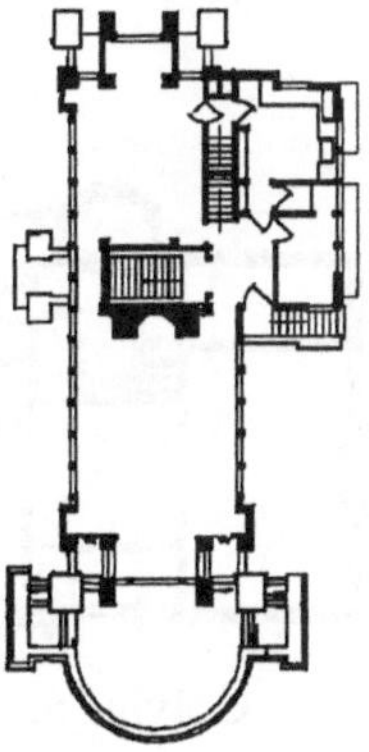   Tomek: 6, 15

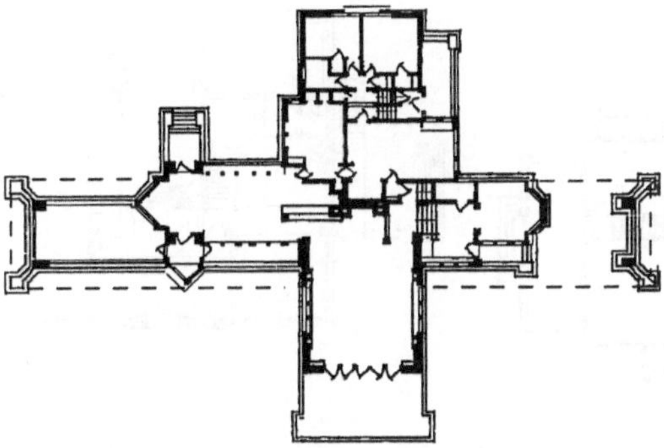

Willits: 5,12 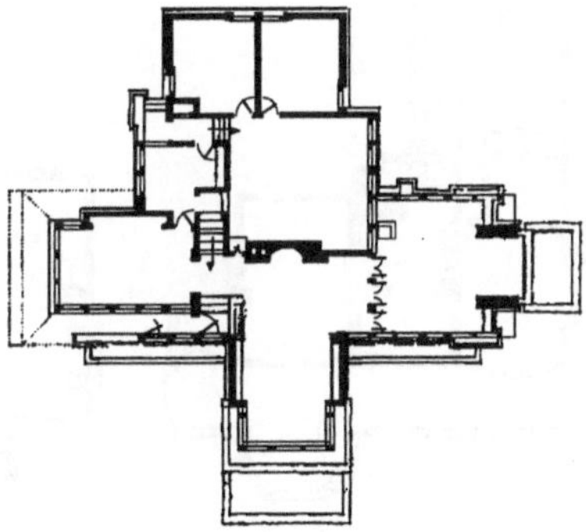   Roberts: 6, 15

Transition Houses: Emmond, Furbeck, Wright (Storrer, 1993)

Emmond: 5, 13   Furbeck: 4, 7   Wright: 4, 7

218

Non-Prairie Houses: Colvin, by George Maher (1916); Jones 5A24 (Jones, 1987);
    Lawson, by Bernard Maybeck (McCoy, 1975); Mallory, by Arthur Rich (Scully,
    1971); Stickley 91 (Stickley, 1982); Winslow, by Frank Lloyd Wright, (Storrer,
    1993)

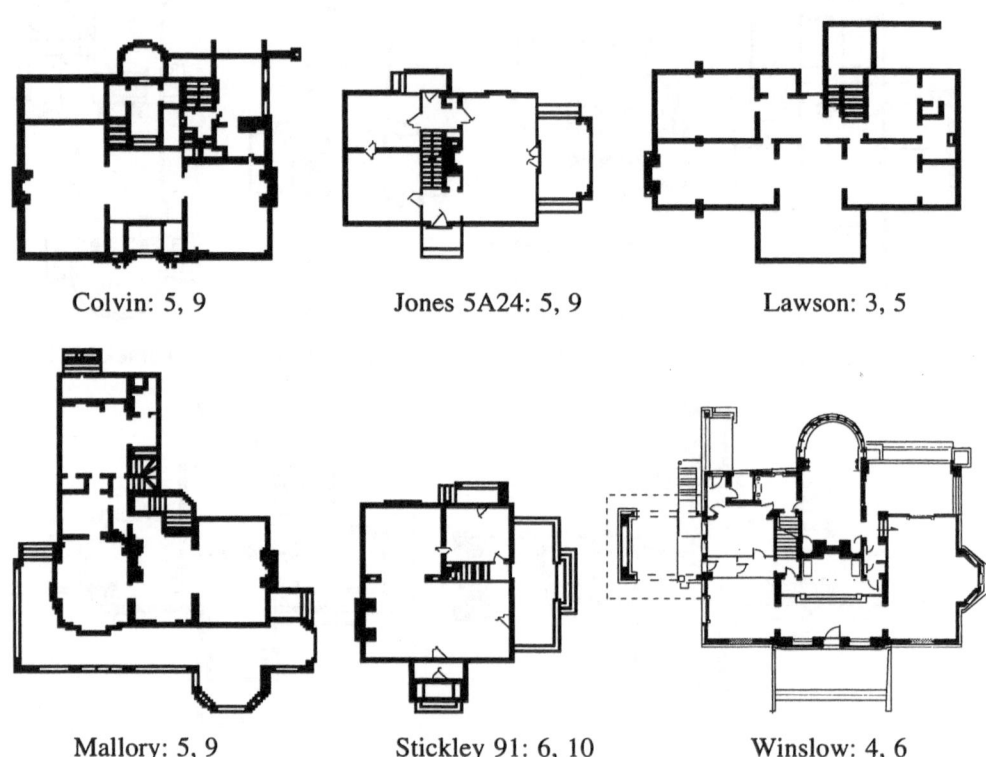

Colvin: 5, 9   Jones 5A24: 5, 9   Lawson: 3, 5

Mallory: 5, 9   Stickley 91: 6, 10   Winslow: 4, 6

Sources of design examples:

(1916) *Architectural Record*, vol. 39, 175.
Jones, R. T. (1987) *Authentic Small Homes of the Twenties: Illustrations and Floorplans of 254 Characteristic Homes*, Dover Publications, New York.
McCoy, E. (1975) *Five California Architects*, Praeger, New York.
Scully, V., Jr. (1971) *The Shingle Style and Stick Style; Architectural Theory and Design from Richardson to the Beginnings of Wright*, Yale University Press, New Haven, CT.
Stickley, G. (1982) *More Craftsman Homes: Floor Plans and Illustrations for 78 Mission Style Dwellings*, Dover Publications, New York.
Storrer, W. A. (1993) *The Frank Lloyd Wright Companion*, University of Chicago Press, Chicago. Also available as *The Frank Lloyd Wright Companion CD-ROM*, Prairie Multimedia, Inc., West Chicago.

MODELING-ASSISTED BUILDING CONTROL

ARDESHIR MAHDAVI
Department of Architecture, Carnegie Mellon University, USA
and
School of Architecture, National University of Singapore

Abstract

The architectural research on provision of computational support for the building delivery process in general and computer aided performance modeling in particular has traditionally concentrated on the building design phase. This paper argues that computational modeling can also successfully apply to the building operation phase. To demonstrate this potential the paper explores a simulation-assisted building control strategy. Specifically, the use of generate-and-test as well as bi-directional inference methods is proposed to derive preferable control schemes and required attributes for control variables based on parametric and iterative simulation runs. The feasibility of the approach is demonstrated *via* illustrative computational examples from the thermal control domain.

1. Introduction

Traditionally, building control systems have operated on the basis of a homeostatic short-term feed back mechanism. For example, thermostatic control of HVAC components involves typical operations (on/off, change in volume and/or temperature of heating/cooling media, etc.) that are essentially guided by temperature sensing in space. More recently, building control systems have become increasingly sophisticated. One of the approaches has been to utilize various methods and tools (including neural nets) to accurately capture the buildings' thermal dynamic characteristics so as to provide a more reliable basis for the control of its behavior (Curtis 1996, Curtis et al. 1993, Mistry and Nair 1993, Osman et al. 1996). In this scenario, control options can be improved ("optimized"), as their past impact on the buildings' dynamic behavior is reflected in the collected information by the sensing system. This paper argues that the above intention, namely to capture buildings' long term dynamic behavior toward enhanced control strategies, can be effectively supported using advanced computational performance simulation routines.

R. Junge (ed.), CAAD Futures 1997, 219-229.
© 1997 *Kluwer Academic Publishers.*

2. The Idea

In the past, performance simulation tools have been mainly used for purposes of building design analysis and evaluation. Less attention has been paid, however, to their potential in view of active control strategies for building service systems (particularly HVAC and lighting devices). In order to realize this idea, a conventional building automation system must be supplemented with a multi-aspect virtual model of the building that runs parallel to the building's actual operation. While the real building reacts "only" to the actual climatic conditions, occupancy interventions, and building control operations, the simulation-based virtual model allows for additional operations:

- The virtual model can move backward in time so as to analyze the building's past behavior and/or to calibrate the program toward improved predictive potency.
- The virtual model can move forward in time so as to predict the building's response (e.g. its hygro-thermal behavior) to alternative control scenarios.

Given the availability of fairly reliable short-term (e.g. three days period) weather prognosis data *via* internet, the predictive capability of the simulation program may be expected to be significantly enhanced, thus providing a useful basis for the evaluation of multiple control options. Obviously, data pertaining to other factors such as the fluctuation of occupancy as well as lighting and equipment use may be provided to the virtual model to further increase its predictive potential. Above and beyond enhancing the effectiveness of dynamic control systems, the suggested approach may yield additional benefits. These include:

- calibration of simulation tools for long-term design and modification feed back,
- prediction of the effects of changes to building hardware and its control systems,
- beta-testing of building control system hardware on simulated data from the virtual building,
- pre-training of neural network and machine learning systems prior to their field utilization using simulation data on building behavior,
- re-training of neural network and machine learning systems to account for the effects of abrupt modifications to building characteristics (e.g. renovation) using simulation data,
- reduction of the number of sensing units necessary for capturing building's real time operational status.

3. Two Approaches

A critical task toward realization of a simulation-assisted building control system lies in the development of a strategy to create a well-defined set of control options as the basis for comparative and/or parametric simulation runs. While there may be numerous methods to derive at such control options, we focus in this paper on two principal approaches (cp. figure 1):

THE GENERATE-AND-TEST METHOD (GAT)

This method involves the rule-based generation of a finite number of discrete control options. Such control options may involve, for example, various on/off timing schemes for intermittent heating/cooling. These schemes are then evaluated and ranked (possibly in view of multiple criteria involving energy, cost, emissions, comfort, etc.) based on the results of multiple simulation runs.

THE BI-DIRECTIONAL INFERENCE METHOD (BDI)

This method (Mahdavi 1993) involves the explicit definition of control and performance variables. An example of a control variable would be the deviation of heating/cooling set-point temperature from the space target temperature. Examples of a performance variable are the annual building energy need, the average cumulative deviation of the maintained space temperature from the set-point temperature, or the average cumulative PPD (predicted percentage of dissatisfied) in a space. Starting from an initial operational state, the bi-directional inference facilitates the derivation of required changes in the control variable(s) based on desired changes in the performance variable(s). This derivation can be accomplished *via* the investigative projection technique (Mahdavi and Berberidou 1995, 1994).

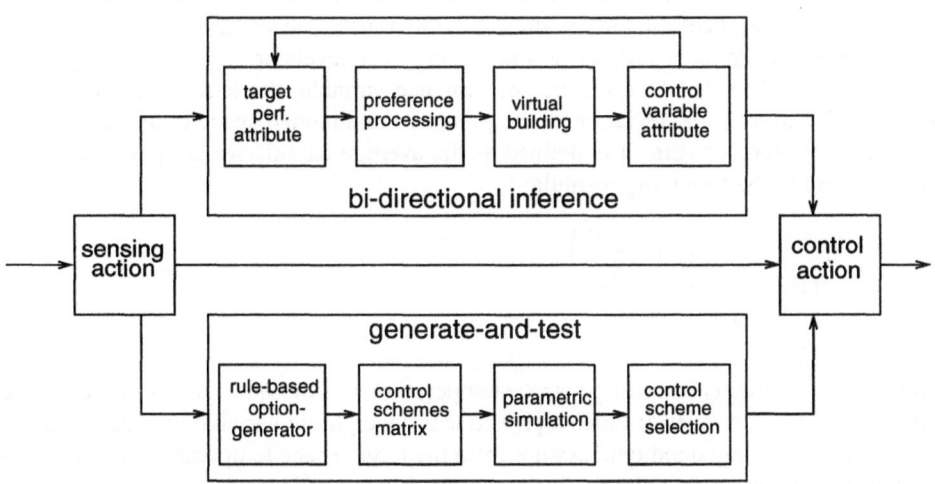

Figure 1. Schematic illustration of the use BDI and GAT for simulation-assisted building control

4. A Demonstrative Example

What follows is a simple computational example to demonstrate the feasibility of the proposed approach. Assume a single-story single-zone quadratic building (6 m by 6 m in plan and 3 m high) located in Pittsburgh, PA, USA. The construction consists of stud-walls with 10 cm fiberglass insulation and 30% single glazing on south and north, light-weight flat roof with 20 cm insulation, and concrete floor slab atop 10 cm insulation. We have fixed these design variables for the specific purpose of the present demonstrative example, since the idea is to emulate an existing structure. If desired, such design variables could also be subject to parametric studies of the kind discussed below. (For example, deployment of shading devices would have a dynamic effect on the energy transmission through the glazing and could be thus modeled as a control variable.)

The problem statement is now to derive a control strategy based on the results of a well-planned set (or sequence) of simulations. We suggested that this "planning" may be accomplished *via* generate-and-test (GAT) and bi-directional inference (BDI) approaches. However, we have to first establish a set of pertinent control and perfor-mance variables. For the purposes of clarity, we limit ourselves in this case to only one control variable and two performance variables. We select as control variable the numeric deviation of the heating/cooling set-point temperatures from the target space temperature (Δt_{sp}), i.e. we specify the extension of the temperature dead-band with our control variable. As our first performance variable we consider the annual total (heating and cooling) energy need of the building q_a. (Obviously, we could define many other similar or related variables such as seasonal energy use levels or separate heating, cool-ing, and electrical loads.) Our second performance variable, the temperature deviation factor (TDF) captures the deviation of the predicted maintained space temperature from the target space temperature. It is defined as the average cumulative temperature devia-tion according to the following formula:

$$TDF = \sum_{i=1}^{n} \frac{w \cdot \left(\frac{|t_d - t_i|}{t_d} \right)}{n} \cdot 100 \qquad [\%] \qquad eq.1$$

In the above equation t_d is the target space temperature, t_i is the space temperature at time step i, n is the total number of time steps, and w is a weighting variable to penalize larger deviations of the maintained temperature from the target space temperature. For the pur-poses of the present study a simple linear relationship is applied:

$$w = |t_d - t_i| \cdot 5^{-1} \qquad eq.2$$

The simulation engine used for the following case studies is the dynamic (heat-balance-based) thermal module in SEMPER (Mahdavi 1996a, 1996b, Mahdavi and Mathew 1995).

USE OF GAT

As mentioned before, this method involves first the generation of a finite number of discrete control options. For the present example five such schemes have been generated for the relevant control variable, i.e. Δt_{sp}. Table 1 represents these schemes labeled A to E.

Table 1: Generated Control Schemes

Control Scheme	Target Space Temp. [°C]	Heating Set-point [°C]	Cooling Set-point [°C]	Δt_{sp} [K]
A	22	21.5	22.5	0.5
B	22	21.0	23.0	1.0
C	22	20.0	24.0	2.0
D	22	19.0	25.0	3.0
E	22	18.0	26.0	4.0

Once these schemes are established, exploratory simulations can be performed immediately. The simulation results provide then a matrix which can be used to organize, rank, and evaluate various control strategies in view of their implications for the relevant performance criteria. For our specific example, such a matrix is given in table 2. This table numerically documents the intuitively expected goal conflict between minimization of energy use on one side and minimization of the deviations of the maintained space temperatures from the target values on the other side. In the present case the resolution of this conflict requires only the definition of the maximum tolerable attribute for TDF. In more complex cases involving larger matrices, well-known methods from the operation research domain can be applied.

Table 2: Simulation Results

Control Scheme	Δt_{sp} [K]	q_a [kWh.m^{-2}.a^{-1}]	TDF [%]
A	0.5	320	0.0
B	1	290	0.1
C	2	240	1.0
D	3	210	3.0
E	4	185	6.1

USE OF BDI

The use of a bi-directional inference mechanism for active convergence support in performance-based computer-aided design has been previously documented (Mahdavi

1993, Mahdavi and Berberidou 1995, 1994). In particular, a preference-based method and an investigative projection technique have been developed to cope with the ambiguity problem inherent to the performance-to-design mapping operation. However, we must now demonstrate that BDI can be also applied toward facilitating simulation-assisted building control processes. Using again the previous building example, and starting from an initial operational state, we would have to map desired changes in a performance variable such as q_a or TDF into a control variable such as Δt_{sp}. Since in the present example all design variables are locked (i.e. factors such as building's thermal mass, glazing area, etc. cannot be changed), only q_a, TDF, and Δt_{sp} can be manipulated *via* BDI. Let us demonstrate this with two illustrative scenarios (cp. the BDI-driven trajectory as illustrated in figure 2):

- The building is in an initial operational state in which is Δt_{sp} is 1 K, the predicted energy consumption is 290 kWh.m^{-2}.a^{-1}, and TDF is 0.1%. The control aim is to minimize TDF without exceeding a q_a value of 255 kWh.m^{-2}.a^{-1}. The gradual decrease of the q_a value is translated *via* BDI in an increase of Δt_{sp} accompanied by increase in TDF. For a predicted energy consumption rate of 255 kWh.m^{-2}.a^{-1}, the control variable Δt_{sp} must be set at 1.7 K to minimize TDF.
- The building is in an initial operational state in which is Δt_{sp} is 3.5 K, the predicted energy consumption is about 200 kWh.m^{-2}.a^{-1}, and TDF is 4.5%. The control aim is to minimize the energy consumption rate without exceeding a TDF of 2.5%. In this case, the gradual decrease of TDF is translated *via* BDI in a decrease of Δt_{sp} accompanied by an increase in q_a. The control aim is realized around a predicted energy consumption of 220 kWh.m^{-2}.a^{-1}, and the control variable Δt_{sp} must be set at 2.7 K.

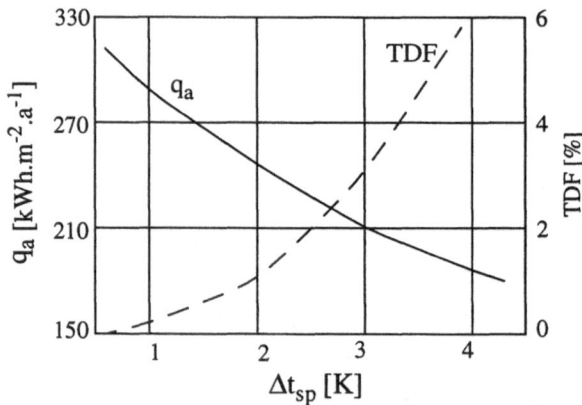

Figure 2. BDI-driven trajectory of performance indicators q_a, TDF and control variable Dt_{sp}

5. A Note on the Early-Design Implications of Control Options

While the focus of this paper was the use of computational modeling to assist the operation of the control systems of actual buildings, the proposed methodology can be also applied to incorporate control issues already in the early stages of the design process. The point is that the availability and characteristics of control options (for HVAC, lighting, etc.) may have significant implications for the decisions made in early design phase. The following case study for the computation prediction of energy use in office buildings allows to better exemplify this point (Mahdavi et al. 1996).

For this case study, building energy use was computed for four buildings with the same volume but different "morphologies" (cp. figure 3 and 4). This difference can be expressed in terms of the "characteristic length" (l_c) of the building which corresponds to the degree of "compactness" of the building and is defined as the ratio of building's volume to the area of its envelope (V/A). Obviously this feature of the building is a direct corrolary of the building geometry and is typically decided upon rather early in the design process (without consideration of factors pertaining to building control options).

The four morphologies considered in this case study are referred to as "square" ($l_c = 9.2$ m), "rectangle" ($l_c = 7.3$ m), "cruciform" ($l_c = 5.5$ m), and "courtyard" ($l_c = 4.6$ m). The simulations were carried out for two different envelope assemblies with corresponding U_m-values of 1.17 $W.m^{-2}.K^{-1}$ and 0.87 $W.m^{-2}.K^{-1}$. Furthermore, two internal load levels were considered, namely 35.6 $W.m^{-2}$, and 40.7 $W.m^{-2}$. The weather data for Washington D.C. was used for the simulations. The results (system energy use) are given in figure 3. From this figure a basic functional relationship between energy use and characteristic length can be implied. However, many factors pertaining to the design and operation of buildings that are not explicitly considered in this diagram, can significantly affect the building's overall energy use. These factors include the magnitude of solar load as affected by total energy transmission through the fenestration, the effects of static and dynamic shading devices, daylighting utilization through continuous dimming of the electrical lighting system, the type and zoning resolution of the HVAC systems, utilization of natural ventilation, etc. To further study this point, additional simulation studies were conducted using the same building morphologies depicted in figure 3 but extending the envelope and systems control options:

- In the previously discussed simulation results the building envelope was only specified in terms of its mean heat transfer coefficient (U_m). The shading coefficient of the glazing system (SC) was assumed to be 0.4 for all of these simulations. In a number of the additional simulations shading coefficient itself is made subject to parametric studies.
- In the previous studies no shading device was modeled. In the new set of simulations an external moveable shading device is among the features considered. The shading operation schedule was derived based on the assumption that the shades would be deployed to avoid direct sunlight incidence on the task surfaces. The shading coefficient of glazing plus shading was assumed to be 0.2.

226

- The previous studies were based on the assumption of a conventional (combined task-ambient) ceiling lighting. In the new set of simulation studies, a daylight-driven continuous dimming option is considered. In this option the lighting can be dimmed down based on the daylight availability in the space.

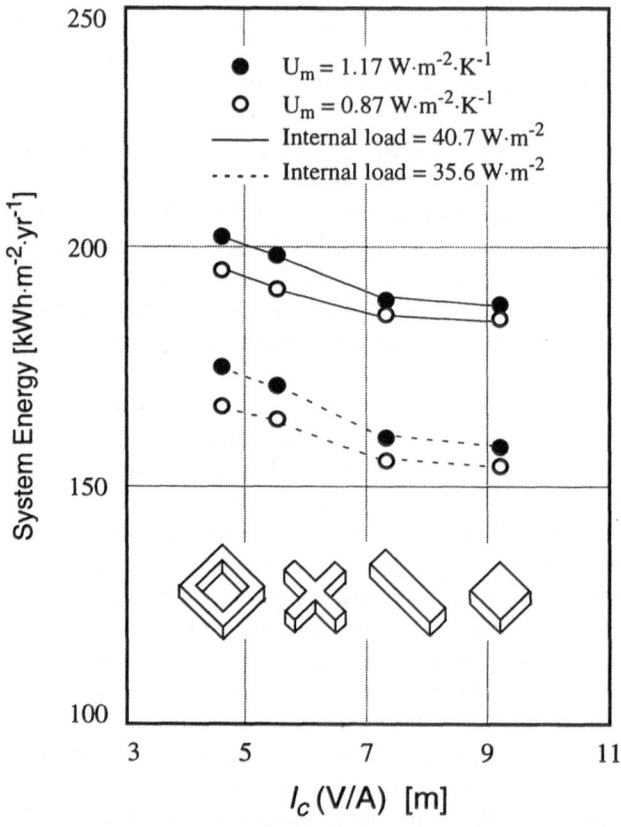

Figure 3: System energy use of four buildings with different morphologies as a function of characteristic length, heat transfer coefficient of the envelope assembly, and the internal load levels

Selected results of these additional simulations (system energy use as a function of characteristic length) are shown in figure 4. Curve *a* in this figure illustrates the "base case" which is reproduced from figure 3 and for which $U_m = 0.87$ W·m^{-2}·K^{-1}, SC = 0.4 (no external shades), and internal loads = 40.7 W·m^{-2}. Curves *b*, *c* and *d* illustrate the results of the additional simulations (for these simulations the U_m-value and the internal load level are the same as in the base case). Curve *b* differs from the base case in that the shading coefficient is assumed to be 0.7. The over-proportional energy use increase for courtyard and cruciform morphologies is probably due to the combined effect of the higher

SC-value and the larger solar exposure. Curve *c* illustrates the results of simulating a moveable external shading device. This dynamic envelope operation modus leads to a rather ambiguous relationship between energy use and characteristic length. Curve *d* demonstrates the simulation results for the aforementioned dimming option. In this case a remarkable reversal of the basic functional relationship of curve *a* can be observed. Building morphologies with more envelope exposure evidently increase the potential for daylight utilization and therefore reduce the overall lighting loads as well as the associated cooling loads.

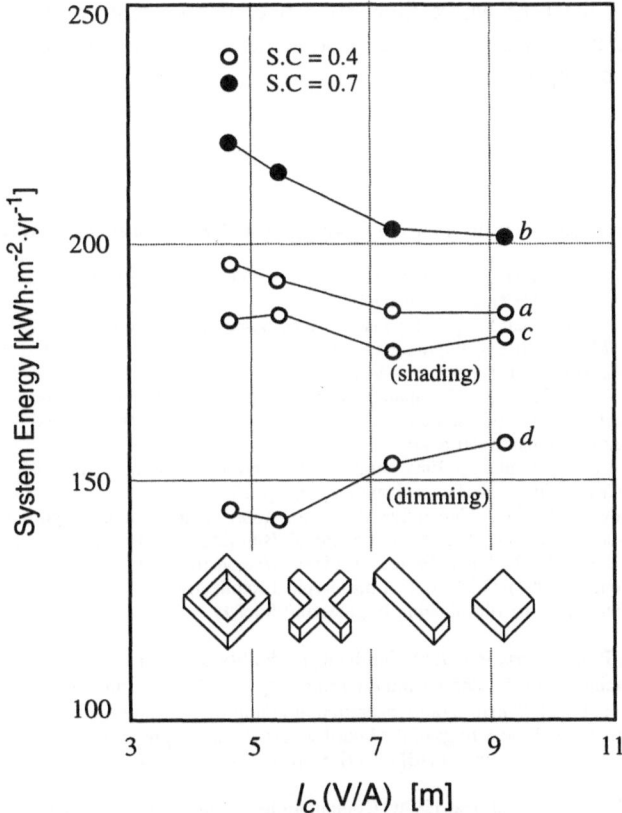

Figure 4:System energy use of four buildings with different morphologies as a function of characteristic length. a) base case, b) alternative shading coefficient option, c) moveable shading option, d) dimming option (see text for more details)

These results clearly indicate that the application of different control technologies may have significant implications for the basic design features such as building geometry. Thus, beyond their potential for real time identification of preferable building control strategies, the above discussed GAT- and BDI-supported methods can also facilitate the "pro-active" consideration of control issues already in the building design phase.

6. Concluding Remarks

To facilitate the understanding of the proposed simulation-assisted building control strategy, we dealt only with examples involving a rather simple building and a relative small set of control scenarios. However, there is no theoretical or methodological reason why much more complex cases cannot be handled using a GAT-based or a BDI-driven control strategy. For example, in the case of BDI, multiple transient control scenarios, dynamic design features, and complex performance criteria may be considered, as long as the corresponding variables and the nature and range of their attributes are non-ambiguously defined. Obviously, in cases where more than one independent control variable may respond to changes in a performance indicator, some form of a weighting or preference methodology must be applied.

References

Curtis, P. S. (1996): Experimental Results from a Network Assisted PID Controller. ASHRAE Transactions. AT 96-21-3.

Curtis, P. S. - Kreider, J. F. - Brandemuehl, M. J. (1993): Adaptive Control of HVAC Processes Using Predictive Neural Networks, ASHRAE Transactions 99(1).

Mahdavi, A. (1996a): Computational Support for Performance-based Reasoning in Building Design. Proceedings of the CIB-ASTM-ISO-RILEM International Symposium "Applications of the Performance Concept in Building". Tel Aviv, Israel. Vol. 1, pp. 4-23 - 4-32.

Mahdavi, A. (1996b): SEMPER: A New Computational Environment for Simulation-based Building Design Assistance. Proceedings of the 1996 International Symposium of CIB W67 (Energy and Mass Flows in the Life Cycle of Buildings). Vienna, Austria. pp. 467 - 472.

Mahdavi, A. (1993): "Open" Simulation Environments: A "Preference-Based" Approach. Proceedings of CAAD Futures '93, CMU, Pittsburgh, Pennsylvania, USA. pp. 195 - 214.

Mahdavi, A. - Berberidou, L. (1995): A Generative Simulation Tool for Architectural Lighting. Proceedings of the Fourth International Conference of the International Building Performance Simulation Association (IBPSA) (Ed.: Mitchell, J.W. - Beckman, W. A.). Madison, Wisconsin. pp. 395 - 402.

Mahdavi, A. - Berberidou, L. (1994): GESTALT: A prototypical Realization of an "Open" Daylighting Simulation Environment. Journal of the Illuminating Engineering Society. Volume 23, Number 2, Summer 1994. pp. 62 - 71.

Mahdavi, A. - Brahme, R. - Mathew, P. (1996): On the Applicability of the "LEK"-Procedure for the Energy Analysis of Commercial Buildings. Energy and Environment, Vol. 31, No. 5, pp. 409 - 415.

Mahdavi, A. - Mathew, P. (1995): Synchronous Generation of Homologous Representation in an Active, Multi-Aspect Design Environment. Proceedings of the Fourth International Conference of the International Building Performance Simulation Association (IBPSA) (Ed.: Mitchell, J.W. - Beckman, W. A.). Madison, Wisconsin. pp. 522 - 528.

Mistry, S. I. - Nair, S. S. (1993): Nonlinear HVAC Computations Using Neural Networks, ASHRAE Transactions 99(1).

Osman, A. - Mitchell, J. W. - Klein, S. A. (1996): Application of General Regression Neural Network (GRNN) in HVAC Process Identification and Control. ASHRAE Transactions AT-96-21-2.

Acknowledgments

The author acknowledges Paul Mathew and Seungju Chang of Carnegie Mellon University for providing simulation data and reference information for this paper.

ON THE PROBLEM OF OPERATIVE INFORMATION IN CAAD

A. MAHDAVI
Department of Architecture, Carnegie Mellon University, USA
V. PAL
School of Architecture, National University of Singapore

Abstract

Computational building performance modeling typically generates large amounts of data. For this data to become operative information, i.e., provide effective feedback to the design process, it must adequately interface with the informational requirements and procedural characteristics of the building delivery process. Toward this end, this paper specifically addresses the potential of aggregate space-time performance indicators.

1. Introduction

Computational models for the prediction of buildings' behavior can generate useful performance-relevant data for building evaluation purposes. Such models have been in use within the framework of both prescriptive and performance-based quality control approaches in building delivery:

The Prescriptive Program: This program involves the derivation of prescriptive design requirements (typically formulated in terms of codes, standards, and guidelines) based on extensive sets of parametric computational modeling studies for typologically classified sets of recurrent building schemes. Although still widely in use, this program has been criticized particularly due to its potentially inhibitory effect on design innovation and will not be the focus of the present paper.

The Performance Program: This program involves *i)* establishing building quality indicators (i.e. building performance variables) and their respective (context-dependent) attributes, and *ii)* the utilization of computational modeling to evaluate the range of such attributes for individual designs.

Although conceptually preferable, the performance program still needs to address the data inflation problem, as advanced (typically dynamic) computational tools for performance prediction commonly generate massive (and sometimes unmanageable) quantities

231

R. Junge (ed.), CAAD Futures 1997, 231-244.
© 1997 *Kluwer Academic Publishers.*

of behavioral data. In this context, there are still many open questions as to how this data should effectively interface with the informational requirements and procedural constraints of the building design and evaluation practice. Various approaches have been suggested to cope with this problem, a few of which are briefly described below:

i) Conservative data transformation: computed performance data are typically numeric and thus less convenient for the development of a global sense of building/space behavior. Creative transformation of such numeric data in visually expressive formats can support the understanding and evaluation of building performance. In this case the intention is not to reduce or manipulate the computed data *per se*, but to use data representations that are more accessible to human information processing.

ii) Non-interpretative data reduction: the vast number of numeric data generated by tools capable of transient simulation may be reduced using simple selective methods or more complex statistical aggregation. Examples of such reductive methods are aggregation of hourly data on building energy performance into "typical day" or "monthly" result formats. In this case, the underlying data base is "manipulated" to create a manageable set of performance attributes. However, the nature of the so generated aggregate indicators (temperature, thermal load, illuminance level, etc.) does not differ from the originally computed simulation results.

iii) Interpretative data translation: physical performance data may be translated into indicators of occupancy evaluation, such as those applied in the thermal comfort domain. In this case, the intention is to support the evaluation and decision making process through the derivation of intuitive measures of occupancy-relevant building quality based on physically defined indicators of building performance.

This paper primarily focuses on approaches to non-interpretative data reduction, arguing that more systematic research efforts are needed to formulate and test methods that allow for meaningful and effective aggregation of space-time performance indicators for simulation-based building evaluation procedures. Such indicators must satisfy two basic requirements. They must strategically reduce the simulated performance base data to the extent that designs can be effectively evaluated, compared, and further developed. At the same time, they must ensure that reductive approach does not eliminate the responsiveness of the indicators to the complexity of and parametric changes in design.

Toward this end, this paper offers contributions in view of aggregate indicators in both space and time domain:

First, some existing and new aggregate thermal performance indicators in time domain are reviewed regarding their potential and effectiveness in decision support. This review includes not only simple statistical (e.g. cumulative) indicators known from the context of prescriptive standards, but also more sophisticated indicators (combined geometric/energetic indicators such as LEK and LEK_{eq}).

Second, a detailed case study from the visual performance area (lighting distribution) is used to demonstrate a new aggregate performance indicator in space domain. Again, the hierarchy of aggregation levels is discussed starting with traditional indicators of light distribution uniformity inside spaces. Next, more sophisticated uniformity indicators are discussed, which address some of the shortcomings of the first generation reductive uniformity indicators, i.e. their unsatisfactory behavior in view of the local deviations of light levels. Finally, a new entropic light uniformity level is introduced to overcome the problems of the second generation indicators, i.e. their indifference toward the various spatial configurations (adjacency relations) of light levels in a room.

2. On Indicators of Thermal Performance

2.1 INTRODUCTORY REMARK

In the domain of thermal performance, prescriptive requirements pertaining to building fabric (particularly building envelope) such as maximum permissible U-values as well as simple and cumulative measures of energy performance (such as area-related or volume-related peak and annual loads) have been in use for a long time. Recent developments on both accounts demonstrate how *a)* the inherently dynamic behavior of building in the time domain, and *b)* certain aspects of building geometry may be reflected in new aggregate performance indicators.

2.2 INTERPRETATIVE CUMULATIVE INDICATORS

Transient simulations typically generate performance results (energy use, space temperatures, etc.) for every time step throughout the simulation period. Where target (desired) levels are known (e.g. a certain space temperature, or zero energy use), secondary (interpretative) indicators may be developed that incorporate the cumulative deviation from those target levels, with or without application of weights to various degrees of deviation. Such performance indicators allow for the comparison of various design alternatives and facilitate code compliance checking. An example of such a cumulative indicator is the temperature deviation factor (TDF) that captures the deviation of the predicted maintained space temperature from the preferred space temperature (Mahdavi 1997). This performance indicator (practically an average cumulative temperature deviation) is defined as follows:

$$TDF = \sum_{i=1}^{n} \frac{w \cdot \left(\frac{|t_p - t_i|}{t_p} \right)}{n} \cdot 100 \qquad [\%] \qquad eq.1$$

Here t_p stands for the preferred space temperature, t_i represents the space temperature at time step i; n is the total number of time steps; and w is a weighting variable to penalize larger deviations of the maintained temperature from the target space temperature.

2.3 TOPOLOGICALLY ENRICHED ENERGY PERFORMANCE INDICATORS

Establishing criteria for the heat transfer through building components is as such a hall-mark of the prescriptive approach to building quality assurance. However, there have been continuous attempts to arrive at related requirements and derivative performance indicators involving a more conclusive aggregation of information needed for building evaluation. For example, a procedure has been proposed and implemented that aggre-gates information on thermal characteristics of the building envelope with a simple descriptor of the building's geometry, namely "characteristic length" l_c, which is the ratio of building volume V_B to building envelope area A_B (Panzhauser 1993, Mahdavi et al. 1996). This has led to the establishment of LEK values (Lines of European k-values, cp. Figure 1) which allow for the definition of the thermal insulation of building envelopes (expressed in terms of envelope's mean U-value) while considering the building enve-lope's geometry (expressed in terms of envelope's characteristic length):

$$LEK = 300 \cdot U_m \cdot (2 + l_c)^{-1} \qquad eq.2$$

Despite this informational enrichment (in view of the concurrent consideration of both thermal characteristics of building envelope and its geometry), LEK still retains a pre-scriptive nature which diminishes its value as a true thermal *performance* indicator. Fur-thermore, LEK obviously does not consider the effects of solar and internal gains. To arrive at an indicator that would achieve this while retaining the benefit of LEK (in view of geometry description), LEK_{eq} has been proposed (Fantl et al. 1996). Once the heating energy need (q_h) of a building and the relevant heating degree days (DD_h) are known, LEK_{eq} can be calculated according to the following equation:

$$LEK_{eq} = 100 \cdot q_h \cdot l_c \cdot (0.024 \cdot DD_h \cdot (2 + l_c)) \qquad eq.3$$

The important point in the above formulation is that q_h itself can be derived based on advanced transient energy simulation. Thus, the single-number indicator LEK_{eq} involves a twofold enrichment in that it includes topologically meaningful geometric information and can embody detailed information on building's energy use derived from sophisti-cated energy simulation routines.

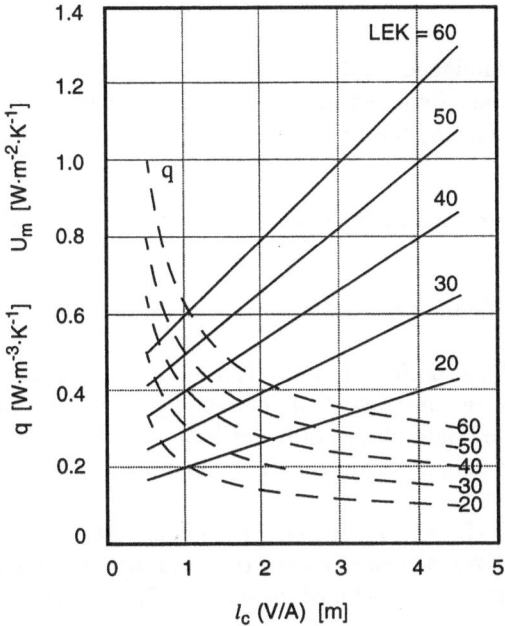

Figure 1. LEK-Diagram (Lines of European k-values) with corresponding volume-specific transmission heat loss values (Mahdavi et al. 1996)

3. A New Light Distribution Uniformity Indicator

3.1 INTRODUCTORY REMARK

We now consider the problem of description of spatial distribution of light, as it represents a particularly good example for the development of a set of indicators that, despite their single-number format, can provide successively higher level of critical information. Uniformity indicators have been in use to describe the degree of uniformity of the illuminance or luminance levels for various applications (e.g. architectural lighting, lighting of sport facilities, road lighting). Numerical attributes of these indicators have been used as *a)* prescriptive definitions of required (or desirable) degrees of uniformity, *b)* simple ("compressed") representations of actual light measurement results, and *c)* descriptors of computer simulation output. We describe in the following (without the intention of exhaustive coverage) examples from three generations of uniformity indicators as space aggregates of visual performance.

236

3.2 FIRST GENERATION

In IES 1993 (pp. 888), illuminance uniformity is discussed in the context of emergency, safety, and security lighting. Required levels of illuminance uniformity are defined in terms of "illuminance uniformity ratio":

$$\text{illuminance uniformity ratio} = \frac{E_{avg}}{E_{min}} \qquad [\text{-}] \qquad eq.4$$

where E_{avg} is the average illuminance and E_{min} is the minimum illuminance.

In IES 1993 (pp. 525-526), a "uniformity ratio" is used to describe requirements pertaining to ceiling luminance in the context of indirect lighting. The uniformity ratio is defined as the "ratio of the brightest area of the ceiling ... to the darkest area of the ceiling, ... in other words, the ratio of the maximum to the minimum".

In DIN and CIE literature (DIN 5044, Hentschel 1982, pp. 196 and 224, Hochstädt and Kuloge 1969, pp. 99A-101A) the length-related (U_1) and the total (U_0) luminance distribution uniformity indicators ("Gleichmäßigkeitszahlen") are used for road lighting design purposes:

$$U_1 = \frac{L_{min}}{L_{max}} \qquad [\text{-}] \qquad eq.5$$

$$U_0 = \frac{L_{min}}{L_m} \qquad [\text{-}] \qquad eq.6$$

where L_{min} is the minimum luminance, L_{max} is the maximum luminance, and L_m the average luminance.

In Hentschel 1982 (pp. 234) illuminance distribution uniformity requirements ("Gleichmäßigkeit") are given for sport stadiums based on recommendations in LiTG 1969 and LiTG 1967. The relevant definition is in this case:

$$\text{uniformity} = \frac{E_{min}}{E_m} \qquad [\text{-}] \qquad eq.7$$

where E_{min} is the minimum (horizontal) illuminance and E_m is the average (horizontal) illuminance.

The above reviewed indicators involve, per definition, numeric values of light levels at a single point. For example, in order to derive the uniformity factor for the illuminance on a horizontal plane (e.g. task surface) in a space according to the above definitions, the minimum illuminance levels must be identified and applied in the computation. The question is, if and how the reliability of the so derived uniformity indicator might be affected due to the uncertainties involved in obtaining the individual (actually measured or computationally simulated) illuminance level at a certain point.

3.3 SECOND GENERATION

In response to the critical point raised above, a number of statistically more elaborate indicators have been proposed. One of these proposals (Mahdavi 1994, Mahdavi et al. 1995) was inspired by the definition of turbulence intensity in ventilation and thermal comfort domain (ASHRAE 1993, Fanger et al. 1987). Turbulence intensity (T_u) is defined as the ratio of the standard deviation of measured air velocities (v_{SD}) during a certain time period to the average value of these measurements (v_m) over the same time period:

$$T_u = \frac{v_{SD}}{v_m} \cdot 100 \qquad\qquad [\%] \qquad\qquad eq.8$$

Mahdavi's proposed new uniformity factor was defined analogous to the above relationship by applying three modifications. First, in the place of velocity in equation 8, a photometric term such as illuminance was used. Second, instead of dealing with the changes of the parameter in the time domain (time-dependent velocity measurements), fluctuations were defined in the space domain (location-dependent illuminance values). Finally, an algebraic modification was applied to insure that uniformity factors can only have values between 0 and 1. The result was the following definition for the illuminance distribution uniformity factor (U):

$$U = \frac{E_m}{E_m + E_{SD}} \qquad\qquad [-] \qquad\qquad eq.9$$

where E_{SD} is the standard deviation and E_m the average value of the illuminance levels.

A comprehensive comparative statistical analysis of this indicator and two conventional uniformity indices (Mahdavi et al. 1995) demonstrated that: a) the proposed uniformity factor has a suitable range for the characterization of a wide range of possible light distribution patterns in architectural spaces (including day-lit rooms) and allows for cross-configuration and cross-space comparisons; b) it is relatively stable despite the uncertainties in obtaining individual readings (or computations) of illuminance or luminance levels; c) it is less affected by the resolution of the measurement/simulation grid; d) it does not "overreact" to minor (locally restricted) variations in illuminance (or luminance) distribution pattern. Furthermore, Mahdavi recently proposed to apply this indicator (eq. 9) toward the solution of the cloud cover description problem in daylighting applications. Recent initial studies involving the measured sky luminance distribution in Singapore appear to suggest an interesting new possibility to specify the sky's cloud cover as a function of the numeric attribute of the sky luminance uniformity indicator.

However, as with all second generation indicators, the uniformity indicator of equation 9 has a clear limitation, despite its obvious advantages compared to the conventional ones. It is indifferent to the specific topological pattern of adjacent illuminance (or luminance) patterns. For example, the fields A and B in figure 2 (which may be thought of as task surfaces with simulated illuminance levels) have obviously entirely different distribution patterns, yet yield identical uniformity attributes no matter if first or second generation indicators are applied.

300	150	300	150
150	300	150	300
300	150	300	150
150	300	150	300

A

300	300	150	150
300	300	150	150
300	300	150	150
300	300	150	150

B

Figure 2. Demonstrative spatial distribution patterns for which a second generation uniformity indicator yields an identical numeric attribute

3.4 THE THIRD GENERATION

In response to this problem and inspired by the entropy notion in thermodynamics, Mahdavi 1996a proposed the concept of a single-number "entropic distribution index" (EDI). To clarify this concept we first provide an illustrative thermal analogy that leads to establishing relative entropic uniformity attributes for a few simple illuminance distribution patterns. Consider the dark-bright illuminance patterns A to E in figure 3. We represent these as low-high temperature zones in a space with an analogous geometry. Assuming adiabatic enclosure, we apply the detailed transient energy simulation tool SEMPER-NODEM (Mahdavi 1996b, Mahdavi and Mathew 1995) to compute for each case the time that is needed to move from the initial state to the thermal equilibrium state.

The conjecture is that the longer the time taken to reach equilibrium, the lower the uniformity of the pattern. To arrive at a uniformity ranking, we first represented these times

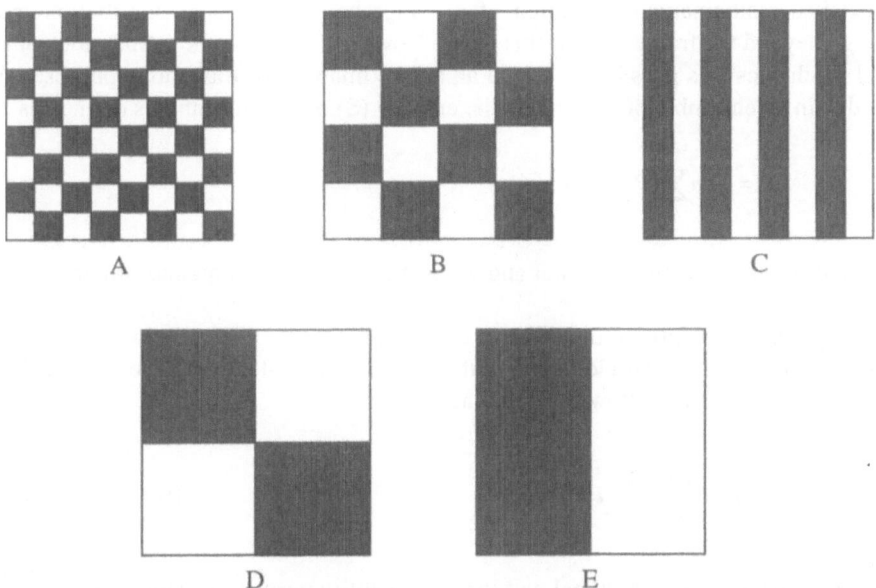

Figure 3. Five demonstrative spatial distribution patterns for which a third generation
uniformity indicator yields different numeric attributes

relative to the shortest equilibrium time among all patterns. The inverse of this relative
time provides an entropic correlate to the visual uniformity of the patterns. We have used
the numeric results of this procedure to rank the patterns A to E according to their degree
of uniformity (cp. table 1). Table 1 clearly demonstrate the sensitivity of the proposed
new entropic light distribution uniformity index (visual EDI) to the topological distribution of individual (simulated or measured) light levels.

Table 1: Uniformity ranking of five demonstrative distribution patterns (cp. figure 3) derived
based on the numeric attributes of Entropic Uniformity Indicator

Distribution pattern	Uniformity rank
A	I
B	II
C	III
D	IV
E	V

The detailed mathematical derivation of a comprehensive entropic distribution index would go beyond the framework of this paper. However, based on the foundations of statistical mechanics it is possible to derive an approximation for illustrative purposes. We know that in an ensemble of n numbers the entropy (S) of the ensemble is defined as

$$S = -kn\sum_{q}(P_q \ln P_q) \qquad\qquad eq.10$$

where k is the Boltzmann's constant and P_q is the probability of quantum state q.

For a rough approximation of the entropic distribution index (EDI) of a two-dimensional orthogonal field with n grid elements (with corresponding illuminance levels E_i), Mahdavi 1996a proposed the following formulation:

$$EDI = \frac{100}{n}\sum_{i=1}^{n}(P_{i,g} \cdot P_{i,l})^{0.5} \qquad [\%] \qquad eq.11$$

where $P_{i,g}$ and $P_{i,l}$ are the global and local probability terms as per the following definitions:

$$P_{i,g} = 1 - \frac{|E_i - E_{m,g}|}{E_{m,g} + E_{sd}} \qquad\qquad eq.12$$

$$P_{i,l} = 1 - \frac{|E_{m,l,i} - E_{m,g}|}{E_{m,g} + E_{sd}} \qquad\qquad eq.13$$

where E_i is the illuminance level at point i, $E_{m,g}$ is the average illuminance of the whole field, E_{sd} is the standard deviation of the illuminance levels of the whole field, and $E_{m,l,i}$ is the local average of the illuminance levels in the immediate neighbourhood of the grid point i.

Using this simplified formulation of EDI, we computed the uniformity levels in a daylit architectural space for 9 different window configurations (cp. Figure 4). The simulation assumptions are summarized in Table 2. Table 3 and Figure 5 include the EDI attributes as well as the corresponding attributes of first generation (U_1 according to eq. 7) and second generation (U_2 according to eq. 9) uniformity indicators.

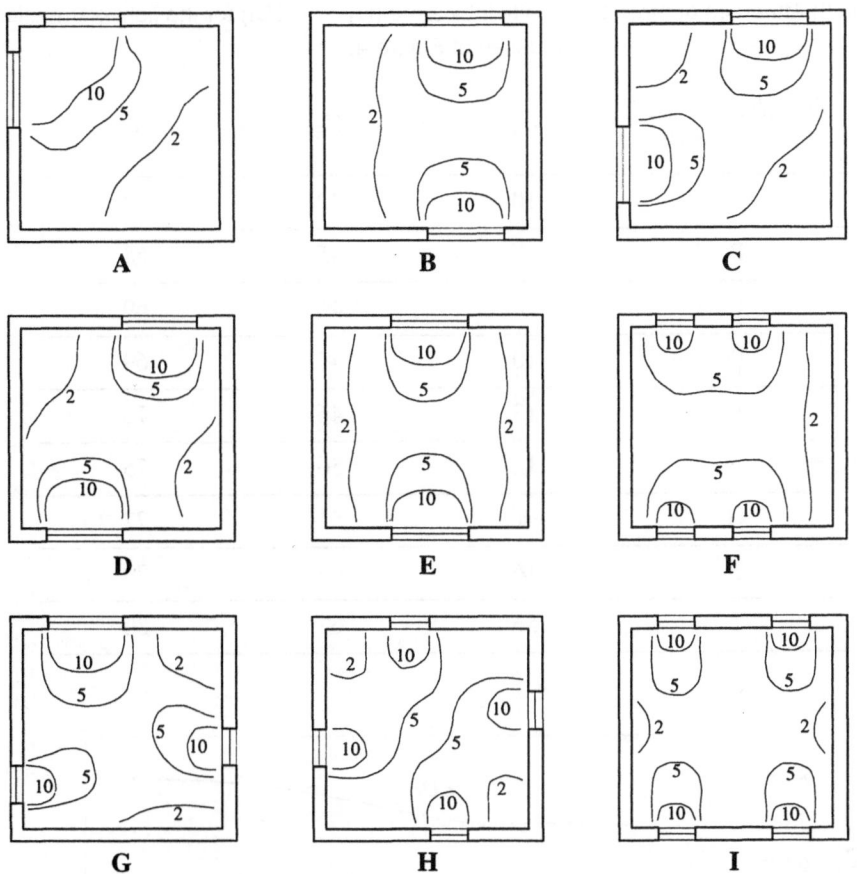

Figure 4: Daylight distribution (expressed in terms of daylight factors) in an architectural space with nine diffrent window configurations.

Table 2: Reflectance (ρ) and Transmittance (τ) assumptions for the simulation of light distribution in an architectural space (cp. Figure 4).

	ρ [-]	τ [-]
Fenestration	0.1	0.8
Ceiling	0.5	NA
Walls	0.3	NA
Floor	0.1	NA

Table 3: Attributes of three Uniformity Indicators (U_1, U_2, EDI) for the light distribution cases of Figure 4.

Distribution case (cp. Figure 4)	U_1	U_2	EDI
A	10	42	62
B	7	45	68
C	4	46	69
D	10	46	69
E	13	46	71
F	12	56	75
G	6	51	77
H	16	63	79
I	56	67	80

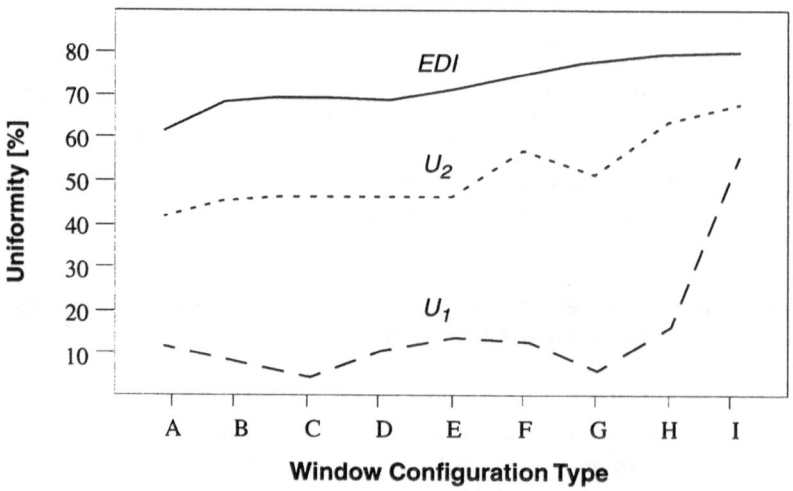

Figure 5: Uniformity Indicators (U_1, U_2, EDI) for the light distribution cases of Figure 4.

These results

- clearly demonstrate the problematic nature of the first generation uniformity indicator (for example, according to the results for U_1, the daylight distribution in case A is more uniform than in case G).
- show that EDI can differentiate distribution patterns even in those cases when the U_2 results yield identical numeric values (cp. cases C and E).

4. Conclusion

In this paper, we specifically discussed some existing and new examples of aggregate space-time performance indicators for simulation-based building evaluation purposes in the thermal and visual (lighting) domains. We demonstrated that such carefully formulated aggregate indicators can strategically reduce the computational performance data, while conserving salient information on buildings' complex space-time behavior. Needless to say, as there is no one unique method to represent and communicate the typically extensive computationally generated building performance simulation results, such aggregate indicators do not represent a substitute for other numeric and graphic means of data representation, but should be seen as a complementary means of provision of operative information in computer-aided architectural design.

Acknowledgement

The authors wish to acknowledge Paul Mathew's contribution in performing the thermal simulations used to generate Table 1 in this paper.

References

ASHRAE (1993): ASHRAE Handbook Fundamentals. American Society of Heating, Refrigerating and Air-Conditioning Engineers, Inc. Atlanta, GA.

DIN 5044: Straßenbeleuchtung, Richtlinien.

Fanger, P. O. - Melikov, A. - Hanzawa, H. - Ring, J. (1987): "Air turbulence and sensation of drought." Energy and Buildings.

Fantl, K. - Panzhauser, E. - Wunderer E. (1996): Der Österreichische Energieausweis für Gebäude, Research Report 1365, Austria.

Hentschel, H. (1982): Licht und Beleuchtung; Theorie und Praxis der Lichttechnik. Dr. Alfred Hüthig Verlag Heidelberg.

Hochstädt, E. - Kuloge, R. (1969): Auswertung von projektierten Straßenbeleuchtungsanlagen hinsichtlich ihrer Leuchtdichtegleichmäßigkeit. Lichttechnik 21, Nr. 9.

IES (1993): Lighting Handbook; Reference & Application. 8th Edition. Illuminating Engineering Society of North America.

LiTG (1967): LiTG-Fachausschuß "Sportstättenbeleuchtung": Beleuchtung für Fußball, Handball und Rugby. Lichttechnik 19, Nr. 9.

244

LiTG (1969): LiTG-Fachausschuß "Sportstättenbeleuchtung": Beleuchtung von Sportstätten für das Farbfernsehen. Lichttechnik 21, Nr. 11.

Mahdavi, A. (1997): Toward a Simulation-Assisted Dynamic Building Control Strategy. Proceedings of the Fifth International Conference of the International Building Performance Simulation Association. Prague, Czech Republic.

Mahdavi, A. (1996a): An Entropic Uniformity Indicator. Report No. DOA-96-08-01. Department of Architecture, Carnegie Mellon University, Pittsburgh, PA.

Mahdavi, A. (1996b): Computational Support for Performance-based Reasoning in Building Design. Proceedings of the CIB-ASTM-ISO-RILEM International Symposium, "Applications of the Performance Concept in Building". Tel Aviv, Israel. Vol. 1, pp. 4-23 - 4-32.

Mahdavi, A. (1994): A New Light Distribution Uniformity Descriptor. CBPD Internal Document, No. 9410-01. Department of Architecture, Carnegie Mellon University, Pittsburgh, PA.

Mahdavi, A. - Mathew, P. (1995): Synchronous Generation of Homologous Representation in an Active, Multi-Aspect Design Environment. Proceedings of the Fourth International Conference of the International Building Performance Simulation Association (IBPSA) (Ed.: Mitchell, J.W. - Beckman, W. A.). Madison, Wisconsin. pp. 522 - 528.

Mahdavi, A. - Brahme, B. - Mathew, P. (1996): On the Applicability of the "LEK" Procedure for the Energy Analysis of Cooling-Dominated Commercial Buildings, Energy and Environment, Vol. 31, No. 5, pp. 409-415.

Mahdavi, A. - Prankprakma, P. - Berberidou, L. (1995): On Numeric Indicators of Light Distribution Uniformity. Proceedings of The 1995 IESNA Annual Conference, New York. pp. 385 - 394.

Panzhauser, E. (1993): Endenergie für Raumwärme, in: K. Masil, K. Fantl, E. Panzhauser, eds., Potential der thermischen Gebäudesanierung in Österreich (Austrian Institute for Economic Research).

On the Evaluation of Architectural Figural Goodness:

a foundation for computational architectural aesthetics

ALEXANDER KOUTAMANIS

Delft University of Technology, The Netherlands

Abstract

The first stage of an investigation into the quantification and computability of architectural aesthetics is reported. Issues considered include the function, sources and role of aesthetic analysis in architecture in the framework of a *descriptive* approach to architectural analysis and design. The main focus is on the applicability of the concept of figural goodness to architectural aesthetics and the derivation of a representation for architectural form suitable to this purpose.

1. Aesthetics, analysis and evaluation

The paper discusses certain issues in the representation of architectural forms which relate to the perceptual coding of images and visual figural goodness. These issues are seen as the foundation of most cognitive processes in architecture, including recognition and memorization. In the present paper they are considered with respect to one specific application, the quantification of aesthetic analysis and evaluation. The importance of aesthetics in architecture and the sensibilities connected to it require a clear statement of the underlying approach and in particular of the hypotheses concerning:

- the function of aesthetic preference in architectural design, as well as in lay interaction with the built environment and its representations;

- the sources of aesthetic preference, i.e. the innate structures and the cognitive reference frames involved in aesthetic analysis and evaluation;

- the role of aesthetics relative to other design aspects in architecture;

- the computability of aesthetic analysis and evaluation.

R. Junge (ed.), CAAD Futures 1997, 245-266.
© 1997 *Kluwer Academic Publishers.*

The intuitive appreciation of aesthetic preference has been one of the hallmarks of architectural design in practice. It has also been one of the main reasons for conflict between the architect and the lay person, as the latter's appreciation of built form and space is less tempered by dominant architectural doctrines and more by the élite that dictates good taste and vogue. As vogue is often at odds with architectural history and criticism, architects have been reluctant to change what they consider to be part of their methodical background. The predominance of the intuitive approach agrees with many types of human interaction with the built environment and its representations. This agreement adds an element of common sense to architectural analysis that may temper indifference to practical problems. However, common sense can be distorted or even refuted by expert opinion and interpretation, especially if the specific human experiences do not involve directly measurable performance criteria. Such distortions and refutations have contributed to the deep dichotomy between form and function in architecture and to the frequent elevation of formal considerations to the highest priority in architectural design, either as a priori norms and canons or as direct and inescapable consequences of functional issues and problems.

Within reasonably well-defined architectural systems formal considerations are in the final analysis part of the constraints of the problem, and specifically of the professional knowledge that constitutes the framework of many design decisions. This has frequently allowed concentration on particular problems and resolution of these problems in ways that satisfied not only the intrinsic formal constraints but also extrinsic performance criteria. In an eclectic period like ours formal considerations lose much of their coherence and consequently much of their value as guiding principles. At the same time their arbitrariness is accentuated by their frequent use as justifications for design decisions that may be otherwise unrelated to the problem. The abuses of aesthetic analysis urge a rigorous, in-depth treatment of aesthetic questions which reveals the true extent and significance of architectural aesthetics and its relations to other design aspects. The basic premises of this treatment are:

- Aesthetics is one of many design aspects that have to be considered in an architectural problem. It is essentially similar and equivalent to functional aspects such as circulation and environmental aspects such as daylighting.

- Aesthetic evaluation is a performance measure and as such it should reflect the interaction between human activities and the built environment that contains these activities. Corollaries of this include:

  - The psychology of perception and cognition are the basic sources of aesthetic analysis and evaluation. This has been common ground in architectural treatises of aesthetics where cognitive psychology has been the main source of explanations for intuitive or conventional aesthetic devices and preferences (Prak 1968; Arnheim 1974; Arnheim 1977; Prak 1979; Arnheim 1988; Holgate 1992; Weber 1995).

  - Aesthetic evaluation should be based on explicit qualitative or quantitative cognitive criteria which replace the often ill-defined architectural norms. In this much can be learned from attempts to quantify aesthetics in general and more specifically in relation to information processing (Birkhoff 1933; Bense 1954; Moles 1968). The prototypical character of these attempts necessitates knowledge transfer from recent research into the quantification of figural goodness, also largely on the basis of information processing. This transfer is the main subject of the present paper.

Equally important to the sources and the relative significance of aesthetic analysis and evaluation is how it should be performed: by which means and towards which goals. Architectural analysis and design (including aesthetic issues) have been driven by normative models belonging to either of the following deontic approaches:

- *Proscriptive*: formal or functional rules that determine the acceptability of a design on the basis of non-violation of certain constraints. Formal architectural systems such as classicism and modernism, as well as most building regulations are proscriptive systems.

- *Prescriptive*: systems that suggest that a predefined sequence of actions has to be followed in order to achieve acceptable results. Most computational design approaches are prescriptive in nature.

As stated earlier, dominance of a specific system or approach in general has been instrumental for the evolution of architecture, as it allowed for concentration of effort on concrete, usually partial problems within the framework of the system and hence supported innovation and the transformation of innovation into global improvement. The strongly eclectic spirit of recent and current architecture reduces greatly the value of normative approaches, as it permits strange conjunctions, arbitrary deviations, far-fetched associations and unconstrained transition from one system to another. In addition, the advent of the electronic era provides through the democratization of computer technologies the means for analyses and evaluations of a detailed and objective nature. These dispense with the necessity of abstracting and summarizing in rules and norms. By this I do not suggest that abstraction is unwanted or unwarranted. On the contrary, abstraction is an obvious cognitive necessity that emerges as soon as a system has reached a stable state, is widely applicable and free from major internal conflicts which may reduce reliability. As a result, one can expect the emergence of new abstractions on the basis of the new detailed, accurate and precise analyses. It is probable that several older norms will be among the new abstractions. This, however, does not imply that we should adopt the top-down strategy followed in many computational studies, i.e. accept unquestioningly current norms and then elaborate these norms in rule-based systems and naïve 'simulations'. Abstraction should occur in a bottom-up fashion that supports new strategies which match the complexity and priorities of today's design problems.

The main characteristic of the new forms of analysis is that they follow an approach we may term *descriptive*. They evaluate a design indirectly by generating a description of a particular aspect comprising detailed measurable information on the projected behaviour and performance of the design. This description is normally closely correlated with the formal representation of the design and therefore permits interactive manipulation, e.g. for trying different alternatives and variations. In short, the descriptive approach complements (rather than guides) human design creativity by means of feedback from which the designer can extract and fine-tune constraints.

In functional analyses it has become clear that most current norms and their underlying principles have a very limited scope, namely control of minimal specifications by a lay authority. They are often obsolete as true performance measures and grossly insufficient as design guidance. The solution presented by the descriptive approach is the substitution of obsolete abstractions with detailed information on functionality and performance, for example abandonment of Blondel's formula of stair sizes in favour of an ergonomic analysis of stair ascent and descent by means of simulation (Mitossi and Koutamanis 1996). The analysis is performed in a multilevel system that connects normative levels to computational projections and to realistic simulations in a coherent structure where the assumptions of one level are the subject of investigation at another level (Koutamanis 1995; Koutamanis 1996).

The approach advocated in the present paper is accordingly an extension of the principles underlying descriptive functional analyses. It is proposed that we should continue to draw the principles of architectural aesthetics from perceptual and cognitive sources and connect these principles to architectural issues but strictly in this order. In other words, rather than starting with ordering the existing architectural aesthetic norms and then proceeding to a search for cognitive relevance and justification, we should attempt to apply general computational models of perception and cognition to architecture. One reason for doing so is the difficulty involved in attempting to form a cohesive system out of disparate cognitive explanations of architectural aesthetic preferences. Another reason is the lack of a suitable theory of architectural perception and visual cognition, i.e. one that links architectural theory with cognitive science and especially with the latter's spectacular advances in the last quarter of our century. Such a theory should not derive from a normative architectural model or system, i.e. should not exhibit any bias towards or against specific approaches. The aesthetic analysis of an architectural object or configuration should potentially accommodate all possible architectural systems (Stiny and Gips 1978). Different systems would correspond to variations in the analysis with respect to the configuration of analytical devices, as well as to (parametric) differences within each device. The common basis of the different systems and of the corresponding analyses is an objective representation of the architectural object, i.e. a description that does not relate to a specific architectural formal system.

The prerequisite of an objective representation brings us to a third reason for commencing our investigation of architectural aesthetics from cognitive science. This reason, which arguably underlies the previous two, is the lack of general agreement concerning the architectural representation of built form and space. Our approach is based on the assumption that the basic components of this representation are the conventional 'solids' and 'voids' of architecture (i.e. building elements and the spaces bounded by the building elements) and their relationships. Lower level (implementation) primitives such as points and lines, as well as Platonic and other abstract prototypical forms, are deemed irrelevant. The main departure from existing related representations, notably the dual graph representation (Steadman 1976; Steadman 1983), lies in that identification and representation of individual components relies less on architectural knowledge and convention and more on visual cognition. Architectural knowledge is treated as a cognitive reference frame against which the perceptual system constructs descriptions of the objects (Rock 1973; Palmer 1989; Rock 1990).

The distinction between the derivation of a description, its interpretation and finally its evaluation is common to computational studies of vision but also of aesthetics (Stiny and Gips 1978). In our case its particular value lies in that it stresses the importance of the representation for aesthetic analysis and thereby of the affinity between figural goodness in perception and the aesthetic appreciation of built form. Figural goodness has been linked to aesthetic response by means of the relation between perceptual arousal and complexity (Berlyne 1960; Berlyne 1971). The working hypothesis of our research is that the representation of architectural objects is subject to coding on the basis of figural goodness. The coding relates to preference for a specific configuration in the percept and hence for a particular interpretation of the object. Architectural aesthetic analysis and evaluation is based on placing this interpretation and coding against a background of cognitive reference frames derived from architectural formal systems. The present paper is a discussion of three basis corollaries of this hypothesis:

- the decomposition of aesthetic analyses into quantifiable factors relating to architectural figural goodness;

- the coding of a representation in relation to figural goodness; and

- the derivation of an architectural representation which supports evaluation and analysis of figural goodness.

2. Aesthetic measures

The first significant attempt to quantity aesthetics was by the American mathematician George D. Birkhoff who effectively established a prototype for subsequent approaches (Birkhoff 1933). In Birkhoff's analysis the aesthetic experience relies on principles of harmony, symmetry and proportion previously stated by among others the Pythagoreans and Leibniz and suggests three successive phases:

1. arousal and effort of attention;

2. the feeling of value or aesthetic measure which rewards the effort of attention; and finally

3. the realization that the perceived object is characterized by a certain aesthetic order.

Birkhoff states that the effort of attention is proportional to the complexity (C) of the perceived object and links complexity, the aesthetic measure (M) and aesthetic order (O) in the basic aesthetic formula:

$$M = O / C$$

Complexity is generally measured by the number of elements in the perceived object. For example, in isolated polygonal forms complexity is measured by the number of distinct straight lines containing at least one side of the polygon, similarly to the gratings of rectangular dissections (Steadman 1983). The measurement of order varies with the specific class of objects to be evaluated but generally takes the form of the sum of weighted contributing elements:

$$O = ul + vm + wn + \ldots$$

where l, m, n, ... are the independent elements of order and u, v, w, ... indices which may be positive, zero or negative, depending upon the effect of the corresponding element. Aesthetic order and consequently the aesthetic measure are relative values which apply to specific classes of objects so restricted that intuitive comparisons of the different objects becomes possible. There is no comparison between objects of different types.

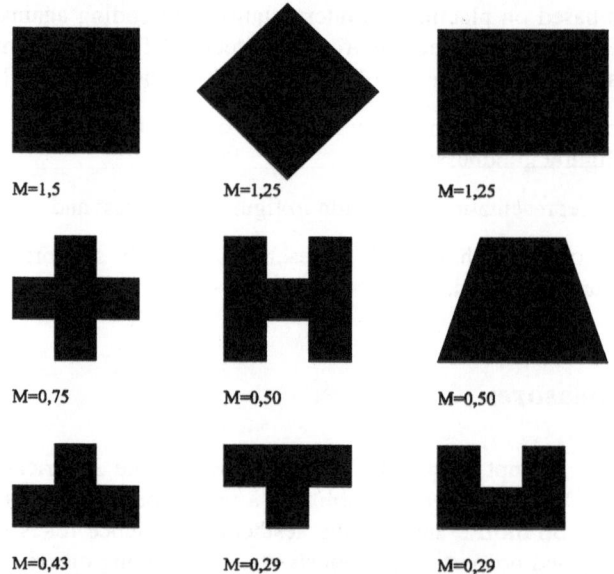

Figure 1. The aesthetic measure of isolated polygonal forms according to
(Birkhoff 1933)

Birkhoff suggests that order relates to associations with prior experience and acquired knowledge that are triggered by formal elements of order, that is properties of the perceived object, such as bilateral symmetry about a vertical axis or plane. Formal elements of order which have a positive effect include repetition, similarity, contrast, equality, symmetry and balance. Ambiguity, undue repetition and unnecessary imperfection have a negative effect. For example, a rectangle almost but not quite a square is unpleasantly ambiguous according to Birkhoff. Also a square whose sides are aligned with the horizontal and vertical is superior to an unnecessarily imperfect square which has been rotated about its centre by 45 degrees "because it would be *so easy* to alter it [the rotated square] for the better" (p. 25).

In the example of isolated polygonal forms aesthetic order is measured by the formula

$$O = V + E + R + HV - F$$

where V stands for vertical symmetry, E for equilibrium, R for rotational symmetry, HV for the relation to a horizontal-vertical network (reference framework) and F for unsatisfactory form. "Unsatisfactory form" encompasses too small distances between vertices or parallel sides, angles too near to 0 or 180 degrees and other ambiguities, diversity of directions and lack of symmetry.

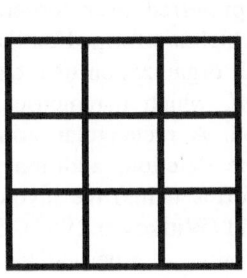 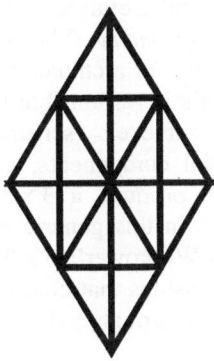

Figure 2. Examples of horizontal-vertical networks according to (Birkhoff 1933)

Ingrained aesthetic prejudices reduce the applicability and reliability of Birkhoff's aesthetic measure. The highest values are achieved with symmetrical forms with the least number of parts. The square with sides aligned to the vertical and horizontal is the clear winner among polygonal forms, followed by the square rotated by 45 degrees and the rectangle with horizontal and vertical sides. Still, the aesthetic measure is important to our investigation for three basic reasons relating more the way the measure is calculated and less to the measure itself. The first is that it equates beauty with order. While this does not hold for aesthetics in general, it is obviously relevant to prescriptive and proscriptive architectural formal systems where conformity to canons and rules, which are often explicitly and paradigmatically expressed, constitutes the usual measure of formal acceptability. The second reason is the factoring of aesthetic order into discrete, independent formal elements with a limited scope each. This too is related to the (didactic) analyses of architectural formal systems. For example, the elements of symmetry and the horizontal-vertical framework in the order measure of polygonal forms has obvious correspondences with the symmetry and taxis levels of classical architecture (Tzonis and Lefaivre 1986). The third reason is the roles of order and complexity in the aesthetic measure and their affinity with information processing and the role of figural goodness in perception. This affinity was not lost on Birkhoff's epigoni who have linked aesthetic measures to information theory (Bense 1954; Moles 1968).

The applicability of Birkhoff's approach to architectural aesthetics is consequently restricted to (a) the analysis of factors contributing to aesthetic appreciation and preference and (b) the evaluation of an object with respect to each of these factors. The benefits of such analyses and partial evaluations lie mainly in the deeper understanding of the nature and relative significance of each factor, especially when the evaluations are founded on perceptual and cognitive models. One example of the correlation between analyses of architectural objects and of perception is the following evaluation of symmetry in classical floor plans.

Tzonis and Lefaivre have described the classical canon as a system of elements, relationships and coordinating devices which constrain rather than direct design decisions (Tzonis and Lefaivre 1986). This system consists of three major levels: *genera, taxis* and *symmetry*. The term genera (preferred over 'orders') denotes the "well-determined sets" of architectural elements which are formed on the basis of fixed local relations. Taxis is responsible for the overall organization of a classical building and contains two sublevels (schemata): the *grid*, which parametrically divides the building into spatial components, and tripartition. A rectangular grid and a simple tripartition schema produce a 3 x 3 pattern. The deletion, addition, repetition and embedding of parts in this generic pattern transforms it into the layout of a classical building, including Wittkower's 5 x 3 Palladian grid (Wittkower 1952). Symmetry is the collection of relationships that constrain the positioning of a particular genus inside the divisions determined through taxis with respect to each other and to the overall structure of taxis.

A similar approach to symmetry is encountered in the work of Stephen Palmer who has considered basic organisational phenomena in perception, such as figural goodness, perceptual grouping and reference frame effects, with respect to local invariance over the group of Euclidean similarity transformations (Palmer 1983; Palmer 1985). He claims that the perceptual system analyses the incoming stimulus information with respect to five fundamentally different but interrelated properties, shape, position, orientation, size and sense. Four of these five basic perceptual properties are intimately linked to simple transformation groups:

- translations along a line (position);

- rotations about a point (orientation);

- dilations (radial expansion and contraction) about a point (size); and

- reflections about a line (sense).

Earlier researchers have linked transformational invariance and figural goodness in evaluations of invariance following eight transformations: central rotations through angles of 0, 90, 180 and 270 degrees and reflections about vertical, horizontal, left diagonal and right diagonal lines through the centre (Garner and Clement 1963; Garner 1974). Palmer extends the notion of invariance under rotation and reflection to other possible transformations, notably dilations and translations, and to reference frame effects (differences in goodness due to different types of symmetries) by treating the transformations as a mathematical group (a set of elements plus a composition operation for putting them together such as that a few properties hold). For each figure there is a set of transformations that leave the figure completely invariant. This set is a subgroup of the initial group of possible transformations and is called the *symmetry subgroup*, as it relates to the mathematical notion of symmetry (Weyl 1952).

The level of symmetry in the classical canon bears close resemblance to the notion of symmetry in the invariance model of figural goodness. The correspondences have been investigated in classical floor plans at three different levels:

- internal symmetry of each space;

- symmetry within each group of spaces; and

- global symmetry of groups in the overall floor plan.

In all cases the taxis schema which determines the global tripartition of the floor plan has been taken as the reference frame for the transformations (Koutamanis 1990). The choice of a reference frame relates to what Palmer terms *the reference frame hypothesis*, i.e. that the effects of the transformations are neutralized by an intrinsic frame of reference which ensures constancy of shape and configuration (as opposed to the invariant features hypothesis which states that shape is represented by detecting those properties of objects that do not change over the relevant set of transformations) (Palmer 1983).

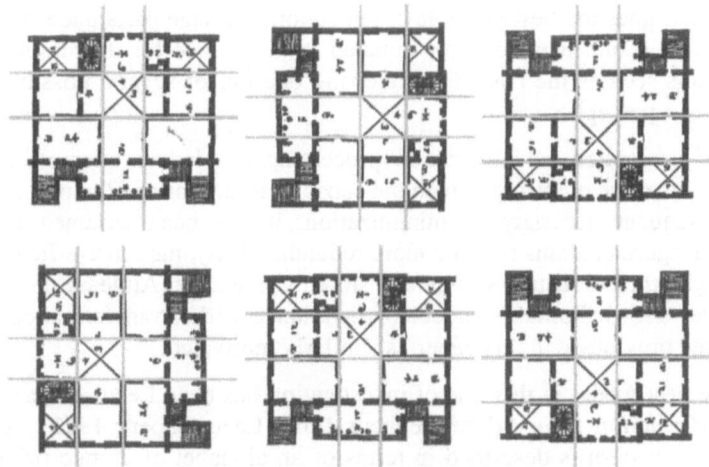

Figure 3. Invariance of a classical floor plan under transformation
(Koutamanis 1990)

The results of the analysis provide a quantitative measure of symmetry for both individual spaces and groups of spaces. Even more significant is that symmetry forms a strong preference criterion for choosing between alternative descriptions of whole floor plans, i.e. different configurations of space groups. Especially in compact floor plans where grouping relationships can be interpreted in different ways symmetry can be the primary criterion for preferring or even accepting a description, as taxis and its containment of space groups is generally unambiguous (Koutamanis 1990).

3. Coding and information

Probably the greatest shortcoming of Birkhoff's approach lies in that it fails to take account of perception, that is, of the processes by which an object elicits a pleasurable reaction. By linking aesthetics to perception we depart from the objectiveness of Birkhoff's measure and adopt an inter-subjective model of aesthetic appreciation which stresses the cognitive similarities that exist between different persons and cultures (Scha and Bod 1993). Inter-subjectivity also allows us to correlate different aesthetic approaches, in our case different architectural formal systems. This is largely due to the reason for such cognitive similarities, the organization of perceptual information.

Gestalt psychologists have formulated a number of principles (or 'laws'), such as proximity, equality, closure and continuation, which underlie the derivation of a description from a percept by determining the grouping of its parts (Köhler 1929; Koffka 1935; Wertheimer 1938). Probably the most important and certainly the most mysterious of the Gestalt principles of perceptual organization is *Prägnanz* or *figural goodness* which refers to subjective feelings of simplicity, regularity, stability, balance, order, harmony and homogeneity that arise when a figure is perceived. Figural goodness ultimately determines the best possible organization of image parts under the prevailing conditions. As a result, it is normally equated to preference for the simplest structure. The principle is seen as the basis for preferring one our of several possible alternative descriptions of a percept.

The view of perception as information processing has led to attempts to formulate figural goodness more precisely. Given the capacity limitations of the perceptual system and the consequent necessity of minimization, it has been assumed that the less information a figure contains (i.e. the more redundant it is), the more efficiently it could be processed by the perceptual system and stored in memory (Attneave 1954; Hochberg and McAlister 1954). Palmer's model of invariance under transformation, which has been discussed previously in this paper, is similarly motivated.

Arguably the best model in this line of investigation has been Leeuwenberg's coding or *structural information theory* (Leeuwenberg 1967; Leeuwenberg 1971). According to Leeuwenberg a pattern is described in terms of an alphabet of atomic primitive types, such as straight line segments and angles at which the segments meet. This description (the *primitive code*) carries an amount of structural information (I) that is equal to the number of elements (i.e., instances of the primitives) it contains. The structural information of the primitive code is subsequently minimized by repeatedly and progressively transforming the primitive code on the basis of a limited number of coding operations:

- iteration, by which the patterns

$a\,a\,a\,a\,a\,a\,b\,b\,b\,b\,b\,b$ (*I* = 12)

$a\,b\,a\,b\,a\,b\,a\,b\,a\,b\,a\,b$ (*I* = 12)

become respectively

6 * [(*a*) (*b*)] (*I* = 3)

6 * [(*a b*)] (*I* = 3)

- reversal, denoted by r [...]:

  $a\,b\,c = r\,[c\,b\,a]$ (*I*= 3)

  Reversal allows the description of symmetrical patterns (Σ):

  $a\,b\,c\,c\,b\,a = a\,b\,c\,r\,[a\,b\,c] = \Sigma\,[a\,b\,c]$ (*I* = 4)

  $a\,b\,c\,b\,a = a\,b\,c\,r\,[\,a\,b] = \Sigma\,[a\,b\,(c)]$ (*I* = 4)

- distribution:

  $a\,b\,a\,c = <(a)>\;<(b)\,(c)>$ (*I* = 3)

- continuation ($\subset ... \supset$), which halts if another element or an already encoded element is encountered:

  $a\,a\,a\,a\,a\,a\,a ... a = \subset a \supset$ (*I* = 1)

The coding process returns the *end code,* a code whose structural information cannot be further reduced. The structural information (*I*) of a pattern is that of its end code.

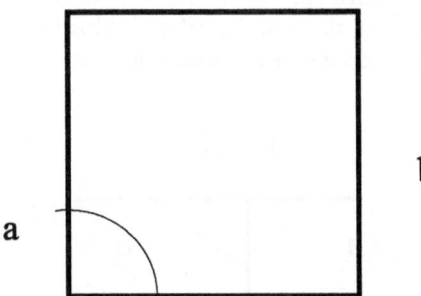

Figure 4a. Coding of square: $a\,b\,a\,b\,a\,b\,a\,b = \subset a\,b \supset$ (*I* = 2)

The structural information of a pattern is a powerful measure of its figural goodness. By equating a figure's goodness with the parametric complexity of the code required to generate it we can both derive the different descriptions an image affords and choose the one(s) that contain the least information. Especially in situations where two or more descriptions are equally acceptable to the human perceiver, as in the Necker cube illusion, measurement of structural information clearly demonstrates that the preferred descriptions are normally equally compact. This suggest that structural information theory is particularly suited to untangling complex, overlapping or intertwined patterns, i.e. situations which are amenable to evaluations of figural goodness by e.g. invariance under geometric transformations only following an initial analysis which segments and disambiguates the image.

With respect to architectural aesthetics we should note the similarities between the measurement of figural goodness on the basis of structural information (including the way an image is coded) and aesthetic analyses by means of shape grammars (Gips 1975; Stiny 1975; Stiny and Gips 1978). For example, the evaluation of floor plans generated by the Palladian grammar (Stiny and Gips 1978) is based on:

- local criteria derived from the floor plans of Andrea Palladio's actual villas; and

- a global aesthetic measure.

The local measures are essentially similar to the generative shape rules of the Palladian grammar and concern either individual spaces or groups of cells in the underlying 3 x 3, 5 x 3 or 5 x 4 grid. The global aesthetic measure is defined as the ratio of the length of the description of a plan (i.e. the sum of the number of cells required for the multicell space types in the plan and of the number of instances of each of these types in the plan) to the length of the information required for its generation (i.e. the number of shape rules required for the generation of the plan).

The local criteria form expressions of Palladian constraints and as such are quite effective in the determining the acceptability of an artificial floor plan on the basis of its similarity to actual designs by Palladio. The global measure, on the other hand, is a measure of the "specificational simplicity" of a plan and therefore a test of the operations and control structures that comprise the Palladian grammar. In other words, the global measure is the equivalent of figural goodness within the framework of a particular shape grammar, even through the relationships between the components of a shape grammar, design decisions and perception can be at times tentative.

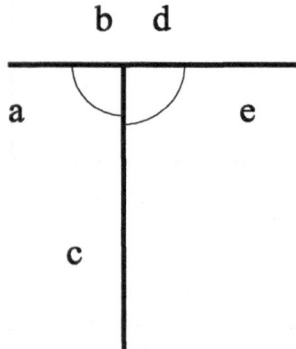

Figure 4b. Coding of branching with bifurcation signs: *a {b c} d e*
(meaning: after *c* return to end of *a* and proceed to following *d*)

Of particular interest to our investigation is the ability of coding to recognize and evaluate alternative groupings of image parts on the basis of basic, purely formal quantitative relationships. These relationships and the resulting group forms are implied in Gestalt theory, as well as in computational architectural representations such as shape grammar and rectangular arrangements. The explicitness of groups in structural information theory and the causal relationship between the configuration of groups and figural goodness satisfy fully one of the basic requirements of our investigation, the correlation of a structured representation with aesthetic evaluation and preference.

4. Architectural primitives

The main problem of theories of perceptual organization, from Gestalt to structural information theory, lies in that they are developed and discussed within abstract domains of simple, mostly two dimensional patterns and elementary primitives such as dots and line segments. Such basic geometric forms should be treated with caution in evaluations of design aspects, as they occupy the lowly level of implementation mechanisms in representations (Marr 1982). The confusion between implementation mechanisms and symbols has been a major obstacle in the evolution of computational design. Moreover, given the highly conventional character of existing architectural representations it is doubtful that the use of implementation mechanisms such as line segments as the basic primitives for architectural aesthetic analyses would reveal much beyond the conventions themselves.

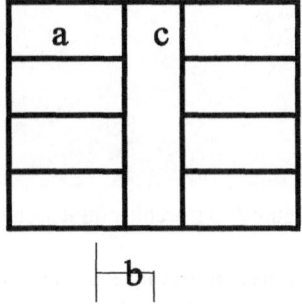

Figure 5. Coding of a floor plan: $a\,a\,a\,a\,b\,c\,b\,a\,a\,a\,a =$
$4 * [(a)]\,b\,c\,b\,4 * [(a)] = \Sigma\,[\,4 * [(a)]\,b\,(c)]$ $(I = 5)$
The end code is a symmetric tripartite configuration of
two space groups flanking a central space.

An extension to the three dimensional forms of the built environment and to the complex two dimensional representations employed in architecture involves the main problem of determining the primitives of these domains. It also invariably increases the complexity of descriptions, as these primitives relate to each other in multiple ways. An initial investigation of the applicability of structural information theory to floor plans has been based on the choice of spaces as the primitives and of formal grouping derived from the chain code as the relationships between primitives (Koutamanis 1990). Even though this investigation has been restricted to establishing preference for one of several previously recognized alternative descriptions of a floor plan in terms of space groups, it made evident that coding efficiency and economy are closely related to intuitive interpretations of architectural figural goodness, also with respect to formal aesthetic devices such as classical symmetry and tripartition.

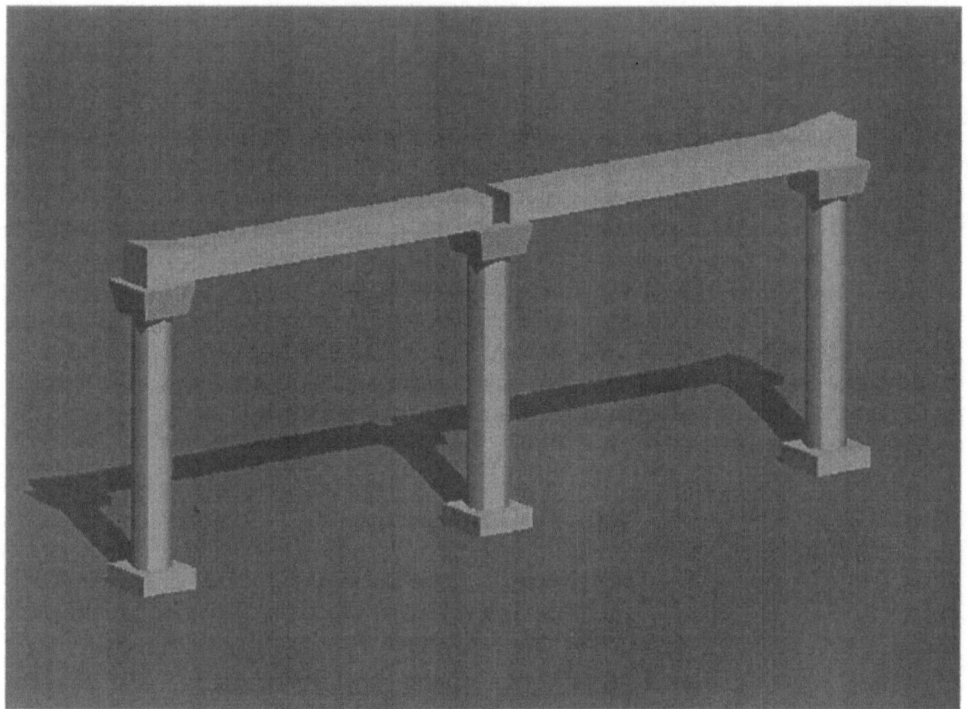

Figure 6. An architectural scene

Many of the problems we encounter in attempts to discover or define the primitives of architectural design are due to a confusion between the real built environment, its architectural representations and the conventions underlying these representations. For this reason we have adopted a sharp distinction between the analysis and manipulation of representations and the perception of and interaction with the built environment. The former rely firmly on architectural conventions and should be accordingly considered from the viewpoint of architectural knowledge. The adoption of building elements and spaces as the primitives of such representations offers pragmatic advantages which should not be neglected. The ability to integrate directly explicit architectural knowledge and the possibility of equally direct correspondences between specifications, regulations or other requirements and the representation of a design form the basis of most design analyses and a responsive background to taking design decisions.

On the other hand, the extension of conventional architectural representations to the perception of built form and space adds little beyond a specialized memory element to general human interaction with the built environment. A preferable starting point is general computational models of perception and recognition which could be enriched with the specialized modes of architecture. These provide a better understanding of perceptual and cognitive devices that also underlie architectural design and analysis. In addition to their direct applicability to the analysis and recognition of realistic architectural scenes they could also ultimately lead to improvements in existing architectural representations.

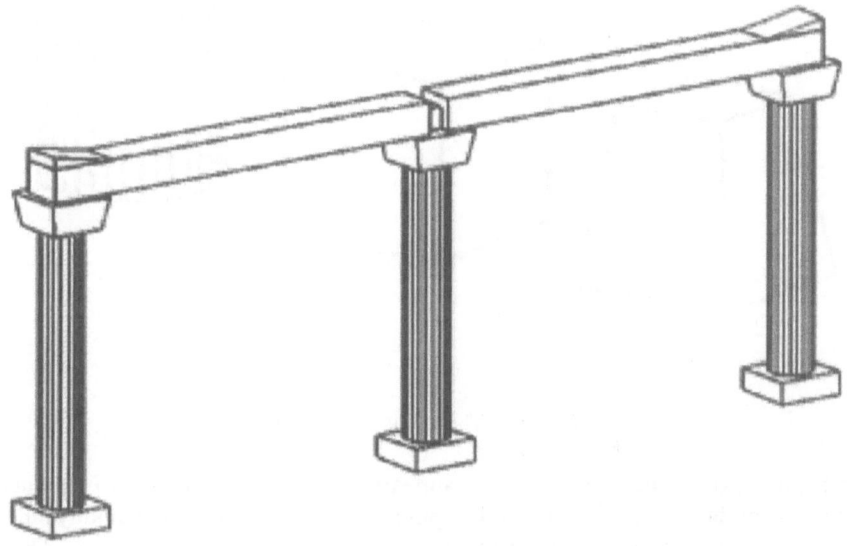

Figure 7. A decomposition of figure 6 into geons

Once low level processing is completed, the first stage in the recognition of a scene is invariably a decomposition of its elements into simple parts, such as the head, the body, the legs and the tail of an a animal. The manner of the decomposition into parts does not depend on completeness and familiarity. An unfamiliar, a partly obscured animal or even a nonsensical shape are decomposed in a more or less the same way by all observers (Biederman 1987). The detection of where parts begin and end is based on the *transversality principle* which states that whenever two shapes are combined their join is almost always marked by matched concavities (Hoffman and Richards 1985). Consequently segmentation of a form into parts usually occurs at regions of matched concavities, i.e. discontinuities at minima of negative curvature. The results of the segmentation are normally convex or singly concave forms.

At first sight one might expect that there is an unlimited number of part types. However, with his *recognition-by-components* theory Biederman has proposed that these forms constitute a small basic repertory of general applicability, characterized by invariance to viewpoint and high resistance to noise. He calls the forms *geons* and suggests that they are only 24 in number (Biederman 1987; Biederman 1995). Geons can be represented by generalized cones, i.e. volumes swept out by a variable cross section moving along an axis (Binford 1971; Brooks 1981). A scene is described by structured explicit representations comprising geons, their attributes and relations derived from only five edge properties.

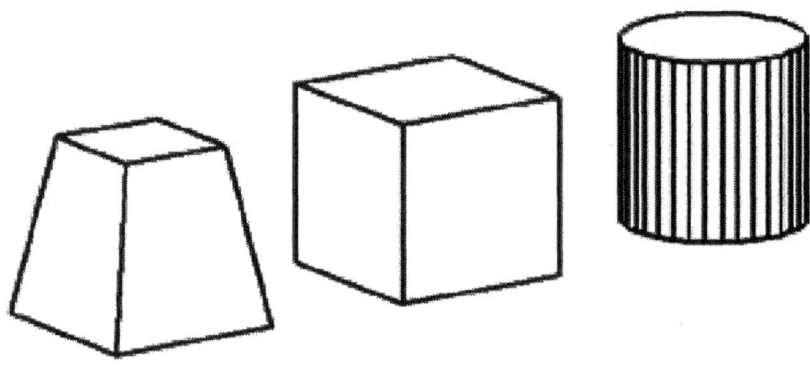

Figure 8. The geons in figure 7

The recognition-by-components theory appears to be as applicable to architectural scenes as to any other scene or object. Decomposition into geons is essentially similar to conventional decomposition into solid building elements and components. The main difference lies in the sensitivity of recognition-by-components to changes in the geometry within what is architecturally a single element. In most cases, however, an element that is decomposed into two or more geons is either a composite element, such as a wall with half columns or pilasters, or a geometrically complex object, such as a T- or L-shaped wall.

A combination of structural information theory and recognition-by-components resolves the problems of both theories with respect to our evaluation of figural goodness. The addition of a compact set of real-world primitives liberates structural information theory from the abstraction of elementary line drawings and extends its applicability to realistic scenes. In Biederman's model coding according to the structural information theory means that there can be grouping of a higher order than local binary relationships. This allows for the development of multilevel representations (hierarchical modular structures (Marr 1982)) which are less complex, better structured and ultimately more meaningful than atomistic relational representations (Koutamanis 1996; Koutamanis 1997). Moreover, the combination of the two theories makes it possible to establish general preference criteria for alternative descriptions on the basis of code compactness which in turn relies on formal grouping principles.

The application of this combination to architectural scenes concentrates in first instance on the definition of primitives and relationships. In that respect, the only deviation from the original theories concerns the relationship that is ignored in coding. In structural information theory this is horizontal alignment. In our investigation we have opted for vertical alignment, in compliance with the general architectural bias for the vertical as canonical orientation. We presume that this bias refers to both a general reference frame reflecting the significance of the vertical in the real world (e.g. gravity) and a specifically architectural reference frame which relates to the interpretation of general orientation preferences in architecture.

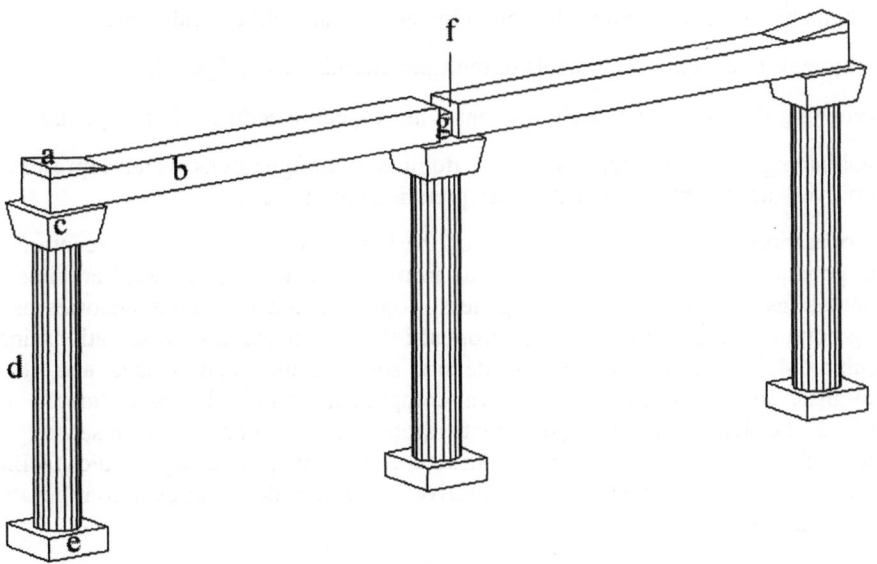

Figure 9. Coding of figure 7

On the basis of the above, the scene of figures 6, 7 and 9 can be coded as follows:

$$a\,b\,\{c\,d\,e\}\,f\,g\,\{c\,d\,e\}\,f\,g\,b\,\{c\,d\,e\}\,a \qquad\qquad (I = 17)$$

$$<(\{c\,d\,e\})><(a\,b)\,(f\,g)\,(f\,g\,b)\,(a)> \qquad\qquad (I = 11)$$

The use of distribution in the second version of the code makes explicit the grouping of the elements comprising the column, as well as the repetition of the group in the scene. This reflects the translational symmetry of the scene (colonnade). The bilateral symmetry that characterizes the total scene is largely lost because of the integrity of the elements and groups in the scene. Bilateral symmetry would be discovered in the code if line segments were used as primitives. This would have meant encoding of the outline of the elements rather than of the elements themselves and would permit splitting of a column into two symmetrical halves with respect to the vertical axis. However, the advantage of discovering and describing explicitly this accidental bilateral symmetry in a repetitive configuration such as a colonnade does not counterbalance the corresponding multiplication of structural information in the primitive code and the initial detachment from the reality of the perceived integral components / geons.

5. The evaluation of architectural formal goodness

Recognition-by-components and structural information theory provide the basis for:

- recognizing and representing the solid elements of an architectural scene;
- grouping the recognized elements in multiple alternative configurations;
- evaluating the alternative configurations with respect to coding efficiency; and
- establishing preference for one or two dominant configurations which represent the intuitively acceptable or plausible interpretations of the scene.

These operations link the representation of the built environment with perception and figural goodness. The necessary deviations from established conventional architectural representations reflect the choice of general cognitive and perceptual theories as the starting point of the investigation. It is proposed that architectural representations and in particular (a) the use of outlines to denote solid entities and spaces and (b) the deterministic decomposition into known components should be reconsidered with respect to the recognition-by-components theory and related vision research. The addition of a memory component to structural information theory would facilitate transition from the basic level of the primitive and end codes to known configurations denoting familiar objects.

The representation of spaces remains a problem that deserves particular attention and further research. The use of outlines, as in figure 5, is the obvious starting point, as it conforms to the way we read floor plans and other conventional representations and to existing computational representations such as rectangular arrangements and shape grammars. This would allow for an exploration of structural information theory and recognition-by-components in the application areas of these representations. From a cognitive point of view, however, the outline of a space in two or three dimensions might not be a relevant or meaningful representation. It has been proposed that surfaces could form a representation level that not only links higher with lower vision (Nakayama, He et al. 1995) but also agrees with the Gibsonian perception of space in terms of surfaces which fill space (Gibson 1966). This view is entirely different from the mainstream Euclidean co-ordinate organization of perceived space whereby the two dimensional retinal image is enriched with depth information derived primarily from binocular disparity. Perception of space in terms of surfaces stresses the biological and ecological relevance of these surfaces as containers of different actions and as subjects of their planning. One example of this relevance is locomotion for the ground surface and related generally horizontal surfaces.

Another issue that requires further consideration concerns the essentially bottom-up character of both recognition-by-components and structural information theory. The addition of a memory component to the system, i.e. a database of geon configurations corresponding to known, familiar entities, would facilitate processing of information at the basic levels and permit rapid transition to the higher levels of the representation. As these configurations would represent compact structures with respect to structural information, we assume that exposure to and recognition of similar or equivalent scenes leads to the transformation of earlier experience into memories which influence our understanding and aesthetic evaluation (Scha and Bod 1993).

The validity of the combination of structural information theory and recognition-by-components for aesthetic analysis is beyond the scope of this paper and subject to further research and empirical analysis. The correspondences between intuitive aesthetic factors, e.g. as formulated in Birkhoff's aesthetic measure, and the coding operations of structural information theory suggest that this approach to perceptual organization is capable of supporting analysis in a coherent manner lacking in studies which employ perceptual organization as an explanation of isolated aesthetic principles.

The representations produced by the combination appear to hold for perception in general and for figural goodness. The transition to aesthetic appreciation is based on the significance of redundancy, as well as to the explicitness of factors which determine our judgements about perceived objects. In architecture the aesthetic contribution of redundancy has been interpreted both positively and negatively. Following the reference frame hypothesis, we may assume that redundancy refers to preferences organized in external (cultural) reference frames which attach different values to complexity and minimal coding. It is conceivable that redundancy is a positive aesthetic factor in one aesthetic system and negative in another. While this is normally only implicit in the system, a closer investigation of the corresponding reference frames reveals it through the significance of coding operations such as repetition and symmetry. The same reference frames reveal other principles which relate to perceptual organization, such as the formalist bias towards "functional expression", i.e. the association of certain elements with specific use types and the signification of concrete affordances in the form of the elements, including the coincidence of different aspects such as formal articulation and structural organization (Holgate 1992).

Finally, it should be noted that the information content of a description relies heavily on the primitives of the representation and hence on the abstraction level that has been chosen for it. In the framework of the multilevel representations of computer vision (Marr 1982; Rosenfeld 1984; Rosenfeld 1990) there is scope for considering the figural goodness and the information content of an object at different abstraction levels. The same applies to architecture, where different abstraction levels not only eliminate details that may be unwanted in e.g. the comparison of a classical and an modernist building while retaining the spatial organization of the designs, but also support correlation of local relationships into coordinating devices which accept interchangeable elements (Koutamanis 1996; Koutamanis 1997).

References

Arnheim, R. (1974). *Art and visual perception. The psychology of the creative eye. The new version.* Berkeley, California, University of California Press.

Arnheim, R. (1977). *The dynamics of architectural form.* Berkeley, California, University of California Press.

Arnheim, R. (1988). *The power of the center. A study of composition in the visual arts. The new version.* Berkeley, California, University of California Press.

Attneave, F. (1954). "Some informational aspects of visual perception." *Psychological Review* **61**: 183-193.

Bense, M. (1954). *Aesthetica.* Stuttgart, Deutsche Verlags-Anstalt.

Berlyne, D. E. (1960). *Conflict, arousal and curiosity.* New York, McGraw-Hill.

Berlyne, D. E. (1971). *Aesthetics and psychobiology.* New York, Appleton-Century-Crofts.

Biederman, I. (1987). "Recognition-by-components: A theory of human image understanding." *Psychological Review* **94**(2): 115–147.

Biederman, I. (1995). Visual object recognition. in *Visual cognition. An invitation to cognitive science. 2nd ed.* S.M. Kosslyn and D.N. Osherson (eds). Cambridge, Massachusetts.

Binford, T. O. (1971). *Visual perception by computer.* IEEE Conference on Systems and Controls., Miami, December 1971.

Birkhoff, G. D. (1933). *Aesthetic measure.* Cambridge, Massachusetts, Harvard University Press.

Brooks, R. A. (1981). "Symbolic reasoning among 3-D models and 2-D images." *Artificial Intelligence* **17**: 205–244.

Garner, W. R. (1974). *The processing of information and structure.* Potomac, Maryland, Erlbaum.

Garner, W. R. and D. E. Clement (1963). "Goodness of pattern and pattern uncertainty." *Journal of Verbal Learning and Verbal Behavior* **2**: 446-452.

Gibson, J. J. (1966). *The senses considered as perceptual systems.* London, George Allen & Unwin.

Gips, J. (1975). *Shape grammars and their uses.* Basel, Birkhäuser.

Hochberg, J. E. and E. McAlister (1954). "A quantitative approach to figural 'goodness'." *Journal of Experimental Psychology* **46**: 361-364.

Hoffman, D. D. and W. Richards (1985). "Parts of recognition." *Cognition* **18**: 65-96.

Holgate, A. (1992). *Aesthetics of built form.* Oxford, Oxford University Press.

Koffka, K. (1935). *Principles of Gestalt psychology.* New York, Harcourt Brace.

Köhler, W. (1929). *Gestalt psychology.* New York, Liveright.

Koutamanis, A. (1990). Development of a computerized handbook of architectural plans., Delft University of Technology.

Koutamanis, A. (1995). Multilevel analysis of fire escape routes in a virtual environment. in *The global design studio*. M. Tan and R. Teh (eds). Singapore, Centre for Advanced Studies in Architecture, National University of Singapore.

Koutamanis, A. (1996). Elements and coordinating devices in architecture: An initial formulation. in *3rd Design and Decision Support Systems in Architecture and Urban Planning Conference. Part One: Architecture Proceedings*. Eindhoven.

Koutamanis, A. (1997). Multilevel representation of architectural designs. in *Design and the net*. R. Coyne, M. Ramscar, J. Lee and K. Zreik (eds). Paris, Europia Productions.

Leeuwenberg, E. L. J. (1967). *Structural information of visual patterns. An efficient coding system in perception*. (Doctoral dissertation, Catholic University of Nijmegen) The Hague, Mouton.

Leeuwenberg, E. L. J. (1971). "A perceptual coding language for visual and auditory patterns." *American Journal of Psychology* **84**: 307-350.

Marr, D. (1982). *Computer vision*. San Fransisco, W.H. Freeman.

Mitossi, V. and A. Koutamanis (1996). Parametric design of stairs. in *3rd Design and Decision Support Systems in Architecture and Urban Planning Conference. Part One: Architecture Proceedings*. Eindhoven.

Moles, A. (1968). *Information theory and esthetic perception*. Urbana, Illinois, University of Illinois Press.

Nakayama, K., Z. J. He, et al. (1995). Visual surface representation: a critical link between lower-level and higher-level vision. in *Visual cognition. An invitation to cognitive science. 2nd ed*. S.M. Kosslyn and D.N. Osherson (eds). Cambridge, Massachusetts.

Palmer, S. E. (1983). The psychology of perceptual organization: a transformational approach. in *Human and machine vision*. J. Beck, B. Hope and A. Rosenfeld (eds). New York, Academic Press.

Palmer, S. E. (1985). "The role of symmetry in shape perception." *Acta Psychologica* **59**: 67-90.

Palmer, S. E. (1989). Reference frames in the perception of shape and orientation. in *Object perception: Structure and process*. B. Shepp and S. Ballesteros (eds). Hillsdale, New Jersey, Erlbaum.

Prak, N. L. (1968). *The language of architecture. A contribution to architectural theory*. The Hague, Mouton.

Prak, N. L. (1979). *De visuele waarnemig van de gebouwde omgeving*. Delft, Delftse Universitaire Pers.

Rock, I. (1973). *Orientation and form*. New York, Academic Press.

Rock, I. (1990). The concept of reference frame in psychology. in *The legacy of Solomon Asch: Essays in cognition and social psychology*. I. Rock (ed). Hillsdale, New Jersey.

Rosenfeld, A., Ed. (1984). *Multiresolution image processing and analysis*. Berlin, Springer.

Rosenfeld, A. (1990). Pyramid algorithms for efficient vision. in *Vision: coding and efficiency*. C. Blakemore (ed). Cambridge, Cambridge University Press.

Scha, R. and R. Bod (1993). "Computationele esthetica." *Informatie en Informatiebeleid* **11**: 54-63.

Steadman, J. P. (1976). Graph-theoretic representation of architectural arrangement. in *The architecture of form*. L.J. March (ed). Cambridge, Cambridge University Press.

Steadman, J. P. (1983). *Architectural morphology*. London, Pion.

Stiny, G. (1975). *Pictorial and formal aspects of shape and shape grammars*. Basel, Birkhäuser.

Stiny, G. and J. Gips (1978). *Algorithmic aesthetics. Computer models for criticism and design in the arts*. Berkeley, California, University of California Press.

Stiny, G. and J. Gips (1978). "An evaluation of Palladian plans." *Environment and Planning B* **5**: 199-206.

Tzonis, A. and L. Lefaivre (1986). *Classical architecture: The poetics of order.* Cambridge, Massachusetts, MIT Press.

Weber, R. (1995). *On the aesthetics of architecture.* Aldershot, Avebury.

Wertheimer, M. (1938). Laws of organization in perceptual forms. in *A source book of Gestalt psychology.* W.D. Ellis (ed). London, Routledge & Kegan Paul.

Weyl, H. (1952). *Symmetry.* Princeton, New Jersey, Princeton University Press.

Wittkower, R. (1952). *Architectural principles in the age of humanism.* London, Alec Tiranti.

HUMAN-COMPUTER INTERACTION AND NEURAL NETWORKS IN ARCHITECTURAL DESIGN

A Tool for Design Exploration

NEANDER F. SILVA
Department of Architectural Design
University of Brasília
Brasília, Brazil

ALAN H. BRIDGES
Department of Architecture and Building Science
University of Strathclyde
Glasgow, United Kingdom

Abstract

Design research has demonstrated that neural networks are able to support creativity. However, there are two main problems with using neural networks in design. One is how you interact with such systems. The second relates to the integration between neural network techniques and other approaches. This paper will describe an integrated model in which those problems are addressed. The resulting system provides an interface in which the neural network output is translated into textual and graphic representations that can play a meaningful role in the design process.

1.   Introduction:

It has been suggested that neural networks yield support for emergence and creativity in design, as demonstrated by Coyne and Newton (1990), Coyne (1991), Coyne and Yokozawa (1992), and Coyne et al. (1993). Those authors argued that support for creativity is achieved by extracting information from implicit knowledge that can be translated as new explicit solutions (Coyne et al., 1993).

However, there are two major problems with using neural networks in design. One is how you interact with such systems. The bare interfaces of neural networks are very poor and passive from the designer's point of view, rendering those systems virtually unusable in design. The second pertains to the integration between neural network

R. Junge (ed.), CAAD Futures 1997, 267-284.
© 1997 *Kluwer Academic Publishers.*

techniques and other approaches or knowledge representations which are essential to the design process such as language and graphic descriptions.

2.   Stand-alone neural networks in design:

An auto-associative neural network design was adopted by Coyne and Yokozawa (1992). After being trained through the exposure to a certain number of examples, represented by different sets of combinations of features, this network becomes 'aware' of what features are mutually excitatory or inhibitory. Coyne and Yokozawa (1992) suggested that if a designer used such trained network by selecting and manipulating features (neurons) on its input layer, the outcome would be not only combinations of features matching examples from the training set, but eventually the emergence of new combinations of features.

For instance, a neural network could be trained with the following set of instances shown in figure 1, which is a an example of a door entrance classification domain composed of nine styles. This domain is obviously simplistic containing only 27 descriptors and night solutions. However, we have chosen to use it because it is just big enough to illustrate the working of a stand-alone neural network, the alternative expert system described later, and the proposed integrated model. Another reason is that it calls for no specialist knowledge on the part of the reader.

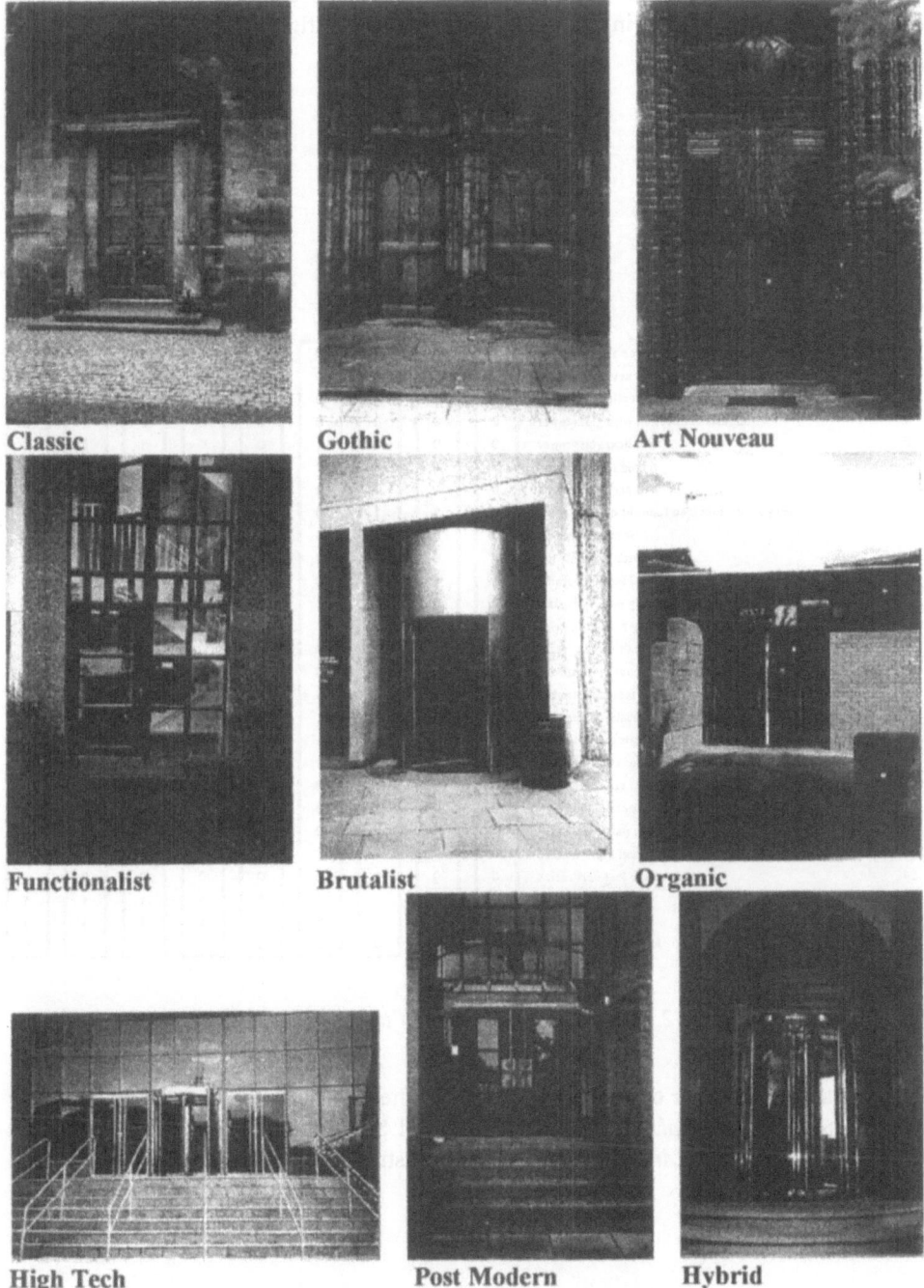

Figure 1. Door entrance domain: illustrations.

The examples could be binary classified according to the presence or absence of certain features, such as those used in the classification shown in figure 2, bellow:

		1 Classic	2 Gothic	3 Art Nouveau	4 Functionalist	5 Brutalist	6 Organic	7 High Tech	8 Post Modern	9 Hybrid Classic + High tech
1	top flat moulding	1	0	0	0	0	0	0	0	1
2	top curved moulding	0	1	0	0	0	0	0	0	1
3	lateral vertical moulding	0	1	1	0	0	0	0	0	1
4	angular connection with glass tower	0	0	0	0	0	0	0	1	0
5	vertical glass tower	0	0	0	1	0	0	1	1	0
6	triangular pediment	1	0	0	0	0	0	0	0	1
7	pointed arch tympanum	0	1	0	0	0	0	0	0	0
8	tracery or steelwork on fanlight or tympanum	0	1	1	0	0	0	0	0	0
9	squared fanlight	0	0	0	0	1	1	0	0	1
10	fanlight with undulate top	0	0	1	0	0	0	0	0	0
11	lateral cylindrical section column	1	1	0	0	0	0	0	0	0
12	Surrounding material: glass	0	0	1	1	1	1	1	1	1
13	Surrounding material: brick	0	0	1	0	0	1	0	1	0
14	surrounding material: smooth stone	1	1	0	0	0	0	0	0	1
15	surrounding material: concrete exposed	0	0	0	0	1	0	0	0	0
16	surrounding material: timber	0	0	0	0	0	1	0	0	0
17	surrounding material: plasterwork	0	0	1	1	0	0	0	0	0
18	surrounding material: metal	0	0	1	1	0	0	1	1	0
19	opening mechanism: swinging door	1	1	1	1	0	1	1	1	0
20	opening mechanism: revolving door	0	0	0	0	1	0	1	0	1
21	leaf type: plain opaque	0	0	1	0	0	0	0	0	0
22	leaf type: paneled opaque	1	1	0	0	0	0	0	0	0
23	leaf type: framed	0	0	0	1	1	1	1	1	1
24	steelwork as leaf decoration	0	0	1	0	0	0	0	0	0
25	leaf material: glass	0	0	0	1	1	1	1	1	1
26	leaf material: metal	0	0	1	1	1	1	1	1	1
27	leaf material: timber	1	1	0	0	0	1	0	0	0

Figure 2. Binary classification of the set of instances.

If after training, the user chooses to make active the input units *'surrounding material: metal'* and *'leaf material: metal'*, the output will be a binary string which represents the solution number 3, *'Art Nouveau'*. This is illustrated in figure 3, bellow.

#	Feature	Input layer	Output layer
1	top flat moulding	○	○
2	top curved moulding	○	○
3	lateral vertical moulding	○	●
4	angular connection with glass tower	○	○
5	vertical glass tower	○	○
6	triangular pediment	○	○
7	pointed arch tympanum	○	○
8	tracery or steelwork on fanlight or tympanum	○	●
9	squared fanlight	○	○
10	fanlight with undulate top	○	●
11	lateral cylindrical section column	○	○
12	Surrounding material: glass	○	●
13	Surrounding material: brick	○	●
14	Surrounding material: smooth stone	○	○
15	Surrounding material: concrete exposed	○	○
16	Surrounding material: timber	○	○
17	Surrounding material: plasterwork	○	●
18	Surrounding material: metal	●	●
19	opening mechanism: swinging door	○	●
20	opening mechanism: revolving door	○	○
21	leaf type: plain opaque	○	●
22	leaf type: paneled opaque	○	○
23	leaf type: framed	○	○
24	steelwork as leaf decoration	○	●
25	leaf material: glass	○	○
26	leaf material: metal	●	●
27	leaf material: timber	○	○

Figure 3. The 'Art Nouveau' entrance description built by the trained neural network.

However, if the user chooses to make active the input units *'triangular pediment'*, *'opening mechanism: revolving door'*, and *'leaf material: glass'*, the output will be the one shown in figure 4, which is a new combination of features, but still represents a sensible solution from the architectural point of view.

		Input layer	Output layer
1	top flat moulding	○	●
2	top curved moulding	○	●
3	lateral vertical moulding	○	○
4	angular connection with glass tower	○	○
5	vertical glass tower	○	○
6	triangular pediment	●	●
7	pointed arch tympanum	○	○
8	tracery or steelwork on fanlight or tympanum	○	○
9	squared fanlight	○	●
10	fanlight with undulate top	○	○
11	lateral cylindrical section column	○	○
12	Surrounding material: glass	○	●
13	Surrounding material: brick	○	○
14	Surrounding material: smooth stone	○	●
15	Surrounding material: concrete exposed	○	●
16	Surrounding material: timber	○	○
17	Surrounding material: plasterwork	○	○
18	Surrounding material: metal	○	○
19	opening mechanism: swinging door	○	●
20	opening mechanism: revolving door	●	●
21	leaf type: plain opaque	○	○
22	leaf type: paneled opaque	○	○
23	leaf type: framed	○	●
24	steelwork as leaf decoration	○	○
25	leaf material: glass	●	●
26	leaf material: metal	○	●
27	leaf material: timber	○	●

Figure 4. New solution emerged from the trained neural network.

The results shown in figures 3 and 4 are consequences of the weights set by the neural network for the connections between each input unit and all output units during the training process.

The above method of using such networks in a direct input layer manipulation, has as main advantage the freedom of testing and of possibly 'forcing' hybrid solutions by picking up pairs of input neurons otherwise considered unusual in the training set.

However, there are disadvantages: the user may get an output that is actually not a stable state, once the process can be terminated arbitrarily. The user may also have to undertake a trial process in which he or she may get lost, once there is no inherent trace facility. Moreover, the user may end up in a dead lock in which the network cannot compute a stable state because the units made active are highly incompatible.

At last, the method does not provide either a plain English interface or a graphic one. In addition, its bare interface is completely passive, that is, all the initiative relies on the user, who must know what he or she wants and also keep track of all events manually.

3. Mustoe's alternative knowledge-based system:

Alternative knowledge-base systems have been devised (Frey, 1986; Mustoe, 1990, 1993). Mustoe (1990) argues that evidently the function of the network of rules in conventional knowledge-based systems is to place a set of individual productions into a correct relationship with a particular solution. Solutions are classified according to their question set, while questions are classified by reference to the solutions they verify.

Therefore, Mustoe (1990, 1993) proposes that a domain knowledge may be better represented through a Boolean classification structure, resembling the table shown in figure 2.

This kind of representation may suggest that we actually went back to a relational database. However, there are substantial differences. The table in figure 2 just illustrates how Mustoe's system, 'Cortex', maps questions into solutions. The solutions are classified independently from each other in a set of binary relationships with the questions that verify them. These relationships are actually encoded in the system as true bit-strings and not as rules. The control system will not operate on them through a query language, but through a direct bit-string manipulation.

Cortex (Mustoe, 1990, 1993) uses a control algorithm that functions by rejecting falsified solutions. This system would verify each of the of the features shown in the table of figure 2 through a series of questions subsequently presented to the user. It will present to the user the currently most frequent question among those still relevant. It will finish either with one or more not falsified solutions, or with a confession that it cannot find a solution. It operates in a process of zeroing-in upon a successively shortened list of still-possible solutions.

The advantages of this kind of knowledge representation are two: firstly, the system will run faster than a rule-based one due to the direct bit-string manipulation.

Secondly, the addition of new solutions will not require a time consuming and expensive process of re-writing 'if... then' statements, as it happens with conventional rule-based systems. The addition of a new solution will imply only the incorporation of another bit-string to the existing set.

A drawback of this kind of knowledge representation is that, in complex domains, a huge number of conditions, or questions, can be generated by this approach. However, this is largely compensated by the control system, as described earlier.

If no inconsistency is necessarily introduced on the knowledge-base already in the system with the addition of new solutions, new horizons are open for building systems in which the knowledge-bases can be consistently expanded. However, as the options must be manually set before hand, Cortex (Mustoe, 1990, 1993) has no inherent knowledge acquisition mechanism as any other knowledge-base system. Nevertheless, its integration with a particular design of connectionist model may provide the answers for this problem.

4. Cortex and neural network's integration: automating and simulating Coyne's model through a 'semi-recurrent' network.

In spite of having quite different control algorithms, there are striking similarities between Cortex and binary neural networks concerning knowledge representation. This similarities call for a potential integration. One way of integrating Cortex with Coyne's model could be the use of the first as the front end of the second. Cortex would act as an intelligent and active interface by filtering user's input, by focusing the user's interest on the most promising solutions and by formatting data for the neural network.

There are two main obstacles to the direct use of the user's answers to Cortex in the input layer of neural networks. The first is related to some differences in knowledge representation. Cortex accepts three possible states for each feature: definitely present, if the answer is 'yes', definitely absent, if the answer is 'no', and neutral, if the answer is 'don't know'. However, a binary neural network input unit has obviously only two states: active and inactive.

The second obstacle is the way in which the neural network is trained and the amount of information provided by the set of answers to Cortex. Very often few of the questions in the question's set of a knowledge-base will be used by Cortex in order to find a solution. Therefore, the user's answers provide limited information, which is rarely enough to produce a stable solution in a neural network.

Both obstacles are overcome if we use a plain feed-forward multi-layered network at training time and a semi-recurrent network, with limited feed-back, at running time. Why do we not use one of this architectures at both training and running time? Firstly, a plain feed-forward multi-layered network is a sufficiently effective architecture to train the kind of knowledge representation that has been described in this paper.

Secondly, the recurrent element that is added at running time, and will be described shortly, is used to provide external control to the process of automating the search for a stable state that would otherwise be undertaken manually. It is not a standard neural network component based in fuzzy logic, but an element that emulates the role played by the user such as seen in Coyne and Yokozawa's (1992) experiment.

If we considered a network design in which each input unit had two attributes, the first specifying if it is active or inactive, and the second specifying if it is 'clamped' or 'unclamped', the user's answers to Cortex could then be used as neural network input. In other words, each input unit would have, besides two possible states (active or inactive), an extra attribute that could have a constant value, if the unit's status is clamped, or a variable value, if the unit's status is unclamped.

A recurrent element between each output unit and its correspondent input unit is then added at running time in which each output is checked against the clamping criteria described above. Figure 5 illustrates this feed-back process.

276

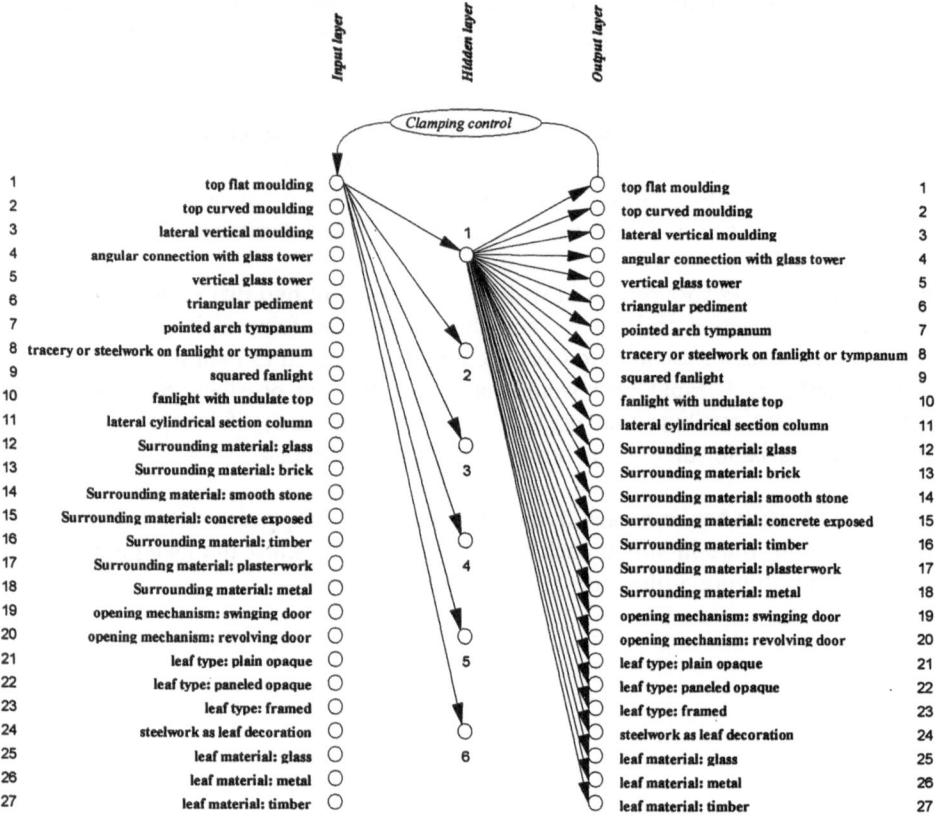

Figure 5. A semi-recurrent network.

In other words, all the output values of 'unclamped' units are accepted and successively re-entered as input. Also, all the output values of 'clamped' units are checked and reset to their initial state (if it has been changed), and successively re-entered as input.

The network could then search for stable states in which reliable solutions could be produced. The process would be terminated in two situations: firstly, when no further changes or mismatches are observed between the input layer and the output layer. Obviously, if the output layer mirrors the input one, there is no reason for re-entering the output values as input any more. Secondly, when the network reaches an infinite loop, that is, the last output equals an output of a previous of stage in the same running process. When these situations happen the process is terminated by the model's algorithm.

This process could lead to three outcomes. Firstly, it could lead to a solution that matches the solution presented by Cortex. Secondly, it could lead to a solution that does not match the solution presented by Cortex. Since the solution found by Cortex is the only one in the knowledge-base able to satisfy the user's input, the network's solution will not match any of the solutions in the original set of examples either. It is thus a new solution. The binary string of such new solutions' can be then easily converted into plain English textual descriptions composed of a list of the features present in the new combinations. From this point onwards the descriptions of precedents can be coupled with graphical illustrations, and the textual descriptions of possible new solutions can be coupled with sets of 3D components without too much difficulty. Therefore, an easy transition from textual information to 3D modelling as design medium may be also provided, in which the first finds a more relevant use and the last is enhanced as an intelligent activity.

5.   Cortex and 'semi-recurrent' network integration: a working example.

Suppose that a user approaches this system. What this user will see on the screen and interact with will be as follows. The system starts by bringing the first question to the screen. This is the question that verifies the most common feature in the set of cases:

Is it supposed to have glass among the surrounding materials?

As the user is not sure yet to which extent glass is to be used around the entrance, he or she answers 'don't know' to this question. The system then brings to screen the next question:

Is it supposed to have a swinging door?

This dialogue continues until the system has tested all the relevant features describing the set of cases. It then brings to the screen the textual description of a first possible solution:

> *The solution may be a functionalist door entrance, with the following features:*
> *-vertical glass tower*
> *-surrounding materials: glass, plasterwork and metal*
> *-swinging door*
> *-leaf type: framed with one or more light cross panels*
> *-leaf materials: non-stained glass, metal*

The solution's description is followed by the graphic illustration of the precedent, in this case in a bit-mapped format:

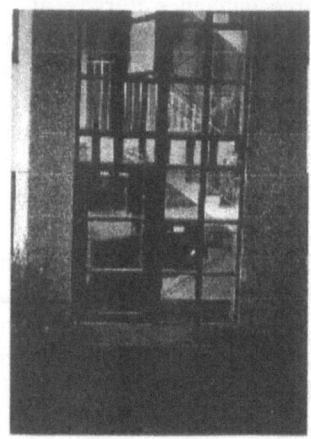

Figure 6. Functionalist solution

The solution's description is followed by the notice:

> *There is an additional possible solution available. To display it press Y, for 'yes'.*

The system then brings to the screen the textual description of a second possible solution:

> *The solution may be a high tech door entrance, with the following features:*
> *-vertical glass tower*
> *-surrounding materials: glass, metal*
> *-swinging door*
> *-revolving door (with four leaves)*
> *-leaf type: framed with one or more light cross panels*
> *-leaf materials: non-stained glass, metal*

The second solution's description is followed by the graphic illustration of the precedent, also in a bit-mapped format:

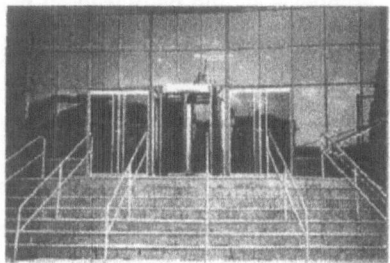

Figure 7. High Tech solution.

The system will then bring to the screen the textual description of a new possible solution, as follows:

A new solution may be possible with the following features:
- vertical glass tower.
- lateral cylindrical section column.
- surrounding materials: glass, metal.
- opening mechanism: swinging door, revolving door.
- leaf type: framed.
- steelworks as leaf decoration.
- leaf materials: glass, metal, timber.

6.   Real world applications:

The small domain described earlier provided a good illustration of the internal procedures of the proposed integrated model. However, its reduced number of precedents (9 cases described by 27 features) did not represent a sample large enough for a neural network learning process. Also, the fewer the number of precedents in the knowledge-base the greater the possibility of retrieving unsatisfactory solutions due to the case-based reasoning representation adopted in Cortex and its closest match searching procedure. Its results are of little use in practical design situations.

However, an 'extended domain', with a much larger number of precedents (122 cases) and descriptors (80 features), describes the problem 'door entrance' in much more detail than in the small domain and provides an environment much closer to a realistic design situation.

For instance, suppose that we are now given the task of designing an entrance door in a non-existing building. We approach the 'extended domain' with the following basic constraints: firstly, the materials should be, as much as possible, industrialised and non hand-crafted. Secondly, it is supposed to be a single door and to give access to a reception desk. Thirdly, an emphasis should be placed on natural heat conservation in the winter.

The system starts by bringing the first question to the screen:

Does the door leaf have a flat top?

Since the presence of a flat door top favours industrialisation, we answer 'yes' to the question above. The system then brings to the screen the next question:

Does it give access to an air lock, vestibule or foyer?

This entrance door is supposed to give access to a reception desk, as mentioned above. In these situations it is usual to have a foyer where people can make enquires and meet other people. Therefore, we answer 'yes'. The system then brings the next question to the screen:

Is it supposed to be the main entrance?

We know that the entrance being designed is an important one. However, as we are not aware yet of the situation in the remaining of the building, the answer is 'don't know'. The system then asks us:

Is it supposed to have glass in the surroundings of the entrance?

We interpret the presence of glass, as a dominant surrounding material, as a means of favouring natural heat conservation in the winter. Therefore the answer is 'yes'. The next question is:

Is it supposed to have non-stained glass in the leaf?

Since the presence of non-stained glass in the leaf can also be interpreted as favouring natural heat conservation in the winter, the answer is also 'yes'. The system then brings to the screen the following question:

Is it supposed to have metal in the leaf?

Since metal favours industrialisation, the answer is once again 'yes'. The system then asks us:

Is it supposed to have metal in the surroundings of the entrance?

Once more, the industrialisation factor drives us to answer 'yes'. The next question is:

Is it supposed to have a Window in both sides?

Since we have decided at an earlier stage that glass was going to be one of the dominant materials in the surrounding of the entrance to provide natural heat conservation, we decide to use a Window in both sides of the door. The answer is therefore 'yes'. The system then asks the next question:

Is it supposed to allow public access?

Although we are aware of the importance of the door, we do not know yet about the type of flow control for this entrance. The answer is thus a 'don't know'. The system then brings to the screen the following questions:

Is it pulled in from the facade?

We did not have a prior opinion on this matter, but we are now inclined to think that the feature above may give some protection from the flow of people in the street and from bad weather. Therefore, the answer to the question above is 'yes'. The next question is:

Does the entrance have a squared fanlight on its upper part?

Since glass is going to be one of the dominant materials in the surroundings of the door entrance, we think of a squared fanlight as a means of increasing heat loss in the summer. Therefore, the answer is 'yes'. The system then brings the next question to the screen:

Does the entrance have one double swinging door?

Considering the initial constraints, which call for a single door, we answer 'no' to this question. The system then brings the next question to the screen:

Is this entrance supposed to have bricks in its surroundings?

Bricks are nowadays industrialised construction materials. However, they do not favour a mechanised construction process to the same extend of their own manufacturing, particularly due to their dimensions and to the need of manual and careful bricklaying. Therefore, we answer 'no' to this question. The system will then bring a solution to the screen:

The most likely solution in the Extended domain is:
The closest match is Case 63, which has the following features:
Flow function:	*main entrance*
Flow control:	*restrict access*

282

Flow connection: *gives access to: air lock, vestibule or foyer*
Formal insertion: *pulled in from the facade*
Door top shape: *flat door top*
Top complements: *squared fanlight*
           flat semi-circular porch

Other complements: *walls supporting porch*
Lateral complements: *a Window in both sides*
Surrounding materials: *glass*
           smooth stone
           metal

Door operation type: *swinging door, one single*
Door leaf type: *framed with three or more light cross panels*
Door leaf materials: *non-stained glass*
           metal

Door handle, if any: *round knob or ring handle*

The following illustration will come up on the screen:

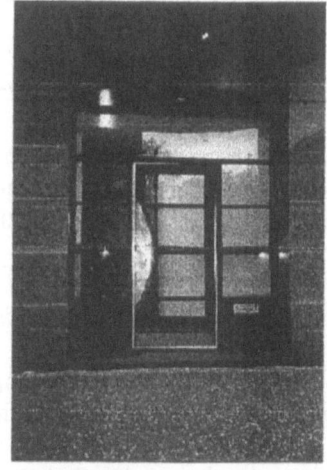

Figure 8. Case 63

The system will then propose the following new combination of features:

A new solution may be possible with the following features:
Flow function: *main entrance*
Flow control: *public access*
Flow connection: *gives access to: air lock, vestibule or foyer*
Formal insertion: *pulled in from the facade*
Door top shape: *flat door top*
Top complements: *squared fanlight*

Other complements:	*vertical glass tower*
Lateral complements:	*a Window in both sides*
Surrounding materials:	*glass*
	metal
Door operation type:	*swinging door, one single*
	revolving door (with four leaves)
Door leaf type:	*framed with one or two light cross panels*
Door leaf materials:	*non-stained glass*
	metal
Door handle, if any:	*long vertical static handle*

The first solution is retrieved because it is the closest match in the case-base. It thus satisfies all the conditions verified by the questions presented to the user. The second solution, which is a result of the neural network computation, also satisfies all the conditions verified by the questions above.

However, it provides some more interesting features regarding natural heat conservation in winter, than the solution retrieved from the case-base. These features are: 'a vertical glass tower' (instead of 'walls supporting porch'), only 'glass' and 'metal' among the surrounding materials (no 'smooth stone'), and the addition of a 'revolving door' (instead of just a 'swinging door: one single').

There may be several possible graphic interpretations of the new solution, but the textual description above already provides some illustration of how the proposed model may augment the designer's creativity. In the example above the 'vertical glass tower' and the 'revolving door' represent features not thought by the designer at the outset of the design task. Yet, they comply with the initial set of constraints, which were that the materials should be, as much as possible, industrialised and non hand-crafted, and an emphasis should be placed on natural heat conservation in the winter.

Several domains could be implemented in the proposed model representing different levels of abstraction for a particular design task. At the end of each section we may accept the proposed solution and move towards its detailing, or decide to go back to the beginning and start another section at the same level of abstraction.

7.   Conclusions:

About 15 domains have been already implemented in the proposed model and they deal with different architectural problems (see Silva, 1995, for a description of the extended domain and experimental data). Each of those domains is being gradually linked to libraries of 3D architectural components. Therefore, in addition the textual descriptions of the new solutions an easy transition to graphic representations is being built.

The main benefits of this hybrid model are: firstly, the system always reaches a stable solution or the most possible stable solution. There is no risk that a search will be terminated at an unstable state. Secondly, there is no risk of the user getting lost in the search for a stable solution, because all the basic choices are made prior to their entering in the neural network, the search is automated and the system itself keeps track of previous actions. Thirdly, there is no risk of getting into an infinite loop since the system filters features by guiding the user through the most promising path and controlling and terminating deadlock situations.

This integrated environment automates the process of searching for stable solution by emulating the user's actions through the semi-recurrent network described earlier.

At last, the system offers a plain English interactive interface throughout the process of solution search. In addition, links to 3D libraries are being developed, which will make neural network output more useful and will enhance 3D modelling as an intelligent activity.

The result of this approach will encourage designers to use intelligent systems in a more enjoyable way and to explore design alternatives towards creativity within meaningful interface and environment.

References:

Coyne, R. D. and Newton, S. (1990) Design reasoning by association, in Environment and Planning B: Planning and Design, vol. 17, pages 39-56.

Coyne, R. D. (1991) Modelling the emergence of design descriptions across schemata, in Environment and Planning B: Planning and Design, vol. 18, pages 427-458.

Coyne, R. D. and Yokozawa. M. (1992) Computer assistance in designing from precedent, in Environment and Planning B: Planning and Design, vol. 19, pages 143-171.

Coyne, R. D., Newton, S. and Sudweeks, F. (1993) A connectionist view of creative design reasoning, in Gero, J. S. and Maher, M. L. (editors) Modeling Creativity and Knowledge-Based Creative Design, Lawrence Erlbaum Associates, Hillsdale, NJ.

Frey, P. W. (1988) A Bit-Mapped Classifier, in Byte, volume 11, number 12, pages 161-172.

Mustoe, J. E. H. (1990) Artificial Intelligence and its Application in Architectural Design, unpublished Ph.D Thesis, University of Strathclyde, Glasgow.

Mustoe, J. E. H. (1993) Cortex User's Guide, Resolution Software, Nottingham.

Silva, N. F. (1995) The use of hybrid technology in the construction of an evolving knowledge-based design system, in Tan, M. e Teh, R. (editors) CAAD Futures '95: The Global Design Studio, proceedings, Centre for Advanced Studies in Architecture, School of Architecture, National University of Singapore, Singapore.

PATTERN-BASED GENERATION OF CUSTOMIZED, FLEXIBLE BUILDING SIMULATORS

JAN PETER RIEGEL, MARTIN SCHÜTZE,
GERHARD ZIMMERMANN
Dep. of Computer Science
University of Kaiserslautern
Germany

This paper describes a domain-specific software development method for the creation of building simulators. The method is based on object-oriented modeling, design patterns and code generation principles. The goal is to provide customizable building simulators that exactly simulate those physical effects an application demands. The numerical accuracy and different algorithms to be used can be tailored to the application's needs. By using object models and preconfigured design patterns, a well-structured simulator model can be created. From this model, the complete product code of a simulator is generated. The patterns help to develop a complete and correct model. Each pattern describes a certain functionality and knows how to generate code to implement this functionality.

1. Introduction

Simulation for performance evaluation of large buildings and their installation is desirable through all phases of the design process. Existing, commercially available building simulators are in most cases large, monolithic systems which neither can be tailored to the simulation needs nor are they easy to use. Worse than that, they frequently require a very detailed description of the building which is not known in early design phases. Therefore, these simulators are used in later stages of the design where the correction of possible errors detected by the simulators is difficult and expensive.

Due to this mismatch between the designer's needs and the simulator's requirements, CAAD system developers try to integrate performance evaluation tools into their systems (e.g., Mahdavi 93). These tools determine important performance indicators during early design phases, sometimes in a two way approach where modifications of the indicators are reflected in possible design modifications (Flemming and Mahdavi 93).

In contrast to the calculation of performance indicators, we propose a 'classical' simulation approach, where buildings are simulated in the time domain. This approach was primarily intended to support building control system engineers with a software test environment, but can also be used for performance evaluation. We do not provide one monolithic, comprehensive simulator. Our software engineering approach to the simula-

R. Junge (ed.), CAAD Futures 1997, 285-298.
© 1997 *Kluwer Academic Publishers.*

tion problem consists of a very flexible simulator generator that creates customized simulators for specific needs and on different levels of abstractions. By this way, the application (CAAD or a control system engineering environment) can be integrated together with the simulator in a design tool (compare Milne 91). The underlying object model of the simulator, as well as the simulated effects, the level of details, and the used physical abstractions can be determined by the user. The modeling of the simulator is supported by a design pattern-based mechanism (Gamma et al. 95/ Pree 95), a concept which originally came from the architectural domain (Alexander, Ishikawa, and Silverstein 77) and found its way to software engineering. When the modeling is done, a code-generator is used to create the product code for an executable simulator.

Chapter 2 describes the simulation principles of our simulators and the different degrees of freedom they allow. The modeling of such a simulator is described in chapter 3. This chapter is divided into three parts, each describing one major step of the modeling. These are: setting up an object model, applying patterns to this model, and transforming a CAD drawing to get a building instance. The generation process is briefly described in chapter 4. The paper concludes with a list of related works (chapter 5) and a discussion of our approach in chapter 6.

1.1 CLASSIFICATION

One application domain where flexible simulators are needed is the development of control systems for buildings. Because it is too expensive and too complicated to test a building control against a real building, simulation is necessary. Here a simulator must act as close as possible like a real building. However, only those effects have to be simulated which influence the or are monitored by the control. Depending on the building control, it might be interesting to test it against a variety of different buildings. The simulator for such control algorithms has to emulate the same timing behavior as a real building, thus real time simulation is necessary.

Another domain where building simulation is useful is Computer-Aided Architectural Design. During the design of a certain building, simulation can be used to verify the performance of construction details. Here, the model of the building is rapidly changing and a simulator must be adapted to simulate exactly the building under development. Simulation should be possible even if only parts of the building are readily designed. Such a simulator is naturally less exact as one using the final design might be, but nevertheless it can be used to detect major design flaws in early phases. By this way, for example, the performance of a central storage heating can be simulated before the entire construction of the building has been completed. If the simulation reveals that the required mass of the heat storage is too big or too small the design can easily be adapted without having to change the complete building.

There are three different categories of simulation environments. *Simulation languages* (e.g. SIMULA, Lamprecht 81) are very universal, but do not provide any help for the development of a domain specific model. To use a simulation language, detailed knowledge of the language and of the problem domain must be present. *Simulation Systems* like GASP (Alan and Pritsker 74) or SMILE[1] are focused on the exact modeling of physical effects. Therefore, they can be used if the physics that should be emulated are well understood and if useful simulation algorithms are known in advance. Complete *specialized simulators* (e.g. TAS[1]) can be used by people which are not experts in the simulation domain, but tend to be static, monolithic systems that cannot be adapted to the application's needs.

The overall goal of our approach is to provide a powerful, but easy-to-use, method for creating building simulators. The user of this method doesn't need to be an expert in the simulation domain. This way, for example, a tool integrator who knows about the application which has to be tested, and is somewhat familiar with software engineering models but who is not familiar with simulation in detail, can use our method. We try to abstract from the implementation of a building simulator by providing simple and understandable simulation patterns that can be used to model the final simulator.

2. Simulation

The building simulator has to be flexible in several ways:

Different physical effects. It should be possible to choose from a variety of effects to be simulated. These effects can be isolated or related with other effects (e.g. humidity is related with temperature). In such a case both effects can be simulated in detail or one effect is emphasized while (for simplicity) the other is only approximated. It must be guaranteed that the simultaneous simulation of more effects leads to correct solutions. This is especially true if discrete effects are mixed with continuous ones.

Different levels of abstraction. Depending on which data are actually available and depending on how accurate the simulation should be, a simulator can be more general or can be modeled specific. Therefore, the 'best' simulation algorithm depends on the current status of the application and on the goal of the simulation (i.e. which effect or behavior should be tested).

Accuracy versus speed. It is a very complex task to simulate physical effects as exact as possible. A customized simulator should be able to trade speed for accuracy in order to get nearly exact results in much shorter time. However, a more precise simulation should also be possible.

Real-time. The simulation should take place in real time or accelerated according to real time, so that the timing behavior of a building can also be simulated. Furthermore, an asynchronous interaction with the simulator is needed to test control systems.

Hardware-in-the-Loop. It should be possible to integrate real hardware into the simulator. This restriction comes from the main purpose of our approach, namely to sup-

---

1. Symbolic Interactive Lotos Execution, LOTOS Tool development group, dept. of Computer science, University of Twente, PO Box 217, 7500 AE Enschede, The Netherlands
1. TAS (Thermal Analysis Software), Environmental Design Solutions Ltd., 13/14 Cofferidge Close, Stony Stratford, Milton Keynes, MK11 1BY, Great Britain

port a control engineer with a powerful tool to test complete control systems including hardware parts. To integrate hardware in the simulator, the simulation must take place in real-time.

To meet all these needs, an application-specific simulator has to be created. For each physical effect to be simulated, we provide one or more simulation components. The interfaces of these simulation components have to be clear and standardized so that single components can be exchanged in order to include a new effect or to chose a different accuracy. Well-defined interfaces also support the process of stepwise refining a simulator. Starting with a rough, imprecise model of the simulator, it could be refined to incorporate more precise algorithms if special effects are to be simulated. This refinement can be hierachical (i.e. starting with an imprecise simulation of a special effect and using more concrete algorithms in later phases to trade accuracy for speed or to change the level of abstraction) or explorational (adding new effects).

The final simulator is built from several small fragments, each providing a partial functionality. In order to make these parts fit together, the simulation must take place on a very local level. Our simulators consist of many different objects which are closely related to „real world" objects. These objects are, for example, walls or rooms. An *object model* describes these objects and their relations. In contrast to other simulation approaches where the calculation is done on a global level (one big system of differential equations is solved with every time step), our simulation objects simulate themselves making the integration of further functionality easy and leaving room for possible local optimizations. For example, in order to calculate the actual room temperature, a room object collects all the heat-flows from its neighboring objects (walls, heating installations, sun radiation, etc.) and uses these values together with the elapsed time interval to compute the new temperature. The room doesn't have to know, how the heat flow was calculated - this is in the responsibility of the wall or heating objects.

The interfaces of the objects and the data they exchange are declared by predefined design patterns. *Patterns* describe an abstract functionality that can be adapted to objects from the object model. Each pattern of our pattern catalog is able to work together with other patterns enabling to model on different levels of abstraction. This way, one simulation object can easily communicate with other objects. If a variant of an existing simulator is to be created, only few patterns need to be exchanged.

As an example, figure 2 a) describes a very simple situation, where a wall only consists of one object. The heat flow through such a wall can easily be calculated depending on the difference of temperatures of both adjoining rooms ($q = (\Delta v \cdot A)/R_\lambda$, with $\Delta v/[K] =$ difference of the temperatures, $A/[m^2] =$ area of the wall, $R_\lambda/[m^2 K/W] =$ thermal resistance of the wall). In figure 2 b) a wall is more complex. Here, it consists of several layers (i.e. the wall object has associated layer objects), each with its own thermal resistance or heat storage capacity. If such a wall is asked about its heat flow, it delegates the question to its layers which sum up the total heat flow. In either case, a room object doesn't have to know about the wall structure - it just uses the well-defined interface to the wall object.

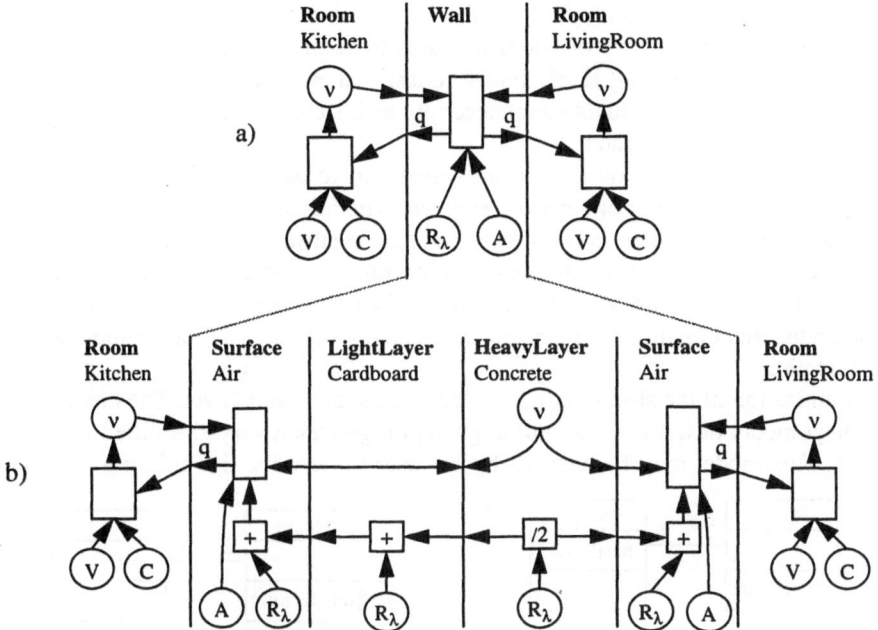

Figure 1. Interfaces of a wall object.
a) simple wall
b) layered wall

3. Modeling the Simulator

The question that now arises is how to specify the simulator without having to deal with the issues of the last chapter in detail.

A software system can be divided into several views. The *static view* describes all parts that do not change while the program runs. Especially data structures belong to the static view. These can be modeled using object diagrams (Rumbaugh et al. 91) or Entity Relationship diagrams. There are several methods how to model data structures. Tools are available to support the modeling and generation of program code out of these models. Static models can be used as an integration platform for the complete system: in these models references to other models describing the functionality or the behavior of a certain object can be stored.

Another aspect of the static view is the instantiation of the object model. That is, the actual data on which the program operates have to be described. We use object models to describe static aspects (see figure 2) and transform data from a CAD editor to instantiate these models, thus building the input to our simulator.

The *dynamic view* on a software system shows how the program behaves during run time. To create dynamic models (e.g. State Transition Diagrams), a software-engineer has to know very much about the problem domain and needs experience in software

290

design. For simplicity, we "hid" all dynamics in special control patterns and in a simulator kernel library. This library provides methods to control the simulation process and can be used on the modeling level. Using control patterns, the dynamic behavior of a simulator can be modeled with abstract descriptions instead of setting up a detailed State Transition Diagram from scratch.

Finally, the *functional view* on a software system has to be described. This is normally the domain of a programmer or software engineer who has to choose the best fitting algorithms and translates them into program code. Because of the limited application domain, we are able to encapsulate possible algorithms in simulation patterns. A pattern generator knows how to implement the pattern's functionality so that the user can concentrate on the functionality a pattern provides and doesn't have to bother about a correct implementation.

The customization of the simulator takes place on the modeling level. The central model is an object model describing the building's topology (see a very simplified example in figure 2). This model may be adapted by the user to describe the building as exact as

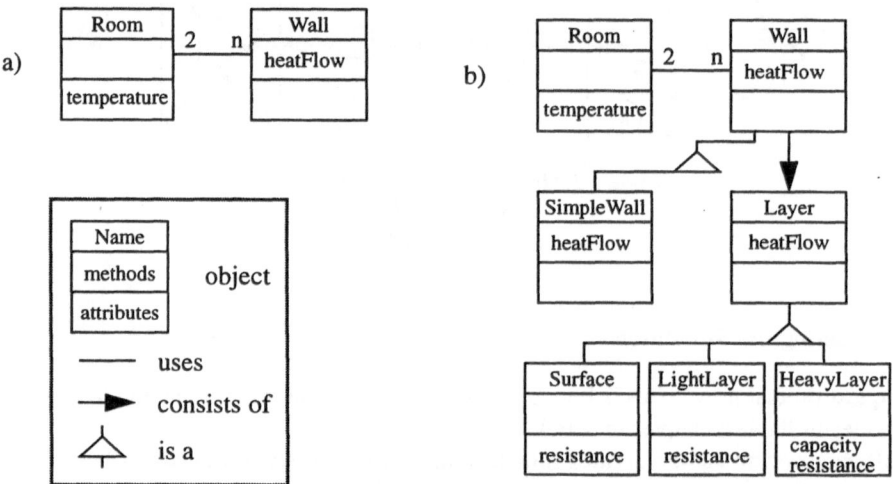

Figure 2. Object models for different simulation methods.
a) direct calculation
b) delegation

needed. For example, the model can be enriched with additional elements describing exact geometry or detailed information about installations and control systems. Figure 2 a) is a small model sufficient to represent simple wall structures (compare figure 1 a)). With a building model like in figure 2 b) more sophisticated simulation approaches as in figure 1 b) can be modeled. In addition to this building model, there are other models describing the simulator kernel objects and the simulator's functionality. To create a special-purpose simulator, these different models have to be linked together. As a 'glue' between the models we use a derivative of design patterns (Gamma et al. 95).

Figure 3 describes the development of a building simulator (see Altmeyer et al. 97). The central model is an object model describing the static view on the simulator. Here, objects like 'Room' or 'Wall' are declared. To describe the behavior of these objects, we use simulation patterns from our pattern catalog. These patterns are 'bound' to the object model by stating how they interact with the objects. A set of generators and transformers is used to generate code from the resulting model.

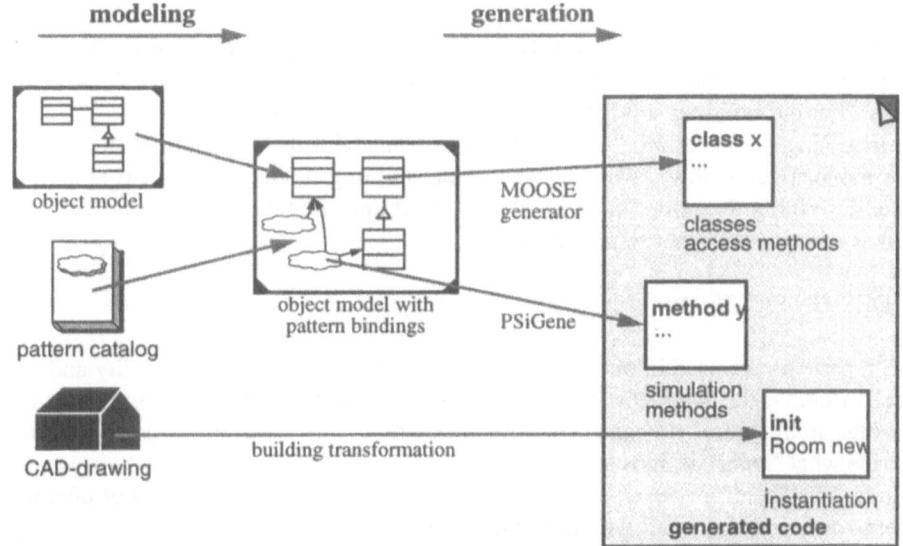

Figure 3. Simulator development with MOOSE/PSiGene

Starting with the adaptation of the building model (i.e. an object model describing a building), design patterns are successively applied to the model in order to get the needed simulator functionality. The result of this modeling step is a refined building model which includes patterns and their bindings to the objects from the model. Encapsulated in each pattern is a code template so that after the modeling step a complete simulator can be generated. The class structure of the simulator program is generated by our software engineering tool MOOSE (Model-based Object-Oriented Software Generation Environment, see Altmeyer, Schürmann, and Schütze 95), whereas the functionality is created using the PSiGene generator (Pattern-based Simulator Generator, Schulz 97). The concrete building instance, which should be simulated, is acquired using a CAD-editor (Speedikon[1]) and transforming the drawing into usable program code. In the following, the three major parts of the modeling process are explained: adaptation of the static models, binding of design patterns, and instantiation of the building. The generation process is briefly explained in chapter 4.

3.1 OBJECT MODELS

A central point of the simulator model is the structural view of the simulation objects. Here, the objects are described along with their attributes and relations to other objects.

---

1. Speedikon X, IEZ AG, Berliner Ring 89, D-64625 Bensheim, Germany

The notation for this description is a derivative of the OMT class models (see Rumbaugh et al. 91 and figure 2).

These models can be edited using MOOSE, a software engineering tool we have written. MOOSE also contains powerful software generators which are capable of generating complete class structures out of these object models including access functions, methods to handle connections between objects, and a persistent storage mechanism. These code generators are available for the programming languages C, C++, and Smalltalk (Visual-Works). For our building simulator, we use the Smalltalk generator.

The modeling of object models goes hand in hand with the application of patterns to these models (see below). We provide a basic building reference model (i.e. an object model describing the static view on a typical building) that can be used as is or can be taken as a basis for further refinements.

3.2 DESIGN PATTERNS

Design patterns are used in our approach to define the simulator functionality and act as 'glue' to define the interactions of objects in the object model. They originally came from the architectural domain. The architect Christopher Alexander used patterns to describe what 'good' designs are and how to obtain them (Alexander, Ishikawa, and Silverstein 77). Therefore, he created a pattern language in the way that a set of interacting patterns describes how and why a building is constructed the way it is. He defines patterns as "a three-part rule, which expresses a relation between a certain context, a problem, and a solution". The idea of patterns has been transformed to the domain of Software Engineering (Gamma et al. 95). Here, each pattern describes a recurring software problem which should be solved, a context in which this problem occurs, and a solution to this problem.

In general, software design patterns give clues how certain problems can be solved using special object structures or algorithms. Design patterns are mainly used to support the modeling and the documentation of software. They should be applicable in many situations, therefore, they are on a rather abstract level and not useful for automatic code generation. In our case, however, we focus on a narrow domain, the domain of building simulation. Therefore, we were able to create specialized patterns that describe small parts of a building simulator. These specialized patterns contain code fragments so that an automatic code generation is possible if such a pattern is used (i.e. bound to an object model).

Each pattern contains a description *when* it could be used, *which* problem it solves, *how* to apply it, *which* underlying object structure it expects, and *how* a generator should generate code to solve the problem. A pattern contains only the solution to a small problem. To solve a complex problem, it often depends on functionality provided by other patterns. For example, to simulate the temperature of a room, the room object needs to know how to calculate the temperature out of incoming heat flows. A pattern *Thermal-Mass* is responsible for the calculation of the temperature and relies for the calculation of the heat flows on functionality provided by other patterns. The parts of a pattern which are defined elsewhere are called 'hot spots' (Pree 95) because the pattern itself only defines the interface making different implementations (by other patterns) possible. Thus

6.1.   Pattern ThermalMass

Intent

The ThermalMass pattern computes the temperature of a mass depending on the amount of heat affecting the mass. ...

Also Known As

Simulation thermischer Masse [1]

Motivation

A volume has to act as a thermal mass to compute its temperature. ...

Applicability

This pattern can be bound to any thermal mass. Typical this is a room or ...

Structure

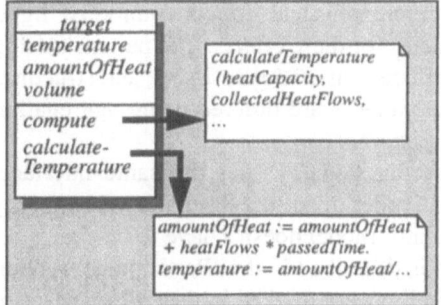

Participants

Objects

• target
  Object to bind pattern to.

Attributes

• temperature
  Last computed temperature.
• volume
  Volume of the thermal mass.

Methods

• compute
  Does the calculation cycle for one time.

Interfaces

• getHeatCapacity
  Determines the storage capacity from the temperature of a mass.
• collectHeatFlows
  Determines the heat-flows from and to a thermal mass.

Collaborations

The computation relies on ...

Consequences

...

Implementation

Smalltalk-Code-Templates

• {init}
  self {amountOfHeat}:= (self {temperature} *
  self {volume} * self {getHeatCapacity}).
  self {timeOfLastComputation}:
      Scheduler simSched simMillisecondClockValue.
• {compute}
  | heatCapacity collectedHeatFlows timeNow |
  timeNow:= Scheduler simSched simMillisecond-
  ClockValue.
  heatCapacity := self {getHeatCapacity}.
  collectedHeatFlows := self {collectHeatFlows}.
  self {calculateTemperature}WithCapacity: heatCapacity
      withHeatFlows: collectedHeatFlows while:
      ((timeNow-self{timeOfLastComputation})/1000).
  self {timeOfLastComputation}: timeNow.

...

Related Patterns

Thermal Exchange

Figure 4. Pattern example (abbreviated)

each pattern includes a description of the interaction with other patterns by using **template methods** which include interface and implementation of a certain functionality and **hook methods** which only describe the interface.

Figure 4 shows parts of one of our patterns (ThermalMass) in detail. Every pattern is structured into several sections to describe the context, the problem, and the solution separately. 'Intent' and 'Motivation' describe the addressed problem using a suitable example from the application domain. 'Applicability', 'Participants', and 'Collaborations' show the context and define how a pattern can be used. 'Structure' and 'Implementation' show the solution. The organization of a pattern is the same as in (Gamma et al. 95), but we have formalized some parts to make code generation possible. The 'Participants' part describes the exact interface of a pattern, i.e., every parameter of the pattern that has to be bound to the object model. In our example, the pattern needs a binding to a target class (e.g. room). This object has to have two attributes *temperature* and *volume*. Also a name for the template method *compute* has to be specified. To calculate a room's temperature, the heat flow into it has to be known. Because there could be many different heat sources and different calculations of the heat flows, the method *collectHeatFlows* is only defined as a hook method. The actual implementation has to be done elsewhere (i.e. with another pattern bound to the wall or to the heating installation). Last not least, the heat storage capacity of the room must be specified. This physical value is depending on the actual room temperature and on the material with which the room is filled (i.e. air). Therefore, this pattern defines only the interface to a method *getHeatCapacity*. With the specification of these six parameters, the pattern is fully bound to the object model and code can be generated.

So far we have collected 14 patterns in a pattern catalog. This small number of patterns is sufficient to model and create building simulators for thermal effects with many variants. We're now extending our catalog to include more physical effects (humidity, light, exchange of air, and others) and different accuracies for these effects. The pattern catalog is divided into three parts concerning physical effects, simulation artifacts, and structural adaptation patterns. Inside these categories, the patterns are ordered using aggregation and inheritance, making it easier to find special patterns.

Each group of patterns that provides a similar functionality uses the same interface. Therefore, it is easy to exchange some patterns in order to build a simulator variant; only local changes are necessary, the rest of the simulator model doesn't have to be changed.

To use a pattern, every interface element which is described in the 'Participants' section must be bound to the object model. Optional bindings are marked in the pattern description. The binding is currently done using a textual description language. We are implementing a graphical editor to be able to perform pattern bindings more easily. Each participant can be bound to an object, an attribute, a method, or to a valid Smalltalk statement, depending on its type.

For example, to simulate the temperature of a room, a 'room' object is needed. The actual temperature depends on heat flows that flow from or to the room through walls or heating equipment. So the room has to be related with its surrounding walls. Therefore there also must be a wall object which is connected to the room. To simulate the temperature of a room, a pattern ThermalMass can be bound to it. With this pattern the room is able to update its temperature using incoming heat flows. The heat flow through a wall depends mainly on the difference of temperatures of its adjoining rooms. A pattern ThermalJunction provides the functionality to calculate these heat flows. Furthermore, the calculation of the room's temperature has to be continuously stimulated so that it is always up to date. The pattern ContinuousComputation binds the room to our simulation kernel library providing just this functionality (see figure 5). Mainly with this three pattern bindings a first, very simple simulator can be modeled.

The binding for the pattern ThermalMass (figure 4) to an object 'Room' is textually described with the following statements:

```
ThermalMass                                    // Name of the pattern
    bind: 'target' to: 'Room';                 // Target class
    bind: 'temperature' to: 'temperature';     // Bind attribute 'temperature'
    bind: 'volume' to: 'volume';               // Bind attribute 'volume'
    bind: 'compute' to: 'calculateNewTemperature'; // Rename template method calculate
    bind: 'getHeatCapacity' to: 'getHeatCapacity';  // Hook method getHeatCapacity
    bind: 'collectHeatFlows' to: 'collectHeatFlows'. // Bind name to hook method
```

Figure 5 shows some sample bindings graphically. Patterns are not only used to provide simulation functionality to one object, they also act as 'glue' between different objects.

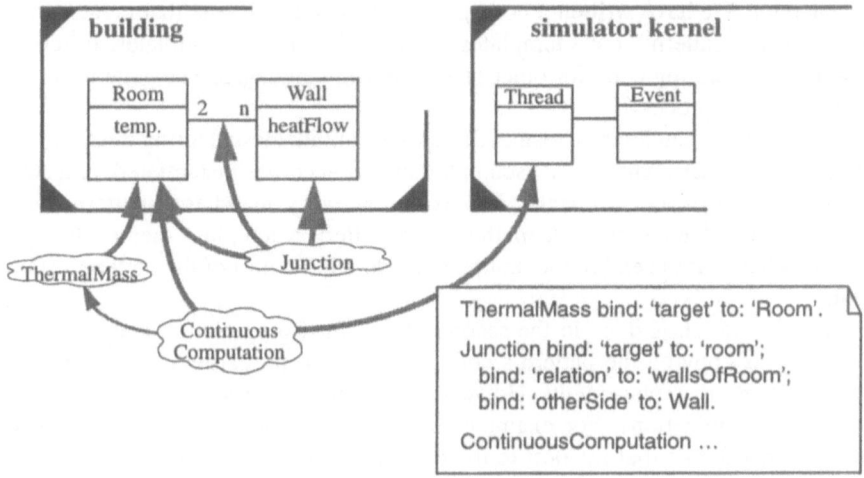

Figure 5. Pattern bindings (sample)

For example, the pattern 'ContinuousComputation' is bound to the Room and to an object 'Thread' from the simulator kernel. This means, that the temperature of a Room should be continuously simulated using a Thread that is controlled by the simulator kernel, 'gluing' a Room with a Thread.

3.3 BUILDING TRANSFORMATION

After the simulator is modeled and generated, it can be used to simulate any building that can be expressed by the building model. But to simulate one special building, the model has to be instantiated. We use a conventional CAD program to draw the plan of the building to be simulated and transform it into code that can be used directly by the simulator. This transformation is done semi-automatic.

We have chosen Speedikon as input to the transformation, because it is object-oriented and operates with objects like 'walls' and 'rooms'. Therefore the transformation to our building model is easy.

Speedikon doesn't handle physical descriptions of construction details like, for example, heat storage capacities of walls. These values have to be entered manually. We use heuristics to get a useful building instance that can be simulated (e.g. by assuming default values for material constants). Manual adaptations are needed to mirror the exact physical constants of the building. By now, the transformation is hard coded for one building model but we will try to use transformation rules in order to be able to handle a greater variety of building models.

4. Generating Simulators

After modeling the building simulator, program code has to be generated to get an executable generator. We have written a code generator called PSiGene that creates Smalltalk code from the patterns' code templates. With different code templates, PSiGene is also capable of generating code for other programming languages. The generation takes place in a two phase process.

Each pattern from our catalog has been coded as a Smalltalk class. During the first generation phase every pattern which was used in the model becomes instantiated. Afterwards PSiGene performs syntactical checks. For example, every mandatory pattern binding must be made, and for every hook method of a pattern a template method has to be defined somewhere. At the end of the first phase, a detailed report of the pattern instances is created.

All the code generation is done in the second phase. In a simple case, code generation can be as easy as copying the pattern's code templates to the target classes while performing simple macro replacements. However, the generator has enough knowledge to do some code optimizations. For example, if more than one code template is given, PSiGene can choose one that fits best to the building model. This is due to the fact that PSiGene does not only know of the pattern bindings, but also regards the building model. Some of our patterns (like StateMachine) use software synthesis techniques because their functionality cannot be fully described using simple code templates.

Figure 6 shows an example of the running simulator. The graphical in- and output consists of several components which were manually adapted to the simulator. However, the whole simulation functionality is generated. With special patterns it will be possible to model and to generate the user interface, too.

So far, we have modeled several small building simulators using patterns. An example we created, is a typical simulator for thermal effects consisting of 24 patterns (11 different types of patterns were used). We suppose that each extra physical effect to be simulated takes about 5 additional types of patterns. The number of used patterns depends on the size of the object model. If many objects should be simulated, more patterns have to be bound to these objects. The object model in this example consists of 15 building object classes and 6 classes from the simulator kernel library. About 5300 lines of code were generated from the building model and the patterns. A building with 9 rooms and several doors and windows consists of about 300 objects.

Comparing the time used to model and generate a simulator using PSiGene with a manually written simulator shows that it is much more efficient using our method even if some patterns which are missing in out catalog have to be created. Building variants of a simulator is a matter of hours (from the concept to the running simulator).

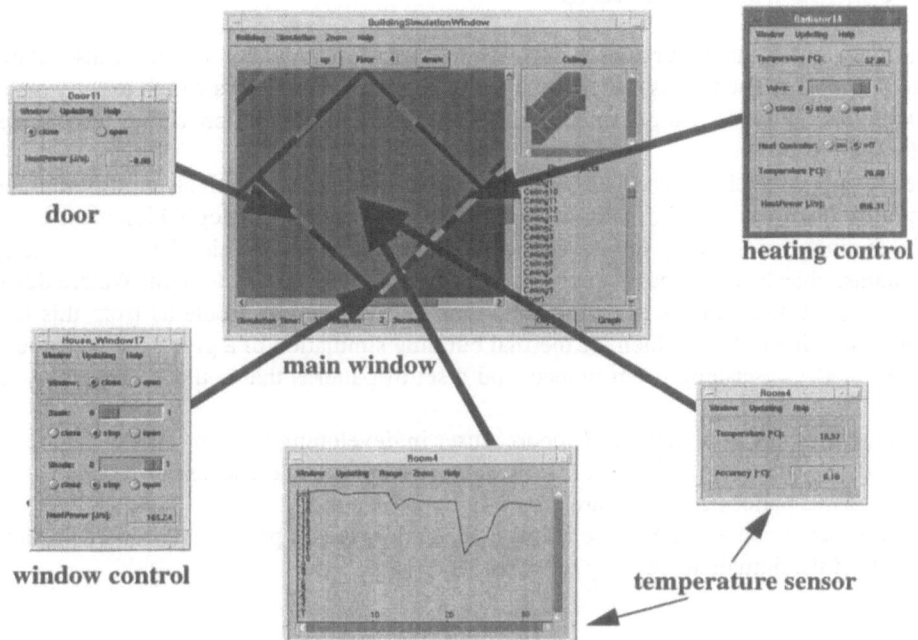

door

heating control

main window

window control

temperature sensor

Figure 6. The running simulator

5. Related works

We are providing a method to model and generate building simulators. To do so, we have combined software engineering methods with application specific knowledge from the simulation domain. The use of design patterns to create software models is a very actual and much discussed topic. The main usage of patterns is during the analysis and design phase (Gamma et al. 95, Pree 95), however, there are some approaches using patterns for software generation (Budinsky et al. 96). Our patterns are small, domain specific pieces that can only be used for software design. The main advantage in this is that we can generate the complete product code. We combined software engineering techniques with domain-specific knowledge and software generators. Related work has been done, for example, by Batory (Batory et al. 94).

The modeling of building simulators is a broad research topic with many facets. Using a simulator during the evolution of a building design is done at the Carnegie Mellon University (Flemming and Mahdavi 93, 95). Here, the simulator is used in a two-way approach: the simulation indicates performance issues of a building which could be used in a reverse engineering step to automatically adapt the building to yield a better performance. This simulator is optimized for this reverse engineering step. Explicit modeling of building simulators has also been done by Filiz Ozel (Ozel 91) who uses object diagrams in conjunction with rules to create a simulator.

6. Conclusion and future works

The main advantage of our approach is that we provide a simulator generator instead of a fixed simulator. The user is therefore capable of creating simulators which exactly match his simulation needs without unnecessary overhead and without providing too detailed information.

We are now collecting more patterns to be able to simulate more effects on different levels of abstractions. All our patterns are collected in a pattern catalog which itself is part of a dictionary describing the domain of building automation. This dictionary includes formulas, objects, and models describing different aspects of the domain. We are developing search mechanisms to extract solutions to given design problems from this lexicon. A solution to the problem of thermal building simulation of a given room can be an object model describing room objects and a set of patterns that could be bound to that model.

Further work has to be done to support a user in developing building models using our method. We are trying to incorporate modeling guidelines and checks for correct pattern bindings in a pattern editor we are currently implementing.

From our experience, we believe that this method increases productivity and helps non-experts of the domain to develop useful tools.

References

Alan, A., Pritsker, B. (1974) The GASP IV simulation language, Wiley, New York

Alexander, C., Ishikawa, S., and Silverstein, M. (1977) A Pattern Language, Oxford Univ. Press, New York

Altmeyer J., Riegel J. P., Schürmann B., Schütze M., Zimmermann G. (1997) Application of a Generator-Based Software Development Method Supporting Model Reuse, in Proc. 9th Conference on Advanced Information Systems Engineering (CAiSE*97), Barcelona

Altmeyer, J., Schürmann, B., and Schütze, M. (1995) Generating ECAD Framework Code from Abstract Models, Proceedings of the Design Automation Conference '95, San Francisco, California

Batory D., Singhal V., Thomas J., Dasari S., Geraci B., Sirkin M. (1994) The GenVoca Model of Software-System Generators, IEEE Software, September 94

Budinsky F. J., Finnie M. A., Vlissides J. M., Yu P. S. (1996) Automatic code generation from design patterns, IBM Systems Journal, Vol. 35, No. 2, (http://www.almaden.ibm.com/journal/sj/budin/budinsky.html)

Flemming, U., Mahdavi, A. (1993) Simultaneous Form Generation and Performance Evaluation: A Two-Way Inference Approach, CAAD Futures '93, Elsevier Science Publishers Ltd., Amsterdam, 161-173

Flemming, U., Woodbury, R. (1995) Journal of Architectural Engineering, Vol. 1, No. 4, 147-152

Gamma, E., Helm, R., Johnson, R., and Vlissides, J. (1995) Design Patterns, Addison-Wesley

Lamprecht, G. (1981) Introduction to SIMULA 67, Vieweg

Mahdavi, A. (1993) Open Simulation Environments: A Preference-Based Approach, CAAD Futures '93, Elsevier Science Publishers Ltd., Amsterdam, 195-214

Milne, M. (1991) Design Tools: Future Design Environments for Visualizing Building Performance, CAAD Futures '91, Vieweg Verlagsgesellschaft mbH, Braunschweig/Wiesbaden, 485-496

Ozel, F. (1991) An Intelligent Simulation Approach in Simulating Dynamic Processes in Architectural Environments, CAAD futures 1991, Vieweg Verlagsgesellschaft mbH, Braunschweig/Wiesbaden, 177-190

Pree, W. (1995) Design Patterns for Object-Oriented Software Development, ACM Press, Addison-Wesley

Rumbaugh, J., Blaha, M., Premerlani, W., Eddy, F., and Lorensen,W. (1991) Object-Oriented Modeling and Design, Prentice Hall, Englewood Cliffs, N.J.

Schulz, S. (1997) PSiGene - A Pattern-Based Simulator Generator, diploma thesis, University of Kaiserslautern

Auditory Navigation and the Escape from Smoke Filled Buildings

Peter Rutherford

University of Strathclyde
Department of Architecture and Building Science
131 Rottenrow, Glasgow. G4 ONG
SCOTLAND

This paper addresses the issue of escape from unfamiliar, smoke filled buildings such as hotels or airports where the scenario of complete visual deprivation may result in occupant death. It proposes that we may be able to apply concise auditory information to the escape procedure, using predictive 'virtual acoustic' techniques in order to assess its feasibility.

Keywords: wayfinding, signal presentation, localization

1   Introduction

As visually orienting creatures, we as humans primarily perceive, explore and navigate our environment through the use of our eyes. The human visual system is therefore often cited as being our primary mode of direct perception. This dependency on visual stimuli has consequently resulted in an innate fear of blindness. We fear, either consciously or subconsciously the prospect of blindness; of being trapped in darkness, debilitated and as a result in a helpless state.

One such scenario of complete visual deprivation is being trapped in an unfamiliar built environment which is on fire and smoke filled, such as a hotel or airport. The occupier is not familiar with the escape exits or indeed the location of the fire itself and is thus exposed to a life threatening situation. In response to this problem, this paper addresses the issue of escape from such buildings, proposing that we might be able to apply concise auditory information to the escape procedure which would guide its occupants to safety. Underlying the whole paper is a description of predictive 'virtual acoustic' models which are being used to evaluate the feasibility of such a proposal. Using psychoacoustic modelling and acoustically 'rendering' the escape routes it is possible to accurately predict occupier ability to navigate solely using the auditory system without exposing them to the dangers of a real fire situation.

R. Junge (ed.), CAAD Futures 1997, 299-304.
© *1997 Kluwer Academic Publishers.*

2   Building egress and wayfinding conditions

Being familiar with the surroundings is greatly needed when a sudden exit has to be made in an emergency situation [1]. If the routes to the various exits are known and if any exits are blocked by smoke then the alternative exits are not so difficult to find. However, where the occupants are not familiar with their surroundings, movement will be slow and wayfinding will prove difficult. It must be kept in mind that, unlike a drawing, no overall view of complete escape is available when a person walks along a corridor. Not only is it impossible to see if the fire is on the other side of a wall or around a corner, but the conditions are repeated on other visually inaccessible levels. Signs to exits may be available, but if the corridor is completely smoke filled then they are in themselves useless.

Additionally, within a smoke filled corridor, the choice of escape action is restricted, resulting in hysteria or panic. Various authors agree [2] that "the association of panic and fires does not hold as long as an escape route appears feasible to the victim". However, panic may be induced if 'there is a limited number of escape routes.....some of these routes are affected by fire and smoke.....some of these routes are as a result blocked'. So how do these people escape? They have limited wayfinding cues, and along with many precipitating factors such as delayed warning and ambiguous messages from co-occupants really do not stand much of a chance of successful escape. In addition, the conventional alarm in and by itself has been shown to be among the most inefficient means to get people to leave a building [3].

There are methods available for aiding navigation such as emergency lighting and luminous escape systems placed on floors and skirting boards, but in the presence of irritant gases such as ammonia based combustion products, their effectiveness is seriously reduced. It is from this, and from psychoacoustic research that this paper originates and argues that our auditory system is sufficiently developed to compensate for severe visual deprivation in order to guide us to a safe exit.

3   Blind navigation and localization ability

There has been much research into people's ability to localize sound in three-dimensional space, some of the earliest of which was in blind navigation where it was found that blind people navigate in space using sound originating from their footsteps, breathing or cane tapping on the ground [4]. The reflected wavefronts informed the subjects of wall proximity, surface materials etc., not unlike that as used by Bat sonar. From this research, it was found that certain signals cannot easily be localized and the purpose of this section is to describe the nature of these signals for presentation within the system.

3.1   *Localization in the horizontal plane*

The most important cues for localizing a sound source's angular position involves the relative difference of the wavefront at the two ears on the horizontal plane. Most importantly the cues for such *lateralization* are interaural time differences (ITDs) and Interaural Intensity Differences (IIDs) (figure 1). These cues are highly frequency dependent. Consider the following example. Source A is straight in front of the listener therefore both the ITD and IID are equal, with path lengths being the same.

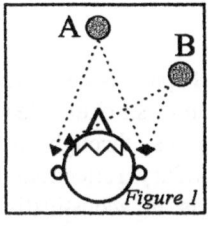

Figure 1

Source B is 60 degrees azimuth to the right of the listener thus the paths are now unequal causing the sound source to arrive later at the left ear than the right. This path difference is the basis for the interaural time difference (ITD) cue, relating to the hearing system's ability to detect interaural phase difference below approximately 1 kHz. There will also be a significant interaural intensity difference (IID) from source B, but only for those waveform components that are smaller than the diameter of the head (i.e. for frequencies greater than about 1.5kHz which are not diffracted). Higher frequencies will be attenuated at the left ear because the head acts as an obstacle, creating a 'shadow' effect on the opposite side. The smaller the wavelength, (i.e. frequency gets higher) the greater the shadow [5].

The previous example can be assimilated to a person *facing* the exit of a room which is engulfed in smoke with a sound signal being presented over the door. They simply have to maintain the binaural balance of this stimuli in order to reach the door, the reassuring signal getting louder as they reach their destination. However, what if the person is in the same situation, only with their face pointing in the *opposite direction*? In this instance, both the ITD and IID are the same as they would be from the front, so what does the ear do to disambiguate front from back? The following section describes this ability.

3.2   *Localization in the vertical plane and from behind*

The ability to disambiguate sources from front to back or from above and below in cases where ITDs and IIDs cannot support this information is described by the role of *spectral* cues in localization. Most significantly, localization is a spectrum dependent effect caused by the absorption and diffraction of sound by the ears pinnae (or outer ear) and torso. The helix of the pinnae (or folds) induce minute timing delays within a range of 0-300 micro seconds causing the spectral content of the sound at the eardrum to be slightly different to that of the source. These microtime delays, resonances, and diffractions can be translated into a mathematical model of the ear known as the Head Related Transfer Function [6] which is different for each position of the sound source. Table 1 illustrates the approximate frequency characteristics of sounds which derive a certain location, i.e. if a signal is presented to a subject from about 6 to 10kHz, (center frequency 8kHz) the subject is more likely to perceive it as originating overhead.

Perceived location	Center frequency (kHz)	Bandwidth (kHz)
overhead	8	4
forward (band#1)	0.4	0.2
forward (band#2)	4	3
rear (band#1)	1	1
rear (band#2)	12	4

Table 1:   Center frequencies and frequency bandwidths for 'directional bands' [7]

4   **Binaural room simulation and signal presentation**

The location of the auditory events in three-dimensional space is only the first step in tackling the navigation problem. The nature of the smoke filled environment itself adversely influences the way sound propagates through it. Elementary filters such as temperature stratification, thermal boundary reflections, smoke absorption and resultant room impulse characteristics such as an acoustically diffuse or focused environment all degrade the occupiers ability to escape, removing vital frequency and temporal information needed to pinpoint the acoustic emitter. In addition, the type of information presented by the system is fundamental to its performance, whether it be pure tones, speech or broadband noise. This section therefore discusses some of these filters through binaural simulation of the sound fields and examines the signals used within its presentation.

4.1   *Binaural room simulation*

Binaural room simulation allows us to authentically expose a person to a certain acoustic environment in order to evaluate its performance. Within this work, the program EASE (Electroacoustic Simulator for Engineers) [8] has been used to calculate the impulse response of the room, and its sister program EARS (Electronically Auralized Room Simulation) to convolve the room response, HRTFs of the ear and an anechoic signal.

Figure 2 illustrates the general concept behind such a system. Firstly, the room or corridor is geometrically modelled (in EASE) for positions, dimensions and orientations of surfaces including the assigning of materials to each surface (which includes absorption / reflectance characteristics). The sound source(s) are then positioned within the space including their directivity (or dispersion) patterns. An audience area is resultantly mapped onto the space of which several listener positions can be taken. Using geometrical acoustic algorithms, namely ray-tracing, (and if needed, the mirror-image method) the propagation of sound from source to receiver is calculated for linear distortions, such as the spectral modifications from absorption in the atmosphere or by the surface materials. As a result, components of the sound field that impinge upon the listener's head, namely direct sounds and reflections of different orders are simulated with their respect to their direction of incidence and arrival time. It must be noted at this stage that two components of the problem, smoke attenuation and influence of heat cannot be calculated with this software and will not be discussed as they are outwith the scope of this paper.

The next stage in the simulation is the inclusion of the listener's ears in the model. The transfer functions (section 3.2) are convolved with each sound ray's direction of incidence on the ear, the respective components from all sources added (within EARS). This essentially gives the binaural impulse response of the room.

The last step of the process is called *auralization*, i.e. transferring this data into audible sound. Here, the dry (anechoic) signal such as the alarm sounding, person screaming and ultimately the navigation beacon are convolved with the room impulse response. As a result, the effectiveness of any signal presented can be derived *virtually*, tested and re-tested in order to optimise its navigation potential. One definite problem with this system is that binaural room simulation usually assumes a static case, i.e. the source or receiver is not moving. In order to assess the viability of any

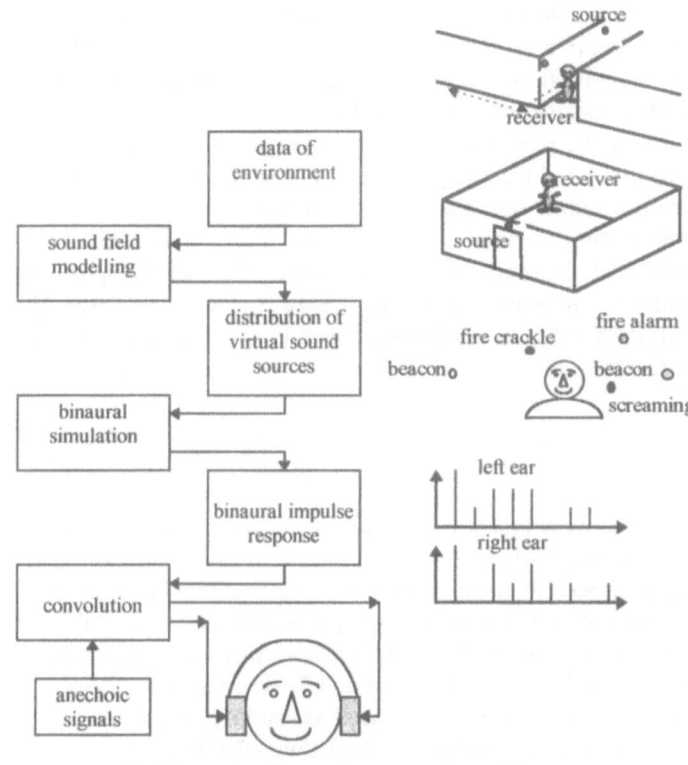

A block diagram of binaural room simulation [9]

localization system, the head must be allowed to move freely to reduce the possibility of front-back confusions which do arise in static cases (the cone of confusion). Additionally, it is difficult to accommodate multiple sources which are presented simultaneously, especially if there is a temporal difference between their presentation. These two points will be described in the conclusion.

4.2   *Preliminary ideas on signal type and presentation*

As can be seen, there are several compounding factors which dictate the type of signal to be presented for navigation. At a primary level, the spectral selectivity of the Head Related Transfer Functions dictates that some form of broadband noise must be used in order to engage all portions of the auditory system's localization ability. However, within this broad range of frequencies, there are certain critical bands which are not used and it may be possible to insert additional signals into these spectral locations. There is one definite reason for inserting signals; broadband sound has no real parallels in our everyday life, or in other words it has no real meaning to anyone. A fire alarm sounding, a baby crying and the roar of a lion all have meaning i.e. warning, comforting etc. This meaning must be incorporated into such a navigation system so that people are not confused with many ambiguous messages. Take for example the use of speech in such a navigation system. Speech has a relatively small bandwidth (1-4kHz) yet it is probably the most informative communication medium available. Unfortunately, speech does not have the necessary spectral components needed for localization and more importantly in a multilingual environment such as an airport, the presentation of evacuation directions in many languages would take too long.

This research as a result concentrates on the use of the broadband messages as described previously and suggests two time-variant approaches:

304

- Presentation of noise which is notch filtered and where pure tones are inserted to form some form of *melody*. As the person reaches the exit, they are reassured by a raising of pitch within the navigation system. If they are close to the fire, the tones undergo random spectral variations denoting complete danger.
- Presentation of noise which as the person reaches the exits gets closer and closer together, i.e. two pulses are presented for every navigation beacon and as the person approaches the exit, these pulses unite into a continuous stream.

As well as attenuating certain frequencies to insert messages, key localization frequencies must be boosted to accommodate the large spectral modifications from smoke and room impulse conditions. There is also the question of masking within the system whereby external *background* sounds also affect the signal presentation (such as the fire alarm).

5   Conclusion

This work is currently very much at the final experimental stage whereby the researcher is close to an optimum solution. As can be seen, there are many difficulties within such research, especially using *virtual acoustic* methods in order to prove them. Firstly, binaural simulation is far from offering real time solutions (taking several hours on a fast Pentium PC for one source) and thus head movement to resolve the cone of confusion cannot be included. Calculating atmospheric attenuation is an additional problem especially when considering that as well as smoke particulates being airborne, they also conglomerate on boundary surfaces thus affecting the absorption characteristics of the materials. The presentation of multiple sources within a corridor also affects the intelligibility of the signal, forming standing waves when pure tones are used within the navigation units. Finally, when using multiple sources, there has to be a time difference between each signal presented utilising another piece of software and many more hours of computation work.

In conclusion, it is the researcher's opinion that we may at some stage in the future see auditory navigation aids being used in the built environment in order to facilitate fire escape. However, the implementation of such systems will be very much technology and legislation driven as the potentially life threatening criticality of building evacuation is not to be taken lightly.

References
[1] **Malhotra, H. L.,** *Fire Safety in Buildings (Report for the Department of the Environment)*, HMSO Publications, 1987, p.57.
[2] **Canter, D.,** (ed.) *Fires and Human Behaviour*, John Wiley and Sons, New York, 1980.
[3] **Proulx, G.,** 1991, "To prevent 'panic' in an underground emergency; why not tell people the truth?", Third International Symposium on Fire Safety, *Science*, Elsevier, Essex.
[4] **Griffin, D.R.**: *Echoes of Bats and Men*. London: Heinemann, 1960.
[5] **Middlebrooks, J. C., Green, D. M.** (1991). "Sound localization by human listeners," *Annual Review of Psychology*. Vol. 42, 135 - 159.
[6] **Wenzel, E.M., Arruda, M., Kistler, D.J., Wightman, F.L.** (1993). Localization using nonindividualized head-related transfer functions," *Journal of the Acoustical Society of America*. Vol. 94, No. 1, 111 - 123.
[7,9] **Blauert, J.,** *Spatial Hearing*. The Psychophysics of Human Sound Localization, MIT Press, Cambridge, Massachusetts, 1997.
[8] **Ahnert, W., Feistel, R.** (1992). "EARS auralization software," In *93rd Convention of the Audio Engineering Society* (Preprint), San Francisco.

I-WALKWAYS

an Exploration in Knowledge Visualization

DANIEL C. GLASER
Department of Architecture
University of California, Berkeley

Abstract. This paper describes a prototype which extends a logic system into a useful design tool to aid in designing pedestrian walkways. A highly interactive program, I-Walkways demonstrates how a logic system can meaningfully aid with design. This technique will allow the designer and the logic system to work harmoniously together to reach a good design solution.

1. Introduction

Encoding non-trivial design knowledge into computerized tools has been a difficult task for researchers in environmental design. One approach for building software to aid the designer is with expert or logic systems. These systems are built by encoding design knowledge into first-order logic expressions. In order for a design to be accepted by a particular logic system it must follow each rule in the system. Thus, it is important that the rules that we specify are compatible with our intentions. In addition, our intentions may change dynamically during the design process. Further, it is improbable to describe all of a person's design intentions. Therefore a logic system must also be flexible enough to allow exploration outside of its rules. I-Walkways demonstrates a technique which provides the instruction of a logic system as well as freedom for the designer to explore his or her design world. This work was inspired by a series of papers by Galle and Kovacs (Galle 1992; Galle 1993; Kovacs 1994).

I-Walkways is a tool written in Java to help facilitate good pedestrian walkway design (Glaser 1997). This tool demonstrates how a designer can harmoniously interact with a logic system to reach a good solution. The system visually encodes two customizable evaluators for pedestrian walkway design. One of the evaluators helps the designer minimize the cost of building their walkway network. The other evaluates how straight a path is to get between two points of the walkway network. Although the usefulness of such evaluations may be debatable, I-Walkways attempts to make these guidelines as transparent to the user as possible. Hence a good visualization system will ensure that the rules that we specify are appropriate. Ultimately the designer has to decide how to incorporate this information into their work. As Jones states, "If a decision maker cannot understand a model nor the solution produced by an algorithm, then the model and algorithm are useless" (Jones 1996).

R. Junge (ed.), CAAD Futures 1997, 305-310.
© 1997 *Kluwer Academic Publishers.*

2. The I-Walkways Program

I-Walkways represents a pedestrian walkway network as a graph on a two-dimensional landscape (Figure 1a). The edges of the graph represent the walkways to be built. A walkway segment has no properties other than where its endpoints are. The endpoints of a walkway segment can either be entry nodes or intermediate nodes. Entry points are places where people can enter the walkway network such as a building or street entrance. They are represented by a square. The designer can add relationships between entry nodes to mark where he or she thinks there will be the most traffic (Figure 1b).

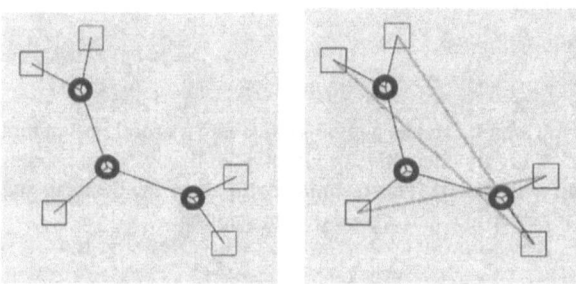

Figure 1. (a) I-Walkways representation of a pedestrian walkway network with five entry nodes and three intermediate nodes (b) after adding predictive main walking lines to the network.

2.1. THE COST EVALUATOR

A few cost evaluators are encoded to help the designer determine how expensive their proposed walkway network is. They represent a few of many potential ways of describing how much a walkway will cost.

At a local level, I-Walkways can help guide a user to place a node such that its neighboring segments have minimal total length. Although there is no analytical solution for this[1], I-Walkways guesses this value through the use of adaptive quad-trees. To convince the designer of its accuracy, there is an option to visualize how I-Walkways arrives at this solution. The system then colors the node according to either its relative (in percentage) or absolute (in pixels) difference from the determined minimal cost point (Figure 2).

---

[1] To find the minimal amount of material to connect an arbitrary number of nodes is NP-Hard. This problem can be reduced to the NP-Hard traveling salesman problem by substituting the cities for the nodes. Hence, at best, we can only find an approximate solution for a reasonable sized problem.

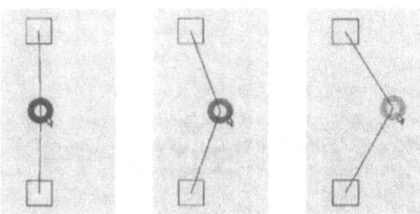

Figure 2. An example of the local cost heuristic being applied to a node being dragged. In the left-most image, the center node starts at a near optimal black state with respect to its absolute walkway cost. In the right-most image it turns cyan to signify that it is at least 25% above its optimal cost.

The system calculates the performance p_{value} of node i by the following formula:

$$p_{value} = \begin{cases} 100 * \dfrac{(C_i - C_o)}{C_o} & \text{if relative performance} \\ \\ C_i - C_o & \text{if absolute performance} \end{cases} \quad (1)$$

where C_i is the current cost and C_o is the optimal cost of node i. The case where C_o is 0 is avoided since a node with only one neighbor is defined to have an infinitely high cost. The current cost of the node, C_i, is computed as the summation of the lengths of all of its neighboring walkways:

$$C_i = \sum_{j=0}^{n-1} \sqrt{\left(x_j - x_i\right)^2 + \left(y_j - y_i\right)^2}$$

where n is the number of neighbors of a node, x_j is the x-coordinate and y_j is the y-coordinate y_j of neighboring vertex j. The color vector p_{color} is thus:

$$p_{color} = \begin{cases} a_{color} & \text{if } p_{value} \leq a_{value} \\ \\ a_{color} + \dfrac{p_{value} * \left(b_{color} - a_{color}\right)}{\left(b_{value} - a_{value}\right)} & \text{if } a_{value} < p_{value} < b_{value} \quad (2) \\ \\ b_{color} & \text{if } p_{value} \geq b_{value} \end{cases}$$

where p_{value} is the cost derived from (1), a_{value} and b_{value} is the user definable cost sensitivity range (Figure 3), and a_{color} and b_{color} are two base colors which signify the minimal and maximal values respectively.

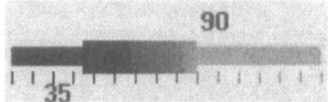

Figure 3. The dynamic cost range bar. Here a_{value}=35, b_{value}=90, a_{color} is black, and b_{color} is cyan. Values 35 and below are black, 90 and above are cyan, and intermediate values are linearly interpolated.

In addition to being a local cost advisor, I-Walkways can help evaluate the cost of larger parts of the network. An intermediate heuristic is used in I-Walkways to assess how costly a path is. In Figure 4, the path connecting the end-nodes which have main walking lines connecting them is colored cyan to signify that it is expensive to build. The coloring of a path is determined similarly as (2) with the exception that the shortest path is used for C_i and the size of the main walking line is used for C_o.

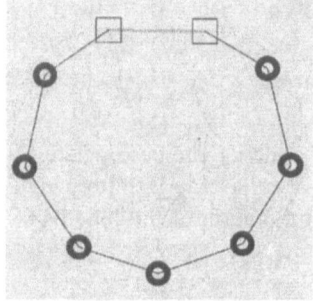

Figure 4. The cost evaluator finds a circumventive path which connects the end nodes of a predictive main walking line. This path is highlighted cyan to indicate that it is an expensive path connecting the main walking lines.

Lastly, I-Walkways reports the global cost for the walkway network. Unlike the previous two modules, this number is not relative to any optimal value. Instead the designer has to rely on comparisons among working and saved solutions to gauge its fitness.

2.2. THE STRAIGHTNESS EVALUATOR

The straightness evaluator helps predict if pedestrians will use the proposed walkways or will take a route off of the designed paths. When modifying a particular node, I-Walkways evaluates how straight the path or paths are which cross a node when modifying it. Although the correlation between straightness and pedestrian behavior "is not without a certain intuitive plausibility, and its unscientific status is hardly atypical of design knowledge" (Kovacs 1993).

Straightness of a node is defined with respect to its local neighbors. I-Walkways draws a set of guide arcs between each pair of neighboring nodes based on two user defined angles, α and β, specified by a range slider similar to Figure 3 (Figures 5a and 5b). This information gets more complex if a node has more than two neighbors (Figure 6). If a node only has one neighbor, a pair of guide lines are drawn to show where the node can be placed and still maintain the straightness quality (Figures 5c and 5d). If the single neighboring node is connected to many other nodes, then I-Walkways draws all the potential straight paths that could be constructed and, if possible, also indicates where the user can place a node to satisfy all the straightness criteria (Figure 6). A coloring scheme similar to (2) is used which compares α to the current node's placement.

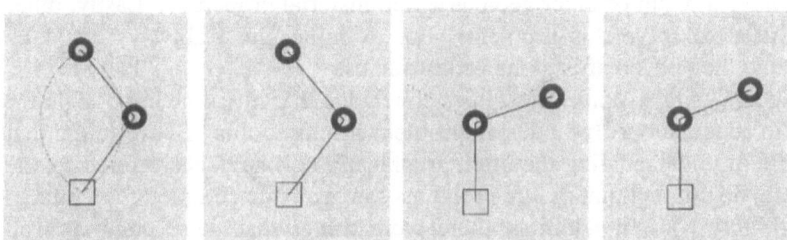

Figure 5. The straightness evaluator for (a) a center node with α =120° (b) α =100° (c) an end-node with α =120° (d) an end-node with α =100°.

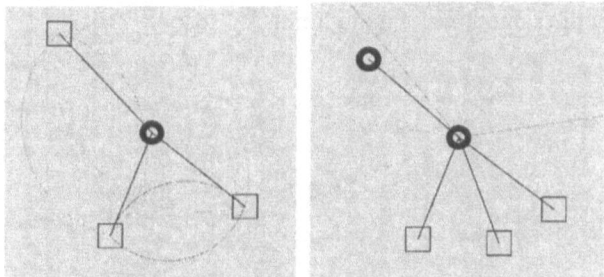

Figure 6. The straightness evaluator (a) advising a node with multiple neighbors (b) advising a node with one neighbor but connecting to many paths. Note the double arc lines in (b) which indicate where all criteria is met.

I-Walkways also checks to see if there is a path among neighboring points to avoid indicating a potential detour where an acceptably straight path already exists (Figure 7).

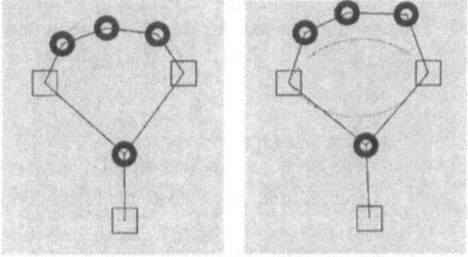

Figure 7. The straightness evaluator applied to the center node (a) with an existing acceptable path and (b) where the existing path between neighbors is worse then the path though the center node.

310

3. Conclusions

There are a number of practical and conceptual limitations for this work. First, the local cost heuristic may help find a local minimum, but it may be detrimental to the overall cost performance. In addition, the underlying model makes many simplifications. For example, the traffic pattern model is very crude, the landscape homogenous, and the walkways do not have a width or other geometric and material properties. Lastly, other methods can be used to solve this problem. For example, the walkways could be created naturally and then paved after some pedestrian use.

While building and using I-Walkways I have concluded that it will be impossible for a generative tool to adequately solve a design problem similar to this. I have found that I was quite capable of understanding the often graphically rich guidance offered by the software and hence, do not feel that design problems can grow too complex for a human to comprehend (Figure 8). In addition, there were things that these computerized evaluators could never be able to help me with. For example, for understanding the implications of a lake nearby. If I was ambitious enough to build a lake evaluator, I could always think of something else which needs consideration for the design. Good CAAD tools should be useful yet at the same time be quite limited in scope.

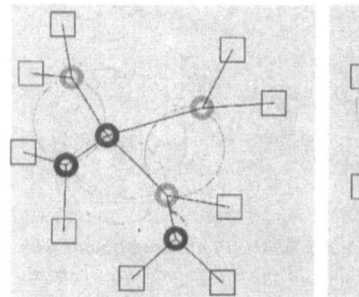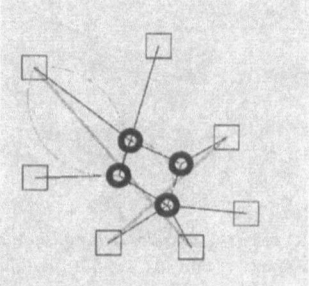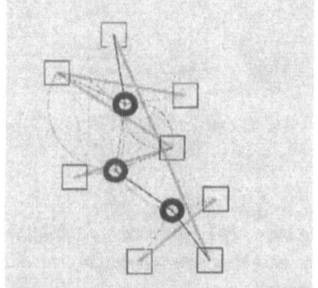

figure 8. Complex walkway networks.

Acknowledgments

I would like to thank Dr. Bruce Land at Cornell for his guidance in the visualization aspects of this project. At Berkeley, Dr. Yehuda Kalay has given me invaluable criticism and support for this work and future directions for it. In addition I would like to thank Dr. Nezar AlSayyad and the CAAD group at Berkeley for their helpful comments.

References

Galle, P., Kovacs, L. (1992) The logic of worms: a study in architectural knowledge representation, *Environment and Planning B*, **19** 5-31.
Glaser, D. (1997) I-Walkways, http://www.ced.berkeley.edu/~dglaser/research/iwalkways/, (10 April 1997).
Jones, C. (1996) *Visualization and Optimization*, Kluwer Academic, Norwell, MA.
Kovacs, L., Galle, P. (1993) The logic of walking: representing design knowledge on pedestrian traffic nets, *Environment and Planning B*, **20** 105-118.
Kovacs, L., Galle, P. (1994) The logic of plaza space: representing design knowledge on shape and function, *Environment and Planning B*, **21** 159-177.

A STUDY OF THE LOCATION OF FIRE EGRESS SIGNS BY VR SIMULATION

NAAI-JUNG SHIH, CHIE-SHAN YAN
Department of Architecture
National Taiwan Institute of Technology
43, Section 4, Keelung Rd.
Taipei, Taiwan, R.O.C.
shihnj@mail.ntit.edu.tw

Abstract

The purpose of this paper is to present a suggestion for the location of fire egress signs along a corridor in a building. The suggestion is made based on a virtual reality simulation of human behavior while rooms are on fire, particularly in a public Karaoka TV entertainment center (KTV). Both the rooms and smoke were modeled to simulate similar situations in which people were asked to find their routes to an egress. Case studies were made of the occurrence of two local severe fire disasters, the official investigation of damages, and related building codes. The simulation concluded that the traditional designation of egress signs at a higher location or just above the door frame may be not function appropriately in indicating the location of exit in case of fire. Since smoke is usually lighter than air and is accumulated closer to the ceiling level, either human vision or egress signs are very likely to be blocked by the darkness of smoke. Vision is additionally restricted because people are suggested to lower their body position to avoid smoke while escaping. Suggestion of alternate location of signage is also made in the research.

1. Introduction

The purpose of this paper is to present a suggestion of the location of fire egress sign along a corridor in a building. The suggestion is made based on a virtual reality simulation of human behavior while rooms are in fire, particularly in a public Karaoka TV entertainment center (KTV). Both the rooms and smoke were modeled to simulate real situation and people were asked to find path to egress. The process can help modeling design environment (Campbell 1996, Stansfield 1996) and facilitate decision-making in design (Campbell & Davidson 1996, Campbell & Wells 1995, Pearce & de Spiller 1995).

R. Junge (ed.), CAAD Futures 1997, 311-316.
© 1997 *Kluwer Academic Publishers.*

While building is getting larger, higher, and complicated, the consequence of fire becomes more severer and the distinguish of fire also becomes more difficult (Building and Fire Research Laboratory NIST 1996, Building Research Establishment 1996). The casualties and damage made to property and life show significant impact to public safety in public buildings in Taiwan. In order to prevent potential fire hazard, related issues have drawn public attention, such as fire prediction, detection of fire, broadcast system, evacuation, fire distinguishing, collapse prevention of structural members, and the number of fire sprinklers or hoses.

Karaoka TV entertainment center (KTV) is very popular in Taiwan. People can sing with the video and original melody from tapes. KTV usually consists of different types of design as in rooms for group of people or in stages for public access. Due to the amount of population simultaneously presented in a KTV and owner's ignorance of fire related facilities, KTV is always considered as a dangerous place and is frequently inspected by related government departments. Many studies were made to local fires occurrences. Survey concludes several factors affect human response in fire and listed all facilities related to fire egress. Official investigation tried to conclude reasons and consequently conduct modification of related building codes to prevent future disaster.

2. Two Cases

Smoke, high temperature, ambiguity of route, difficulty to rescue, and improper management of fire protection devices are among the main factors causing casualties when fire occurs. Two recent cases have drawn public attention and are discussed in terms of security design of fire protection. Investigation shows that:

- equipment which caused fire was placed on a corridor which was the main passage route during evacuation
- equipment which caused fire was placed near main exit
- interior was decorated with combustible materials
- exterior walls lacked openings for occupants to escape and smoke to ventilate naturally
- automatic systems to distinguish fire and to control smoke were not installed

In general, the problems related to fire protection are listed as follows.

- ignoring the principle of two-way corridor design
- mismanagement of passage ways and exits
- lacking the guidance of building managers, guards, or staffs
- incorrect movement or behavior of occupants
- problems in zoning and space planning
- missing clarity of exit signage in case of fire

Based on the classification of building codes, KTV and restaurant are rated as the second and third priority in building inspection (see Table 1). Significant casualties were caused in the two cases, although most of the design already met the requirements specified in building codes. Most of the casualties were caused by smoke for the reasons listed below:

- building codes do not differentiate the requirements based on the height of building and the amount of interior combustible materials
- current building codes lacks a complete regulations for smoke control and differentiation for levels of security in zoning and automatic smoke control
- current building codes do not explicitly specify the demand for the space of escape and do not differentiate situations based on floor area, density of population, amount of combustible materials, smoke control, etc.

As a result, most of the building codes only specify the minimum requirements. The design of fire control will never reach a sufficient level of security if general codes are applied. In order to reduce casualties, the codes should be modified based on current situations, materials, and customs.

Table 1. Inspection priorities of building types

Priority	Building Types
First	hotels (five stars), department stores
Second	theaters, movie theaters, MTV, KTV, hotels
Third	large restaurants, ball rooms, clubs, sauna, entertainment centers, etc.

3. Factors affect Evacuation and Planning

Evacuation represents the movement of occupants away from hazard region when building is on fire. Evacuation behavior is related to the movement of occupants who usually have several characters, such as being oriented by light, referring to familiar routes, and blindly following crowd. There are many factors are influential when fire occurs, such as fire sources and combustibility, fire spread and smoke flow, detection and fire fighting, escape and protection of occupants, and the capability and response of fire fighting from outside.

Based on the previously mentioned characters, routes for evacuation need to be designed properly:

- simple and clear
- more than one direction should be provided for refugee
- no interruption or dead end occurs to a passage way: both ends of a passage way should lead to a safe place, looped corridor usually are safer
- behavior in emergency should be considered
- applying windows, balconies, and facilities to evacuate occupants
- include area or floor for refugee in space planning

4. Test

Since the route for evacuation and the signage of exit are important factors in fire protection. A test was conducted, recorded, and analyzed accordingly.

314

The test conducted in this research was based on several restrictions:

- the simulation of egress route considers a floor area as the basic region of classification
- the planning of egress route is mainly horizontal, instead of vertical
- vertical stairs is the only traffic core safe for evacuation
- elevators cannot be used in fire
- the simulation of fire egress route is the main concern, starting location of fire and burning types are not included
- smoke, which generated from burning materials and fire, covers top one third of room height
- evacuation speed is one meter per second

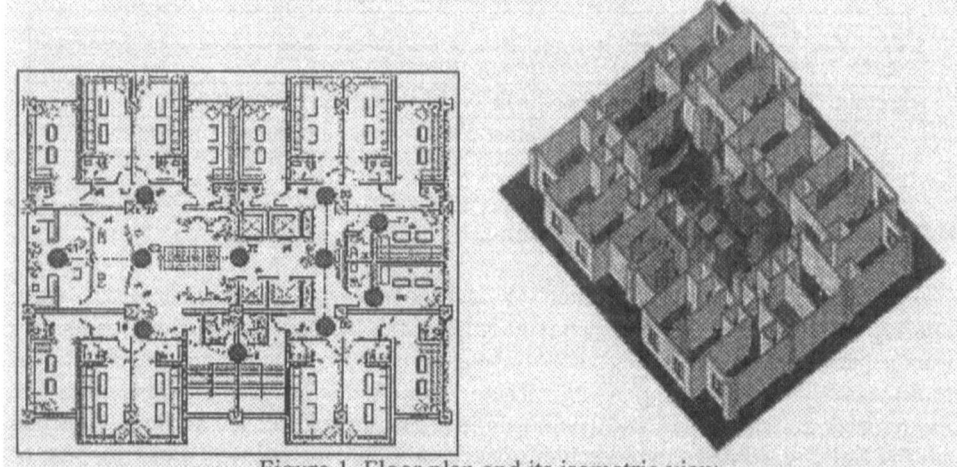

Figure 1. Floor plan and its isometric view

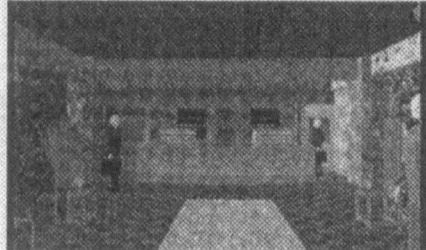 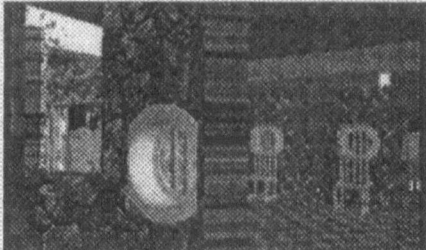

Figure 2. Interior views

The simulation applied QuickTime VR through which eleven scenes (see Fig. 1) were built to match the intersection points of corridors in plan (Bouman 1996). The scenes were built and rendered by ArchiCAD (see Fig. 2) and played by testees with VR Player. Each scene allows testees to view the surrounding environment in 360 degrees. A scene can be connected to the next by following the indication of an arrow. By clicking the arrow, a testee will be moved to the center of next scene, which simulates the escape from a node (an intersection of two corridors or a hallway) to another. Testees can choose directions in a scene to determine which one is the appropriate way to escape, without clear indication of egress signs. Each testee was demonstrated with the manipulation of cursor which corresponds to the movement in a scene first.   In the beginning, each testee was led by a waiter from entrance (the eleventh scene) to a room which represents the seventh scene where the simulation starts by informing her/him of the fire.   Once fire condition is given, testee tried to escape though the manipulation of cursor.   In the mean time, tester recorded the number of scene and the period of time that each testee had traveled before reaching exit.   Escaping route was drawn and shown to each testee.   Whole manipulation process of each testee was recorded by camcorder.

This test was an interactive manipulation which is different from traditional animation with defined route.   However, the connection between scenes was conducted by quick transfer from one to another, instead of continuous movement.   Due to the limits of computational capacity of computer and the demand for more memory, only eleven scenes were installed, instead of one scene at each interval.   Not many escaping routes could be selected.   Additional location of fire, which has significant influence to the judgment of routes, was not taken into consideration.

Future tests would include the improvement of the construction of computer models and manipulation process.   Tests would allow continuous movement between any two locations on plan, demonstrate the influence of the speed of smoke spread, specify the height of signage in various design, compare the difference between the existence of sign, and add more scenes in representing real environment.

4.5 FINDINGS

The test process was recorded and evaluated based on traveled distance and required period of time (see Fig. 3).   Distance is checked by the sequence of scenes that a testee has traveled in order to calculate the number of path segments.   The total period of time is measured in order to check if a certain limit is exceeded, by comparing to a human's escaping speed as one meter per second.   Recorded data show that testees tended to escape by taking the route that they came from.   The number of traveled scenes varies from three, which is exactly the shortest path to the entrance, to nine.   In general, time is proportional to the distance traveled.

The simulation concluded that traditional designation of egress sign at higher location or just above the door frame may be not functioned appropriately in indicating the location of exit in fire.   Since smoke is usually lighter than air and is accumulated closer to ceiling level (1/3 of room height), either human vision or egress sign is very likely to be blocked by the darkness of smoke color.   Vision is additionally restricted because people are suggested to lower their body position while escape for the presence of smoke.   The design of egress indication should be integrated with corridor design or the building components within the passage way, in contrast to exit indication only. The indication should provide egress direction at the first place when occupants are in corridors. The building components can be skirts, wall finishes, floor finishes, or any additional signage which have clear indication. Detailed suggestion of the location and the shape of signage is yet to be tested in future research and confirmed in real situation to prove its feasibility.

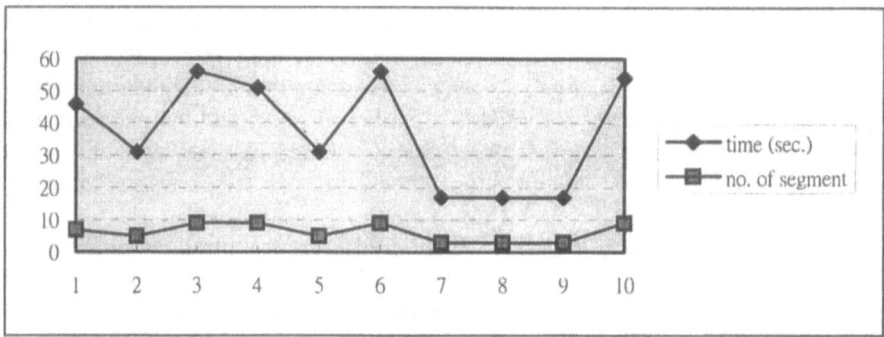

Figure 3. Relation between distance (no. of path segments) and time required

5. Conclusion

A simulation was conducted to confirm the influence of location of fire egress sign along a corridor in a building, through an image-based virtual reality simulation. Based on the findings of simulated result, the test concluded that traditional designation of egress sign at higher location or just above the door frame may be not functioned appropriately in indicating the location of exit in fire. Detailed suggestion of the location and the shape of signage is yet to be tested in future research and confirmed in real situation to prove its feasibility.

References

1. Bouman, O. (1996) Real Space in Quicktimes, Rotterdam: NAi Publishers.
2. Building and Fire Research Laboratory, National Institute of Standards and Technology (1996) Interactive Virtual Environments for Fire Simulation Project.
3. Campbell, D.A. (1996) Design in Virtual Environments Using Architectural Metaphor: A HIT Lab Gallery, Unpublished master's thesis, University of Washington, Seattle, Washington.
4. Campbell, D.A. and Davidson, J.N. (1996) "Community and Environmental Design and Simulation: The CEDeS Lab at the University of Washington." Daniela Bertol (Ed.) Designing Digital Space: An Architect's Guide to Virtual Reality. New York: John Wiley & Sons.
5. Campbell, D.A. and Wells, M.A. (1995) Critique of Virtual Reality in the Architectural Design Process, HITL Technical Report No. R-94-3. Seattle, WA: HIT Lab, University of Washington.
6. Fire Research Station, Building Research Establishment (1996) Using Virtual Reality to visualise buildings & fires Project.
7. Pearce, M. and de Spiller, N. (1995) Architectural Design [Profile No. 118, Architects in Cyberspace], 65(11-12).
8. Stansfield, S. (1996) "Virtual Environment for Architectural Walkthrough: Three Applications," The International Journal of Virtual Reality, Vol.2, No.3, pp.14-16.

A KB CAAD SYSTEM FOR THE PRE-CONCEPTUAL DESIGN OF BIO-CLIMATIC AND LOW ENERGY BUILDINGS

DR. ABRAHAM YEZIORO AND PROF. EDNA SHAVIV
Faculty of Architecture and Town Planning
Technion - Israel Institute of Technology, Haifa, Israel

This work discusses the structure of knowledge base CAAD systems for the design of solar and low energy buildings, along with the presentation of the different knowledge bases required for such systems. The general discussion is followed by presenting a KB-CAAD system PASYS, that was developed as a tool for determining thermal comfort design strategies in the pre-conceptual design stage. PASYS is based on a knowledge base which stores the existing information concerning thermal comfort rules of thumb and accurate procedural calculations, which facilitates defining thermal comfort design strategies that suite best the local climatic conditions and the specific constrains of the design problem at hand.

1. Introduction

Significant data and knowledge concerning bio-climatic and energy conscious design has been compiled during the last decade. However, the information has not been utilized widely by the designers due to its inaccessibility. When the initial design is based on correct climate- conscious principles, it may guarantee that later design stages will conform with the above principles and save expensive modification. The implication of changes at later design stages may sometimes be a redesign of the entire project. It is vital, therefore, to understand already at the pre-conceptual design stage what are the correct bio-climatic and passive solar design strategies which satisfy the requirements of the local conditions and which enable to achieve thermal comfort conditions, while conserving non renewable energy.

Shaviv and Kalay (1990) suggested that the process of designing energy-conscious buildings can be viewed as a sequence of decisions made at different levels of abstraction, each successive level more detailed and specific than the former one. These levels of abstraction correspond to discrete design stages, which include (Cross, 1977, Jones, 1970, Mitchell, 1977):

a. Briefing: statement of user needs.
b. Pre-conceptual design: feasibility study and determination of detailed program requirements.
c. Conceptual design: exploring different schematic design alternatives that agree with the programmatic requirements. This stage is concerned primarily with geometry and orientations, without considering material compositions.

R. Junge (ed.), CAAD Futures 1997, 317-330.
© 1997 *Kluwer Academic Publishers.*

d. Preliminary design: determining material compositions and building
                          details.
e. Detailed design: exploring different detailed design alternatives. This
                          stage deals with structure and material composition
                          considerations.
f. Design documentation: preparing building documents.

Shaviv and Kalay (1990) proposed a methodology for supporting computationally all
stages of an energy-conscious design and evaluation process, by partitioning the design
process into discrete stages and identifying the energy-conscious characteristics of each
stage. The methodology is based on merging procedural simulation and knowledge-based
heuristic methods into one integrated system. The knowledge base contains heuristic
rules for the design of passive solar buildings. Whenever possible, the knowledge-base
guides the designer through the decision making process. However, if the rules of
thumb are not valid for the particular design problem, the system guides the architect by
means of the procedural simulation model (Shaviv and Peleg, 1990).

Nevertheless, the heuristics used by expert systems of this kind are formed by
generalizations based on average cases. They may be totally inappropriate when applied
to a particular, nonstandard case. Also, heuristic methods are suitable for evaluating the
expected energy performance of typical design alternatives at the very early design stages
but are not suitable for evaluating the actual expected performance of detailed particular
design solutions.

Consequently, although procedural and heuristic methods are applicable and useful at
particular stages of the design process, neither one can support the entire design stages
on its own. Shaviv and Kalay suggested therefore to use through out the different design
stages a combined heuristic and procedural methods as is shown in Figure 1. In this
paper we further elaborate this approach by presenting the structure and the different
elements of such KB-CAAD system.

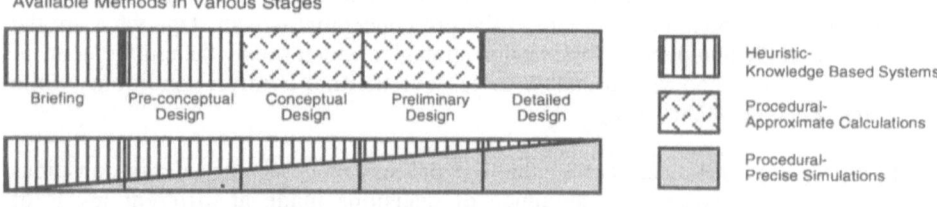

Figure 1: Schematic comparison of available vs. proposed energy evaluation
tools in different design stages (from Shaviv and Kalay, 1980).

2. The Design Process

2.1 THE ITERATIVE DESIGN PROCESS

The design process is characterized by being an ill defined problem, which means that
while searching for solutions, a better understanding of the goals and constraints might

occur. Therefore, we present the design process as an iterative process (see Figure 2). An iterative process allows to advance securely from stage to stage, but at the same time enable the possibility of returning to previous stages. As the design advances, it should acquire a greater level of knowledge and details. However, the availability of the required knowledge at the right time, and specially during the early design stages, may reduce the necessity for too many iterations, or the need to go back to a very early design stages, after an advanced design stage has already been achieved.

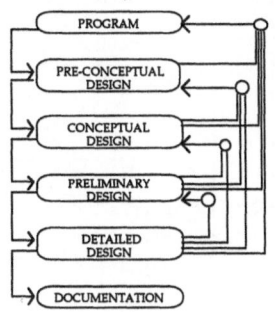

Figure 2: The Iterative Design Process

2.2 THE DESIGN ACTIVITIES

Each design stage, as was presented in figure 1, is composed from four activities (Maver, 1971); analysis, synthesis, appraisal and decision (see figure 3). We shall define these activities as follows:

a. Analysis: Defining the goals, relations between the design parameters and possible conflicts as well as arranging the random by collected data in a reasonable way.

b. Synthesis: developing partial and/or full design solutions.

c. Appraisal: comparing the performance of the solution with predefined goals and constraints, and verify the consistency of the solution.

d. Decision: choosing the best solution.

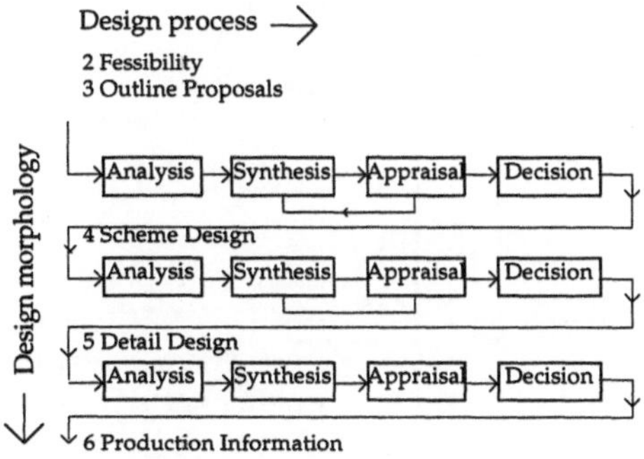

Design process →

2 Fessibility
3 Outline Proposals

Design morphology

→ Analysis → Synthesis → Appraisal → Decision

4 Scheme Design

Analysis → Synthesis → Appraisal → Decision

5 Detail Design

Analysis → Synthesis → Appraisal → Decision

6 Production Information

Figure 3: Basic scheme of a design process and its main activities (from Maver, 1971)

320

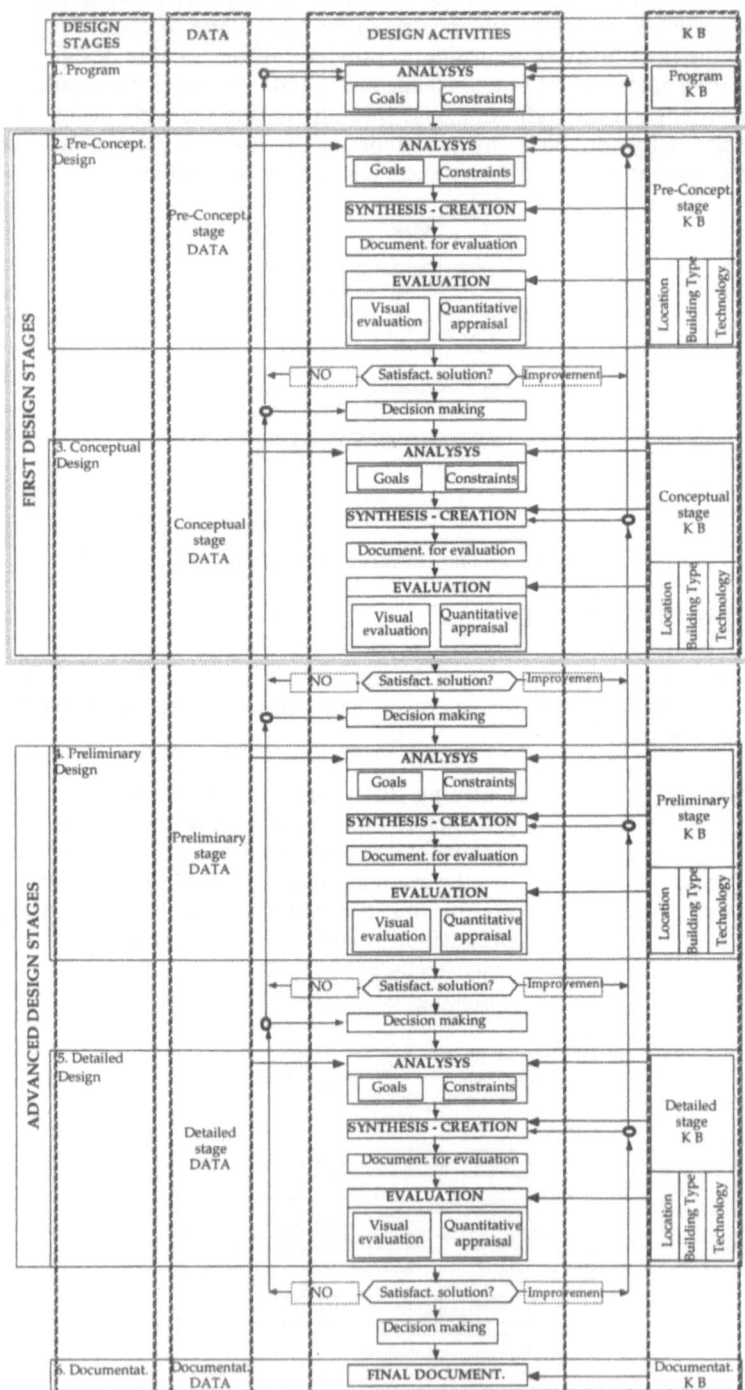

Figure 4: Basic scheme of a design process, main activities, knowledge bases and data

2.3 KB AND THE DESIGN PROCESS

To perform each activity at each design stage, one needs to accumulate the proper data and knowledge. The following question arises: should the data and/or knowledge base be related to the different design stages, or to the different design activities? Through the comprehensive development of a KB-CAAD system for the pre-conceptual and conceptual design stages (Yezioro, 1994), we found that the same data and knowledge bases are required for all activities in each design stage, but different data and knowledge bases are required for each design stage. We have summarized this conclusion in a chart that extend the design process suggested by Maver to include the knowledge bases as part of it (see figure 4).

Let us emphasis the fact that the knowledge, in KB-CAAD systems, should be available at the right time, but only the relevant knowledge should be presented, to avoid confusion due to excessive information. On top of it, the system should identify the level of accuracy required at each design stage. Redundancy in information can be as bad as not giving information at all, as the non energy expert designer might not be able to judge what to choose.

2.4 TYPES OF KB

Knowledge bases can be classified according to three main types (see figure 5):

a. Location: according to general geographic location (climatic zone) or specific location, for example, Tel Aviv or Jerusalem.
b. Type of building: according to recommendations for different types of buildings, for example, residential buildings, offices, schools, etc..
c. Technology: according to the technological possibilities required to realize a solution.

The knowledge bases can also be divided into two types according to:

a. Fixed KB: there is no need or possibility to change it
b. Dynamic KB: should be changed according to the specific case.

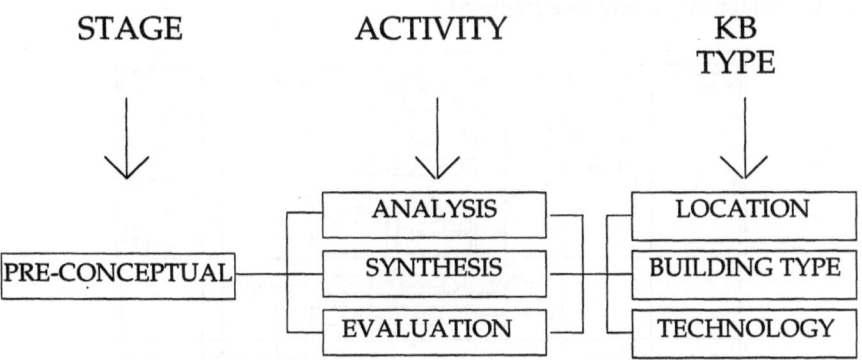

Figure 5: Types of knowledge base

3. PASYS - a Computer Aided Passive Solar Design System

The suggested outlines for the structure of knowledge base CAAD systems described above, is based on a thorough development of a computer-aided Passive Solar Design System (PASYS) that we have carried for the pre-conceptual and conceptual design stages. PASYS enables us to understand the relations between the different elements of the KB-CAAD system, and it serves as a case study.

PASYS behaves as an intelligent expert and is based both on a knowledge base which stores the existing information concerning solar-climatic construction in the form of rules of thumb, and on precise procedural models which enable finding solutions suited for the local climatic conditions. This knowledge can be retrieved and is available upon request. It also stores examples and explanations to allow the designer to better understand the different available design possibilities and in this way leads the designer to the best passive and low energy buildings in given climatic conditions.

The role of PASYS during the pre-conceptual design stage is to analyze the climatic conditions. On the basis of this analysis and the comfort chart, stored as a knowledge base, it suggests all possible combinations of passive design strategies that best fit the given location.

In the second stage, the conceptual design one, the role of PASYS changes and it presents existing passive systems for the summer and winter that comply with the suggested design strategies, as well as different examples of built solutions and principles, including explanations about each system and various heuristic rules to guild the designer. The passive solar systems can be determined on the basis of the required size of south glazing and the actual existing one. PASYS carries out a consistency check of the required passive systems for winter and summer, and recommends only those with components that suit both the cooling and heating requirements.

In each design stage PASYS deals with the various activities: analysis, synthesis, documentation, evaluation and decision making. The system is capable of proposing solutions corresponding with the particular design stage. These solutions take into account the constraints determined both by the designer and by the system itself, owing to the knowledge bases it contains. This work is limited to the presentation of the pre-conceptual design stage only (see figure 6).

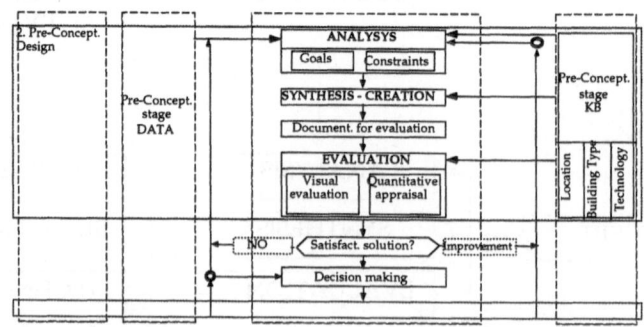

Figure 6: Basic scheme of the main activities in the pre-conceptual design stage

4. PASYS as a Design Tool for the Pre-conceptual Stage

Most Computer-Aided design systems are aimed to serve as evaluative tools at the advanced design stages, after a design solution has already been established by the architect. There are few CAD tools to help the designer in selecting the best suited design strategies at the very early pre-conceptual one. The decision about the preferred design strategies also determines the form of the solution and building and its performance.

Bio-climatic comfort charts and psychrometric charts have been proposed by different researchers as a method to analyze the local climatic conditions at a given location (Olgyay 1963, Milne and Givoni 1979, Arens et al 1980, Szokolay 1986, ASHRAE 1989). These charts help the designers to find whether the local climatic conditions fall in the range of what is defined as the thermal comfort zone and to find out what passive and low energy design strategies best suite these conditions. We know however, that the thermal comfort depends heavily on the type of clothing (CLO) and level of activity (MET) and therefore, adaptation of the thermal comfort zone and the different passive design strategies zones on these charts should be corrected accordingly (see Fig. 7) . Such a correction is not an easy task, especially if the whole range should be corrected in order to see the complete picture. Additionally, the designer who is not an expert in bio-climatic design, does not always know the kind of constraints he should select for his current problem. Furthermore, such charts include an enormous number of possibilities, and there is no guarantee that the desired climatic design strategy will be selected.

We present here a new KB-CAAD system for determining thermal comfort design strategies that overcomes the difficulties presented above. This system was developed according to the lines describes above. It presents the user with a dynamic and sensitive chart adapted to the level of activity, the type of clothing and defines the suitable constraints for the specific type of building. These adaptations can be performed automatically according to the kind of problem at hand, or as requested by the designer. The system is based on a knowledge base, which stores the existing information concerning thermal comfort rules of thumb, and precise procedural calculations, which enable defining very easily the thermal comfort design strategies that best suite the local climatic conditions and the specific constrains of the design problem.

In the process of designing solar and low energy buildings, we can define the activities for the pre-conceptual design stage as follows:

4.1 ANALYSIS

In the pre-conceptual design stage the analysis activity is the most dominant one. At this stage first adaptations between project demands, as is dictated by the program, specific constraints, specific conditions of the location and the available design strategies are taken place. In the analysis activity local climatic conditions are verified and checked against goals in order to establish the proper design principles.

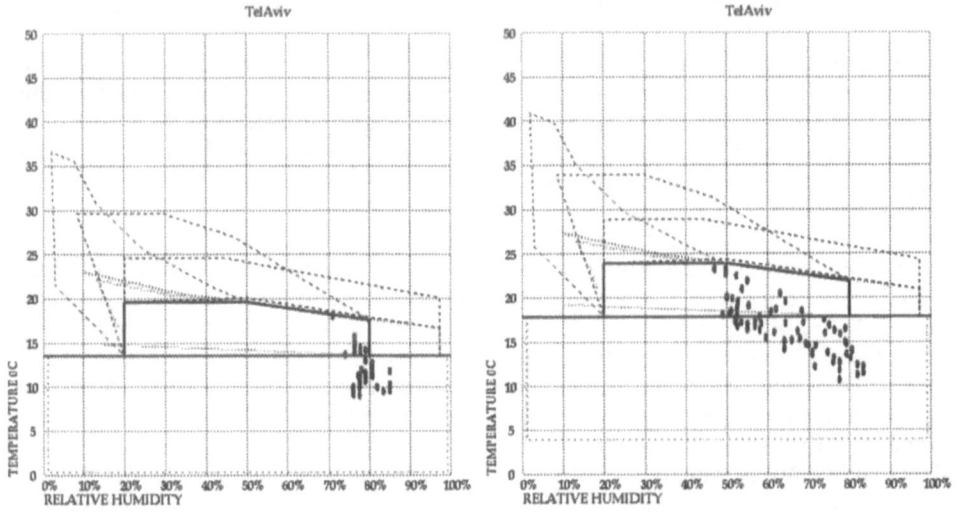

(a) Winter-Night: 3.0 clo 0.8 met

(b) Winter-Day: 1.0 clo 1.6 met

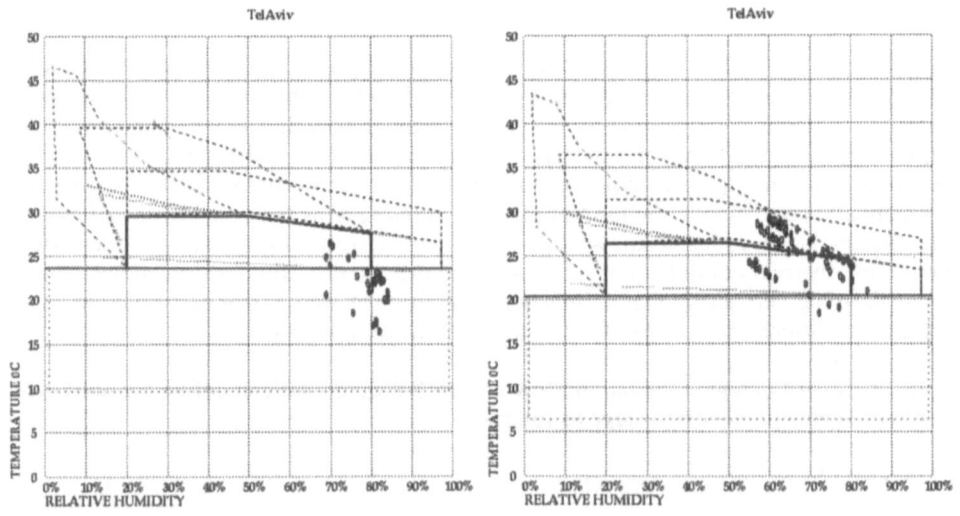

(c) Summer-Night: 0.5 clo 0.8 met

(d) Summer-Day: 0.5 clo 1.6 met

Figure 7. Dynamic bio-climatic chart for residential buildings.

4.1.1 Goals

From the climatic point of view the main goals for this stage are:

(I). To achieve a thermal comfort solution that will require minimal use of non-renewable energy. In other words, to look for energy conscious design solution on one hand, and to use, as far as possible, passive solar and low energy cooling strategies on the other hand.

(II). To reach goal (I) by using minimum thermal comfort design strategies, which means that a smaller diversity passive systems for heating and cooling will be required in the building. Goal (II) will ensure that less different building elements should be added to the building in order to achieve the required thermal comfort, thus obtaining a simpler and more economical solution.

4.1.2 Constraints

The constraints for this stage are: the level of activities, the level of clothing during winter and summer, day and night, as well as the maximum allowed wind, radiation and moisture in the building.

4.1.3 The knowledge base

The pre-conceptual design stage is based on a wide knowledge that presents the designer with the different comfort zones, the thermal comfort design strategies and the reasonable constraints. The knowledge bases can be classified according to three types, presented above, according to the location; like Jerusalem, Tel Aviv etc., building type; like residential buildings, commercial buildings, etc., and technology; like the different passive solar systems: direct heat gains, sunspace etc., or the different passive cooling systems, like thermal mass, night ventilation, natural ventilation radiative cooling, etc..

The knowledge base fixes the appropriate level of activity and type of clothing during day and night, for winter and summer and fixes the appropriate range of values for the different constraints. For instance, if the chosen activity is 'residential', than the system will allow higher wind speed in the room (see Fig. 8) than if the chosen activity would have been 'offices' (see Fig. 9, where the allowed wind speed in summer is only 1 m/s, in order to avoid papers being blown away). Also, it will assign proper values for clothing (CLO) and activities (MET), both for day/night and winter/summer. As for example in residential buildings this values for winter will be 1 CLO for day time and 3.0 CLO for the night, and 1.6 MET for the day and 0.8 MET for night time (see Fig. 8), while in the office building it will consider these values only for day time and these values are 1 CLO, and 1.4 MET (see Fig. 9). Moreover, the knowledge base recommends an appropriate comfort zone. For residential buildings a wider zone is applied (see Fig. 8) based on the chart suggested by Milne and Givoni [1979], while for an office building a narrower one is proposed (see Fig. 9) based on the chart suggested by Arens et al. [1980]. In this way an inexperienced designer has to deal with only few variables, while ensuring the consistency of the solution. All this is achieved just by clicking 'ADVICE ON' on the vertical menu (see Figures. 8 and 9). On the other hand the expert designer can choose 'ADVICE OFF' and examine any design variable by assigning different values to it. Thus he can "feel" the influence of the different design parameters on the space of possible solutions. For example, he can learn that just by dressing more heavily in winter or less in summer, thermal comfort may be achieved without adding mechanical means.

4.1.4 The data

The data required is according to the location, i.e. the local climatic conditions; temperature, relative humidity, wind velocity, etc. and the type of building; residents, office buildings etc..

4.2 SYNTHESIS - CREATING SOLUTIONS FOR PASSIVE COOLING AND HEATING DESIGN STRATEGIES

Since our design tool is a computerized one, it can carry out an hourly analysis of the climatic conditions in any specific location during the whole year. This is done after all the constraints are set up, either by the expert system of PASYS, or by the designer. The analysis is performed with respect to these constraints. The examination of the data consists of calculating the percentage of time during a month, a season, and the whole year, that the point, presenting the hourly climatic condition (temperature and humidity), falls inside the borders of the thermal comfort zone, or any thermal comfort design strategy. In such a case it is added to the total number of times such an event occurred.

4.2.1 Simple Solutions

First "simple" solutions are proposed instantly by the system (see right side of Fig. 8). These simple solutions are composed of only one cooling strategy and one heating strategy. The evaluation of each simple solution is performed and presented graphically and numerically for each month, season and for the whole year.

Each simple solution is represented by four parts as follows:

a. The first part indicates the comfort zone (CZ) i.e. it shows the percentage of time that the points presenting the climatic conditions (temperature and humidity) fall inside the boundaries of the comfort zone.

b. The second part is the passive solar heating strategy (PSH) i.e. the percentage of time that there is a need to provide heating in order to achieve comfort conditions. There is only one heating strategy defined in the thermal comfort charts and it represents in general the requirement for any passive solar system.

c. The third part represents one of the four possible cooling strategies i.e. the percentage of time that there is a need to provide a passive cooling strategy, in order to achieve comfort conditions. The possible cooling strategies are natural ventilation (NV), artificial ventilation (AV) using basically fans and ventilators, high thermal mass with night ventilation (HTM) and evaporative cooling (EC).

d. The fourth part is the amount of time that we need to provide the building with a non-renewable energy in order to achieve thermal comfort conditions. The need for non-renewable energy occurs when the sum of the above three parts is not 100%, otherwise its value is zero.

The overall evaluation for the simple solutions takes on the SGI Indy workstation less than one second. This means that the evaluation of any change that the designer makes on any constraint value or on any design parameter, are presented simultaneously on the screen and can be appreciated by the designer. If the designer is not satisfied with the results of the simple solutions he may ask the system to generate and present him "complex" solutions.

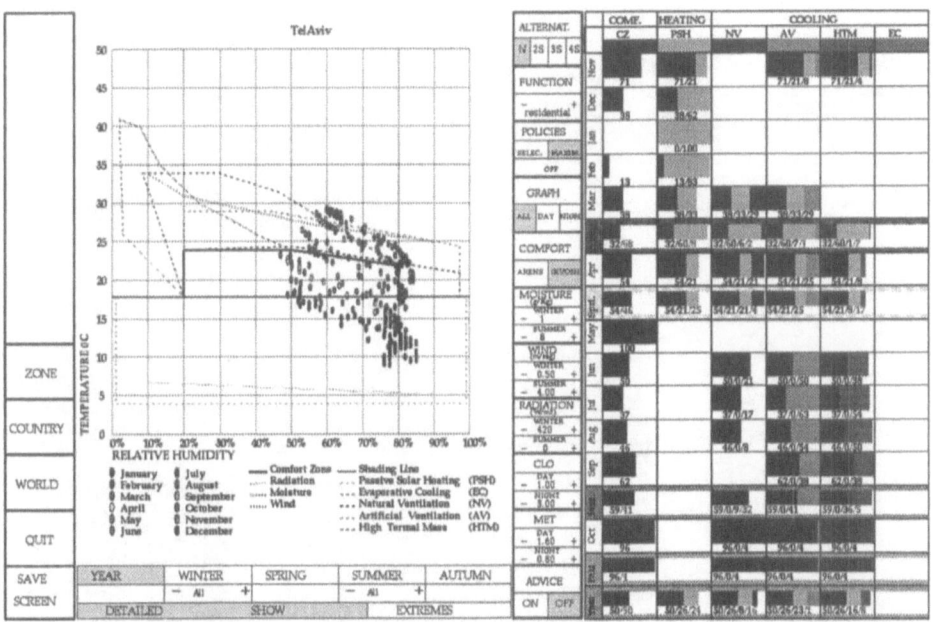

Figure 8. The bio-climatic chart for residential buildings and the space of simple solutions for the thermal comfort design strategies.

4.2.2 Complex Solutions

Contrary to the simple solutions that consist of only one passive cooling design strategy, complex solutions consist of more than one. (See the top part of Fig. 9). Let us mention that the bio-climatic chart includes only one zone that presents passive heating design strategy (without describing specific passive solar system). However, it shows more than one possibility for selecting a passive cooling design strategy. This means that the designer has the freedom to select more than one passive cooling strategy at the same time and that there exist different combinations to achieve a complex solution. When the designer selects 'ADVICE ON', the system takes care that any solution that includes more cooling design strategies than the former created solutions, should be better, in terms of thermal comfort and minimal use of non-renewable energy, than the previous created solutions. Such an evaluation, which is performed automatically while creating new solutions, guarantees that the two main goals presented before are satisfied (i..e. using as much as possible passive solar energy and achieving a low energy solution by using less different building elements). On the other hand the designer can choose the 'ADVICE OFF' option in order to create and examine any possible design solution he wishes to test.

The overall evaluation of two strategies takes less than a second on an SGI Indy workstation. Evaluation of three strategies takes less than five seconds and of four strategies about 10 seconds only. In any case the designer may choose the number of combinations he wants.

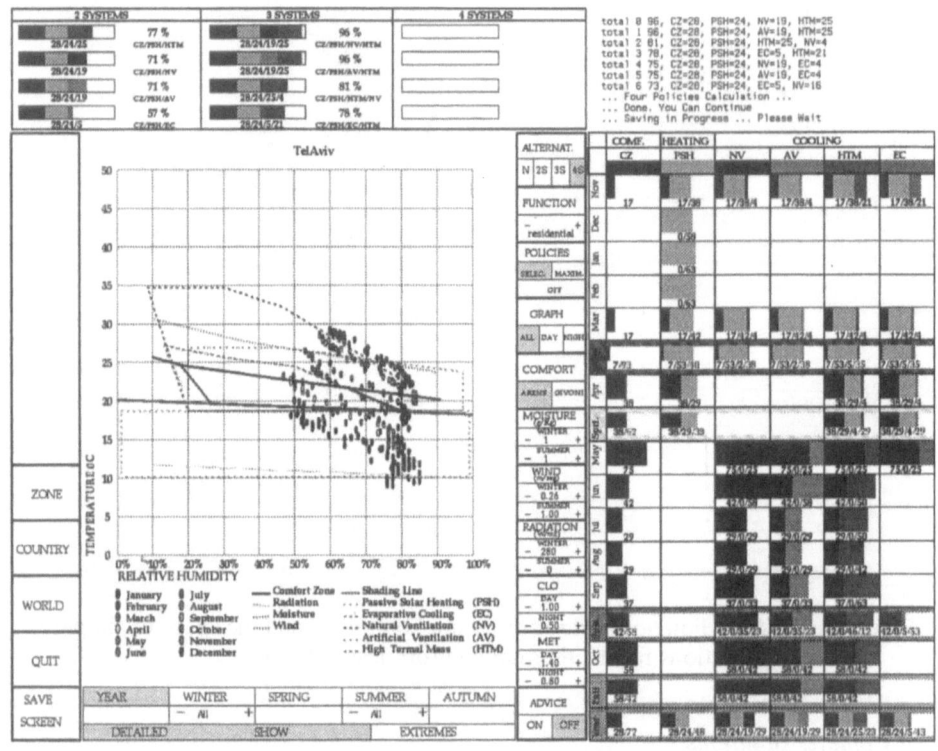

Figure 9. The bio-climatic chart for an office building and the space of simple and complex solutions for the thermal comfort design strategies.

4.3 DOCUMENTATION FOR EVALUATION

In the pre-conceptual design stage the documentation activity is very simple and is presented as schematic graphs (see fig. 9 up). It shows the types of required strategies, the energy saving by each strategy and the required backup system (the white spot). The documentation allows to evaluate quantitatively and visually the different possible solutions.

4.4 EVALUATION

In order to be able to distinguish between the performance of the different design alternatives so that one can make the right decision, it is important to include the evaluation activity in all the design stages including the pre-conceptual one. Such evaluation may prevent the necessity to perform too many iterations throughout the design process as well, although it does not guarantee it. The evaluation is of two types:

a. Visual evaluation. It receives values by means of the senses or feelings. In our case the color and size of the spots in the schematic graph shows visually the potential saving obtained from each strategy and the potential saving obtained from all the strategies acting together.

b. Quantitative evaluation. The solution is graded by means of calculations or measurements. In our case the amount of non-renewable energy consumption needed in order to achieve thermal comfort conditions is calculated as well and shown for each design alternative. This value can guide the designer in choosing his/her preferred solution.

4.5 DECISION MAKING

In this activity the designer examines the different available alternatives for selecting passive design strategies in order to choose the one that best fits the problem in hand. The complex solutions presented on the top part of the screen (see Fig. 9 up) turn to be an active menu where the designer may choose any of these solutions as the preferred one just by pointing at it. The preferred solution will be evaluated more thoroughly in the next step, which is the conceptual design stage and in which the exact passive solar systems and their preferred size are determined. The description of choosing the best passive heating system, which is not embedded directly in the bio-climatic chart, is out of the scope of this paper and will be presented elsewhere.

In order to maintain the flow of the process and allow the iteration design process, the active menu, which is composed from the complex solutions, created in the pre-conceptual stage, remains throughout the whole design process. This active menu, allows the designer, should it become necessary, return and choose different complex solution for farther development.

5. Summary and Conclusions

In this work general outlines for structuring KB-CAAD systems are presented along with an example of developing a KB-CAAD system (PASYS) for the design of solar and low energy buildings. The system is very heavily in use by the students in the faculty of architecture and town planing. It allows them to understand better how to design energy conscious buildings in the different locations in Israel and the entire world, without the need to really be in the location of the project (like the situations in international competitions). Even us, the expert in the field, found the system very helpful and we use it in all our consulting jobs. On top of it, we use this KB-CAAD system as researcher, by running it for the different locations in Israel, with the aim to establish design guidelines for selecting the best design strategies in different locations in Israel (Shaviv, Capeluto and Yezioro, 1996).

References

ASHRAE. (1989) "Fundamentals Handbook SI Edition." American Society of Heating, Refrigerating and Air Conditioning Engineers, Inc. Atlanta, GA.

Arens, E., P. McNall, R. Gonzalez, L. Berglund, L. Zeren. (1980) "A New Bioclimatic Chart for Passive Solar Design." Proceedings of the 5th National Passive Solar Conference, American Section of the International Solar Energy Society.

Cross N, (1977) "The Automated Architect." Pion Ltd., London UK.

Jones J.C, (1970) Design Methods: Seeds of Human Futures, John Wiley and Sons, New York NY.

Maver, T. W. (1971) "Building Services Design - A Systematic Approach to Decision- Making." RIBA Publications Ltd., London.

Milne, M., B. Givoni. (1979) "Architectural Design Based on Climate." in D. Watson (ed.), Energy Conservation Through Building Design, McGraw-Hill, Inc.

Mitchell W.J, (1977) Computer-Aided Architectural Design, Van Nostrand Reinhold, New York NY.

Olgyay, V. (1963) "Design with Climate - Bioclimatic Approach to Architectural Regionalism." Princeton University Press. Princeton, New Jersey.

Shaviv, E., Y. E. Kalay. (1990) "Combined Procedural and Heuristic Method to Energy Conscious Building Design and Evaluation." in "Evaluating and Predicting Design Performance" Y E. Kalay (ed). John Wiley & Sons, Inc.

Shaviv E., U. Peleg. (1990) "A Knowledge Based Computer-Aided Solar Design System." ASHRAE Transactions, V. 96, 2.

Shaviv, E., I. G. Capeluto, A. Yezioro (1996) "Climatic and Energy Aspects of Urban Design in Hot-Humid Region of Israel. Part I: Principles for Climatic and Energy Design in Hot-Humid Climate and Determination of Design Strategies." Center for Research and Development in Architecture, Faculty of Architecture and Town Planning. Technion - Israel Institute of Technology, Haifa, Israel. Ministry of National Infrastructures, Division of Research and Development, Jerusalem, Israel.

Szokolay, S.V. (1986) "Climate Analysis Based on the Psychrometric Chart." International Journal of Ambient Energy, Vol. 7, No. 4. Ambient Press Ltd.

Yezioro, A. (1994) "Form and Performance in Intelligent CAAD Systems for Early Design Stages in Solar Building Design." Ds.C. Thesis. Supervisor: Prof. E. Shaviv. Faculty of Architecture and Town Planning. Technion - Israel Institute of Technology, Haifa, Israel.

ALGORITHM AND CONTEXT: A CASE STUDY OF RELIABILITY IN COMPUTATIONAL DAYLIGHT MODELING

K. P. LAM
School of Architecture, National University of Singapore

A. MAHDAVI
Dept. of Architecture, Carnegie Mellon University, USA

V. PAL
School of Architecture, National University of Singapore

Abstract

The systematic use of reliable modeling data is believed to improve the building design quality. The key term here is "reliability". There is general agreement that reliability in the context of modeling-assisted CAAD depends on the accurate description of both contextual parameters (climate, site, etc.) and building features (geometric and non-geometric properties) as well as the validity of the underlying simulation algorithms. In this paper, we specifically address the importance of detailed contextual information and computational algorithms for the reliability of the daylight modeling results.

1. Introduction

It has been argued that computational prediction of building performance can contribute to the improvement of design quality. This potential, however, depends on the reliability of computational results, which is in turn affected by the degree of accuracy in describing the contextual parameters (e.g. microclimate, sky conditions) and the validity of simulation algorithm.

In this paper, we specifically explore the implications of such factors in the domain of daylighting. Toward this end, we *a)* compare six alternative sky models in view of their performance in reproducing the "actual" sky luminance distribution in the tropical context of Singapore; *b)* investigate the implications of sky model selection for the detailed computational prediction of daylight distribution in an architectural space; *c)* address the implications of using a simplified calculation procedure instead of a detailed simulation model.

R. Junge (ed.), CAAD Futures 1997, 331-344.
© 1997 *Kluwer Academic Publishers.*

2. Comparative Study of Six Sky Luminance Models in a Tropical Context

2.1 MOTIVATION

Reliable prediction of illuminance and luminance distribution in daylit interiors necessitates an accurate model of the daylighting source, i.e., the sky. In this section we present a preliminary comparative assessment of six sky luminance models in view of their applicability in the tropical context, using measured data in Singapore.

2.2 THE SKY MODELS

The six sky luminance distribution models considered are:

1) CIE standard overcast sky: This sky as defined by the CIE has a non-uniform, isotropic luminance distribution, increasing from the horizon to the zenith (Hopkinson et al. 1966).

2) All weather sky model: Developed by Perez et al. (1993) on the basis of sky luminance data collected for predominantly clear skies, this model uses a clearness index and a sky brightness factor to compute the sky luminance distribution.

3) ASRC-CIE model: This model is a linear combination of four skies - the CIE or Kittler clear sky, the Gusev turbid clear sky, the intermediate sky and the CIE overcast sky. The coefficients of linear combination are computed using the sky clearness and the sky brightness factors (Littlefair 1994).

4) Brunger's model: This model describes the sky luminance distribution by parameterizing insolation conditions as functions of the ratio of global to extraterrestrial irradiance (Brunger 1987).

5) Kittler's model: Sky luminance distribution is described by parameterizing the insolation conditions in terms of the illuminance turbidity which is given as a function of direct normal extraterrestrial illuminance and direct normal ground illuminance (Kittler 1986).

6) Perraudeau's model: A theoretical formulation of a cloud ratio is used for computing sky luminance distribution (Perraudeau 1988).

The above models (with the exception of the CIE standard overcast sky) base their sky luminance distribution computations on the basic parameters of solar altitude, solar azimuth, diffuse horizontal irradiance, global or direct normal irradiance, and extraterrestrial solar irradiance. Kittler's model needs the ground reflectance as well.

For the purpose of the present study, the solar altitude, azimuth, and extraterrestrial solar irradiance are computed using Sol-Aris (Mahdavi and Lam 1994, Mahdavi et al. 1994).

Values of diffuse horizontal irradiance and global irradiance are obtained through measurement (see Section 2.3). The direct normal irradiance is computed using the diffuse and global irradiance values and the solar altitude.

2.3 MEASUREMENT

2.3.1 *Measurement Setup*
The Daylight Measurement Station at the National University of Singapore was upgraded from a General Class Station to a Research Class Station with the installation of a sky scanner in June 1996. The sky luminance distribution measurements are augmented by diffuse and global irradiance and illuminance measurements which are being used to validate the sky luminance distribution models.

The sensors for the measurement of the illuminances and irradiances are placed on the roof of the lift shaft of a four storey building sited on the top of a ridge to maximize unobstructed views of the sky. The following measurements are taken and data stored on an hourly basis from 7 a.m. to 7 p.m. daily.

(a)   Global irradiance on the horizontal and on the north, east, south and west facing vertical surfaces.
(b)   Diffuse irradiance on the horizontal.
(c)   Global illuminance on the horizontal and on the north, east, south and west facing vertical surfaces.
(d)   Diffuse illuminance on the horizontal.

The maximum, minimum, average and instantaneous hourly values are stored.

Shadow rings are used to provide the diffuse horizontal illumination and radiation values. The sensors for measuring illuminance and irradiance on vertical surfaces are screened from ground reflections by a circular ring and four diagonal fins within the circular area, painted matt black.

The sky scanner is installed on the roof of a lift shaft in the same building complex where the above equipment is placed. It is connected to a data logger placed in a nearby room. It measures both luminance and radiance at 145 points of the sky by scanning the sky dome. Scans are taken every half hour. The measurement points are shown in Figure 1.

2.3.2 *Data Collection Period*
Data collected for 41 days during the months of June, July, September, and November 1996, were used for the study presented.

334

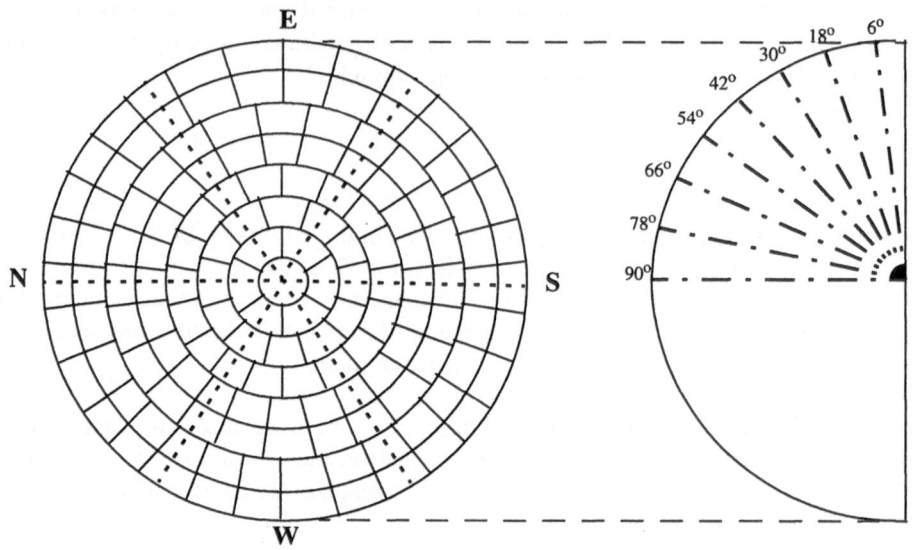

Figure 1: The measurement points for the sky scanner.

2.4 RESULTS

2.4.1 *Comparison Based on Correlation*
A total of about 65000 pairs of measurement-prediction values were used to derive the degree of correlation (r) between the measurements and the predictions were determined for each model. The results are summarized in Table 1.

Table 1: Correlation Between Measured and Predicted Values

Model	*correlation (r)*
Perraudeau	0.64
Perez	0.62
Kittler	0.57
Brunger	0.57
ASRC-CIE	0.36
CIE	0.17

2.4.2 *Comparison Based on Relative Error*

To obtain an intuitive sense of the deviations between measurements and predictions, mean relative error was used according to the following definition

$$RE_m = \left[\exp\left(\frac{1}{n} \cdot \sum_{i=1}^{n} \ln \frac{L_{p,i}}{L_{m,i}} \right) - 1 \right] \cdot 100 \; [\%] \qquad \text{eq. 1}$$

where
$L_{m,i}$ is the measured sky patch luminance
$L_{p,i}$ is the predicted sky patch luminance

For n = 1, this formulation obviously yields the "classical" definition of relative error (RE):

$$RE = \frac{L_p - L_m}{L_m} \cdot 100 \; [\%] \qquad \text{eq. 2}$$

The results are given in Table 2. Figure 2 shows the mean relative errors together with the deviation ranges which contain 70% of all relative errors. Figure 3 shows the mean relative errors of the models as a function of the sun position (expressed as the angular distance of the relevant sky patch from the sun position).

Table 2: Mean Relative Errors of the Predicted Values

Model	Relative Error [%]
Brunger	-3.0
CIE	20.8
ASRC-CIE	-23.0
Perez	26.5
Kittler	29.0
Perraudeau	33.6

2.4.3 *Comparison Based on Linear Regression*

Figure 4 shows the measurement vs. prediction regression lines for all models.

336

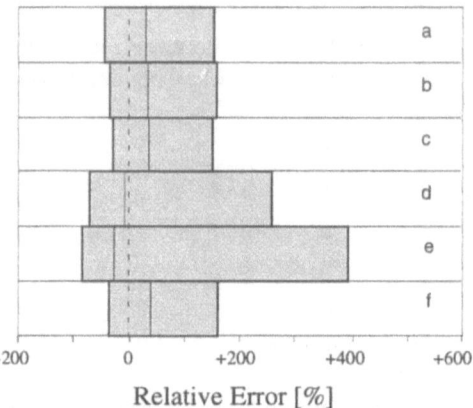

Relative Error [%]

Figure 2: Mean Relative Errors with the corresponding Deviation Ranges containing 70% of the relative errors. (a: CIE, b: Perez, c: Kittler, d: Brunger, e: ASRC-CIE, f: Perraudeau)

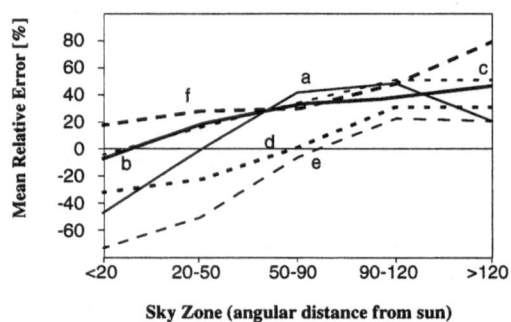

Figure 3: Mean relative error as a function of the angular distance of the relevant sky patch from the sun. (a: CIE, b: Perez, c: Kittler, d: Brunger, e: ASRC-CIE, f: Perraudeau)

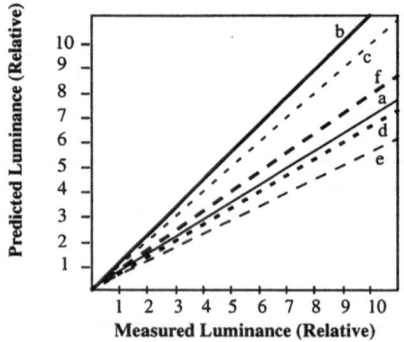

Figure 4: Measurement vs. Prediction regression lines for the five sky models. (a: CIE, b: Perez, c: Kittler, d: Brunger, e: ASRC-CIE, f: Perraudeau)

2.5 DISCUSSION

The rather short period of data collection and the resulting small set of data may not warrant a final judgement as to the relative performance of the five sky models in terms of their general reliability and in particular their applicability in the tropical context. However, if we assume that the currently available empirical data represents a good sample of the prevailing sky conditions in Singapore, the following provisional ranking among the models can be derived based on both relative error and correlation data:

Table 3: Provisional Relative Overall Performance of the Five Sky Models with Regard to the Measured Singapore Data

Model	Rank
Perez, Kittler	I
Brunger, Perraudeau, CIE	II
ASRC-CIE	III

None of the models appear to be ideal for the prediction of Singapore's sky luminance distribution. The question is, however, to what extent the differences among the models may have a significant impact on the reliability of prediction of daylight levels in architectural spaces.

3. The Implications of Sky Model Selection for the Prediction of Daylight Distribution in Architectural Spaces

3.1 PROBLEM STATEMENT

In the previous section, we compared the predictions of six sky luminance models with actual luminance measurements obtained with a sky scanner in a tropical context. In this section, we explore the implications of the sky model selection for the prediction of illuminance distribution inside architectural spaces. Specifically, we seek to answer the following questions:

a) What variations in the simulated illuminance levels for a typical architectural space is to be expected as a result of using different sky luminance models?

b) Is the extent of this divergence significant in view of other uncertainties involved in the design and evaluation process (e.g., building description, validity of the daylight simulation program, etc.)?

c) What can be said of the relative performance of various sky luminance models given the available measurement results?

d) What is the implication of using a simplified calculation procedure instead of a detailed simulation program?

3.2 APPROACH

3.2.1 *The Sky Luminance Models*
The six models that were considered for this study are described in Section 2.2.

3.2.2 *Simulations*
The program DSM (Mahdavi and Pal 1996, Mahdavi and Berberidou 1994) is used for the simulation of illuminance distribution in the test space.

DSM utilizes the three component approach (i.e., the direct, the externally reflected, and the internally reflected daylight component), to obtain the resultant illuminance distribution in buildings. The direct component is computed by numerical integration of the contributions from all of those discretized patches of the sky dome that are "visible" as viewed from reference receiver points in the space. External obstructions are treated by the projection of their outline from each reference point onto the sky dome and replacement of the relative luminance values of the occluded sky patches with those of the obstruction. A radiosity based approach is adopted for computing the internally reflected component. An earlier version of this program was compared to three other prediction tools and was found to provide reliable results (Mahdavi et al. 1993).

Simulations were performed for a fixed grid of points on the floor of the test space (cp. Figure 5). For the same time period, simultaneous measurements of indoor and outdoor illuminance levels were obtained.

3.2.3 *Simplified Calculations*
The BRS Split Flux method was used for simplified hand calculations (Hopkinson et al. 1966).

3.2.4 *Test Space*
An actual space (a living room in a multi-storey building in Singapore) was selected for simulations and measurements. A schematic plan of this test-space is given in Figure 5, which also includes the simulation/measurement grid.

Reflectances (ρ) of the surfaces of this room as well as the transmittance (τ) of the fenestration were obtained based on spot measurements and were used as the input parameter for the simulation model (cp. Table 4).

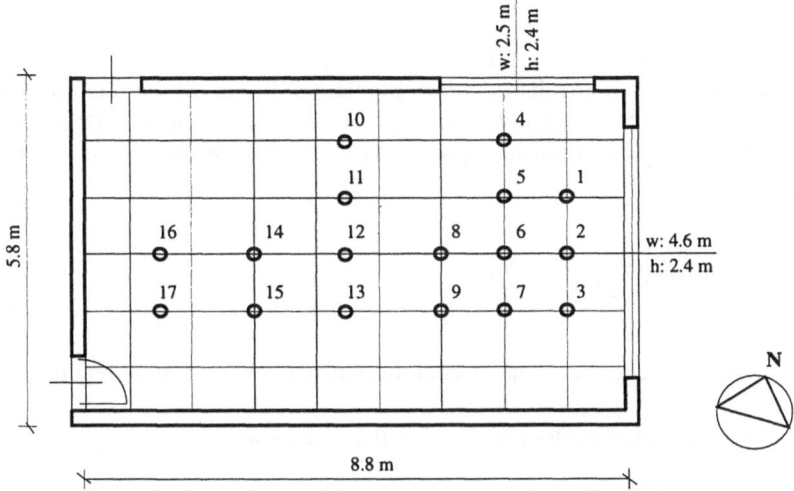

Figure 5: Schematic illustration of the test space and the location of the 17 reference points.

Table 4: Input Parameters for Simulation Model

	ρ [%]	τ [%]
Fenestration	0.10	0.70
Walls	0.60	NA
Floor	0.70	NA
Ceiling	0.65	NA

3.2.5 Measurements

Indoor Measurements. Continuous indoor measurements were performed over a period of three days for a selected number of points in the test space (cp. Figure 5). A total of 119 time-averaged (hourly) illuminance readings were used for the purpose of this study.

External Measurements. To "calibrate" the sky luminance models for the simulation procedure, certain measured data (diffuse and global horizontal irradiance, as well as diffuse horizontal illuminance) are needed. These were obtained from the radiometric and photometric measurement station at the National University of Singapore (Ullah 1996).

340

3.3 RESULTS

3.3.1 *Comparison of the Sky Models*
Indoor illuminance levels for the reference points (cp. Figure 5) were computed for all six sky models using the aforementioned daylight simulation program (DSM). Based on these illuminance values, mean daylight factors (DF_m) were derived according to the following equation:

$$DF_m = \frac{1}{n} \cdot \left(\sum_{i=1}^{n} \frac{E_{p,i}}{E_{e,i}} \right) \cdot 100 \ [\%] \qquad \text{eq. 3}$$

where

$E_{p,i}$ the simulated indoor horizontal illuminance level for point p at hour i

$E_{e,i}$ the simulated outdoor horizontal illuminance level for point p at hour i

n the number of (hourly) illuminance readings for point p

The results are shown in Figure 6, together with the daylight factors derived from illuminance measurements for the same points.

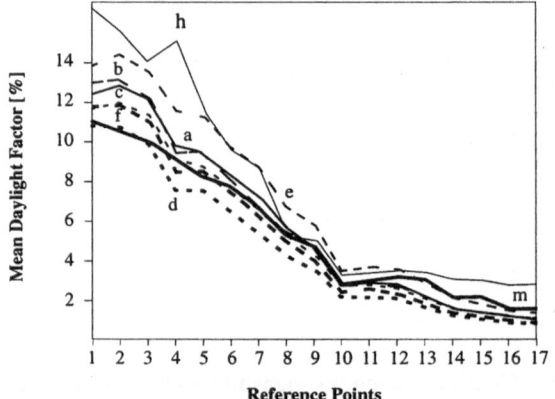

Figure 6: Mean simulated daylight factors for 17 reference points in a test room (cp. Figure 1) using six different sky models (a: CIE, b: Perez, c: Kittler, d: Brunger, e: ASRC-CIE, f: Perraudeau, h: Hand calculations, m: Measured values).

Figure 7 includes the daylight factors averaged over all six models (together with the corresponding range of standard deviations), as well as the measured results.

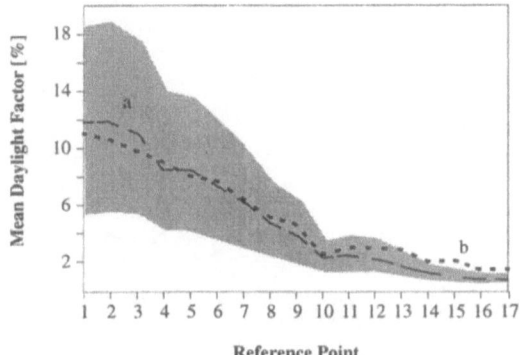

Figure 7: The predicted daylight factors (curve a) averaged over the six sky models with the corresponding standard deviations (the shaded zone). Also given are the measured average daylight factors (curve b).

3.3.2 *Comparison of the predictions and measurements*
The relative error (RE) for the predictions of each sky model was calculated for each reference point according to the following equation:

$$RE = \frac{DF_{m,s} - DF_{m,o}}{DF_{m,o}} \cdot 100 \ [\%] \qquad \text{eq. 4}$$

where

$DF_{m,s}$ is the simulated mean daylight factor

$DF_{m,o}$ is the measured mean daylight factor

The results are shown in Figure 8. Figure 9 shows the mean relative errors with the corresponding standard deviations.

3.3.3 *Summary*
Table 5 ranks the sky model and algorithmic options considered in this paper according to their performance in predicting the daylight levels in the test space.

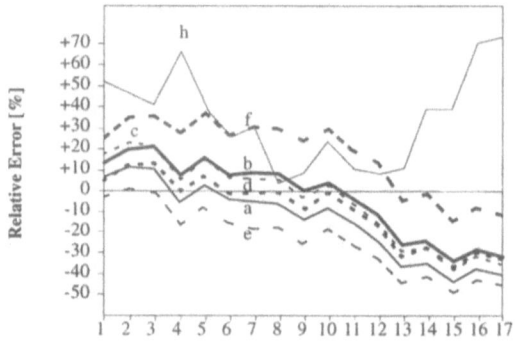

Figure 8: The relative errors in predicting the daylight factors for the 17 reference points in the test room using DSM with various sky models (a: CIE, b: Perez, c: Kittler, d: Brunger, e: ASRC-CIE, f: Perraudeau) and using hand calculations with CIE sky (h).

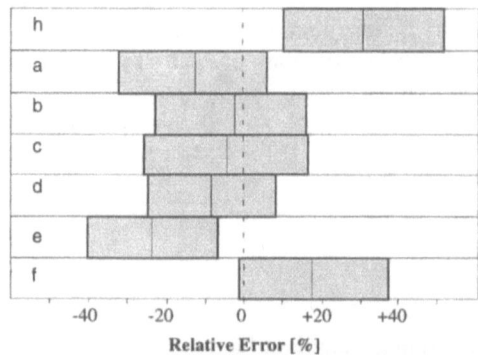

Figure 9: Mean Relative Errors (averaged over the 17 reference points in the test room) with the Corresponding Standard Deviations. (a: CIE, b: Perez, c: Kittler, d: Brunger, e: ASRC-CIE, f: Perraudeau, h: Hand calculations).

Table 5: Ranking of the Predictive Reliability of the Combined Algorithm and Sky Model Options

Rank	Sky Model	Algorithm
I	Perez, Kittler, Brunger	DSM
II	CIE	DSM
III	ASRC-CIE, Perraudeau	DSM
IV	CIE	BRS Split Flux

3.4 DISCUSSION

The following discussion of the results must be seen in the context of the following limitations and constraints:

- While the space selected for this study is quite typical for residential buildings, we do not imply that the results obtained for this space unconditionally apply to all spaces, independent of their geometry, photometric properties of the building components, and the building orientation.
- Parallel continuous measurements of indoor and outdoor illuminance levels could be carried out only for a rather short period of time.
- While DSM has shown to be a rather reliable prediction tool, factors such as uncertainties in the determination of the input data (geometry of space and apertures, photometric properties of the building components and elements) may lead to inaccuracies in prediction of illuminance levels. Based on prior experience, errors of up to ±12 % must be expected.
- The main purpose of this paper was to explore the divergence of predictions based on various sky models relative to each other. The data in this paper on the prediction-measurement differences is not meant to be seen as the ultimate judgement concerning the performance of the sky models.

Having established this understanding, we proceed to answer the four questions stated in the problem statement (cp. Section 3.1):

a) While the sky models considered for this study display a remarkable agreement in terms of their overall trends, they clearly lead to divergent predictions of indoor illuminance levels in absolute terms (cp. Figure 6).

b) This divergence in predicted illuminance levels for the six sky models is not negligible. As Figures 6, 7, and 8 clearly demonstrate, the choice of a specific model does significantly affect the predicted illuminance levels. Here, "significance" is meant to imply that the errors caused by the prediction of a specific sky model may occasionally be of higher magnitude than those caused by the inaccuracies of the model itself (algorithmic limitations) or the model input data (building description).

c) DSM's prediction of the trends in daylight distribution in the test space generally match the trend established by the measurements, independent of the sky model selected. However, the absolute value of predictions do deviate to various extents from the measured results. According to the currently available measurement results, DSM predictions are closer to the measured values (particularly in the front portion of the test space) when Perez, Kittler, or Brunger model are used.

d) As it can be seen from Table 5, the use of simplified calculation procedures is not desirable, as it leads to high errors. This is particularly significant in the context of the present study, as the errors may not be attributed to the underlying sky model (CIE), whose predictions are not significantly more error prone than the other sky models.

In conclusion, we would like to note that the combination of an advanced simulation model and a detailed sky model that would closely match the local sky luminance distribution patterns is very likely to provide the designer with lower errors in prediction of the daylight distribution in architectural spaces. More specifically, it is likely that this error in predictions of a detailed simulation tool and a reasonably accurate sky model (with a better local fit than the six models considered in this study) may not be significant if seen in the context of the uncertainties involved in the description of the geometry, materials, furnishing, and use patterns of interior spaces in the design phase.

Acknowledgment

The authors wish to acknowledge the contributions of Dr. M. B. Ullah (Associate Prof., School of Building and Estate Management, National University of Singapore) in providing the measured outdoor irradiance and illuminance data used for this study.

References

Brunger, A. P. (1987). "The Magnitude, Variability and Angular Characteristics of the Shortwave Sky Radiance at Toronto". University of Toronto, Ph.D. Thesis.

Kittler, R. (1986). "Luminance Models of Homogeneous Skies for Design and Energy Performance Prediction". *Proceedings, 2nd. International Daylighting Conference*, Atlanta, GA: American Society of Heating, Refrigerating and Air-Conditioning Engineers.

Hopkinson, R. G., P. Petherbridge and J. Longmore (1966). "Daylighting." London : Heinemann.

Lam, K. P., A. Mahdavi, M. B. Ullah, E. Ng, V. Pal (1997). "The Implications of Sky Model Selection for the Prediction of Daylight Distribution in Architectural Spaces". *Proceedings, 5th. International IBPSA Conference*, Prague, Czech Republic.

Littlefair, P. J. (1994). "A Comparison of Sky Luminance Models with Measured Data from Garston, United Kingdom". *Solar Energy* Vol. 53, No. 4, 1994: pp. 315-322.

Mahdavi, A. and V. Pal (1996). "The Characteristics of DSM". Internal Report DOA-01-11-96, Department of Architecture, Carnegie Mellon University, Pittsburgh, PA, USA.

Mahdavi, A. and L. Berberidou (1994). "GESTALT: A Prototypical Realization of an 'Open' Daylighting Simulation Environment". Journal of The Illuminating Engineering Society. Vol. 23, No. 2. pp. 62-71.

Mahdavi, A., L. Berberidou, K. P. Lam, Z. Li (1994). "Computational Support for the Concurrent Design of Natural and Electrical Lighting Systems in Buildings". *The 1994 IESNA Annual Conference*, Miami, FL. pp. 721-730.

Mahdavi, A., and K. P. Lam (1994). "A Computer Application for the Consideration of Solar Radiation in Architectural Design Process". Internal report 94-03 (Pittsburgh, PA: Center for Building Performance and Diagnostics, Department of Architecture, Carnegie Mellon University).

Mahdavi, A., L. Berberidou, P. Mathew, K. J. Tu (1993). "Prediction of Daylight Factors in 'Realistic' Settings: A Demonstrative Case Study". Journal of The Illuminating Engineering Society. Vol. 22, No. 1. pp. 40-44.

Perraudeau, M. (1988). "Luminance Models". *National Lighting Conference and Daylighting Colloquium*, Cambridge, England.

Perez, R., R. Seals, and J. Michalsky (1993). "All-Weather Model for Sky Luminance Distribution - Preliminary Configuration and Validation". *Solar Energy* Vol. 50, No. 3, 1993: pp. 235-245.

Ullah, M. B. (1996). "International Daylighting Measurement Program - Singapore Data I: Quality of Data Gathered Over a Long Period". International Journal of Lighting Research and Technology, CIBSE Series B, Vol. 28, No. 2, pp. 69-74.

A COMPUTER-AIDED EVALUATION TOOL FOR THE VISUAL ASPECTS IN ARCHITECTURAL DESIGN FOR HIGH-DENSITY AND HIGH-RISE BUILDINGS

THOMAS S. P. LI
Department of Architecture
The University of Hong Kong
Hong Kong

BARRY F. WILL
Dean
Faculty of Architecture
The University of Hong Kong
Hong Kong

Abstract

The field of view, the nature of the objects being seen, the distances between the objects and the viewer, daylighting and sunshine are some major factors affecting perceived reactions when viewing through a window. View is one major factor that leads to the satisfaction and comfort of the users inside the building enclosure. While computer technologies are being widely used in the field of architecture, designers still have to use their own intelligence, experience and preferences in judging their designs with respect to the quality of view. This paper introduces an alternative approach to the analysis of views by the use of computers. The prototype of this system and its underlying principles were first introduced in the CAADRIA '97 conference. This paper describes the further development of this system where emphasis has been placed on the high-rise and high-density environments. Architects may find themselves facing considerable limitations for improving their designs regarding views out of the building under these environmental conditions. This research permits an interactive real-time response to altering views as the forms and planes of the building are manipulated.

1. Introduction

"Visual aspects" are here defined as the responses perceived by users while viewing out of a building. The field of view, nature of the objects being seen, the distances between the objects and the viewers, and the daylighting and sunshine are some key factors affecting these aspects. Unlike many other engineering criteria such as structural

345

R. Junge (ed.), CAAD Futures 1997, 345-356.
© 1997 *Kluwer Academic Publishers.*

loading, wind loading, and dimensions of escape routes, etc., the "view" being a qualitative entity presents difficulties in being measured using conventional mathematical tools but it is probably one of the major factors that leads to the psychological satisfaction and comfort of the users inside the building enclosure. Yet little research has been carried out on the visual aspects from the users' point of view whilst most concentration has been placed on the exterior appearance of the building and its external relationship to the built environment. Generally in addition, the results of the limited studies performed have been presented in descriptive forms or in drawings that can be subjectively interpreted by human observers but not by computers.

A "window" here is defined broadly as any opening that permits users to view out of the interior of a building. Obscure or translucent openings are not considered here, as they are not related to the visual aspects of the user interaction. Positions and orientations of window openings are crucial to the visual performance of a building as windows respond critically to daylighting, sunshine and view, of which each process in its own way can be decisive to the synthesis of a successful architectural design. The primary goal of this study is to find a way to assist architects in designing "visual efficient" buildings. Designers can use this evaluation tool in the early stages of the design process to evaluate the performance of their designs concerning the visual aspects. The evaluation results are shown on the computer screen in real-time. The designers can then modify or refine their designs according to the results. By going through an evaluation and modification loop an excellent architectural design with the visual aspects related to other design criteria can be achieved. In order to use this evaluation tool in the early design stages, a modelling tool has been developed to allow the designer to input the design geometry of the building to the computer in a user-friendly manner. As the result is in real-time, changes are immediately incorporated and evaluated.

The prototype of the modelling tool and the underlying principles and methodologies of the evaluation tool were first introduced in the CAADRIA '97 conference (Li and Will, 1997). This paper describes the further development of this system. The system development is concentrated on high-rise and high-density environments because of the fact that architects face many more limitations in improving their designs for better quality of view in such environments than in any other circumstance.

2. Summary of the previous development

This research relates to a single module of the Interactive Optimization Tools for Architects system (IOTA) which is currently being developed to assist architects in the design process. The modules of the IOTA are plug-ins and can be used individually or in concert. Obviously the more plug-ins that are switched on the greater the demand on computing power and the slower the overall process but the advantage is that more realistic and rational decisions can be made. This particular module can be divided into two parts – the modelling tool and the evaluation tool. In the early stages of the design process, most decisions about the design are not yet made nor is the geometrical information of the design stabilized. On that ground most of the existing commercial computer-aided design software, such as AutoCAD, ArchiCAD, etc., are not suitable applications in these early design stages. They all require a certain amount of information about the design geometry in order to build the computer model. They do not respond easily to amendments of the computer models once the geometrical data are input. This modelling tool is built so as to make the use of the evaluation tool at the earliest design stages easy. For the sake of evaluation and of presentation of the results, the shapes of the buildings are approximated by 3m-cubes. A cube represents a unit of space which may not necessarily be empty. There may be rooms, corridors, staircases or internal partitions penetrating a cube. As a cube is a basic test unit it can neither be too large nor too small. Setting the cube too large would lead to a rough approximation of the building form while setting it too small would increase the number of iterations for testing and thus slowing the interactive process. A cube of 3m dimension was therefore chosen since it is approximately the typical height of a floor and its centroid plays an important role in averaging eye heights.

An architect can start designing the building form by specifying the number of floors, location and dimensions of the building on the site. As a default value a building of rectangular form is assembled accordingly as a stack of cubes. The designer can then remove unnecessary cubes from the rectangular block to form the shape that is required. When the basic 3D building form is established, the designer can then insert different elements such as slabs, windows, cores and internal partitions into the massing model.

As the building form is undergoing changes the corresponding performance of the design related to the visual aspects is being tested in real-time by the evaluation tool. The view seen by a person standing inside the preliminary building is simulated and evaluated based on the geometrical information of the building design and its surrounding environment which has been input previously. A basic test unit of the building is a cube of which the person is assumed to be standing in the middle of it. The eye level of the person is assumed to be 1.5m above the floor, that is, the eye is exactly at the center of the cube. (Future developments will include standing and seated observers and averaging iterations.) The program determines the view by emitting rays from the center of each cube in a 360° array horizontally and a 90° array vertically. By checking the obstacles hit by the rays the computer can know what objects can be seen by the person, how far they are and what the views are composed of. These data are then fed to a fuzzy system for computing the expected Visual Aspect Performance Index (VAPI) value. The final product of the Evaluation Tool is a building composed of colored blocks for the sake of easy visualization of the performance.

The fuzzy system model developed consists of five fuzzy sets and forty-five fuzzy rules. The five fuzzy sets are the quantity of plants (green area) in the view, the average distance between the observer and the surrounding buildings, the number and size of buildings in the view, the proportion of sky in the view and the proportion of sea in the view. These factors are chosen based on Markus' comprehensive studies on the psychological aspects of views (Markus, Brierley and Gray, 1972; Markus and Gray,

1972). It was stated clearly by Markus that the first four factors were highly correlated with people's general satisfaction with their living environment. The last factor which is specific to Hong Kong is added by the authors. It conforms to Markus' findings that people prefer more open space and fewer buildings in the view.

3. Improvements on the Modelling Tool

In order to make the 3D model of the site environment realistic without increasing the loading on the computational power, texture mapping is incorporated in the modelling tool. The advantage of texture mapping is that photo-realistic visualization can be achieved easily without having to increase the complexity of the 3D model. The demand on extra computational power is low if there is a hardware texture mapping device. As a result real-time photo-realistic 3D computer models can be visualized on the computer monitors when the designers are making their buildings using the modelling tool.

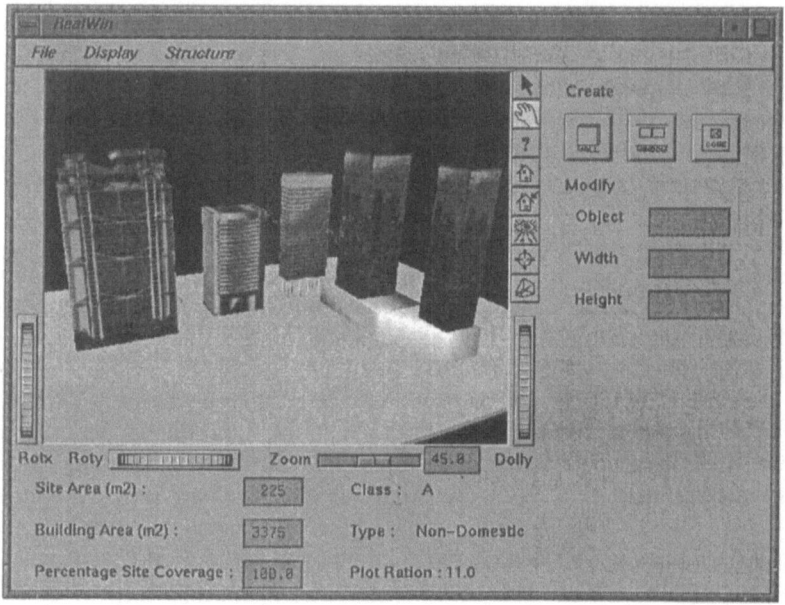

Figure 1

In the previous version of the program, as the form of a building was only approximated by 3m-cubes, flaws tended to appear when building models with curved faces were introduced. The resultant jagged faces were considered unacceptable and improvements were deserved to be made. One way of improvement is to decrease the dimensions of each cube so that the jagged faces can be smoothened. But as a block is a basic testing unit for the evaluation tool decreasing the dimensions of each block only results to the increase in the number of testing units and thus an increase in the time needed for the testing iterations. Another way is to deform the outermost blocks to fit into the building form exactly but this may require the manipulation of complicated Bézier surfaces in the worst case which will certainly degrade the performance of the computer. The technique employed in this modelling tool is to approximate the form by polyhedrons because the number of faces of a polyhedron is not much more than that of a rectangular block. The overhead required on the computing power is not significant compared with the Bézier surfaces. The outermost blocks can be deformed to any polyhedrons so that the building form can be approximated accurately. The mechanism of deformation is that the designers are allowed to add and remove vertices on the block. The designer can change the positions of these vertices by dragging the vertices interactively. The blocks can then become polyhedrons. The user interface of the modelling tool provides for the designer the ability to do the deformation from six directions of the rectangular blocks, such as top, bottom, left, right, front and back. Figure 2 to Figure 5 show three major operations – movement, addition and deletion of the vertices on the blocks. Multiple blocks can be deformed at a time in order to speed up the deformation process.

Figure 2 shows the TOP view of the block. Two vertically overlapped points are selected and moved simultaneously to form a trapezoidal prism.

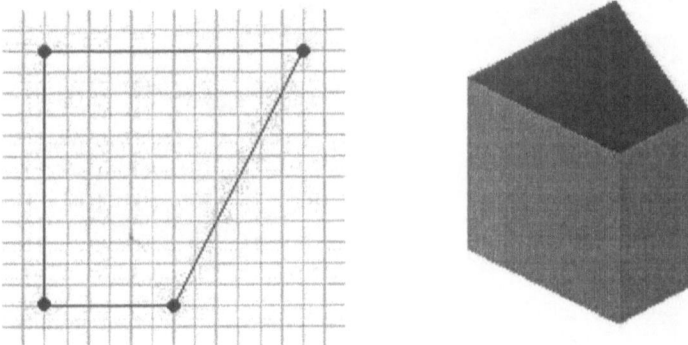

Figure 2

Figure 3 shows the TOP view of the block. Two vertices on the TOP face are moved inwards to form a trapezoid.

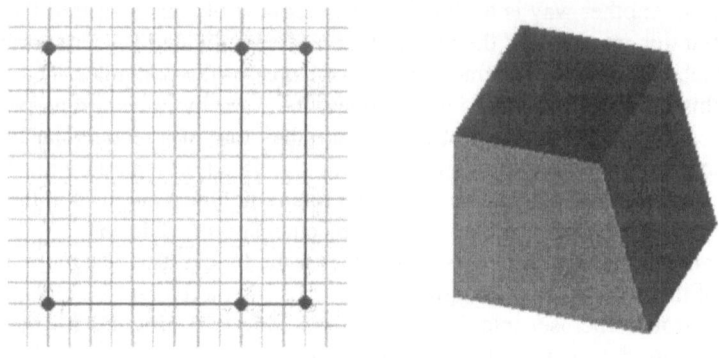

Figure 3

Figure 4 shows the TOP view of the block. Two vertically overlapped points are selected and deleted to form a triangular prism.

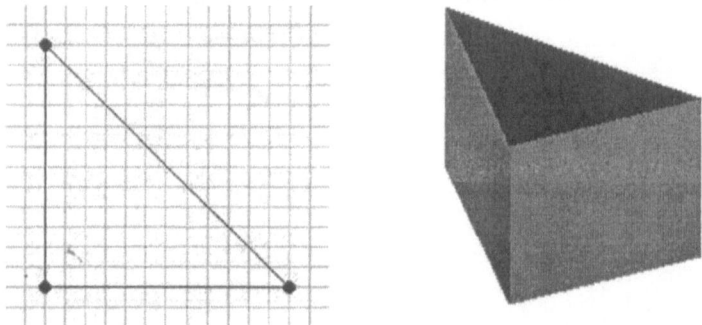

Figure 4

Figure 5 shows the TOP view of the block. Two vertices are inserted and moved simultaneously to form a pentagonal prism.

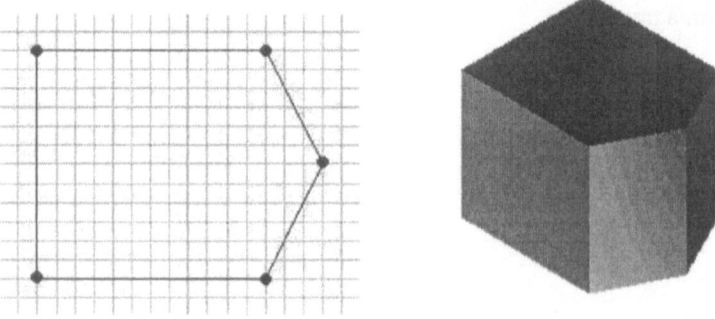

Figure 5

Furthermore, in order to prepare for taking sunlighting into account in the solar evaluation module, some features are added to this modelling tool to enhance its ability in modelling complex shapes and its ability to adopt sunlight into the considerations. Firstly, the building models can be moved freely within the site boundary so that the designer can choose the best position for the building to receive more sunlight in winter and less sunlight in summer. Secondly, the form can be extruded upward or compressed downward to form a tall and slim building or a short and flat building so that the building can be changed to other shapes easily to receive or escape from sunlight whilst keeping the plot ratio constant.

4. Improvements to the Evaluation Tool

In the previous development of the fuzzy system model, interviews were conducted to investigate the correlation between the five factors and the users' general satisfaction with the views. Fifty pictures of some common scenes in Hong Kong were shown to the interviewees who were requested to rank these pictures with respect to their preferences. These fifty pictures were taken in different areas in Hong Kong, from urban areas to rural areas, from construction sites to wasteland. In the previous studies, those fifty pictures were taken without any criteria, but were randomly taken in different areas in Hong Kong. As a refinement some factors other than those five fuzzy sets affecting people's general satisfaction can be considered. For example people may prefer a well landscaped and complete built environment to a construction site.

A few amendments have been made on the evaluation tool so as to improve the accuracy of the fuzzy system model. In this study only an urban area is considered not only because of the reasons set above but also owing to the fact that designing a building with a good quality of view in a high-rise and high-density environment is the most challenging for an architect. It is hoped that this system can help designers to achieve a better quality of view in their designs under such conditions. The number of rules is raised from 45 to 100. In this study 105 photos taken in the urban areas in Hong Kong were shown to interviewees to rank according to their order of preferences. The photos show nearly all types of views that can be seen in the urban areas in Hong Kong. Two examples of the photos are shown in Figure 6. They exemplify some of the differences that landscape can make in an urban setting.

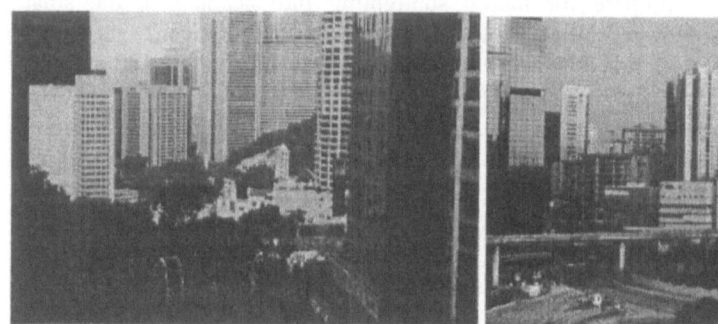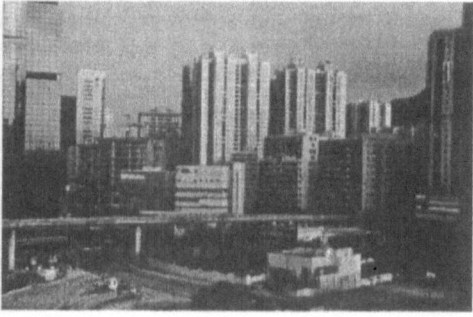

Figure 6

Interviewees, besides being asked to rank the photos, were also asked to evaluate the five factors (included in Table 1). For example they might be asked if the proportion of green plants in a photo was "no", "low", "average" or "high". The verbal comments and the average ranking of the 100 photos were used to build the rule base of the fuzzy model. The average ranking of these 100 photos was treated as the dependent variable – the Visual Aspect Performance Index (VAPI). The five fuzzy sets and their corresponding subsets are shown in Table 1. Obviously these photos take into account only daytime views but in future the model will also consider comparable night views.

Fuzzy Set	Fuzzy Subset
Average Distance between the Observer and the Surrounding Buildings	Near distance (N) Medium distance (M) Far distance (F)
Proportion of Buildings in the scene	No building in the scene (N) Low proportion of buildings (L) Average proportion of buildings (A) High proportion of buildings (H)
Proportion of Sky in the scene	No sky in the scene (N) Low proportion of the sky (L) Average proportions sky (A) High proportion of sky in the scene (H)
Proportion of Green Plants in the scene (including mountains, trees, grass and artificial planting)	No green plants in the scene (N) Low proportion of green plants (L) Average proportion of green plants (A) High proportion of greens plants (M)
Proportion of Sea in the scene	No sea view (N) Low proportion of sea in the scene(L) Average proportion of sea view (A) High proportion of sea in the scene (H)

Table 1

The other five photos were used to verify the accuracy of the fuzzy system model. They were pixelated using PhotoShop. Different elements were outlined for the ease of counting the proportions in the pictures. The actual distances between the observer and the observed buildings were measured from a map to compute the average distance. Table 2 shows the parameters for the picture shown in Figure 7.

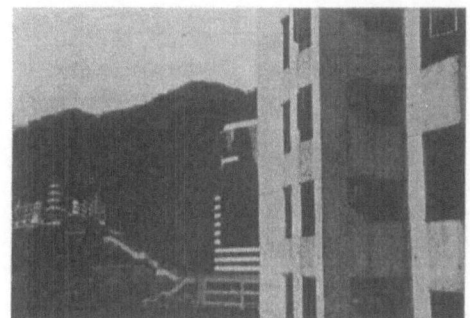 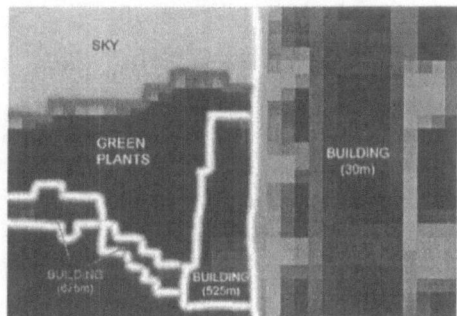

Figure 7

	Percentages of the elements in the Picture			
Distance of Buildings	Buildings	Sky	Plant	Sea
134m	53%	15%	32%	0%

Table 2

The membership functions of all fuzzy sets were amended to give a better approximation of the real world. A membership function converts real-world data of a certain element to a degree of membership of the fuzzy subset. The degree of membership is a real number in the interval [0, 1]. "1" indicates strict belonging while "0" indicates a certain element which has no relation with the subset. A value between 0 and 1 indicates partial belonging (Zadeh 1965). The new membership function of the Average Distance between the Observer and his Surrounding Buildings is shown in Figure 8.

354

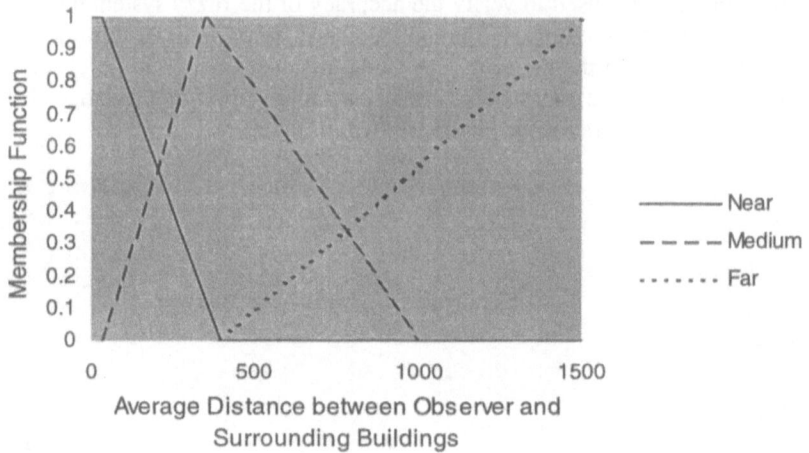

Average Distance between Observer and
Surrounding Buildings

Figure 8 : Membership function of Average Distance between Observer and
Surrounding Buildings

Table 3 shows the membership functions of the fuzzy subsets for the picture shown in
Figure 7.

Average Distance of Buildings			Proportion of Buildings in the scene				Proportion of Sky in the scene				Proportion of Plants in the scene				Proportion of Sea in the scene			
N	M	F	N	L	A	H	N	L	A	H	N	L	A	H	N	L	A	H
0.72	0.33	0	0	0	0.88	0.12	0	0.75	0.25	0	0	0.4	0.9	0	1	0	0	0

Table 3

For example, by matching the fuzzy subsets with those in the rule shown in Table 4, the
degree of belonging to the rule can be computed by the popular max-min operation.

$$\min [0.72, 0.88, 0.75, 0.4, 1] = 0.4$$

Average Distance of Buildings	Proportion of Buildings	Proportion of Sky in the scene	Proportion of Plants in the scene	Proportion of Sea in the scene
N	A	L	L	N

Table 4

Applying the same process to each rule, we can derive a set of values. The expected
VAPI is computed by the center of gravity method.

Using the parameters computed from the computer model of the building created with
the Modelling Tool and applying the fuzzy system described above, the VAPI of the
design can be predicted by aggregating the VAPI from the rule base of the fuzzy model.
A general picture of the VAPI allocation in each part of the building can be established.

The VAPI value of each cube is mapped to a color which is assigned as the color of the cube. The final product of the Evaluation Tool is a building composed of colored blocks. Figure 9 shows an example of a building with color mapping. The color goes from red, to yellow, to green and then to blue. Red indicates good quality while blue means poor quality. Each block interacts with its adjacent blocks so that blocks on the exterior of the building shade those inside. Progression through the blocks reduces the view quality assigned. Horizontal slabs also affect the available views for the inner blocks because of the reduced field of view.

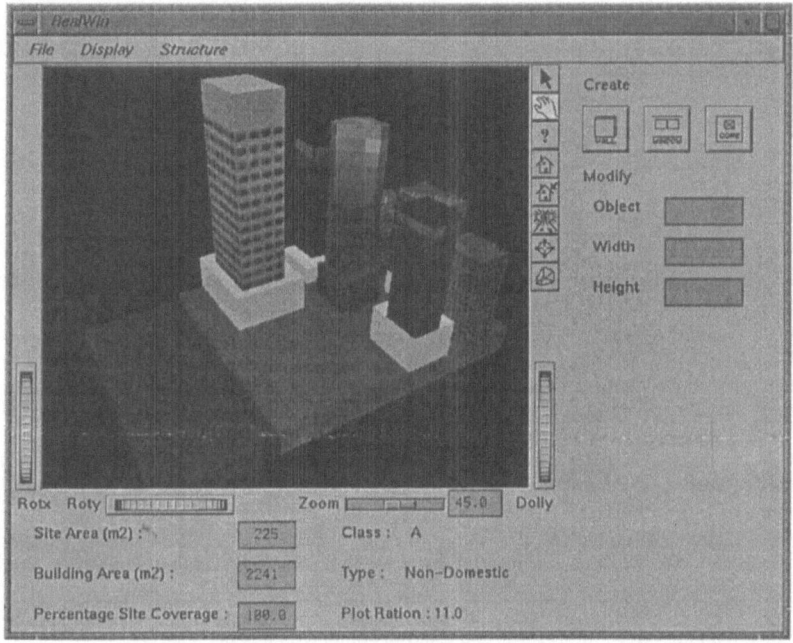

Figure 9

5. Conclusion

This paper not only demonstrates the way of using computers in helping designers to make better decisions on the visual aspects of buildings, but also illustrates the possibility of using computers in analyzing other parameters. Traditionally, designers can only predict the outcomes of their designs concerning such aspects as utilization of sunlighting, quality of view, complying with the building regulations, etc. by way of their own experiences, knowledge, and personal preferences. Often luck plays an important role in the success of a design solution. Computers, instead of making decisions for architects, can provide them with supporting information, analyses and comparisons so that architects can give solid evidence to support their decisions. This modelling tool is envisaged as one element of a package of design decision supports

(IOTA) that can interactively respond as the architect manipulates his building in a creative and free manner. Future inclusions in this tool will permit the masking of an exterior wall to represent variable window openings, sill and head heights and column sizes and spacing. Light reflectors and refractors or shading devices will also be accommodated. These elements will respond directly to the inputs to the facade design mask which will interactively appear on the 3D model as it is manipulated in form and position. Alternative solutions can thus be objectively compared as the design process progresses and this is particularly relevant where qualitative assessments, such as the users' overall satisfaction with views from the building, vary with the changing design. This process can be carried out simultaneously with the external facade design of the building and can be related to the solar impacts from the sunlight tool already developed by Wong and Will (1996).

Reference

Lam, W. M. C.: 1977, *Perception and Lighting as Formgivers for Architecture*, McGraw-Hill, pp14-30.

Li, T. S. P. and Will, B. F.: 1997, A Computer Based Evaluation Tool for the Visual Aspects in Window Design, *Proceedings of the CAADRIA '97 Conference* (To be published), Taiwan

Markus, T. A., Brierley, E. and Gray, A.: 1972, *Criteria of Sunshine, Daylight, Visual Privacy and Viewing in Housing,* Building Performance Research Unit, Department of Architecture and Building Science, University of Strathclyde, Glasgow.

Markus, T. A., and Gray, A.: 1972, Windows in Low Rise, High Density Housing – *the Psychological Internationale de L'Eclairage Conference: Windows and Their Functions in Architectural Design,* Istanbul.

Wong, W. C. H. and Will, B. F.: 1996, An Analysis of Using a Digital 3D sundial as a Design and Decision Support Tool, Proceedings of CAADRIA '96 Conference, Hong Kong.

Zadeh, L. A.: 1965, Fuzzy Sets, Information and Control, vol. 8, Orlando, pp. 338-353.

A MOBILE AGENT ORIENTED METHOD OF SIMULATING THE INTERACTION BETWEEN A BUILT ENVIRONMENT AND THE OCCUPANTS' ACTION

HARUYUKI FUJII
Izumi Research Institute, Shimizu Corporation, Tokyo, Japan

AND

SHOICHI NAKAI
Department of Architecture, Chiba University, Chiba, Japan

Abstract. *The thermal comfort of a built environment in question and the energy efficiency of the building providing the built environment is one of the aspects that plays an important role in the decision making in architectural design.. However, it is not easy to deal with the interaction between a built environment and actions of occupants that change the environment in a conventional way of architectural environment simulation. Focusing on the interaction, the authors propose a method of mediating programs, which evaluate the quality of a building or simulate the performance from different aspects, in a Mobile Agent Oriented Community, so as to compose a module of the design support system.*

1. Introduction

This research aims to explore the development of an architectural design support system dealing with multiple disciplines employed in architectural design process. Focusing on the interaction between a built environment and actions of occupants that change the environment, the authors propose a method of mediating programs, which evaluate the quality of a building in question or simulate the performance from different aspects, to compose a module of the design support system.

The quality of a built environment is evaluated from diverse aspects such as egress, energy efficiency, etc. One of aspects that plays an important role in architectural design process is the thermal comfort of a built environment in question and the energy efficiency of the building providing the built environment. They could be predicted by simulating the thermal behavior of a building, its occupants' actions that bring about the changes in the built environment, and the interaction between the thermal behavior and the actions. The interaction should be considered since a built environment affects its occupants' action while an occupant's action brings about changes in the built

357

R. Junge (ed.), CAAD Futures 1997, 357-372.

environment where the action occurs. Architectural design process is formulated as a goal-oriented, constrained, decision-making, exploration, and learning activity that operates with in a certain context (Gero 1990). The evaluations of the quality of a building in question and the simulation of the building performance facilitate the decision-making process and the exploration process (Maher 1995) by giving information with which the participants in the design process can formulate a less ill-formed design problem and find the corresponding solution from an ill-defined problem.

To support the decision making process and the exploration process, an architectural design support system is required to incorporate application programs evaluating the quality of a designed building and/or programs simulating the building performance. The authors propose to enable a simulation program of human action and that for numerical analysis of thermal behavior to interact with each other, as if a human interacts with an environment through actions. It is true that many application programs for numerical analysis of the thermal behavior of a building have been implemented. However, to the authors' knowledge, the programs don't deal with the interaction between the thermal behavior and its occupants' actions. The actions bringing about changes in thermal behavior are just given as a preset data like a climate data. On the other hand, the achievements in artificial intelligence research provide formal methods of simulating human actions. The world, represented in a symbolic method, changes only when an action, which brings about the change, is done. Therefore, it is not easy to let such programs interact with each other. In addition, most of the programs have been implemented independent of one another. One might want to claim that it could be an alternative way to implement a new program. However, we should take advantage of existing application programs, even though they are written in different programming languages and have their own I/O convention and internal data structure. It is not realistic or desirable to require that those programs are (revised to be) implemented in one programming language or system. The programs should be mediated without affecting each program's external and internal representation of a design.

The authors have proposed a translation approach to the mediation of the existing programs (and newly implemented programs), and the conception of a mobile agent-oriented community (MAOC) and the conception of an agent interchange format (AIF) as a method of the translation approach. In MAOC, mobile agents carrying a design problem interact with evaluation programs and problem solvers to solve the problem. AIF is a representation format to exchange mobile agents and information between modules in distributed systems like MAOC through a computer network. An AIF expression carries information about the tasks to solve a given program and about the domain in question.

This paper is organized into seven chapters. Following a brief review of MAOC and AIF in chapter 2 and the problems in the task of implementing a simulation program of the indoor climate-human action interaction in chapter 3, chapter 4 describes how MAOC mediates a simulation program of thermal behavior and a simulation program of human action to simulate of interaction between a built environment and individual actions. Where, the simulation program of thermal behavior is developed by using thermal circuit network and a simulation program of human action is developed on the basis of the principles in artificial intelligence and philosophy of action. Chapter 4 also mentions the conception of a universal language to share information about a building in question among the mediated programs. In chapter 5, a small built environment in which two occupants try to control room air temperature by operating HVAC or by opening or closing a window, thermo-start actions, and its transition are simulated as examples by the programs mediated by MAOC. Chapter 5 refers to the results of the simulation, too. Chapter 6 mentions a role of the indoor climate-human action simulation as a tool for researches in architectural environment. In chapter 7, it is concluded that the interaction between transition of a built environment and its occupants' actions can be simulated by mediating a simulation program of thermal environment and a simulation program of human actions in MAOC, though these programs are implemented independently. It is also concluded that the simulation of such interaction will possibly facilitate the decision-making and the exploration in architectural design process.

2. Mobile Agent-Oriented Community

The authors have proposed a translation approach to the mediation of the existing programs and newly implemented programs, and the conception of a Mobile Agent Oriented Community (MAOC) and an agent interchange format (AIF). A detailed discussion of MAOC is presented elsewhere (Fujii, H., Nakai, S., et al., 1996). Here, a brief overview of the approach is given (figure 1).

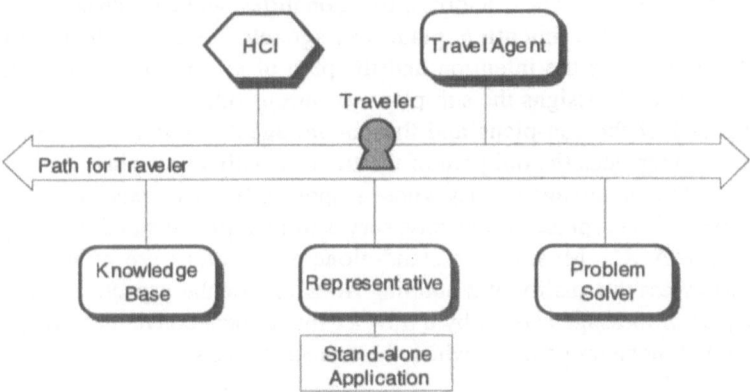

Figure 1. The Conception of a Mobile Agent Oriented Community.

We define a mobile agent-oriented distributed design support system to be one that is composed of a set of autonomous modules that we call on-site agents, mobile agents, and a set of paths, a computer network, for mobile agents to be sent back and forth among sites. We call such a system a Mobile Agent-Oriented Community, or MAOC.

A mobile agent takes responsibility for its own problem solving. A mobile agent that is assigned a problem creates a plan to solve the problem and executes each step of the plan by interacting with agents, on-site agents, that solve a part of the problem. A mobile agent is an object in a certain class when it is in a program module called site in which agents perform their activities, while it is a packet of information when it is outside the site. That is, a mobile agent in MAOC is in the form of text information containing domain knowledge and task knowledge. In order to represent this knowledge, the authors have proposed a concept of language and system independent Agent Interchange Format (AIF) which we believe a unique feature of the MAOC approach.

An on-site agent is an object that mobile agents interact with through a speech act to solve a problem carried by the mobile agent. On-site agents stay inside the sites and wait for mobile agents. The communication between on-site agents is mediated by mobile agents. An on-site agent responds to a mobile agent's speech act with respect to the type of the speech act and helps a mobile agent to solve a problem. It is not assumed that there is a single on-site agent that maintains control of the whole community.

Traveler class, from which a traveler is instantiated, is a subclass of mobile agent class. In the initial stage of problem solving, each mobile agent called traveler begins with a problem and a vague and incomplete plan to solve the problem. The plan is then decomposed into a sequence of well-organized sub-plans that are relatively concrete and complete through the interaction with on-site agents called travel agents. Ideally, the entire community finds a complete plan.

An instance of HCI class, an HCI, is a human computer interface, that facilitates the interaction between the user and MAOC. An HCI displays the surface language expressions of a design in question and the interaction between the user and MAOC.

Each instance of travel agent class, a travel agent, maintains explicit information about other agents, i.e., id, class, address, functionalities, and can retrieve an incoming traveler's information, i.e., intention, plan, message, etc. A travel agent performs the following tasks: (1) gets the intention and the plan of a traveler, (2) decomposes the plan into sub-plans, (3) assigns the sub-plans to various other on-site agents, (4) makes a plan composed of the sub-plans and the on-site agents to which the sub-plans are assigned, and (5) replaces the old plan of the traveler with a new plan. Specialist class is a subclass of the on-site agent class whose responsibility is to execute a task within a particular aspect. The representative class serves to incorporate stand-alone application programs into MAOC. Most of the stand-alone applications are numerical analysis programs that assess the quality of a building from a particular aspect. A representative translates a part of message carried by a traveler into a local convention used as the I/O data for the stand-alone program for which the representative serves.

3. Problems in Simulation of A Thermal Environment and Actions

Leaving a discussion of MAOC aside for a moment, let us turn to a discussion concerning simulation of the interaction between a built environment and human actions. The problems in the simulation are worth a mention since they can be solved by MAOC rather than traditional approaches.

The thermal comfort of the environment, indoor climate, and the energy efficiency provided by a designed building is one of the important aspects from which the quality of a designed product is evaluated. The author (Fujii, 1989) claims that they should be predicted by considering the interaction of the thermal behavior of a building and the occupants' actions causing the changes in a building that bring about the changes in the thermal environment.

To the authors' knowledge, although many indoor climate simulation programs have been developed, the programs don't deal with the interaction between the thermal behavior and its occupants' actions. The reason for this, the authors believe, is the following. A building and the thermal behavior of the building are represented simultaneous differential equations. It is predicted by numerical analysis on the basis of the equations. On the other hand, the formal methods of simulating human actions, which are provided by artificial intelligence research, require not numeric but symbolic representation of a world. In addition, the world, represented in a symbolic method, changes only when an action, which brings about the change, is carried out. These facts prevent us from dealing with the two different types of simulation methods to let them interact with each other. As a result, the actions bringing about changes in thermal behavior are just given to an indoor climate simulation program as a preset data like a design climate data and no interaction of changes in the indoor climate and the occupants' action is handled.

3.1 REPRESENTATION OF THE INTERACTION

The environment inside or outside a building and actions performed by occupants in the building interact with each other. An occupant perceives the environment. If the occupant feels uncomfortable with the environment, he/she may perform actions to change the perceived environment into the environment that he/she wants. As a result of the actions, the environment changes and brings about a certain consequence. The authors assume that the interaction of the environment provided by a building and the occupants can be regarded as a dynamical system composed of an environment and agents as shown in figure 2.

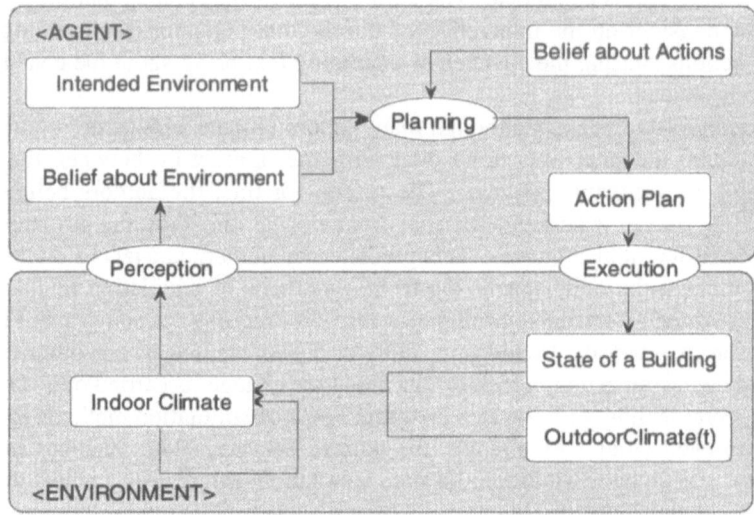

Figure 2.   Interaction of an Agent and an Environment.

A state of the environment, the indoor climate and a state of building, is regarded as the input to an agent. An agent interprets the input from the environment and updates and maintains the belief about the state of environment. If the agent desires to change the state into another state, namely goal state, the agent makes a plan of actions to get the goal state and carries out the plan. The actions executed are the output from the agent and the input to the environment. When an action is carried out a state of the environment changes to another state. For example, if an agent opens a window that is closed before the action, the window is open after the action. The change in a part of a state brings about the changes in other parts of a state. When a window is opened, the ventilation ratio of the room, which the window is installed, changes. This incident may affect the indoor climate.

3.1.1 Representation of Thermal Behavior of a Building

From an aspect of the thermal environment, a building consists of a number of thermal storage concentrated mass, points, and a number of heat flow paths between two of the points in parallel. A building design is represented as a thermal circuit network. A thermal circuit network is a graph whose vertices and edges represent the thermal storage points and the heat flow paths, respectively. Design drawings of a building are translated into a thermal circuit network representation before the thermal behavior of the building is analyzed.

The balance of heat flows in a thermal circuit network is represented as equation 1.

$$M\frac{dx}{dt} = C \cdot x + C_0 \cdot x_0 + R \cdot g \qquad (Equation\ 1)$$

Equation 1 is a matrix representation of simultaneous differential equations representing the balance of the thermal properties, the thermal state of the building in question, and the state of weather elements and the other input. Where, M is a thermal capacity matrix describing the thermal capacity of each of the thermal capacity masses, C and C_0 are thermal conductivity matrices between two of the masses, x and x_0 are the vectors describing the temperature of the masses, R is a heat absorption ratio matrix, and g is a vector describing heat gain or heat loss of the masses. C, C_0, R and g are subject to change corresponding to changes in a state of the building whose thermal properties are represented by the equation. x and g are subject to change corresponding to a state of weather elements, a state of the building, and occupants.

Equation 2 is the matrix representation of the simultaneous forward difference equations derived from equation 1. The temperature of the thermal capacity masses of the building are calculated by equation 3.

$$M \frac{x_{t+\Delta t} - x_t}{\Delta t} = C_t \cdot x_t + C_{0,t} \cdot x_{0,t} + R_t \cdot g_t \quad \textit{(Equation 2)}$$

$$x_{t+\Delta t} = M^{-1} \cdot \left(C_t \cdot x_t + C_{0,t} \cdot x_{0,t} + R_t \cdot g_t \right) \Delta t + x_t \quad \textit{(Equation 3)}$$

3.1.2 Representation of Changes in an Environment by Actions

Artificial intelligence studies have been providing the ways of formally representing human actions. An intelligent agent is one of the sophisticated concepts and methods. An intelligent agent is a software object that has certain basic capabilities. For example, an agent should perceive the environment where an intelligent agent is and should deliberately react to events or changes in the environment in a goal-directed manner. The environment is described in the internal representation accessible to the agent in a state-based manner (Müller, 1996).

Many methods employ symbolic representation of the environment such as first order logic to represent a state of affairs of the environment. Let STATE be the type of a state of affairs of the environment. Information about a transition from a state of affairs of a designed product to another state of affairs of the product is represented as an ordered pair of a description about the state before the transition and a description about the state after the transition. Therefore, the type of a description of transition can be expressed by a pair <STATE, STATE>. The authors have been taking advantage of this symbolic representation of the actions and the state changes (Fujii, H., Katukura, H., and Nakai, S., 1996).

3.2 QUESTIONS IN SIMULATION OF THE INTERACTION

Some questions arise about the implementation of a simulation program of the interaction between a built environment and the occupants' actions. They are:

1. How to mediate a numerical analysis program for simulation of the thermal behavior and a symbol processing program for the simulation of agent's actions?
2. How to deal with the different types of the representation forms of the same environment?

The next chapter explains MAOC implementation as an answer to the first question and a translation approach as an answer to the second question.

4. MAOC Implementation and a Translation Approach

We discuss how MAOC and a translation approach answer the questions arised in the previous chapter.

4.1 MAOC IMPLEMENTATION

MAOC can be implemented as a combination of a client program, which a user interacts with, and a number of server programs, each of which provides a part of solutions to a problem from a certain aspect. Both the client and the server programs need to be capable of exchanging an agent represented in an AIF format. The authors have tried two different approaches for implementing the MAOC concept. The first one has its base entirely on the Web mechanism, meaning that each Site of MAOC is composed of a combination of a Web server and CGI applications. One of the drawbacks of this approach has been found that the communication between the sites is quite limited due to the simplicity of the HTTP protocol used in the Web mechanism. The second approach for implementing the MAOC concept, which is discussed in this section, is to develop a new client-server system combined with the Web mechanism.

It is no doubt that the Web has already become a de facto standard of interfacing the network. Since almost everyone is comfortable with using this popular service, it is reasonable to use a Web browser as a client program, i.e., HCI, of MAOC. However, as the underlying HTTP protocol lacks the flexibility of expanding functionalities, we will limit the use of the Web mechanism only to the interface part. The rest of the system, a group of server programs (or, sites), is implemented as a number of independently running programs that communicate with each other over the network. Java has been used as the implementation language.

Figure 3 shows a schematic illustration of the implementation for the indoor climate-human action simulation. There are two kinds of components in figure 3, i.e., an HCI and sites. The HCI consists of a Web browser and a Java applet that interacts with the user for obtaining information on the problem under consideration and for providing information on the solution returned from the sites. The rest of the systems are sites where a Travel Agent, Representatives, and a Specialist wait for a Traveler to visit. All of these sites are instances of Agent Server class that provides an environment for both mobile and on-site agents to perform their tasks (Rodley, 1996).

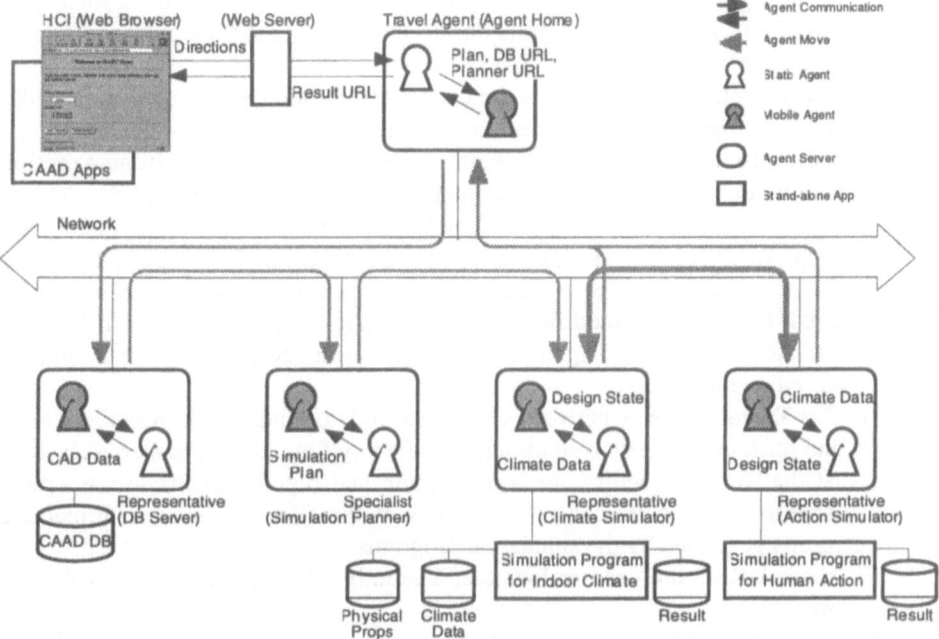

Figure 3. MAOC Implementation of Indoor Climate-Human Action Simulation.

In the case of the indoor climate-human action simulation, mentioned in the previous chapter, problem solving proceeds in the following way.

1. The user first provides information on what he/she would like to do through the interaction with a Java applet on a Web browser. This information gets transferred to a Travel Agent that stays in a site called Agent Home, where a Traveler gets created with an overall plan for problem solving, a URL of a DB server and a URL of a simulation planner.

2. According to this initial plan, the Traveler is sent to a site called DB Server, where he obtains CAD data for a building under consideration. This is done through the interaction with a Representative of a CAAD database server.

3. The Traveler is then sent to a site called Simulation Planner, where he talks to a Specialist of indoor climate-human action simulation and receives a detailed plan for a simulation task.

4. On the basis of the simulation task, the Traveler is first sent to a site called Climate Simulator, where he gives a current state of the design, such as the location of doors and windows along with their conditions (open/close), to a Representative that is responsible for the indoor climate simulation. This Representative interprets the data given by the Traveler and generates input data to a stand-alone indoor climate simulation program. The program performs a simulation task for a single time interval and computes the new state of indoor climate based on the input data and databases for physical properties of materials and climate data for environmental simulation. This

result is given back to the Traveler and the current state of indoor climate gets modified. Intermediate results during the computation are stored in a local result file.

5. The Traveler is then sent to a site called Action Simulator, where he provides with a current state of the indoor simulation, such as the temperature of a room, to a Representative that is responsible for the human action simulation. This Representative interprets the data given by the Traveler and generates input data to a stand-alone human action simulation program. The program performs a simulation task for a single time interval and computes the new state of human action such as 'open a window' or 'turn on an air conditioner'. This modifies the current state of the design which is given back to the Traveler. Again, intermediate results during the computation are stored in a local file.

6. The Traveler moves between Climate Simulator and Action Simulator back and forth until a certain condition specified by Simulation Planner is reached. At that time, he returns to Agent Home and reports results to the Travel Agent. The results consist of URL's of the files that store intermediate results generated by the simulation programs.

7. These URL's are used to locate and examine the simulation results by the user.

In the current implementation, exchange data in the AIF format, which are transferred between the sites, consists of Java class data of Traveler class and plans and states represented in a text format.

4.2 A TRANSLATION APPROACH

Another crucial issue involved in the implementation of the indoor climate-human action simulation that requires the mediation of different types of application programs is how to represent the domain and the task under consideration. An application program developed for a particular purpose uses the data structure and the I/O convention suitable for the purpose. It is not realistic to redefine the data structure and the I/O convention of the existing programs. Instead, the authors have been investigating a translation approach. The explanation of the conception of the approach consisting of surface and universal languages is give in elsewhere (Fujii, H., Nakai, S., et al., 1996: Fujii, H. and Nakai, S., 1997). The conception is inspired by Chomsky (1986) and Coyne et al. (1990).

In this section, the conception related to the indoor climate-human action simulation is mentioned (figure 4).

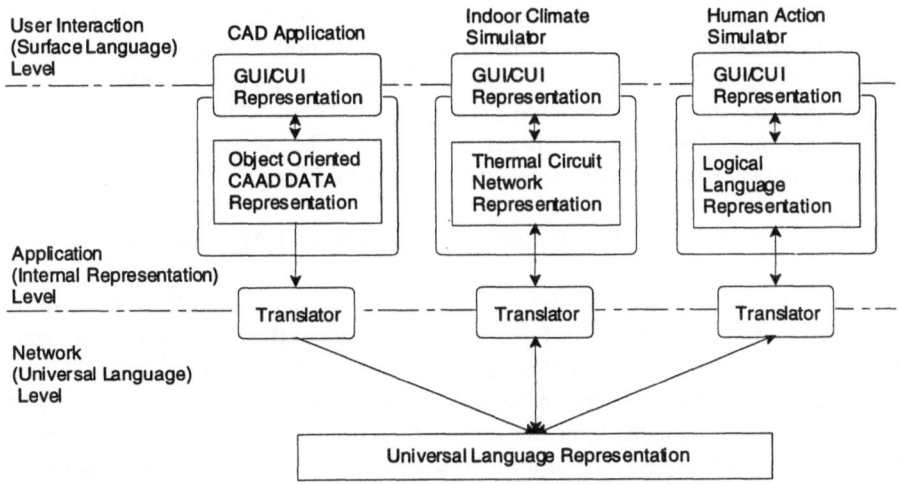

Figure 4. Relations between Surface Languages and a Universal Language.

Information required to the indoor climate-human action simulation is conveyed from an on-site agent to another on-site agent in the form of a universal language. An CAAD data represented in an object oriented manner of a building in question is translated into the universal language representation when it is carried by a mobile agent. When the mobile agent arrives at the indoor climate simulator, the universal language representation is translated into the thermal circuit network representation of the building. The indoor climate simulator predicts a state of the indoor climate of the building on the basis of the thermal circuit network. The thermal circuit network, together with the state of the indoor climate predicted, is translated into the universal language representation and given to the mobile agent waiting for the result. The interaction of the indoor climate simulator and the human action simulator is executed by the mobile agent conveying the universal language representation. The mobile agent conveys the universal representation to the human action simulator and the logical language representation in a state-based manner is generated from the universal representation. It is intended that the logical language representation is compatible with the logical framework proposed by Brazier et al (1994; 1995). When the human action simulation is done, the result, the changes in a state of the world in the building, is translated into the universal language representation.

For the purpose of the efficiency of the translation occurring many times, we assume that the part of a state of affairs that has not been changed by an application program remains the same.

5. Example

An example that tells us how the indoor climate-human action interaction simulation, which is incorporated with a CAD system in a Mobile Agent Oriented Community, supports exploration (Maher 1995) in a design problem solving process. The example is not a real situation but is likely to arise in architectural design. A part of an architectural design process is regarded as a problem solving process composed of the generation process and the verification process (Akin 1986). The current state of the design is evaluated from some aspects to verify if the design is capable of the performance that is expected to be provided by the design (Gero 1990).

5.1 A DECISION MAKING PROBLEM IN ARCHITECTURAL DESIGN

Suppose that an architect needs to define the detail of the exterior walls of the building whose planning layout is shown in figure 5. It is assumed that the building is located in Tokyo. The architect wants to choose one from two types of wall details, namely, the outside insulation type and the inside insulation type (figure 6). The architect plans to investigate, from the aspects of the indoor climate and the energy efficiency, the difference between two types. In addition, the definition of problem in an architectural design process is subject to change with respect to the feedback from the verification process.

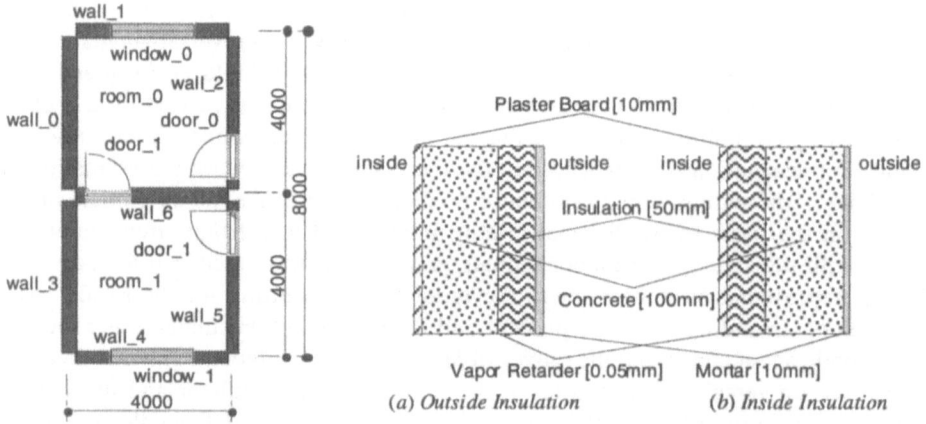

Figure 5. Planning Layout   *Figure 6. Wall Details*

The task described above is composed of sub-tasks each of which is carried out by different types of application programs, i.e., a CAD system, a program for numerical analysis of heat transfer in a building, and an artificial intelligence program simulating actions of intelligent agents. MAOC takes care of mediation of the programs.

We assume that two occupants (agents) are staying the building described in the previous section. Agent_0 is in room_0 and agent_1 is in room_1. Both agent_0 and agent_1 feel comfortable when the air temperature of the room where they stay is between 26.0 degrees C. and 28.0 degrees C. They feel hot when the temperature is above 28.0 degrees C. and feel cold when the temperature is below 26.0 degrees C. The action patterns of the agents are described below.

When agent_0 feels hot and believes that the outside air temperature is lower than the air temperature of the room agent_0 stays, agent_0 opens a window. Before opening a window, agent_0 turns the air conditioner, which is installed in the room, off if it is working. When agent_0 feels hot and doesn't believe that the outside air temperature is lower than the air temperature of the room, agent_0 turns the air conditioner on if the air conditioner is not turned on, or agent_0 operates (power_up) the air conditioner so as to increase the heat extraction by the air conditioner if the air conditioner is already turned on. Before operating an air conditioner, agent_0 closes windows if they are opened.

When agent_1 feels hot, agent_1 operates (power_up) the air conditioner, which is installed in the room where agent_1 stays, so as to increase the heat extraction by the air conditioner. When agent_1 feels cold, agent_1 operates (power_down) the air conditioner so as to decrease the heat extraction by the air conditioner.

5.2 SIMULATION RESULTS

The duration of the simulation is 10 days with the time interval of 5 minutes. The design air temperature of Tokyo for the estimation of air conditioning cooling load is used as the outside air temperature. The solar input, radiative heat transfer, the heat gain or heat loss from ceilings and floors are ignored for simplicity. The heat gain from the occupants in the building is taken into consideration.

Table 1, 2 and figure 7, 8 show the interaction between the indoor climate and the occupants' action. The results shown in the tables and figures are those of ninth day. Table 1 and 2 show the thermal comfort perceived by two occupants and actions performed by the occupants. Figure 7 and 8 are graphs showing the fluctuation of air temperature and air conditioning cooling load in the case of the outside insulation type and the inside insulation type, respectively. Table 1 and figure 7 are in the case of the outside insulation type. Table 2 and figure 8 are in the case of the inside insulation type. We can read the interaction of the indoor climate and the occupants' actions from the tables and figures. For example, agent_0 feels hot and opens window_0 at 03:35 (table 1). After the event, the air temperature of room_0 decreases (figure 7). Agent_0 feels cold and turns aircon_0 off at 17:20 (table 1) and the air temperature of room_0 increases (figure 7). We can read the difference between the transition the air temperature of room_0 and that of room_1. We can relate the difference to the difference between the action pattern of agent_0 and that of agent_1.

In this example, the simulation results tell us that there is no significant difference between the outside insulation and the outside insulation with respect to the transition and the air temperature in the architectural spaces and the air conditioning cooling load. The architect could not determine which design is better with respect to the thermal comfort and the energy efficiency of the building. Therefore, the architect could choose one of the designs focusing on the other aspects such as economy, durability, etc. He/she may redefine a design decision making problem on the basis of the information acquired before and after the simulation in the example.

Table 1. Perception and Action of Agents (Outside Insulation)

Time	T_outside [C]	T_r0 [C]	Agent_0 Percpt	Agent_0 Action	T_r1 [C]	Agent_1 Percpt.	Agent_1 Action
9:00:55	27.625	26.015	neutral	-	25.999	cold	aircon_1 power_down
9:01:00	27.6	25.999	cold	aircon_0 power_off	26.215	neutral	-
9:03:35	27.025	28.005	hot	window_0 open	27.996	neutral	
9:03:40	27	27.858	neutral	-	28.003	hot	aircon_1 power_up
9:05:30	26.9	27.19	neutral	-	25.999	cold	aircon_1 power_down
9:07:05	28.208	28.006	hot	window_0 close & aircon_0 power_on	27.843	neutral	-
9:07:20	28.533	27.556	neutral	-	28.033	hot	aircon_0 power_up
9:10:40	32.233	28.021	hot	aircon_0 power_up	28.007	hot	aircon_1 power_up
9:17:15	31.375	26.007	neutral	-	25.993	cold	aircon_1 power_down
9:17:20	31.3	25.971	cold	aircon_0 power_off	26.187	neutral	-
9:18:00	30.7	28.119	hot	aircon_0 power_on	27.031	neutral	-
9:18:05	30.642	28.031	hot	aircon_0 power_up	27.094	neutral	-
9:19:10	29.883	25.997	cold	aircon_0 power_off	27.240	neutral	-
9:20:00	29.3	28.074	hot	aircon_0 power_on	27.114	neutral	-

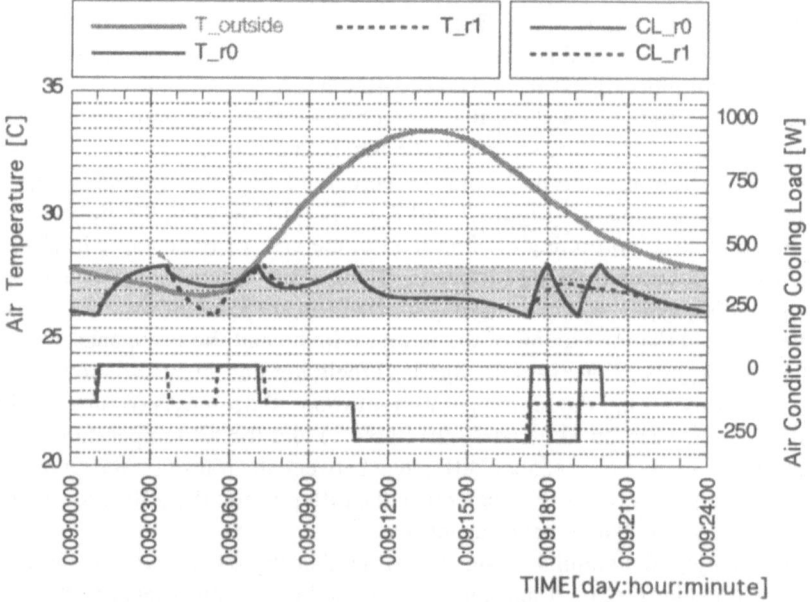

Fugure 7. Air Temperature and Cooling Load (Outside Insulation)

Table 2. Perception and Action of Agents (Inside Insulation)

Time	T_outside [C]	T_r0 [C]	Agent_0		T_r1 [C]	Agent_1	
			Percpt	Action		Percpt.	Action
9:00:45	27.675	25.999	cold	aircon_0 power_off	25.987	cold	aircon_1 power_down
9:03:15	27.025	28.003	hot	window_0 open	27.996	neutral	-
9:03:20	27.1	27.881	neutral	-	28.016	hot	aircon_1 power_up
9:05:25	26.883	27.214	neutral	-	25.993	cold	aircon_1 power_down
9:07:05	28.208	28.037	hot	window_0 close & aircon_0 power_on	27.851	neutral	-
9:07:20	28.533	27.600	neutral	-	28.044	hot	aircon_0 power_up
9:10:30	32.1	28.022	hot	aircon_0 power_up	28.020	hot	aircon_1 power_up
9:17:10	31.45	25.968	cold	aircon_0 power_off	25.967	cold	aircon_1 power_down
9:17:50	30.85	28.091	hot	aircon_0 power_on	26.942	neutral	-
9:17:55	30.775	28.003	hot	aircon_0 power_up	27.011	neutral	-
9:19:05	29.942	25.924	cold	aircon_0 power_off	27.202	neutral	-
9:19:55	29.358	28.034	hot	aircon_0 power_on	27.112	neutral	-

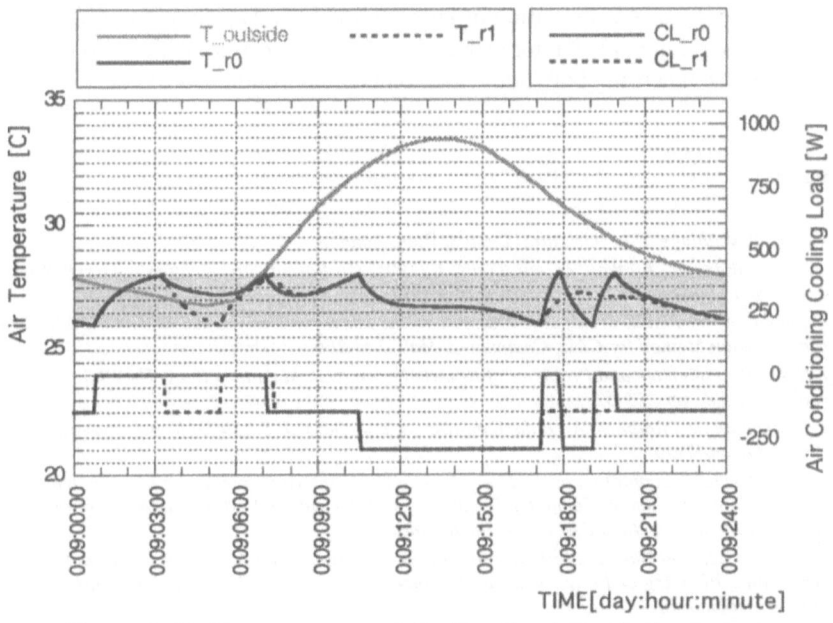

Figure 8. Air Temperature and Cooling Load (Inside Insulation)

6. Representation of the Interaction as a Research Tool

In addition to the role as a tool of predicting the quality of a designed building for a design support system, the representation of the interaction between the indoor climate and the human actions has another role. The role is to explain phenomena described by the empirical data concerning the indoor climate of a building which is occupied. Even though the empirical data is affected by human actions, it is often the case, in researches in architectural environment, that the human actions are not described or taken into consideration in a formal way. Therefore, the assumptions on the human actions in a built environment are hardly verified in a logical manner. The authors believe that our representation of the indoor climate-human action interaction helps the researches in architectural environment to deal with the human actions in a logical manner.

7. Conclusion

The authors may conclude that it is possible to simulate the interaction between transition of a built environment and its occupants' actions by mediating a simulation program of thermal environment and a simulation program of human actions by MAOC, though these programs are implemented independently of each other. The authors may also conclude that the simulation of such interaction will possibly facilitate the decision-making process and the exploration process in architectural design process by providing a tool for (A) simulation of indoor climate and calculation of heat load/loss including effects of human behavior caused by environmental transition and (B) analysis of empirical data concerning both human action and environmental transition.

References

Akin, Ö. (1986) Psychology of Architectural Design, Pion, London.

Brazier, F., van Langen, P., Ruttkay, Zs. and Treur, J. (1994) On Formal Specification of Design Tasks, in Gero, J. S. and Sudweeks, F. (eds.), Artificial Intelligence in Design 94, Kluwer, 535-552,

Brazier, F., van Langen, P, and Treur, J. (1995), A Logical Theory of Design, in J. S. Gero (ed.), Advances in Formal Design Method for CAD, Chapman & Hall, 243-266.

Chomsky, N. (1986) Language and Problems of Knowledge, in A. P. Martinich (ed.), The Philosophy of Language, 2nd Ed., pp. 509-527.

Coyne, R., Rosenman, M., Radford, A., Balachandran, M. and Gero, J. S. (1990) Knowledge-Based Design Systems, Addison-Wesley.

Hujii, H. (1989) Dynamic Simulation of the Interaction between Interior Environment and Individual Behavior, 12th Symposium on Computer Technology of Information, System, and Application, Architectural Institute of Japan, 427-432. (in Japanese).

Fujii, H., Katukura, H., and Nakai, S. (1996) Formal Representation of an Agent who Plans and Acts, Journal of Architecture, Planning and Environmental Engineering, No. 482, Architectural Institute of Japan. 249-258. (in Japanese).

Fujii, H., Nakai, S., et al. (1996) A Mobile Agent-Oriented Approach to a Distributed Design Support System, Gero, J. S. and Sudweeks, F. (eds.) Artificial Intelligence in Design '96, 485-504.

Fujii, H. and Nakai, S. (1997) On Formal Representation of Multi-Disciplinary Design Process, Maher, M. L., Gero, J. S., and Sudweeks, F. (eds.), Preprints Formal Aspect of Collaborative CAD, 343-368.

Gero, J. S. (1990) Design Prototypes: A Knowledge Representation Schema for Design, AI Magazine, 11(4), 26-36.

Maher, M. L. (1995) Formalizing Design Exploration as Co-Evolution : A Combined Gene Approach, in J. S. Gero (ed.), Advances in Formal Design Method for CAD, Chapman & Hall, 3-30.

Müller, J. P. (1996) The Design of Intelligent Agents - A Layered Approach, Springer-Verlag, Berlin.

Rodley, J. (1996) Writing Java Applets, The Coliolis Group.

MUD: EXPLORING TRADEOFFS IN URBAN DESIGN

MARK D. GROSS*, LAURA PARKER*, AND AME M. ELLIOTT[†]

*Sundance Laboratory for Computing in Design and Planning
College of Architecture and Planning
University of Colorado, Boulder, CO, USA 80309-0314

[†]College of Environmental Design
University of California, Berkeley CA, USA 94720
{mdg@cs, parker@ucsu} .colorado.edu, aelliott@ced.berkeley.edu

Abstract

The design of cities and neighborhoods involves multiple stakeholders with various agendas, each comprising multiple criteria. Any design proposal will rank differently against each stakeholder's agenda, and effective participatory design requires that stakeholder interests are mutually understood and negotiated. We describe a program to promote this understanding and negotiation among stakeholders, called MUD, that enables each stakeholder to articulate their criteria for judging designs, to make design proposals, and to score designs against the criteria. By enabling stakeholders with different values and different areas of expertise to exchange design proposals and agendas we hope to foster understanding and stimulate negotiation.

1. Introduction

Efforts to design successful built environments must be negotiated among stakeholders who hold diverse and often conflicting values about the built environment (Branch, 1985). Because of this multiplicity a platform for comparing design solutions is required. We are developing MUD, a decision support tool for teaching and learning about multiple objective design problems. Our MUD program enables stakeholders to score designs against various agendas, providing a flexible, quantitative, and visual means to compare designs in order to facilitate further negotiation.

Designing for a community requires balancing diverse viewpoints and values. Neighborhood planning meetings, for example, include not only architecture and planning professionals but also individuals and groups advocating their own agendas. Any design product will perforce make tradeoffs among the various values. Therefore the design process should offer opportunities for participants to explore and understand these tradeoffs. We describe a computer based tool called MUD (Multi-user Urban Design) that supports this exploration and attempts to promote a better understanding of design tradeoffs by urban design stakeholders. We plan to employ the MUD program as

R. Junge (ed.), CAAD Futures 1997, 373-387.

a learning tool in planning and design courses at the University of Colorado at Boulder to teach students about tradeoffs among stakeholders in multi-criteria design.

We define a 'stakeholder' as any person, group, or entity that is affected by changes made to the built environment or has some vested interest in those changes. For example, an individual resident, a business group, a city engineering department, and the mayor may all be stakeholders in an urban or neighborhood design. Each stakeholder holds values about the built environment, by which they judge proposed designs. We use 'agenda' to mean the expression of these values as a set of specific criteria and weights that allows a stakeholder to assign a numerical score to any proposed design. We define a criterion as a quality of the physical environment that can be measured as present or absent.

Using the MUD program entails three conceptual phases. In agenda-setting a stakeholder expresses a set of design objectives using a formal language of elements, relations, and preferences. In design exploration, a stakeholder uses a simple two-dimensional CAD program to lay out a design alternative. In evaluation, a stakeholder examines the performance of the design alternative with respect to various agendas, including his or her own. MUD is highly interactive so the three phases are not rigidly sequenced. The designs and agendas of other stakeholders are available and a stakeholder can test a design against the agendas of other stakeholders or test designs made by other stakeholders against his or her own agenda.

Multiple criteria is the hallmark of any design problem, even those few design problems that involve only a single stakeholder. And, in real-world design multiple stakeholders are the norm, not the exception. Therefore we believe that MUD offers a model for teaching and learning about multi-criteria, multi-stakeholder design in any domain, and is not limited to urban or neighborhood design. However, urban design appears to be a particularly easy domain in which to understand the concept of diverse stakeholders who each have various agendas.

A great deal of work has been done on solving multi-criteria design problems, based on weighted optimization methods and pareto optimality (Gero, 1985). In such systems, a designer enters all the various constraints, objectives, and preferences, and using numerical methods the computer program produces a solution. Optimization is useful when the designer(s) can make accurate and precise assessments of the constraints, objectives, and preferences. However, this is often not the case in design problems. Rather, as the late design methodologist Horst Rittel observed, "understanding the problem is identical with solving it." (Rittel, 1972) In other words: Understanding tradeoffs among objectives in a design problem and reaching agreement among stakeholders about the relative merits of objectives precisely enough to express them as a set of weights is a most difficult part of design, which typically requires a great deal of exploration. We therefore emphasize that MUD is not a system for design problem solving; *it is a system for exploring design tradeoffs.*

The language-based approach we follow in MUD, providing a vocabulary for developing design evaluations, complements Kalay's P3 model (Kalay, 1997) (process, product, and performance), which deals with informing the public and soliciting feedback, aiming to construct a shared understanding by helping participants come up with a consistent design vocabulary.

Rittel proposed to support the exploration of tradeoffs in community planning processes through what he called "issue based information systems" (Rittel & Kunz, 1970), instances of which have been built by McCall, Conklin, and others (Conklin & Begeman, 1988; McCall, 1989). Such systems support deliberation about a design by providing a structure for representing issues, answers, and the pros and cons of each alternative and sub-alternative. This type of system is invaluable for keeping track of the design argument among participants, but alone an issue-based system is unhappily divorced from the act of design. Systems that integrate argumentation in a design environment have been built by McCall and his colleagues, including the Janus-Crack design environment for kitchens, and the PHIDIAS-II HyperCAD system for the design of space based habitation (Fischer, McCall, & Morch, 1989; McCall, Bennett, & Johnson, 1994). However, neither of these systems emphasized evaluation in the context of design, nor did they allow end-user programming of evaluative criteria. In Janus-Crack, for example, evaluation was limited to hard coded critics that detected specific conditions in the design. On the other hand, evaluation of design performance has been explored in a number of computer aided design systems (Kalay, 1992; Rutherford, 1993). Some of these systems support multi-criteria design; few, however also support multiple stakeholders. Combining these approaches, our MUD program entails both evaluation of designs and the design of evaluations by multiple stakeholders.

The remainder of the paper is structured as follows: In Section 2 we outline an example of an urban design project in Boulder, Colorado, USA. Analysis of the project points out some features needed in a computer system to support urban design. In section 3 we describe the nature of the relationships between stakeholders, their agendas and the criteria that comprise those agendas. We present the elements of the MUD program, including the MUD-L language for expressing urban design criteria. Section 4 describes the MUD interface: how designers specify agendas and lay out site plans. They can then test their site plans against their agendas, and against the agendas of other stakeholders. Section 5 outlines the implementation of the MUD program. Section 6 concludes with a discussion and some directions for future work.

2. Real Urban Design: Boulder Future Employment Rezoning Project

We examined a real urban design project, the City of Boulder Future Employment Project (BFER), that was conducted in January and February, 1997. We wanted to observe the design process and methods employed by urban planners to ascertain the consistency of our approach with current planning practices. BFER's goal was to enact a comprehensive rezoning of the city to balance the number of projected jobs with available housing. As subsidiary goals, BFER also aimed to reduce traffic congestion and to retain the small scale character of the city.

BFER was broken down into eleven community zoning districts. Each district held an open house meeting for community members to discuss and negotiate proposed changes.

We focused on district #5, the Boulder Junction area. The program for the Boulder Junction area called for changes to four land use zones: the Main Street/Mixed-use zone, the Industrial/community business zone, the Mixed density residential zone and the Mixed-use zone. In each district a new zoning requirement was set for 17 different 'use definitions' that spell out physical requirements, or criteria. Use definitions used in the Boulder Junction area were: minimum lot area, maximum building height and minimum distance from building to road. These measurable design qualities correspond to the relationships we provide in MUD (see section 4).

From the analysis of the Boulder Future Employment Rezoning project, we found that four key features are needed in a computer system to support participatory urban design. These features are as follows:

2.1. PRELIMINARY DESIGNS: A STARTING POINT FOR NEGOTIATION.

In the BFER project, city planners began negotiation by presenting a preliminary design to the public. Presentation of a design for consideration by other stakeholders is the primary means of communication in planning and must be incorporated into a system that supports design with multiple stakeholders. The BFER project held public information displays around the community, distributed information packets and maintained a World Wide Web site with a description of preliminary design phases. Open house workshops were held in each of the proposed areas of change to provide a forum for presenting and discussing designs. The BFER web site contained descriptions and diagrams of proposed changes in each district.

Presentation of preliminary design can also be construed as the schematic design phase. This phase is necessary for negotiation before more detailed plans are proposed. The schematic design presentation must remain accessible to stakeholders so that changes can be easily made. For example, in the BFER project, maps were displayed showing alternative zoning solutions to the same problem, thus allowing for manipulation of the design on a conceptual level using schematic representations of possible solutions.

2.2. AGENDA SETTING IS PART OF THE URBAN DESIGN PROCESS.

Stakeholders have agendas, whether they express them or not. By formally stating one's agenda, stakeholders articulate their needs and the criteria they wish to see met. Agenda setting was conducted in BFER through various means. Rezoning the city to balance jobs and housing was the agenda of the City Council. The agenda of The Downtown Alliance (a group of citizen boards including the Downtown Management Commission, Downtown Design Advisory Board, Landmarks Preservation Advisory Board, Planning Board) was to develop the downtown area in a manner that maintains the livability and "feel" of downtown, and to protect the downtown's historic character. Other nonprofit downtown stakeholder organizations participated in BFER such as Downtown Boulder Inc., Historic Boulder, and representatives from several neighborhood groups. The various stakeholders worked together to implement changes that incorporated successful trade-offs between the interests of each group. Trade-offs in a project like this can be expressed in terms of the agendas articulated by stakeholder groups.

2.3. DELIBERATION IS A MEANS TO DESIGN EXPLORATION

A computer system for supporting multi-stakeholder urban design must be able to accommodate the discourse and deliberation that occurs in design exploration. In BFER, the design exploration phase was conducted through open house sessions with the community, presentation of preliminary summaries of proposed zoning changes, development of 'use and bulk' charts, and overviews of current zoning in the city of Boulder. The open house sessions were used to explore the tradeoffs between the objectives of various user groups and City Council. Deliberation occurred during the open house sessions through analysis of the maps of proposed changes and 'use charts' and 'bulk charts'. Use charts specify which uses are permitted, prohibited, or conditional, or require use reviews for each of the City's zoning districts . Bulk charts include building height limitations, lot coverage requirements, building setbacks or yard requirements, usable open space requirements, minimum lot sizes, and density requirements. Preliminary designs developed by professional planners were presented to the public by City Council and comment and feedback from the public was solicited. The feedback was used in considering the next phase of changes. Summaries of proposed changes were placed as information displays in the Public Library and in the Courthouse.

2.4. INCREMENTAL SPECIFICATION OF DESIGN CRITERIA

Design can be seen as a process of incremental refinement. In BFER the initial exploratory design phases led to attempts to specify the initial project objectives in physical terms. The overall project objective was to balance housing and jobs. After several iterations of preliminary design, the objective was defined more explicitly. The objective became for City Council and other participating stakeholders to rezone commercial areas into mixed use and residential areas. Again, the objective was defined in more specific terms: Rezone commercial areas in the Boulder Junction area to attain a ratio of 3:1 residential to commercial. Thus the larger higher-level objective of balancing jobs and housing was ultimately redefined in terms of specific land use practices.

3. Stakeholders have agendas made of weighted criteria

In urban design, as in many complex design problems, (almost) everything is connected to everything else in a complex chain of causality. Design thus entails a game of attempting to solve one problem while not disturbing previously reached solutions, and trying to understand the effects any design decision will have on other decisions. Because there are many different stakeholders with diverse agendas, it is difficult to keep in mind the decisions that are important to others. Methods for partitioning decisions into closely related clusters address this problem analytically (Alexander, 1966; Owen, 1970). Here we explore a simpler approach of providing stakeholders with a visual reference to the agendas of other stakeholders. MUD provides this visual reference with gauges that we call 'urbometers' that graphically display the performance of a design with respect to a given stakeholder's agenda.

3.1. STAKEHOLDERS HAVE AGENDAS

All stakeholders in an urban design project have various objectives they wish a design to meet. We call the set of a stakeholder's objectives an 'agenda'. An agenda can consist of a single objective: an advocacy group works to keep a potentially polluting factory from being built in their community. The importance of a single objective is paramount. But an agenda can also consist of many sub-objectives with varying degrees of importance. For example, a city council wishes to control growth, promote business, and attract tourism. The various issues may have different levels of importance.

Urban design projects are often cast under an overarching rubric. For example, the Boulder Future Employment Project focused on balancing jobs and housing. However, there are many means to accomplishing the objective. Stakeholders do not necessarily focus on this top level goal; rather they are concerned about the possible effects of proposed changes with respect to their own agendas. A group of business owners may not have a vested interest in Future Employment, yet they have a strong interest in the quality and character of the downtown area. Negotiation between stakeholders with differing agendas (which may also contain divergent criteria) must occur in a framework for comparison that can provide an analysis of how a specific proposal affects one's own and other stakeholders' agendas.

The agendas of stakeholders often deal with political and economic objectives rather than directly with physical ones, and it is a challenge to translate these 'higher level' objectives into specific criteria and decisions about physical form. For example, business owners may wish to increase the number of customers; this may be realized concretely by providing additional on-street parking within 100 meters of the business. Neighborhood residents may wish to increase the safety of a street; this may be realized concretely by reducing the width of the street to slow traffic, or by installing a stop sign at every corner.

We concentrate here on the physical decisions that must ultimately be made to realize a design, and leave the translation from 'higher level' goals to their 'lower level' physical implementations as a problem for the user. That is, we require MUD users to state their criteria and agendas in terms of the urban design elements and spatial relationships that make up a plan, rather than higher level objectives that ultimately drive their decision making. We leave for a future project translation from political and economic goals to the physical agendas in MUD (but see Koile, this volume, on the relation between physical form and non-physical design characteristics (Koile, 1997)).

3.2. AGENDAS ARE MADE UP OF CRITERIA

An agenda comprises a set of criteria. Each criterion expresses a single goal, or desired property of the design. Here are some criteria:

- "I want a park"
- "I want a two-to-one ratio of office space to commercial space."
- "I want a park within a distance of 100m of every block of housing."

- "I want two parking spaces for every 100 m^2 of housing."
- " I want no more than 1000 m^2 of commercial space."

From these examples it is clear that a criterion involves one or more design elements, ('park', 'housing', 'commercial space'), properties of these elements ('area of park', 'number of parking spaces'), and constraints and relationships on these properties ('no more than 100 m^2 of', 'within a distance of', 'a ratio of'). We provide a limited palette of design elements, which correspond to the basic elements typically found in urban design site plans. For example, our palette includes park, commercial, street, parking, and housing elements.

Some properties involve a single element ('the length of this park'). Other properties are collective over all instances of an element class ('area of housing'). We concentrate on the properties of area (square meters of housing, commercial, park, etc.), number-of (there must at least two commercial areas), and linear dimensions. Some relationships are unary; they involve only a single design element or class of design elements. For example, 'I want a park', 'no more than 1000 m^2 of commercial space' involve only a single element or element class. Other relationships are binary; they involve two elements or classes: 'two parking spaces per 100 m^2 of housing', 'distance from housing to park'.

3.3. EXPRESSING CRITERIA IN THE MUD-L LANGUAGE

We have constructed a language for discourse about urban design in which stakeholders can express their values. We call the language MUD-L for Multi-User Urban Design Language (pronounced "muddle"). MUD-L aims to capture and convey the sorts of criteria described above in a fashion that both a human and a computer can read and interpret. MUD-L has a simple Lisp syntax to express elements and relations. For example, the criterion "I want a park" is expressed in MUD-L as (Exists Park). The criterion "I want a ratio of two-to-one office space to commercial space" is expressed as (Ratio Office Commercial 2/1). MUD-L statements are Lisp predicates, so evaluating a design with respect to criteria expressed in MUD-L is as simple as calling Lisp's 'eval' function on each criterion.

The stakeholder must determine the importance of each criterion, that is, its weight in relation to other criteria in the agenda. For example, for the criterion (Exists Park), the stakeholder must decide "How important is it that a park exists in this area?" Thus, the stakeholder assigns each criterion a numerical weight to express its relative importance. The user specifies a numerical weighting value for each criterion on a scale 1-10 (10 = very important); MUD normalizes the weights so that the total of the normalized weights equals 1.0. An agenda thus comprises a set of criteria and their associated weights. For example:

```
8 (exists park)                        ; I want a park, weight 8 (very important!)
5 (min-area housing 10000)             ; Area of housing ≥ 10,000 m², weight 5
3 (max-distance housing park 100)      ; distance housing to park≤100m, weight 3
```

4. The MUD Interface

MUD runs as a single-user system on a personal computer (Macintosh), though several users can run MUD on separate workstations and collaborate by sharing designs and agendas stored on a central server. Using MUD entails three distinct activities, which we describe below.

4.1. AGENDA SETTING

Each stakeholder constructs an agenda using a simple dialog-menu interface to construct statements in MUD-L, the simple urban design language we have constructed for MUD. The MUD-L language comprises nouns, predicates, and quantifiers. Nouns describe the elements of an urban design, for example units of housing, commercial building, parks, and streets. Predicates describe characteristics of design elements and relations among them, for example maximum-number-of, area-ratio, maximum-distance-between. Quantifiers allow specific numeric values to be added to the criteria: numbers of units, areas, and distances.

Figure 1 shows the interface dialog for constructing an agenda. The dialog contains a table of relationships (left side, first column), a table of basic urban design elements (left side, second column), and a type-in box for specifying numeric values. A stakeholder constructs each criterion by selecting a relationship, one or more urban design elements, and numeric quantifier values. For example, from the relationships column, the stakeholder chooses 'exists'; then from the elements column, 'Park'; and finally enters the weight '8' to construct the weighted criterion ((Exists Park) 8). The stakeholder also selects a weight for each criterion, and adds the weighted criterion to the current agenda. The list of criteria in the agenda appear in the table at the right. A name for the agenda can be entered in a typein box at the bottom.

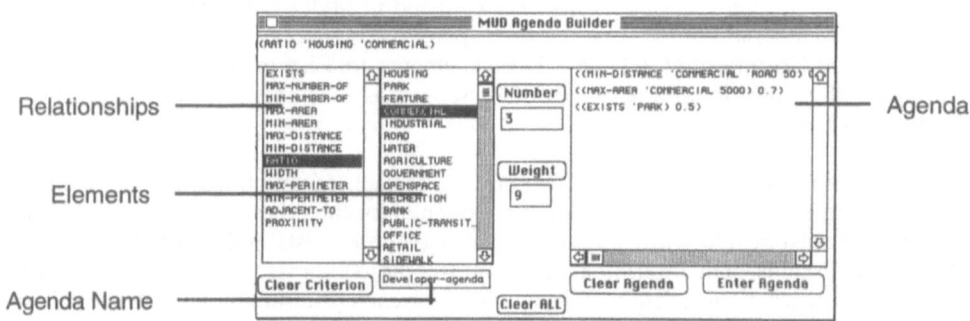

Figure 1. Stakeholders select elements, relationships, and weights to construct the criteria in an agenda.

When the stakeholder finishes adding criteria to construct the agenda, the system produces an 'urbometer', a gauge that displays the score of design alternatives according to the newly constructed agenda (figure 2). The black horizontal bar indicates the current score—the extent to which the current design satisfies the agenda. The criteria and their (un-normalized) weights appear below (scale 1-10) . The bullet (black circle) to the right of a criterion indicates that the criterion is satisfied in the current design state.

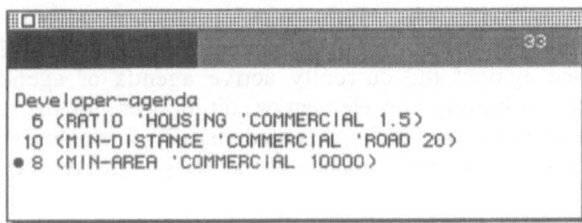

Figure 2. Urbometer displays score of design with respect to an agenda.

4.2. LAYING OUT SITE PLANS

Traditionally urban designers create plans by laying out a site plan and iteratively producing preliminary designs and evaluating them with respect to design requirements. MUD supports this process with a simple palette-based drawing program. Figure 3 shows the palette and work area.

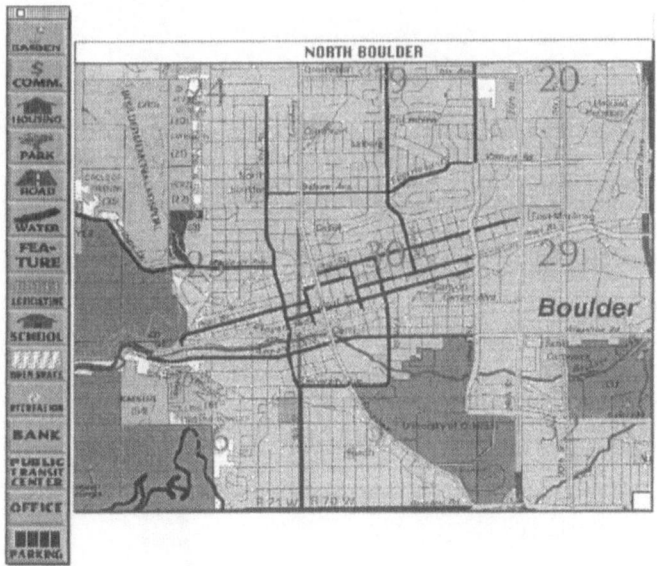

Figure 3. MUD tool palettes and work area with an underlay of the site.

The work area displays a site plan underlay, typically a map or an aerial photograph scanned and imported into MUD. On top of this underlay the stakeholder-designer selects elements (park, housing, commercial blocks) from the tool palettes at the left and assigns them locations and sizes. The elements appear as color coded rectangles, lines, and graphic symbols similar to traditional urban design representations.

4.3. EVALUATION: WATCH THE URBOMETERS

As a stakeholder lays out urban design elements in the work area, MUD scores the design performance against the currently active agenda or agendas. Scoring is incremental, updated with every new element or editing operation the stakeholder makes to the design. The agenda scores are represented visually on a collection of urbometers, which indicate graphically the performance of the design with respect to the criteria in that agenda.

Figures 4 and 5 show alternative designs and the scores of these alternatives according to three different agendas. The three agenda urbometers are presented together in a single window to make it easier to compare them. The three agendas each have three criteria. They differ both in the criteria they require and their weights. As elements are added to the design the urbometers fluctuate, gauging to what extent each agenda has been satisfied.

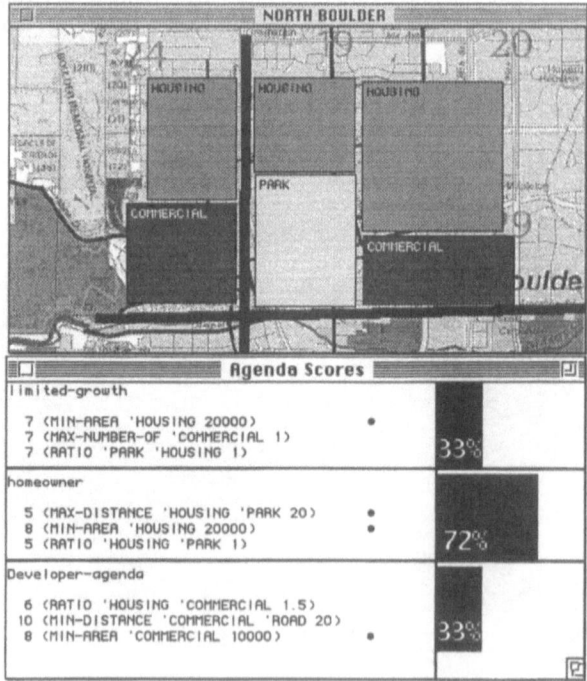

Figure 4. Design alternative #1 and scores according to three agendas. A black dot (bullet) indicates a satisfied criterion; the black histogram bar indicates the design's score according to the agenda.

Figure 4 shows a design with two commercial areas (dark gray) located along the east-west road (thick black line) and a park at the center of town, with housing zones to the north. The first agenda gives a low score to the design, because only one of its criteria is satisfied (minimum area of housing 20,000). The weights on the three criteria are equal, so in this case the score is 33%, or 1/3 of the criteria met. The second agenda gives a better score to the design, because the minimum area of housing is met as is the maximum distance between housing and park. Only the 1:1 ratio of housing to park is not met. Finally, the third agenda scores the design poorly as only the minimum commercial area is met.

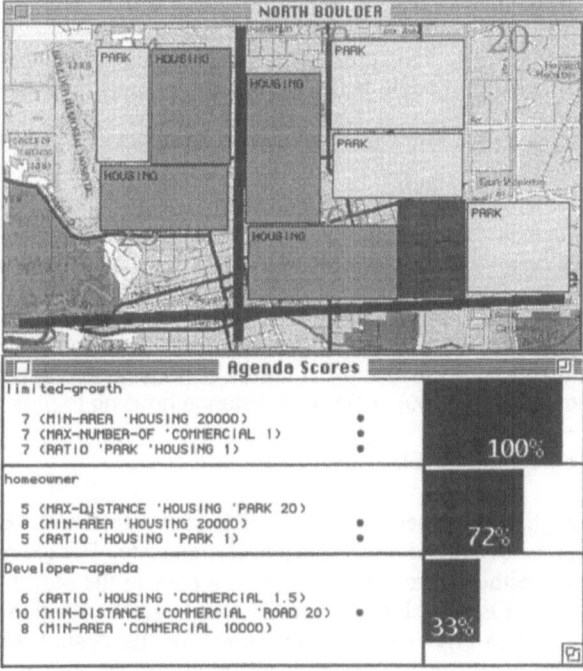

Figure 5. Design alternative #2 scores differently according to the same agendas as above.

Figure 5 shows a second design alternative, in which housing is clustered around the north-south road and parks are arranged at the outlying edges. A single commercial zone is located toward the east end of the east-west road. The scores according to the same agendas as in figure 4 are shown below. The first agenda is entirely satisfied (100%): the design meets its minimum area of housing, its limit of one commercial area, and its ratio of 1:1 park to housing. The second agenda is partly satisfied: the maximum distance criterion from housing to park is not met (the housing zone at the crossroads is too far from the nearest park). The third agenda has only one satisfied criterion: the maximum distance from the commercial zone to the road. The score is slightly greater than 33% because that criterion is weighted more heavily than the other two.

5. Implementation

5.1. DATA STRUCTURES FOR URBAN DESIGN ELEMENTS

MUD was begun as an extension of a 2-D draw program, written in Macintosh Common LISP as a student project for a programming class. A main display list keeps track of each element added to the design. Each element datum includes the element type, a particular name (if the designer chooses to specify a name), and its location coordinates (in pixels) needed to draw it on the screen. For example:

```
(housing pruitt-igoe ((100 100) (200 150))
```

describes a block of housing called "pruitt-igoe" that is a rectangle located at coordinates top left = (100,100) to bottom right = (200, 150).

5.2. DATA STRUCTURES FOR CRITERIA AND AGENDAS

The data structure for each agenda consists simply of an agenda name and a list of weighted criteria in which the weights have been normalized to sum to 1.0. For example, the agenda described above would appear normalized as follows:

```
(agenda park-housing
    ((exists park) .5)                        ; I want a park, weight 8 (very important!)
    ((min-area housing 10000) .3125)          ; area of housing ≥10,000 m², weight 5
    ((max-distance housing park 100) .1875))  ; distance housing to park ≤ 100m, weight 3
```

5.3. EVALUATION OF DESIGNS

Evaluation of a design with respect to an agenda is straightforward. Each criterion is a predicate, which can be evaluated on the current design state. For example, (exists park) is either true or false; either there is or there isn't a park in the current design. If there is, the criterion's weight is added to the agenda score. Each criterion is thus evaluated and the weights added to arrive at the total score for the design with respect to the agenda. The maximum score is one; the minimum, zero. For example, if (in the current design) there is a park, and there is more than 10,000 m^2 of housing, but the maximum distance from housing to park is greater than 100m, then the design will receive a score of $(.5 + .3125 + 0) = .8125$.

6. Discussion and Further Work

We have described the current version of the MUD program. It provides a visual environment for proposing urban designs and evaluating them according to multiple agendas, each consisting of multiple criteria. The MUD-L language provides a means for stakeholders to express their objectives in terms of specific physical design characteristics. The MUD program provides an environment for collaborative work in urban design that aims at consistent recognition of all stakeholder agendas during design. It provides stakeholders a means to compare and evaluate various designs according to different values. It allows designer-stakeholders to quickly see the effects of a design on the agendas of other stakeholders. Expressing urban design goals in a language that can be understood by designers and also quantified begins to provide a framework for argument, evaluation, and decision making.

6.1. MUD IN THE CLASSROOM AND STUDIO

In interviews in a third year (architecture) design class, students found it very difficult to articulate specific spatial relationships that enhance or detract from neighborhood quality. Instead, their remarks were directed at experiential qualities (safety, noise level, privacy) and not at specific, measurable, qualities of the built environment. Yet specifying physical characteristics is the essential job of designers. When pushed, students resorted to the ineffable: 'when I draw it just comes to me', 'I can't describe it but I know it when I see it'. We began to realize that explicit translation from experiential qualities to physical decision making is an important but neglected skill in our curriculum, and one that our MUD program might help students to develop.

Also, we have found that students of architecture and planning find it nearly impossible to deal with multiple stakeholder design problems. Typically a student designer will either give up and ignore all but one of the stakeholders' objectives, or try to synthesize *a-priori* a compromise agenda, in order to proceed with the designing. Compromise is inevitable, but students tend to adopt a compromise agenda before they have fully understood the tradeoffs entailed in diverse stakeholder objectives.

It seems difficult for novice designers to apprehend the fact that a given design may at once be good from the perspective of one group of users and bad with respect to another. Therefore, we plan to employ MUD as a teaching tool in a course on design methods in hope of making students aware of the complexities of multiple agenda, multi-criteria design and the attendant need for negotiation.

In this classroom-studio setting we will provide a design scenario and assign students different roles. For example, we will ask one student to represent the developer who wishes to build commercial buildings; another to represent young parents with children, and so on. Each student must construct an agenda that corresponds to their assigned role. We will ask them to make a design that maximizes performance with respect to their agenda. They will post their agendas and designs in a shared electronic bulletin board on the class web page. Then we will ask them to use their own agenda to score designs made by other students and to score their own design against the agendas made by others. We hope by these means to begin a classroom dialogue among designer stakeholders that promotes collaboration and negotiation, using urban design plans and agendas as concrete learning tools: 'objects to think with'.

6.2. FROM PHYSICAL FORM TO EXPERIENTIAL CRITERIA

We've insisted in this version of MUD that stakeholders express their design agendas in terms of specific physical criteria by which a design can be measured. We believe (as we mentioned above) that this is a valuable exercise for stakeholders: to translate experiential criteria such as "I want a place for my children to play" into physical and directly measurable statements like "I want a park within 100 meters of my house". However, we would also like to provide computer support for making these translations from experiential to physical decisions. This could make it easier for stakeholders who are not skilled designers to express their values in terms of 'higher level' goals and then use the program to evaluate designs. This would entail the construction of a higher level language on top of MUD-L, in which experiential qualities such as safety, quiet, good for children could be described in terms of physical criteria such as street width, presence of traffic lights, and distance from house to school.

Providing a higher level language of experiential qualities opens the door to more innovative problem solving. When design agendas are stated in terms of physical criteria, it remains only to evaluate and compare designs. Our MUD program provides a simple and visual means to do this. But there may be several ways that the environmental quality of safety (for example) can be provided; a skilled designer can choose among these alternative methods to provide desired outcomes. For example, one stakeholder may want on-street parking and therefore designs a wide street; another wants safety and therefore designs a narrow street (to slow down traffic). At the lower level of these physical decisions the difference in objectives seems irreconcilable. But, reasoning at the higher level of experiential goals, a stop light will resolve the apparent conflict. Thus enabling stakeholders to express values in terms of experiential qualities may allow more innovative design solutions.

6.3. THE REASONS BEHIND AGENDAS

We mentioned earlier the decision support methods of Rittel, McCall, and others who employ an 'issue based information system' to support argumentation, negotiation, and deliberation among various parties to a design decision. The issue base constitutes a repository of facts and opinions about a design and its desired outcome, and can be valuable both during a design process to support group and community decision-making, as well as afterward, to record design histories and rationale. However, these decision recording tools omit a crucial part of the process: construction and evaluation. Conversely, our MUD program, which provides exactly this type of support, fails at archiving the argument, negotiation, and deliberation that naturally surrounds construction and evaluation by a group of differing stakeholders. It makes sense to put these two approaches together: Consequently, our ideal architecture for multi-user, multi-criteria design decision making would employ both kinds of support tool a construction kit for making designs, the means to set agendas for evaluating designs and to evaluate designs according to agendas, and the integrated means to record these events for later perusal in a structured database. An issue based information system such as McCall's PHIDIAS-II program would seem suited to these activities, and we plan to explore using this or similar technology to support design negotiation in future versions of MUD.

Acknowledgments

The CAD Research Group at the College of Environmental Design, UC Berkeley offered support and valuable discussions to Ame Elliott. Laura Parker received support for this project from the University of Colorado Undergraduate Opportunities Program (UROP).

References

Alexander, C. (1966). *Notes on the Synthesis of Form*. Cambridge: Harvard University Press.

Branch, M. C. (1985). *Comprehensive City Planning: Introduction and Explanation*. Washington DC.: Planners Press.

Conklin, J., & Begeman, M. (1988). gIBIS: a hypertext tool for exploratory policy discussion. In *Conference on Computer-Supported Cooperative Work*, (pp. 140-152). Association for Computing Machinery.

Fischer, G., McCall, R., & Morch, A. (1989). JANUS - Integrating Hypertext with a Knowledge-based Design Environment. In *Hypertext '89 Proceedings* (pp. 105-117).

Gero, J. S. (Ed.). (1985). *Optimization in Computer Aided Design*. Amsterdam: North Holland.

Kalay, Y. (1997). P3: An Integrated Environment to Support Design Collaboration (CAD Research Group Working Paper). Center for Environmental Design Research, University of California, Berkeley.

Kalay, Y. E. (Ed.). (1992). *Evaluating and Predicting Design Performance*. New York: Wiley.

Koile, K. (1997). Design Conversations With Your Computer. In R. Junge (Eds.), *CAAD Futures '97* (this volume).

McCall, R. (1989). MIKROPLIS: a hypertext system for design. Des. Studies, 10(4), 228-238.

McCall, R. J., Bennett, P., & Johnson, E. (1994). An Overview of the PHIDIAS II HyperCAD System. In A. Harfmann & M. Fraser (Ed.), *ACADIA (Association for Computer Aided Design in Architecture)*, (pp. 63-74). St. Louis, MO.

Owen, C. L. (1970). DCMPOS: An Algorithm for the Decomposition of Directed Graphs. In G. L. Moore (Eds.), *Emerging Methods in Environmental Design* (pp. 133-146). Cambridge, MA: MIT Press.

Rittel, H. G. (1972). On the Planning Crisis: Systems Analysis of the 'First and Second Generations'. Bedriftsøkonomen(8), 390-396.

Rittel, W., & Kunz, W. (1970). Issues as elements of information systems Working Paper 131. Center for Planning and Development Research, University of California, Berkeley.

Rutherford, J. H. (1993). KNODES: Knowledge Based Design Decision Support. In U. Flemming & S. van Wyk (Eds.), *Computer Aided Architectural Design Futures '93* (pp. 357-374). Amsterdam: North Holland.

AN INTEGRATED COMPUTING ENVIRONMENT FOR COLLABORATIVE, MULTI-DISCIPLINARY BUILDING DESIGN

LACHMI KHEMLANI and YEHUDA E. KALAY
Department of Architecture
University of California at Berkeley

Abstract. The increasing complexity of the built environment requires that more knowledge and experience be brought to bear on its design, construction and maintenance. The commensurate growth of knowledge in the participating disciplines—architecture, engineering, construction management, facilities management, and others—has tended to diversify each one into many sub-specializations. The resulting fragmentation of the design-built-use process is potentially detrimental to the overall quality of built environment. An efficient system of collaboration between all the specialist participants is needed to offset the effects of fragmentation. It is here that computers, with their ubiquitous presence in all disciplines, can serve as a medium of communication and form the basis of a collaborative, multi-disciplinary design environment. This paper describes the ongoing research on the development of such an integrated computing environment that will provide the basis for design and evaluation tools ranging across the many building-related disciplines. The bulk of the discussion will focus on the problem of a building representation that can be shared by all these disciplines, which, we posit, lies at the core of such an environment. We discuss the criteria that characterize this shared building representation, and present our solution to the problem. The proposed model has been adapted from geometric modeling, and addresses explicitly the difficult problem of generality versus completeness of the represented information. The other components of the integrated environment that are under development are also described. The paper concludes with some implementation details and a brief look at two evaluation tools that use the proposed building representation for their task.

1. Introduction

If an opinion poll were to be conducted among people about their approval of the built environment, most would express their dissatisfaction with some aspect or another of their living or working conditions. Undeniably, modern buildings are far from perfect in their ability to satisfy the physical, social, cultural, and economic needs of the people who are affected by them. The need for an improvement in the overall quality of our buildings is consistently felt in all sections of our society.

What does an "improvement in quality" really entail? It would mean raising the standards of all the individual processes that a building "lives" through in its lifecycle—its design, construction, and maintenance—as well as the overall quality of the combined effects of these processes. For they are not discrete processes; rather, they are highly interdependent. Hence, coordinating them while striving to achieve greater overall quality is highly critical to the building process.

As our society becomes increasingly more complex, a proportional increase in the complexity of the built environment is inevitable. Buildings must fulfill a host of diverse criteria, abide by innumerable codes and rules, and an ever-increasing list of

389

R. Junge (ed.), CAAD Futures 1997, 389-416.
© 1997 *Kluwer Academic Publishers.*

constraints. A direct consequence of this increased complexity has been an enormous growth in the number of diverse professionals who need to be involved in the design and construction of a building. Gone are the days when this responsibility rested almost entirely on the shoulders of the "master builder." Today's built structures incorporate so many aspects—social, aesthetic, technical, financial, and legal, to name a few—that it is impossible for any single discipline to deal effectively with each and every one of them. Specialization has become inevitable as the only way in which the intricacies of each aspect of the building can be dealt with, and this trend will only continue to grow in the future.

Specialization makes the already involved process of building design even more intricate and time-consuming. All the individual design and construction specialists do not work together on a design. Not only are they physically located in different places, they are also not usually working on the same design model. The architects may do a preliminary design and send it over to the structural engineers and the HVAC engineers; the structural engineers may demand certain changes, but by this time the HVAC engineers may have already done their job, which they would have to repeat because of the changes. Meanwhile, the architects might have a chat with their clients and make some more changes, which, by the time are communicated to all the concerned participants might result in much wasted work. Since every project has a budget constraint, and consequently a limited time for "finalizing" the design, such delays in communication can be disastrous. Often, there are also gross errors in transmitting information, and in interpreting it. All these can lead to costly mistakes, which may not be rectifiable. Naturally, they can all end up severely compromising the quality of the design.

Over and above the communication issue, however, there is yet another serious problem that specialization brings in its wake. It is very difficult, if not impossible, for the specialists to have a clear vision of the "overall goodness" of the project. Due to their limited view-points, each specialist tries to optimize the design for his/her own discipline, which quite conceivably may come at the expense of other disciplines. Eliminating windows on the west side of a building to save energy is a case in point: it might also deprive the inhabitants of a fabulous view.

What is really needed to bring the various specializations together into a coherent whole is an *effective system of collaboration*. This calls for the development of an environment within which all the building design professionals can work together, so that there are no delays, inconsistencies, miscommunications, and other errors. This environment also has to provide a means to negotiate partial solutions, to trade them off against each other so the overall result is improved. Essentially, the environment has to specifically recognize the over-riding significance of collaboration in enhancing the quality of building design, and be geared towards actively fostering and providing for that collaboration.

It is in the development of such an environment for collaborative, multi-disciplinary building design that computers may play their most important role in the field yet. The use of computers in enhancing the quality of building design, as well as the efficiency of the design process itself, has been the subject of much research for more than three decades. In some areas, these efforts have met with remarkable success. A wide variety of sophisticated computer tools is available for design visualization in the form of drafting, 3D modeling, rendering and animation software. These have become an inseparable part of the "toolkit" of each and every one of the professionals involved in the design and construction of buildings. In sharp contrast to the widespread availability and use of visualization software, computer tools that can provide "intelligent" assistance in the actual design process in the form of analyses, criticism, evaluation, prediction or generation, are noticeably absent. Yet, with the increasing need for collaboration in building design, such computer aids are now needed more than ever before. In fact, in today's building scenario, a collaborative design environment could be made possible only through computers, since this is the only medium both common yet powerful enough to serve as the communication channel.

This paper describes our research on such a computerized environment for collaborative building design. More specifically, it details our solution to the problem of a shared building representation for the envisioned collaborative design environment, which we see as one of the most critical components of the sought environment.

2. The Problem of Collaboration

We have seen that a building is a complex artifact with manifold aspects to it, each of which is attended to by professionals who specialize in that aspect. Not surprisingly, these professionals are concerned with quite varied sets of information about the same building. The architect is mostly concerned with spaces (i.e. voids), the structural engineer is mostly concerned with walls, columns, and floor slabs (i.e. solids), the fire safety engineer is concerned primarily with openings, and so on (Figure 1.)

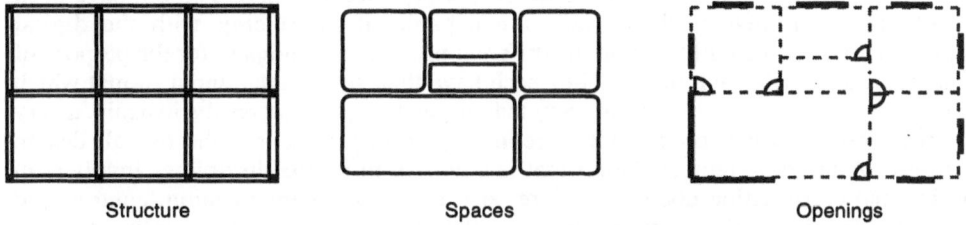

| Structure | Spaces | Openings |

Figure 1. Multiple "viewpoints" of the same building.

In fact, we might say that each one of the specialists—architects, engineers, construction managers, facilities managers, even building owners and end-users—seem to have a different "world view" of the building. In general, architects often emphasize the quality of artifacts over their function, purpose, and the processes of making them. Engineers tend to emphasize the function or purpose of the artifact, placing less emphasis on the process of making, and still less on its formal qualities. Construction managers are interested mostly in the process of making, whereas facilities managers are interested in the process of maintaining the building. Owners and end-users are usually not interested in their environment, as long as it does not impinge on their activities and interfere with the achievement of their personal or institutional goals (i.e., how well the place supports the education of students, the fabrication of goods, or the healing of patients).

All these unique world views have been developed by the different professionals through their education and practice. They are trained to understand the world in a particular way, and they develop a unique language to facilitate representation, communication and comprehension within their own sub-culture. Yet, all these professionals are undeniably working on *one* project, and they need to think in terms of optimization of the "whole" versus the individual "parts" they are each concerned with.

The difficulty also arises when these professionals have to collaborate, as they must do. How do they communicate and exchange information with one another? How does each profession interpret and understand the information presented to it by another profession? Currently, there is no firmly established protocol of communication, much less a means to explicate that which is communicated.

2.1. REQUIREMENTS OF A COLLABORATIVE DESIGN ENVIRONMENT

Considering the problem of multiple viewpoints, the most crucial aspect of a collaborative environment would be a *building representation* that can be shared by all disciplines, and can become the basis of a *shared understanding* of the overall improvement of the project. If all the discussions, feedback loops, and design decisions related to the building could be based on one and the same representation, there would be less room for cross-disciplinary miscommunication, delays and errors. The situation would be analogous to the "ultimate" collaborative scenario where all the "experts" are in the same room, working together off the same table, fulfilling their respective roles as and when needed, making decisions and resolving conflicts as and when they arise by talking to each other, clarifying doubts and bargaining trade-offs.

Over the years, a host of stand-alone computer tools for supporting different aspects of building design, such as energy, cost, fire safety, circulation, and so on, have been developed. But few of these tools are capable of interfacing with the digital representations of the building that the different professionals prepare for the purpose of communication with each other. They each have their own unique input format which is often non-graphical and therefore very tedious, and output their results in again, a very specific format, which is difficult to redirect gainfully back into the overall design process. A building representation that is common to all the disciplines involved in design and construction could, therefore, also facilitate communication, control, and effective decision-making by bringing together all the individual discipline-specific design tools and aids into one, integrated environment.

In short, we propose that what is needed for effective computer-aided collaborative design is a unified building representation which is accessible and comprehensible to all the professionals in the building team, which not only allows the sharing of *information* but also the sharing of *understanding*, and which facilitates the development of *design tools* for different aspects that can be "plugged" into it.

2.2. CRITERIA FOR A "GOOD" BUILDING REPRESENTATION

How can the suitability of a proposed building representation for collaborative, multi-disciplinary building design be assessed? We posit that a "good" representation will be distinguished by four important characteristics:

2.2.1. *Well-formedness*
The most basic requirement of a building representation is that it must have semantic integrity. Windows must be within walls rather than floating in space. If a wall is moved, all the spaces that the wall is part of must change in configuration accordingly. This calls for a representation far more semantically "intelligent" than the one, for instance, used by popular CAD softwares which are severely limited in their ability to provide adequate support for design tools.

2.2.2. *Completeness*

The ideal building representation would undoubtedly be one which is as close as possible to the actual building itself. This would mean that all the architectural elements—spaces, walls, doors, windows, staircases, and so on—would be represented along with all their attributes, not just individually, but also in their relationships with each other. Likewise, all the structural, mechanical, security, and other elements need to be represented explicitly. This calls for a building representation that is *complete* in all respects, leaving nothing to be inferred (or possibly—mis-inferred) by the users.

2.2.3. *Generality*

A building representation, to be of universal value, should be general-purpose enough to be able to describe any building, irrespective of its function. No doubt, it might be over-ambitious for a representation to aim to model *every* kind of building possible; there will always be some configuration far too complicated for it to handle. For instance, the Sydney Opera House might be difficult to represent accurately compared to another, structurally less complicated, building of the same function. Thus, it might be reasonable for a representation to target possibly 60-70% of architecture as within its capability to model, with the deciding factor being the complexity of the building rather than anything else.

2.2.4. *Efficiency*

An efficient building representation is one that is computationally fast and compact. Yet, these two criteria will almost always conflict with each other. Speed can be enhanced by redundancy, because the sought information will be immediately available without the need to compute or search for it. Redundancy, on the other hand, will make the space requirements enormous, and is a potential source of errors in both the creation and the manipulation of the building-related data. A good building representation has to find an efficient balance between speed and non-redundancy.

2.3. DIFFICULTIES IN DEVELOPING A COLLABORATIVE DESIGN MODEL

The main difficulty in deriving a building representation for collaborative, multi-disciplinary design arises precisely due to the fact that all the different "world views" of the individual specialists have to be integrated in one shared representation. Partial models, which deal only with the domain of one specialist—such as space planning, structural engineering, or energy evaluation—are relatively easier to accomplish, and many such models are already becoming available commercially (Eastman and Siabiris 1995). However, they typically support only one or a few tasks, usually at the expense of other tasks. Thus, a model designed to support the structural analysis of buildings, for example, is less capable in the area of habitability analysis, which requires representation of intended activities more than building components. Therefore, while such "focused" models provide support for some common design, prediction and evaluation utilities for the individual specialist, they are unable to support the involved and open-ended range of design and evaluation activities that characterize the overall process of building design.

On the other hand, there are also models which are aimed at supporting a broad range of applications, but these typically sacrifice completeness for the sake of generality. A good example is a system like Autocad, which can represent almost any geometrical feature of a building, but lacks the semantic knowledge of building parts such as walls, windows, kitchens, and so on.

From the point of view of collaborative design, neither type of model is acceptable: the narrow-range but detailed models can support only a few applications, and are therefore not useful when a broad range of applications is desired. The broad-ranged but insubstantial models can only partially support all applications, lacking the ability to represent the semantics of the design. Our efforts have been focused on developing a model that can do both: have a considerable, if not exhaustive knowledge of the semantics of the designed building, while at the same time have the capability of supporting a wide and open-ended range of applications from different disciplines.

2.4. OTHER ATTEMPTS

The importance of a collaborative design environment for the effective application of computer aids in enhancing the quality of building design has, by now, become well established within the CAAD research community.

Two general directions have emerged: one which provides a low-level, shared information exchange platform, augmented by separate disciplinary models; the other which provides a comprehensive common database that can be used by all the participating disciplines. The first direction has been demonstrated by efforts such as STEP, COMBINE and IBDE, and is predicated on the belief that each discipline knows best what it needs, hence should be entrusted with developing its own disciplinary model. While this solution is, in many ways, more practical and easier to accomplish, it suffers from a lack of shared understanding, and from a reduced level of semantic communication. The other path, demonstrated by efforts such as OXSYS, ICADS, EDM and BDA, has recognized these potential deficiencies, and attempted to develop more comprehensive design databases.

In this section we briefly review some of these individual research efforts from the point of view of the four characteristics of "goodness" defined earlier—well-formedness, completeness, generality, and efficiency. A more comprehensive description of these systems can be found in Galle's recent survey of current research into computer modeling of buildings (Galle 1995).

The focus of STEP is on data exchange, rather than the actual representation of the building. In fact, it is concerned with defining a standard for the representation and exchange of "product information" in general, not just building data (Bjork and Wix 1991). Thus, STEP seems to be too general-purpose to provide an efficient and workable solution to the problem of building representation, and too shallow as far as semantic content is concerned.

The COMBINE project, working with the STEP standard, deals with a neutral file exchange of data among various application programs—primarily energy and HVAC (Amor et al 1995). It does have, at its core, an "integrated data model" for the building description, which gets translated to the applications. However, this data model was developed by synthesizing the individual data schemas of the various design tools it wanted to integrate, so it is not a complete model. Each time a new tool has to be integrated into the system, the data model has to be re-worked.

Both the OXSYS and the IBDE systems are too specialized to provide the solution to a generic building representation. The OXSYS system was developed for the design of hospitals and ancillary buildings in accordance with a particular style of construction and relies on a predefined "kit of parts" to cover a large portion of the design and construction process (Richens 1977). The IBDE system consists of seven stand-alone, knowledge-based design tools, each dealing with a specific aspect of high-rise office-building design, integrated into one system (Fenves et al 1994).

The focus of the ICADS system is not on the detailed representation of a building, but on developing a controller for multiple rule-based expert systems that it brings together to collaborate in evaluating a design (Pohl and Myers 1994). Only a limited number of architectural objects are recognized by this system, so it is not very versatile.

The EDM and BDA building representations come closest to our definition of "suitability" for collaborative design, based on the four criteria discussed earlier. EDM was developed to address issues in the design of buildings with wide ranges of design abstractions, construction technologies, and variations in building use. These are represented by the constructs BOUNDED_SPACE, CONSTRUCTED_FORM, and ACTIVITY, respectively, which in turn, form the three main components of the Generic Building Model, the basic framework of the EDM (Eastman and Siabiris, 1995). It is meant to be open-ended, so that additional descriptions can be added. The EDM thus aims to be general-purpose as well as complete. As far as efficiency is concerned, it stores both space and structure explicitly which might lead to considerable redundancy. It is in the issue of semantic integrity, however, that the EDM differs most from our approach. The EDM separates the geometric representation of the building from the integrity constraints that endow it with semantic validity. The semantics are computed as and when needed. Our model, on the other hand, aims to have as much of the semantic integrity built into the topological representation of the building itself.

BDA (Building Design Advisor) is an integrator of a host of existing (and yet to be developed) evaluation systems such as DOE-2, Superlite, etc. (Papamichael et al 1996). As such, it provides a unified interface and a common data representation and exchange format. Much effort has been devoted to rigorously define the values that are passed from one application to another, and capture the semantics, origin, and nature of the exchanged data. BDA does so by providing custom sockets into which the evaluation systems plug into the BDA environment. Nonetheless, while BDA recognizes the need for a common database, which is the repository of the values that are produced and are required by the various applications, it assumes that the design of the building will occur elsewhere (even though BDA provides a simple graphical editor). This assumption is evident in the structure of the database, which seems to be more a passive repository of objects than a dynamically updateable one. In other words, while there is a "place" to store objects and the relationships between them, maintaining the database's semantic integrity is a task that must be assumed by some intelligent agents that are not part of BDA system itself. These agents are assumed to be very capable, since the structure of the database does not, in and of itself, enforce well-formedness.

3. Our Solution

Our approach to the development of a collaborative design environment is different from many of the research efforts just discussed. Of most significance is the fact that we are not starting out with any specific existing application programs and developing a building representation based on what information they need and how they need it. Instead, what we are asking is: "How should a building be represented so that it is understood as a building, so that the underlying data structure has been created to represent only a building and nothing else." We seek a building representation that can give us any kind of information about a building, from *any* disciplinary point of view. A few examples are:

- What spaces does it have? (architecture)
- Where are the columns and walls on each floor located? (structural engineering)
- For a given external wall, how many separate internal surfaces does it have? (energy evaluation)
- In which direction does a particular door open? (architecture and construction)
- What is the total area of wall surface? (cost analysis, construction and maintenance)

We are not concerned with exactly which specific application programs will be using this information. Rather, we would like to ensure that such information is easily available from the representation, simply because it *is* the representation of a building! In other words, we accept the need for specialization, but at the whole building, or project level, rather than on the disciplinary level. As such, our representation may not be suitable for mechanical engineering, or for VLSI design, as many of the current CAD systems are. In return, we gain semantic knowledge, because the representation is familiar with such terms as "wall," "window," or "HVAC system."

3.1. THEORETICAL FRAMEWORK

We present in this section, the theoretical underpinnings of our solution. First, the building representation that we have developed at the core of our proposed model will be described in detail. Thereafter, the overall schema of the integrated building environment, including the representation, will be presented.

3.1.1. *Basic Requirements of a Building Representation*
All buildings are essentially assemblies of similar components, such as walls, slabs, doors, windows, and so on. Where one building differs from any other is the manner in which these components are put together—the combinations are potentially infinite. In other words, it is the *assembly* that makes every building unique. Therefore, as far as a building representation is concerned, it seems appropriate to make a distinction between the components themselves and the manner in which they are assembled. This has the added, and very important advantage, of re-usability: the detailed information about the various components would be common across various building projects and types, and does not have to be developed or maintained for each building project separately.

The building representation, therefore, gets naturally partitioned into two logically separate components:

1. A Project database
2. Several Object databases

Project Database (PDB). This database links all the elements of the building into the appropriate hierarchy, and stores information about the geometrical location of elements with respect to each other. It is unique to each building project. Detailed information about any element can be obtained from the Object Databases. Because of the hierarchical organization of the elements, much of the semantic content of the building representation will be inherent in the Project Database; for example, a space will necessarily belong to a floor (i.e. level), an opening will necessarily be contained within a wall or slab, and so on. Hence, it is in this component of the overall representation that the semantic integrity of the information is maintained.

Object Databases (ODBs). These databases will be commonly shared by all building projects. They will contain detailed information about every possible component used in buildings, not just for physical elements such as doors, windows, and so on, but also for more abstract elements such as spaces. The information would include material composition (for physical elements), as well as attributes, behavior, dimensional limits, and so on.

The relationship between the PDB and the ODBs is depicted in Figure 2. It shows how a PDB can draw on several ODBs, and how the ODBs can contribute to several projects at the same time.

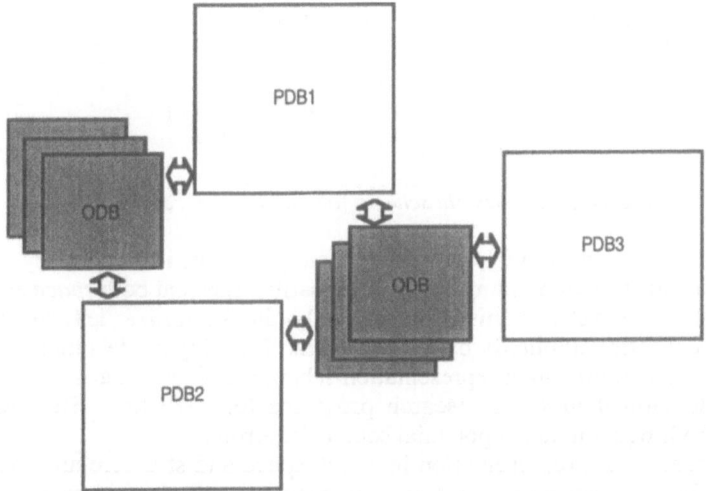

Figure 2. The relationship between the PDB and the ODBs.

Both these components of the building representation will now be discussed in detail. The PDB is more extensively presented since it is the data structure we have developed underlying the PDB that is most unique about our proposed building representation. We believe that this data structure satisfies all four criteria of a suitable building representation listed earlier, particularly that of *generality* and *completeness*— which we have seen are the criteria that other systems fail to accomplish adequately.

3.1.2. The Problem of "Completeness": Duality of Space and Structure
In developing a data structure for the representation of a building, it might be very natural to arrive at a structure similar to that shown in Figure 3, where the building has been divided into its two most basic components: the *spaces* that are contained within the building, and the *structure* of the building. Each one of these components is then expanded further. This method of organization would seem to work, except that it does

398

not take into account the close coupling between space and structure. For it is the *structure* that defines the space; alternatively, a *space* also determines the structure that bounds it. Thus, when either entity is explicitly represented, the other is also defined; moreover, one entity cannot be modified without affecting the other.

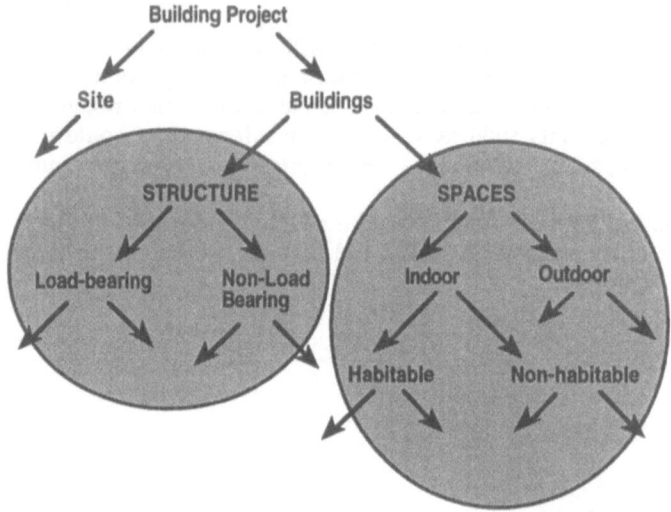

Figure 3. A possible schema for building representation.

A semantically-rich model, which can support multi-disciplinary design and performance evaluation of buildings, must necessarily represent both *space* and *structure*. Most commercial CAD systems represent only the *structure*, leaving the *space* it encloses to be inferred implicitly by the designers. Building models that represent both entities separately, following a representation scheme similar to that shown in Figure 3, have been developed in some research programs (e.g., EDM). Such redundancy, however, is both wasteful and a potential source for errors.

What is needed is a representation in which space and structure are both explicitly represented, but in a non-redundant, inter-dependent, complementary way. We have developed such a representation, which includes both structure and space, using a data structure modeled after the well-known, geometric, winged-edge model.

3.1.3. *The Winged-Edge Data Structure*
The problem of duality between *space* and *structure* in architecture is analogous to that of the duality between *edges* and *faces* in (planar) polyhedra. A polyhedron consists of a number of connected faces bounded by shared edges. Moving an edge should automatically affect both the faces that share the edge (as well as the other faces that share the corresponding vertices). Considerable research was devoted to the development of an appropriate representation scheme that could facilitate such operations efficiently. In 1972, Baumgart came up with the "Winged-Edge" data structure (Figure 4), which has since then been successfully implemented in several modeling systems such as GLIDE, BUILD, WORLDVIEW, and DESIGN (Baumgart 1972, Baer et al 1979, Kalay 1987, Kalay 1989).

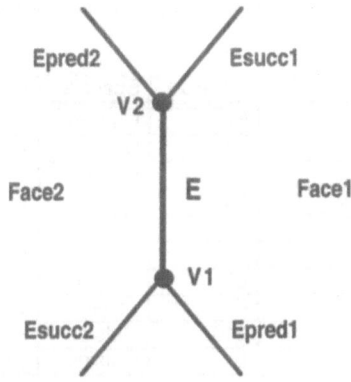

Figure 4. The Winged-Edge data structure.

This data structure is centered upon the representation of the EDGE. The boundary of a solid can be completely defined by a data structure in which each edge is referenced by two preceding edges (Epred1 and Epred2), two succeeding edges (Esucc1 and Esucc2), two faces (Face1 and Face2), and two vertices (V1 and V2). It then takes only the specification of the geometric coordinates (X, Y, and Z) of each vertex for the complete definition of the polyhedron itself. This is a compact, non-redundant, consistent, and efficient data structure.

Considering the faces and edges of a polyhedron as analogous to the spaces and walls of a building, respectively, we can use a similar data schema to solve the problem of duality of space and structure in building representation. As a simple example, consider the three-room apartment shown in Figure 5. The topological skeleton of this design is fully captured in the EDGE table, depicted in Table 1.

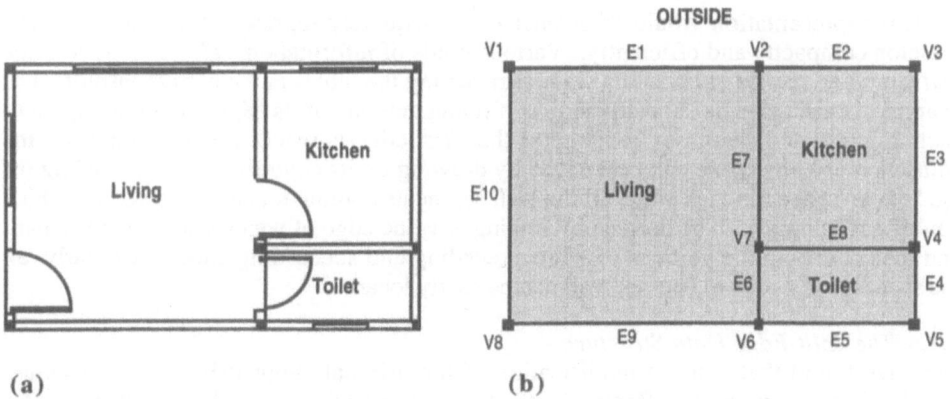

(a) (b)

Figure 5. (a) The actual plan of a studio apartment; (b) its topological skeleton with the wall-edges and vertices explicitly marked (openings have been ignored for now).

Table 1. The Edge table representing the building skeleton shown in Figure 5b, based upon the Winged-Edge data structure.

Edge ID	Space 1	Vert 1	E-Pred 1	E-Succ 1	Space 2	Vert 2	E-Pred 2	E-Succ 2
E1	Living	V1	E10	E7	Outside	V2	E2	E10
E2	Kitchen	V2	E1	E3	Outside	V3	E3	E1
E3	Kitchen	V3	E2	E8	Outside	V4	E4	E2
E4	Toilet	V4	E3	E5	Outside	V5	E5	E3
E5	Toilet	V5	E4	E6	Outside	V6	E9	E4
E6	Living	V7	E7	E9	Toilet	V6	E5	E8
E7	Living	V2	E1	E6	Kitchen	V7	E8	E2
E8	Kitchen	V4	E3	E7	Toilet	V7	E6	E4
E9	Living	V6	E6	E10	Outside	V8	E10	E5
E10	Living	V8	E9	E1	Outside	V1	E1	E9

It is now easy to see how more information can be added to this representation, in the form of additional tables, or pointers to attribute records. For example, it is possible to add geometry, simply by adding X, Y, Z coordinates to the vertices, as depicted in Table 2. Likewise, information can be added about the spaces and the walls, by referencing the appropriate object in one of the accompanying ODBs.

Table 2. The list of X, Y, and Z values for the vertices of the building skeleton shown in Figure 5.

Vertex ID	X	Y	Z
V1	0	0	0
V2	12	0	0
V3	20	0	0
....	....	....	....
....	....	....	....

This representation would then suffice to adequately represent the given building skeleton compactly and efficiently. Various kinds of information, *related to both space and structure*, can be obtained by simply traversing the tables in the correct fashion. For example, a space can be "constructed" by finding any one of its edges, and moving on to each appropriate "successor" edge until that first edge is reached again; similarly, the adjacencies of any space can be derived by drawing up its edge-list and then reading off the "other" space for each edge; all the wall segments coming together at a vertex (which could be a column) can be derived by finding any one edge of which that vertex is a part, and then checking the vertices of all its preceding and succeeding edges. Similarly, all the spaces that share an external wall can be easily located.

3.1.4. *The Split-Edge Data Structure*

We have found that a slight modification of the original Winged-Edge data structure enables it to adapt more effectively to the representation of walls and spaces of a building. Instead of a single edge representing a wall segment, we now have a *pair* of half-edges as shown in Figure 6. This is known as the Split-Edge data structure, which has been adopted for geometric modeling as well, instead of the Winged-Edge model (Kalay 1987, Kalay 1989).

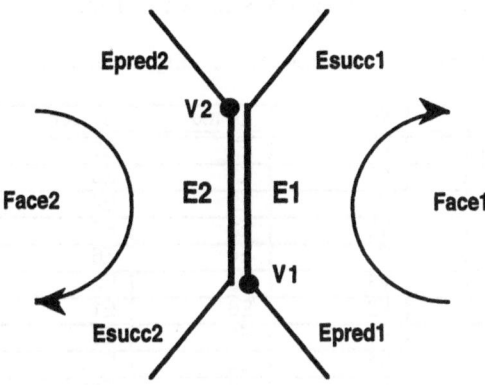

Figure 6. The Split-Edge data structure.

In the Split-Edge model, every half-edge is associated with one preceding half-edge, one succeeding half-edge, one vertex, one face, and finally, a corresponding *opposite* half-edge. Translated into the building data structure, this means that each physical wall is represented by two wall *segments*, each of which is associated with only one space— the space that it adjoins or faces. Thus the two wall surfaces are defined separately; this enables the explicit specification of the finishes and other properties of each surface— information that will be crucial in evaluations such as lighting and energy consumption.

Taking the example of the same three-room apartment shown earlier, its topological skeleton using the Split-Edge Data Structure would be as shown in Figure 7.

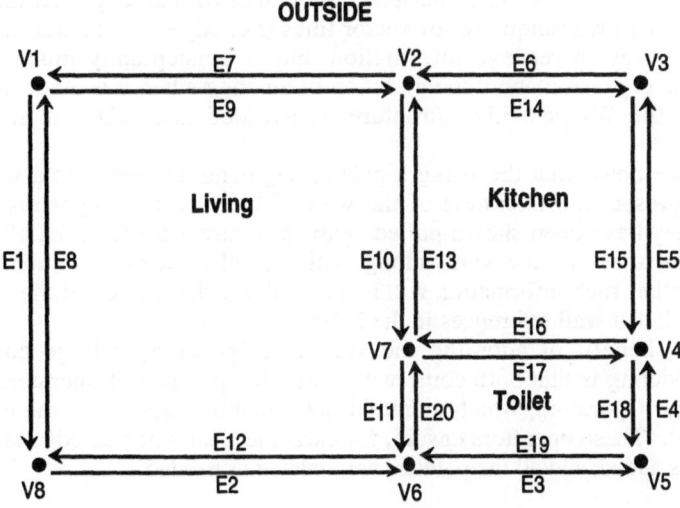

Figure 7. The topological skeleton of the studio plan shown in Figure 5a, now represented using the Split-Edge model.

The Edge table for this topological skeleton is shown in Table 3.

Table 3. The Edge table representing the building skeleton shown in Figure 7.

Edge ID	Space	Vertex	E-Pred	E-Succ	E-Opp
E1	Outside	V1	E7	E2	E8
E2	Outside	V8	E1	E3	E12
E3	Outside	V6	E2	E4	E19
E4	Outside	V5	E3	E5	E18
E5	Outside	V4	E4	E6	E15
E6	Outside	V3	E5	E7	E14
E7	Outside	V2	E6	E1	E9
E8	Living	V8	E12	E9	E1
E9	Living	V1	E8	E10	E7
E10	Living	V2	E9	E11	E13
E11	Living	V7	E10	E12	E20
E12	Living	V6	E11	E8	E2
E13	Kitchen	V7	E16	E14	E10
E14	Kitchen	V2	E13	E15	E6
E15	Kitchen	V3	E14	E16	E5
E16	Kitchen	V4	E15	E13	E17
E17	Toilet	V7	E20	E18	E16
E18	Toilet	V4	E17	E19	E4
E19	Toilet	V5	E18	E20	E3
E20	Toilet	V6	E19	E17	E11

The Half-Edge table is twice as long as the Winged-Edge table, but has fewer columns, so the space requirements are only marginally larger. It has the advantage that there is now only one vertex and one space associated with an edge. At the same time, each space has its own unique set of vector lines (i.e. edges). This not only enables a more efficient way to retrieve information, but is conceptually much simpler and "cleaner". The confusion about the direction of an edge when it is associated with two spaces, as in the Winged-Edge structure, is avoided altogether in the Split-Edge structure.

It should be noted that the using a pair of segments to represent a wall is not an attempt to represent the thickness of the wall. Thus, the two segments are actually coincident; they have been shown pulled apart in Figure 7 for the sake of clarity only. The actual thickness of the wall, along with its other attributes such as U-value, finishes, and other such information will reside with the description of the wall object in the ODB which that wall references in the PDB.

The added benefit of adopting the Winged-Edge or Split-Edge concepts from geometric modeling is that both come with a full complement of operators, commonly known as Euler Operators, which maintain the semantic integrity of the data structure (Wilson 1985). These operators ensure, for example, that when an edge is split in two, a new vertex is added, and all the pointers are updated as needed.

3.1.5. *The Data Structure of the Project Database*
We have implemented a working version of the project database, based upon the Split-Edge data structure. It is schematically depicted in Figure 8. Although far from complete, it is able to represent a building skeleton quite comprehensively, including multiple levels (floors), as well as doors and windows.

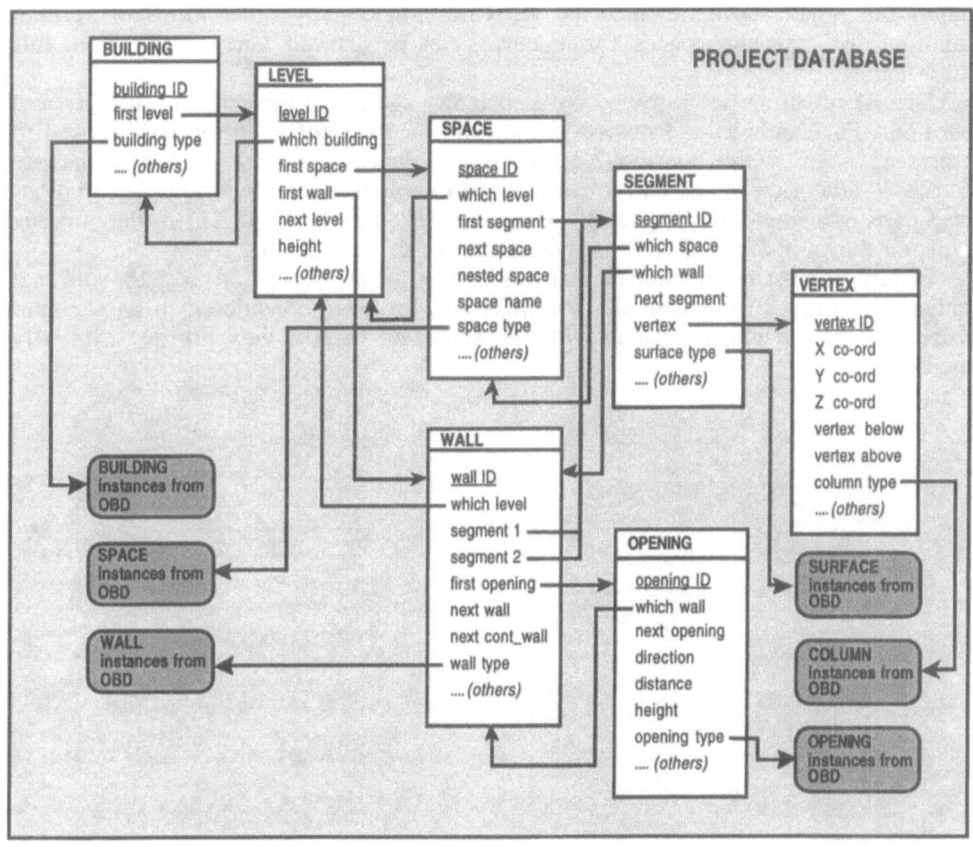

Figure 8. The data schema of the project database.

The PDB is a relational database comprising, for now, seven "entities" related to each other as shown in Figure 8. (The backward pointers are not essential but are provided to improve the efficiency of search operations.) Where the entity is also an architectural element (e.g. space, wall, opening, etc.), it is linked to the corresponding instance from the ODB by having a field denoting the *type*, which holds more detailed information about that element.

It is not mandatory for a space to be bounded by a physical wall; in other words, all spaces do not have to be rooms. Two or three spaces can come together to form an actual physical "room," for example, a living+dining+kitchen in a house or an apartment. In this case, the individual spaces can be modeled as being bounded by "virtual walls"—walls that do not physically exist. However, it would be necessary for that wall to be represented in the WALL table, with a flag to indicate its type.

As far as vertical relationships between elements are concerned, one option was to apply the Split-Edge data scheme in the same manner in which it was used to capture horizontal relationships between elements. However, we found this to be complex, inefficient, and wasteful, considering the fact that most spatial relationships in architectural design are conceived of horizontally, in plan. Hence, we chose to express vertical continuity simply by storing the contiguity of vertices in the form of the vertex below and vertex above fields of the VERTEX entity. Any other kinds of vertical relationships, between spaces for instance, can be derived fairly easily from this information.

Special situations are resolved by associating additional attributes to the different entities. For example, the nested space field of the SPACE entity can be used to represent "holes" within a space (e.g., courtyards, ducts, atria, etc.). Similarly, the next cont_wall field of the WALL entity can be used to collate all the individual, contiguous wall parts of a single, physical wall. For example, edges E10 and E11 of the building skeleton shown in Figure 7 are contiguous parts of the same wall.

Finally, it must be mentioned that this data structure is currently 2.5 dimensional rather than fully 3 dimensional. Volumes are, for now, considered to be standard extrusions of the planar spaces with the specified height; they are not explicitly modeled.

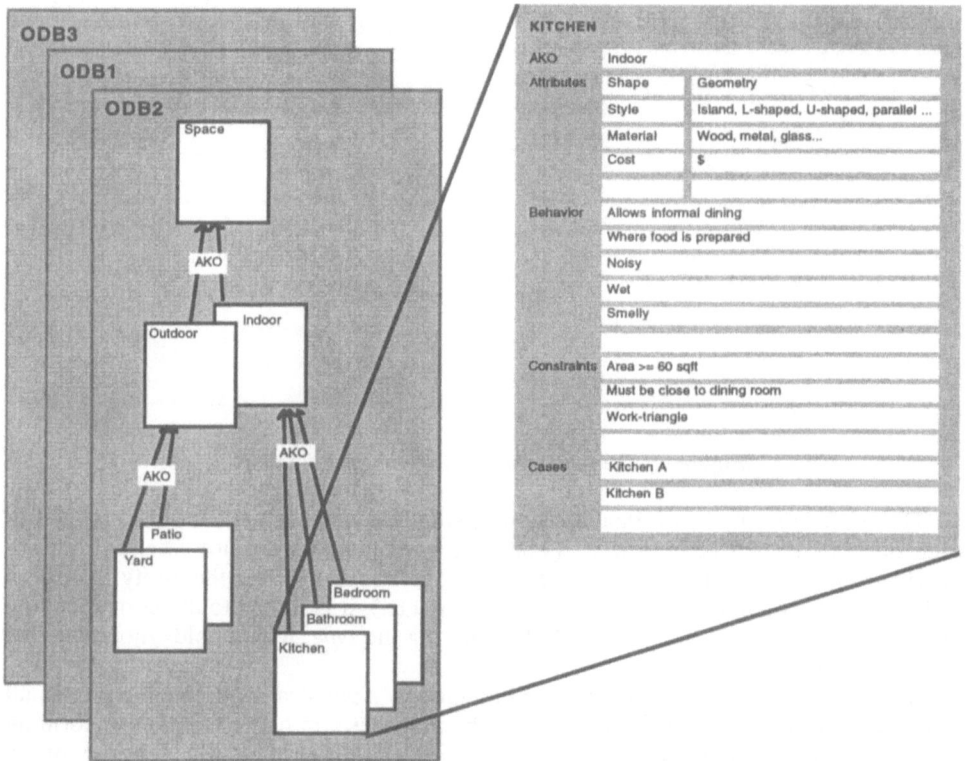

Figure 9. An example of an object database.

3.1.6. *Object Databases (ODBs)*
Each one of the ODBs is an object-specific knowledge base. It contains information needed to generate and to evaluate objects of the kind it represents. This information includes:

1. *Attributes*, such as the object's name, its form (shape), the materials it is made of, and other properties (e.g., manufacturer, cost, etc.).
2. *Classification* relationships, which define an inheritance path for properties this object shares with other, more generalized objects of the same type (e.g., that the kitchen is a kind of indoor space).
3. Information that describes the *behavior* of the object (e.g., that the kitchen is where food is prepared).
4. Integrity *constraints*, which are logical propositions associated with attribute values, and define generic expectations from the object (e.g., that the work triangle must be within certain dimensional limits).
5. *Cases*, in multimedia form, which provide anecdotal information about the object.

Objects are stored in hierarchical classification structures, which facilitate more compact, non-redundant representation—a method that has now been popularized by object-oriented programming languages. An example of a KITCHEN object and its inheritance hierarchy is depicted in Figure 9. This object shares many attributes with other indoor spaces, such as bedrooms and living rooms, in that it is enclosed by walls, has indoor thermal properties, and so on. The classification hierarchy stores shared attributes at the lowest level in the tree that is still shareable by all the objects that can benefit from this information. An inheritance mechanism makes higher-level information available to objects that are stored in lower levels of the hierarchy. Figure 10 depicts a few more of such classification hierarchies.

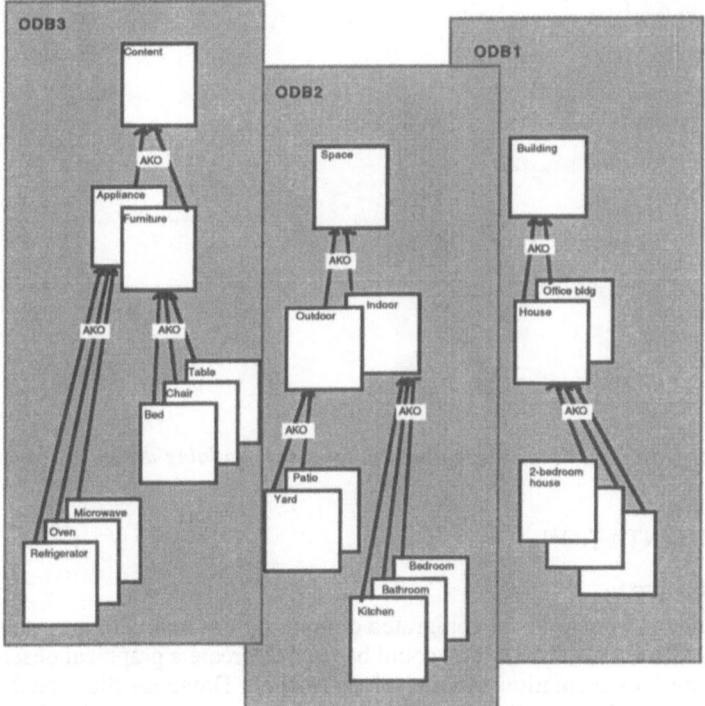

Figure 10. Some examples of possible object databases.

406

3.1.7. *Integrated Building Design Environment*

The building representation just presented forms the basis of an integrated computing environment for collaborative, multi-disciplinary building design, the overall scheme of which is illustrated in Figure 11. It includes the PDB, the ODBs, as well as the design and evaluation tools for the many building-related disciplines that will interface with the building representation for the purpose of query and update. We refer to these tools as "operators" or "IDeAs" (for "Intelligent Design Assistants"). Several IDeAs can come together to collaborate on one building project, and they can also partake in several other projects at the same time. We envision that one of the IDeAs can act as the "project manager" for each project. Many of the IDeAs would be similar to the existing stand-alone applications for energy evaluation, costing, structural analysis, and so on. However, many others would also have to be conceived from scratch to work with the new collaborative paradigm, possibly as coordinating and mediating tools between existing applications in the integrated environment.

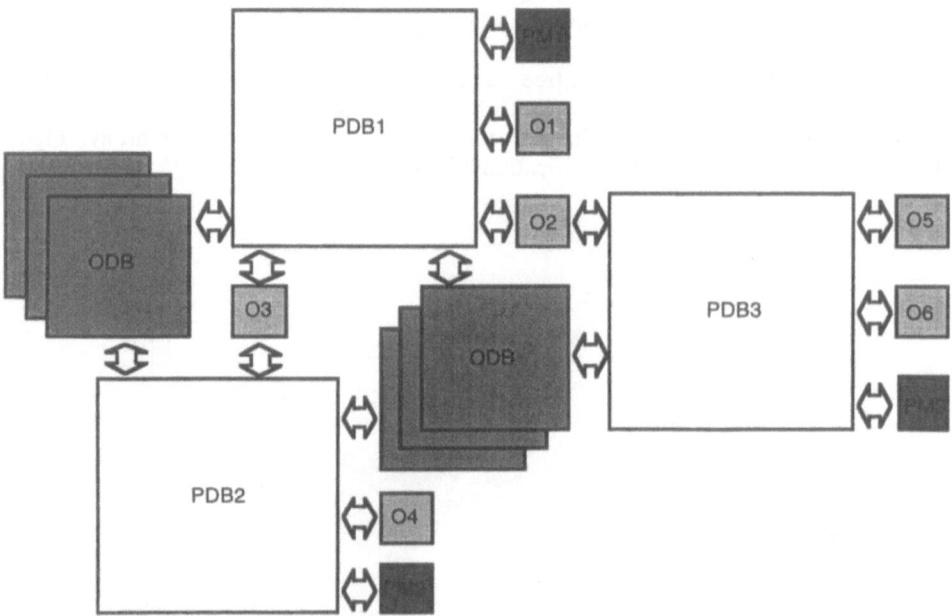

Figure 11. The overall schema of the integrated building design environment.

3.2. IMPLEMENTATION

3.2.1. *Graphical Editor*

An important component of the integrated computing environment that we felt the need to develop was a CAD editor which could be used to create a graphical description of the building in the representation scheme of the Project Database illustrated in Figure 8. While this may not be as sophisticated as a full-blown commercial drafting or modeling software, it had to be reasonably efficient and user-friendly. More importantly, it needed to have the facility to carry out a variety of semantically-meaningful operations that currently available general-purpose CAD softwares do not provide. Taking a simple example, deletion of the peripheral room of a building should automatically update the walls of the adjoining rooms from being "internal" to "external" walls.

This task has been completed in our customized editor to the extent to which a proposed building project can be quickly and comprehensively described, including the connections between the PDB and the ODBs. Consider the sample plan shown in Figure 12 which took less than 5 minutes to generate. This, essentially, is the Project Database. The links to the various Object Databases are made by associating "types" to each of the PDB entities—levels, spaces, walls, surfaces, columns, and openings. Figure 13 shows this link being made through a dialogue box that opens up by clicking on one of the wall entities.

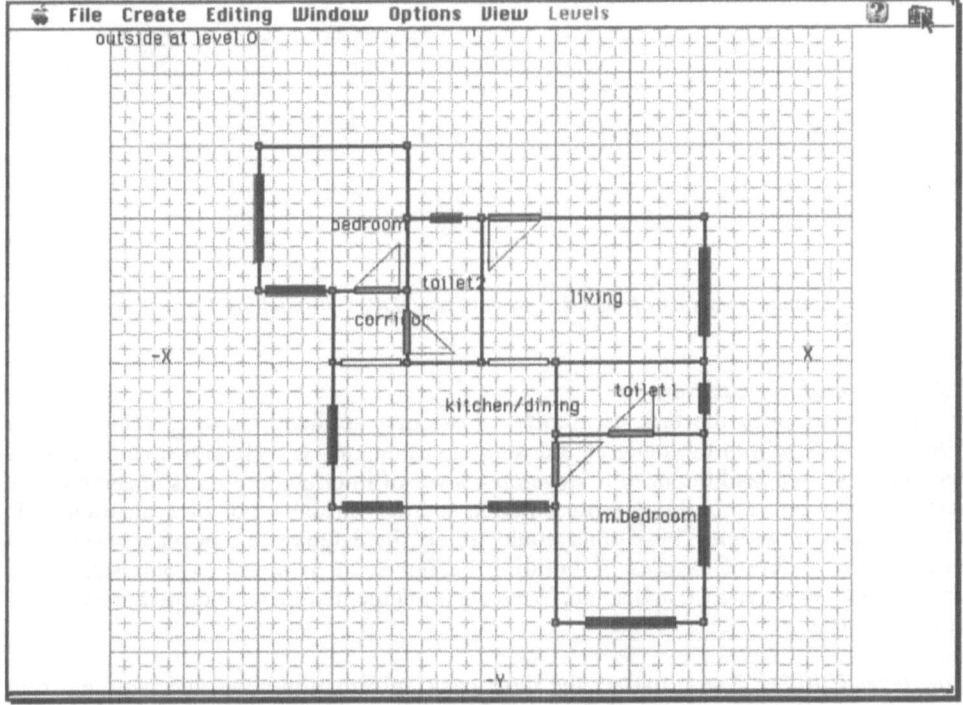

Figure 12. The customized graphical editor implemented to describe the PDB of a building.

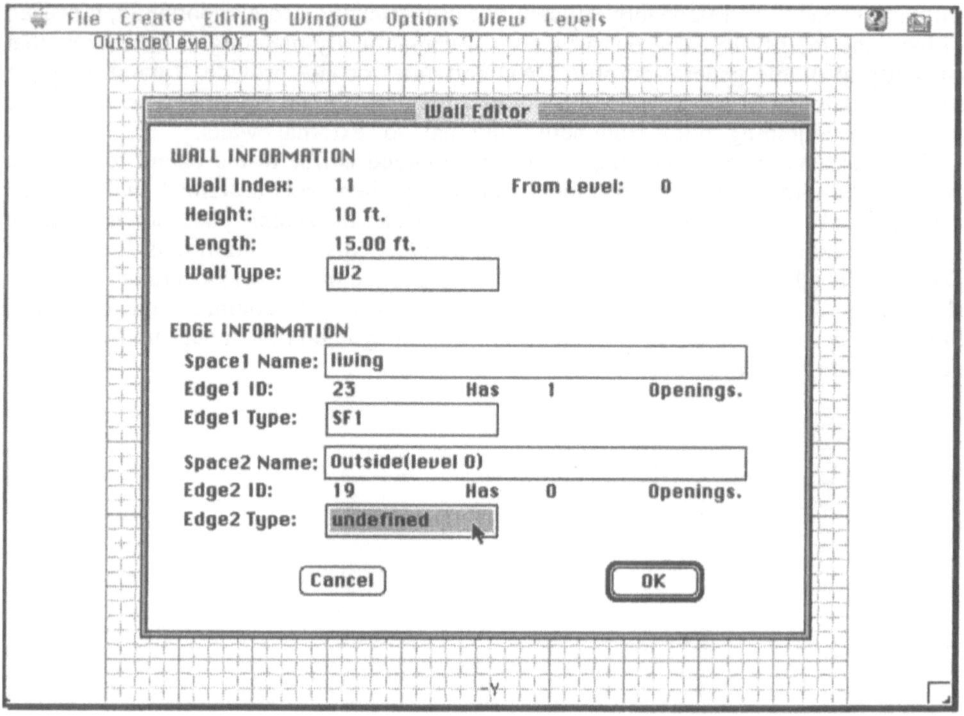

408

Figure 13. Linking elements of the PDB to various ODBs by attaching "types" to entities.

3.2.2. *Intermediary Utilities*

In order to simplify the task of modifying existing applications or developing new IDeAs to work within the integrated design environment, as well as to avoid needless repetition, we realized the need to develop a layer of intermediary tools to serve as the interface between the building representation and the various kinds of evaluation tools that will interact with it. These tools would "filter" the database and provide customized "views" for different aspects of building design such as space layout, structures, construction, energy, cost, and so on. We have implemented these intermediary Utilities at two levels:

Low-level Utilities. These extract the most basic information about the building from its implemented representation. A few examples are given below:
- The sequence of wall surfaces that make up a space.
- The area of a space.
- The two wall surfaces that make up a wall.
- The length of a wall.
- The number of doors or windows in a given wall, and their individual as well as collective areas.
- The geometrical location of a given vertex (column) or wall.
- The position of an opening within a wall and its width.
- The object specifications of any entity which has been linked up to an ODB.

High-level Utilities. These make use of the low-level functions to provide some higher-level information about the building. A few examples follow:
- A list of all the spaces within a particular level.
- A list of all the external walls.
- An adjacency matrix, showing which spaces lie adjacent to each other.
- A connectivity matrix, showing which spaces are connected to each other.
- The shortest path to get from one space to another.
- Contiguous vertices in multiple level building projects.
- All instances of an entity with specified object characteristics.
- Extent of visibility of one space from another.

3.3. TESTING

Two independent research interests have been supported so far by the integrated design environment described in this paper, serving as a much-needed testbed for its further development. The first was an evaluator for the spatial layout of two bedroom apartment units; the second was an evaluation of house designs specific to the Korean lifestyle and culture. For both evaluators, the building plans to be analyzed were first drawn using the graphical editor described in Section 3.2.1. Then, the data files generated were run through the Intermediary Utilities described in the preceding section. The output from the latter process served as input to the evaluator functions.

A brief description of each evaluator is given next, along with an evaluation summary generated by it for two test examples so that a comparison can be drawn.

3.3.1. *Apartment Evaluator*
The attempt of this tool was to simulate the "multi-disciplinary" nature of building design by a "multi-criteria" evaluation of a design problem. Accordingly, the apartment design was tested for several criteria—optimality of space, proximity, privacy, circulation, and climatic comfort—not just at the level of the whole unit, but at the level of each individual space of the apartment. The actual evaluation of the fulfillment of these criteria was based on well-established standards of good apartment design in architectural practice. The purview of the evaluation extended primarily to performance aspects related to layout and functionality. Aspects like structure, construction and materials, cost, energy consumption, and so on, were not considered.

Evaluation Methodology. The evaluation was carried out by determining the performance measure (i.e., "quality") of each room of the unit separately, by aggregating its performance measure for each of the five aspects mentioned earlier—optimality, proximity, privacy, circulation, and comfort. This, in turn was determined by aggregating the raw scores of all the individual criteria that provide a measure of that aspect. A *raw score* could be either 1 or 0, depending upon whether the criterion was satisfied or not. (Ideally, many intermediate values should be possible to bring out the finer shades or subtleties of differences in performance, but for the sake of simplicity here, we dealt with just the two extreme cases.)

All of the measures for individual spaces were then aggregated with certain aspects which applied to the apartment as a whole to give its total performance measure. The aggregation, at each level, was done by assigning appropriate weights to the various criteria.

Test Example 1. The graphical representation of the plan created using the customized CAD editor is shown in Figure 14.

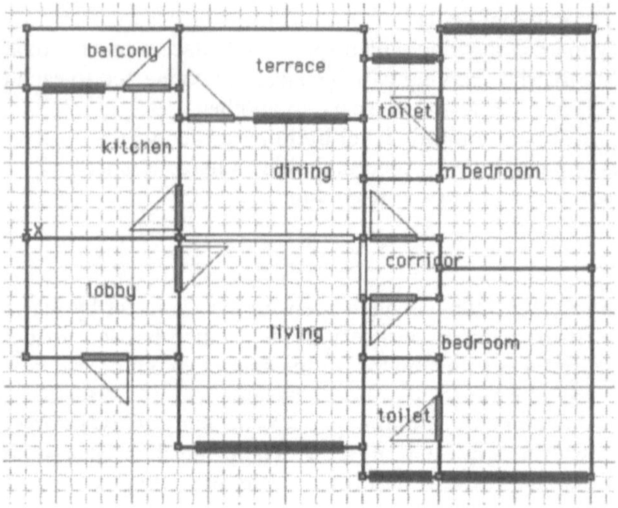

Figure 14. Test Plan 1 for the Apartment Evaluator.

The output from the Evaluator for this design, giving the results of the evaluation methodology described earlier, is reproduced in part below. (The entire listing, showing the performance criteria for all aspects, and for all spaces, is too lengthy to present here.)

```
**********        TWO-BEDROOM APARTMENT EVALUATOR          **********

**********         EVALUATION SUMMARY OF ASPECTS           **********
```

	Performance Aspect	Weight	Score
LIVING ROOM QUALITY	Optimality of Space	15	100.00
	Proximity Requirements	25	100.00
	Efficiency of Circulation	20	30.00
	Privacy	25	70.00
	Climatic Comfort	15	40.00
	TOTAL (Weighted)		69.50

... and so on until ...

OVERALL APARTMENT QUALITY	Quality of Outdoor Spaces	30	55.00
	Other Adjunct Spaces	15	40.00
	Overall Circulation Pattern	35	100.00
	Economy and Serviceability	10	0.00
	Miscellaneous	10	100.00
	TOTAL (Weighted)		67.50

```
**********           FINAL EVALUATION SUMMARY            **********
```

Performance Aspect	Weight	Raw Score	Weighted Score
Living Room Quality	17	69.50	11.82
Dining Area/Space Quality	11	65.00	7.15
Kitchen Quality	14	50.50	7.07
M. Bedroom Quality	17	88.00	14.96

Bedroom 2 Quality	17	80.00	13.60
Toilet 1 Quality	5	94.00	4.70
Toilet 2 Quality	5	74.00	3.70
Quality of Other Aspects	14	67.50	9.45
FINAL TOTAL	100		72.45

FINAL SCORE OF UNIT =   72.45 out of 100.00

**********   END OF EVALUATION PROGRAM   **********

Test Example 2. Contrast Test Plan 1 with the plan shown in Figure 15, obviously a poorer one to an experienced architect. The evaluation results (only the summary is given) reflect the difference in quality.

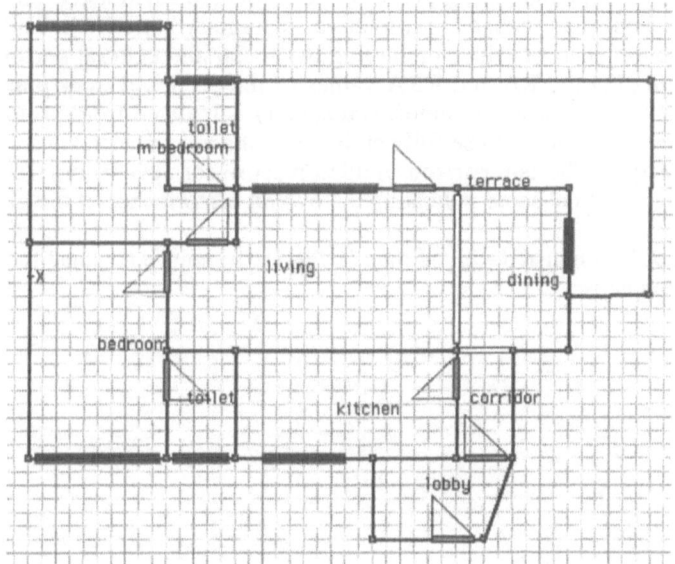

Figure 15. Test Plan 2 for the Apartment Evaluator.

**********   FINAL EVALUATION SUMMARY   **********

Performance Aspect	Weight	Raw Score	Weighted Score
Living Room Quality	17	36.25	6.16
Dining Area/Space Quality	11	85.00	9.35
Kitchen Quality	14	60.50	8.47
M. Bedroom Quality	17	70.50	11.98
Bedroom 2 Quality	17	62.50	10.62
Toilet 1 Quality	5	94.00	4.70
Toilet 2 Quality	5	59.00	2.95
Quality of Other Aspects	14	53.50	7.49
FINAL TOTAL	100		61.72

FINAL SCORE OF UNIT =   61.72 out of 100.00

**********   END OF EVALUATION PROGRAM   **********

412

3.3.2. *Korean House Evaluator*

The objective of this tool was to evaluate a given house design to determine how well it would function according to Korean lifestyle and culture (Lee 1997). The unique social and cultural factors that were consistently identified by analyzing a range of traditional and modern Korean house designs were used to define the goals and the criteria of the evaluation.

Evaluation Methodology. Korean houses are characterized by high connectivity between a central courtyard (the center of the house), and the kitchen and certain other key spaces. The evaluation of the design proceeds from the notion that the topological morphology of the floor plan expresses the social and cultural characteristics of a dwelling. This notion has been developed by Hillier and Hanson (1984) into a quantifying method that converts the topological diagram of a floor plan to numeric data. It accounts for both the connectivity of the underlying graph (called "relative asymmetry", or RA), and for the size of the floor plan (called "real relative asymmetry" or RRA). This method was employed here to generate values predicting the social and cultural performance of the house by calculating the RA and RRA values of the proposed floor plan. These were then compared to a set of ideal values, statistically calculated from prototypical house designs that were considered to be fully embodying the unique social and cultural factors of Korean lifestyle. The comparison yielded a score for several criteria, which were aggregated to give a final score indicating the overall performance of the given design.

Test Example 1. The house design shown in Figure 16 comes reasonably close to being a "good" Korean house.

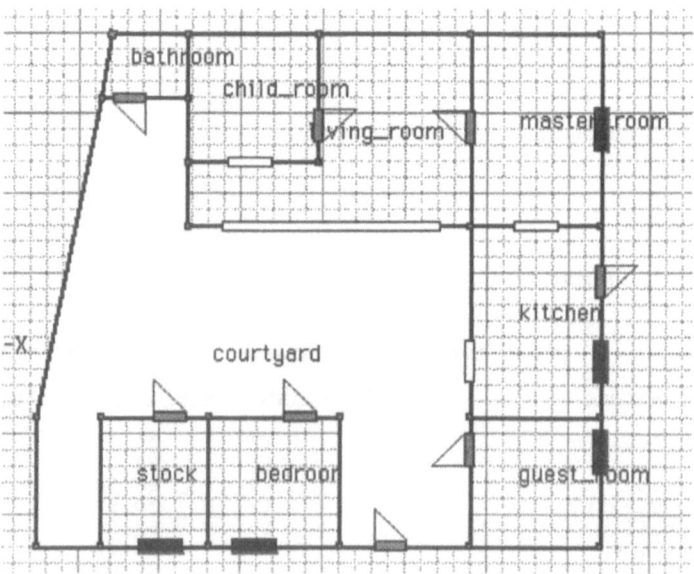

Figure 16. Test Plan 1 for the Korean House evaluator.

Some excerpts from the evaluation follow. After deriving the adjacencies and connections, the shortest path from each space to all the other spaces was determined as shown below.

```
* Shortest Paths between Spaces

From Space:   0 ( outside of building )
     To Space              : Nodes : Path
outside of building   0    :    0 :  0
child_room            1    :    3 :  1 <-- 3 <-- 9 <-- 0
master_room           2    :    2 :  2 <-- 4 <-- 0
living_room           3    :    2 :  3 <-- 9 <-- 0
kitchen               4    :    1 :  4 <-- 0
guest_room            5    :    2 :  5 <-- 9 <-- 0
bedroom1              6    :    2 :  6 <-- 9 <-- 0
stock                 7    :    2 :  7 <-- 9 <-- 0
bathroom              8    :    2 :  8 <-- 9 <-- 0
courtyard             9    :    1 :  9 <-- 0

Sum of Nodes of Space 0 = 17
```

... and so on. The shortest paths, in turn, were used to derive the RA and RRA values for each space.

```
* Relative Asymmetry Values

       R.A. value of space 0 = 0.222222 : outside of building
       R.A. value of space 1 = 0.388889 : child_room
       R.A. value of space 2 = 0.305556 : master_room
       R.A. value of space 3 = 0.166667 : living_room
       R.A. value of space 4 = 0.194444 : kitchen
       R.A. value of space 5 = 0.277778 : guest_room
       R.A. value of space 6 = 0.277778 : bedroom1
       R.A. value of space 7 = 0.277778 : stock
       R.A. value of space 8 = 0.277778 : bathroom
       R.A. value of space 9 = 0.055556 : courtyard

* Real Relative Asymmetry Values

       R.R.A. value of space 0 = 0.726216 : outside of building
       R.R.A. value of space 1 = 1.270879 : child_room
       R.R.A. value of space 2 = 0.998548 : master_room
       R.R.A. value of space 3 = 0.544662 : living_room
       R.R.A. value of space 4 = 0.635439 : kitchen
       R.R.A. value of space 5 = 0.907771 : guest_room
       R.R.A. value of space 6 = 0.907771 : bedroom1
       R.R.A. value of space 7 = 0.907771 : stock
       R.R.A. value of space 8 = 0.907771 : bathroom
       R.R.A. value of space 9 = 0.181554 : courtyard
```

Much of the final evaluation was based on these RA and RRA values. It is summarized below.

Evaluation Criterion	Score	Weight
• Difference of Real Relative Asymmetry (RRA) values from standard values	94	1
• Hierarchy of space, evaluated by comparing Depth Values from Entrance with standard ones	80	1
• Frequency of Contact (Degree of Communication), evaluated by Depth Values of Public Spaces— Living_Room, Corridor, Dining	75	1

	Score	Weight
• Openness to Neighborhood, evaluated by Depth Values of the first Open space from Entrance	100	1
• Degree of Privacy, evaluated by comparing RRA Values of Private Spaces with standard values	97	1

<u>FINAL SCORE</u>   <u>89</u>

Test Example 2. On the other hand, the design shown in Figure 17 does not perform as well.

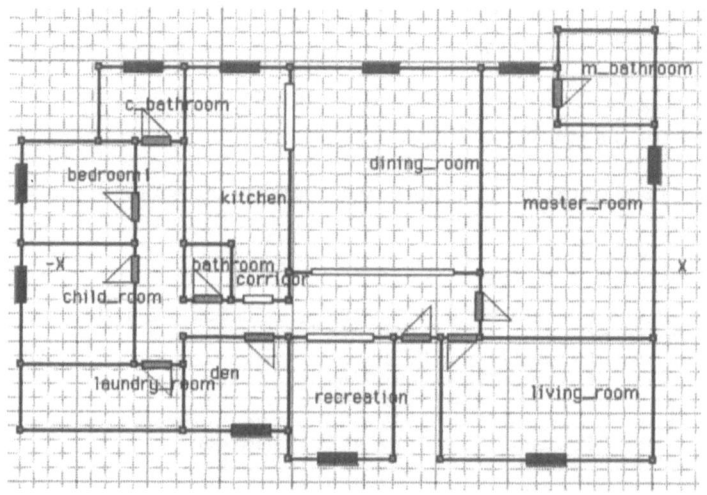

Figure 17. Test Plan 2 for the Korean House evaluator.

The evaluation summary for this design, generated by the evaluator, is reproduced below.

<u>Evaluation Criterion</u>	<u>Score</u>	<u>Weight</u>
• Difference of Real Relative Asymmetry (RRA) values from standard values	77	1
• Hierarchy of space, evaluated by comparing Depth Values from Entrance with standard ones	70	1
• Frequency of Contact (Degree of Communication), evaluated by Depth Values of Public Spaces– Living_Room, Corridor, Dining	55	1
• Openness to Neighborhood, evaluated by Depth Values of the first Open space from Entrance	50	1
• Degree of Privacy, evaluated by comparing RRA Values of Private Spaces with standard values	87	1

<u>FINAL SCORE</u>   <u>67</u>

4. Conclusions and Further Work

We have shown that the proposed Split-Edge data structure can serve to construct a suitable building representation that can potentially support the needs of the varied professionals involved in the design and construction of buildings. It is *complete* because it can represent both space and structure, it is *general* because the type and use of the building being modeled is immaterial, and it is *well-formed* by virtue of the hierarchical organization and coupling of the entities. We are optimistic that it is also *efficient,* but we will wait until we have tested it for a very large building project before we claim that with absolute certainty.

We have already tested the representation with two evaluation tools for architectural design that were described in this paper. These tools are, of course, just a precursor to more design and evaluation tools for other building-related disciplines that we intend to develop. We see the tool-building process as one of the best ways to ratify and further develop the underlying data structure for the PDB. To this end, we plan to invite various disciplinary experts to use our proposed representation as the basis for the development of their design aids.

At the same time, we will continue to develop further the database itself; it is still in a very rudimentary state. As we proceed further with its design development, there are several issues that need to be investigated and resolved: How can spaces come together to form specific groups or clusters, in addition to the pre-defined grouping mechanism of levels? Do floor/ceiling slabs need to be represented separately, or should they be inferred from the LEVEL entity? Should spaces on the same physical level but with discontinuous floor slabs be grouped under one level? How can inclined or even more complicated roof structures, and walls that are not strictly vertical and have irregular heights, be represented? What about columns that lie within a space rather than at a vertex—do they need a separate means of representation? How can special, but very common elements like staircases, be compactly represented? How can the data structure be made extensible so that additional information for any element can be readily added as and when needed?

Extensive further work is also required on the other components of the integrated building environment. The ODBs, and their interface with the PDB, is another substantial component of the research project. We also intend to focus on the communication and control issues, knowledge representation issues, and user interface issues as the research progresses.

Finally, we plan to use the new and powerful communication medium of the Internet within which to situate the integrated environment, and exploit this medium's potential for universal access of any information to facilitate the desired collaborative design paradigm. Accordingly, all current and future development work will be carried out in the Java programming language. Our ultimate vision is to have the building representation as well as all the desired tools and evaluators located in "cyberspace", from where they can be accessed by any member of the building team, from any place, at any time, as and when needed.

Acknowledgments

It is obvious that this project is very large, and involves many individuals, too many to list. We wish to thank, in particular, Dr. Jin Won Choi from the Department of

Architecture, Ajou University, Korea, for his significant contribution towards developing the PDB and the customized editor. Thanks are also due to all the PhD and MSc students in the CAD Research Group of the Department of Architecture at Berkeley, particularly Choong-Hoon Lee, who developed the Korean House evaluator described in this paper.

References

Amor, R., G. Augenbroe, J. Hosking, W. Rombouts, and J. Grundy. (1995) Directions in Modeling Environments, Automation in Construction 4(3), 173-187.

Baer, A., C.M. Eastman and M. Henrion. (1979) Geometric Modeling: a Survey, Computer-Aided Design 11(5), 253-271.

Baumgart, B. (1972) Winged Edge Polyhedron Representation, Technical Report CS-320, Stanford Artificial Intelligence Laboratory, Palo Alto.

Bjork, B.C. and J. Wix. (1991) An Introduction to STEP, Technical Report jointly published by VTT Technical Research Center, Laboratory of Urban Planning and Building Design, Finland, and Wix McLelland Ltd., England.

Eastman, C.M. and A. Siabiris. (1995) A Generic Building Product Model incorporating Building Type Information, Automation in Construction 3(4), 283-304.

Fenves, S., U. Flemming, C. Hendrickson, M.L. Maher, R. Quadrel, M. Terk, and R. Woodbury. (1994) Concurrent Computer-Aided Integrated Building Design, Prentice-Hall, Inc. Englewood Cliffs.

Galle, P. (1995) Towards Integrated, 'Intelligent', and Compliant Computer Modeling of Buildings, Automation in Construction 4(3), 189-211.

Hillier, B. and J. Hanson. (1984) The Social Logic of Space, Cambridge University Press, New York.

Kalay, Y.E. (1987) WORLDVIEW: An Integrated Geometric Modeling/Drafting System, IEEE Computer Graphics & Applications 2(7), 36-46.

Kalay, Y.E. (1989) Modeling Objects and Environments, John Wiley & Sons, New York.

Lee, Choong-Hoon. (1997) Evaluation of Social and Cultural Performance of Modern Korean Housing Design, Masters Thesis, University of California, Berkeley.

Papamichael, K., J. LaPorta, H. Chauvet, D. Collins, T. Trzcinski, J. Thorpe, and S. Selkowitz. (1996) The Building Design Advisor, ACADIA '96 Conference proceedings, Tucson, Arizona.

Pohl, J. and L. Myers. (1994) A Distributed Cooperative Model for Architectural Design, Knowledge-Based Computer-Aided Architectural Design (G. Carrara & Y.E. Kalay, eds.), Elsevier Science Publishers, Amsterdam.

Richens, P. (1977) OXSYS-BDS Building Design Systems, Bulletin of Computer-Aided Architectural Design 25, 20-44.

Wilson, P.R. (1985) Euler Formulas and Geometric Modeling, IEEE Computer Graphics & Applications 5(8), 24-36.

Human - Machine - Design Matrix

A Model for Web-based Design Interaction

DANIEL E. TSAI,   *Research Associate, Harvard University*
SUNGAH KIM,   *Harvard University*

A model of human-machine design interaction is presented. The model is based on synchrony of actions [Mitchell95a] as it relates to designing, presenting, and discussing a design object over an electronic medium. The model descriptively accommodates existing technologies and areas of CAD research. The model prescriptively illuminates future CAD vis. the Web. The model is based on 3 factors : synchrony, presence, activity; and 2 players : human and machine. Future technologies are considered in terms of the a shifting role of the machine (the computer and the net) from server to agent to actor.

1. Starting Point : Synchrony and Presence

In The City of Bits, Mitchell identifies and contrasts common forms of person to person communication in terms of physical versus virtual presence, crossed with synchronous versus asynchronous actions. The emerging world of the Web is at the center of a heated transformation from physical to virtual. Electronically mediated counterparts to conventional physical forms of communication are taken to a new extreme when considered in the context of the World Wide Web (the Web).

	Physical	*Virtual*
Synchronous	[1] Physical Presence, Synchronous (*Face-to-face meeting*)	[3] Virtual Presence, Synchronous (*telephone, video conference, electronic meeting, talk radio*)
Asynchronous	[2] Physical Presence, Asynchronous (*Leaving messages, mail*)	[4] Virtual Presence, Asynchronous (*email, fax, electronic bulletin board*)

Figure 1 : Matrix-1 - The starting point from [Mitchell95a].

417

R. Junge (ed.), CAAD Futures 1997, 417-430.
© 1997 *Kluwer Academic Publishers.*

2. Design Extensions

The application of virtual presence, synchronous and asynchronous actions have been put into practice. Virtual design studios [Mitchell95b][Wojtowicz95] have shown how existing technologies can be applied to design studios that are spatially and temporally disjunct. Email, Web pages, remote file transfers, video conferencing, whiteboards [Bly88] and other existing communication technologies have been used to show how collaborative designing can be extended via the internet.

This paper extends the idea of synchrony and presence, first as a descriptive framework for design activities and second as a prescriptive model of how the machine (i.e. networked computational machines) can play a greater role in design activities vis. the Web.

Synchrony and presence has been related specifically to architectural design in "Aspects of Asynchronous and Distributed Design Collaboration" by Wojtowicz, Papazian, Fargas, Cheng and Davidson in [Wojtowicz95], switching 'local' and 'global' for 'physical' and 'virtual':

	Physical	*Virtual*
Synchronous	[1] (*Face-to-face meeting*) Conventional design studio	[3] Design teleconference, videoconference
Asynchronous	[2] Digital design correspondence, pinup board	[4] Networked design studio

Figure 2: Design collaboration matrix - [Wojtowicz95].

The above description includes general design related activities. It may be useful to further articulate distinct (if not iterative) acts within the design process. The purpose of this subdivision is to identify distinct differences and implications for where machine role and human purposes can be advanced.

Presuppose that designing is subdivided into acts of creation, presentation, and discussion of a design component (i.e. the representational object - the artifact). The initial four quadrants of Matrix-1 expand into 3 sets of 4:

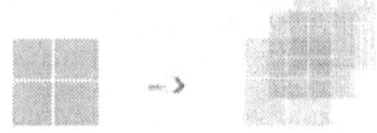

2.1. DESIGNING THE COMPONENT.
The term *designing* is used here as an act of creating or modifying a representation of an (architectural) object with the aid of tools. Matrix-1 simply translates to :

A. Designing	Physical	Virtual
Synchronous	[1A] Physical Presence, Synchronous *designing*	[3A] Virtual Presence, Synchronous *designing*
Asynchronous	[2A] Physical Presence, Asynchronous *designing*	[4A] Virtual Presence, Asynchronous *designing*

Figure 3: Design Matrix

[1A] Physical presence, synchronous designing occurs when a design team works within the same physical space. This is the traditional studio situation.

[2A] Physical presence, asynchronous designing occurs when drawings, models, notes and other physical media are worked on and exchanged between designers without direct contact. This is also part of the traditional design studio situation when designers have different schedules.

[3A] Virtual presence, synchronous designing occurs when the computer is used as an interactive tool by more than one designer. At the low end of the complexity spectrum, is the electronic white-board, where designers can draw on a common electronic document. At a higher level of complexity, designers could be working on a 3D model simultaneously in real-time. Ultimately, this becomes real time multi-user CAD.

[4A] Virtual presence, asynchronous designing occurs when the computer is used by a group of designers that do not see each other's changes instantaneously and are not constrained by other's actions. If a design starts from a common state, then proceeds asynchronously, multiple versions are produced. If such versions are brought together, then discrete actions, i.e. changes, are serialized and made into a unified model. This is serialized or multi-version CAD [Katz86][Kim97].

A. Designing'	Physical	Virtual
Synchronous	[1A] Conventional Team Designing	[3A] Multi User CAD
Asynchronous	[2A] Design Exchange	[4A] Multi Version Multi User CAD

Figure 4: Design Matrix re-codified

2.2 PRESENTING THE DESIGN COMPONENT

Presentation is defined here as fundamentally uni-directional. A story is told, and objects are presented without viewer interaction with the presenter. Unlike the prior case where the focus is on creating or modifying a design object, the purpose here is to communicate the design object - in whatever state it is in.

B.Presentation	Physical	Virtual
Synchronous	**[1B]** Physical Presence, Synchronous *presentation*	**[3B]** Virtual Presence, Synchronous *presentation*
Asynchronous	**[2B]** Physical Presence, Asynchronous *presentation*	**[4B]** Virtual Presence, Asynchronous *presentation*

Figure 5: Presentation Matrix

[1B] Physical presence, synchronous presentation occurs during the course of conventional presentations in front of an audience. This is the conventional presentation or lecture.

[2B] Physical presence, asynchronous presentation occurs when a designer leaves a physical model or a drawing for someone to review. An exhibit is a more public example. In all cases, the design object stands by itself without its designer to present it in person.

[3B] Virtual presence, synchronous presentation occurs during a guided tour through a design by the designer, via tele-presence. This case includes live broadcasting / live viewing and one way video link.

[4B] Virtual presence, asynchronous presentation occurs when a design object is put on virtual display. As with case [2B], the design object has to stand by itself -- as a drawing, a HTML page published onto the World Wide Web [HTMLa] [HTMLb][HTTP], a virtual reality model [Ames96], an animation, or a video taped presentation. This can be extended to include a virtual museum and virtual places (such as VRML worlds).

B.Presentation'	Physical	Virtual
Synchronous	**[1B]** Lecture/Presentation	**[3B]** Tele-Presentation
Asynchronous	**[2B]** Physical Exhibit	**[4B]** Electronic Publishing, Virtual Museums, Virtual Worlds

Figure 6: Presentation Matrix re-codified

2.3 DISCUSSING THE DESIGN COMPONENT

Discussion is the bi-directional extension to presentation. A means of response, and a sense of participatory equality is necessary in this scenario. Once again, it is distinct from designing because although they are both bi-directional, the purpose in the discussion is to communicate and inquire about the design (object). Although comments may lead to altering the design object, its immediate purpose is discourse.

C. Discussion	Physical	Virtual
Synchronous	[1C] Physical Presence, Synchronous *discussion*	[3C] Virtual Presence, Synchronous *discussion*
Asynchronous	[2C] Physical Presence, Asynchronous *discussion*	[4C] Virtual Presence, Asynchronous *discussion*

Figure 7: Discussion Matrix

[1C] Physical presence, synchronous discussion is the conventional design review or critique.
[2C] Physical presence, asynchronous discussion occurs when drawings, models, data are passed back and forth and returned with comments. This is a physical form of 'Request for Comment'.
[3C] Virtual presence, synchronous discussion is a tele-review with shared media. Video conferencing, internet audio, white boards are the technologies that can be applied to this area.
[4C] Virtual presence, asynchronous discussion is the 'Request for Comment'. An exchange of email on a design is a simple example. A discussion group on an electronic form is another example. This case can also be embodied in Web site that allows comments to be made with a running commentary complete with images and models.

C. Discussion'	Physical	Virtual
Synchronous	[1C] Meeting / Critique	[3C] Tele-Critique
Asynchronous	[2C] Physical requests for comment	[4C] E-forum

Figure 8: Discussion Matrix C Re-codified

2.4 SUMMARY

Summary A-Designing B-Presenting C-Discussing	Physical	Virtual
Synchronous	[1A] Team Designing [1B] Lecture/Presentation [1C] Meeting / Critique	[3A] Multi-User CAD [3B] Tele-Presentation [3C] Tele-Critique
Asynchronous	[2A] Design Exchange [2B] Physical Exhibit [2C] Physical RFC	[4A] Multi-Version Multi-User CAD [4B] Electronic Publishing, Virtual Museums, Virtual Worlds [4C] E-forum

Figure 9 : Summary

3. Web/Machine Roles

The areas distinguished by the model so far do not specify how a globally networked computing system (i.e. the Web) is significant. This can be approached by considering the role of the machine (i.e. the networked computational system). The role of the machine can be described in 4 discrete role-levels, ordered in increasing machine involvement :

(1)The machine as a communication medium between persons (medium);
(2)The machine as a processor and store of information (server);
(3)The machine as that acts on the behalf of a person (agent) [Kay84][Maes84][Maes93] and ultimately,
(4)The machine as an intelligent, autonomous cohort (actor).[i]

The computer aided design implications for treating the computer within these capacities are now evaluated in relation to the multi-dimensional matrix developed so far.

Synchrony	x	Presence	x	Activity	<=>	MachineRole
synchronous asynchronous		physical virtual		design presentation discussion		medium server agent actor
Starting matrix (4) Design Extensions (12) Machine Roles (48)						

Figure 10 : Synchrony, Presence, Activity and Machine roles

Up to this point, there has been an implicit assumption that the parties participating in the articulated activities {designing, presenting, discussing} are human - i.e. the role-players are human, and the tools are physical (pen, paper, wood, saws...) or electronic (computer, telephone, fax).

An abstract (producer-consumer) data flow model [Pressman87][Yordon78] [DeMarco79] will now be used to articulate a more precise relationship between players, and in particular, identify roles that both human and machine do and can play.

Starting with a simple a model of information flow between 2 entities. One entity is the producer of information. The other entity is the consumer of information. This is a uni-directional model of information flow.

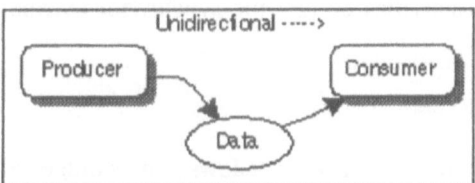

Figure 11 : Producer and consumer of information.

If both entities are information producers and consumers, then a bi-directional elaboration follows:

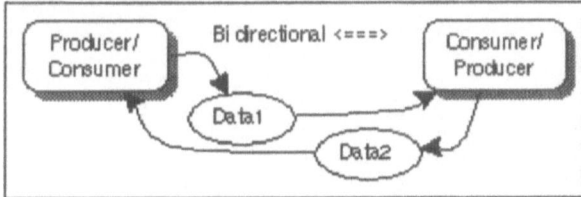

Figure 12 : Bi-directional information flow.

So far, the model is abstract and does not specify by what means the information passes between the two parties. This is the point where the role of the machine needs to be defined. The omission of the machine as an intermediary reduces to scenarios of traditional physical person-to-person interaction. Given that the focus is the involvement of the machine (networked computational machine), only cases involving the machine will now be explored.

3.1 MACHINE AS COMMUNICATION MEDIUM

The simplest use of the machine is as a communication medium between parties. This is a storage-less (secondary, not primary or cached) use of the machine, exemplified in the telephone, the video teleconferencing devices, and fax machines.

424

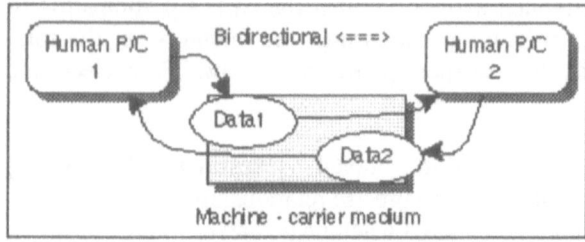

Figure 13 : The machine's role is as carrier medium of information.

In relation to the activities of designing, presenting and discussing designs, the machine as communication medium is exemplified by design whiteboards, presentation broadcasting, discussion tele-conferencing. Only synchronous activities are possible without the use of secondary storage. These activities occupy the synchronous-virtual quadrant of the matrix.

3.2 MACHINE AS INFORMATION PROCESSOR/SERVER.

The introduction of (secondary) storage into the machine allows for storage of information. Information produced does not have to flow directly to consumers. It can be stored in the system, processed and served to consumers on demand.

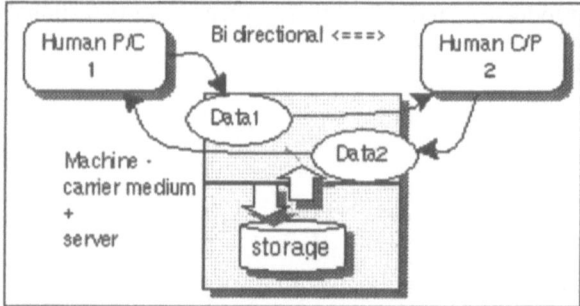

Figure 14 : The machine's role is expanded to storage and basic processing of information.

The role of the machine as information processor and server is well known today. All computational machines have some form of primary and secondary memory, and therefore have the capacity to store and process information. What is really of interest in using this incredibly simple model, is how it relates to the prior model of synchrony and presence.

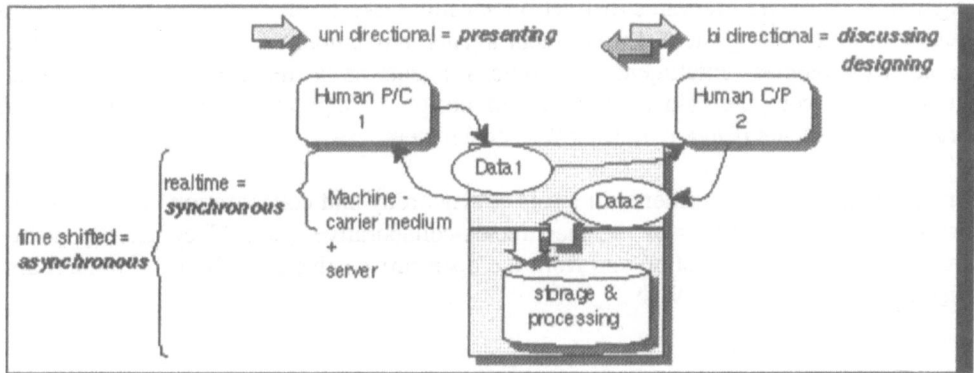

Figure 15 : Information flow and machine role are mapped to synchronicity and activity.

Synchronous activities in principle do not require secondary storage - these are real time activities. Asynchronous activities, however, require secondary storage - these are time shifted activities. For example, real time streaming of digital audio over the internet [Audio] can be done via peer-to-peer connection between the talker (the audio producer) and the listener (the audio consumer). In contrast, communicating between 2 (or more) parties by posting information on a Web site or a electronic bulletin board is an asynchronous (time-shifted) activity. The producer of information (the Web author) relies upon the machine's storage system to store and serve the information to consumers (readers/surfers).

The direction(s) in which information flows helps to distinguish in machine terms, what designing, presenting, and discussing requires (not what they are). Presenting is in essence a uni-directional activity. The lecture, the video taped or animated movie presentation, and the broadcasting show are all not interactive - information flows from the presenter (the information producer) to the audience (the information consumers). In contrast, interaction requires bi-directional flow of information. Collaborative designing, and discussion of a design requires a feedback loop [Bly88]. In the first case, the design object is being created collaboratively, and in the second, it is being exhibited and discussed.

3.3 MACHINE AS (DESIGN/PRESENT/DISCUSS) AGENT

Up to this point, the machine's role is first as a carrier of information, and then as storage vessel, conventional processor and server of information. The players have always been the human designers. The next step is to consider the networked computational machine as an agent - specifically as a design-agent, presentation-agent, and discussion-agent, within the context of synchrony and collaboration.

426

Agents are software programs that act on specific tasks on the behalf of others, with a degree of autonomy [Maes84]. Intelligent agents may employ reasoning or knowledge-based technologies. Intelligent autonomous agents is a intense area of research, including in computer aided architectural design - e.g. intelligent navigation and presentation agents [Engeli95], sans Web enablement.

Presently, agents are considered in terms of synchrony and collaboration. The following are proposed articulations of the machine-as-collaborative-agent. They directly derive from the outlined codification and from the open prospective availability of human and machine resources on the Web..

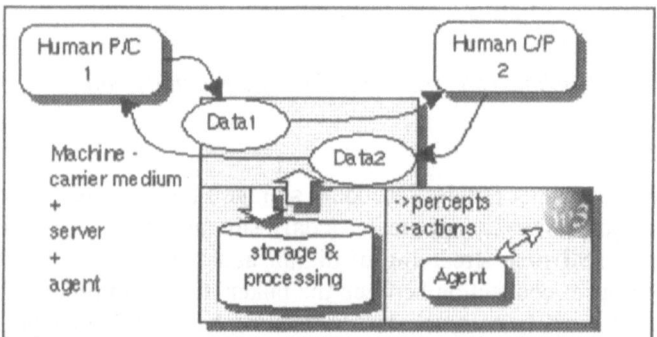

Figure 16 : Information Agents add a further level of processing capability to the machine's role.

3.3.1 Proposed Web / Machine Roles - Synchronous Collaboration Agents
Agents can engage designers/presenters/discussants on synchronous media (e.g. video conference, whiteboards, etc.) on demand. The collaboration agent is synchronous because the participants are connected by synchronous (real-time) media. For example, if a critic is needed to discuss a design, a collaboration agent could search for an available candidate and hook that person up. If someone was needed to help design a component, the collaboration agent could find an available designer to hook up to the other side of a whiteboard or CAD system.

Area	Possible Uses
Synchronous-design-agents :	-- agents that connect up designers via video-conference based on needs or requests. -- agents that generate design alternatives on the fly, as the design collaboration takes place -- agents that add information to the design as it is developing (without constraining it)
Synchronous-presentation-agents :	agents that find an audience to hear a presentation or to find a speaker on a topic.

| Synchronous-discussion-agents : | agents that find critics to discuss a design. |

Figure 17 : Possible use of synchronous agents.

3.3.2 Proposed Web / Machine Roles - Asynchronous Collaboration Agents
Asynchronous collaboration agents can be differentiated from synchronous counterparts by capitalizing on time-shifted availability and independence of Web resources and participants. Agents can act to match asynchronous requests for and submissions of information between parties. This scenario is the electronic design marketplace, where the design agent finds a suitable match to your request for a component or tool. The marketplace is asynchronous because the market participants (designers, presenters, critics) contribute or make requests at different times, autonomously from other parties. It then becomes the agent's task to *'make the market'* , or broker exchanges, by searching, analyzing and matching producers and consumers of (design) information. The response from the agent may be real-time or time-shifted.

Area	Possible Uses
Asynchronous-design-agents :	-- agents that manage a marketplace for design components and tools. -- agents that attempt to reconcile and put together designs as they develop
Asynchronous-presentation-agents :	-- agents that create a marketplace for pre-recorded presentations and audiences. -- Presentation agents can also be used to guide the audience through the material in a custom fashion
Asynchronous-discussion-agents :	-- Agents that match people to discussions in an electronic forum. -- Agents that comb through designs and present critiques and commentary from a specific viewpoint (perception rules).

Figure 18 : Possible use of asynchronous agents.

One can imagine, that the advent of a collaborative design environment with designing/presenting/discussing agents, would upon submission of a design object or online creation of a design object, get responses back from both human collaborators and agents. Given that human participants and agents could be selective (private, by invitation only), or public (as much of the Web is still), design feedback, spontaneity and serendipity could flourish.

428

Fantasy design log :
> - creating my design online…
> - a materials agent encounters my design and gives me a suggestion to some interesting textures encountered at http://…
> - I go away for java…
> - I check the project web site and my design has been rendered and integrated into the master VRML model…not bad but what until the crit-agents get a handle it!
> - working late…and I get blasted into this a video conference on nurbs, gotta go!
> - end log 97.04.14:02:02

Figure 19 : Fantasy design log in the era of design agents.

3.4 MACHINE AS (DESIGN, PRESENTATION, DISCUSSION) ACTOR

Agents act on the behalf of another. Actors are first class participants. Whereas agents are asked to perform a defined task for a design participant, such as finding a design component, an actor is a collaborator - a designer, a critic, an intelligent guide. The role of the machine then becomes that of participant - collaborator. In the producer/consumer model, the machine assumes the (formerly solely human) role. This ultimate scenario is the design example of a Turing test [Turing50], where design collaborators over the Web do not know whether the design collaborator, the presenter or the audience, or the design critic is a person or a machine.

As agent technologies develop and diversify into a myriad of embedded specializations it is plausible to extrapolate - perhaps in the conceptual flavor of Minky's Society of Mind [Minky85] - to the level of intelligent electronic actors.

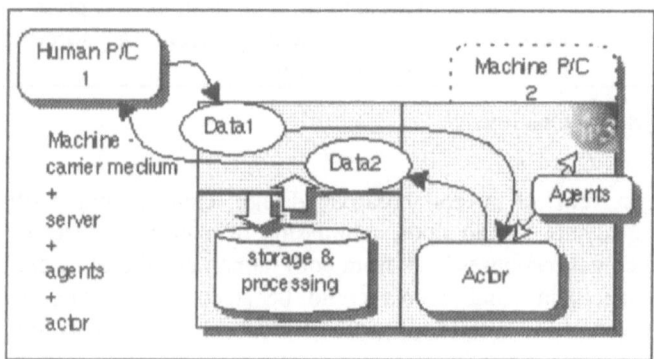

Figure 20 : The machine as Actor becomes a first class participant.

4. Concluding Remarks

The advent of the Web does not change or alleviate many design computation problems. It sets a globally shared virtual environment in which to test and solve problems. The Web is an electronic playground where participants across the ether can be known or unknown, singular or countless, man or machine. Actions and requests for information can occur in synchrony or asynchrony with collaborators. When the networked computational machine's role is expanded beyond that of carrier medium or processing engine, or information server, to that of agent and ultimately of participant, new possibilities for design collaboration appear.

References

[Ames96]   Andrew Ames, David Nadeau, John Murphy. (1996) The VRML Source Book, John Wiley, NY.
[Barrett92]   Barrett, E. (ed.) (1992) Sociomedia : Multimedia, Hypermedia, and the Social Construction of Knowledge. MIT Press. Cambridge
[Bly88]   Bly, S. (1988) "A use of drawing surfaces in different collaborative settings." In Proceedings of the Conference on Computer-Supported Cooperative Work (CSCW '88), p250-256, Portland, Oregon, September 26-28, ACM Press.
[Bush45]   Bush, V. (1945) "As We May Think" Atlantic Monthly, July
[Booch94]   Grady Booch. (1994) Object-oriented Analysis and Design with Applications, Benjamin/Cummins, Redwood City, Calif.
[Engeli95]   Engeli, Kurmann, Schmitt (1995) A New Design Studio, Intelligent Objects and Personal Agents in Virtual Environment, in Acadia 1995.
[Katz86]   Katz, R., E. Chang, R. Bhateja. (1986) "Version Modeling concepts for Computer-Aided Design Databases". Submitted to ACM SIGMOD Intl. Conf. On Management of Data.
[Kay84]   Kay, Alan (1984): Computer Software. Scientific American 251(3, September), 53-59.
[Kim97]   Kim, SungAh (1997) Version Management in Computer-Aided Architectural Design. Thesis, accepted by Harvard University Graduate School of Design, Feb. 1997.
[Maes84]   Maes, Pattie (1994): Agents that Reduce Work and Information Overload. Communications of the ACM 37(7, July), 31-40.
[Maes93]   Maes, Pattie; Kozierok, Robyn (1993): Learning Interface Agents. In: Proceedings of AAAI '93 Conference Washington, D.C. July 1993. 459-465.
[Minsky85]   Minsky, Marvin (1985): The Society of Mind. Simon & Schuster, New York. 339 pages.
[Mitchell95a]   Mitchell, W. J. (1995) The City of Bits. MIT Press. Cambridge
[Mitchell95b]   Mitchell, W.J., McCullough, M., (1995). Digital Design Media. Van Nostrand Reinhold New York:.2nd ed.
[Russell95]   Russell, S., Norvig, P. (1995). Artificial Intelligence, Prentice Hall, NY.
[Simon95]   Simon, A. R., (1995). Strategic Database Technology, Morgan Kaufman Publishers, San Francisco, CA.
[Turing50]   Turing, A. M. (1936). Computing Machinery and Intelligence. Mind, 59:433-460.
[Wojtowicz95]   Wojtowicz, J.(ed.) (1995). Virtual Design Studio, Hong Kong University Press. Hong Kong

WWW References
[Audio]   http://www.cnet.com/Content/Reviews/Hands/012797/ra30.html ; also see other internet audio streaming at
                http://www.yahoo.com/Computers_and_Internet/Software/Internet/World_Wide_Web/Browsers/ Plug_Ins/Sound_and_Video
[HTMLa]   http://www.ncsa.uiuc.edu/General/Internet/WWW/HTMLPrimer.html
[HTMLb]   http://www.w3.org/pub/WWW/MarkUp/
[HTTP]   http://www.w3.org/pub/WWW/Protocols/

---

[i] The actor-agent-server paradigm is used by [Booch94] to differentiate software objects by their complexity : an actor is an object that can operate on other objects but is never operated upon by other objects; an agent can both

operate upon other objects and be operated upon by other objects. An agent acts on the behave of an actor or an agent; a server never operates upon other objects by is only operated upon by other objects. A server provides services. This is congruent with more general uses of the terms. Servers provide a formal set of services (communications, database management, repository services, global naming services, etc.) [Simon95]. An agent is generally described as something that perceives and acts in an environment [Russell95]. Agents contain knowledge, can work autonomously, work on specific tasks on the behalf of others.

Shared Virtual Reality for Architectural Design

LUCA CANEPARO, *Design Network Lab, Dipartimento di Progettazione architettonica, Politecnico di Torino, v.le Mattioli 39, 10125 Torino, Italy, e-mail media@centauro.polito.it*

The paper presents the implementation of a system of Shared Virtual Reality (SVR) in Internet applied to a large-scale project. The applications of SVR to architectural and urban design are presented in the context of a real project, the new railway junction of Porta Susa and the surrounding urban area in the city centre of Turin, Italy.

SVR differs from Virtual Reality in that the experience of virtual spaces is no longer individual, but rather shared across the net with other users simultaneously connected. SVR offers an effective approach to Computer Supported Collaborative Work, because it integrates both the communicative tools to improve collaboration and the distributed environment to elaborate information across the networks.

1. Introduction

Our group, the *Design Network Lab*[1] in the Department of Architectural Design, School of Architecture of the Polytechnic of Turin, in 1996 began to implement an information system to support the design and management of a large-scale project.

The project concerns a central part of Turin, the new intermodal transport system of Porta Susa and the surrounding urban area. The new Porta Susa railway station is becoming the fulcrum of the overall system of exchange between high-speed and local trains, surface transport, both public and private, and the future underground.

The information system is the work environment implemented to support the comprehensive analytical, planning, design, building and managing aspects of the Porta Susa project.

The present implementation is based on the shared virtual environment technology. Shared Virtual Reality (SVR) integrates and co-ordinates the work of the public administration, the contractors, the firms, the suppliers, the building companies, etc. Capillary and precise communication implies a different organisation of the overall project through the reorganisation of the work and the interaction between individuals, groups and organisations (Greenbaum and Kyng 1991). Besides "professionals", a further main aim of the SVR is communication and interaction with public transport users and citizens.

---

[1] The scientific heads of the Laboratory are Prof. Anna Maria Zorgno and Prof. Pio Luigi Brusasco.

R. Junge (ed.), CAAD Futures 1997, 431-442.
© 1997 *Kluwer Academic Publishers.*

1.1 PORTA SUSA PROJECT

The Porta Susa project is a major and long-term investment of the Municipality and National Railway Company, and is going not only to change a central urban area (Figure 1), but also to renew the overall system of communications of the city.

The Porta Susa project intends to innovate the Turin communication system towards an integrated system of exchange between different means of transport. The present main railway station, Porta Nuova, is not suitable for high-speed trains because it is a railhead, like most of nineteenth and twentieth century central stations.

The project is increasing the importance of the Porta Susa railway station as the lines will be quadrupled and moved underground. The Porta Susa area is becoming the fulcrum of the integrated system of exchange between trains and aeroplanes by means of a direct train link to the airport, and between the future underground and the private and public surface transport. To provide for these increasing requests and functions, a new railway station will be built at the lowered track level, a few hundreds meters away from the actual one.

Due to the roofing of the lines, the area, no longer occupied by the railway junction, will be available for new purposes and activities. The result will be a spacious boulevard, which will join two, at present separated, parts of the city. On this area the Municipality is planning an international architectural contest, perhaps one of the first contest based on Internet.

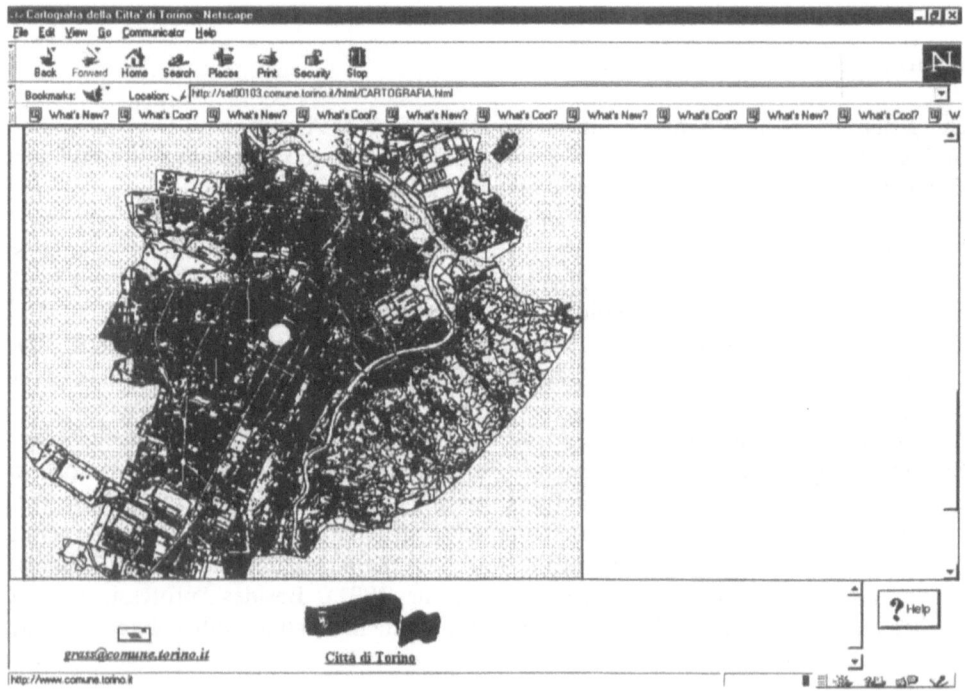

Figure 1. Turin and the Porta Susa area (white point).

1.2 COMMITMENTS OF THE INFORMATION SYSTEM

The information in a large-scale project is made up of hundreds or thousands of documents: drawings, drafts, blueprints, pages of reports and technical specifications, letters, manuals, etc. This large body of documents is made and updated continuously by the large number of people from all fields working to the project.

The information system is the environment committed to improving the work of documenting the project and to facilitate the various tasks which as a whole constitute the information-work: storing, retrieving, sharing, adding, modifying and managing documents.

As considered in a previous paper (Caneparo 1997), the main commitments to the information system are distributed access, flexibility, scalability and simplicity of use.

The *distributed access* permits every firm, contractor, supplier and other trading partner to store information on their servers. Meanwhile the users can access the overall information in a transparent way, and are no longer required to know if a document resides locally or remotely. For example, a user can visualise a drawing or a report consisting of parts residing on various servers (Figure 2). Download - upload changes the meaning, the users can browse documents and drawings as if the local disks are a Web site or explore Web sites as if they are local disks. The overall documentation of the project is location-independent and so, thanks to Turin's fiber-optic network, the access of a remote server could be as efficient as a local one.

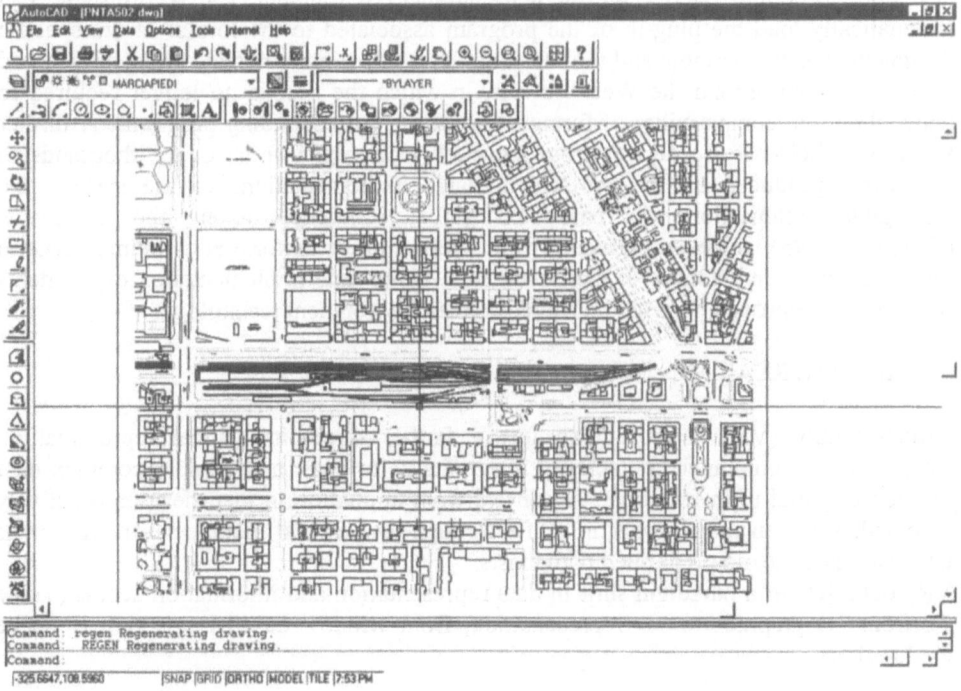

Figure 2. Drawing made of parts residing on different servers.

Flexibility is the capacity to work with the presently used formats of documents, as well as with futures ones, not yet foreseen in the current implementation. Flexibility in the processing of different formats of document is obtained by means of the Multipurpose Internet Mail Extension (MIME) of the Web, which allows the processing of various document formats, hence not only HTML documents. This potentiality of the HTTP protocol allows one to associate to each specific type of document the application necessary to visualise, modify, print or save it. Flexibility is understood, as well, as the capacity to dynamically redefine working-groups according to the ongoing task. Flexible definition of work groups requires not only more efficient communication, but also tools to support dynamic redefinition of the information flows.

The *scalability* is the possibility of integrating further individuals or groups into the project, dealing with their computer systems and networks. This should be made possible, either for the entire time of the project or, instead, for a limited time, e.g. to gather specific know-how.

Effective *simplicity of use* comes from the machine understanding what the user wants to do. Most of present available programs do not take into consideration real human functions, but instead offer tools to automatize information work. What we have tried to do with the Porta Susa information system is to integrate the tools the users are actually familiar with, in a networking-distributed environment. Present simplicity of use derives from the integration of the different applications (CAD, word processor, spread sheet, etc.) with the Web browser. This is particularly true with the most recent applications developed to be integrated in an Internet - intranet environment. The Web browser can automatically load the plug-in or the program associated to the format of the specific document. Today's graphic and multitasking OSes allow one to create an integrated and coherent system around the Web browser, in which the user is no longer required to know about the compatibility of formats and of the corresponding programs. A further key to simplicity of use derives from the intelligible organisation of the thousands of documents produced by the project. In early implementations of the system, the intelligible organisation was based on global coherence among the documents (Caneparo 1995), achieved by means of the definition of uniform and common criteria for the creation and storage of the documents. Up-to-date implementation experiments with another approach, focused on shared virtual environment technology.

1.3 VIRTUAL REALITY

Virtual reality (VR) is more than just a further three-dimensional representation. Because of the computer interactivity it permits us to enter and explore complex data through a spatial representation. VR differs from animation in that the user is actively involved: s/he can move -walk in or fly through- the virtual model, the movement is unconstrained, i.e. no predefined path exists.

VR could deliver a paradigm shift in data representation and information access (Table 1), from 2D graphics to 3D representation, from window based interaction to space exploration.

	Time-sharing	Desktop	Networks
Decade	1970 -'79	1980 -'89	1990 -
User	Specialist	Single	Group
Interface	Text	2D / 3D	VR / Multimedia
Interaction	Read and type	Paint and click	Navigate and communicate

Table 1. Paradigm shift in data representation and information access.

In the Porta Susa project not only designs and prototypes, but the whole documentation is accessible by means of VR. Through Internet the numerous and various protagonists of the project have the immediate opportunity to inspect the work in progress.

CAD 3D models are converted to Virtual Reality Modelling Language (VRML) and then uploaded to the server. Several programs and plug-ins for WWW browsers are available to explore VRML models. Since the conversion from CAD to VRML models is automatic, the representations accessible through the Internet are easily kept in sync with the design at its current stage.

Spatial representation is crucial for exploring architecture and buildings, but VR should not be limited to this task. VR enhances the understanding of interferences between flows of different transport systems, i.e. passengers of the high-speed trains, local trains and underground. VR improves the storage of a single document among the several thousands of the overall project and then its easy retrieval by "walking in" a 3D representation of the project with the relevant documents associated to the objects and buildings.

2. Shared Virtual Reality

The exploration and critique of a VR model across the net is essentially an individual process, because each person, independently from others, downloads the model and examines it. Shared Virtual Reality (SVR) differs from VR in that the experience of 3D spaces and objects is no longer individual, but rather is shared among several users across the Internet. SVR opens *virtual places* in Internet, *cyberspaces*[2], where people can enter, meet and communicate with others connected simultaneously.

SVR produces a further paradigm shift merging the capacity of network communication with three-dimensional representation. As *avatars*[3] people can meet in virtual spaces representing any physical or symbolic places whatever (Figure 3a, 3b).

SVR is a new medium, a different way to achieve real-time 3D interaction, different from the Web wheel of authoring - publishing - browsing pages.

For architects and planners SVR means a new way to communicate their ideas, to *broadcast* a design, to meet clients and contractors directly in the 3D model. As soon as

---

[2] According to the definition of William Gibson (1984), cyberspace is the total digital network, a place of meeting and communication.
[3] In Hinduism an avatar is the Terrestrial incarnation of a god or goddess. In SVR an avatar is the user "incarnation" in cyberspace.

436

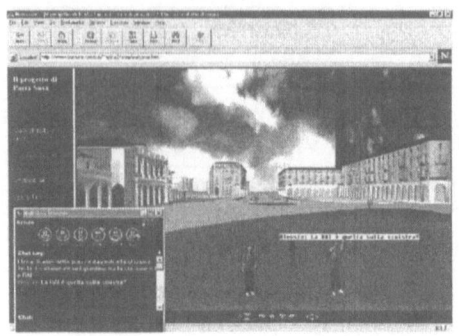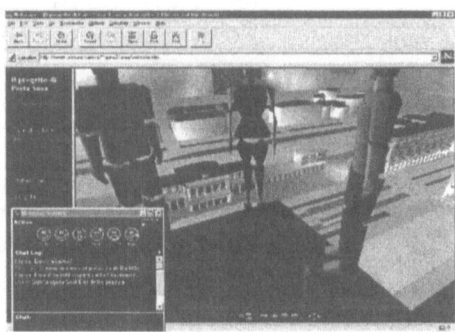

Figure 3a, 3b. Avatars meeting in cyberspace.

early stages of the design are modelled, they can be explored and discussed making everyone feel at home, even if they are sited in front of a computer at the office. Virtual tours and meetings about the project can be arranged to consider ongoing problems and decisions among small or large groups. Virtual meetings can be scheduled in advance or, instead, arranged ad-hoc among people working on the same task.

2.1 SHARED VIRTUAL REALITY PLUG-IN

The tool to gain access through the Internet to SVR is a plug-in for the main WWW browsers. The plug-in allows us to explore three-dimensional scenes by means of the Virtual Reality Modelling Language (VRML)[4].

The plug-in is automatically loaded when the HTTP protocol defines the corresponding 3D format. To the WWW browser the plug-in adds the tools for exploring 3D space, for visualising other users who are simultaneously connected and for communicating with them.

At present the plug-in runs with Windows 95 and NT on Intel CPUs, while porting to other platforms is expected.

2.2 EXPLORING SPACES

Moving a virtual observer controls exploration of virtual spaces and designs. The users define the virtual observer's movements and this is an important aspect of VR interaction. The method of interaction depends on the available hardware. With two-dimensional devices (e.g. mouse, trackball) the user defines a plane on which, by means of the pointing device, s/he traces and modifies on the screen a vector defining the path of the observer in the environment. With three-dimensional devices (e.g. spaceball) the user gains continuous control of the movements in 3D.

The SVR plug-in simulates the physical properties of objects, for example, it does not permit one to pass through objects, and the movements of the observer are restricted according to physical laws such as acceleration and gravity. The simulation of these

---

[4] For VRML specification, see http://vag.vrml.org/ .

properties slows down the interaction with the world considerably, in that a powerful CPU and a 3D graphics accelerator card considerably improves smooth movements and interaction feedback.

2.3 DATA FORMAT OF SVR

The SVR plug-in adopts the Virtual Reality Modelling Language (VRML), version 2 [5]. The VRML is a file format for describing three-dimensional interactive worlds and objects, conceived to be used in conjunction with the WWW. Since May 1997 the VRML is recognised as an ISO standard, and defines a worldwide accepted language.

In essence, the VRML is a plain ASCII language to describe 3D worlds. As HTML is a language to format documents for the Web, VRML is a language to create spaces and objects. Interpreters (browsers) for VRML are widely available from several companies and institutions for many different platforms.

VRML defines the syntax to describe a scene, consisting of lights, geometries and their properties. The properties attributed to geometries are colour, texture, animation and hyper-links to HTTP documents in Internet. Version 2 adds the possibility of animating objects, and improving VR worlds with sounds and the visibility option, that is fog.

The SVR plug-in implements the VRML 2's multi-user capacities and makes use of Java programming.

Java adds further action to VRML worlds. Java applets can be associated to polygons and objects to perform specific actions based on events, time or user's interaction. For example, it' s possible to use a wall of a building as a blackboard or sketch book, or define an interactive environment to assemble objects from primitive solids.

Multi-user capacity is inherent to the nature of VRML. The SVR plug-in implements several options to make VRML worlds effective cyberspace communities of people across the Internet. To gain access to a SVR world the user has simply to follow the appropriate link from her/his WWW browser. After a while (depending on the size of the model) s/he will be projected into the shared virtual reality world.

After the user has selected from her/his WWW browser the link to an SVR world, the WWW server defines the MIME type that the Web browser associates to the SVR plug-in. The Web browser loads the plug-in and transfers to it the address of the VRML model to be downloaded. The plug-in downloads and interprets the VRML language in a virtual 3D world. If the world contains the definition of a SVR server, the plug-in loads the multi-user module. The multi-user module is the client, which requests, over the net, the defined SVR server. The reply from the SVR server specifies the number and position of the other users connected. If there are other users in a defined *aura*[6] of interaction, the plug-in downloads from the WWW server the geometry and description of the appropriate avatar. After the download is completed, the avatar is displayed in the proper position, while the plug-in regularly transmits the position of the user and in the meanwhile receives the positions of the other users.

---

[5] For VRML version 2 specification, see http://vag.vrml.org/VRML2.0 .

[6] The aura is the predefinible radius of visibility and interaction among avatars.

2.4 COMMUNICATION IN SVR WORLD

The multi-user module integrates a chat program. The chat allows people in SVR worlds to exchange brief written messages. The plug-in forwards the message to the SVR server, which redistributes it to every user, connected (Figure 4). The ongoing experiment focuses on vocal messages, so, as a result, the communication becomes more friendly. A main drawback of vocal messages is that the communication is not duplex, a certain time lapse between the submission and the reception exists, and single packets can arrive delayed or out of sequence. Individuals in wide spread groups can receive the message at different moments, making effective interaction difficult.

The integration of the SVR plug-in with the main WWW browsers permits one to use other tools of communication. To gain audio capacity, RealAudio™ or a conference program can run concurrently with the SVR plug-in.

3. Design Applications of Shared Virtual Reality

3.1 SHARED VIRTUAL PROTOTYPES

One of the primary aims of VRML is the sharing of three-dimensional models across the net. For architects and engineers VRML makes it possible to share models on the basis of a platform independent standard. No matter what the CAD and modeller programs are

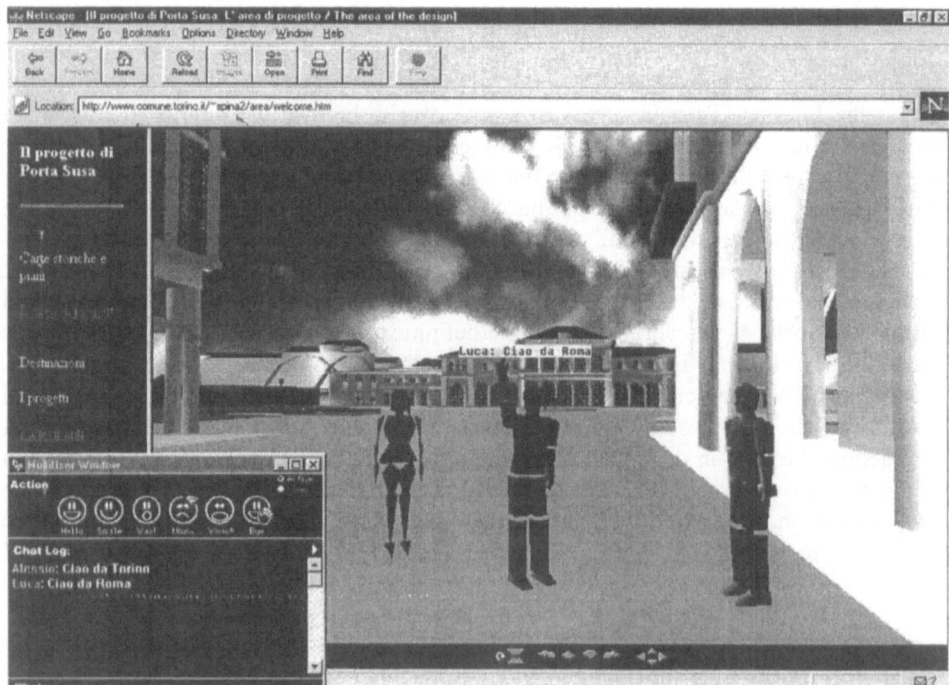

Figure 4. Chat active among users simultaneously connected.

439

used, the model can be exported to the VRML format and shared across Internet.
Buildings, structures, infrastructures can be closely examined before they are built. The different participants in the project can explore design and technological solutions in depth and in detail by walking in and flying through. The ability to explore the model is important in order to examine the interrelations between the parts. The use of colours can produce either realistic representation, for example to evaluate the environmental impact, or a symbolic representation, for example to highlight the integration and the interference among parts and objects. Because the virtual spaces are shared, persons from different places can meet in the virtual prototype to examine a design, to discuss problems and to take decisions. Decisions made in early design phases cost less than those made later since changes requested during the building process, if they are not impossible, will result in increased costs.

Design stages, as soon as they are computer modelled, are made available across the net. It could be possible to have no intrinsic gap between conception, simulation, examination and decision. Moreover, due to SVR the decision making process can involve every participant, no matter what time-schedule-appointment. Everyone, in front of her/his workstation, laptop or notebook, can examine the shared virtual prototype and join the meeting (Figure 5a, 5b).

Porta Susa is a large-scale project involving several teams working on different aspects, sometimes distinct, but more often interrelated. Sharing 3D data among the teams is necessary but not sufficient, because it is essential to make and keep these data coherent. The coherence is based on both modularity and extensibility. VRML 2 standard has been conceived with modularity and extensibility in mind. VRML is based on blocks, which can be loaded separately and then easily combined. Each block can be upgraded or substituted and new blocks can be added. In the Porta Susa project a taxonomy of VRML blocks has been defined, of which the most primitive elements are the building parts. These parts are available on the server, and shared among the various participants. The blocks from CAD or VRML can be easily combined into more complex objects, as far as whole buildings.

3.2 THREE-DIMENSIONAL DATA BROWSING

It has been previously mentioned that we expect the VR to be a paradigm shift. VR on the Internet, i.e. VRML, consists of a new way of interacting not only with objects and

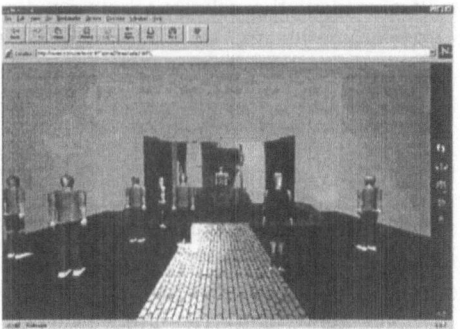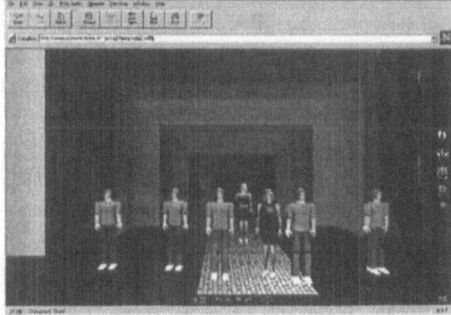

Figure 5a, 5b. Meeting in the virtual auditorium.

buildings, but with information in general.

VRML worlds can be modelled to represent not only designs, but also the interrelation between the objects in the design and the overall information of the project.

The Porta Susa project consists of thousands of drawings, drafts, blueprints, pages of reports and technical specifications, data, historical changes, etc. A primary aim of an information system on the project is storing and retrieving such a vast amount of heterogeneous documents, which varies and modifies during the life of the project itself. SVR offers the information system two paradigms: the model of buildings and the model of relations among documents. The two models are presented separately, but in fact they are closely interrelated, because, by means of VR, it is possible to move from one model to the other and back again.

3.2.1 Model of Buildings

The model of buildings is a simplified three-dimensional representation of the overall Porta Susa area. To store a document (whatever its medium), the user walks into the model up to the object/building which that document relates to. Then s/he just links the document to the related object or group of objects. To retrieve a document, the users explore the VR model up to the object to which the document relates and then with the cursor s/he explores the surroundings by selecting HTTP hyper-links to documents. The VRML and HTTP permit the building of a "web" of interrelated links among documents on the Internet, allowing users to view different and remote servers as ubiquitous data sources. Moreover different media are managed transparently by the MIME protocol, freeing the user from most concerns about data formats and their compatibility. VR has proved a powerful tool to manage building information, which is intrinsically related to a spatial paradigm.

3.2.2 Model of Relations among Documents

Instead of representing the physical relations between objects, the model of relations among documents presents the links connecting documents, directories and servers. The single documents are the nodes and the hyper-links are the arrows connecting them. Further objects represented are the directories, grouping document-nodes, and the servers, storing the directories.

The logical structure of hyper-links is more easily understood as a graph. The graphical representation, whether two or three-dimensional (Figure 6a, 6b), highlights the organisational structure of the project, and makes evident the relations between the documents accessible through the net. The user can explore the overall documentation of the project "flying through" the model of the relations between the documents, and can browse the interrelated documents by clicking on (Figure 6b).

3.3 COLLABORATIVE ENVIRONMENT

The Porta Susa information system is the operative tool conceived and created to support and improve the collaboration between the protagonists of the Porta Susa project. As noted in previous studies (Benford 1991) (Garlegher, Kraut and Egidio 1990) (Greenbaum and Kyng 1991) (Schuler and Namioka 1993) (Tan and Teh 1995),

Computer Supported Collaborative Work (CSCW) is not based on a single factor, but rather on the integration of three primary aspects: memory, process and collaboration.

Memory is the knowledge of the project and design from its early stages, to the present state. Active memory provides both the long-term storage of information and the representations necessary to deal with the huge amount of heterogeneous documents. The memory consists of drafts, drawings, reports, letters, technical specs created, exchanged and updated during the project. In a previous paper (Caneparo 1997), I have considered the forms of representations used to deal with memory (primarily precedents and indexing schemes) and how they relate to the ISO 9000 standards.

Process is the capacity to work with the information of the project. This capacity consists of the user's ability to access, update, modify both single, and bodies of, documents. Often the process spans the project both horizontally, between users with similar tasks, and vertically, between persons with different roles. A single task, e.g. modifying a drawing, could involve several draftsmen, architects, contractors, public administrators, etc. Work progresses by means of both concurrent and independent processes. The information system should be closely coupled to the data processed and flexible enough to support a dynamic environment, gradually evolving according to the tasks faced.

Collaboration, the users do not make explicit distinction between working cooperatively or individually. Several persons are required to deal with various tasks in the large-scale project. This can be done by traditional hierarchy, by subdividing responsibilities and tasks, or in a more flexible way, by setting up dynamic, ad-hoc, groups to deal with a specific job. Dynamic definition of groups requires more efficient communication in order to effectively exploit the pool of human resources and to share the knowledge and skills of the single members.

SVR offers a unitary approach to the different aspects of CSCW. The approach adopted is the *virtual yard*, where people involved in the project from his/her computer can connect to and "walk in" to retrieve information, to meet someone, to discuss a topic and to take a decision regarding an aspect of the project.

The virtual yard is not the model of a real, physical, place, but instead a symbolic representation of the project. The participating contractors, subcontractors, firms, suppliers and other trading partners manage their own virtual stands in the area of the

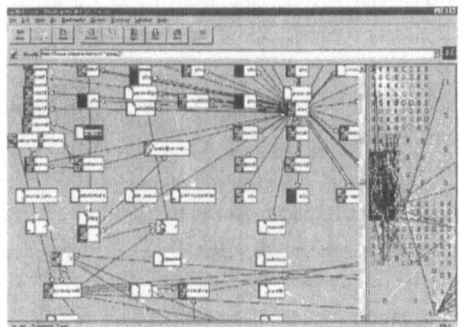 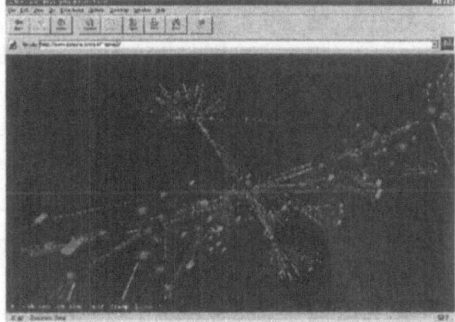

Figure 6a, 6b. 2D and 3D navigable maps of the hyper-links between documents in the Porta Susa project.

project on which they are working. These stands represent the distribution of know-how, duties, responsibilities and jobs and provide hyper-links to relevant information and to the people who are working on it. The cluster of these stands constitutes the nucleus of the virtual yard. Around this nucleus there is an auditorium, where participants in the project can meet each other and discuss by means of the SVR tools (cf. 2.4). There is also a library into which one can enter to gain access to the 3D models of the logical structure of hyper-links between documents (cf. 3.2.2).

Acknowledgements

Participating to the project are: Councillor for special projects, Turin Council; State Railway Company; AutoDesk Italia SpA.; Kinetix Inc.; Texas Instruments Italia SpA. For the ongoing cooperation my thanks go to Giorgio Emprin, Elena Girotto, Alessio Gotta, Tecla Livi, Maria Luigia Priore. Parts of this research have been funded by the Consiglio Nazionale delle Ricerche, Progetto Finalizzato on Problems of complex design, coordinator Prof. Edoardo Benevenuto.

References

Further information on the project is available on-line at:

http://www.comune.torino.it/~spina2

http://sat00103.comune.torino.it/

Benford, S. (1991) "Requirements of activity management", in Bowers, J. and Benford, S. (eds.) *Studies in Computer Supported Cooperative Work: Theory, Practice and Design*, Elsevier Science, Amsterdam, pp. 285-298.

Caneparo, L. (1995) "Coordinative Virtual Space for Architectural Design", in Tan, M. and Teh, R. (eds.), *Proceedings of CAAD Futures '95*, Centre for Advanced Studies in Architecture, Singapore, pp. 739-748.

Caneparo, L. (1997) "Shared Information System for Urban and Architectural Design", in Coyne, R. Ramscar, M. Lee, J. and Zreik, K. (eds.) *Design and the net. Proceedings of the Sixth International EuropIA Conference*, Europia Productions, Paris, pp. 39-52.

Garlegher, J. Kraut, R. and Egidio, C. (eds.) (1990) *Intellectual teamwork: social and technological foundations of cooperative work*, Lawrence Erlbaum Ass., Hillsdale.

Gibson, W. (1984) *Neuromancer*, Ace Books, New York.

Greenbaum, G. and Kyng, M. (eds.) (1991) *Design at Work*, Lawrence Erlbaum Ass., Hillsdale.

Schuler, D. and Namioka, A. (eds.) (1993) *Participatory design: principles and practices*, Lawrence Erlbaum Ass., Hillsdale.

Tan, M. and Teh, R. (eds.) (1995) "Co-operative Design", in *Proceedings of CAAD Futures '95*, Centre for Advanced Studies in Architecture, Singapore, pp. 637-771.

THE MULTI-USER WORKSPACE FOR COLLABORATIVE DESIGN

SUNGHO WOO, EUNJOO LEE AND TSUYOSHI SASADA

Department of Environmental Engineering,
Osaka University, Japan

Abstract : The design process requires a collaboration between organizations and individuals. This paper considers recent research in collaborative design system, it is called the multi-user workspace. The group-oriented multi-user workspace is a design environment where the collaborative work progresses smoothly between individuals or organizations in architecture design process. This paper describes the recent research concerning the multi-user work space with inter-university collaboration. The design is processed in the multi-user work space where participants interact with each other. This paper aims to prove the availability of the Multi-user workspace and to understand its problems.

Keywords : collaborative design, multi-user workspace

1.Introduction

The architectural design requires collaborative work among various participants who are architects, clients, engineers and various agencies in stages of the design process. The Sasada laboratory has conducted the collaborative architectural design through various projects. We found some important points in the process of projects. The face-to-face meeting is periodically held in real space (for example, conferencing room) and the presentation of the project is carried

443

R. Junge (ed.), CAAD Futures 1997, 443-452.
© 1997 *Kluwer Academic Publishers.*

444

out in the design process. The participants fixed the date for presentation using e-mail, telephone and fax. It's very difficult for all participants to be present in the same place, so the projects were performed in a asynchronous fashion.

Sometimes the participants need to synchronously access the virtual space, what we call the multi-user work space, where they review the design for solving the problems quickly in the design process. In the multi-user work space, the presentation data exist in multi-media format, such as text files, images, movies and 3D objects. The participants access the multi-user work space, review the design while communicating with other participants, and they solve problems by collaborating with each other.

This paper will describe the recent research concerning multi-user work space with inter-university(Osaka University in Japan and KyungHee University in Korea) collaboration. The design is processed in the multi-user work space where participants interact with each other. This paper aims to prove the availability of the multi-user workspace and to understand its problems.

2. Multi-user work space

The Sasada laboratory has carried out various projects with collaboration. According to those projects, design system is composed uniquely to facilitate the collaboration. These projects recently carried out, let us to know the necessity of new media.

2.1 Some models of the collaborative work

The WWW mediated projects classify into 3 types as follows.

· Collaborative design between designers and public
· Collaborative design between designers
· Collaborative design between designers and clients

2.1.1 Collaborative design between designers and public (Maruyama Project)

On the Island of Awaji in Osaka Bay in Japan, the Maruyama fishery harbor is located. In accordance with the master plan, we designed facilities of harbor one bye one. The design process is opened to the public from the early stage using its home page. The home page for this project is composed of texts, 2D images(including sketches, CG generated still images and photos), and 3D objects. For the purpose of operating the 3D objects at high speed on Internet, the data is composed of the model of the plan without the environmental data.

The public could not review the design completely under these conditions. Communication between designers and public was only through e-mail, and the information flow was unidirectional.

2.1.2 Collaborative design between designers (Ayuyagawa Project)

The Sumoto City planned to throw the big bridge over the Ayuya river, and requested the design of it to the Sasada laboratory. This case is processed between designers who have various specialties, and the feedback among the participants is very quick. The BBS records the comments and questions with image automatically, so the bidirectional communication is accomplished.

2.1.3 Collaborative design between designers and clients (Mikata Project)

The Mikata-cho in Fukui Prefecture is famous for the vestiges of Jomon Age. The Mikata-cho planned to make Jomon Park including the museum, recreation facilities and research center. The clients could review the design through 2D images of project home page and operate the VRML of the plan from all viewpoints.

In this project the interactive bulletin board is used for bi-directional communication. The clients download the image of project home page, and write his or her opinion on it and attach the image with e-mail. The agent program automatically separate the text and the image, and upload it to project home page. All of attach file is 2D image.

2.1.4 Lessons from these projects

Through these projects we found the necessity of new media with some conditions for smooth design review, that is synchronous, real and interactive media. We begun to research new collaboration system in 3D virtual space like any place in reality. If this 3D virtual space is the same as real space, the participants, who are clients, or specialists, or designers, can perceive the things as their own way.

2.2 Concept of multi-user work space

Like any place in reality, the street is subject to development in Multi-user work space. Users can build their own small streets feeding off of the main one. They can build buildings, parks, signs, as well as things that do not exist in reality, such as special neighborhoods where the rules of three dimensional space-time are ignored. The events happen in a virtual three-dimensional world in a computer-driven alternate universe where people assume avatars, converse, travel, plan, and build.

Multi-user work space consists of shared virtual space and private virtual space. Shared virtual space allows multi-users to interact on the Internet. On the other hand, private virtual space is not shared for private work. The shared virtual space allows many kinds of tools in Multi-user work space to be shared, and changes in Multi-user work space such as motion, data modifications and sound are transmitted to the users immediately. In this case, the problem of having a seamless environment in time and space is solved. The private virtual space has a function in a separate space (not shared). The design process needs to set limits to the public, and needs to progress in private space until the design is partially completed. Thereafter, the shared Multi-user work space accommodates changes. In addition, the people concerned in the design process are free from the restraints of being present at the same time and same place. This can be compared to using a bulletin board for communication. The asynchronous function improves the quality of communication or design.

The method to express the representation of multi-user in the virtual space is very important. An avatar is the embodification or representation of a user's awareness and identity within a multi-user computing environment. For virtual worlds, an avatar is an audiovisual representation of a human user that facilitates communication with other users in cyberspace. The purpose of the avatar is to give other users in a multi-user distributed VR simulation a proxy or surrogate for the purposes of simplifying and facilitating multi-user interactions on a social level. Thus, the representation of many human traits are important in the avatar representation of the user. For example, emotion and gesture may be vital for creating optimal virtual conversation interfaces for users.

Virtual space is built by Project server. Project Server consists of a property server, an avatar server and Web Server. The property server manages all the data related to the building and objects in the world. The avatar server communicates avatar traffic through the world. In addition, a web server is required for storing the art work for the world. This web server can be anywhere on the Internet. The client who has their own browsers can access the project server.

3. Inter-University Project (KyungHee Project)

KyungHee University in Korea will establish the graduate school of architecture engineering for training design professionals. The established education system was given too much of logical reasoning, so the curriculum of the graduate school of architecture engineering is made centered upon practical experience. The building is composed of CAD room, computer room and lecture room with slide, audio and video facility. For the design of the building, the Inter-University project between KyungHee University in Korea and Osaka University in Japan has begun.

In the early stage of the project both of universities set up the project home page and shared the data. KyungHee University presented a topographical map, the picture of the building around the site, and the drawings in their home page. Osaka University provided various tools such as groupware calendar, bulletin board and web robots database to support multi-user workspace.

3.1 Objects making with Data base

KyungHee University produced 3D models of the buildings around the site, and presented those images and dxf files in their home page. Osaka university downloaded the dxf files from KyungHee home page, converted it into VRML for design review at any time on the Internet, and represented it to Osaka University home page. The multi-user work space is made by arranging the 3D objects after the 3D model is produced. Moreover, we could review and refine the design in the multi-user workspace.

3.1.1 Standard format for sharing 3D model data

We have different modelers(ArchiCAD, formZ, AutoCAD), so the 3D model data is shared by using dxf and VRML format. In the early stage of the project, KyungHee University made 3D model while Osaka university made it into VRML. As time went by KyungHee University understood how to make VRML and QTVR, and finally they made 3D model and VRML. In this process, we communicated how to use internet technology.

3.1.2 Data arrangement with web robots database

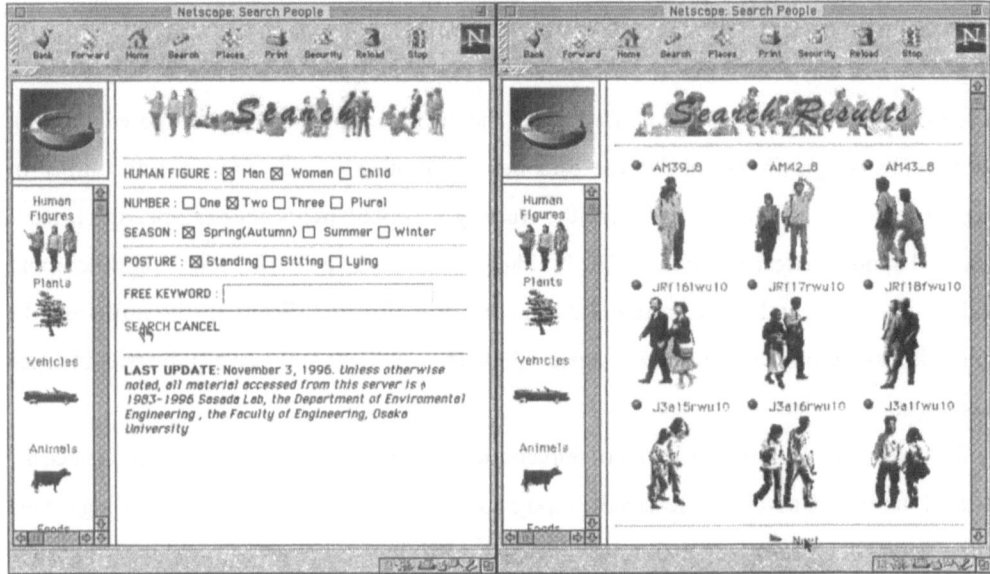

Figure 1 Web Robots

We used Web Robots Database for data arrangement on the virtual space. Web Robots Databases are used to manage and effectively search the distributed data on the internet. The List of Active Robots has been changed to a new format. This format allows more information to be stored, allows faster updates, and information to be more clearly presented.

The Web Robots Database is composed of robot, intelligent agent and a search engine. A robot is a program that automatically traverses the Web's hypertext structure by retrieving a document, and recursively retrieving all documents that are referenced. Intelligent agents are programs that help users with things, such as choosing a product, or guiding a user to fill forms, or even helping users find things. A search engine is a program that searches through databases of HTML documents gathered by a robot. (Figure 1).

3.1.3 communication with BBS

The BBS is used for communication between participants, and records the problems, questions and answers with multi-media data including text, image, VRML and movie. User post message with attached file, such as image, VRML and movie, BBS represent the contents user posted with inlined multi-media data (Figure 2).

posted with inlined multi-media data (Figure 2).

The representative communication tools have its own advantages and disadvantages. The characteristics are as follows(Table 1). Comparing these communication tools, BBS is most useful in this project.

Table 1. Comparison of communication tools

	data format	records	comments
e-mail	text, image, movie, VRML	user's machine	record in participation time
BBS	text, image, movie, VRML	Project Server	record all history of project, whenever user reference the history, categorized by the questions

450

chat	text	not save, or user's machine	record in participation time
news group	text	News Server	once user read, never can't read, tree architecture

3.2 Design review in multi-user workspace

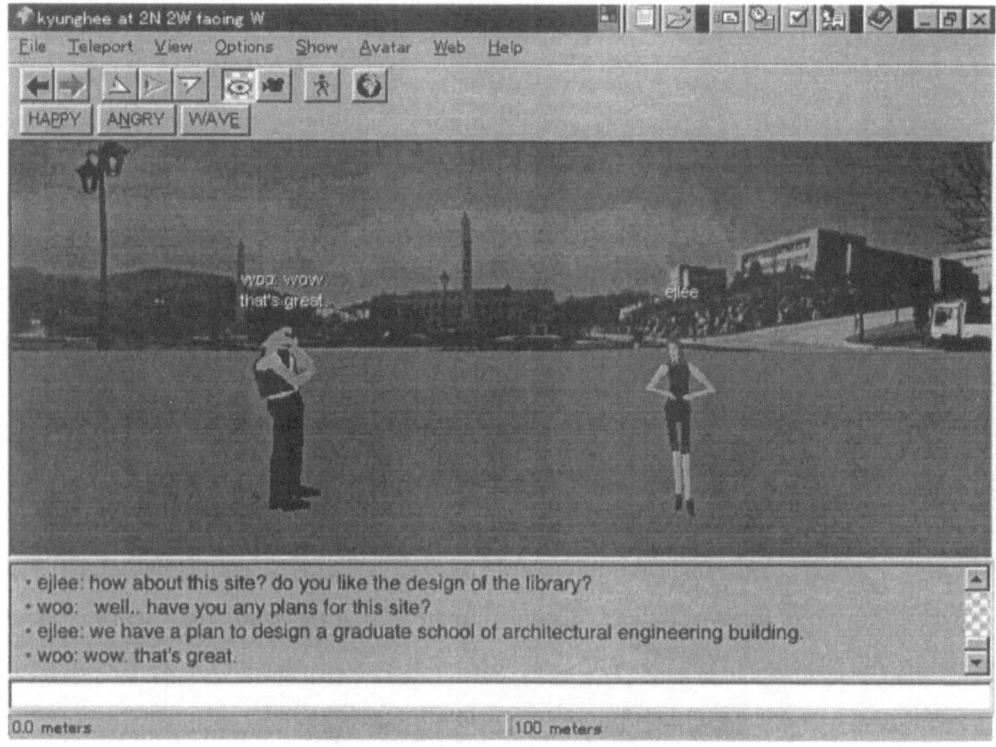

Figure 3 Multi-User Workspace

The server of multi-user workspace is setup after the background image of the site is completed. The server of multi-user workspace is composed of property server, avatar server and web server. Both universities made 3D models the buildings around the site, and placed it in multi-user workspace. The 3D model data produced by various modeler made into DXF or VRML format, and finally converted into RWX format(renderware format). These RWX data are saved in project server, and the participants can use it via their own browser. The available data are buildings, facilities, 2D images of human figures or trees, texture materials, 3D vehicles and street furniture.

The participants reviewed the design using their browsers whenever they wanted. They could simply arrange, move, and transform the objects in the multi-user workspace. The participants could recognize the avatar as a remote representation of human identity and existence, and express the emotion through the avatar. It's possible to communicate precisely through this synchronous communication.

After the environmental data is located in multi-user workspace, the participants reviewed each of the design alternatives while communicating with each other. The communication tool was BBS and chat, and the contents of the communication was recorded. As the result of the design review, 3 of the design alternatives were accepted (Figure 3).

4. Conclusion

Architecture design is the result of the teamwork of multi-disciplinary groups including designer, client and committee. The design process includes the exchange of information between individuals and organizations, and discussions among professionals of the team.

In this study, professor and students enter into a shared virtual space which they made, and review the design interactively. This system is an available environment for participants to access data made by collaboration on the web at anytime. Each participant can see what he or she wants using the interactive function, such as walkthrough and examiners (review tool of rotating the object) and non-interactive functions, such as still images and animation, while he or she communicates with another participant.

Each student can review the design in the virtual space made by the professor and students. It is easy to use an authoring tool for making data, gathering opinions of participants on the bulletin board, VRML viewing and application analysis of the shadows from a distributed area. All of them are composed componently. This design environment makes design review possible at any stage of design in Multi-user work space on the web.

In this study, the Multi-user work space makes it clear that there is the possibility of open architecture with progressive communication between professor and students in various locations, with the Kyunghee project.

452

Reference

http://www.worlds.net:80/alphaworld/aw-about.html

http://206.79.196.34/avatar/Avtdefn.html

http://sap.mit.edu/dsof/research/creative_design.html

Sungho Woo, Yoshihide Takenaka, Tsuyoshi Sasada (1996) Architgectural Virtual Space in Design
    Education, CAADRIA96, 27-34

Sungho Woo, Claude Comair, Tsuyoshi Sasada (1997) Architectural Virtual Space in collaborative
    design, IFIP97, 65-76

Tsuyoshi Sasada (1995) Computer Graphics as a Communication Medium in the Design Process, CAADFutures,
3-5

Milad Saad, Mary Lou Maher (1995) Exploring the Possibilities for Computer Support for Collaborative
Designing, CAADFutures, 727-738

Tsuyoshi Sasada (1991) Open Design Environment, Proceedings of the 4th International Conference on
Computing in Civil and Building Engineering, 57-64

Dean Taylor, Kevin O'connor (1997) Experiences with remote collaboration for concurrent engineering,
IFIP97, 29-48

Mao-Lin Chiu (1997) Representations and Communication channels in collaborative architectural design, IFIP97,
77-96

John Danahy, Rodney Hoinkes (1995) Polytrim:Collaborative setting for Environmental Design, CAADFutures,
647-658

James Rutherford (1995) A Multi-User Design Workspace, CAADFutures, 673-686

Thomas Kvan (1995) Fruitful Exchanges:Professinal Implications for Computer-mediated Design, CAADFutures,
771-776

John S Gero, Mary Lou Maher (1996) Current CAAD Research At The Key Center Of Design Computing
University Of Sydney, CAADRIA96, 35-52

Claude Comair, Atsuko Kaga, Tsuyoshi Sasada (1996) Collaborative Design System with Network Technoogies
in Design Projects, CAADRIA96, 267-286

Miltan Tan (1996) Design Thinking and the need for open access to Multimedia Sources, CAADRIA96, 99-108

A CONCEPTUAL NETWORK FOR WEB REPRESENTATION OF DESIGN KNOWLEDGE

RIVKA OXMAN, ANAT SARID, SHOSHI BAR-ELI,

AND RUTH ROTENSHTREICH

Faculty of Architecture and Town Planning

Technion, Haifa, Israel, 32000

abstract:

The nature of the Internet as a medium for the representation, storage and accessing of design knowledge is explored and various research issues were introduced. The appropriateness of certain characteristics of the medium as a potential environment for a new interactive way of doing design by exploring design ideas are investigated. Considerations of the Web as a collaboratively constructed and maintained design resource are explored. Cognitive models are proposed in order to support cognitive behaviors in search , browsing and concept expansion. Our particular approach utilizes the ICF (issue-concept-form) as a conceptual network for design knowledge bases. Finally, a report is given on a pilot program demonstrating how the exploitation of the ICF model structured around design chunks can support the construction and maintenance of shared design resources.

R. Junge (ed.), CAAD Futures 1997, 453-473.
© 1997 *Kluwer Academic Publishers.*

1. Introduction

The increasing use of the internet and availability of design subjects has opened up opportunities for the representation of design knowledge. New distributed and interactive tools, virtual reality on the net, multimedia, networked digital video, and Java applications are providing new and excitting media for the construction of virtual design spaces (Schmitt, 1996). Despite these developments, the nature of the net as a medium for the representation, storage and accessing of design knowledge does not yet provide direct support for the enhanced content of ideas and concepts which are part of design thinking. The cognitively rich behavior of design can be supported, but how? If we are interested in the development of cognitively powerful knowledge representations, we require accepted models behind the organization and presentation of design knowledge.

In this paper, we report on research in which the Internet is considered an environment for the representation, storage and retrieval of design knowledge. The research demonstrates that the embedding of design conceptual content in Web-based representations provides a powerful advantage for expanding the cognitive richness of design content. It further contributes to expediting the processes of search and retrieval in design knowledge bases.

Our approach considers strengthening conceptual content through structuring the representation of information as a key to the advancement of Web-communication and utility. Information structure, the way information is displayed and organized, has implications for functionality in navigation, browsing and search. We demonstrate that appropriate structuring can also emulate and support cognitive behaviors. Stucturing also provides the added advantage of enhancing the possibility of collaborative development of shared sources of knowledge. In order to move in this direction, we require conventionalized structures and models behind the organization and presentation of design information.

Three components: design descriptions, underlying information models and the intrinsic potential of Web technology provide the theoretical and methodological foundations for Web-based Design Information Systems. Web technology can contributes to new forms of interactivity, communication, presentation and collaboration by the integration of a variety of tools. However, without the introduction of knowledge models capable of cognitive level performance, we cannot advance beyond the presentation level of design information. Thus, our long-term objectives are: to strengthen the cognitive content (particularly conceptual richness) of design representations; to exploit Web utilities optimally in design information resources; to conventionalize models of design knowledge in order to support the shared construction and usage of open Web sites; to enhance forms of search.

How can design be represented in a way that encodes the conceptual content of design knowledge ? We demonstrate that ìstructured conceptual representationsî provide, beyond the improvement of functionality, potential for supporting conceptual linkages. This behavior appears to emiulate the important cognitive phenomenon of concept expansion, a form of the evolution of design concepts which is characteristic of creative thought in design.

In the system we demonstrate how enhanced conceptual representations of design concepts can support search through a body of design ideas in an associative manner. It is the complexity of this Web of associations which emulates concept formation in creative design in the human designer. In our approach, current emphasis has been placed upon the exploration of cognitive models as a methodological basis for the representation of both typological and design precedent knowledge. We present certain of the theoretical and research issues in a cognitive approach to Web-based design information bases, report on classes of design conceptual structures which have been implemented in the system, and consider the implications of our current approach as a basis for shared, open sites.

The paper reports on an educational experiment, and the resulting information base which has been developed by a group of architectural students. The formalization of this conceptual knowledge (Regionalism in Architecture) on the Internet has proved to be extremely challenging as an educational medium. This attempt at the formalization of conceptual knowledge in an information base has helped to identify issues for further research and development. Among these, we place emphasis upon shared, or conventionalized, knowledge structures. These are considered a precondition for collaboratively constructed Web-based design resources.

The work demonstrates the potential of future collaborative development of shared design sites. We consider selected theoretical and research issues related to this powerful concept. As a design community it would be of great potential benefit, if we could focus collaborative efforts on the structures, norms and procedures required for the development of shared environments. We believe that only through large-scale collaborative efforts will it be possible to formalize domain knowledge in design. Such knowledge can also eventually contribute to domain-specific search engines.

In the following section, issues related to the Net as a shared Web-based design resource are presented. Net attributes are reviewed with respect to their potential contribution to the exploitation of the Net as a design resource. On the basis of this review of characteristics, we propose a set of issues which must be addressed in order to achieve a shared Web-based representation for design knowledge.

2. Shared Design Web-Space

The Web is a large and continuously extendible information resource. As such it has certain characteristics which are problematic relative to the objectives outlined in the previous section:

-       information structure and organization

The current unstructured nature of the medium limits the potential for joint projects and shared activities which require a higher level of structuring in the representation of information. Beyond collaboration, structuring is a subject which also has implications for the functionality of Web sites, in general, as well as for information bases. Information structure, the way information is displayed and organized, has further implications for navigation, browsing and search. Navigation is dependent upon the structure of linkages. Linkages are currently hand-crafted and are dependent upon the individual views of the developer. If we as a professional community wish to promote collaborative development of shared knowledge, we require conventionally accepted models behind the presentation and organizational structure of design information.

Among the issues for development which are raised by these problems are the definition of the presentational and conceptual content of designs and the way in which the organizational structure of information affects utility and cognitive performance

-       domain dependent vs domain independent information bases

Given the quantity of information now available on the Web and the open introduction of information without guidelines or restrictions, finding and exploiting material becomes extremely inefficient. One potential solution appears to be the employment of domain specific organizational media such as indices based upon domain specific conceptual vocabularies. An important methodological problem is how to acquire this knowledge and how to construct domain vocabularies.

-    textual vs graphical representation

Today it is possible to make references, or to access information, primarily through textural means. Current Net tools such as Map-Edit support graphical indexing, but in a limited way. The possibility of graphical indexing is obviously of great significance to the performance of design information bases which generally contain much graphical content. Graphical indices might enable search for similarity based upon graphic content, or some code of graphic content. Conventions for the graphical coding of designs must be developed along with appropriate computational tools for graphical search.

-    passive vs active presentation

The Web has so far operated as a medium for disseminating information to distributed locations. Interaction with information is still primarily static, or passive. The Internet supports the presentation of information, either in 2d or 3d, in animation, video, etc., but does not adapt and respond to specific requirements of users. With the development of new programming languages such as JAVA, there exist possibilities for the inclusion of programs and algorithms within sites which enable dynamic operative characteristics. This raises the general issue of how to effectively support the dynamic and individualized activation of diverse forms of information rather than the presentation of static documents.

We have reviewed four broad categories of Net characteristics which appear to have significant implications for Net-based design resources:

a. structure and organization in shared sites and cognitive performance;

b. domain specific indices enabling more efficient search and utility;

c. graphical indexing for more intuitive design search

d. interactive knowledge-bases.

In the following sections we consider these problems with respect to our current research.

3. Cognitive Models as a Basis for Web Representation of Design Knowledge

We have proposed a model for a Shared Design Web-Space. The model is based upon a general approach to the representation and structuring of design information. Related to the model, we also propose a basis for establishing domain-specific indices. A concept of employing the graphical content of design schema as graphical indexing for designs.

We include in the general model a cognitive model of design representations (Oxman, 1996). That is, beyond the graphical and basic textual presentation of the design itself, as is characteristic of current Web presentation, the proposed approach would provides underlying conceptual knowledge. Beyond the presentation of general information, a design information base can provide knowledge which would support the understanding, reasoning about, and re-use of a design precedent.

In recent years we have attempted to develop formalisms for the representation of cognitive-based representations of design knowledge which can be exploited in design and design reasoning processes (Oxman, 90, 94). Among the forms of design knowledge representation which are as relevant candidates are the following:

- precedents (related to the computational paradigm of case-based design): this provides a presentation of the holistic solution as well as decomposition of the solution description as a set of separate design ideas (chunks). It includes a rationale for the re-use of the solution (in CBR terminology: case-adaptation).

- typologies (related to the computational paradigm of prototype-based design): in addition to the presentation of designs it is possible to provide a representation of the schema, or prototype, which underlies the specific solution. This representation might also include the refinement process, or the steps of development of a final solutions from the source schema.

  These two forms of representation, along with their rationale for re-use of information, would enhance the possibilities for active rather than passive representation of knowledge.

- indexing through design schemas can potentiall support graphical indexing for references of similar designs.

4   The Web as a Design Case-base

CBD (Case-based Design) (Maher et al., 1995) has developed as a domain-based application of CBR (Case-based Reasoning) (Kolodner, 1994). CBR is a cognitive-computational theory which has been applied to the modeling of design. A foundational research priority in CBR is to model human reasoning in re-use of past solutions. Since the re-use of the experience of prior solutions in all of its complex phenomena is characteristic of design, CBR is a highly relevant technology for design fields. In order to differentiate CBD, we refer to it as Precedent-based Design, since design precedents are prior design solutions which have something to contribute to current design problems.

In the behavior of Web users as well as in the construction of sites, we can observe case-based phenomena. For example, indexing, browsing, down-loading and re-use of material are all phenomena which have an analogous relationship with CBR (Oxman, 1996). This affinity of the Net as a resource-base and its relationship with Case-based Reasoning suggests possibilities for a new form of development of design resources within the Net. CBR has developed a rich experimental and applicative technology which has recently resulted in CBD systems in different tasks in design. In constructing design resources within the Net, we can derive much experience from approaches in CBR including search and adaptation paradigms.

With the development of Network technology CBD appears to be entering a phase of new developmental potential. The Net already functions as a large case-based system (Oxman, 1996). What is required in order to enhance the performance of the Net as a world-wide CBD system? What foundational work is necessary in order to enable collaborative activities towards this end. Should we entertain the idea of an international research program to develop a global Shared Web Design-Space? If this is a desirable objective for our community, should this effort be conceived of as a collaboratively developed global case-based site?

Certain issues are relevant to advancing this possibility :

- the Web as a global design case-base: large scale case bases, their function and performance requirements must be defined;

- collaborative development and usage of design cases: the possibility of a jointly constructed and collaboratively maintained design resource site requires development of conventionalized models, norms and specs;

- standardization of the representation of information in a format which provides more than the presentation of visual information: this requires the acceptance of a cognitive model;

- design search engines based on semantic taxonomies for design: how to develop semantic, or conceptual, taxonomies to support linkages and browsing; can semantic Nets of design issues and design concepts furnish convenient category sets for a global indexing schema?

- design search engines based on graphical design ontologies: graphical indexing and browsing potential and the indexing of designs according to a graphical design ontology require development;

- automation of indexing and linkage: the process of index creation can be automated;

- case acquisition in the Net: how are new cases acquired and introduced;

- legal issues: legal problems related to the re-use of intellectual property in the form of designs, design representations, etc.;

- interactive design: if interactivity within the site is possible, how can it be achieved?

- adaptation engines: how can we achieve the dynamic and interactive activation of information, for example, in the provision of rationale for the re-use of design knowledge in current design problems.

5. ICF as a Formalism for Shared Design Work Space

5.1 BACKGROUND

The term design precedent has been employed in our work on design knowledge representation (Oxman, 1994) in order to designate that particular design case conceptual knowledge which has something to contribute to current design problems. One of the distinctive problems in representing designs is the richness and complexity of their descriptive content. Each design contains many related chunks of information which are often difficult to describe or to decompose. Furthermore, not all of the information embedded in complete and exhaustive records of existing designs may be immediately relevant for aiding in current design problems. A design chunk is defined as an original annotation of an entity of conceptual content which characterizes the uniqueness of a specific part of the design solution in a design precedent.

We have attempted to identify the kinds of knowledge which exist in design precedents and to formalize their representation. The representation formalism which we have developed was termd ICF (Issue-Concept-Form). A typical ICF formalism provides explicit linkages between design issues of the problem, a particular solution concept, and a related form description of an element of the design solution. In our research we have developed the ICF formalism in order to make these conceptual linkages of design ideas explicit. The ICF model addresses these problems of representation. It is based upon a decomposition of holistic case knowledge into separate chunks of design knowledge. Each design chunk represents the linkage between design issue, concept and form in a design precedent. The chunks are defined as components in a structured memory according to a semantic network of issues, concepts, and forms.

Design issue: the design issue is domain-specific semantic information related to goals and issues of the problem. It may be a point related to the design task which is deliberated by the designer. Such points may be formulated by the programmatic statement, the intrinsic problems of the domain, or by the designer himself.

Design concept: the design concept is a domain-specific formulation of a design idea in relation to an issue. It is a form of ideation related to the design task. It is a verbal statement of a solution principle. The way to a solution, rather than the explicit physical description.

Design form: the form is the specific design artifact which materializes the solution principle. It is a physical design, particularly related to the structure of design objects.

This tri-partite schema has significant implications for memory organization, indexing, and search. It contributes to an indexing system which supports issue and concept-related search in the design precedent representation. Furthermore, this approach enhances the capability of browsing within memory and can support cross-contextual indexing by exploring the net of concepts, and following it back to appropriate structural descriptions.

5.2   ICF ON THE WEB

The ICF formalism has been exploited to promote a conventionalized and structured representation for design information in the development of a shared site of design projects in the Web. Beyond the simple objective of presenting design objects, the formalism provides a structure of representation which emphasizes the encoding of design rationale and method as well as the formal and functional elements of the design. Given the structured nature of the representational set, and the significantly developed relationship between textual and graphic presentation in ICF, this ontology preserves a high level of specific representational flexibility within sites while providing the essential structure to support queries, search and exploration of ideas through the networked linkages between sites.

In the following chapter, we provide a description of an experimental case study. In our experiment, a set of domain-specific semantic information related to design issues is explicated and employed as part of the indexing system for cases in the Web.

Design concepts employed in a collection of design precedents exemplify conceptual approaches to the conceptual resolution of these issues in selected examples of regional architecture (Work by Anat Sarid, Ruth Rotenshtriech, Shoshi Bar-Eli).

5.3   CASE STUDY: REGIONALISM IN ARCHITECTURE

In the current system precedents have been drawn from case studies of the Mediterranean House, a sub-class of regionalism for which the case-base provides a collection of important precedents as represented by their conceptual contributions. Within the case-base specific formal elements are identified which provide structural, or conceptual, solutions for the significant issues.

Among typical issues are: topography, controlling the view, content of elements, content of vernacular architecture, regional attention to natural physical forces such as light, spatial institutions such as contained private space, the spatiality of the wall, materials and technology, etc. The collection of examples which constitute a case-base for the Mediterranean House provide, through the convention of the ICF model, a means to address certain of the theoretical issues in constructing structured, shared design Web-space sites.

The model also addresses other issues which were outlined above such as domain-specific indexing. That is, the contribution of the multiple independent site designers working within the convention of ICF can contribute to a body of taxonomic content which constitutes a conceptual vocabulary for the specific domain, Regionalism in Mediterranean Architecture. This vocabulary can then function as a domain specific index and support search for relevant design ideas. The index is automatically updated as the site is expanded.

5.4 SYSTEM DESCRIPTION

We have implemented a system which contains knowledge which constitutes a precedent base for the Mediterranean House, and for its design. Figure 1 illustrates the current implemented version of the system. The system currently consists of a set of formatted Web pages. The format includes four frames. Each frame contains a specific representational formalism or presentational media. The frames of a typical page are:

a.   Title window

The title window is at the upper right corner of the page and contains the title of the current subject and its related content described by text. For example, Web page b. presents in the title window the name of a selected issue and provides a textual description of its related concepts. In figure 1. the title of the selected issue is: "relation to local architecture".

b.   Top-down representational window

The *top-down representational window* is located at the upper left corner of the page. It contains a relevant part of the ICF complete semantic map which supports browsing linkages from a higher level of abstrcaction to a lower one. For example, issue no. 2 - "relation to local architecture" is linked to its related concepts: concept no.2.- "controlled use of local elements" and : concept no.3. "planting an element from local vernacular architecture". Each top down window contains its own explanation box which is the description window below the net diagram.

467

c.   Bottom-up representational window

The *bottom-up representational window* is located at the bottom right corner of the page. It contains a relevant extracted part of the ICF complete semantic net which supports browsing linkages from a lower level of abstraction to a higher level. For example, concept no. 3 - "planting an element from local vernacular architecture" is linked back to its related issues such as "relation to local architecture" and "building and site relation".

d.   Description window

The description window contains textual and graphical illustrations. It is located at the bottom left part of the page. It is usually linked to presentation pages by textual and/or graphical references. At each level, the description page contains the relevant material. The explanation window contains a textual explanation of the related abstraction. For example, an explanation of concept no.3. "planting an element from local vernacular architecture" is the following: "Using elements of local vernacular architecture in design is an act of quotation: - copying an element derives from the need to preserve, or respect, the importance of local buildings, methods and elements that were developed through the years. By quotation, or emulation, the architects creates an immediate connection to local architecture that can easily be traced and understood". Key words in the text can also be employed as mouseable indices.

5.5   SYSTEM OPERATION

In contrast with a system with highly structured networks of objects and linkages from design object to design object (coarse level of indexing), we experimented with conceptual indexing (finer level of indexing) in order to support associational search and browsing exploiting the principle of semantic nets. The system has provided a means to test certain of our theoretical assumptions regarding both cognitive behaviors in information bases and the conditions for shared design Web sites. We are continuing development of a large scale system (The Modern Chair) for the continuing development of the approach and the validation of research assumptions.

An example of browsing and exploration modes in the ICF system is illustrated in figures 1 and 2. The first step is the selection of an issue from the set of all currently exisiting domain issues. Once an issue is selected, the second page appears with related concepts. The user selects relevant conceptual linkages by activating the top-down window. Related concepts appear and cross-contextual linkage can be established. At this stage the partial ICF map of concepts related to forms appears on the screen. In the same fashion, the linkage which has been discovered may introduce the discovery of another design concept associated with the form element of another precedent. Similar concepts may activate several designs. By activating a top-down window which connects concepts to forms the user may explore how a similar design concept may be realized by different form elements in two designs.

Browsing supports the exploration of cross-contextual linkages in both directions. By activating the top-down and the bottom-up windows there exists the possibility to search and browse the whole semantic network. The user can go forward by selecting the top-down window or go back to a previous page by activating the bottom-up window and select any relevant domain category. Top-down navigation is illustrated by the arrow on the left side of the figure. Bottom- up navigation is illustrated by the arrow on the right side of the figure. For example, a selection of issue no. 2 - "relation to local architecture" activated its related concepts. Further selection of one of the concepts, for example, concept no. 3 "planting an element from local vernacular architecture" provides a link to its related forms. The selection of one of the forms, e.g., form no. 1 "roofs" links to its related precedents. The selection of one of the precedents links to the specific precedent pages. For example, the selection of precedent no. 1 "the Bonet House" links to its precedent presentation page.

A bottom-up navigation process can be supported at any stage. By selecting a specific abstraction at the bottom right-hand side of the screen, the system links the page which describes the selected abstraction. For example, the selection of issue no. 2 - "relation to local architecture" of the concept page "planting an element from local vernacular architecture", links back to other concepts of the same issue, to provide for a selection of another concept such as - "controlled use of local elements".

5.6 COMPUTATIONAL TOOLS

In the following section we provide a preliminary review of the tools which we have employed in the pilot program as well as some comments on current technological developments and constraints.

In our current work we have employed the following Internet tools: an HTML language for textual descriptions, and the construction of HTTP references. We have employed Map-Edit for the editing of forms and image maps in order to achieve graphical indexing and interactivity. This tool employs geometrical entities such as squares, rectangles, polygons etc. for the editing of specific objects and image maps. The disadvantage of Map-Edit is that it requires a re-identification of objects. What we might expect from a more flexible referencing tool is the ability to identify a certain object or grouping of objects interactively. This probably requires a link with a VRML tool which is employed for visual presentations and graphical descriptions. We are currently considering the use of JAVA for achieving dynamic icons and animation processes. The installation of more complex process-related descriptions, such as typological schema and algorithms for both refinement of schema and adaptation of cases are still in a research stage.

470

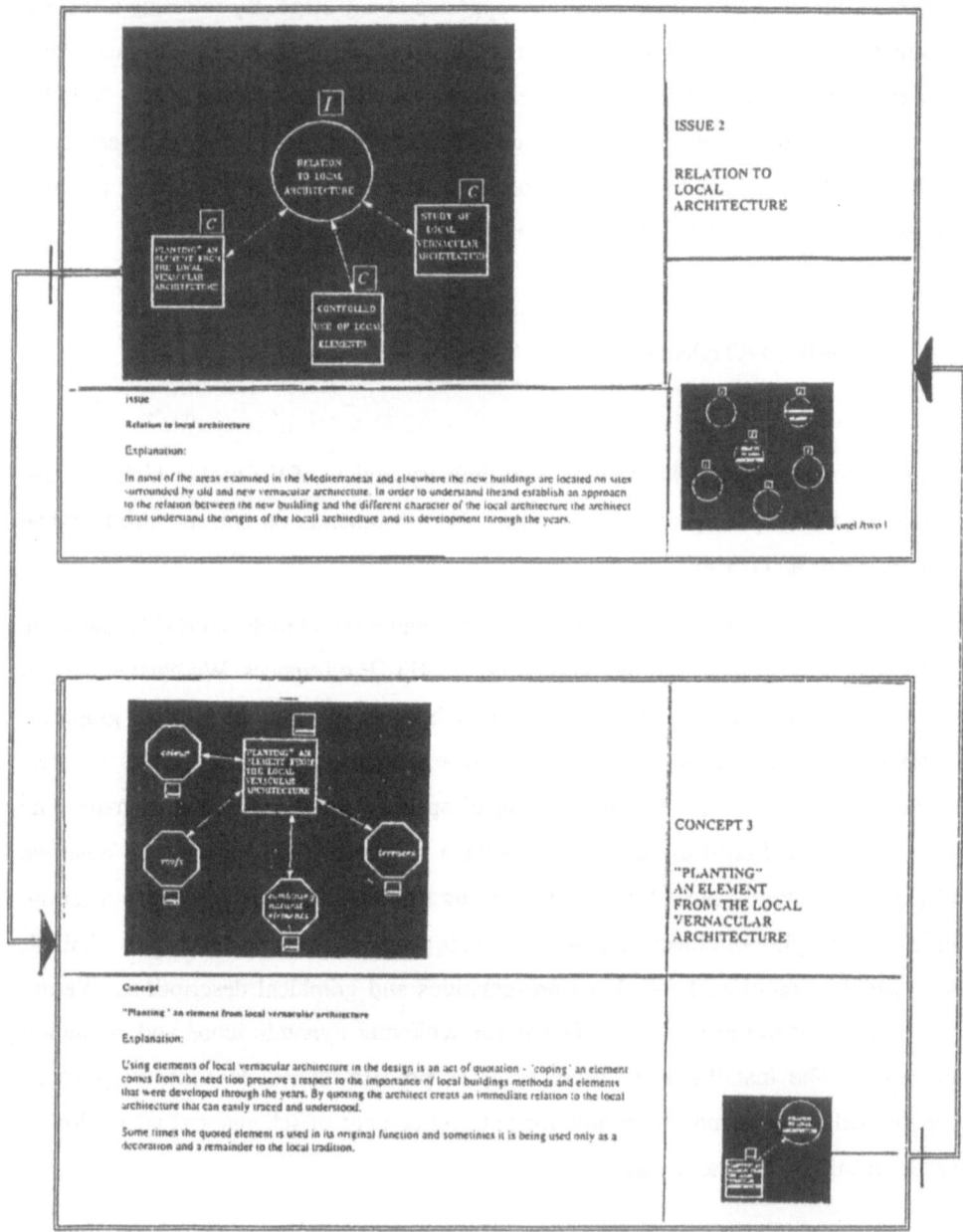

Figure 1. System operation of top-down window and bottom-up window for issues and concepts

471

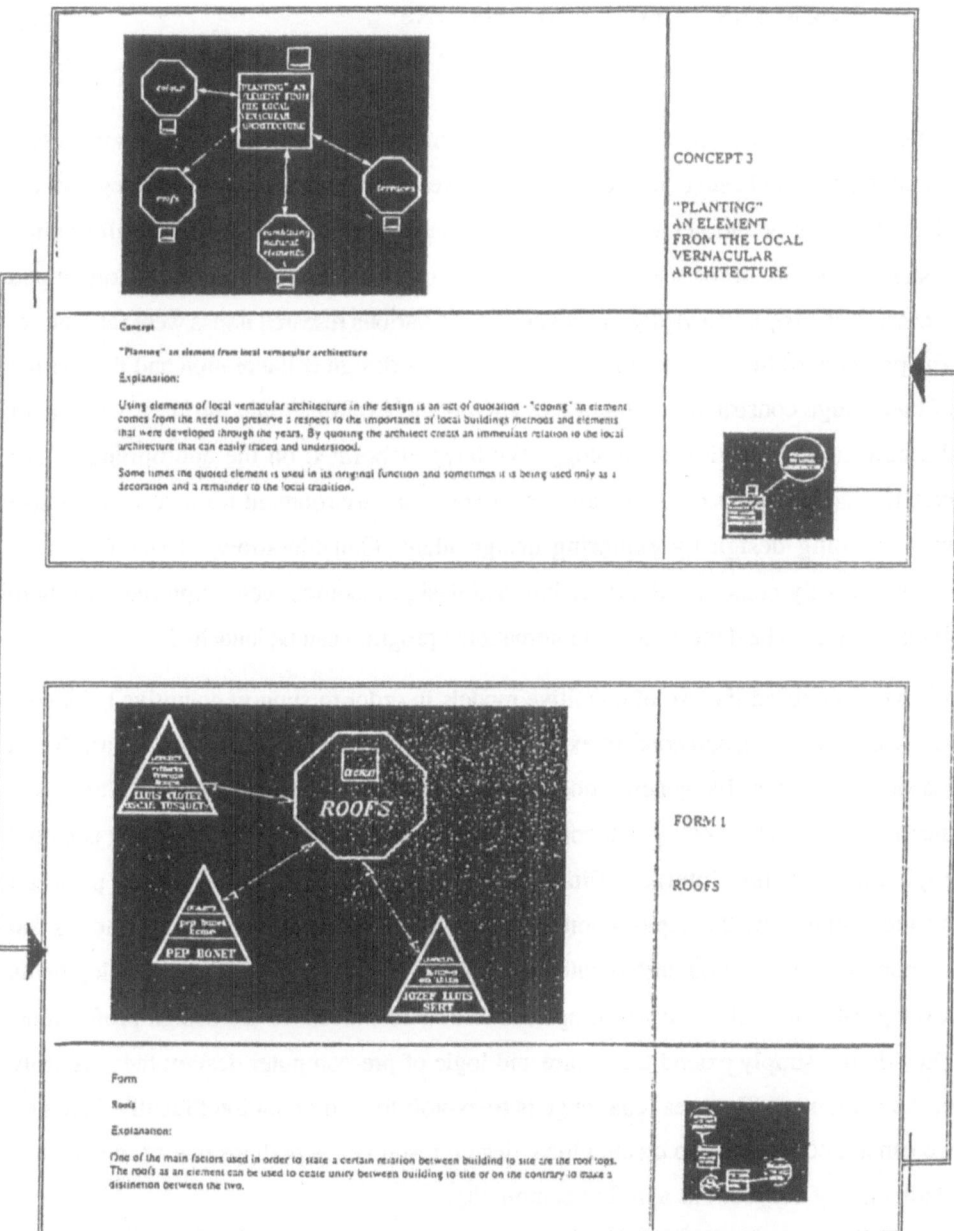

*Figure 2. System operation of top-down window
and bottom-up window for concepts and forms*

6 SUMMARY AND CONCLUSIONS

In this paper we have considered the Web as an environment which provides resources for developing and using design knowledge bases. It is an approach which integrates the characteristics of the net with certain problems and potential of design information systems. The nature of the Internet as a medium for the representation, storage and accessing of design knowledge was explored and various research issues were introduced. The potential of this new medium as a resource for design collaboration and the sharing of knowledge construction imperatives attempts to exploit certain intrinsic potential of the new communication technology. We have elaborated on the appropriateness of certain characteristics of the medium as a potential environment for a new interactive way of doing design by exploring design ideas. Considerations of the Web as a collaboratively constructed and maintained design resource were explored, but there remains work to be done before an international program can be launched.

We have proposed the use of cognitive models in order to support cognitive behaviors in search , browsing and concept expansion. Our particular approach utilizes case-based reasoning as a possible general model for design knowledge bases. We have extended the inter-relationships between theoretical issues in case-based design and their potential application in the Internet. Finally, a report was given on a pilot program demonstrating how the exploitation of the model structured around design chunks can support the construction and maintenance of shared design sites. These developments show great potential for the development of new resources for the design professions. They do not simply extend the nature and logic of pre-computer design; they are truly of the computer. The great challenge is to exploit this potential intelligently. Whether we can act collectively to create global design resources is perhaps one of the greatest challenges of our work as a design community.

Acknowledgments

This research was supported by the Technion Fund for the Promotion of Research at the Technion. This work is part of an on-going research program to explore theoretical and implementational issues of the shared construction of Web resources and the cognitive performance of design knowledge bases. As it is a long-term project, certain of the ideas have been previously partially presented, and are here up-dated by our most recent findings and developments.I wish to acknowledge the contributions of my teaching assistants Anat Sarid and Dina Guez.

Bibliography

1. Kolodner, J. , (1994), Cased Based Reasoning, Morgan Kaufmann, New York

2. Maher M. L., Balachandran, B. and Zhang, D.M. (1995) Case-Based Reasoning in Design, Lawrence Earlbaum Associates, Hilsdale, NJ

3. Mitchell J.W. (1995) City of Bits: Space, Place, and the Infobahn, The MIT Press,

4. Negroponte N. (1995) Being Digital, Alfred A. Knopf N.Y.

5. Oxman Rivka E. (1990) "Prior Knowledge in Design, A Dynamic Knowledge-Based Model of Design and Creativity" Design Studies, Butterworth-Heinemann, Vol. 11, No. 1, pp. 17-28

6. Oxman Rivka E. (1994) "Precedents in Design: a Computational Model for the Organization of Precedent Knowledge", Design Studies, Butterworth-Heinemann, Vol. 15, No. 2, pp. 141-157.

7. Oxman Rivka E. (1996) "Design Scape - Future Directions for Web/cas-Based Design", Position Paper in International Workshop on Case-Based Design Systems, AID'96 Fourth International Conference on Artificial Intelligence in Design. Workshop Proceedings, Stanford, U.S.A

8. Oxman Rivka E. (1996) "Shared Design-Web-Space in Internet-Based Design" International Conference, Ecucation for Practice, Lund, Sweden

9. Schmitt G. (1996) Architektur mit dem Computer, Vieweg

An Advanced Groupware Approach for an Integrated Planning Process in Building Construction

CHRISTIAN MÜLLER / Universität Karlsruhe (TH), ifib
cmueller@ifib.uni-karlsruhe.de

1 Abstract

Increasing complexity of today's buildings requires a high level of integration in the planning process. Common planning strategies, where individual project partners cooperate mainly to exchange results, are not suitable to jointly develop project goals and objectives. Integrated planning, a more holistic approach to deal with complex problems, is based on a high degree of communication among team members and leads to a goal oriented cooperation. This paper focuses on the application of an advanced groupware approach suitable to support efficiently an integrated design process in construction. First an appropriate planning process model will be presented, which differs from common product model approaches and takes into account the great importance of team- and goal orientation in integrated planning. Then the idea of an open CSCW-platform is proposed, which basic structure and containing elements are based on the defined planning model. Appropriate cooperative planning scenarios can then be ad-hoc modeled and configured dynamically on this CSCW platform according to the requirements of the specific project. For the participants of the planning process, the resulting groupware approach represents an integrated computer based working environment. This environment allows a kind of immersion into the project. Finally a prototypical implementation of this approach will be shortly discussed.

2 Motivation

In almost every area of planning the demand dealing and solving problems following a more holistic approach is growing, which means to take into account aspects outside the own point of view. This is especially true in the area of architectural planning, where the common planning strategies result into many problems. Some of the reasons are: Most buildings are one of a kind products, big and heterogeneous consortia, individual customers and the dynamic in goal definition and requirements. It can be expected, that many of the resulting problems are avoided by following an integrated planning approach.

To combine a growing specialization of the individual and a holistic view, the application of cooperative and team oriented method of working seems to be suitable.

R. Junge (ed.), CAAD Futures 1997, 475-480.
© 1997 *Kluwer Academic Publishers.*

So the individual competence can be used as best as possible and at the same time it can be integrated efficiently into the overall context.

Therefore our efforts towards an integrated planning process focus on establishing an higher level of cooperation and goal orientation. The application of team-oriented working methods requires changes in management and information-technology. The research area of CSCW addresses the basic theoretical principles of computer support of cooperative work. The expression *Groupware* is used to denote systems, which implement some of the concepts of CSCW. Those systems contain implicitly aspects of cooperative working methods. Therefore the application of groupware systems without appropriate planning methods gives away a big chance for improvements and could sometimes also lead to the opposite due to frustration by not coming up to the high expectations. I will cover both aspects in this paper - the planning model and the groupware approach.

3 Integrated Planning Model

3.1 BACKGROUND

As mentioned above we need a planning model, that focuses on cooperative and integrated methods. Figure 1 shows the focus of traditional and future planning models.

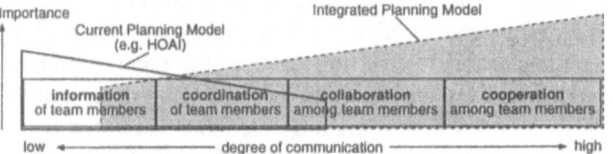

Figure 1. Focus of different planning models

Whereas today the focus is still on passive methods like being informed, the new planning model emphasizes the active role of the planning participants as cooperating team-members.

There have been several approaches to model design processes by formalizing logical and physical dependencies on the level of design objects and linking communication related aspects to them. This leads to complex deterministic product models, which are hard-coded in solution strategies. They are only valid for projects with very specific and restricted requirements. Common product models can be used as database models for particular CAD tools. However, this approach seems not to be suitable to support an integrated design process with dynamically changing requirements.

Unlike the concept of a product model, which is focusing interactions on a generic geometrical model of the planning object, the presented idea is approaching the problem by viewing and modeling goals and requirements. From this point of view a project structure matching the planning task

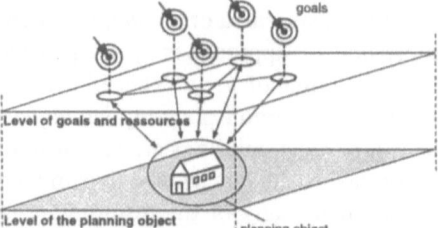

Figure 2. Two level approach of the planning model

is derived, which can be adapted dynamically by detailing the goal and task definitions. The two different approaches can be figured out by defining two levels looking onto the planning process (Figure 2).

3.1.1 Level of project goals and resources

At this level it is possible to structure a project in any meaningful kind through providing planning scenarios. After that it transforms itself to a platform allowing a continuous and dynamic structuring and organization of the cooperation process. This concerns e.g. the definition of goals, tasks, information infrastructure, tools, methods, roles, and the allocation of resources in every respect.

This level offers every participant of the planning process the context for his individual contribution, and is therefore a system for the metaplanning ("planning the planning"). This makes it possible to integrate knowledge and experience over all stages and in all tasks of the project.

3.1.2 Level of design objects

Generic problem areas such as architectural design require a high degree of freedom of choice and the ability to compromise on certain subjects in collaboration with others. Because of this, the variety of solution strategies to implement design objects can not be predetermined and is therefore only limited by the ability of team members to innovate new solution strategies. Various partial models of the planning object can be found on this level, which are different in abstraction, aspects and representation. They will be integrated according to the commitments on the other level.

3.2 THE PLANNING MODEL

The planning process model is based on the idea, that the complete planning task can be described as a net of interdependent task bundles (Figure 3.). These task bundles will be handled collectively in a team-oriented manner and will be indicated in the following as *nodes*. A node is determined by the goal or task definition in his center. The definition of the tasks is

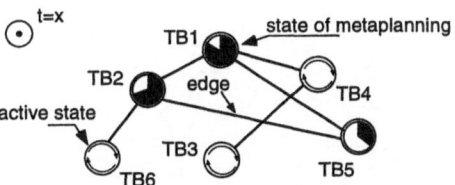

Figure 3. Project represented as a net of interdependent nodes

derived in a iterative and holistic process based on the goal definitions by the team members of a node. This is a fundamental difference to common planning procedures.

The particular task bundles can have logical and informational dependencies on each other. These dependencies are called *edges*. This means that for integrated handling of each task bundle, information and resources from others are necessary. At the beginning of a project, edges are qualitative descriptions at any level of detail of dependencies between collective tasks. Based on the so described dependencies, information flows are growing in different ways along these edges during the project. Every edge has an administrator. He must be a node team member connected by the edges. He

478

guaranties implicitly initialization, control, and coordination of information flows be-

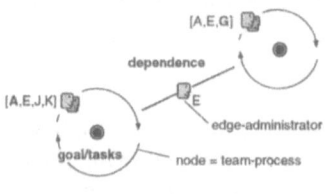

Figure 4.

tween nodes (Figure 4). This role must be filled permanently during the project. The process to obtain a meaningful system of nodes and edges will not be discussed here due to the conciseness of the paper. Beside the trivial final state, every node can have two states, which are taken on in this order at least one time: *State of metaplanning* and the *active state*. The function of node state of „metaplanning" is planning the planning process with the goal to achieve the ability of work of this node as soon as possible. This is defined as the complete definition of all aspects linked to the node for the specific planning task. These are:

- Definition of the tasks based on the goal definition
- Setting up the team: roles, competence
- Selection of the methods , tools and partial models to be used
- Setting up and configuring the information infrastructure
- Allocation of resources

After completion of the metaplanning state the node switches into the active state, where the task oriented cooperation with the goal of achieving the task linked to that node begins. If there is any change in goal or requirements, the node switch back to the metaplanning state trying to reach as soon as possible the active state again. This is going on until the goal is completely achieved (final state).

Unlike the common way of structuring the architectural design process, e.g. the German HOAI, project related phases are not taken into account for the concept of the planning process model. The structure of the planning process model is identical for all project phases and sees the project as a system of nodes, which must be dynami-

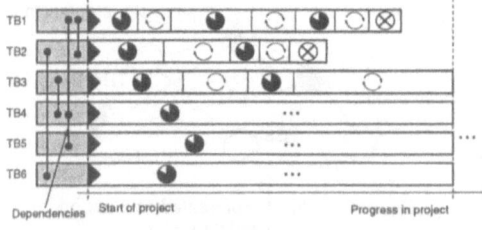

Figure 5.

cally configured and worked out (continuous switching between the two node states). Because of that and the importance of early project phases, all task bundles start jointly at the beginning. Figure 5 shows that all nodes start at the same time with their node state of metaplanning, but switch differently into the active state. It is not necessary to define constraints of time because it can be seen as a resource or results implicitly from the logical and informational dependencies. In summary, the planning model can be characterized as follows:

- Linked multi-team concept; best benefit of the human resources
- Project structure and -management follows dynamically the requirements of the planning object
- Planning model is restrictive on the level of goals and tasks
- Dependencies between task bundles are especially taken into account

4 Advanced Groupware Approach

4.1 CHARACTERISTICS

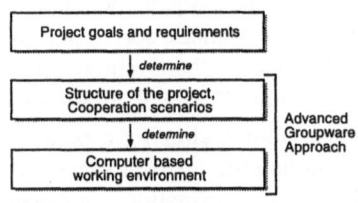

Figure 6.

Carrying out a project following the presented planning model results in high requirements regarding dynamic and flexibility, so that it can be achieved only with an appropriate computer support. The idea is to merge project structure and working environment (Figure 6). This enables a kind of immersion into the project through the working environment, which always represents the current project state.

Common concepts and methods in the area of CSCW research or characteristics of existing groupware systems will not be discussed here. The following will summarize the additional aspects of an „advanced" groupware approach, to fulfill the requirements of presented planning model.

4.1.1 Interactive modeling of the planning environment

An important aspect of the planning model is the phase of „metaplanning". That means, that a continuous planning of planning and specification of the working environment is also a project planning task. It must therefore be possible to model the net of nodes and edges as well as the cooperation scenarios within a team-process (node) while the project is running. This includes definition of tools, team members, their roles, information-management , etc.

4.1.2 Integration by the context „node"

To be aware of the context is a very important aspect to achieve an effective cooperation. In our case, a node represents a context, where all the resources are collected, that are necessary to achieve the goal linked to him. Therefore a node is the main structuring element. Everything involved is part of the resources: task definitions, team members, tools, etc. as well as the complete information, created in or flowing into the node. An imported conclusion is that the information management always follows the project structure. For example, tools needed for the tasks in the specific context e.g. can be collected in a toolbox linked to the node and offered to the team.

4.1.3 Extended awareness

Modeling, the integration by context and navigation in a project makes great demands on the graphical user interface to represent and visualize the project and the working environment. It seems to be suitable for that handling metainformation, which can be used for the representation of the objects on the GUI.

4.2   PROTOTYPE „P3"

P3 is a Client/Server based software study, based on the idea of providing meta-information and references (URLs) of a project. This meta-information can then be used to configure the particular working environments of the users. The project models are managed by a RDBMS. The *P3* client allows users to access the database server and to model virtual project scenarios. *P3* points to the basic structure of the System: projects consist of (team-)processes involving a number of persons. Figure 7 shows the client project window, with a list of processes and persons. Double clicking a process topic in the list opens a process window with icons representing the team incorporated members. Attached to every process is a specific toolbox, an information container containing URLs (ftp-protocol) and a window with roles, that can be given to the team members. The third window at the bottom shows information about the particular working environment of the user like his communication possibilities, existing tools, his photograph and URLs to WWW-based information. The project scenario can now be modeled easily by drag&drop operation of the elements.

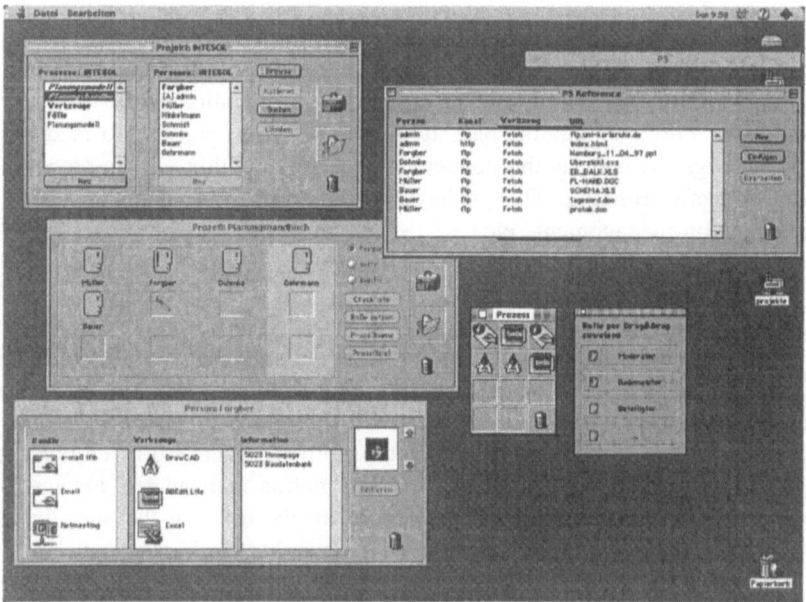

Figure 7. Screenshot of the prototype P3

5   Conclusion

The idea of a new integrated planning model and an appropriate groupware approach was shortly discussed in this paper. The presented aspects are part of various research projects and my doctoral thesis currently undertaken in at the „Institut für Industrielle Bauproduktion", Prof. N. Kohler at the University of Karlsruhe, Germany.

KNOWLEDGE, AGENCY, AND DESIGN INFORMATION SYSTEMS

JEFFREY HUANG, SPIRO N. POLLALIS
Harvard University Graduate School of Design

This paper addresses CAAD from an organizational point of view. We employ recent developments in organizational economics to model the organizational processes in building design. Based on an analysis of (i) the cost of transferring knowledge, and (ii) agency cost in existing design organizations, we propose a framework for redesigning organizational processes and for developing appropriate design information systems. The paper describes work on a larger on-going research project at the Harvard Design School on intra- and interorganizational design information systems.

Keywords: computer supported cooperative design, knowledge transfer, process modeling, organization theory, agency problems.

1. Introduction

The increasing fragmentation and growing specialization in the building industry has led to a division of design activities and responsibilities across organizational boundaries, and resulted in communication and contractual barriers among the participants. Recent developments in computing and the emergence of networking phenomena such as WAN, EDI and Internet provide a promising situation for overcoming these problems, while necessitating in research a shift from isolated desktop computing thinking to a social computing approach [Mitchell 1995]. Recent research in computer supported cooperative design (CSCD) addresses this shift. However, most of the research efforts in these areas focus on the technical aspects of cooperative design, such as standards for communication, product and information models, and user interface systems, while the underlying organizational and contractual processes have been widely neglected.

This paper addresses these organizational problems. It is based on our belief that an understanding of the organizational processes in building design will lead to a better and more effective development of appropriate design information systems. The remainder of this paper is organized as follows. Section 2 defines the problem of design organizations as a problem of collocating knowledge and decision right. In section 3 the components of organizational costs are described: the cost of knowledge transfer and

481

R. Junge (ed.), CAAD Futures 1997, 481-488.

agency cost. Finally, section 4, discusses the implications of the proposed organizational model for design information systems.

2. The Problem of Design Organizations

Designers have limited knowledge at two levels. First, there is the feasibility frontier of technology, which depends on the knowledge available at any given time [Pollalis and Bakos 1994]. Second, there are limitations specific to each individual, which March and Simon termed "bounded rationality". Accordingly, humans' sensory and mental faculties are a scarce resource with limited storage, processing, and input/output capabilities [March and Simon 1958]: knowledge possessed by any individual designer is limited to a small subset of the existing state-of-the-art knowledge known to humanity.

This limitation of the individual's mental and sensory faculties makes it impossible for the designer (in the following referred to as she) to make every detailed design decision in the building design process personally. In particular, with the growing specialization and globalization of the profession, she must increasingly rely on the specialized and local knowledge of other participants in the field, such as the engineer, the local subcontractor or the manufacturer, to make detailed design decisions. As a result she faces the following choice. Either she gathers as much information as she can from these specialists to make design decisions herself, or she delegates decision authority to the participant who possesses the appropriate knowledge.

Design situations which requires knowledge that is transferred at low cost, such as price and quantity information, calls for the first solution. In situations, however, in which specific ("sticky") knowledge is involved, such as how to solve a complex detailing problem, the second solution, delegating decision rights to the specialist, is appropriate, as the cost of transferring this type of knowledge is high [von Hippel 1990].

The problem of design organizations is thus defined as a problem of collocating decision rights with the relevant specific knowledge; and the structure of design organizations as an efficient response to the structure of its information costs [Jensen and Meckling 1991]. According to this definition, a change in information cost (e.g. through the introduction of new design information systems), must parallel a change in organizational structure.

3. Knowledge Transfer and Agency Cost

If all information could be shared at zero cost, and if there was no divergence between the goals and objectives of the design actors (as is assumed by traditional micro-economic models) then the problem of collocating decision rights and knowledge would assume no importance. However, the transfer of knowledge is costly, and design actors have divergent interests resulting in agency cost.

3.1. COST OF TRANSFERRING KNOWLEDGE.

A designer attempting to gather the requisite knowledge to make every detailed decision herself will pay the cost of gathering and moving every single piece of knowledge necessary to make the decision. Further, she is likely to commit errors due to her bounded information processing capability.

We use knowledge here to denote all knowledge required to make design decisions; i.e. not only internal knowledge held by the designer (what is typically labelled "design knowledge" in traditional CAAD research), but also, and in particular, external knowledge not possessed by the designer due to her bounded rationality. Further, by transferring knowledge we mean effectively transferring knowledge and not just passing information. The simple borrowing of a CAD programming manual, for example, is not sufficient to transfer programming knowledge to the recipient.

The cost of transferring knowledge depends, to a large extent, on the nature of knowledge transferred. Figure 1 illustrates the continuum of the different types of knowledge involved in design decision making. At one end of the spectrum is explicit knowledge which is transferred at low cost. As knowledge becomes more tacit, the cost of transferring knowledge increases.

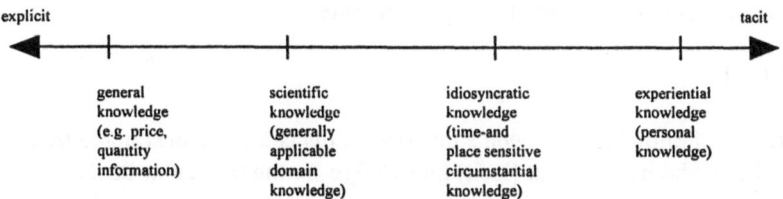

Figure 1. The spectrum of knowledge in building design ranging from explicit to tacit knowledge.

We distinguish between general, scientific, idiosyncratic, and experiential knowledge which are described in more detail below. While general knowledge denotes the most explicit type of knowledge and is transferred at low cost, the three other types of knowledge, scientific, idiosyncratic, and experiential are located in the continuum between explicit and tacit knowledge and are costly to transfer. In design, the different types of knowledge are:

(1) General Knowledge. Examples of explicit knowledge are prices and quantities -the manufacturer can easily tell the designer the price of standard building products or she can look them up herself in a product catalog. However, even this explicit knowledge is not transferred completely without cost. The input of data into product

catalogs is time-consuming and, because of the time-sensitive nature of the data, requires continuous update [Pollalis 1994]. In practice, the process of putting data into catalogs resembles often a Sisyphus task: the data changes faster than can possibly be updated.

(2) Scientific Knowledge. Scientific knowledge is costly to transfer to nonscientist or scientist in different fields because of its implicit references to rules about cause and effects known only within its particular scientific domain. For example, the technical properties of a structural steel member (the forces, inertia, and so forth) which might be crucial for deciding a facade system are not easily communicated to the architect, the owner or the electrical engineer.

(3) Idiosyncratic knowledge. Idiosyncratic knowledge includes knowledge of specific local site conditions, skills and preferences of individuals, the peculiarities of specific machines and particular design opportunities. This information is often place sensitive and perishable, and thus very difficult to transfer. For example, knowing about the availability of an attractive vacant site is only useful if it is acted upon immediately. If this information is not used immediately it may become useless.

(4) Experiential Knowledge. Experiential knowledge is acquired through personal experience. It combines intuition, judgment, common sense, and the capacity to do something without necessarily being able to explain it. The exercise of skills such as sketching, using a CAD program, and craftsmanship are examples. Experiential knowledge requires costly learning processes to transfer.

3.2. AGENCY COST

From this description of knowledge, it seems that design decisions are best made by the participant in the field who has the best (tacit) knowledge for the particular decision. This would be true if the objectives and goals of the design participants were perfectly aligned. However, in a realistic setting, assigning decision rights is not so simple. The designer confronts a problem deriving from what is commonly called agency costs [Jensen and Meckling 1976].

An agency relationship is an agreement under which one party, the principal, engages another party, the agent, to perform some service on the principal's behalf. In the context of the design organization, agents are consultants, engineers, manufacturer, and surveyors, who are engaged to exploit the economics of specialization.

Agency costs are the costs that result from the incentives agents have to take actions that increase their well-being at the expense of the principal. Agency theory argues that this occurs (1) because of the divergent (self-) interests of principal and agent, and (2) because the principal cannot perfectly and costlessly monitor the actions and information of the agent (information asymmetry).

Consider a structural engineer to whom design decision right is delegated. Let us assume the decision right involves the authority to determine the diameter of a column. In this example, the designer is the principal and the structural engineer is the empowered agent. Given the choice between two possible diameters for the column, a larger one, which the structural engineer can recommend from the top of his head, and a thinner one, which corresponds better to the original design intent (e.g. "an elegant solution"), but which would require additional calculations, he is likely to select the first, because of his natural interest to avoid risk and maximize profit. We assume here that making a "safer" choice will involve less calculations, thus increasing the profitability of the agent's service provided. Further, we assume that there are no personal or professional profits to be made by recommending a "daring" solution. In order to obtain the elegant solution, the principal (the designer) must incur additional costs which corresponds to the agency costs.

More generally, the components of agency costs are:

(1) Monitoring costs incurred by the principal to observe the agent's actions. Monitoring refers to direct observation, such as on-site supervision of subcontractors' activities, or the establishing of appropriate control mechanisms, such as shopdrawing processes, which ensures the designer that the manufacturer's details or materials correspond to her design intent.

(2) Bonding costs incurred by the agent to make his or her service more attractive. The agent might buy insurance premiums to guarantee that he will not take adverse actions or to ensure that the principal will be compensated if he does so. In today's practice the bonding capacity of agents, such as engineers, specialist engineers, manufacturers is often an essential part of their service offerings.

(3) Residual loss. Usually it does not pay to resolve the incentive conflict completely. Residual loss is the effective loss that results despite the bonding and monitoring costs incurred.

4. Implications for Design Information Systems

We have defined the problem of design organization as a problem of collocating decision rights with relevant knowledge, and have described its components in detail: the cost of transferring knowledge, arising from the "stickiness" of scientific, idiosyncratic and experiential knowledge, and the costs due to the opportunistic behavior of the agent, the agency costs, composed of monitoring costs, bonding costs and the residual loss. Acknowledging the existence of these organizational costs provides us with a framework for an understanding of design organizations which takes into consideration the limited rationality, the self-interests, and the natural conflicts of design participants. This framework implies several things for the design, development and implementation of design information systems.

Firstly, it gives us a theoretical explanation for why existing cooperative design information systems have not been taken up so readily by the building design industry: the existence of high agency cost and knowledge transferring cost undermines their effective use, despite their availability and maturity. This is consistent with results from our field studies which suggest that the main barriers for electronic design collaboration are not technical, but organizational: reluctance and "lack of incentives" of participants to share knowledge, and "impossibility to explain matters on the screen" [Huang 1995].

Secondly, it indicates that the development of design information systems is not simply a technical task to be left to systems analysts, programmers and software developers. It demands an integrative approach that takes into account the organizational costs. Design information systems should aim at either reducing the cost of knowledge transfer, the opportunistic behavior due to agency behavior, or both. In our research unit at the Harvard Design School, we have designed the following three example design information systems, which address these issues.

(i) A shared project-centered database which tracks data entries and status of documents. The benefits of this "Project Information Center" are increased overall transparency and reduced monitoring costs. The shared database further cuts bonding costs, by encouraging a shift from multiple, individual insurance coverage to a single wrap-up (project) insurance policy. This is less costly, because it eliminates redundant, overlapping coverage; further increased transparency makes risks better observable to insurance companies, who in turn are ready to lower insurance fees.

(ii) An on-line three-dimensional CAD model with integrated groupware/discussion databases. This "3D Groupware" provides a platform for concurrent schematic design and design development. Communication among engineers, architects, manufacturers, and owners are facilitated by discussion databases directly attached to three-dimensional building parts. Specific knowledge of participants are thus transferred to a shared model on a "middle ground". The tacit knowledge of these specialists are referred to by semi-structured searchable pointers. This keyword-like knowledge index points to the corresponding parts of the three-dimensional geometric model, and is referred to from relevant discussion database entries. It enables quick access to relevant specific knowledge for particular design decisions.

(iii) Finally, an "Electronic Marketplace" on the World Wide Web for buying and selling building products and materials, and for exchanging building design expertise [Bakos 1991]. Open to a prequalified "trusted" community, the electronic marketplace employs market coordination mechanisms for "automatically" collocating decision rights with the relevant knowledge/information (Figure 2).

For a more detailed description and discussion of these design information systems see [Huang 1997].

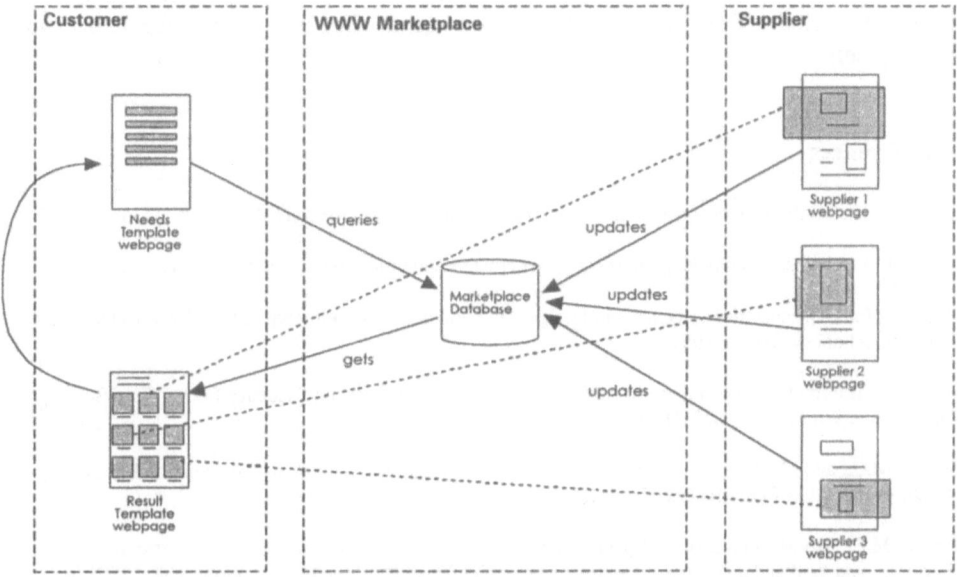

Figure 2. Architecture of the WWW electronic marketplace developed at the Harvard Design School. The efficiency of the marketplace design derives from the combination of optimized distribution of critical product/expertise data and the use of semi-formalized templates.

Thirdly, it gives us guidelines for designing future computer-supported organizations. An example application in this direction is the "process handbook" for building design we are currently developing. The process handbook is a tool and methodology for redesigning organizational processes systematically based on coordination theory [Malone et al. 1993]. We decompose the building design process into its core activities and dependencies, and analyze how these dependencies can be managed by alternative, computer supported coordination processes borrowed from other industries, such as the automobile, the aerospace, the consulting and the software design industry. Our framework of knowledge and agency costs provides us with valuable selection criteria for choosing among the alternative coordination processes and for adapting those to building design.

Finally, we believe that in future computer-supported design networks, replacing human agents with intelligent or autonomous computer agents will not reduce, but augment the

488

agency costs described in this research (at least the first generation of computer agents). These agency costs arise primarily because users are not likely to trust computer agents immediately and completely, and will thus incur necessary monitoring and bonding costs. Thus our organizational model provides us with a framework not only for present human participants, but offers guidelines for an effective management of future computer agents. We are confident that the designer will play a leadership role in the emergent networked design practice, if he or she succeeds to leverage distributed specific knowledge from human and computer agents effectively, and delegate decision rights wisely.

5. References

Bakos, Y. N., "A strategic analysis of electronic marketplaces," MIS Quarterly, 15, 3 (1991), pp. 295-310.

Hayek, F.A., The use of knowledge in society, American Economic Review, 35, September 1945, 1-18.

Huang, J. Inter-Organizational Systems in Design. Doctoral dissertation in preparation. Harvard University, Graduate School of Design, 1997.

Huang, J. "Information Technology in Building Design and Construction: Evidence from a Field Study." Unpublished Working Paper. MIT Sloan School of Management, 1995.

Jensen, M.C., and Meckling, W.H., Theory of the firm: managerial behavior, agency costs and ownership structure, Journal of Financial Economics, 3, 1976; pp. 305-360.

Jensen, M.C., and Meckling, W.H., Specific and General Knowledge, and Organizational Structure, in Werin, L., and Wijkander, H. (eds.), Contract Economics, Basil Blackwell Ltd., Oxford, U.K., 1992; pp. 251-274.

Malone, T.W., Crowston, K., Lee, J., and Pentland, B., "Tools for inventing organizations: Towards a handbook of organizational processes," Working Paper, Cambridge, MA: MIT Center for Coordination Science, 1992.

Mitchell, W.J., "CAD as a Social Process," in: Tan (ed.), Proceedings of CAAD Futures '95, September 24-26, Singapore, 1995.

Pollalis, S.N., "Construction Technology in the Electronic Architectural Studio," In Proceedings, 2nd International Conference on Design and Decision Support Systems, Vaals, The Netherlands, August 15-19, 1994.

Pollalis, S.N., and Bakos, Y.J., "Technology in the Design Process," Journal of Architecture and Planning Research, 1994.

Von Hippel, "Sticky Information, and the Locus of Problem Solving: Implications for Innovation," Management Science, 40, no.4, pp. 429-239, 1994.

DATA EXCHANGE IN DESIGN/REALISATION PROCESS IN BUILDING TRADE, AN EXPERIMENTATION WITH WOOD-FRAME PANELS

SAHNOUNI Y.[a], BIGNON J.C.[ab] and LEONARD D.[a]
CRAI (Research Center of Architecture and Engineering)
[a] *School of Architecture of Nancy, France.*
[b]*School of Architecture of Strasbourg, France.*

Abstract

Exchange of computerized data is today at the centre of interest for most of trade partners and authorities of standardization. The aim is to set up continuous and cooperative processes of exchange, without "re-modeling" or loosing information.

The paper presents our research on building modeling at design stage and its application to data exchange during the design/construction process. An experimentation about data transfer between two software is presented. The first one realizes the design process according to the developed data model (arTec[1]). The second one is a trade software for the design of wood-frame panels (Woodpecker[2]).

1. Introduction

The current AEC industry is comprised of diverse disciplines. Each using its own processes and terminology to accomplish a common goal which is the realization of a building. Over the last 15 years, use of CAD tools in the building trade evolves through different main stages. During the eighties, tools are developed to meet the particular requirements of the partners in order to increase productivity. Furthermore, data exchanges between the diverse disciplines are limited to graphical data, without any semantical description. It often leads to "re-model" information, because it is not adapted to the own terminology of the diverse users. Then, during the nineties, productivity means integrating the different processes into a single one from design to realization. The hardware and software technology improvments and the several works on standardization make possible this interoperability. We go from data exchange file format problems to a logic of data sharing, which involves semantical description of design objects.

---

[1] Architecture Technology. A data model developed at the CRAI, France, 1992.
[2] Woodpecker, developed by CastorProduct society, Gerardmer, France.

R. Junge (ed.), CAAD Futures 1997, 489-500.

1.1 BUILDING MODELING

The design / construction process can be approched through different aspects, wich reveal the great segmentation of the process and the difficulty to model it in a continuous way.

- *different partners and multiple views* : client, architect, engineer, etc. Each one performs a specific role and has one or several views of the building. They have different primary interests for the same project and use a different description of the building [1]. It is difficult to find a standard semantic description. The same building component can have various definitions when used by different trades, with their specific practices and ways of working.

- *independent participants and interpretation of information* : during the life cycle of a building, the different participants are not part of the same organization. Even they work in a cooperative way, information is contained in different sources. Building programs, drawings, cost estimates, etc, are all independently stored in different formats. Information is then converted, interpreted, and modified, which often leads to "redesign".

- *use of CAD software* : over the last 15 years, it has evolved through different stages : drawing with lines, creating elements based on lines, grouping elements into templates for repetitive drawings, connecting text attributes to templates, and connecting templates to relational data bases to retrieve needed non-graphic information. In all cases, the graphic representation was separate and independent of the non-graphic data storage and affected exchange formats definition.

Modeling the building along the life cycle needs developing conceptual models which specify data to describe work along its life cycle, where each partner identifies his specific information. These models must also support data transfers preserving the "building design objects" features.

1.2 ISO SPECIFICATION OF BUILDING CONSTRUCTION

The objective of standardization is to provide a neutral mechanism able to describe product data throughout the life cycle of a product, independently of any particular system. The nature of this description makes it suitable for neutral file exchange, and makes up a basis for implementing and sharing product databases.

ISO technical committee for industrial automation systems and integration proposes two models [2] : a product model (static), and a process model (dynamic). The product model (product object) identifies all the entities and relationships which constitute the object. It is organised around three concepts : facilitie (ex:factory), physical part (ex : wall) and space (ex:premise), described by their properties (shape, size, thermal properties, etc). The process model describes the life cycle of the building (inception, design, production, use, decommissionning). Object design evolves through states, with the instanciation of its attributes, with different values along the process [3].

The standard STEP[3] provides the ressources and methods for describing data related to the products used in the design/realization process , as a basis for CAD editors. It is organised as a series of parts : geometric and topological representation (part 42), product explicit shape representation (part 225), product life cycle (part 208), etc.

This standardization brings a new approach of the building life cycle modelling. Common ressources aim to increase interoperability between AEC sectors. These generic ressources are then specialized to the partner needs, to develop application tools. However, we think that there is a lack in the model of ISO about the state chaining in the product life cycle. It is specified as a set of states, each one of them is clearly defined. But the mechanism with which a product evolves from a state to another is not specified. Though, design activity consists of transforming the building along the process. Choose a material, calculate pemissible load of a bearing structure, etc are design actions. We can consider that it is not important in data exchange because we transfer static states of the project, but standardization is also elaborated for CAD editing, which performs design activity.

1.3 A MODEL FOR AIDED TECHNICAL DESIGN

Objective is to draw up the building technical design process according to technical definition of the building, and to the products which compose the building elements, with their "making use of" rules. The arTec model fits into the "rational anticipation stage"[4] of the building life cycle. Through levels of design, information is progressively transformed from an abstract and geometrical state to a technical defined state, by a mechanism using design operations [3].

We aim to identify generic paradigms in order to define technically the information and cover a great number of building elements (walls, floors, etc) and the most common constructive technologies (concrete, wood, steel, etc) as a shared basis usable by several partners.

The paper is organized into two main sections. In Section 2, we present the concepts that the building data model is based upon. We emphasize on integration of technical features in building description.Then, we describe design operations used to evolve through the design/construction process. In Section 3, we describe an experience of data exchange between our prototype and a trade software for wood-frame panel design,

---

3 STandard for Exchange of Product model data, ISO norm 10303.
4 The design process begins when the project reaches a sufficient definition level to be transferred by conventionnal figuration tools.

using own formats of exchange. We propose finally to use the neutral format of STEP as another way to bring semantical features of building design objects.

2. arTec building data model

2.1 CONCEPTS

We describe the technical design process of a project using :

- *design objects* (physical parts) : there are all the components of the building (walls, floors, windows, columns, beams, etc). ArTec design objects have a geometrical representation and are described by a set of semantical attributes (function, material, etc) at each level of design. We introduce relations between physical parts as virtual design objects, necessary to model the evolution of information through the building life cycle (for example : choosing a technical solution at the junction of two walls must be modelled and described) .

- *design levels* : the work evolves during the process through levels of design : volumic level (spatial figures), logical level (technical choices) and elementization level (decomposition into products, realization documents) [4]. Design objects evolve from a level to another, with progressive technical semantization, using design operations.

- *design operations* : design operations settle the transition from a state to another [5] and define semantically the design objects. These operations, which are described later, are primitives and can be combined to define complex operations.

2.2 PROCESS MODELLING WITH LEVELS OF DESIGN

We use object oriented methods to represent the model [6]. We think it is partially adapted to model the building structure. We can represent the design objects by classes, with concepts as composition (a building is composed of physical parts), sort of (a wall, a window, a beam are "sort of" physical part). However, we have more difficulty to describe relations between design objects. We must represent a relation as a class which contains references of the objects relied. Furthermore, we cannot model object evolution along the life cycle, by transforming it from a class to another, corresponding to the design levels of arTec. In order to specify the dynamic evolution of the building, we have introduced the concept of generalization/specialization by *alternative* [7]. The same entity can change its caracteristics and has differents states through it's life cycle.

2.2.1 arTec object representation

The model is organised around object classes (*Figure 1*). Each physical part is an " object class", defined by a set of geometrical and technical attributes.

We distinguish planar and linear entities according to their geometrical representation, by modelling their boundaries. The planar entities are modelled with a profile[5], the linear entities with an axis (two points). In both of them, boundaries brings technical information. We consider that several technical problems are arised and resolved at the boundaries of the design objects.

Relations between physical parts are of two kinds : chaining and positionning. Chaining links design objects which have common boundaries. It is graphically represented by connectors[6]. Positionning is a topologic relation which means for instance that two walls are parrallel or belong to the same network.

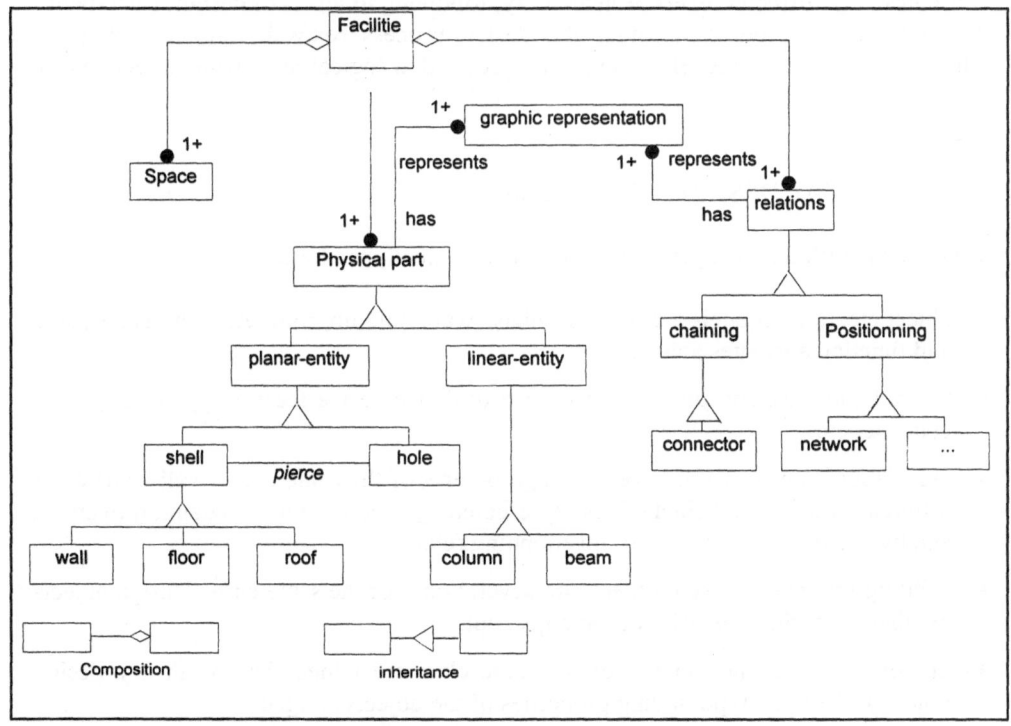

Figure 1: arTec model : design objects and relations.[7]

---

[5] A set of outlines representing the object boundaries. Each outline is a set of segments.
[6] Graphic representationof technical unspecification between objects.
[7] The diagram is designed with OMT method (Object modelling technic, J.Rumbaugh 1991)

494

2.2.2 *arTec object through levels of design*

ArTec object is alternatively at a functionnal stage (volumic), technological stage (logic) or technical description stage (elementization). Three classes are defined (*Figure* 2).

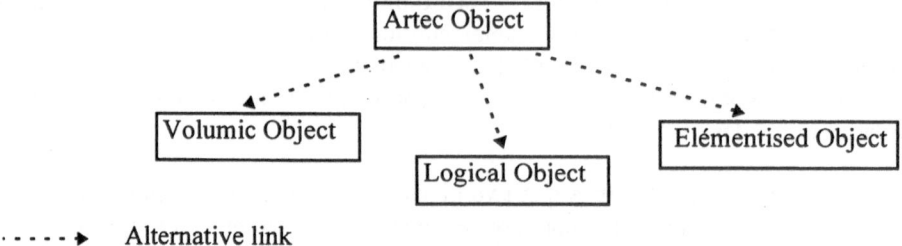

· · · · · ➤   Alternative link

Figure 2: three levels of design.

Attributes have different levels of incertainty, following to levels of design. Uncertainty decreases as we progress through the design process (ex:wall thickness is given arbitrary at the volumic level (20 cm), and precised at logical level when a technology is chosen (16cm).

2.3 TECHNICAL DESIGN OPERATIONS

We have identified basic operations to define technically the building :

- object creation : models a designed object (wall, column, roof, etc) with shape, size and function attributes attached.

- technological instanciation : affects to a design object a technology (wood, steel, concrete, etc).

- association : links two or several design objects of the same class (wall-wall) or of different classes (wall-floor). It is represented by a connector for chaining relations and by a network or an axis for positionning relations.

- splitting-up : cuts a design object into several ones of the same class. Design objects obtained save the properties of the object cut.

- merging : two design objects of the same class are joined. The result is a design object of the same type, saving properties of the objects merged.

- re-sizing : modifies design object size (thickness, section, length, etc). It consists of moving object boundaries (point, segment).

- re-positionning : modifies object design position. It can be a combination of translation and rotation.

- re-formalizing : modifies object design shape. We can transform round section of a column into a square one.

- boundary caracterization : affects to design object boundaries technical caracterization according to the technology chosen.

Design operations are not only geometric ones. They attach semantic features to objects. When a wall is split-up, the result objects have a common boundary which brings joint information.

These generic operations are specialized to each physical part (resizing a wall or a column is not the same operation). Design operations allow feed-back mechanism. The design - realization process is not linear and top-down. A technical option at the logical level can involve to change entity properties and go back to the volumic level.

2.4 DATA TRANSFER AT EACH LEVEL OF DESIGN

ArTec gives a greater place to design-construction viewpoint. It doesn't perform all the views necessary to building design like structural design, thermal design, etc. Nevertheless, it can provide information needed by applications for other views at a given state in the design process. Each level defines a coherent state of constraint satisfaction. Information about the building can be extracted to be used by another partner (ex : structural engineer must calculate the section of columns). ArTec can retrieve data to carry on the technical design.

3. Experimentation : data exchange between arTec and Woodpecker

Data transfer is carried out between the application which performs the arTec data model and Woodpecker software, which realizes the "building product composition technic"[8] of wood-frame panels. According to arTec model, Woodpecker performs the elementization level. Experimentation is realized on walls and windows, which are common to both of these applications.

3.1 SOFTWARE DESCRIPTION

3.1.1 arTec prototype

We have developped the prototype with Autolisp language on Autocad software,mainly used as a graphical interface. Walls are defined at the volumic level by a set of

---

[8] Calepinage in french

496

outlines : an outside one and a set of inside ones, representing holes. Outline is a closed set of coplanar segments, which represent wall boundaries. Thickness is the sum of two half ones on both sides of the outline. As all the arTec entities, the wall has an identifier which is the same along the process. The graphical entity generated in Autocad is represented in a 3D system coordinates relative to the wall entity.

3.1.2 Woodpecker Prototype

Woodpecker is a trade software running on Macintosh system. Walls are wood-frame panels defined in 2D system coordinates according to vertical panel plane. For each panel, origin (0,0) of the system is at the down left on the outside face of the panel. Panels are modelled with simple geometric primitives (rectangles, triangles, trapeziums) or their association. Technical caracteristic on boundaries are specified, in order to design the wood-frame solution.

3.1.3 Data exchange

ArTec data project is stored in the own format of the prototype, in ASCII mode. A program converts arTec format into Woodpecker format. This procedure avoids the panel geometrical modelling stage in Woodpecker.

3.2 AN EXAMPLE OF MODELLING PROCESS

The following project illustrates the process : modelling the volumic design objects, defining them at the logical level, transferring data to Woodpecker for panel design.

3.2.1 Modelling the project at the volumic level

The project is modelled (*Figure 3*) and default thickness (20 cm) is assigned to all the walls. Connectors are represented automatically when the system meets wall common boundaries.

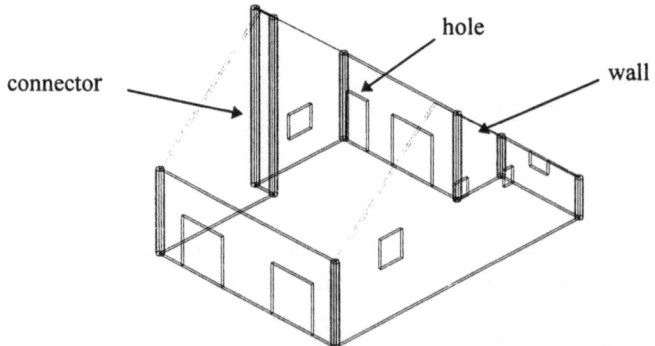

Figure 3 : Building at volumic level.

497

3.2.2 Project instanciation at logical level

The " wood-frame panels " technology is assigned to all the components of the project, which implies automatic operation of thickness resizing (16 cm) according to the technology constraints.

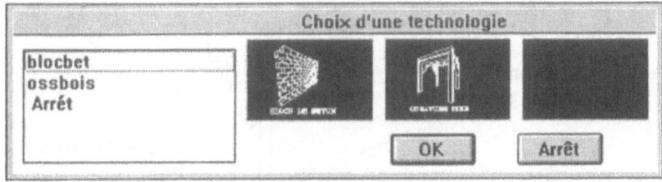

Figure 4 : technology instanciation operation.

3.2.3 Chaining resolution

All the connectors are resolved by choosing a technical solution to the junction of the walls (Figure 5). This operation implies resizing and repositionning the walls (these operations are made automatically).

Figure 5: connector resolution.

3.2.4 Boundary caracterization

Object boundaries are defined in the geometrical representation of the physical part. Technical caracterization operation consists of choosing a boundary type (*Figure 6*) and attach it as an attribute to the boundary (described as a segment in the wall) (*Figure 7*).We have added to the prototype the caracteristics used in Woodpecker to define panel boundaries.

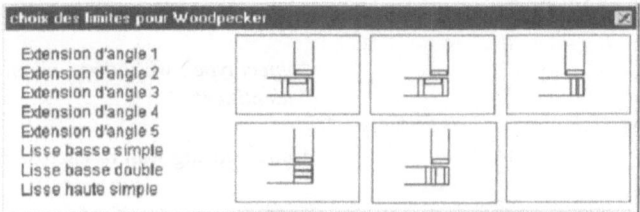

Figure 6 : choosing boundary caracteristics for the wall.

498

In wood-frame technology, choosing an angle extension type is important to settle cladding.

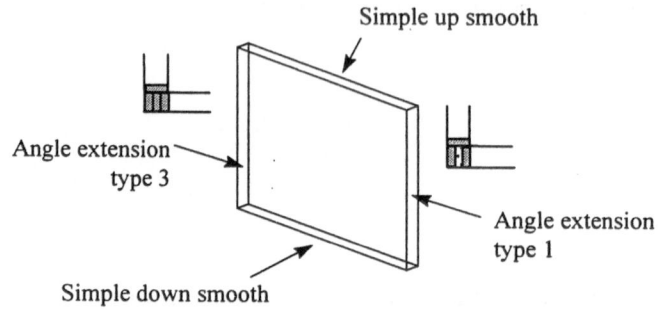

Figure 7: wall boundary caracterization operation.

3.2.5 Wall cutting for prefabrication

Wood-frame panels are prefabricated in a workshop and transported to building site to be assembled. Panels which have too large dimensions or a complex shape, are decomposed into simpler ones, in order to facilitate prefabrication and transport. This activity is performed by a combination of design operations (ex : *Figure 8*) : splitting-up (1) into two panels, window repositionning (2) and boundary joint caracterization(3).

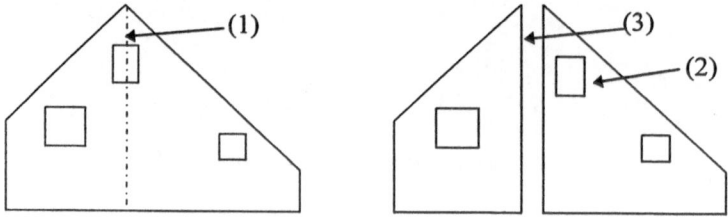

Figure 8 : an example of design operations.

3.2.6 Data transfer to Woodpecker

Data project has been saved in arTec format and translated in Woodpecker format. Object is described by its geometry and technical caracteristics of its boundaries.File format is written as follows :

```
PAV                              /*object type : wall, floor, window, ...*/
num            8                 /* identifier*/
thickness 10.0          /* in cm*/
carc           wall              /*load-bearing wall or enclosure*/
points
          num    coord           /*num : point identifier*/
          1      x y z            /*coord : wall system coordinates */
```

lims /* boundary list */

num	np1	np2	nol	carl
1	1	2	9	LBS (simple down smooth)
...				

/* num : boundary identifier; np1,np2 : identifiers of segment points; nol : identifier of chaining connector; LBS : technical caracteristic*/

holes

17 18 /* list of hole identifiesr */

tech "wood-frame-panels" /*technology attribute*/

3.2.7 Realization of the wood-frame panels

Each wall is designed separately. Technical solution is designed according to the boundary types specified in arTec. Working drawings (*Figure 9*), wood quantities (classified according to section and length) are determined. Computerized data is then used by an automatic machine for wood conversion. Panels are realized in workshop and finally assembled in building site.

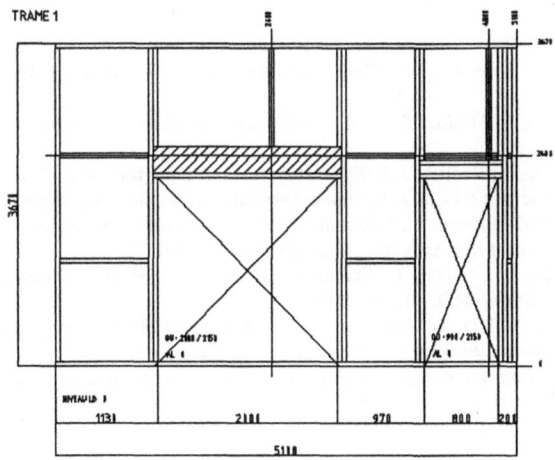

Figure 9 : working drawing for a panel.

3.2.8 Assessments

Data exchange experimentation proves that design-realization process can be continuous and cooperative. Design operations perform this evolution. However, two problems emerge. The first one concerns panel orientation for distinguishing inside and outside, which is important for cladding design. The solution proposed is to attach a *normal* vector to wall profile, oriented towards outside. The second one is the expression of designer wishes. There is a kind of information that cannot be expressed with data (ex : particular section of a wood pieces) which is not chosen at the logical level. We propose to add in the file exchange format a commentary field where the designer can express his particular requirements and wishes.

3.3 DATA EXCHANGE USING STEP

We said in the first section (section 1.2) that there is a lack in both of ISO and STEP to model transformation from a state to another in the life cycle of the building. In data exchange, we don't need to transfer the history of designed objects, describing the object and all the operations performed to reach a given state. We need to transfer static state of objects. If it is necessary to go back and modify the object, task must be performed by the related application. Data is exchanged at a given state in the building design process. So, STEP format of exchange (part 21) seems relevant, since it can bring semantical features in the design object description. It also avoids to write a format converter adapted to each exchange.

4. References

1. Van Nederveen, G.A., Tolman, F.P. (1992) *Modelling multiple views on buildings*, Automation in Construction 1, 215-224.
2. ISO Technical Report, (1993) *Classification of information in the construction industry,*ISO/TC 59/ SC 13.
3. Bignon, J.C., Leonard, D., Sedille, J. (1992) *Modelisation des transferts d'information techniques en C.A.O.* Rapport de recherche du C.R.A.I., Plan construction et architecture, report, 108p.
4. Bignon, J.C., Leonard, D., Piquee, Y., Sahnouni, Y., (1995), *ArTec : un modele du cycle de vie des objets batiment en cours de conception technique,* EuropIA 95, 437-450.
5. Boudon, Ph., Deshayes, Ph., Pousin, F., Schatz, F. (1993*) enseigner la conception architecturale, cours d'architecturologie;* Editions de la villette, 316p.
6. Masini, G., Napoli, A., Colnet, D., Léonard, D., Tombre, K. (1989)*Les langages à objet, langages de classes, langages de frames, langages d'acteurs,* InterEditions, 584p.
7. Leonard, D. (1989) *Conception et representation des objets a comportement complexe,*Inforcid congress, Nancy, France.

DESIGN AND MODELLING IN A COMPUTER INTEGRATED CONSTRUCTION PROCESS – THE BAS•CAAD PROJECT

ANDERS EKHOLM and SVERKER FRIDQVIST *CAAD, School of Architecture, Lund University, Sweden.*
E-mail: Anders.Ekholm@caad.lth.se and Sverker.Fridqvist@caad.lth.se

Abstract

A new approach to product modelling in a design context is proposed. CAD-software must not only enable product modelling, but must also support product design. This is not fully achieved in the traditional 'enumerative' approach to product modelling. We discuss how product design and modelling can be based on a 'facetted' approach to information modelling, and how a data model that supports the design process can be based on a framework for system information. The background for our research is the current development in the construction industry towards a computer integrated construction process. A first prerequisite for this is the use of computer based models. Another prerequisite is that CAD-software can support the design of the results of the construction process, including construction works, user organisations, and the production and facility management processes. A third prerequisite is that computer based models are built with standardised concepts and terminology to enable exchange of information between different actors and computer systems during different stages of the construction process. Principles for organising frameworks for user organisation and construction works information are presented in an appendix.

1 Introduction

1.1 DISPOSITION OF THE PAPER

The introduction of the paper discusses prerequisites for a computer integrated construction process; of specific interest in the design context is the shift from computer aided draughting to computer based modelling. The second part of the paper discusses principles for structuring product model information and presents principles for structuring frameworks for user organisation and construction works information. The third part of the paper presents some requirements on a data model for a product design tool.

R. Junge (ed.), CAAD Futures 1997, 501-518.
© 1997 *Kluwer Academic Publishers.*

1.2 PREREQUISITES FOR A COMPUTER-INTEGRATED CONSTRUCTION PROCESS

The current development towards a computer-integrated construction process is characterised of both an increasing use of computers and an integration of the different stages of the process (Björk 1995:12). The use of computers considerably increase the possibilities of handling the huge amounts of information that characterise the construction process. A computer-integrated construction process will enable a faster, safer and more complete transfer of information between actors and stages in the process. For example information from the brief development and the design proposal stages will more easily be made available and utilised during the production and the facility management stages.

The realisation of a computer-integrated construction process depends on a multitude of factors beside the development of computer technology. Many obstacles depend on lack of theoretical knowledge of both the building design process and the development of CAD-programs for design. Three main prerequisites for the realisation of a computer-integrated construction process are here considered as a background for the BAS•CAAD project.

A *first* prerequisite for a computer integrated construction process is the use of computer based models, so called product models, of buildings and other objects of design. Computer based models enable simulations during different stages of the design process, for example for cost- and quantity analysis, energy calculations, evacuation studies, and studies of activities during production and use of the building (Eastman 1991).

A *second* prerequisite for a computer integrated construction process is that CAD-software can be used not only as a product representation tool but also as a product design and modelling tool. This would enable the use of CAD already in the earliest stages of the design process. CAD-software must enable design and modelling of all the systems and processes that are formed, not only the building, but also the building user organisation, the production process, and the facility management process (Ekholm et al 1994).

A *third* prerequisite for carrying out a computer integrated construction process is the establishment of standardised and scientifically well-founded principles for structuring information in product models. During the design process a multitude of models are developed and different actors and computer systems must be able to interact and exchange information. In order to enable communication among actors and computer systems and to ensure conceptual consistency and compatibility among models, it is necessary that the models are built as well with scientifically well-founded concepts as by use of common classification systems and technical standards (ISO 1994a).

1.3 CAD – FROM COMPUTER AIDED DRAUGHTING TO COMPUTER AIDED MODELLING

An essential factor in realising a computer integrated construction process is a design process that develops and utilises computer based models, so called product models. CAD-software is mainly developed for a traditional design process where it is used as a draughting tool for producing production drawings in the later stages of the process. Currently research and technical development within CAD-software are directed towards modelling tools, where data in the computer can be structured to build a computer based model of an object (Galle 1995).

The introduction of building product models represent a revolution to information handling in the construction process. Information can be exchanged between computer systems and different actors without the need for human interpretation, thus accelerating the speed of information transfer and eliminating sources of human misunderstanding. Among the major research questions that have emerged, and which also have been given the most attention, are those concerning the structure of building product models and the transmission of building product data between different actors and computer systems throughout the construction process, e.g. GARM (Gielingh 1988), RATAS (Björk 1989), STEP (ISO 1994a), and COMBINE (Augenbroe 1995).

This research has concentrated on questions regarding buildings and other construction works, while additional systems and processes, that also result from the construction process, so far have been given less attention. Among such results are the building user organisation and the building management process. That also these results can be represented as computer based models is essential in order to achieve the overreaching goal of a computer integrated construction process.

The earliest stages of the construction process, the brief and the design proposal stages, deal not only with the building but also with the user organisation as an object for design. Every construction project has a brief-formulation stage which includes a description of the user organisation and its requirements on the building. The design proposal stage includes spatial layout and co-ordination of the user organisation and the building. CAD-software that could represent the user organisation would be useful not only in the brief-stage but also to facility management.

Today's practice overlooks the fact that architectural drawings provide very little representation of the building's occupants or the environmental context: "One complete half of the 'ensemble' is almost completely missing" (Steadman 1979:184). Neither commercially available CAD-software for building design, nor software for product modelling developed in a R&D context provide explicit representation of the user organisation and consequently have limited use at the earliest stages of the design process.

Another essential issue which is not yet really approached in the building research community is the question of product design. Until now, work within building product modelling has presupposed that the essential properties of the building are determined before constructing the product model. The development of software to be used as an environment for both product design and modelling is still in its infancy and represents a challenge both to construction research and commercial software development (Junge 1995).

1.4 THE BAS•CAAD PROJECT

This paper gives an overview of some of the results of the BAS•CAAD research project at the division of Computer Aided Architectural Design, School of Architecture, Lund University. BAS•CAAD is an acronym for Building and User Activity Systems Modelling for Computer Aided Architectural Design. The BAS•CAAD project has the overall aim to contribute to the development of tools for computer aided design in the early stages of the construction process including the brief development stage.

Today there is a lack of knowledge about how CAD-software can support product design and modelling. The BAS•CAAD project addresses this problem by dealing on the one hand with principles for structuring product information, and on the other hand with requirements for a computer based information system for product design and modelling.

In the project we are developing a basic data model, a generic framework, from which different domain specific frameworks and product models can be built. This will ensure conceptual consistency and enable exchange of information between different models. The data model is based on a philosophically and scientifically well-founded property theory developed by Mario Bunge (1977 and 1979).

A prototype CAD-software that enables product design and modelling is currently being developed within the project. This prototype shall also be developed to enable spatial design and co-ordination of a building and a user organisation. The approach concerns questions of architecture and building science including questions of IT.

The results from the BAS•CAAD project are expected to be applied in different ways. The systemic principles for the data model may be of general interest for conceptual modelling in information systems development. The CAD-software prototype for product and process modelling may be further developed in applications, e.g. for building design and architectural programming. The theoretical foundation in the project may be of interest in other contexts like classification and standardisation.

2 Principles for structuring product model information

2.1 FRAMEWORKS, THEORIES AND MODELS

A representation of a thing, that resembles the represented thing in some way, is in everyday language called a model. A concrete model is a thing that physically resembles the modelled thing, while a conceptual model is a mental representation of a thing, made up of concepts that represent the real thing. A computer based model is a concrete model built into a computer, generally with the help of modelling software. A concrete model, be it in a computer or in clay, is a representation of a mental model, it is not a direct "footprint" or a mould of reality. We do not model things directly "as they are" but as we "see" them in everyday praxis, expressed in a scientific theory, or in an artists vision. The consequences of this for computer software for design will be discussed further on.

In the context of computer based information handling a "conceptual model" is also called an "information model" (Schenck and Wilson 1994:10). Referring to computer based models, e.g. product models, we normally mean the information model and not a concrete representation in the computer.

It is of interest in this context to distinguish between frameworks, theories and models. A framework, or context, loosely defines the key concepts of a domain (Bunge 1983:323). A theory is more precisely defined, it is a logically organised context (ibid:331). A model concerns a specific member or group of members of a domain and is based on either a framework or a theory. If the model should represent the dynamical behaviour of the represented thing it must be based on a theory and not a framework. The fact that the term 'model' is often used both for models, frameworks and theories gives rise to an ambiguity in terminology. For a discussion see (Björk 1995:8) and (Ekholm 1996:2).

A computer based information system consists of a conceptual schema, an information base and an information processor (ISO 1985). The *conceptual schema* is a generic conceptual representation of the part of the real or formal world we are interested to handle information about. This part of the world is also called the *universe of discourse*, UoD. The conceptual schema is either a framework or a theory. If the modelling purpose is to simulate the behaviour of a product, then the specific relations among entities are of interest and the schema must be a theory. On the other hand, if the modelling purpose is only to transfer information about entities, then it is sufficient that the conceptual schema is a framework. The *information base* in the information system describes the state, i.e. the values of the attributes, of the UoD at a certain time. The *information processor* is a software tool that makes it possible to query and update the conceptual schema and the information base. A specific information model consists of a conceptual schema together with the attribute values in the information base.

In a traditional information system a *computer based model,* e.g. a product model, is defined on the basis of a predefined conceptual schema. The information processor allows the user to assign values to the attributes in the schema. An information system for product design is different in that the information processor also enables design operations on the conceptual schema in order to develop a conceptual model for a new and beforehand unknown product.

2.2 THE NEED FOR STANDARDISED AND SCIENTIFICALLY WELL-FOUNDED FRAMEWORKS

A product information model or, for short, *product model* is defined as "an information model which provides an abstract definition of facts, concepts and instructions about a product" (ISO 1994a). In the context of computer based information transfer it is necessary that product models are based on scientifically wellfounded standardised frameworks in order to fulfil the requirements for unambigousness and exchangeability. Of interest to this study are three main abstraction levels of frameworks:

1. The ontological level, concerning very generic properties like object, property, thing, system, space and time common to every description of reality.
2. The domain level, concerning properties common to a certain technology or trade, e.g. construction or shipbuilding.
3. The application level, concerning domain specific artefacts, e.g. buildings, masts or dams.

That these levels are relevant for standardisation of frameworks is confirmed by the work within the STEP Project (Standard for Exchange of Product Model Data). STEP is an international standardisation activity within the International Standardization Organization, ISO; the objective of STEP is to enable "unambiguous representation and exchange of computer-interpretable product data throughout the life of a product" (ISO 1994a). For example both the formal language EXPRESS, which is used for developing conceptual schemas, and the so called "Integrated resources" in STEP, that are common to all or most product information models, belong to the ontological level. Models for different kinds of domain specific artefacts like high-rise buildings, steel structures or curtain walls, belong to the "Application protocol" level (ibid). The domain level is not formally distinguished in STEP, but the Building Construction Core Model, BCCM, within the AEC-domain is a typical example of a framework at the domain level (ISO 1996).

Standardisation has a long standing international tradition in the construction context, for example through the international applications of the Swedish SfB-system[1]. Currently a framework for classification within the construction industry is under development as an ISO International Standard based on the Technical Report TR 14177 (ISO 1994b). The purpose of this standard is to be a basis for the development of national and international classification systems.

Among the objectives of the BAS•CAAD project are to develop principles for structuring frameworks as a basis for standardisation of information on the ontological level, especially the data model for the design tool, and on the domain level for buildings and user organisations.

2.3 AN ONTOLOGICAL FRAMEWORK FOR DESIGN

In order that frameworks and standards shall be mutually compatible and stable it is necessary that they have a sound scientific basis. In this study such a foundation is provided by a scientific[2] ontology developed by Bunge (1977 and 1979). A short presentation of the concepts of thing, property and system in Bunge's ontology follows here. A slightly more extended presentation is given in Ekholm (1996).

A *thing* is a concrete object, among its most basic properties is its existence (Bunge 1977:160). In order to exist, the thing must interact with its environment and therefore a thing has environment and structure (Bunge 1979:6), see Figure 1.

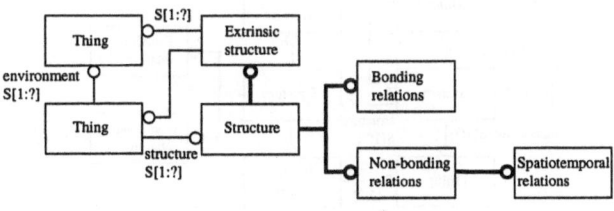

Figure 1. Thing[3]

---

1 SfB is an acronym for "Samarbetskommittén för Byggnadsfrågor", a Swedish organisation that developed the SfB-system for building classification (ByggAMA 1950).

2 In his "Treatise on Basic Philosophy", Mario Bunge discusses different views on what ontology is about. His own view is that the ontologist should "stake out the main traits of the real world as known through science, and that he should proceed in a clear and systematic way" (Bunge 1977:5).

3 The framework diagrams in this paper are developed in EXPRESS-G, the graphical counterpart of EXPRESS, a formal language which was developed in its present form to be used within STEP (Schenck and Wilson 1994). In the EXPRESS-G schemas the circle shows the direction of a relation towards the entity pointed at by the circle/arrowhead. A broader line indicates a subtype relation. The cardinality of a relation is indicated by the figures within the square brackets and expresses the number of entities that occur in a relation (ibid:316). The syllable letter S expresses that the entity at the arrowhead is a set in which the order of the instances is unimportant and that each instance may only be related once (ibid:145).

The environment is the set of things with which the thing interacts. A thing does not necessarily have parts, but it has an external structure which is the set of relations between the thing and its environment. A relation is a mutual property of the related things. For example the property to leave a graphite trace is a mutual property of a lead pencil and a paper. A mutual property is often described as an attribute of each of the related things; the term used for this attribute in information modelling is "role", e.g. the role of the pencil is "marker", and the role of the paper is"marking surface". Relations may be bonding and non-bonding; bonding relations affect the state of the related things, while non-bonding relations do not. Relations to reference frames like spatial and temporal relations are non-bonding. The state of a thing is the values of its properties at a specific time.

A *system* is a complex thing with composition, environment and structure, both extrinsic and intrinsic, see Figure 2. In order to be a system it must have integrating, bonding, relations among its parts otherwise it is just an *aggregate* (Bunge 1979:6). The structure is the set of all the systems relations, both intrinsic and extrinsic. Intrinsic relations hold among the parts of a system; they are mutual properties of the parts. Intrinsic properties are basic to external properties, e.g. the pencil mechanism must hold and feed the lead in order that the pencil shall fulfil the role as a marker.

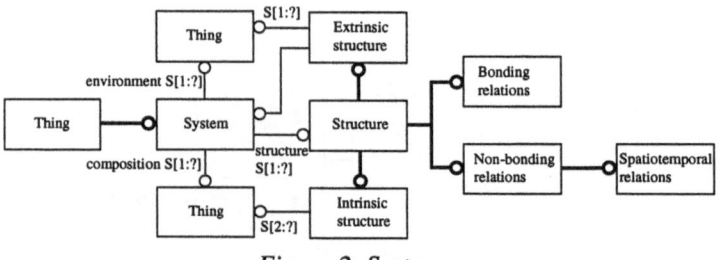

Figure 2. System

3 Frameworks for user organisation, construction works and construction space information

In order to accomplish a computer integrated construction process, principles for standardised computer based representations must be developed. In this section is shown how such principles can be based on the systems framework. The schemas in this section also illustrate how the basic representation of a system with composition, environment, and structure can be extended to represent domain specific systems by attaching domain specific system objects and property objects. In a CAD-program these domain specific objects would be found in a property library.

3.1 A FRAMEWORK FOR USER ORGANISATION INFORMATION

An organisation is normally described and organised in a planning process for administrative, business or production activities. The purpose of such a description is to understand the functioning of the organisation. If the organisation needs a building for its activities, the organisation's relations to the building are described in a building brief. The brief design stage starts with a description of the user organisation concentrating on those aspects that are of importance in relation to the use of buildings. Together with other information, e.g. requirements emerging from town planning, this description is used as a basis for defining the requirements on the building.

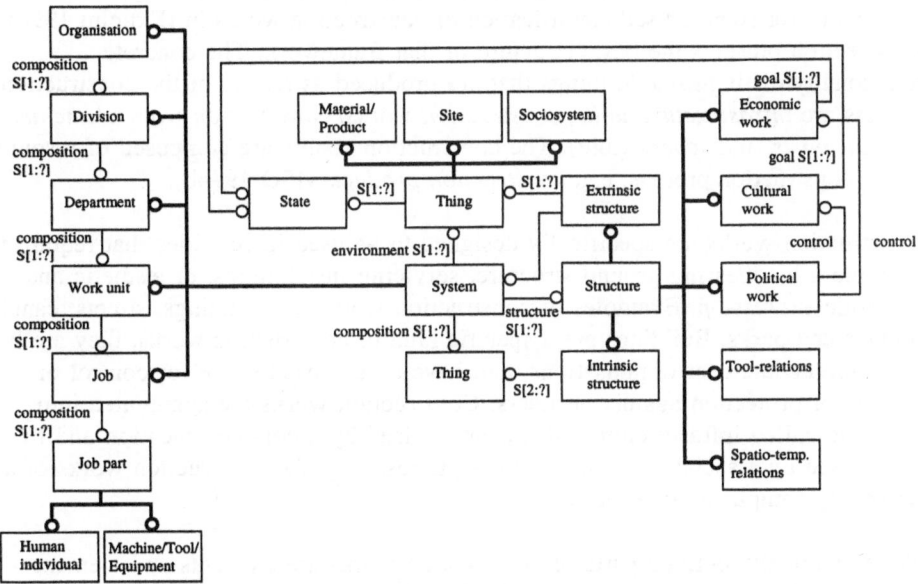

Figure 3. A framework for organisation information

In Ekholm and Fridqvist (1996) we define an *organisation,* see Figure 3, as a system with:

1. A *composition* of organisation units. The composition has a level structure with a lowest level of jobs. A job is composed of human individuals, and/or machines, tools or other equipment. Jobs combine into higher levels like work units, departments and divisions (Child 1984:85). Among the subsystems are an economic or cultural system, and a control system that manages the former (Bunge 1979:189).
2. An *environment* of things with which the system interacts and which is partly transformed by its work. To the environment belong the material resources transformed by work and the resulting products. Also sociosystems towards

which the organisation directs its cultural work belong to the environment. Finally the organisation has a physical site, for example a building or the natural environment.

3. A *structure* including an intrinsic social network based on work, with tool-relations to the artefacts in the system. Some members of the organisation manage others with respect to their work in the organisation. To the extrinsic structure belong the economic and cultural work as well as relations to the site. To the structure of an organisation belong its spatiotemporal properties.

3.2 A FRAMEWORK FOR CONSTRUCTION WORKS INFORMATION

The framework for system information has been applied for development of a framework for standardised classification of construction works in (Ekholm 1996). This section presents the basic structure of that framework. The concrete functionally distinguishable things that are produced as results of the construction process are *infrastructure units, construction works, construction work elements, element parts*, and *spaces* (ibid). The construction results are composed of resources of the construction process, e.g. *construction products* (ISO 1994b).

Construction works are specifically designed to be used in activities that require for example a loadbearing ground structure, servicing installations, or æsthetic and symbolic expression. Examples of construction works are buildings, streets, canals, bridges and parks. Buildings are a specific kind of construction works, they are built to accommodate user organisations, and have an enclosed space for control of climate or protection against intruders. Construction works are aggregated into larger so called infrastructure units, characterised by a common location and by being used by a social organisation for a purpose, e.g. the construction works of a university campus or an airport.

The functionally defined parts of construction works are elements and element parts. An element like a gypsum wall with a space enclosing function may be composed of element parts like loadbearing scantlings and enclosing gypsumboards. An element part does not have the complete function required to be classified as element. An element part is composed of one or many assembled and transformed construction products, and constitute the lowest level of construction artefacts, they do not have the complete function required to be classified as elements (Ekholm 1996).

The internal bonding relations between the parts of a building can be caused by gravitation or by fixture devices. Among the external relations are the functions to the users and the transfor-mation-relations to the site, both of which are bonding relations. Among the non-bonding relations are the spatial relations, and indirectly, the interpretation-relations to those who experience the building as a system and a

511

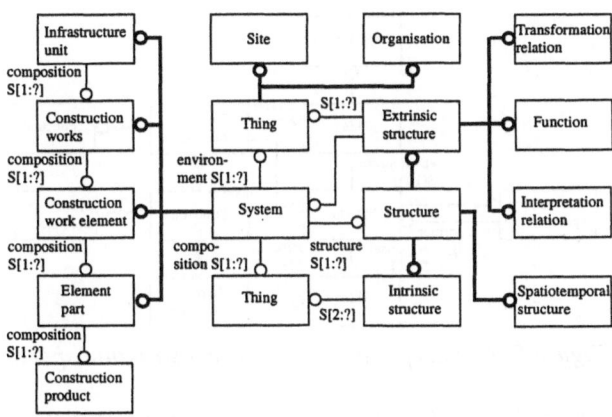

Figure 4. A framework for construction works information

sign, in order to appreciate its architecture and history. These concepts are related in a framework for construction works information, see Figure 4.

3.2.1 A framework for construction space information
In the construction context, space is treated as a factual material thing. For example, a space function program developed in architectural programming contains requirements on the buildings spaces, e.g. for surface materials, fire resistance and sound reduction levels. The properties of spaces in buildings are designed for occupancy of users, machinery and equipment. Spaces are classified by their basic function in relation to the users and other agents, for example office and communication spaces, and climate- and fire-zones (ISO 1994b).

In order to develop a framework for space in the construction context, a generic definition of space is necessary. In Bunge's Ontology, a spatial relation is a mutual property of things; spatial relations are non-bonding relations. Using the convention of naming things after their properties, the term *space* denotes a collection of spatially related things (Bunge 1977). The referent of the concept *space* are things, and the concept represents their spatial properties. This concept of space is practical in construction, but contrasts to our perception of space as an intangible object enclosed by surfaces of things.

According to this definition any collection of things may be regarded as a space; both a constellation of stars, an alley of trees and a room of building elements are spaces. In the construction context spaces are constituted by construction works and their parts. Of specific interest to the users are spaces that can be used for occupancy and that have enclosing functions. Ekholm (1996) has proposed the following definition of space in the construction context: "A space in the construction context is an aggregate of construction works, their parts and other

512

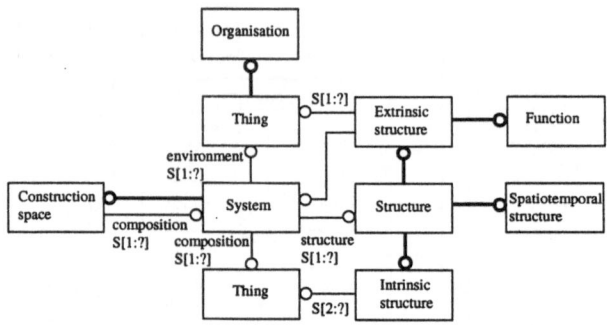

Figure 5. A conceptual schema for construction spaces

things defined only with regard to their materially or experientially enclosing properties". See Figure 5.

The user organisation that uses the building's spaces belongs to the environment of the spaces. The question of which parts belong to the construction spaces and which belong to the organisation depends on the context. For example, in one case a bookshelf may be considered a building element belonging to a construction space, while in another case the bookshelf is considered part of the organisation.

Spatial layout design deals with the spatial co-ordination of spaces and the user organisation located within the construction spaces.

4 A facetted approach to design and modelling

4.1 ANALYSIS AND SYNTHESIS TOOLS

When discussing problems in technical and scientific research, Bunge recognises two main types of problems, analysis and synthesis: "The analysis problem is: Given a system (i.e. knowing or assuming its composition, environment, and structure), find its behaviour. The inverse problem is that of synthesis: Given a behaviour, find or design a system that realises it" (Bunge 1983:274).

This fundamental difference between analysis and synthesis is reflected in the design tools that can be used to assist these processes. In (Fridqvist and Ekholm 1996) we suggest a division of design tools into two groups: analysis tools and synthesis tools. *Analysis tools* are used for evaluating designs, i.e. problem solutions obtained through synthesis. Examples of analysis tools are software for cost estimation, stress analysis and energy balance calculation. Typical for analysis is that once the system is known, the task of finding solutions to problems regarding the system will follow given rules. Thus analysis tools can be highly automated, provided the necessary data is at hand in a suitable form.

Bunge's definition of synthesis has as a consequence that properties must be possible to define and describe separately from the things that have them. A *synthesis tool* is used for creating designs. In (Fridqvist and Ekholm 1996) we propose that a synthesis tool has three characteristics:

1. It assists the designer initially to describe the desired behaviour or functions of the thing to be designed.
2. It supports composing and describing things that have these functions.
3. It presents data in ways that are suitable both for analysis tools and for the realisation of the design.

A consequence of the first characteristic is that a synthesis tool not only must be able to represent the domain objects, but also the designer's way of thinking about the objects.

4.2 DETERMINATION OF PROPERTIES IN THE PROCESSES OF CLASSIFICATION AND DESIGN

In a design process the main objective is to find a solution consisting of parts that together satisfies the requirements on the resulting whole. During the process the understanding of the requirements on the whole as well as the necessary properties of the parts develops gradually and the conceptual model gets increasingly more specific (Lawson 1990:91). However, the process does not unfold simply by subclassing from generic to specific, instead it is a search process characterised by addition and removal of attributes from the conceptual representation . For example in building design an internal wall may initially have the function loadbearing, but later in the process the loadbearing properties may be removed and finally the wall properties may be reduced to a separation function held by a textile curtain.

A starting point for the BAS•CAAD project is that a design tool must enable a free combination of properties. How such a system would be organised may be conjectured by making an analogy to principles for classification applied in the context of information retrieval. Two different approaches to classification can be distinguished, the enumerative and the faceted (Hunter 1988).

The principle behind an enumerative classification system is that all relevant classes, simple and compound, are listed. An object is classified, according to its properties, in one of the available classes in the classification system. A problem with this kind of classification is that objects that have new kinds or new combinations of properties, which are not represented in the classification, cannot be classified properly. In a design system based on similar principles as the enumerative classification, the designer would have to choose among objects with a predetermined selection of properties. The advantage of this is that the design system can be optimised for handling well known combinations of properties, as for example concrete slab or steel truss. The drawback is that only things corresponding to the predetermined classes can be modelled.

The principle behind a facetted classification system is that a thing is classified by a combination of properties collected from different facets. A facet is a set of properties of similar kind like characteristic function, material, shape etc. To design according to this principle would mean that the designer determines the properties of an object by combining properties from different facets. For example the object may be determined to have a certain 3D-shape listed in a shape facet. The shape can be further specified through the interface of a geometric modeller. Among other properties it would be possible to decide whether the object is composed or homogeneous, or to determine its material, etc.

By enabling a "free" combination of properties the designer would be able both to start on a high abstraction level and to create new kinds of parts. In a practical design system a combination of these two principles would be favourable. Sometimes it would be necessary to build a combination of properties and at other occasions it would be practical to choose a part with a predefined compound set of properties from a library.

4.3 CLASS-CENTERED AND OBJECT-CENTERED APPROACHES TO DATA MODELLING

The enumerative and the facetted approaches to product classification and design are seemingly equivalent to the class-centered and object-centered approaches to data modelling, respectively, discussed by Garrett and Hakim (1994). In the following, information modelling refers to the development of conceptual representations of the UoD, while data modelling refers to the implementation of the conceptual model into objects in the database, see Björk (1995:26) for a discussion of this distinction.

In a *class-centered* approach an information model is implemented as computer based classes. Commonly a representation of a specific domain object is created as an instance of one such class. For example, the concept of physical parts of the building with a loadbearing wall function is implemented as the class "loadbearing wall". A representation of a specific loadbearing wall is created as an instance of the class.

A problem arises when the class-centered approach is used for implementation of a modelling tool for design. Since an instantiated object inherits all the characteristics (attributes and methods) of the class, addition of a new attribute may not be consistent with the class definition and may require a transfer of the instance to another class. For example a design change from loadbearing wall to non-loadbearing wall would require a reclassification of the instantiated wall-object. Prob-

515

lems of pre-determined classification of product model objects has been discussed
by Junge (1995), Garrett and Hakim (1994) and in papers concerning different
software systems developed for general representations of design data like EDM-2
(Eastman et al 1995), and Feature based modelling (van Leeuwen et al 1993).

The class-centered approach does not comply with the requirement that a synthesis
tool must allow a free specification of properties, separate from the things that have
them. For example to be able to represent the design of a brick wall and its change
from loadbearing to non-loadbearing, an object would at a certain time be associated to attributes like "enclosing", "vertical plate", "loadbearing", and "brick", and
at a later stage have "loadbearing" removed. In analogy with the facetted approach
to classification this may be called a facetted approach to design.

According to Zhao and Biliris, such a functionality would be achieved by an object-
centered approach to data modelling (Zhao and Biliris 1993). Furthermore, object-
oriented data models can be grouped into two main categories: class-based and
prototype-based. Class-based models focus on classes while prototype-based models
focus on objects. The *object-centered* approach proposed by Zhao and Biliris
combines these two methods and "achieves dynamic object evolutions, flexible
object classification, and strong typing".

4.4 THE BAS•CAAD DATA MODEL

The object-centered approach is developed within the context of computer science.
In the BAS•CAAD project we have chosen to build a data model that enables a
facetted approach to design, on the basis of Bunge's property theory. The reason for
this is our hypothesis that the data model of the design tool and our approach to
information modelling must be mutually consistent.

In order to build a data model that can be used for product design, we need to
implement a view of the world that conceptually separates the individual from its
properties. Such a view is needed in both science and design, for example in
science the question is "what are the properties of this thing?", and in design the
question is "what properties must this factually possible thing be given?". This
separation makes it possible to question the properties of a thing without
questioning its existence.

According to modern science, the world consists of things characterised by their
properties. In this view there are no "bare" things without properties and there are
no "free" properties. In spite of this, it is in both science and design convenient to
assume the existence of bare individuals that can be equipped with different kinds
of properties. Such an approach is supported in Bunge's ontology, where a thing is
understood as a substantial individual with substantial properties (Bunge 1977:162).

This reflects the way a designer thinks during the design process. Although the properties of a designed object change, the designer may still treat the transformed object as having the same identity through the design process. An example is the situation when a wood-carver works on a piece of wood to make a bowl; the original piece of wood gradually transforms into a bowl, but to the wood-carver it keeps its identity through the transformation.

According to this view it is possible to at the same time maintain the identity of a thing and be uncertain of its properties. In the BAS•CAAD data model this is achieved by introducing the class of individuals possessing the "existence" property. The members of the "individual object" class are "bare" individuals without other properties than a feigned existence; the members of "property" classes are attributes, i.e. concepts representing real properties of things.

A short presentation of the basic objects in the BAS•CAAD data model is made here as a summary from Fridqvist and Ekholm (1996). The data model distinguishes between three main object classes: System object, Property object and State value object.

System objects alone represent only the identity and existence of a concrete thing. System objects have the attributes *composition, structure* and *environment*. The structure is the set of all the systems relations, both internal and external. The composition and internal structure are basic to the intrinsic properties of the system, while the system, its environment and the external structure are basic to the extrinsic properties of the system. The properties of a system are modelled by associating a Property object with a System object.

Property objects represent properties of concrete systems. A Property object is instantiated only once, to ensure compatibility between different modelled objects. Consider, for instance, the property 'mass': In the real world there is only one kind of mass. This should be reflected in a model, so that there is only one instance of property object that represents 'mass'. A Property object always has a state function with a state space. The *state space* represents the possible states of the associated systems with respect to the property.

State value objects represent the actual value of a property's state function with respect to a specific system object. For example the property 'colour' has a state function for traffic lights that can take the state values 'red', 'yellow' and 'green'.

Property objects may be grouped in domain characteristic property libraries, for example a property library for building design will hold properties specific to buildings. The development of property libraries will have to be guided by analyses of the different design domains the tool is intended to be used within.

System objects, Property objects and State value objects are to be determined by design operations through a user interface. Characteristics of design operations and design moves have been discussed in an earlier paper by Fridqvist and Ekholm (1996).

The actual implementation in the BAS•CAAD project of System object, Property object and State value object is made in Smalltalk, an object-oriented development environment. The principles for the design and use of the model design tool are to be further developed and tested in a prototype currently under development. This testing will include the development of a small building model and a model of a user organisation. The aim is to enable a spatial representation and co-ordination of the systems as in a lay-out process.

References

Augenbroe G. (1995). Combine 2. Final report. Delft: Delft University of Technology.

Björk B.-C. (1989). Basic structure of proposed building product model. *Computer aided design* Vol. 21, No 2, pp. 71-78, 1989.

Björk B.-C. (1995). Requirements and information structures for building product data models. Espoo: Technical Research Centre of Finland, VTT Publications no 245.

Bunge M. (1977). Ontology I: The Furniture of the World, Vol. 3 of Treatise on Basic Philosophy. Dordrecht and Boston: Reidel.

Bunge M. (1979). Ontology II: A World of Systems, Vol. 4 of Treatise on Basic Philosophy Dordrecht-Boston: Reidel.

Bunge M. (1983). Epistemology and methodology I: Exploring the world. Vol. 5 of Treatise on Basic Philosophy. Dordrecht: D. Reidel Publishing Company.

ByggAMA 1950 (1950). ByggAMA 1950, Allmän material och arbetsbeskrivning för husbyggnadsarbeten. Stockholm: A V Carlsons Bokförlags AB.

Child J. (1984). Organization: A guide to problems and practice. London: Paul Chapman Publishing Ltd.

Eastman C. M. (1991). The evolution of CAD: Integrating multiple representations. *Building and Environment*, Vol. 26, No 1, pp. 17-23, 1991.

Eastman C. M., Assal H., and Jeng T. (1995). Structure of a database supporting model evolution. In *Modelling of buildings through their life-cycle*. Proceedings of CIB workshop on computers and information in construction (eds. Fisher M., Law K., and Luiten B.) Stanford University, Stanford, Ca, USA, August 21-23.

Ekholm A., Fridqvist S., af Klercker J. and Ljunggren N.-O. (1994). Building and user activity systems modelling for computer aided architectural design. Paper presented at the IT-BUILD conference in Lund 1994. Lund: CAAD, School of Architecture, Lund University.

Ekholm A. (1996). A conceptual framework for classification of construction works. *Electronic Journal of Information Technology in Construction (ITcon)* Vol. 1. Stockholm: Royal Institute of Technology. URL: http://itcon.fagg.uni-lj.si/~itcon/.

Ekholm A. and Fridqvist S. (1996). Modelling of user organisations, buildings and spaces for the design process. In *Construction on the Information Highway*. (Ed. Ziga Turk). Proceedings from the CIB W78 Workshop, 10-12 June 1996, Bled, Slovenia.

Fridqvist S. and Ekholm A. (1996). Basic object structure for computer aided modelling in building design. In *Construction on the Information Highway*. (Ed. Ziga Turk). Proceedings from the CIB W78 workshop 10-12 June 1996 in Bled, Slovenia.

Galle P. (1995). Towards integrated, "intelligent", and compliant computer modeling of buildings. *Automation in Construction*. Vol. 4, No 3, pp. 189-211, 1995.

518

Gielingh W. (1988). General AEC reference model. External representation of product definition data. Doc. no 3.2.2.1, TNO-report BI-88-150, Delft, The Netherlands.

Garrett Jr J. H. and Hakim M. M. (1994). Class-centered vs. Object-centered Approaches for Modelling Engineering Design Information. Proceedings of the IKM-Internationales Kolloquium über Anwendungen der Informatik und der Mathematik in Architektur und Bauwesen, pp. 267-272, Weimar, Germany, March 16-18, 1994.

Hunter E. J. (1988). Classification made simple. Hants: Gower.

ISO (1985). Concepts and terminology for the conceptual schema and the information base. ISO/DTR 9007 (TC97), also SIS teknisk rapport 311. Stockholm: SIS.

ISO (1994a). Industrial automation systems and integration - Product data representation and exchange - Part 1. ISO 10303-1:1994(E). Geneva: International Organization for Standardization.

ISO (1994b). Classification of information in the construction industry. ISO Technical Report 14177:1994(E). Geneva: International Organization for Standardization.

ISO (1996). Building construction core model, BCCM, ISO 10303. ISO TC184 SC4 WG3 N496. Geneva: International Organization for Standardization.

Junge R. (1995). Aspects of new CAAD environments. *Modelling of buildings through their life-cycle.* Proceedings of CIB workshop on computers and information in construction (Eds. Fisher M., Law K., and Luiten B.) Stanford University, Stanford, Ca, USA, August 21-23.

Lawson B. (1990). How designers think. Second edition. Oxford: Butterworth Architecture.

Leeuwen J. P., Wagter H. and Oxman R. (1995). A feature based approach to modelling architectural information. In *Modelling of buildings through their life-cycle.* Proceedings of CIB workshop on computers and information in construction (Eds Fisher M., Law K., and Luiten B.) Stanford University, Stanford, Ca, USA, August 21-23.

Schenck D. A., and Wilson P. R. (1994). Information modelling: The EXPRESS Way. Oxford: Oxford University Press.

Steadman P. (1979). The evolution of designs. Cambridge: Cambridge university press.

Zhao H. and Biliris A. (1993). An Object-centered Data Model for Engineering Design Databases. In *Proceedings of the Third International Symposium on Databases for Advanced Applications.* Daejon, Korea, April 1993, pp 133-140.

LIFE CYCLE MODELS OF BUILDINGS - A NEW APPROACH

NIKLAUS KOHLER, BERTRAM BARTH, SANDRO HEITZ,
MANFRED HERMANN
ifib- Institut für Industrielle Bauproduktion.
Universität Karlsruhe, D-76128 Karlsruhe

Keywords:

Life cycle costs, Life cycle impact assessment, Product models.

1. Life cycle costs

The idea of life cycle cost was developed a quarter of a century ago. A wide dissemination of the term was given through a report for the US Secretary of Defense "Life
Cycle Cost in Equipment Procuration" [LMI65]. This report was followed by a series of
guide lines in the defense field and later on in other government activities. The basic
definition of life cycle costs is: "The sum of all costs incurred during the lifetime of an
item, i.e. the total of procurement and ownership costs." [DHI89].
The primary uses of life cycle costs are:
   - comparing competing projects
   - long range planning and budgeting
   - selecting among competing bidders
   - controlling an ongoing project
   - comparing logistic concepts
   - deciding on the replacement of an aging equipment.
There are several life cycle costs models available in literature. [DHI89] distinguishes
between general, non specific models and specific models developed for a particular
application.
In the building field attempts have been made to introduce the notion of life cycle costs
mainly through building surveys and for public owned buildings [TRE75], [BEK80].
Recorded data of construction, refurbishment and maintenance costs of buildings show
that over a 50 year period the total costs amount to approximate twice the investment
costs (without financial costs).

R. Junge (ed.), CAAD Futures 1997, 519-531.
© *1997 Kluwer Academic Publishers.*

520

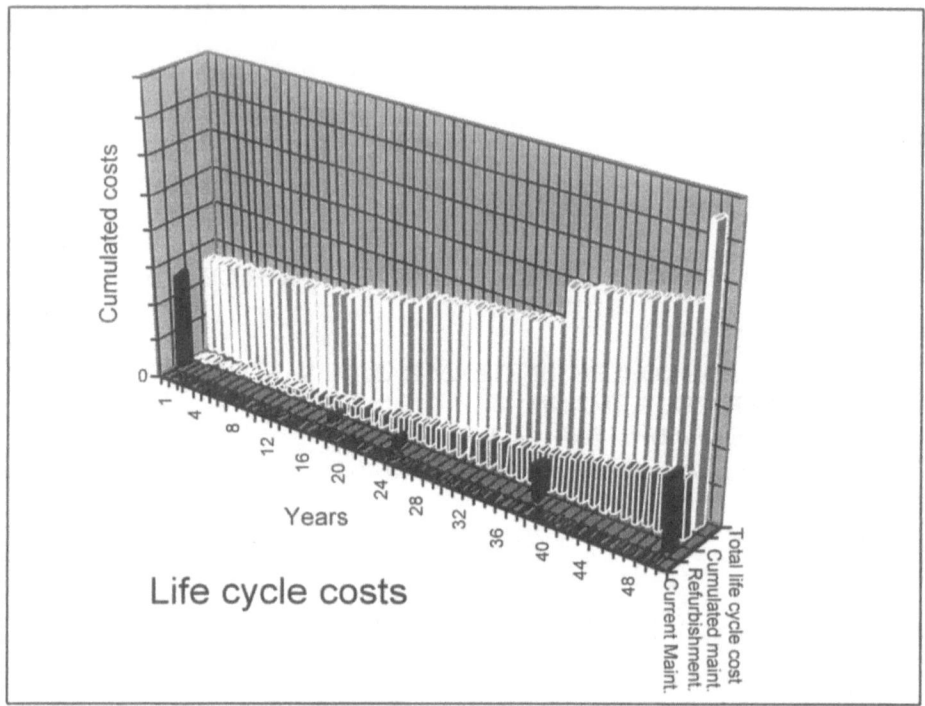

Figure 1. Total costs of a swiss appartement building over 50 years [BFK93]

All existing life cycle cost models reflect the objectives of the equipment's owner. Even if they take into account all costs occuring during the life cycle, they generally only consider "real costs", which have to be paid. The environmental issue leads automatically to an extension of this traditional life cycle costing. The real overall cost of any type of item is much larger than the costs actually paid (including interests) if we take into account the social costs associated with production, use and disposal. For objects with large resource consumption (energy, land, mass) and emissions, this aspect becomes predominant. These costs are considered "external" costs in economic theory [KAP51],[HOH88], because they are not charged to the user of a piece of equipment but are accounted for by society (this is why they are also called "social" costs). The idea of adding at least the energy and direct emissions (pollution) aspects to general life cycle costs in the building field is quite recent [KOH87].

The calculation of life cycle costs as well as mass flows has to take into account the specificity of buildings compared to current industrial products. First of all buildings are very complex products. This complexity is hidden by the fact that we all live in buildings and consider them to be simple and obvious things. But the planning process is long and complicated, because a great number of technical, social, economical, ecological and cultural factors have to be considered. Buildings have a life time, which is much longer than current industrial products and the conditions of use change frequently and are difficult to predict. Through their long life time and their link to the site, buildings have been one-of-a-kind products. Even if techniques of industrial mass production have been introduced to parts of buildings, the one-of-a-kind nature of the whole object remains. However planning will be considerably improved by new computer based techniques.

Concerning the global environmental impact, the building stock induces the largest energy and mass flows of all production sectors. It represents the largest financial, physical and cultural capital of the industrialized societies. In the future it will also become the largest raw material resource for new buildings.

Taking these facts into account, the classical life cycle costing approach must be considerably enlarged in two directions:

Time system limits: A building starts with the expression of a social need and it ends with the physical disposal of the construction materials.

Domain system limits: A building creates, during its whole life time, an ongoing extraction of materials and energy from nature and a large and ongoing flow of emissions back into nature. The ecosphere becomes the final system limit.

2. Life cycle based criteria

2.1 SYSTEM LIMITS

The basic modeling approach for the life cycle of buildings does not start from geometry but from system theory and in particular from ecological system theory. For Odum "the environment has organisms, chemical cycles, water, air, humans, machines, soil, cities, forests, lakes, streams, estuaries, and oceans; and connecting them all are flows of energy, including that associated with matter and information" [ODUH83]. The environment can be described in a system language, which is basically an energy circuit language. In this system all occuring phenomena are accompanied by energy transformations. The energy language keeps track of flows of potential energy from sources going into storages or into transformation (work) and finally into degraded forms leaving the system as heat sinks. Pathways of the energy language are pathways of energy flow.

The rate of flow of energy into useful work is defined as power.

In this processes the first (conservation) and the second (entropy) laws of thermodynamics apply. Odum adds a third important principle, which has been observed in natural systems: the feedback of energy from storage stimulates the in-flow pathways as a reward from receiver storage to inflow source. By this feature the flow values developed reinforce the processes that are doing useful work. Feedback allows the circuit to learn. Lotka formulated the maximum power principle, suggesting that systems, which develop designs that maximize the flow of useful energy prevail.

522

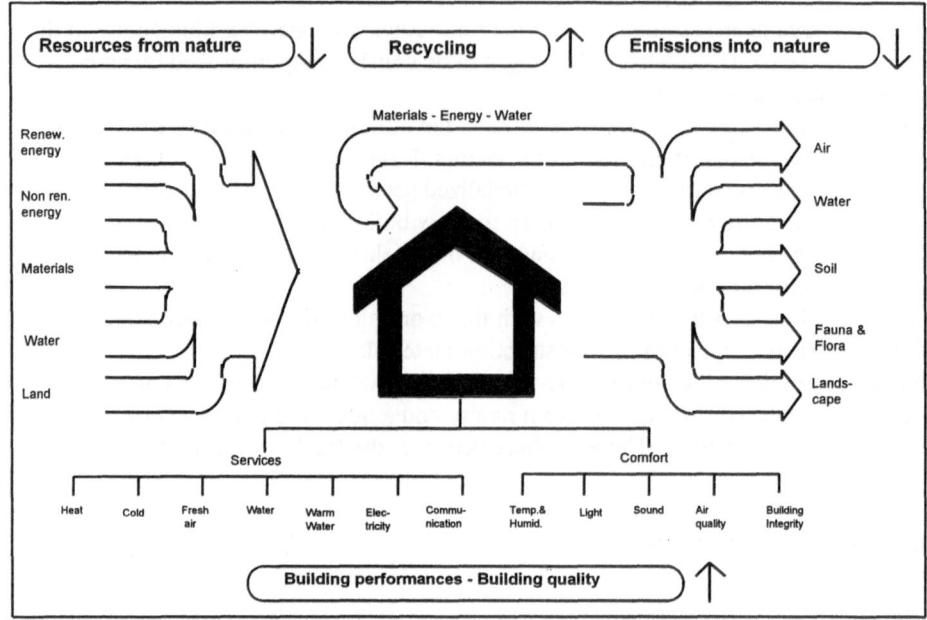

Figure 2. Resource and emission flows of a building [REG96]

The concept of system limits refers basically to the limits in time and space. This approach has been developed in a qualitative [ODUE71,ODUH83] and quantitative [LOT25] way by system ecology [DEL92] and has become the basis of life cycle impact assessment (LCIA). The experience from life cycle impact assessment shows that all results depend strongly on the chosen system limits. This is particularly true for the economic and ecological aspects of the building process.

2.2 ENERGY AND MASSFLOWS

A building can be represented during its life time as the superposition of different flows and activities along the different phases of its life time:

Physical flows:
- Material (building material, water).
- Energy (embodied and operation energy).
- Waste (building materials and waste from use).
- Emission (waste released into the air or soil).

Financial flows:
- Internal costs
- External costs

Information flows:
- Documents
- Communication in all forms

Money is an exchange medium that flows as a countercurrent to materials, energy and information flows. Financial flows can therefore be associated with all of the physical flows and activities. This allows to identify the internal as well as the external costs. All planning and managing activities include documentation and communication as well as data processing in many forms. These are all information flows.

In order to be able to establish the life cycle costs, the flows of materials and the construction operations of a building must be known for its entire life cycle. This knowledge exists to a large extent today, but it is dispersed and in very different formats. All building operations (planning as well as construction and use) can be evaluated according to mass flow, energy flow, information flow, resource use, financial flows and environmental impact criteria. The techniques of life cycle impact assessment (LCIA) [SET91,93] show that it is very important to separate the quantitative data from the evaluation. The evaluation of the flows from different points of view and their assessment in a larger context must always be possible. The goal of the evaluation is to enable the decision maker (designer, owner, politician, producer) to make conceptual, political, constructive and economic choices.

2.3 FUNCTIONAL UNITS

The notion of functional (or reference) unit is used in LCIA as well as in the traditional cost planning techniques like cost targeting. The performance based approach in planning is entirely dependent on a coherent choice of functional units. By taking into account several life cycle phases, the choice of common functional units becomes decisive for the possibility to practice feedbacks and learn from past experience.

2.4 TIME CONSTANTS

Buildings can be considered as composed of elements and parts with very different time constants varying from nanoseconds for light propagation to hundreds of years for the replacement of bearing structures.

Nanosecond	Light
Second	Sound
Minute	Air movement
Hour	Temperature
Day	Energy
Week	Use cycle
Month	Cleaning
Year	Maintenance
10 Years	Refurbishment
100 Years	Life time

Figure 3. Time constants in buildings

Furthermore several time scales are superposed: planning time, use time, building life time; they all have historical character and are not reversible. Furthermore according to the second principle of thermodynamics buildings always contribute to the raise of entropy, only the speed of the process can be influenced by planning and use techniques.

2.5 PROCESSES

The life cycle approach leads to a dynamic way of understanding buildings. This is opposed to the predominantly static way of describing buildings, essentially by pre-ferring the geometric, static way of actual CAD systems' descriptions. At present there is a shift from the geometric modeling based CAD approach to a work flow based modeling approach. Because of their work flow character it is difficult to integrate cost planning, scheduling, specification, refurbishment and facility management into the exsting CAD systems. By analyzing the work flow process we realize that they finally all relate to energy and mass flows from and to nature. These flows will therefore become the fundamental process unit in the description of the building process during its life time.

The recent attempts to apply ecological models to computing [HUB88] offer new perspectives for more complex conceptual modelization and scientific computing (simulation, computer experiments). New modeling and computation techniques allow to simulate decentralized, distributed systems supporting complex simulation as well as new cooperative planning techniques through the use of networks.

3. Process modelization

3.1 PRESENT BUILDING MODELING

The need to explicit modeling of buildings emerges from the recognition of the limits of existing geometry based CAD systems [IWC89]. There has been an agreement that only a common semantic model could be a basis for different applications and views. The ongoing modeling discussion is generally agreeing on a distinction between rooms (functions) and building elements (construction elements and systems). For certain applications like cost planning a hierarchical approach is appropriate [BED92], for other aspects like topology, other models are necessary. There is still a large discussion about the way to introduce object oriented techniques on a modeling and software engineering level. First attempts to use agent based techniques are recorded. The STEP approach integrated the life cycle perspective from the beginning [GIE88]. The present discussion around the building application protocol of STEP is still not finished, no clear standard solution is in sight.

3.2 DESIGN AND BEHAVIOR

The modeling of the design process has as a target the object, which has to be constructed. In this stage the object has no existence and therefore no behavior yet. The dynamic aspects related to the building process and building use are not modeled in detail. If we look at buildings during their life cycle, we look mainly at existing buildings, at the building stock. The diagnosis of the temporal behavior of existing buildings becomes the main interest in modeling, the center of the models. New construction becomes a special case of maintaining buildings [KOH94].

3.3 THE PERFORMANCE BASED APPROACH

The idea of a performance based approach in building comes from the attempts to develop open industrialized building systems. The implementation of solar energy related research and energy conservation strategies could not succeed as long as there were no clear performance standards for all phases of the design, planning and use process. Cost planning techniques, above all cost targeting, also used performance approaches. The application of life cycle impact analysis has given an additional importance to the performance aspect by introducing a whole series of new targets and functional units. In all cases the determination of the reference unit is very important. By analyzing the performance standards, above all in life cycle impact assessment the questions became more and more fundamental. Taking into account the new system limits and the considerable mass and energy flows induced by the very first decision to build, the new question was: do we really need a building, or can we solve the same problem by other means, which have less impact? Often non-building solutions (e.g. through the application of information and communication techniques) prove to be preferable to a building.

4. The common building model

4.1 THE "BUILDING AS BUILT"

The basic idea is that the common model for the life cycle of building can only be the "building as built". Several authors came to the same conclusion [BJÖR92]. The "building as built" is the starting point of the life time of a building and of its induced mass, energy, work and monetary flows. All planning steps, which precede the "building as built" can be considered as a temporarily uncompleted building or as not yet instantiated structure.

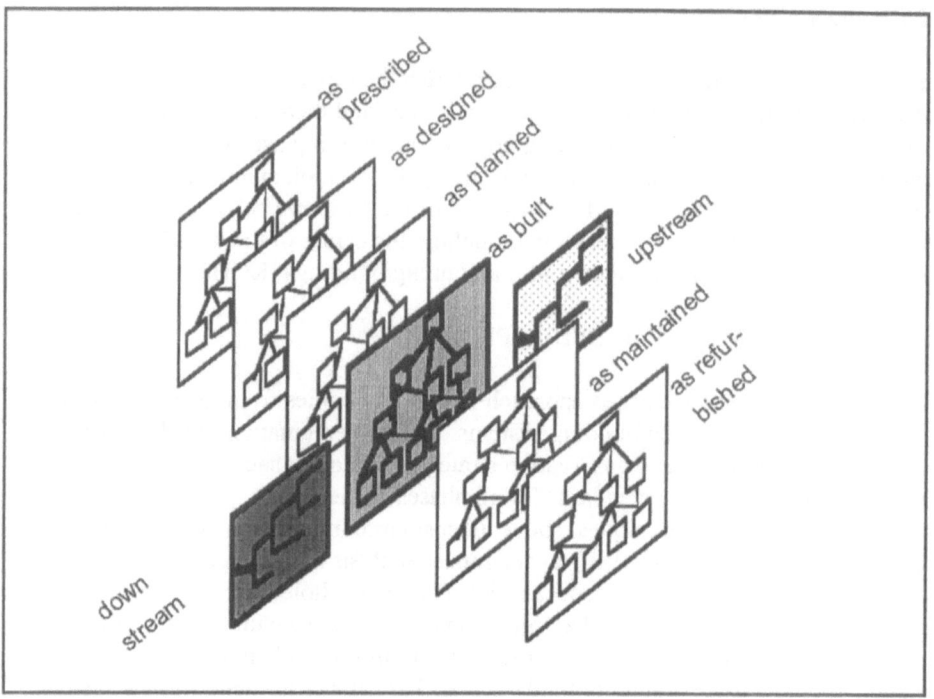

Figure 4. Overview of the life cycle model

The design process reveals the building (it discovers the underlying structure). The questions of functional units is very crucial because as long as a functional unit has not been given a specific value through a planning decision, it must take a default value, which can be the average value of a large number of similar buildings. This allows to produce a large number of simulations of possible design outcomes, which are of course not exact, but which are plausible.

The basic assumption is that buildings of a certain function (housing, office buildings, hospitals, factory etc.) are much more similar than we generally think. Their cost and environmental impacts during their life time can already be determined during the design brief and through performance specification by associating performances and functional units. It also implies that simulation techniques can be used very extensively to verify if the performance targets are reached during the ongoing planning phase.

The impacts of the building during the life cycle phases after construction (building as maintained, refurbished and demolished) can be simulated the same way, taking into account the upstream and the downstream processes.

The question of how to integrate time into the life cycle models is crucial. As well as there are different time constants in the building construction and use, there are different uses of time during the planning process and the life time of a building. It is important

that they all refer to a common scale. This is fundamental for several planning problems:
- the management of planning time, above all the versioning problem
- the management of construction time (scheduling)
- all simulation referring to energy, environment, construction
- facility management
- replacement strategies

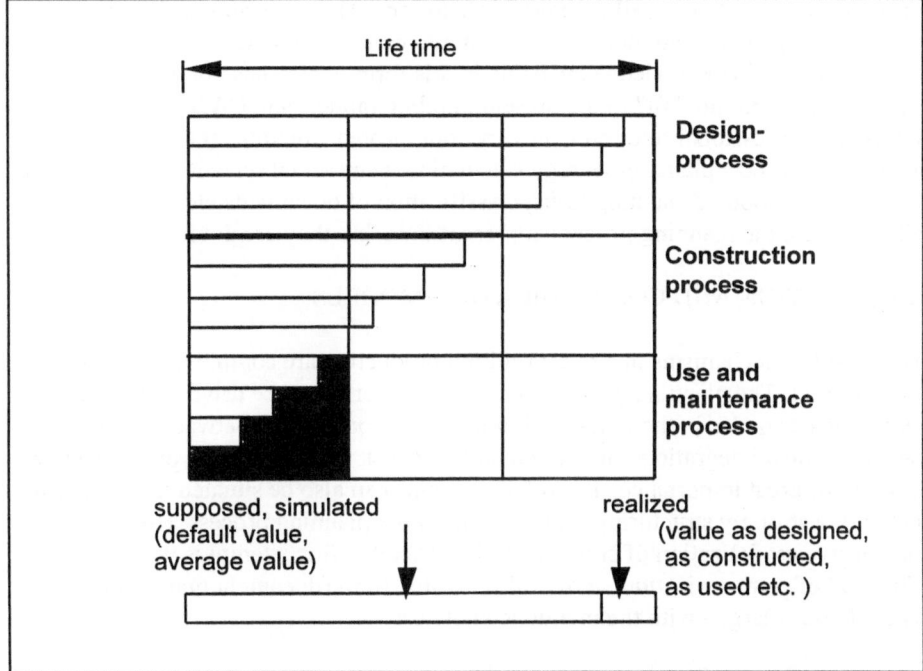

Figure 5. The planning process as combination of realized and supposed parts

4.2 THE PLANNING PROCESS

If we consider the "building as built" as the central representation then the planning and use process of a building can be considered as the gradual replacement of average or default values by actually realized values. In the beginning a building is therefore described by 99 % of average (default) values and 1 % really planned (realized) values. This principle can be applied through the use of different (common) functional units. The advantage is that the whole building is considered and nothing is forgotten. It is therefore possible to represent buildings as combinations of planned and not planned parts, of realized and supposed parts, of real and virtual parts, of past, present and future parts.

5. Implementation

5.1 CATALOGUES AND STRUCTURES

A large part of information and knowledge about buildings is contained in catalogues. Catalogues of building systems (they can vary from small parts of buildings to whole buildings) establish relations (rules) between the different parts, defining in this way a coherent solution. The whole building can therefore be considered as situated in an n-dimensional space containing all the possible solutions. The same approach can be made with specifications and their aggregation into construction elements [HEI95]. There is a considerable knowledge available in this form, which has not been taken into account because it did not "fit" to the present building models and CAD systems. Standards and professional rules contain very complicated structures (element decomposition for cost planning, energy calculation, contractual schemes, procurement methods, specification of planning tasks, classification of documents etc.). They actually structure the planning process at present time, but they are not integrated.

5.2 COOPERATION AND COMMUNICATION MODELS

The cooperation mechanisms in a one-of-a-kind production are complex, to control them is essential. The planning process being mainly performance driven, the issue of reference units as nodes in the horizontal integration (cooperation between planners) as well as the vertical integration (simulation and feedback over the phases of the planning process) are of great importance. The reference units can also be situated in time giving this network a strongly structuring aspect of the whole planning process. However, the comprehensive modelization of cooperation is difficult, a first attempt is the modelization of communication, above all in written form (document management), which could be enlarged with the mentioned network.

5.3 BUILDING REPRESENTATION

A building in general can be most easily described as composed of:
- a <u>particular building</u>, which has functions, is situated on a particular site and has a time of existence. It is composed of
- <u>rooms</u>, which have functions and requirements. Their reference units refer to functions (e.g. 1 m^2 of classroom of an elementary school) and requirements (e.g. light level, resulting temperature). All information related to rooms is contained in the so called room book, which evolves during the life cycle of the building. Rooms are delimited by construction elements and contain system elements
- <u>elements</u>, which have a physical reality and performances. They are either con-struction elements (separation and structure functions) or system elements (technical functions, converting or transporting different media). In addition to their compositional information they have other characteristics like their status (degree of realization: average, planned, new, refurbished etc.), their time tag and

their data status (form of description, origin, quality, contract status). Their reference units refer to their composition (material layers, duct characteristics), performance (e.g. acoustic absorption, efficiencies) and use of resources (cost, materials, land, time).

- use processes of rooms, which are dynamic (evolution of temperature during a day, change of function after 10 years).Their reference units relate to the requirements (e.g. internal loads in W/m^2). They consume resources and create emissions.
- construction and maintenance processes, which are dynamic and which refer to elements (set up time, probable life span). Their reference units are those of the specifications. They consume resources and create emissions.
- resources of different types. Their reference units are related to mass, energy and time flows.

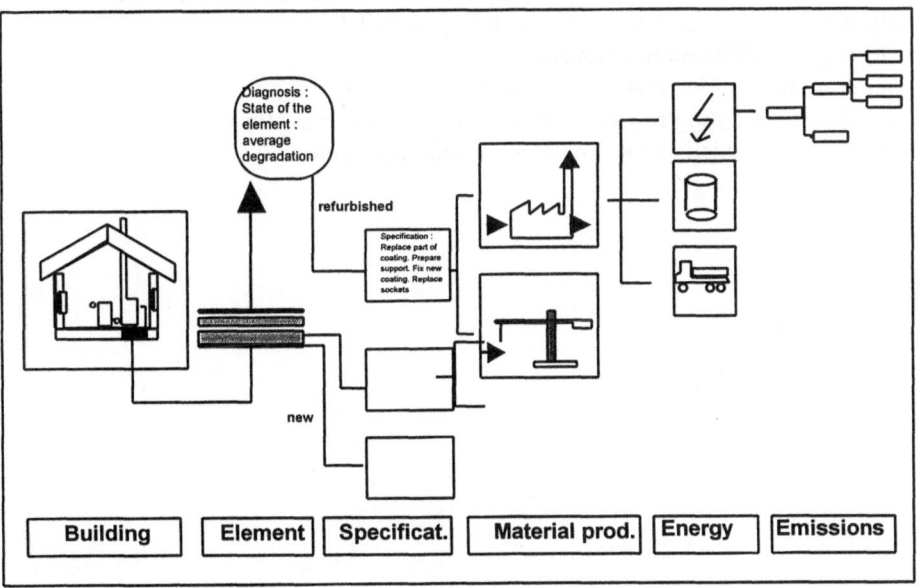

Figure 6. Building representation based on elements and specifications

There are complex topological relations between rooms and elements. Rooms are defined by surfaces of construction elements and contain technical system elements. Technical systems have their own compositional structure (generally hierarchical). The reference units are interrelated through the composition of the building. They can generally be recomposed by generalization from the "building as built" to earlier stages. In the other direction this is not possible without additional information.

The "building as built" can be considered as the most complete representation of a building. All other stages can be derived (by generalization for the early design phases, by simulation of the use and deterioration for the life cycle). The largest part of the information about a "building as built" is contained in the complete specifications.

There are catalogues of specifications and catalogues of elements composed of specifications, which allow to describe most buildings. At present time, the most complete general description of a building is given by the structure and standard specifications, which are used in cost estimation [CRB91].

6. Conclusions

The life cycle approach is not an additional view to the actual, mainly design and geometry based representation of a building. It is a different structural principle, which has the following characteristics:
- it has large and explicit system limits
- it is process based and not object or geometry based
- it uses reference (functional) units to link construction and performance
- the central modeling level is the "building as built"
- it integrates different time levels
- it allows to combine real and virtual components
- it integrates existing construction, cost, energy and mass flow data
- it is based on existing data contained in catalogues etc.

References:

[BED92] Bedell, J.R., and Kohler, N. (1992): "A Hierarchical Model for Life Cycle Costs of Buildings", Proceedings of the Computers in Building W78 Workshop, May 1992.

[BEK80] Bekker, P.C.F. (1980): "Life time theory of dwellings", in CIB Symposium Quality and Cost, in Building. EPFL Lausanne 1980

[BJÖR92] Björk, B.-C. (1992): "A Unified Approach for Modelling Construction Information", Building and Environment, special issue on databases for project integration, 1992.

[CRB91] CRB (1991): "CCE Cost Classification by Elements", CRB Swiss Research Centre for Rationalization in Building and Civil Engineering, Zurich 1991.

[DEL92] Deléage, J.-P. (1992): "Histoire de l'Ecologie", Paris 1992

[DHI89] Dhillon, B.S. (1989): "Life cycle costing", New York 1989

[GIE88] Giehling,W.F. (1988): "General reference model (GARM) ", TNO Report BI-88-150, 1988

[HEI95] Barth, B.; Eiermann, O.; Haida, A.; Heitz, S.; Hermann, M.; Kukul, E.: "Life cycle modeling of buildings", in EuropIA'95, Lyon 1995, Hermes, Paris 1995

[HOH88] Hohmeyer (1988): "Social costs of energy consumption", Berlin 1988

[HUB88] Hubermann, I.A.: "The ecology of computation", Elsevier, Amsterdam. 1988. computing

[IWC89] International Workshop on Computer Building Representation, CH-Chexbres 1989, EPFL-Lausanne 1989

[KAP51] Kapp, W. (1951): "The social costs of private entreprise", Harvard 1951

[KOH87] Kohler, N. (1987): "Energy Consumption and Pollution of Building construction", International Congress on Building Energy Management 87, EPFL Lausanne 1987

[KOH91] Kohler, N. (1991): "Life cycle costs of buildings", Buildings and the Environment, Proceedings of the Forum at the University of British Columbia, Vancouver BC, March 1991

[KOH94] Kohler, N. et al.: "Energie- und Stoffflußbilanzen von Gebäuden während ihrer Lebensdauer", Schlussbericht Forschungsprojekt BEW, Ifib - Universität Karlsruhe 1994

[LMI65] Logistic Management Institute (1965): "Life cycle costing in equipment procurement", Washington D.C. 1965

[LOT25] Lotka, A.J. (1925): "Elements of mathematical biology", New York 1925

[ODUE71] Odum, E.P. (1971): "Fundamentals of Ecology", Philadelphia 1971

[ODUH71] Odum, H.T.: "Environment, power and society", New York 1971

[ODUH83] Odum, H.T. (1983): "System Ecology", New York 1983

[ODUH87] Odum H.T.; Pillet G.: "Energie, écologie, environnement", Geneva 1987

[ROE71] Georgescu Roegen N. (1971): "The entropy law and the ecomomic process", Cambridge Mass. 1971

[SET91] SETAC (1991): "A Technical Framework for Life Cycle Assessment", Smugglers Notch Workshop Report, Washington D.C., USA 1991, 134p.

[SET93] SETAC (1993): "A conceptual framework for Life-Cycle Impact Assessment", mars 1993, 146 p.

[TRE75] Trenton, H.P. (1975): "Terotechnolgy: the right lifespan", Building, April 1975, London

[BFK93] Impulsprogramm Bauliche Erneuerung: "Unterhaltskosten von Gebäuden und Bauteilen", Bundesamt für Konjunkturfragen, Bern 1993

[REG96] Final report REGENER Project, Ecole des Mines, Paris 1996

SOME THOUGHTS ON THE EXISTENCE OF A GENERIC BUILDING OBJECT

JAMES A. TURNER
College of Architecture and Urban Planning
The University of Michigan

The purpose of this paper is to propose a new universal data structure, called a Generic Building Object (GBO), to support the reinvention and re-implementation of a building data base application called PLAN. The paper reviews various building models as presented explicitly and implicitly in the writings of other computer-aided building design researchers.

1. Introduction

In the late 1970s researchers at the Architecture and Planning Research Laboratory, College of Architecture and Urban Planning, The University of Michigan, developed an application for the interactive creation of a building data base. The development of the program, named PLAN, was sponsored in part by the Construction Engineering Research Laboratory at Champaign, Illinois as part of its long-term CAEADS (Computer Aided Engineering and Architectural Design System) project [Mitchell et al. 1978].

PLAN [Turner et al. 1982, 1983] had the feel of a "smart" drafting system—one which knew about walls, doors, windows, floors, furniture and rooms, instead of points, lines, hatching and text. The program was unique for its time since it allowed various architectural analysis applications to "share" its flat-file data base. Development was steady for three years and versions of the software were used by U.S. Army Corp of Engineers offices and by our own students.

Many things have changed in the computer world in the last 15 years: PLAN's user interface, although modern at the time, was written without an underlying windowing system, and PLAN's internal data base was supported with standard FORTRAN fixed-size arrays. For good reason, most modern applications have as much time invested in graphic user interface design as in software design, and most applications are written in an object-oriented programming language.

2. Why a generic building object is desirable

Modern programmers like to think in terms of "objects," aggregations of data structures and operations which act together to mimic real-world things The many advantages of object-oriented programming or object-oriented design will not be presented in this paper.

R. Junge (ed.), CAAD Futures 1997, 533-552.

Reason enough is that once committed to object design and implementation, there are many useful "classes" available (as there were traditionally many function libraries available) to support the writing of new applications. For this project standard class libraries will be used to support the windows interface, fundamental geometry data structures, transformation matrix operations, and scene composition and viewing [Turner 1996].

A collection of generic building objects would support building-related CAD applications allowing programmers to develop applications quicker and perhaps provide an easier opportunity for integration between applications. Programmers will adopt a generic building class library if: 1) class methods are available for storing and retrieving the various building objects; 2) methods are provided for drawing and measuring all class data structures; 3) class data structures are rich enough to support most geometric and non-geometric building entities; and, 4) the class is a good abstraction of building objects; that is, the data structures and methods are complete enough to emulate the different types of building components.

3.   Features of a generic building object—lessons learned from PLAN

The goal of this project is to determine if it is possible to create a single class, a building object meta-model, from which all other building components can be derived. Fundamental questions to be addressed are: Will the proposed object be too simple to be of any good. That is, will its have to be so basic that it includes only administrative data such as those found attached to files—create date, last change, owner, permit status, type, etc. Or will it necessarily be too complex, with data available to support all perceived sub classes. The goal of the study is not to provide a mechanism to support all complex data structures and algorithms necessary for applications. It is assumed that a generic building object would provide sufficient methods for mapping its data into other more sophisticated application data structures.

The following functionality and data structures are required:

- provide a complete set of **geometric data structures**. PLAN used the geometric data structures provided in [ARCH:LIBRARY 1988] and used by [GEDIT 1993]: 2D network (called NW2-SETs) for storing wall data, 2D polygons (PG2-SET) for storing rooms, 2D hybrids (HY2-SETs, combination of lines and polygons) for furniture, corner and opening symbols. PLAN was inherently two-dimensional, but a complete generic building object will need to support two- and three-dimensional geometries.

- support symbol **instantiation**. Much of building design is the choosing and locating of catalog parts. A mechanism must be provided to store library part definitions and their transformed instances. PLAN provided for the instantiation of doors, windows, furniture and corner symbols which were chosen from pre-defined, **project independent libraries**. Hopefully, some day there will be available a complete collection of machine-readable catalogs and libraries in a consistent format.

• support **standalone applications** such as energy, cost, structural analysis. PLAN provided a function for reading the entire flat-file data structure. Analysis programs only needed to include the data structure declarations (also stored in a file) and call the file reading function to gain access to a building project. PLAN also provided a set of functions (such as a function for getting a room polygon) for retrieving the various components from the data base. But for the most part analysis programs needed to get at the data directly by accessing values from PLAN's numerous arrays—a tricky, error prone task even for the authors of PLAN.

• support all building object **relationships**. Seldom do building components exist in a data base without attachments or links to other components: rooms and walls are members of a floor; walls are connected to other walls; furniture must be in a room; openings must be on enclosures. The building objects must also be able to adapt to an application's overall data storage design (super-imposed building model); that is, an object must be able to be stored in an hierarchical, network, or relational data model.

• support change through a **dependency mechanism**; for instance, when a wall is moved, modified or deleted, all components related to it must be notified so that they may react. PLAN accomplished this through fixed, algorithmic dependencies; that is, when a wall was modified, the two adjacent room polygons were re-generated, and any openings were moved or removed.

4.   Review of Generic Building Models

Prior to the design of PLAN there was only a small amount of literature related to computer-aided design available. Specifications for CAD systems which influenced its development were given in [Coons 1963] (a very early—if the not the first—mention of the acronym "CAD"):

"A computer system, to work in partnership with a designer, must have several clearly definable capabilities. It must be able to accept, interpret, and remember shape descriptive information introduced graphically... Coupled to this graphic facility must be a computational facility for unraveling and performing all of the mathematical analysis and computations that pertain to the design process. The computer should be able to furnish information about standard parts, standard materials, and standard processes. This is essentially an operation of catalogue storage and retrieval."

and [Eastman 1975]:

"If the building itself is described – say, as a large set of polygons with attributes affixed and each subsystem appropriately linked – then not only could analysis be made without any coding of data, but any kind of drawing could in theory be produced of any part of the total building and its components. ... The 'official' building description is stored in the computer. Only one (comprehensive) model is required for a total project. ...   In the long run, of course, both machine-encoded building descriptions and longitudinal integration are expected to become part of the philosophy of most Computer Aided Building Design designs."

and [Mitchell 1979] (paraphrased):

"In order to develop a systematic method for symbolic description of a building, it is necessary to conceive of the building as divisible into a collection of discrete element, and to devise a notation for identifying and describing the relevant properties of each of these elements. ...to decide which attributes should be described. Attributes might be geometric, physical, economic or of any other kind relevant to the applications. ...it may also be necessary to describe various relations which exist between elements. ...Buildings typically include many standard elements. ..it would be more appropriate to store one detailed description, plus a list of location instances... Different types of geometric description can be distinguished according to: type of geometric element, topological attributes and relations, level of detail, and geometric assumptions."

There has been much written in the last ten years regarding data structures to support computer-aided building design. The following is a review of recent articles which contain suggestions for characteristics of a generic building object.

4.1. RATAS

[Björk 1989] presents basic characteristics of a building product model developed under a Finnish study called RATAS. Although this is an early paper by the author, it gives a clear explanation of a proposed building product model.

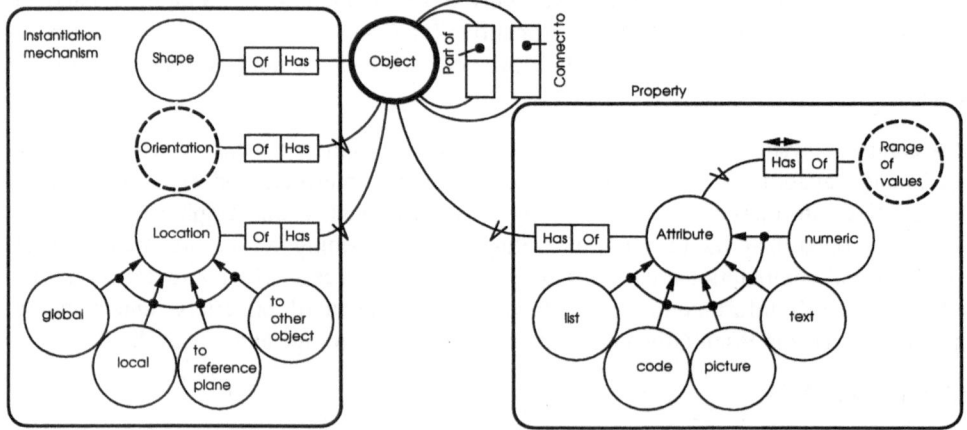

Figure 1. NIAM [Nijssen 1989] RATAS Model

The RATAS object includes *attributes* of various types and allows numeric and text attributes to specify a range of acceptable values. These restrictions may be set by local building regulations or client demands, or may be as simple as: integers greater than zero, or the colors RED, GREEN and BLUE. Explicit data structures for topological relationships are provided in the form of "part-of" and "connect-to" relationships. The author warns that without hierarchical and network relationships between objects, the building description would only be a "catalogue of the constituent parts of the building." It is assumed that an object (such as a wall segment) can be connected to and a part of many other objects.

The author also states that *instantiation* features of shape, location and orientation are important, but does not elaborate on how these characteristics would be supported. He instead suggests that these will be based on data structures found in commercial CAD systems or as specified in the STEP standard. Also mentioned is the need to provide a mechanism for *views* of the building data to allow sub-setting based on attributes and relationships.

4.2. Engineering Data Model

[Eastman 1991] presents a data model named EDM which is based on the assumptions that design knowledge is fundamentally different than traditional data bases. This is because design knowledge is incomplete through most of the design process, must be in the form of multiple representations, and includes procedural as well as declarative knowledge. EDM was developed with the following goals: to represent function as well as form; to support multiple levels of abstraction for various phases of product life cycle; to represent (the semantics of design and engineering) information sufficient for all uses; to allow application specific views of information; and to allow for extensible views of information for new evaluations, technologies and generalizations. In addition, EDM supports constraints in the form of expressions involving attributes (properties). EDM considers the management of relations to be a "central issue in design information management."

The primitive structures of EDM (Figure 2) are *domain* (a named data type and a range of possible values), *aggregation* (property—a collection of variable/value pairs), and *constraint* (a relationship between variables). Eastman distinguishes between constraint definitions and constraint calls.

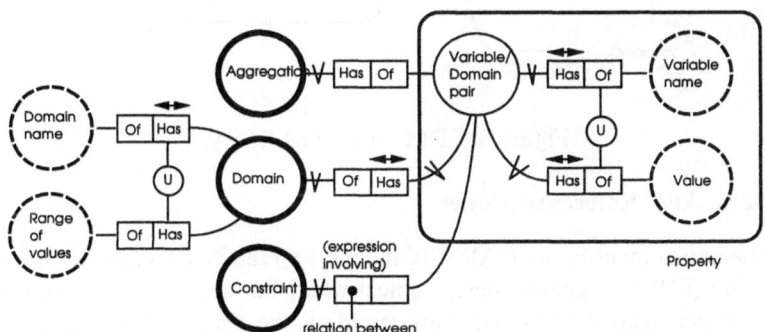

Figure 2. EDM Basic Elements

The basic knowledge structure in EDM (Figure 3) is the Functional Entity (FE) which can be embedded within other FEs; can contain a set of *properties* (such as shape models and

538

material properties) and *constraints*; can support *abstraction* (subsets of all properties or relaxed constraints—can be used as an application-specific or design phase *view* of the FE); can be *accumulated* with other FEs according to specified conditions (a set of constraints satisfied). Accumulations can be further grouped together to form a *composition*. An *abstraction* mechanism is supported through specification of a set of properties.

EDM [Eastman 1995] also supports the relationship between spaces and walls (and ceilings, floors, ...):

"EDM represents both the solid constructed portion of a building and also its spatial form, in a manner allowing (both) to be manipulated and analyzed, while maintaining consistency between the two sets of descriptions."

Geometric data structures are also stored as Functional Entities. Basic FEs such as vertices and lines are combined into higher order FEs such as boundary representation solids. Constraint satisfaction guarantees the correctness of the geometry. For example, lines of a face can be forced to be coplanar, two-connected, and non-intersecting.

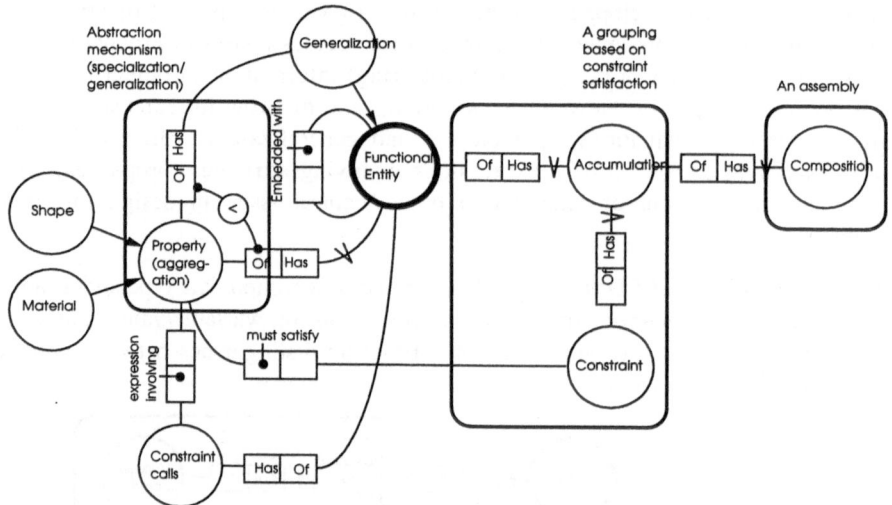

Figure 3. EDM Functional Entity.

4.3. General AEC Reference Model

The fundamental element in the GARM [Gielingh 88][van Nederveen 1993] is a Product Definition Unit (PDU) - a generic design object which has application to many disciplines and building specialization types. An entire building, the structural system, a window or a single brick can be a PDU. A PDU can also exist at different *levels of detail*: generic— perhaps a parametric object; specific—an object with fixed parameters; and occurrence—an instance of a specific object. A PDU may be decomposed into other PDUs, and information is recorded as *characteristics* at the level of the PDU. The

GARM reflects the fact that design information becomes more concrete as the building process moves through its different stages towards completion.

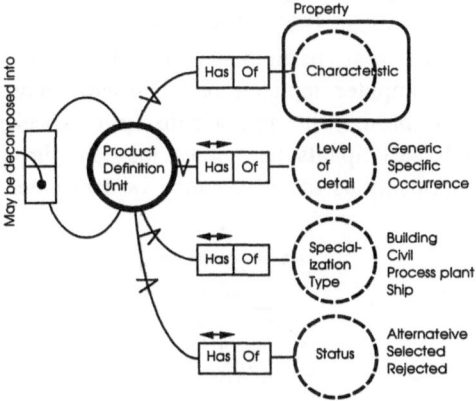

Figure 4. Product Definition Unit

One clever feature of GARM is the Functional Unit (FU) which is a *required* PDU. For a design to move forward, one or more *proposed* Technical Solutions (TS) must be found whose characteristics satisfy the FU requirements. Archiving Functional Units along with Technical Solutions allows for the remembering of design object requirements, not just final *specified* solutions. GARM demonstrates that objects must play different roles in support of the design process depending on level of detail (GENERIC, SPECIFIC and OCCURRENCE), and status (ALTERNATIVE, SELECTED, REJECTED).

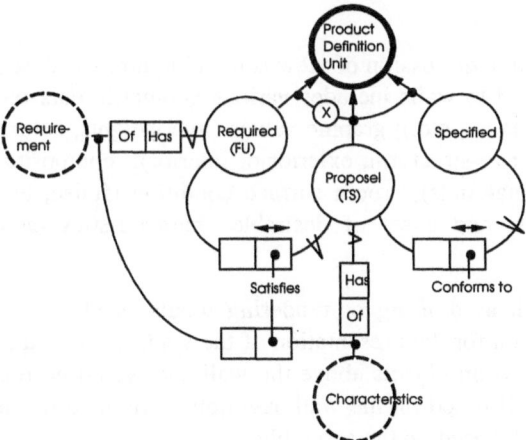

Figure 5. Functional Unit - Technical Solution

The GARM gives a detailed model of geometry representation which include 3D wire-frame, surfaces, and solids, CSG and boundary representations, and 2D drawings.

540

4.4. Actor—Application View Mechanism

The very important concept of allowing different *views* of neutral design information is described clearly in [van Nederveen 1993]. The mechanism allows for different views of design information for different participants in the design process, whether the participants are people or computer applications. The view may simply be a sub-setting of all information available through filtering or a discipline-specific transformation of the information. For human participants the views may be through standard data base queries; for computer applications the views may be facilitated through function calls (methods).

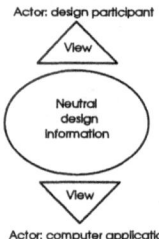

Figure 6. Actor - View Model

[Junge et al. 1995] also states the need for different views of design information:

"A problem arises with the multi-disciplinary nature of many system components playing a role in several partial model for systems. ... This reflects the human way, as various participants of the design process have different conceptualizations of the design object."

The usual example in a discussion of view is the ubiquitous wall segment. Typical design information attributed to walls includes various geometric data (single line, double line, solid, room surface faces, etc.); graphic values (line weights, color, fill patterns); various types (structural or non-structural, exterior or interior); composition of assembly (order, type, thickness of materials); room surface condition (finish, texture, color); opening types and locations; and a set of desirable characteristics (acoustic, fire, structural, privacy, security).

An application such as drafting or rendering would need to extract only graphic and geometric information for the presentation of the wall. A structural engineer would need loading information from objects above the wall and would need to derive the structural properties of the wall based on the wall assembly. An acoustician would need different properties of the wall based on the assembly.

4.5. Time Dimension

[Smeltzer 1993] suggests that, since design is only up-to-date for short periods, an important addition to design information is a time dimension. This *temporal property* would allow for variations on design stage such as design versions, design variants, and design studies. To compound the problem: "There may be several design variants of a certain design version at the same time." The authors offer these notions as abstract and ask the question: "What temporal aspects are relevant to the description of architectural objects and to the recording of design information ...?"

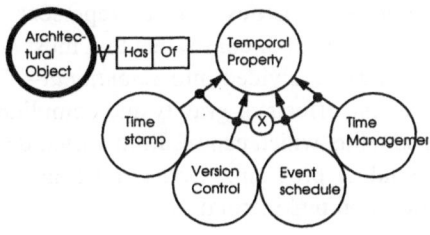

Figure 7. Temporal Property

4.6.   Multi-modal representation

[Carrara et al. 1992] makes a distinction between information and knowledge, and that to represent knowledge we must represent the process that acquires, interprets and applies the knowledge. Design knowledge comprises: *descriptive knowledge* which represents the objects and concepts, their function and relationships; *normative knowledge* which represents goals and constraints; and *operational knowledge* which represents strategies for selecting or generating objects and methods for assigning values. These three "modalities" must be able to represent the profession's shared experiences, the designer's personal experiences, and the circumstances of a specific design project.

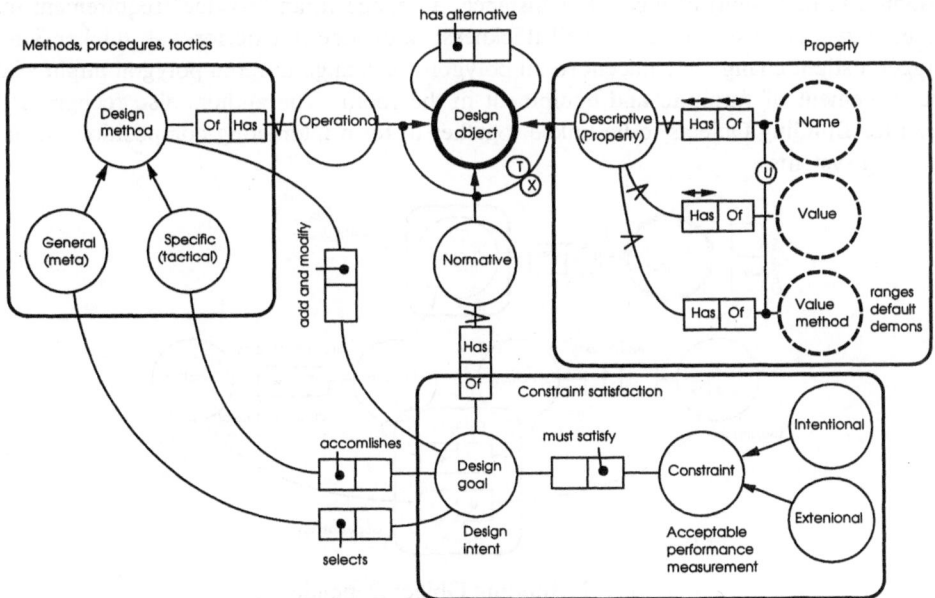

Figure 8. Design Object Model

542

The authors propose using a frame structure to store object *properties*. Frames provide a nice mechanism for storing ranges of acceptable values and instructions for handling missing values (a typical situation during early design stages) through default values or demons. Normative knowledge is defined as either *goals*—properties which represent design intent—or *constraints*—properties which represent acceptable performance requirements. A goal consists of a set of constraints that must be satisfied before the goal is satisfied. Constraints are further divided into *extentional*—"hard" constraints that are either satisfied or not such as the effects of gravity and compliance with building codes— and *intentional*—"soft" constraints which can result in varied degrees of satisfaction, such as those specified in the building program. Goals and constraints provide a context for tradeoffs and judgmental decision making in design.

The authors suggest that constraints can be of the form, **constraint (value | range)**, such as:

> illumination level (150 fc)
> lighting system (ambient)
> area (125 sqft, 150 sqft)

4.7. Object Dependency

[Bhat et al. 1993] defines Building Objects (Figure 9) as having structural and behavioral attributes. Structural attributes describe real-world objects and behavioral attributes (as functions) are design support tools. The notion of *dependency* is suggested by allowing structural and behavioral attributes to be linked to create actions (change in state) in response to (a design) change. For instance, a change in an "r-value" requirement may necessitate a change in exterior wall thickness; a change in exterior wall thickness may trigger a slight change in adjacent room polygons; a change in room polygon might affect the placement of furniture and equipment in the room The authors also recognize the need for Building Objects to be linked together to form interacting compositions such as building systems.

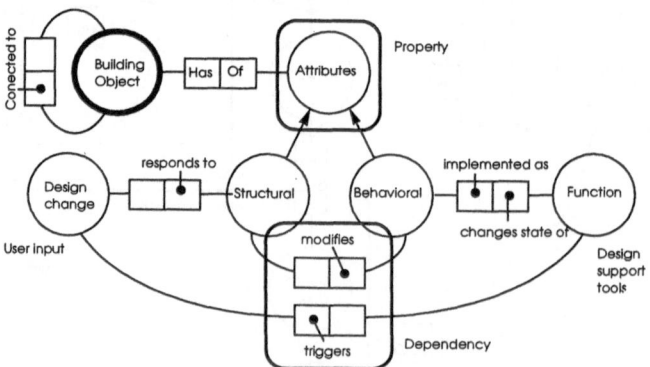

Figure 9. Building Object Dependency

[Tsou 1992] presents a "generic cost control model" (Figure 10) which was implemented in Smalltalk and used to support his dissertation research. An object has *properties*, which represent facts or internal states and can be used for analysis. An object has *constraints*, which represent functional requirements, preferences, safety requirements, building code requirements, and architectural design considerations. Tsou warns that only some types of constraints can be transformed into a "computer-readable" form. An object has *methods* which can either be analysis or input/output methods. The *view* of an object is simply defined as an output method.

The most interesting feature of this generic model is its *dependency* mechanism. When a change occurs, the changed object will "broadcast a change event" to a list of objects who are dependent upon it.. The change event will then trigger each dependent object to react to the change, perhaps through an "update" method. The example Tsou gives is: Increasing the bay spacing of the floor system will force re-selection of the floor system and increasing the floor depth; which will increase the floor-to-floor depth; which will increase the building height; which will increase the total exterior wall area; which will increase the construction cost.

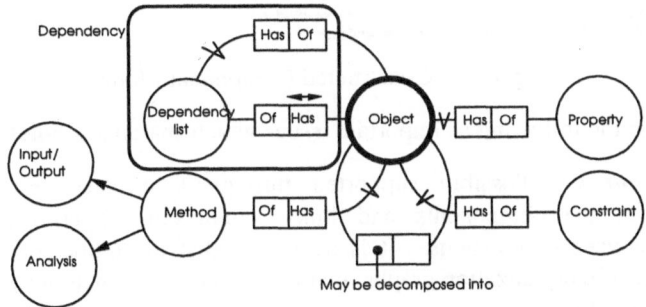

Figure 10. Dependency List

[Junge et al. 1995] re-enforces the need for supporting object dependency: "If one design object has been modified, the system [must be able to] keep track of the consequences to other design objects."

4.8. Constructed Component

[Zambian et al. 1991] defines a constructed facility as a collection of components with each component having *attributes*. Attributes are either functional or descriptive, with descriptive attributes as either spatial—geometry and topology, including shape, size, adjacencies—or non-spatial—all other physical properties. The authors also stress the need to provide different *functional views* of component attributes based on discipline-specific needs. The view mechanism must be able to respond to spatial and non-spatial queries such as: "find all architectural rooms supported by at least one structural beam whose grade of steel is the same as that of the structural column located at the northwest corner of the HVAC zone K3."

544

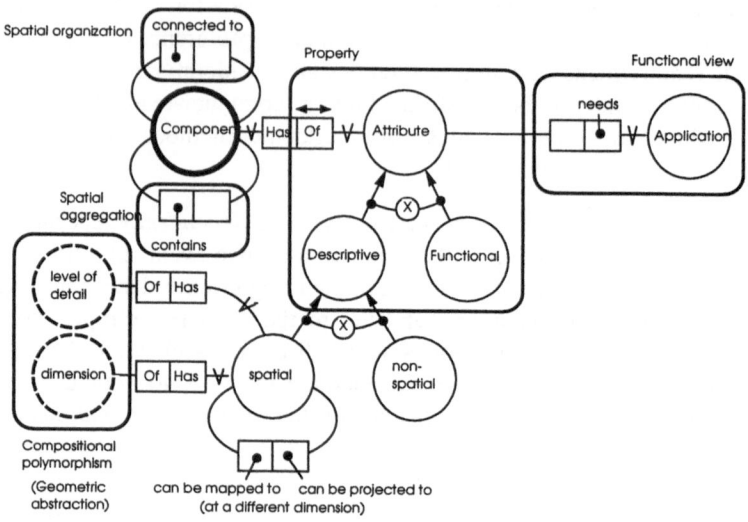

Figure 11. Constructed Component Model

The authors give the following as important issues which spatial attributes must support:

spatial organization: Possibly supported through a directed graph with nodes corresponding to the components and links (connected to) corresponding to the relationships between components. It is suggested that discipline-specific views may need to have specific organization graphs; that is, a single component may be part of more that one graph structure.

spatial decomposition and aggregation: Components are usually organized in an hierarchical fashion (contains); that is, objects subdivide into smaller objects. It is noted that disciplines must be able to aggregate and decompose objects according to specific needs.

compositional polymorphism: The ability to retrieve spatial information at various levels of detail or geometric abstraction. Different representations, such as geometries of different dimensions or at different levels of detail, can co-exist. The authors propose that if a 3D representation of an object is needed by an application, but only a 2D representation exists (or vice-versa), functions should exist for generating one representation from the other.

4.9. Design Intent

[De La Garza et al. 1994] presents an argument for including *design intent* (as "implicit" design knowledge) information along with typical "explicit" design knowledge. This stored rational for design decisions is essential for construction firms to be able to produce detailed design drawings which are "consistent with the overall function, performance and quality requirements of the project." The authors provide the following classification of design knowledge:

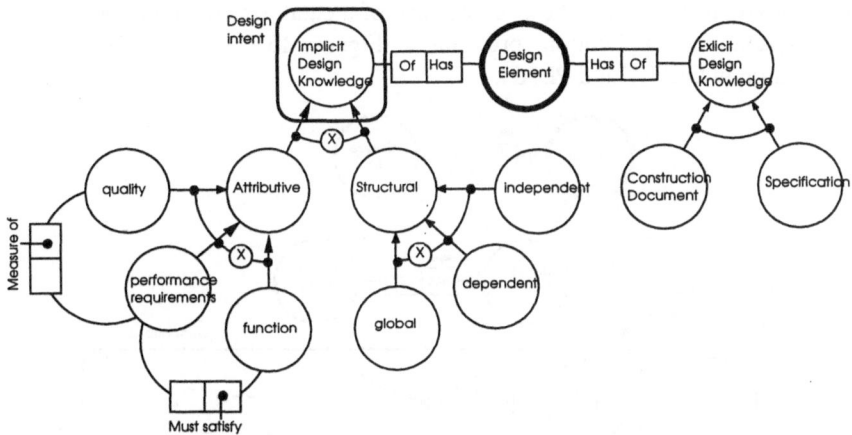

Figure 12. Design Intent Classification

Function refers to the purpose of a design element, while *performance* refers to the requirements it needs to satisfy. The function of a beam is structural while its performance is to provide support for the floor above. *Quality* is a measure of how a design element performs and meets its requirements. Independent design intents are the requirements an element must satisfy independent of other elements. Dependent design intent is due to the interaction of elements. Global design intent affects a collection or class of elements. The authors give no hint as to how design intent should be stored or used.

4.10. Part Libraries and Object Instantiation

The need for building component instantiation from project independent parts catalogs is stated succinctly in [ISO CD 13584-31 1995]: "Products are often made of parts." [Wittenoom 1995] stresses the need for other types of external information sources:

"Information about projects is rarely created completely afresh for each project. The process ... will usually draw on manufacturers' product information, building regulations, design codes and standards, marketplace perceptions, current architectural or engineering design practice, 'accepted industry practice' and solutions developed in previous projects."

ISO 13584 hints at the type of information necessary in a "computer-sensible representation and exchange of parts library data:"

"Each global shape description shall be associated with a set of numeric-typed, string-typed or Boolean-typed parameters whose set of values characterize each part of the part family. logical specification of an interface between a parts library system and a product modeling system."

546

Not much advise is given as to how an instance of a catalog part should be maintained by a CAD system or application program. Figure 13 borrows from instantiation implementation in GEDIT and PLAN.

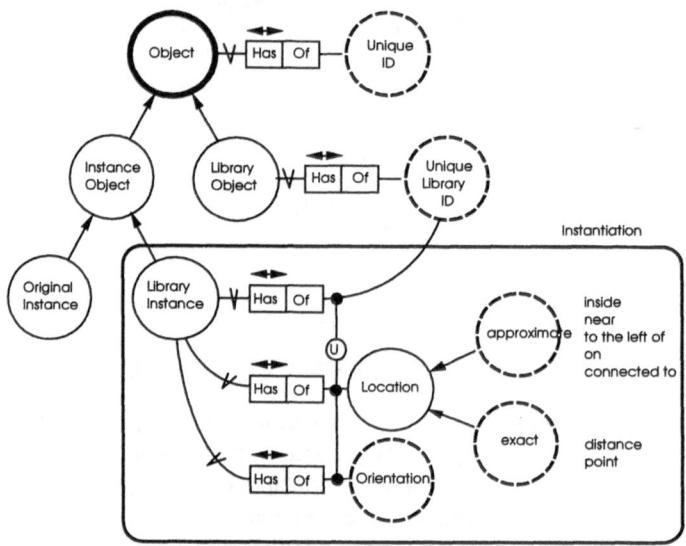

Figure 13. Object Instantiation

As Figure 13 shows, an Object must be able to represent both instances of objects and library objects—those from "parts libraries," which can be instanced. The "Original Instance" (perhaps a bad name) represents the one-time objects which are not instances. These include project-specific objects like rooms, wall segments, and building site. Necessary instantiation data includes the ID of the Library Object to be instanced, locational data, and orientation data. It can be argued that since most manufactured parts come in specific sizes, a scaling transformation should not be included.

5.   Proposal for a Generic Building Model

Many features were presented in the papers reviewed. Some features are obvious and were part of many models. For instance, a generic building object must have geometric and non-geometric properties, and must be able to combine with and be part of other objects. Good suggestions were given for storing non-geometric property values and geometries. Almost no hints were given for storing and using more qualitative values such as constraints and design intent.

Figure 14 declares the Generic Building Object to have a *name, unique ID* and a set of *properties* and a set of *geometries*. Hierarchies are supported by a *part of* relationship; Groupings are supported by a *combined with* relationship; and network links are supported by a *connected to* relationship. An intermediate "node" object is necessary to support the need for a single object (such as a wall segment) to be part of, combined with, and connected to many other objects.

It can be argued that these relationships should be maintained by the application program and not be a fundamental part of a GBO.

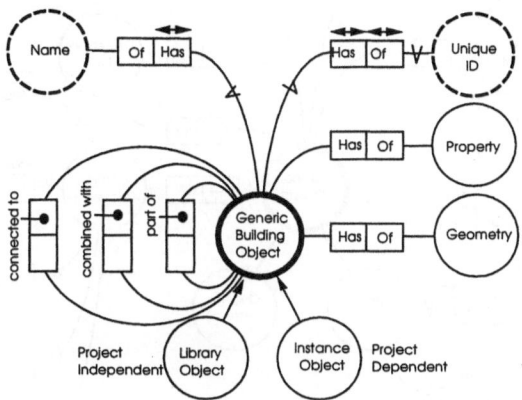

Figure 14. Proposed Generic Building Object

As shown in Figure 15, a Library Object has at least a *product identifier* and a *generic type*: Its geometric representation(s) and properties are inherited from the GBO.

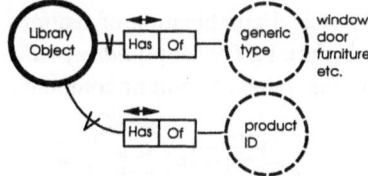

Figure 15. Library Object

Figure 16 shows that there are two types of instance objects: a *library instance* and an *original instance*. A library instance refers to a transformed copy (*location, orientation*) of a library object (*library object ID*). *Constraints* (as suggested in [Carrara 1992] and *design intent* are stored as properties.

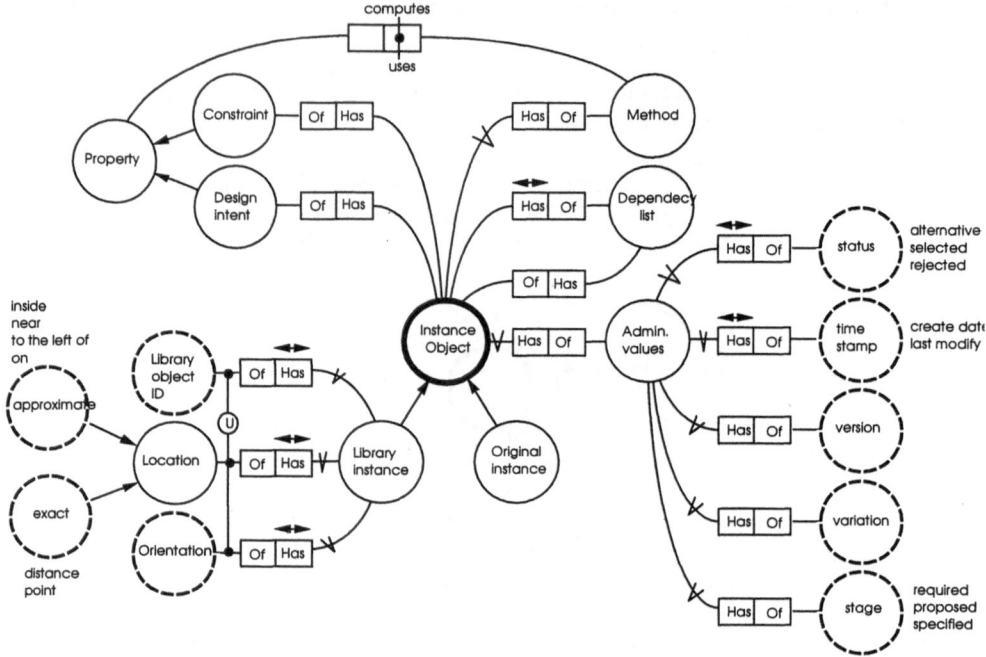

Figure 16. Instance Object

The property object (Figure 17) has a combination of features from EDM, RATAS, and Carrara. Each property must declare its (variable) *name*, *value type*, *range* of acceptable values, and *method* for determining value (default or computed).

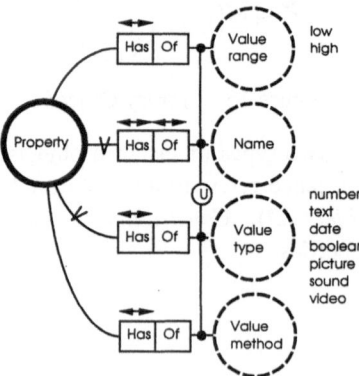

Figure 17. Property Object

A geometry class is not defined. An in-house class [Turner 1996] will be used to support the test implementation. Figure 18 shows that geometry needs a set of methods for input/output (drawing, sketching) and analysis (area, volume, centroid, nearness). The model also shows that is desirable to support different geometric representations of a single GBO, including geometries at different levels of detail and at different dimensions.

For instance, a furniture catalog object may have a generic 2D and 3D symbol, a 2D drafted image, and more complete 3D geometries at various levels of detail.

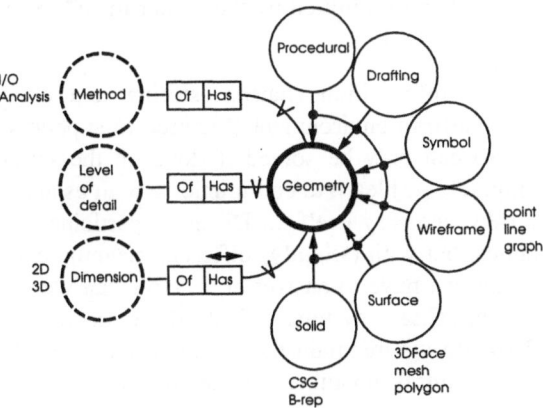

Figure 18. Geometry Object

6. Conclusion

The next step in this project is to validate (or abandon) the proposed models (and subsequent class structures) through a test implementation. It is hoped that the following questions can be answered before a complete re-write of PLAN is attempted:

Object linking. Is the management of the potentially large number of building objects and their relationships the responsibility of the application's data structures (how much should an application depend on the linking and grouping mechanisms in the proposed model?) For instance, the relationship between walls and rooms is a difficulty one to support since the same geometric information can possibly be stored twice: the edges of room polygons coincide with the faces of walls. The problem arises not in the storage of the edges, but in the initial drafting and further editing of the edges. Does the application allow the wall to be moved and have the room polygon follow (room polygon dependent on walls), or is the room polygon allowed to be modified and the walls re-configured (walls dependent upon room polygon)? It is a difficult task to allow the modification of both the walls and the room polygons and to keep a consistent data structure. The strategy adopted in PLAN allowed the user to <u>only</u> modify the walls of a floor (similar to the various double line tools in AutoCAD, MicroStation and form•Z). Each wall line knew its type—and therefore the wall thickness and right and left offsets—and each line knew if it represented an interior or exterior wall. The room and perimeter polygons were generated automatically [Turner et al. 1985] from the resultant double line network. Links between wall lines and polygons were kept so that rooms knew which walls were used in their generation, and walls knew which (GEDIT 1993) rooms they shared.

Semantics. Is a standard dictionary of terminology essential for applications to communicate with each other and with a data base of objects. The communication between the PLAN data base, standard parts libraries, and application programs was made easier by using a common vocabulary. This is a difficult task involving much translation for independent applications. (The original set of 83 definitions in the "AEC Attribute List" was taken from PLAN and submitted by the author to [IGES 1990].) The need was stated very well by [Tolman 1997]:

> "The problem of open information sharing, storing and exchanging in a multi-application and multi-platform environment (required to support Computer Integrated Construction and such) can only be solved if done on the semantic level. What is needed [for] our computers [to] learn to communicate with us and with each other is the terminology of the B&C industry itself. ... Different participants in the B&C industry use the same semantics, but with (slightly) different meaning and attribute definitions. The main questions that we have to answer are: 'How can we provide semantic inter-operability between the CAxx systems of all those people, if they use different semantics?', and, 'How do we maintain the semantic integrity of a project database?'. We should develop a 'neutral' unambiguous set of B&C semantics, including some basic attribute definitions... ."

Methods. Can abstraction and view mechanisms be supported by methods? Can generic methods be provided to applications to handle their object sub-setting and retrieval needs?

What kind of generic methods must exist for retrieving values of properties. What other kinds of generic methods are needed? It is assumed that an application view would be implemented as an object method. That is, each generic building object may have to provide a fixed set of methods for extracting necessary data. Non-standard applications may have to rely on generic `"get_property_value()"` functions.

How will methods handle objects whose properties and geometries are only vaguely specified ("the kitchen needs a stove", "the room is to have a south-facing window")?

Must all derived objects provide a standard set of building related methods for analysis such as `"compute_cost()`, `compute_weight()`, `count_components()`, `compute_volume()?"`

Constraints. Constraints are shown in the model as being part of instance objects. Do library objects come with preset constraints such as "A bathtub should be in a bathroom"? That is, should the constraint be attached to the GBO instead?

Dependency. How can circular dependencies be identified and resolved? Can a generic method support dependency?

References

ARCH:LIBRARY, 1988, A collection of functions and subroutines for the manipulation of 2D and 3D geometries. It also contains 2D and 3D matrix and vector manipulation routines and a set of routines to aid in sketching and creating parallel and perspective wire-frame and hidden line free drawings., Initiated in 1974, *Programmer's Manual,* Architecture and Planning Research Laboratory, Ann Arbor, MI

Bhat,R.R., Gauchel,J., Van Wyk,S., 1993, "Communications in Cooperative Building Design,". in *Proceedings*, CAAD Futures '93, Pittsburgh

Björk,B.-C., 1989, "Basic structure of a proposed building product model," *Computer Aided Design,* 21(2), 71-78

Carrara,G., Kalay,Y.E., Novembri,G.,1992, A Multi-modal representation of design knowledge", *Automation in Construction 1* (1992) 111-121, Elsevier

Coons,S.A., 1963, "An Outline of the Requirements for a Computer-Aided Design System," *Proceedings of the Spring Joint Computer Conference* , Detroit, MI

De La Garza,J.D., Oralkan, G.A., "An Object Space Framework for Design/Construction Integration," *Automation in Construction 3* (1995), 283-304, Elsevier

Eastman,C.M. 1975, *Spatial Synthesis in Computer-Aided Building Design* , John Wiley & Sons.

Eastman,C.M, 1991, "Use of Data Modeling in the Conceptual Structuring of Design Problems", in *Proceedings*, CAAD Futures '91, ETH Zürich

Eastman,C.M, Siabiris,A.,1995, "A generic building product model incorporating building type information", *Automation in Construction 3* (1995) 283-304, Elsevier

GEDIT, 1993, A geometric modeling and visualization program, Initiated in 1976, *User's Manual,* The College of Architecture and Urban Planning, The University of Michigan, Ann Arbor, MI

Gielingh,W., 1988, General AEC Reference Model (GARM), ISO TC184/SC4, TNO-IBBC.

The Initial Graphics Exchange Specification (IGES) Version 5.0, 1990, U.S. Department of Commerce, NIST, Center for Building Technology, Gaithersburg, MD, NISTIR 4412

ISO CD 13584-31, 1995, "Parts Library Standard", ISO TC 184/SC4/WG2

Junge,R., Liebich,T., 1995, "New Generation CAD in and Integrated Design Environment: a Path towards Multi-Agent Collaboration,"in *Proceedings*, CAAD Futures '95, Singapore

Mitchell,W.J., 1979, *Computer-Aided Architectural Design,,* Van Nostrand Reinhold, NY

Mitchell,W.J., Oliverson,M., 1978, "Computer Representation of Three-Dimensional Structures for CAEADS", Technical Report P-86, Construction Engineering Research Laboratory, Champaign, Illinois

Nijssen,G.M, Halpin,T.A, 1989, *Conceptual Schema and Relational Database Design: A Fact Oriented Approach*, Prentice Hall, Englewood Cliffs

Rumbaugh,J., Blaha,M., Premerlani,W., Eddy,F., and Lorensen,W., 1991, *Object-Oriented Modeling and Design*, Englewood Cliffs, NJ: Prentice Hall

Smeltzer,G.T.A, Dijkstra,J., 1993, "A Time Dimension for Computer-Aided Architectural Design Systems," in *Proceedings*, CAAD Futures '93, Pittsburgh

Tolman,F., 1997, On-Line Workshop Series: ISO TC184/SC4/WG3/T12, AEC - STEP Building & Construction Group, from e-mail message dated Tue, 04 Mar 1997

Tsou,J.-Y., 1992, *Using Conceptual Modelling and an Object-Oriented Environment to Support Building Cost Control During Early Design*, Doctoral dissertation, College of Architecture and Urban Planning, The University of Michigan, Ann Arbor

Turner,J., Borema,P.L., Hall,T., 1982, "Soft Drafting: Sketch Input for Computer-Aided Building Design", Architectural Research Laboratory, Ann Arbor, MI

Turner,J., Borema,P.L., Hall,T., 1983, *ARCH:PLAN Version 2.0, Architectural Plan Sketching Program* , Architectural Research Laboratory, Ann Arbor, MI

Turner,J., Hall,T., 1985, "The Automatic Generation of Room Polygons from a Weighted and Directed Planar Graph of Wall Lines", Architecture and Planning Research Laboratory, Ann Arbor, MI

Turner,J., 1996, "A Case Study of Object-Oriented Design and Implementation of a Generic 3D Geometry Viewer", http://www-personal.umich.edu/~turner/Abstracts.html, The College of Architecture and Urban Planning, The University of Michigan, Ann Arbor, MI

van Nederveen,S., Bakkeren,W., Luiten,B., 1993, "Information Models for Integrated Design", in *Proceedings*, CAAD Futures '93, Pittsburgh

Wittenoom,R., 1995, "Use of Draft ISO 13584 PARTS LIBRARY Standard for Design Decision Support", presented at the CIB78-95 Workshop, Stanford University , http://www.wt.com.au/~ausstep/stepdemo/readings/raw/cib7895.html

Zamanian,M.K., Fenves,S.J., "A Framework for Modeling and Communicating Abstractions of Constructed Facilities," in *Proceedings*, CAAD Futures '91, ETH Zürich

SORTS : A CONCEPT FOR REPRESENTATIONAL FLEXIBILITY

RUDI STOUFFS
Architecture and CAAD, Swiss Federal Institute of Technology Zurich
RAMESH KRISHNAMURTI
Department of Architecture, Carnegie Mellon University, Pittsburgh

This work is based on the recognition that there will always be a need for different representations of the same entity, albeit a building or building part, a shape or other complex attribute. This exigency ensues, formally, to define the relations between alternative representations, in order to support translation and identify where exact translation is possible, and to define coverage of different representations. We consider an abstraction of representations to model sorts that allows us to define algebraic operations on sorts and recognize algebraic relationships between sorts, providing us with a method for the analysis of representations, and the comparison of their coverage. We present the basis of support for a multi-representational environment.

1. Motivation

As computers become more powerful, we envision new domains, situations and environments where the computer can play an important role as a facilitator. This requires the integration and manipulation of increasing amounts of knowledge and information. In a collaboration in space and time, knowledge may be combined from multiple domains and exchanged between multiple disciplines; information may be provided and requested by all participants in this activity process; and information may change during the life-cycle of a designed artifact and because of advancements in knowledge and technology. Such collaboration creates additional impediments to effective computer support. The transfer of knowledge and information between multi-disciplinary partners and the interpretation of this data are non-trivial tasks.

Exchange of Information. A key problem is the loss of information in data exchange between representations supporting different views or abstractions of the same artifact. Within the domain of geometric modeling, many different representations have developed over the past thirty years, based on a variety of models, associated operations and underlying concepts. At the same time, much effort has been spent in order to arrive at a single representational standard for design and engineering data/ tasks. The IGES (Smith et al., 1988) and STEP (Bloor, 1991) projects are prime examples. However, no single geometric representation exists to solve every problem. Over time, new design and evaluation methods develop that require new information

R. Junge (ed.), CAAD Futures 1997, 553-564.

and extended representations. For example, the ever increasing concern for performance issues in building or engineering design necessitates the development of extended representations and manipulations of design geometries, including form and material properties, specified at various levels, in a hierarchical manner with dynamic relationships.

Multiple Representations. We strongly believe that there will always be a need for different representations of the same entity, albeit a building or building part, a shape or other complex attribute. Rather than provide specific applications for the translation between alternative representations that may serve the same or similar purpose, the need exists, formally, to define relations between alternative representations, to define coverage of different representations and to define where exact translation is possible. Our objective is to offer general translational support that recognizes the coverage of and relations between different representations, and to support different representations within a single, but, possibly locally or temporally disconnected environment (Stouffs et al., 1996). Such an environment must be able to identify when and where exact translation is possible so that the data-flow can be assessed and data-integrity can be monitored.

We conceive an algebraically based formalism that provides a handle for dealing with, operating on, and interrelating representations. This abstraction, termed *sorts*, defines formal operations on sorts and recognizes relationships between sorts, providing us with a method for the analysis of representations and the comparison of their coverage. Sorts are distinguished by their component sorts, their compositional relationships, and their assigned names. The manner in which sorts relate depends on the manner of their composition. Alternative compositions of the same component sorts give rise to alternative views of the same data, and can be derived from one another. Algebraically, sorts incorporate all embedded views, and contrary to current CAD approaches, sorts do not impose an object/geometry-centered view. For instance, by specifying a compositional hierarchy of the component sorts using a dependency relationship, the top component naturally becomes the focus of the information set. By altering the dependency relations, representations can be restructured to reflect any particular view of the information, adapting the organizational structure to the informational purpose.

Alternative Concepts and Ontologies. Probably the first and foremost problem associated with an all-encompassing representational model is the imposition of a fixed frame of reference. Instead, different disciplines adopt different ontologies; the same term may prescribe different meanings, the same concept may define alternative representations. Such semantic incompatibilities provoke data communication problems that may prevent collaboration at its fullest potential. The concept of sorts only provides for a common syntax, allowing for different vocabularies and languages to be created, and providing the means to develop translational facilities between these. For example, a point may be specified with any number of coordinates depending on its dimensionality, its coordinates may constitute integers, reals or rationals, these may be bounded in space, etc. Sorts can be defined accordingly, compared and related, and translational support provided for. There is no imposition of concepts beyond the purely syntactical, and the alphabet of building blocks for the vocabulary definition can be readily extended at all times.

Extensibility and Adaptability. No language thus created ever needs to be static. A vocabulary may be extended from the existing alphabet or using newly developed building blocks. Augmented representations that provide support for extended information and technological advances can be achieved by combining sorts into a new composition or by specifying additional component sorts to an existing composition. Far from having to redevelop not only the data but also the applicative operations, the concept of sorts aims to provide almost continuous support to evolving representations, providing for an environment that supports exploration and trial, even with respect to the representation. Data can be readily converted to new and extended (or condensed) representations, procedural operations may remain applicative if written with flexibility in mind. This extensibility provides ample support to explore extended representations for such purposes as building performance evaluation or the inclusion of design history and design intent into an artifact's design information.

2. Definition

Conceptually, a sort specifies a set of similar models that can be described in terms of a single set of equations. We denote the system of equations for a sort its *characteristic individual*. The models of a sort constitute its *individuals*; a *form* is any group of individuals of the same sort. Given any system of equations, the corresponding set of all models described by these equations defines a sort. Then, each individual of the sort is specified by assigning a value to each of the equational parameters. For example, a point is specified by its tuple of coordinates. From a purely conceptual point of view, sorts can be related by comparing their systems of equations, in a mathematical manner. For instance, since each equation constraints the values its parameters may adopt, a sort *subsumes* another sort if its equations form a subsystem of the other sort's equations. Note that while a sort commonly specifies a continuum of models, the extent of this continuum can easily be altered with respect to any conceptualization or purpose, by adding to and/or removing from its system of equations.

In practice, we specify the characteristic individual of a *primitive sort* as an elementary data type. Primitive sorts combine to *composite sorts*, using the formal operations to specify the compositional relationships. For instance, the *attribute* operator provides for (recursively) subordinate compositions of sorts using an object-attribute relationship in both a one-to-one and a one-to-many instantiation. For example, a sort of labelled points is specified as a sort of points, with one or more labels assigned to each point in the data form. The operation of *sum* allows for co-ordinating, disjunctive compositions of multiple sorts, under many-to-one and many-to-many instantiations, where each sort may -but does not have to- be represented in the data form. For example, a shape rule has both a *lhs* (left-hand-side) and *rhs* (right-hand-side) shape, either of which can be omitted. Finally, naming sorts provides for a semantic-like differentiation of sorts that may otherwise be syntactically identical, e.g., *lhs* and *rhs* denote equivalent -not identical- sorts.

While an analysis of the representational structure may be sufficient for a one-way translation, in order to provide for continuous support for data exchange in a multi-representational environment, some domain-specific knowledge is additionally needed. Consider the conversion and subsequent re-conversion of some data between two sorts. At least a canonical representation for individuals and forms is required in order to control information-loss in a strict sense. If the data is also altered between conversion and reconversion, additionally, appropriate ways of manipulating the data within the conceptual framework of the sorts are needed to achieve the same level of control. While certain other purposes or cases may demand even further domain-specific knowledge, providing uniform ways of handling different and a priori unknown data is a necessary condition for correct translation.

As an example, consider the association of building performance data to design geometries. The behavior of these data as a result of alterations to the geometries can be expressed through a number of operations chosen to match the expected behavior. When data is presented into a collaborative environment, the applicable manipulative operations are provided as well, allowing any application to properly interpret, manipulate and re-present this information without unexpected data loss. Therefore, the definition of a sort includes a specification of the operational behavior of its individuals for common arithmetic operations. The foundation of this approach is an algebraic model for shapes (Stouffs, 1994) that offers a uniform and consistent approach for dealing with geometries of mixed-dimensionality and non-spatial attributes, using common arithmetic operations of sum, difference and product (intersection) (Stouffs and Krishnamurti, 1996, discuss the adaptability of the algebraic model to domain-specific functionality). Extending this model to sorts, these serve as an abstraction of representations that enable us to compare multiple representations in terms of their coverage and thus support translation.

Notation. In the sequel, we adopt the following notation for sorts and their definition. We reserve the letters *a* through *h* to denote sorts (i.e., their names) and *i* through *l* for characteristic individuals (as elementary data types). A primitive sort is specified by a name, a characteristic individual and, optionally, a number of arguments dependent on the particular characteristic individual (see section 4 for some examples). We write $a : [i]$ or $b : [j](x)$. In some cases, a primitive sort specifies multiple aspects, each of which is assigned a name and considered a sort. The actual primitive sort only exists as a linking construct. We write $(c, d) : [k](y, z)$. The definition of a composite sort similarly consists of a name and an expression in terms of its component sorts using the symbols '^' and '+' to denote the respective attribute and sum operations. Parentheses provide for the nesting of definitions. We write $e : b + c$ and $g : f \, \hat{} \, e$, $g : f \, \hat{} \, (e : b + c)$, or $g : f \, \hat{} \, (e : (b : [j](x)) + c)$.

3. Matching

Compositions of sorts can be compared and potential matches classified according to their similarity, in rough terms, as equivalent, similar (as composed of equivalent sorts) and convertible. A more detailed classification is provided in table 1, using a numerical ordering system, termed *levels*, to differentiate the matches computationally. Different integral levels specify different matching types, decimal levels allow for finer comparisons. Equivalent as well as similar sorts guarantee correct conversion of the data without information loss, except semantically. Incomplete or partial conversions, either through augmentations (by adding sorts under the attribute operation) or diminutions (by removing sorts under the attribute operation) always result in data loss. Whether data-loss occurs in complete conversions, i.e., through rearrangements of the component sorts (by inverting attribute relationships), depends on the behavioral categories of the constituent sorts. It also depends on conversions between primitive sorts that share the same or similar characteristic individuals, but differ in one or more constraints or arguments (i.e., equations). Finally, the operation of sum specifies a subsumption relationship on sorts, where one sort may match a part, under sum, of another sort, providing an additional (perpendicular) grading scheme.

level	match	interpretation	example
= 0.0	identical	semantic equality	$a \leftrightarrow a$
< 1.0	equivalent	semantic derivability	$a \leftrightarrow b : a$
< 2.0	strongly similar		$a \,\hat{}\, b \leftrightarrow a \,(c : b)$ $a + (d : b + c) \leftrightarrow$ $(e : a + b) + c$
< 3.0	weakly similar	syntactic equality	$a : [i] \leftrightarrow b : [i]$
< 4.0	convertible	syntactic convertibility	$a \,\hat{}\, b \leftrightarrow b \,\hat{}\, a$ $a : [i](x) \leftrightarrow b : [i](y)$
< 5.0	incomplete • augmentation • diminution	partial convertibility	• $a \rightarrow a \,\hat{}\, b$ • $a \,\hat{}\, b \rightarrow a$
= 5.0	incongruous	no valid conversion	

Table 1. Integral levels of sort matching.

This approach allows us to monitor data-integrity during the design process, at all times, for a large variety of data. Specifically, we know that data-integrity is maintained for each data sort under any of the formal operations on data forms within

558

this sort; the coverage of data sorts can be compared; data can always be moved from more-restrictive to less-restrictive sorts without data loss; and data loss can be measured when moving data in the opposite direction. Active control over which conversions should and should not be allowed or considered is presented to the user in the form of a level tuner (see table 2): three user-defined levels specify level intervals of predefined handling behavior. Additional control is envisioned using rules of exception.

level	handling
$< l_1$	sorts are considered equal, conversion is performed without notification
$< l_2$	user is notified of conversion
$< l_3$	user's approval is requested for conversion
$\geq l_3$	conversion is not allowed, unless upon user's specific initiation

Table 2. Level tuner.

4. Exemplar Sorts

With the arrival of the world-wide web into the architectural domain, (virtual) architecture has conquered new territories, presenting the challenge to support, newly, any and all data types that can be envisioned. Rather than supplying slots for generic entities in a database-like fashion, sorts provide the ability to select the appropriate operational behavior, resulting in a more intelligent integration of the entities in the information environment. Currently, we consider characteristic individuals for geometric entities (i.e., points, lines, line segments, planes, plane segments and volumes), attribute entities (e.g., labels, weights, identifiers and signs), relational entities (e.g., properties) and hypermedia entities (e.g., images, text). Below, we describe some of these characteristic individuals, with their behavior, in detail.

Sorts of Labels and Points. A sort of labels provides the simplest example. A label, a string of characters, can be created, deleted or, as a combination of both, altered. When creating/adding a label to a form of labels, it becomes an additional element in the set that represents this form. When deleting/subtracting a label, the element is removed from the set. The algebraic operations behave as set operations; an empty set specifies an empty form. The resulting behavior is termed *discrete*. Discrete behavior ensures that two individuals are combined into one, only if these are identical. Another example of discrete behavior is a sort of points. Though a point can and may be considered as a composite sort, i.e., corresponding to a tuple of coordinates, specifying a characteristic individual for points trivializes consistency checking and generally simplifies data management and retrieval. For example, a one-step creation ensures that each point has the exact number of coordinates. The conversion of such a point into any other description, e.g., as a composite sort of coordinates, can be added easily.

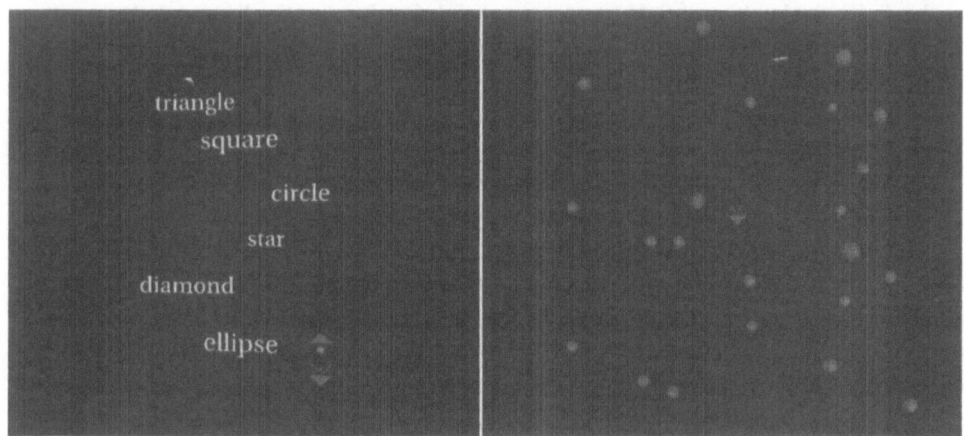

Figure 1. Visualization of forms of labels and points in VRML. Bige Tunçer.

Sorts of Images. An example of where such conversion is already provided, while it is both commonplace and necessary, is a sort of images. The characteristic individual for images, referencing an image, e.g., filename, accepts an argument specifying the image type, e.g., GIF, JPEG and TIFF. This argument to the sort's definition provides the ability to create separate sorts for gif-images, jpeg-images and tiff-images. In notational form, these become:

```
gif-images  : [Image](GIF)
jpeg-images : [Image](JPEG)
tiff-images : [Image](TIFF)
```

Provided that an application(s) or procedure(s) is available for the bytecode-wise conversion of images between different types, the characteristic individual for images specifies a conversion operation that can be invoked automatically. For example, in order to convert a form of images to "web-able" images, it suffices to define sorts of

```
images : gif-images + jpeg-images + tiff-images
web-images : gif-images + jpeg-images
```

and convert the form of images to a form of web-images.

Sorts of Properties. More complex representations often include relationships between data entities, for instance, for the purpose of referencing library entities, or for specifying a relational system between information aspects and artifacts. For example, a boundary representation for solids specifies vertices, edges, faces and solids, with edges linking vertices, edges bounding faces, and faces bounding solids. A characteristic individual for properties provides similar capabilities to sorts. A *property* is a two-way link between two individuals of given sorts. In order to prevent information-loss when handling properties and their associated individuals, an individual must know of any properties that link it to other individuals. All properties of the same sort thus assigned to an individual make up an attribute form with respect to the individual. Appropriate sorts must be defined accordingly under the attribute relationship. For example, a polygonal boundary representation with vertices and edges may define the following sorts:

```
points : [Point]
line-segments : [LineSegment]
(endpoints, segments) : [Property](line-segments, points)
vertices : points ^ segments
edges : line-segments ^ endpoints
```

An edge is defined as a line segment with (two) endpoints; a vertex is a point with any number of segments originating from it. The operational behavior of properties, termed *relationship*, is an extension of the discrete behavior, incorporating appropriate data-management procedures and ensuring data-consistency. For example, when deleting or otherwise modifying either individual associated to a property, this property is deleted or modified correspondingly.

5. Behavioral Categories

Most computer aided design applications, currently, adopt an object-behavioral approach. This approach specifies that once an entity is created (as an object), it remains unaltered except through explicit user intervention. Such behavior closely resembles the discrete behavior. The latter specifies that two individuals combine into one, only if these are identical. An object-oriented behavior can be achieved for any sort, by combining this with a sort of (unique) identifiers under the attribute relationship. The resulting data form is akin to a database of individuals, where each individual has a unique key assigned.

Often, a sort may profit from a more complex behavior that induces additional strengths. For instance, under the discrete behavior, a conscientious decision is required from the user on any change to the individual. This does not readily support creativity and novelty which rely on the concept of emergence, i.e., the recognition of information components and structures that are not explicitly present in the information and its representation, and on the restructuring of information (Stiny 1993; Krishnamurti and Stouffs, 1997).

Computationally recognizing emergent structures requires determining a transformation under which a specified similar structure is a *part* of the original structure (Krishnamurti and Earl, 1992). This part relationship can be freely defined, as long as it constitutes a partial order relationship. The algebraic model (Stiny, 1991; Stouffs, 1994) is based on such a part relationship. Under the algebraic model, a form is specified as an element of an algebra that is ordered by a part relation and closed under the algebraic (sum, difference and product) operations and (affine) transformations. Fundamental to the algebraic model is that under the part relation, *any* part of a form is a form. As such, a form specifies an infinite set of (sub)forms that are each part of the original form, and users can deal with forms in indeterminate ways. The *maximal element representation* (Krishnamurti, 1992; Stouffs, 1994) captures this notion. Figure 2 illustrates this definition of a form, under the part relation, with a form of line segments. The six maximal line segments can be grouped to three by three to form two triangles. However, other interpretations are possible and up to five triangles can be recognized in the shape and manipulated as such.

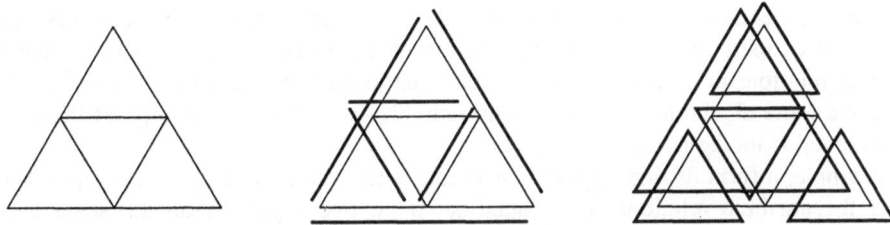

Figure 2. A shape with six maximal lines, specifying five different triangles.

Interval Behavior. The maximal element representation for line segments specifies an individual as an interval segment on an infinite line carrier, and a form as a (minimal) set of such individuals. A form is termed maximal if no two segments on the same carrier touch or overlap, i.e., these must be disjoint. Any two non-disjoint individuals are combined into one. Then, an individual is a part of another individual if its segment is embedded in the other segment on the same carrier. The algebraic operations on forms differ in their behavior from the discrete behavior in that individuals interact, not only if these are identical, but also if these overlap or touch.

Ordinal Behavior. Stiny (1992) explores the application of the maximal element representation to geometries with weights as attributes. Weights may be considered to denote thicknesses for points and lines, or tones for planes and volumes. A behavior for weights becomes apparent from drawings: a single line drawn multiple times, every time with different thickness, appears as it was drawn once with the largest thickness, even though it assumes the same line with other thicknesses. Thus, unlike behaviors described above, weight individuals from the same form always combine into a single individual. This (maximal) individual has as weight value the least upper bound of all the individuals' weight values, i.e., their maximum value. This behavior is termed *ordinal*; using numbers to represent weights, the part relation on weights corresponds to the less-than-or-equal relation on numbers.

Behavior of Compositions. A composite sort inherits its behavior from its component sorts in a manner that depends on the compositional relationship. Under the operation of sum, the behavior is that of the component sort for each component. Forms from different component sorts never interact, the resulting form, termed *metaform*,

562

corresponding the composite sort, is the group of forms for all component sorts. When adding an individual to a metaform, this individual gets added to the appropriate component form. When an algebraic operation applies to two metaforms of the same sort, the operation instead applies to the respective component forms. A metaform is empty only if all component forms are empty.

The attribute operation on sorts specifies a dependency relation on the sorts in a composition, where each component, except the first, defines an attribute sort to the previous component. That is, a corresponding form consists of individuals of the first component sort each of which has, as attribute, a form corresponding to the sort as a composition of all but the first component, in a recursive manner. Thus, the behavior of such a sort is defined by the behavior of its first component sort. Specifically, when an algebraic operation applies to two forms of the same composite sort (under the attribute relationship), identical individuals merge and their attribute forms combine under the same algebraic operation. Any individuals that have an empty attribute form are removed from their respective forms.

For example, a form of line segments with attributes, corresponding to a composite sort under the attribute relationship, is maximal if no two segments on the same carrier overlap and any two segments that touch have non-equal attribute forms. Then, an individual is contained in a form if a discrete classification of this individual into component individuals can be found such that each component is a part of an individual in the form, and the respective attribute forms adhere to the appropriate part relation.

Singly-associated versus multiply-associated sorts. Behaviors play an important role when assessing data-loss in information exchange between different sorts. Reorganizing component sorts under the attribute relationship into a different compositional hierarchy may alter the corresponding behavior and trigger data-loss. Consider a sort of weighted points, i.e., a sort of points with attribute weights, and a sort of points of weights, i.e., a sort of weights with attribute points. A form of the former sort is a set of non-coincident points, each of which has a single weight assigned. These weights may be different for different points. The resulting behavior of the form is discrete. A form of the latter sort is composed of a single weight with an attribute form of points, and has ordinal behavior. In both cases points are associated with weights. However, in the first case different points may be associated with different weights, whereas, in the second case all points are associated with the same weight. In a conversion from the first to the second sort, data-loss is inevitable.

Sorts with ordinal behavior are also denoted *singly-associated* sorts. A form of a singly-associated sort necessarily contains only a single individual, unless empty. In contrast, forms of *multiply-associated* sorts can contain any number of individuals. Sorts with discrete, relationship or interval behavior are all multiply-associated sorts. The previous exposition always applies to compositions of both singly-associated *and* multiply-associated sorts under the attribute relationship.

6. Discussion

Our intention is not to create an all-encompassing representation as a new standard that would be "up to date" as to current research and usage. Given the sheer variety of representations available and used, any single standard, however flexible, will always typify a rather subjective evaluation that is indifferent to the purposes of the different representations. Moreover, as technology evolves and knowledge increases, any standard would become quickly outdated. Indeed, solid modelers with new capabilities are being developed, such as the parametric solid modelers found in Pro-Engineer™, SDRC's IDEAS™ and Autodesk's DESIGNER™ for which no adequate translators exist.

Instead, we propose an alternative approach that is not restricted to currently known representations and that may take away the need for standards altogether. Conceptually, we distinguish the data classes and subclasses that constitute a representation and consider a multi-way communication system using these data classes as the vocabulary elements (see also Stouffs et al., 1996). This vocabulary is not a priori limited and may be extended at all times by any application. Using sorts to achieve such a communication system, new applications that adopt sorts as their representational framework may synthesize any or all of the sorts' strengths into their functionality. How and whether these strengths are presented to the user depends on this functionality. Redeveloping an existing application to profit from sorts, without altering the original functionality, only extends the application with readily accessible communicational capabilities, but does not provide a glimpse of the flexibility that is intrinsic to sorts. How such flexibility can be presented to the user, in the best and easiest manner, makes for an interesting question on its own. Currently, we are developing database support for persistent information, data visualization tools in VRML, and a JAVA-based interpreter/interface.

Sorts were conceived in order to respond to and deal with issues of data-loss in information exchange, mainly between existing applications. These, primarily, profit from a framework for data communication constructed on sorts. When dealing with a substantial number of applications and corresponding representations, all of which may not be known at the time of development, it becomes inefficient to develop a tool for every (possible) channel linking two applications. Instead, consider a support system for data exchange consisting of a node for facilitating data transfer and translation, and a query language for communication between the node and each application. Such a system only requires from each participating application a (single) communication front-end that can pose and answer queries to and from the node. For an example, consider the domain of solid modeling. Whereas different boundary representations specify different relational systems to link the boundary entities, the entity classes of vertices, edges, faces and solids are common to the domain and, in some variations, to the representations. Using a domain-specific query language that understands these concepts, and the node's ability to explore different entity relationships, a multi-way communication system can be established that supports solid representations (Stouffs et al., 1996, present a theoretical analysis of solid representations for this purpose).

Additional non-geometric data can be treated as attributes to the geometric entities and communicated as such.

Sorts present a method for the analysis of representations, and the comparison of their coverage. A multi-way communication system based on sorts provides the ability to identify when and where exact translation is possible. While we cannot guarantee complete and correct translation, at least, data-flow can be assessed and data-integrity monitored.

Acknowledgments

The first author wishes to thank Kuk-Hwan Mieusset and Bige Tunçer for their contributions to the current developments, respectively, database support and data visualization. He also wishes to thank Gerhard Schmitt for his undaunted belief and support.

Bibliography

Bloor, M.S. (1991) STEP-standard for the exchange of product model data, Standards and Practices in Electronic Data Interchange, IEE Colloquium, 2/1-3, The Institution of Electrical Engineers (IEE), London.

Krishnamurti, R. (1992) The maximal representation of a shape, Environment and Planning B: Planning and Design 19, 267-288.

Krishnamurti, R., and Earl, C.F. (1992) Shape recognition in three dimensions, Environment and Planning B: Planning and Design 19, 585-603.

Krishnamurti, R., and Stouffs, R. (1997) Spatial change: continuity, reversibility and emergent shapes, Environment and Planning B: Planning and Design 24.

Smith, B., Rinaudot, G.R., Reed, K.A., and Wright, T. (1988) Initial Graphics Exchange Specification (IGES), Version 4.0, SAE/SP-88/767, Society of Automotive Engineers, Warrendale, Pa.

Stiny, G. (1991) The algebras of design. Research in Engineering Design 2, 171-181.

Stiny, G. (1992) Weights, Environment and Planning B: Planning and Design 19, 413-430.

Stiny, G. (1993) Emergence and continuity in shape grammars, CAAD Futures '93 (eds., U. Flemming and S. Van Wyk), 37-54, North-Holland, Amsterdam.

Stouffs, R. (1994) The Algebra of Shapes, Ph.D. dissertation, Departement of Architecture, Carnegie Mellon University, Pittsburgh, Pa.

Stouffs, R., and Krishnamurti, R. (1996) The extensibility and applicability of geometric representations, 3rd Design and Decision Support Systems in Architecture and Urban Planning Conference, Architecture Proceedings, 436-452. Eindhoven University of Technology, Eindhoven, The Netherlands.

Stouffs, R., Krishnamurti, R., and Eastman, C.M. (1996) A formal structure for nonequivalent solid representations, IFIP WG 5.2 Workshop on Knowledge Intensive CAD II (eds. S. Finger, M. Mäntylä and T. Tomiyama), International Federation for Information Processing, Working Group 5.2.

THE DISTRIBUTED ARCHITECTURAL MODEL FOR CO-OPERATIVE DESIGN

Shun WATANABE and Kiichiro KOMATSU
Institute of Policy and Planning Science
University of Tsukuba
1-1-1 Tennodai, Tsukuba-shi, Ibaraki
JAPAN

Collaborative design has become one of the most significant topics in the field of design science and computing. Many studies have been made on proposing methods of collaborative design computing from various points of view. In this paper, the latest technological approach in the field of computer science is taken to illustrate future design systems. The distributed architectural model is proposed to support collaborative and concurrent design.

I will begin by discussing existing methods for design collaboration, and I will also mention the CORBA (Common Object Request Broker Architecture) specifications for the framework of the distributed computing environment. The semantic/presentation split basis is introduced as the essential for developing distributed applications, and the strategy for adapting AKM (Architectural Knowledge-representation Model) to this basis will also be considered. Then I will introduce the sample implementation of our distributed architectural model in the Distributed Smalltalk environment, and also explain IDL (Interface Definition Language) interface of architectural objects.

Keywords: architectural model, collaborative design, distributed computing

1. Introduction

Currently, architectural design projects have been becoming bigger and more complicated. It is impossible for a single designer to command whole parts of the projects. They involve many specialists and many sub-projects must be performed concurrently within the project with various constraints related to each other.

On the other hand, since the world-wide computer network has been becoming more popular because of the Internet, collaborative design has become one of the most significant topics in the field of design science and computing. Common technologies and popular applications for the Internet communication have also been developed, and many studies have been made on proposing methods of collaborative design computing from various points of view. Some papers describe how to communicate architectural design information by transferring CAD data files through this network, and another describes how to discuss the design problems by using teleconference systems in the virtual design studio.

R. Junge (ed.), CAAD Futures 1997, 565-570.
© *1997 Kluwer Academic Publishers.*

These researches usually describe how to modify existing design systems and applications to adapt to the network environment and then use them as substitutions for tools of collaborative design so that everyone can easily give common consent to the design. However, it will be difficult enough to support collaborative design projects by modifying existing design systems since they are essentially designed and implemented to be operated by a single user and are not supposed to be handled by multiple users in the concurrent operations at one time.

It is indispensable for the ideal system of collaborative design to support these concurrent operations to a single data (object). In architectural design projects, many designers and engineers investigate the same architectural design from their various views of expert knowledge and make alterations to the relations and attributes of the design elements throughout the design process. The system should be able to inform all related staff of these alterations simultaneously since they all will always require the latest information about the design.

2. Communication of Design Information

Methods to communicate architectural design information to other designers through the computer network can be classified roughly into the following three levels from the view of information processing technology:

2.1 FILE TRANSFERING

The easiest method of communicating architectural design information is to transfer data files and make copies to the remote client through the computer network. FTP may be the most popular and effective application for distant communication. The WWW can be regarded as an extension of file transfer technology in which the graphical user interface is appended to link other documents. Supplying data of 3-dimentional space by VRML and QTVR on the web service is also regarded as a further extension. At this level, the remote client can browse through the data about the design information transferred to it, but the remote client cannot alternate original data on the server directly. Even though alterations can be made on the copies individually, it is impossible to merge them without creating problems. In addition, there is no guarantee that the data sent to the remote client is the latest design information since an action of transferring files is brought by the client.

2.2 FILE SHARING

Another method of communicating architectural design information is to share files between clients through the computer network. NFS may be the most popular application and it is effective for local collaboration. At this level, a client can browse through and alternate original data about design information on the server file system directly. File sharing is well functioning on the network environment where remote file systems are mounted seamless and permanent for end users, but sometimes just functioning as an alternative method for file copy as is usual with the non administrated systems. In addition, the same design information cannot be browsed through simultaneously by several users since an individual file is the minimum unit for exclusive access control.

2.3 MEMORY SHARING

The ideal system for communicating architectural design information would be to share memory space and to transfer up-to-date information to remote clients through the computer network simultaneously. This system would be composed of multiple processes distributed over the network, and the exclusive controls of them should be very complicated since there is a possibility that some concurrent processes would access the same place of shared memory at that time. The local system cannot control remote systems or the networks that connect them. Currently, the distributed computing environment has been the most significant topic in the field of computer science and many technologies are under development. To enhance design collaboration, future design systems should employ this technology so that it can be handled by multiple users.

3. Distributed Computing Environment

A distributed application generally consists of the client system which requests services and the server system which provides these services. The OMG (Object Management Group) proposes that the standard specification called CORBA (Common Object Request Broker Architecture) should realize these interfaces through the common ORB (Object Request Broker). That is, the CORBA can be said to be enhanced technology for object-oriented programming.

In distributed object-oriented systems, when a request to create an object in remote memory is noticed, a surrogate to refer to the remote object is also created in local memory. In the local system, since messages cannot be sent to the remote object directly, they are alternatively sent to the surrogate. The surrogate by itself cannot process the request so the ORB gets involved to intercept and forward the message appropriately. The ORB on the remote system translates a service request from a requester through the local language, such as C, C++, or Smalltalk, to the implementation-neutral IDL (Interface Definition Language), and forwards it to the ORB on a server system where a correct provider object is located. The server's ORB then translates an incoming request to the local language and forwards it to the provider object for processing. Therefore, as long as the applications conform to the CORBA standard, it is possible to communicate between them even if they are written in different local languages.

The AKM (Architectural Knowledge-representation Model) was proposed as the framework for developing our knowledge-based CAAD system in 1989. In the AKM, the constraints between architectural design components, such as floors, rooms, columns, and beams are mainly concerned with the view according to cognitive science. Details of the AKM are described in the reference paper titled, "Knowledge Integration for

Architectural Design." The AKM is totally object-oriented and the future perspective to shift it to the distributed computing environment has been considered from the beginning of its development. So, our work of adapting the AKM to the distributed computing environment is to split the functions of the AKM objects into client parts and server parts to optimize performance, then translate critical methods into IDL and register them to the interface repository.

4. Semantic and Presentation

A certain architectural object has different expressions in different design specifications according to its characteristics. For example a single column which constitutes architectural design is presented in different appearances between a plan and a section, and a drawing and a table. The different appearances are no more than fragmentary descriptions about the particular attributes of the architectural object according to the rules for expressing the design specification. Therefore, there is a need for architectural objects that can be accessed simultaneously by several designers and engineers at the same time so that the objects' relations and attributes can be altered throughout its design process.

Similar discussions have been made about the field of distributed computing. To optimize performance of distributed applications, the semantic/presentation split is proposed to permit users to have simultaneous views of objects. Within distributed computing, one logical object is split into two functional parts, with the presentation object handling the bulk of user interaction, and with the semantic object holding an object's persistent state. The presentation objects and semantic objects are linked in a many-to-one relationship. The views of the presented objects may be different in appearance but reflect the same underlying state (information) of the semantic object.

Here, we have to concern the difference between the architectural model, such as the AKM, and the graphic model such as a surface model or a solid model. The architectural model represents the relations and attributes of architectural design components, whereas the graphic model represents geometric features which can also be one of the attributes of the architectural design component. The semantic function of architectural objects is to generate, hold, and maintain the objects' relations and attributes, while the presentation function of architectural objects is to produce and render a graphic model for visualization respectively according to the objects' attributes and relations. Methods which should be translated into IDL are functions, to transfer the geometric parameters of architectural objects between semantic and presentation objects, that make presentation objects produce an up-to-date graphic model.

5. Implementation of the Distributed Architectural Model

Our sample implementation of the distributed architectural model is written in the Distributed Smalltalk environment. In this environment, further services specified by OMG for developing distributed applications, such as the naming policy, the event notification, and the object lifecycle, are provided in addition to the ORB. Our implementation also follow these specifications.

Now, essential classes of the AKM are transferred into the distributed architectural model. For example, the class of "Column" in the AKM is split into two classes, which are "ColumnSO" and "ColumnPO." "ColumnSO" represents a semantic of columns and "ColumnPO" represents a presentation according to the naming policy of Distributed Smalltalk. These class definitions follows the column's definition in the AKM, but the methods are classified into two classes according to the semantic/presentation split basis. IDL statements to change the position of a column of the distributed architectural model are described as follows:

```
// Column

module Column   {

        interface ColumnPO : Presentation {

                #pragma    selector setPosition setPosition:
                void setPosition (in Point newposition);
        };

        interface ColumnSO : ApplicationSO {

                #pragma    selector setPositionBy setPosition:by:
                void setPositionBy (in Point newposition, in ColumnPO po);
        };
};
```

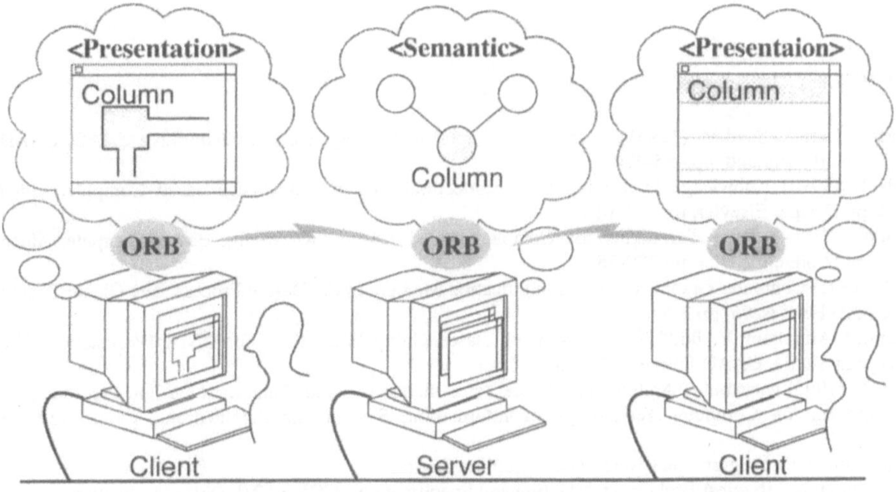

Figure-1 Relations of the semantic/presentation model

where "interface" defines a factory (class) and ":" means its inheritance. The symbols which follow the "selector" are the Smalltalk method name, and the corresponding function is written in the next line.

Once a designer tries to change the position of a column from the client view, the request message of setPosition:by: is once sent to the surrogate of aColumnSO of the client, and it is translated into the IDL through the ORB, then it is forwarded to aColumnSO of the server. The corresponding method in Smalltalk is to reassign the instance variable for position of aColumnSO to aPoint, and to broadcast a setPosition: message to every presentation as follows:

```
setPosition: aPoint by: aPO
        position := aPoint
        self presentersExcept: aPO do: [:po | po setPosition: aPoint]
```

Therefore, every client view will be rendered again according to the latest information about a column position just accepted by aColumnPO.

6. Conclusion

Throughout this paper, I have explained about the principle technology and the implementation of the distributed architectural model which is necessary for the system of collaborative and concurrent design. The model is composed of semantic parts and presentation parts, and enables several designers and engineers to collaborate on an architectural design project from different locations simultaneously. In the current implementation, every processing of clients and server is handled in Smalltalk. In the next implementation, geometric processing, which is the function of presentation objects on the client, will be handled in OpenGL to improve performance.

References

Maher.M.L., Gero.J., Saad.M. (1993) Synchronous Support and Emergence in Collaborative CAAD, CAAD Futures '93, North-Holland, pp.455-470

Watanabe,S. (1994) Knowledge Integration for Architectural Design, Knowledge-Based Computer-Aided Architectural Design, Elsevier, pp.123-146

Mitchell,W.J. (1994) Three Paradigms for Computer-Aided Design, Knowledge-Based Computer-Aided Architectural Design, Elsevier, pp.379-388

Sasada,T. (1995) Computer Graphics as a Communication Medium in the Design Process, The Global Design Studio, CAAD Futures '95, pp.3-5

Morozumi,M. Murakami,Y. Iki,K. (1995) Network Based Group Work CAD for UNIX Workstation, The Global Design Studio, CAAD Futures '95, pp.637-646

Watanabe,S. (1995) Distributed Architectural Model for Collaborative and Concurrent Design, Proceeding of 18th Symposium on Computer Technology of Information, System and Applications, pp.127-132 (in Japanese)

Distributed Smalltalk Programmer's Reference, ParcPlace Digitalk

The Common Object Request Broker: Architecture and Specification - CORBA2.0, OMG Publications

PRODUCT DATA MODEL FOR INTEROPERABILITY IN AN DISTRIBUTED ENVIRONMENT

RICHARD JUNGE
Faculty for Architecture, Professorship for CAAD
Technical University Munich

DR. THOMAS LIEBICH
CAB Research & Consult
Munich

Abstract

This paper belongs to a suite of three interrelated papers. The two others are 'The VEGA Platform' (Junge, Koethe, Schulz, Zarli, Bakkeren. this volume) and 'A Dynamic Product Model' (Junge, Steinmann, Beetz. this volume) These two companion papers are also based on the VEGA project. The ESPRIT project VEGA (Virtual Enterprises using Groupware tools and distributed Architectures) has the objective to develop IT solutions enabling virtual enterprises, especially in the domain of architectural design and building engineering. VEGA shall give answers to many questions of: what is needed for enabling such virtual enterprise from tht IT side. The questions range from technologies for networks, communication between distributed applications, control, management of information flow to implementation and model architectures to allow distribution of information in the virtual enterprises.

This paper is focused on the product model aspect of VEGA. So far modeling experts have followed a more or less centralized architecture (central or central with 'satellites'). Is this also the architecture for the envisaged goal? What is the architecture for such a distributed model following the paradigm of modeling the 'natural human' way of doing business? What is the architecture enabling most effective the filtering and translation in the communication process. Today there is some experience with 'bulk data' of the document exchange type. What is with incremental information (not data) exchange? Incremental on demand only the really needed information not a whole document.

The paper is structured into three parts. First there is description of the modeling history or background. the second a vision of interoperability in an distributed environment from the users coming from architectural design and building engineering view point. Third is a description of work undertaken by the authors in previous

R. Junge (ed.), CAAD Futures 1997, 571-589.
© *1997 Kluwer Academic Publishers.*

project forming the direct basis for the VEGA model. Finally a short description of the VEGA project, especially the VEGA model architecture.

Introduction

Having a close look at the daily work of architects, it becomes obvious that a very considerable portion of their time and efforts is spent with organizing the collaborative design process and communication with the client, official bodies construction companies and other engineers. The pressure coming from these duties is rising, and special project managers are employed not only for large projects. By the same time the responsibility and influence of the architect on the whole building process, which does not end with the completed design drawings, is reduced, part of their traditional scope of the process will be lost.

Considering the roots of the profession an architect was the „master of builders", as the Greek origin of the name shows. There were no special means necessary to allow him to communicate with his fellows, all were on site and design specifications where often drawn directly on stone or earth. Sketches merely served as schematic overviews. The first change of this paradigm appeared during the quattrocento, when the Renaissance architects introduced new design methods and were separated from the building site. Detailed and scaled design drawings now became part of the contract and had to be communicated with the builders. With the increasing complexity of buildings, and the parallel process of further division in work, far more detailed design specifications have to be communicated and coordinated with a growing number of participants. The question here is, whether the developing technology can provide the appropriate means to allow architects to maintain their central role or whether they will become design professionals amongst others being coordinated by project managers working directly for the client or main contractor.

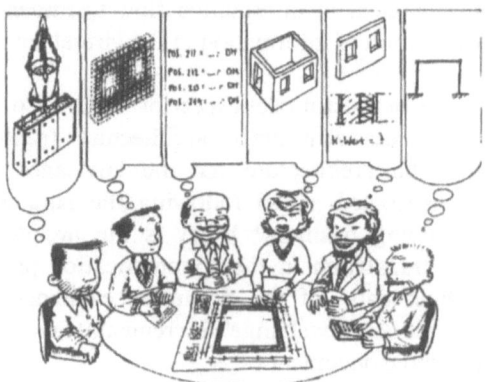

Figure 1 Collaboration today

On the technological side of the problem there is the overall impression that efforts where mostly assigned to the development of individual software as for drawing and designing, both within commercial sector and research while the communication aspect was almost neglected. At least the communication aspect was understood as an integral part of architects professional work In consequence the often quoted „islands of information", or islands of construction computing, have to be acknowledged. How to bridge those and to allow design programs to collaborate on behalf of the designers as natural as designers do? What could be expected from such a technology and what are the main components?

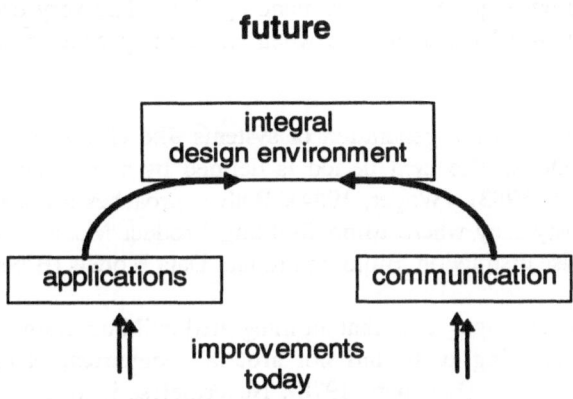

Figure 2 Improvement of IT

Basic Technology

Such a 'future integral design environment' modeled after the human reality of design and communication needs at least two basic technologies. First software has to understand the semantic of the objects it is dealing with. The solution is a 'semantical kernel' as a basis for applications, a product data model. The second is a 'transport system' for information, not just data, that allows direct and flexible communication between, again following the human reality, distributed applications and their product data models. The technology for this part is COAST (COrba Access to Step models) which is an extension of CORBA (Common Object Request Broker Architecture). This paper will focus on the product data model part, COAST and CORBA technology is being explained in 'The Vega Platform' (this volume).

Product Modeling is nowadays a widely accepted methodology for the development of advanced software systems in engineering domains, such as architecture and building. The methodology improved a lot especially during recent years through European projects such as COMBINE, Neutrabas, ATLAS; COMBI; PISA and since 96, VEGA as well as a lot more from other then the AEC domain. The progress however is not based on the fact that these projects are all EU funded but rather on the fact that all these projects where following the basic idea: using and at the same time improving the technologies and methodologies of the STEP environment. As a result there is nowadays a powerful basic technology of STEP- CASE- tools available as well as a widely accepted baseline towards product modeling.

The modeling approach or architecture in these projects was different partly, certainly, because projects had different goals. These developed from integration of applications in specialized domains to the development of a platform for interoperability through distributed product models. Another reason for different model approaches certainly was the state of knowledge at the corresponding time. The very origins of product models are to be found back at a time when the term product model wasn't even invented.

A starting point surely is the generation of systems like GLIDE and BDS (Richens, 1978) and many others. The next period is defined by a first generation of expert systems (Landsdown, 1983), (Wager, 1984). Both approaches are using data models. Today one would say they where using Building Product Models, or to follow B-C Bjoerk's more precise distinction, Building Product Data Models (Bjoerk, 1995).

A certain degree of consensus, e.g., that an integrated building design system has to be structured around building parts and not around geometrical objects was already reached in the early years (Eastman, 1978). Nevertheless it took more then a dozen further years until the structures of building product models have become clearer (Junge, 1994).

The stream to which above mentioned projects belong to started with the development of two generic models. The GARM (General AEC Reference Model) (Gielingh, 1988) was developed for the STEP project (Standard for the Exchange of Product model data). The GARM provides a first basic concept by introducing the 'Product Definition Unit' (PDU) and its subtypes 'functional unit' and 'technical solution'. The Building Systems Model (Turner, 1990), developed as an input to STEP, also shows a top down strategy to model a building. It therefor introduces functional systems as, e.g., enclosure, structural, mechanical, etc. and their entities.

With the development in the IMPACT project very intensively studied fundamental modeling principles as specialization, discrimination and orthogonalisation as well as implementation principles as extension and instance. The model introduced the technique of 'layered models (IMPACT 1991). This term was used for a concept of a structure with a number of models of different levels of abstraction on different 'layers within one larger model structure. This as a significant architecture for the design of models with a larger scope.

The ESPRIT project Neutrabas (Neutrabas, 1991), which is in the ship building domain, used concepts developed in IMPACT at the same time and extended product modeling technique further. It is one of those projects that started to deal with the question, respectively solutions, for objects, belonging to more than one system and are playing different roles in them. The discussion on this problem is still open today.

This open discussion leads directly to the question how a building product model that would cover the needs of all parties involved in the process of design, construction, maintenance and management has to be structured. Today the belief is that there is no way leading to a homogeneous single central building product model is accepted nearly unanimously. In a number of current projects one finds various terms used for strategies dividing the universe of discourse into partial models. Terms used to describe such partial models are 'topical model' (van Nederveen and Tolman, 1992). The ESPRIT project ATLAS has 'view type models' (ATLAS, 1994) and ESPRIT project COMBI uses 'partial models' and 'application models' (Junge et al, 1994). Although the solutions are slightly different, a way has been found that leads to practical building product data models. Currently, the discussion focuses on the question, how to interconnect these partial models?

The discussion is on strategies where these partial models should be compatible with each other. The idea is that there is a "central" part of the modeling domain that would be shared by all partial models (Luiten et al., 1991). These central parts are often called kernel or core models. Again one finds this concept in ATLAS and COMBI, as LSE Project Type Model or as Central Neutral Model The main question still remains: What is the focus of a core model? Is it the sum of parts used in at least one application or domain? Or should it be a small but generic kind of 'overhead model' which aims at staying stable when the scope of the overall model expands? A proposed solution for the first is e.g. the minimal kernel model as used in the NICK project (Tarandi, V. 1993)

The Building Construction Core Model (BCCM, ISO STEP Part 106) (Tolman and Wix, 1995) is based on the opposite concept. Here the basic assumption is that all objects that are used by more than one system should go into the core model. This at first glance simple concept implies that the core model will explode into a kind of central model. The question is how many objects do not belong to more than one system? This seems to be a concept that bears the danger in it that the model will 'explode' because of the shear number of entities it will have to deal with and it will not be a stable core for the suite of models it is intended to serve until all these models are being developed themselves.

It seems that the BCCM is following a certain 'One Model Mystique', everything that is used in more than two apps has to go into the model. Such a model for Computer Integrated Construction (CIC) could easily reach a number of some thousand entities. Even if most of the attributes would go into the connected 'domain models', thus leaving itself to be a semantically flat model, it seems very questionable if such a model would be manageable. A comparable approach, the IPIM, was already discussed in STEP during '88 and '89 and it was given up.

Period	Focus of development	Example Model
1975-85	'pre-product models'	
1975-80	basics of building models	GLIDE
1980-85	models for expert systems	
1985-	building product data models	
1986-90	generic product models	GARM, BSYSM
1990-95	aspect and layered models	COMBI, ATLAS
1995	'consolidated models'	O.P.E.N

Figure 3 Model history

One could draw a kind of time table showing the main focuses of research and development of product modeling technology. A variation of such a time table presented by Bjoerk (Bjoerk 1993) is shown in fig 3. The concept of 'consolidated models' is introduced into the discussion in this paper for the first time.

Model layer	Example data structures	Example models
Fundamental data model	objects, attributes, relationships, generalisation- specialisation, rules, methods, messages	Relational data model Entity-relationship -model The object as in objectoriented programming The frame
Generic product description model	Data structures for describing aggregation- decomposition, type objects, versions, shape and location information, abstraction levels etc	STEP Resource Parts EDM GARM OOCAD
Building product data model kernel	Generic object classes which are particular for a specific design or construction discipline and or phase in the construction process	Building Systems Model GSD Model RATAS Framework KBS MODEL NICK
Aspect models	Detailed object classes which are particular for a specific design or construction discipline and or phase in the construction process	COMBINE IDM RATAS quantity take off prototype Structural Steel Model CIMsteel LPM
Aplication models	Detailed object classes which constitute the conceptional model of one particular application	Not subject for this framework

Figure 4 The layered Framework Architecture

In 1994 Bjoerk presented a structure of how the different product modeling concepts very briefly and incomplete described above can be grouped. He called it a 'Layered Framework Architecture for grouping data structures which are used in building product data models' (Bjoerk 1994). Bjoerk writes: 'The set of data structures that make up a fictive comprehensive building product model is partitioned using an onion-like overall architecture. ....a decomposition into five layers would seem appropriate'.

Following Bjoerk all these layers combined would form a comprehensive building data model. The models quoted in the framework where more or less focused on the necessary exploration of only one the layers. From today's level of knowledge it is obvious that none of those 'layers' as standalone solution can provide the necessary basis for a comprehensive Building Product Data Model.

The authors therefore propose the 'Consolidated Model' as the attempt to combine the pieces developed in previous projects to an architecture mature for an commercial implementation of the paradigm of the 'Building Product Model'.

Steps towards the 'Consolidated' Model

The Consolidated Model architecture emerged out of a more or less consequently and gradually development in a series of projects in which the authors where involved during the last five years. These are COMBI, an ESPRIT project, NextCAAD I, NextCAAD II, both are industry funded development projects and VEGA, again an ESPRIT project. One common basis for all these projects was an as much as possible and consequent application of so called 'STEP methodologies' and the search for a model architecture flexible and comprehensive enough to serve as a basis for commercial applications in architecture and engineering.

COMBI

The COMBI project (1993-95) developed a prototype environment for cooperative design focused on the domain of structural engineering. It envisions an intelligent environment based on the idea of 'integration by communication' (Junge 1991). The first focus is the development of four application tools for structural design. These tools form a chain beginning with beginning with soil mechanics and foundation design to structural analysis end ending with reinforcement design. The second focus is on integrating these four application tools, while the third focus is on linking tools which are developed on the basis of a product model with an external traditional CAD system. For these three tasks a product model framework is being developed.

This framework has been created under consideration of the evolutionary multi stage and multi agent nature of the design process and for unified integration approach. The architecture of the framework can schematically be represented by four hierarchically structured levels (Figure 5). It can be envisioned as a network of computers and users, where communication and coordination is achieved through a shared information medium and a control mechanism which provides facilities for the integration of loosely coupled design tools.

The application tools, that support the work of the individual designers shown on the bottom level have their own application dependent product model, which need not necessarily conform to a standardized product model specification. Thus, application data are processed and stored only within their application domain, which helps to avoid information explosion and unnecessarily complicated data structures. The three main reasons for this decision had been:

- the application data structure must be organized according to the needs of the application methods in order to achieve maximal run time performance,
- existing tools must be integrated without internal modifications, and
- specific data extensions needed by the application methods are often only for temporary use and can be generated automatically, e.g. by a finite element mesh generator, as it is the case in the COMBI prototype.

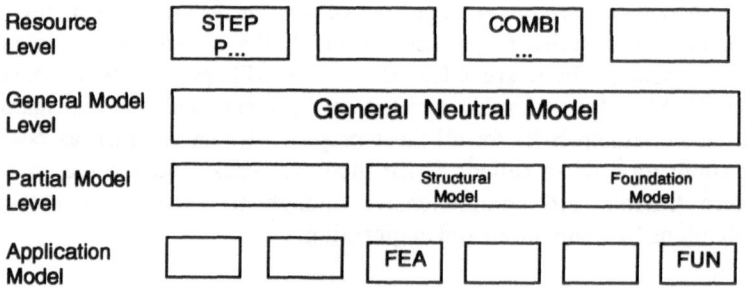

Figure 5. COMBI Framework

The two middle levels are forming the common kernel of the integration framework. They are divided into a central neutral model and several COMBI partial models designed to serve specific engineering domains. Communication within one domain is supported by the corresponding partial model, communication between different domains need additionally to their partial model the Central model and thus increasing the need as well as the complexity of the necessary mapping mechanisms. The uppermost level serves as a common source, containing resource parts from STEP as well as small COMBI proprietary resources. The framework can be classified as a layered architecture using a central kernel and partial models.

COMBI's application models contain all possible output/ input entities and their attributes from the corresponding application tools. With the help of an intelligent center, a tool capable of filtering and translating the incoming information for use in the receiving system, however, the desired communication can be achieved in a more flexible way. The model translation/ conversion technique used by the COMBI Communication Manager (CCM) knows the origin of the entities respectively their attributes within the sending systems. Therefore the CCM performs the translation between different specification forms of design objects and attributes within the participating systems. The rules that describe the input and output data format and the transactions are realized and executed in KEE. The CCM establishes the exchange of instances between files using STEP physical file format between different models.

Figure 6 COMBI's CCM

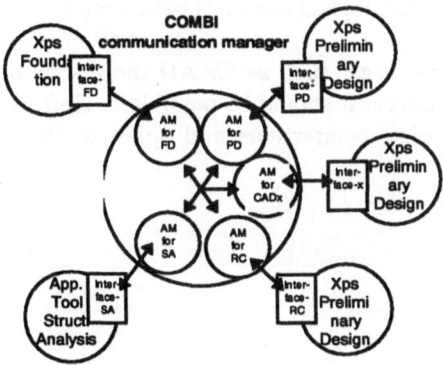

NextCAAD I

The objective of the project is to develop a CAD system that is not primarily oriented towards the graphical and geometrical processing as current systems are. the basic idea is that the architectural design object is considered as an logical entity within the product model. Architectural design objects can be rooms, spaces, building components of different complexity, grids, etc.. Parts of the objects' definitions are dealing with its physical shape, physical properties, others with its representation, which is always an abstraction of the real design object.

The existence of an object is not based on the representation of its geometry on the screen respectively the object's definition in geometry data. This is the fundamental step, which creates the basis for a consistent application of the product model philosophy. An object is a more abstract entity that can be viewed from different viewpoints. Thus it has different representations, e.g. different geometric representation, characterized by different sets of attributes.

580

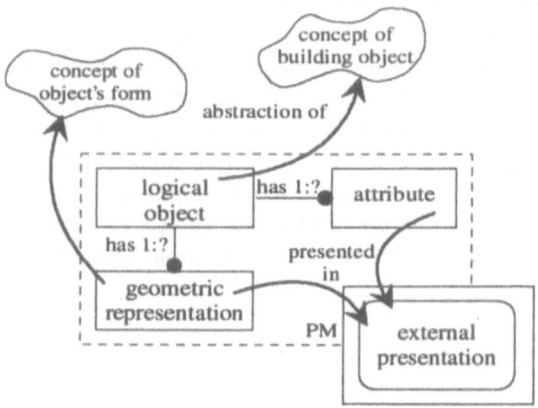

Figure 7 The logical object as basic entity

An application covering a whole area, such as CAAD the discipline of architecture, has to be dynamic by its own nature. It should be possible to shift the focus from, e.g. the spatial layout, to the building components and again to the bill of quantities.

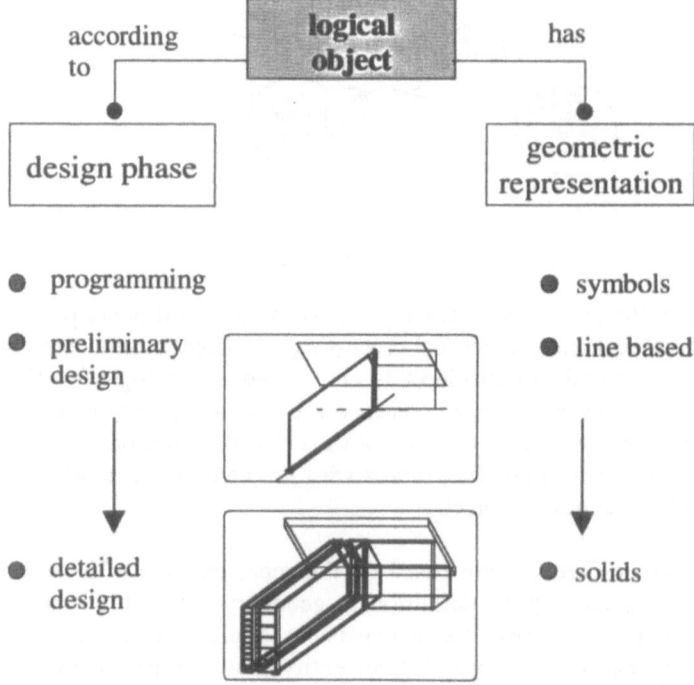

Figure 7 Different geometric representations of an design object

Therefore the modules of the system, and accordingly the partial models of the product model, should be interconnectable in a dynamic way. Thus, the underlying product model is structured as a web of interconnected modules, in opposite to the layered approach, e.g., in STEP.

The next generation CAAD has to deal with many phases of an object's life cycle, from programming phase to detailed design. Objects may occur and disappear during the life cycle, usually they are transformed or refined many times. In particular the geometric representation change from vague and fuzzy symbolic descriptions to sketch-like two dimensional geometry and again to exact two- and three dimensional geometric entities. Thus, multiple geometric representation is of special concern for the project.

Design is a highly dynamic process with many diverse particular activities and constantly shifting focuses between them. Translated into a product model this means that there are many individual focuses. This could be modeled as a huge singular model. Disadvantages are discussed in the introduction. Even if it where theoretically possible to model such a thing, practically it is not feasible, because one has to have particular results in a foreseeable range of time. All this leads to modular models. The conceptual idea for bringing the dynamic nature of the design process into a product model was that of a number of modules that could be connected together in dependence of the momentary context requirements of the CAAD system. This is the idea of a web of dynamically interconnected modules.

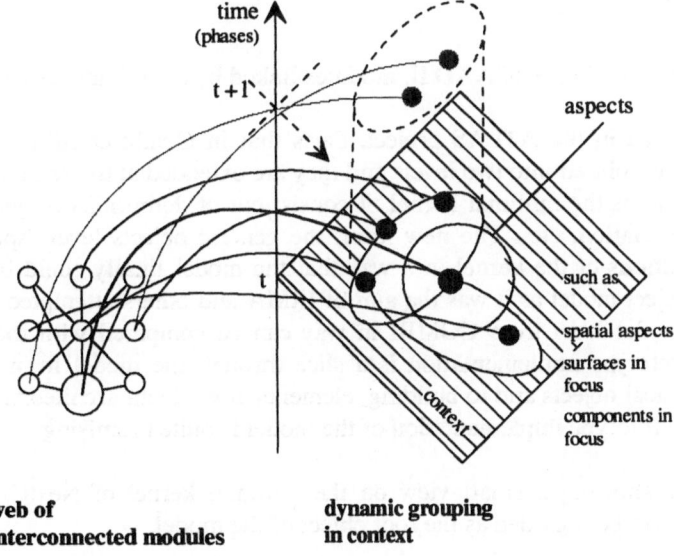

**web of
interconnected modules**

**dynamic grouping
in context**

Fig. 8 dynamic grouping of interconnected modules

582

The concept of this model architecture could not be realized for many reasons. The resources of the NextCAAD I project where reduced and it was focused to architectural design only. This goal could be reached with much simpler model architectures that or based on more proven ground. Secondly it still is doubtful if it would have been possible to implement such an architecture with the implementation technique of that time and project. Today with developments from the VEGA project a realization would be more likely

THE NextCAAD II PROJECT

The model architecture of the NextCAAD II project in a certain way is a step back. However the model is built upon experiences from the COMBI project as well as on

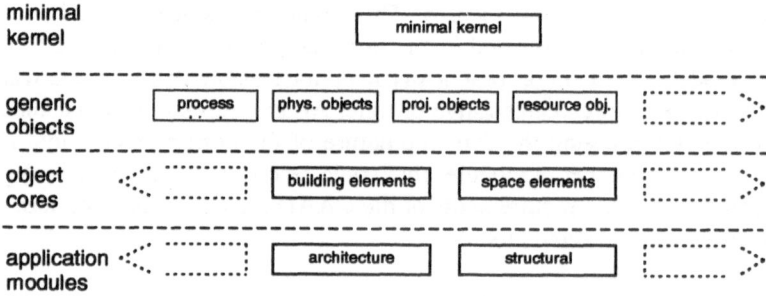

Figure 9 NextCAAD II, modules linked by a minimal kernel

concepts realized in the ATLAS project. Parts that in Combi constitute the 'general model' are now splitted into two levels and they are extended at the same time. A small bur generic part is the minimal kernel. it consist out of definitions of generic objects, attributes, association, etc.. The new level, the generic objects level specializes and extends definitions of the kernel in a way that the model finally could be used as an integrated project model as it was the aim of IRMA and others attempted for example. The partial model level from COMBI in way can be compared with the object_core level. The prototype implementation is a slice through the model form the minimal kernel to physical objects and to building_elements down to an architecture application module. The prototype implementation of the model is quite promising.

Figure 10 is showing a small view on the minimal kernel of NextCAAD II. The design object can be regarded as the root object of the model.

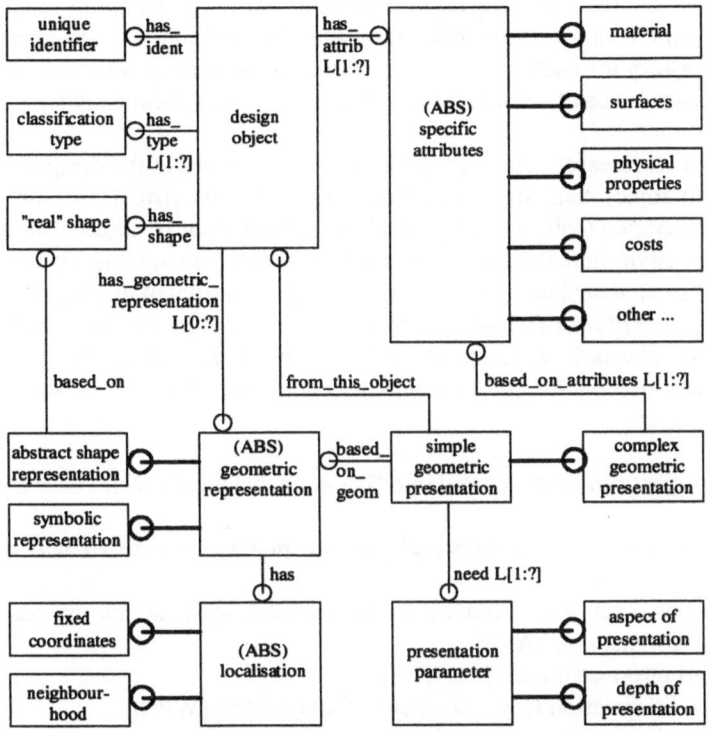

Figure 10 Design object as root of the minimal kernel

The ESPRIT project VEGA

The author is convinced that the only computer based environments really excepted and freely integrated in daily working habits will be those that are working in a similar manner as the human experts of a design and/ or construction team do. Such a team consists of various experts from different domains and expertise. They are highly dependent on each others working results as basis for own work. A strong interactive communication using all kinds of documents and the spoken word is a necessity. The accessibility of information as basis of each others work is based on interpretation which transforms the received information into the experts domain world. The experts are working on the same building elements but having very distinct understandings, semantics of those elements.

One could say all those experts are having different 'Brain Implemented Models' (BIM) compared with the models we are used to talk about. There is no person existing with an above all central knowledge, each is understanding their own domain often (or normally?) using different 'expert languages'. See example of architect and structural engineer talking about a concrete column or load bearing wall. In addition to their own

BIM they have to have a filtering and translation knowledge enabling them to find the mutual implications of a certain decision on their tasks. Without such 'mechanisms' it would be necessary for each and every time to communicate in a lecture style manner, surely not a useful type of communication between experts during a design task.

Transferred to an research issue in the IT world this picture of the design/ construction team leads to topics that are very actual today: IT for virtual enterprises, CSCW (Computer Supported Collaborative Work) and distribution of objects ore data models in an network environment to name only a few. Solutions enabling virtual enterprises collaboration in an distributed product data model environment are the objective of the ESPRIT project VEGA (Virtual Enterprises using Groupware tools and distributed Architectures). The task structure of VEGA gives an answer to the question which ingredients are needed for an implementation of the 'BIM' idea in an computer environment.

The VEGA project can be structured into five main technical tasks. These are:

1. Conceptual model architectures and product models for a distributed environment (PDM)
2. Technologies enabling communication between applications in an distributed model environment (COAST)
3. Distributed information services (DIS)
4. Work- and information flow management and control (WFL)
5. Implementation architectures and techniques

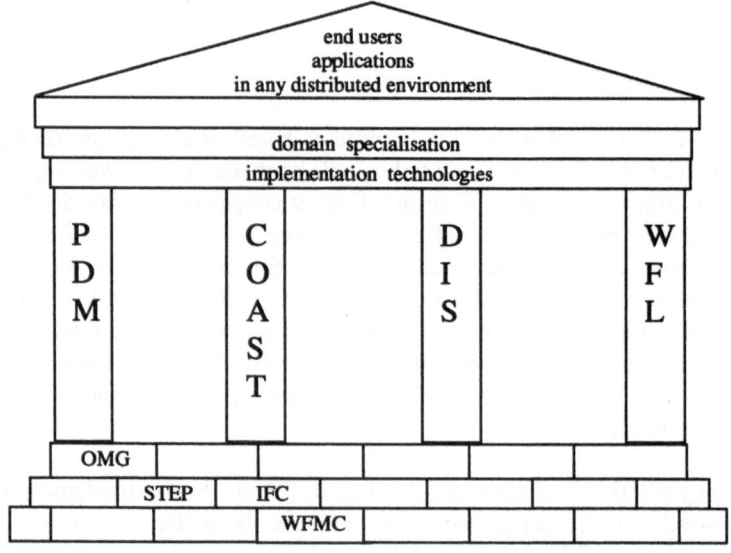

Figure 11 The structure of the VEGA project

To use a picture the first four main tasks are forming columns standing on a basement and bearing an architrave and a gable. The basement bearing the columns are standards like ISO's STEP, OMG's CORBA, WFMC's Workflow interface definitions, IAI's IFC and the Internet with HTML and VRML. And it is not only that these standards or supporting these tasks, there is also a counter reaction back into the standards. OMG has accepted a proposal for extensions of CORBA coming out of VEGA's COAST, there are influences into the workflow interface definitions as well as into IFC's architecture. 'The VEGA Platform' (this volume) is a more detailed description of COAST, DIS and WFL.

The architrave is formed by implementation technologies and domain specialization. These are consisting of two parts: first a generic implementation environment for EXPRESS product data models schemata, the Dynamic Product Model. This PDM among other features allows dynamic evolution of a schema under runtime of the application. The second is a generic product data model which is the necessary task for a specialization of the PDM for the domain of architecture and engineering. 'A Dynamic Product Model' (this volume) is a detailed description of that task. The following is dealing with the 'VEGA model'.

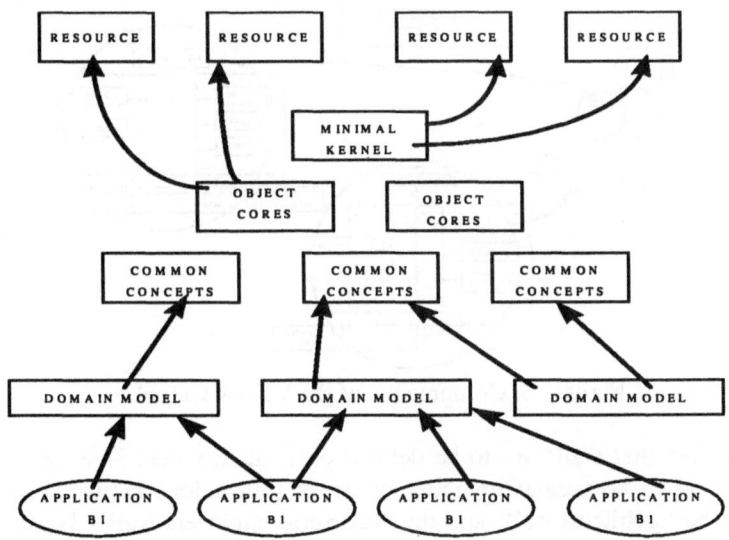

Figure 12 The VEGA model architecture

The architecture of the VEGA model is building again upon COMBI and especially extending NextCAAD II model prototype capabilities. As in the COMBI Framework 'recourses' are used. These can be 'borrowed' and/ or adapted ones, for example from STEP, or native ones, developed for the project. The idea simple is that parts used in several places of the schema should be referenced and not copied into the parts.

586

The minimal kernel is quite identical with its forerunner, but off course results from prototype implementation are introduced. The two levels which where named generic objects and object cores are now being coupled together under 'object cores'. It is always a little bit fuzzy to distinct one level from the other or to clearly identify the need for a 'level' to be introduced. So one could discuss this separation or not separating of the two. The objective however is similar as before: to extend and specialize definitions of the minimal kernel, which are held very generic into the direction of domain model objects.

A level introduced in the VEGA model architecture is that of 'common concepts'. The idea behind this is, using the example of a wall for explanation, is the following: In most building product definitions of building elements like wall are being found at the level of aspect or domain models. This results in the fact that there are different wall definitions with respect to their domain requirement or aspect. The level atop those domain models end with definitions of the general building element of which wall is a

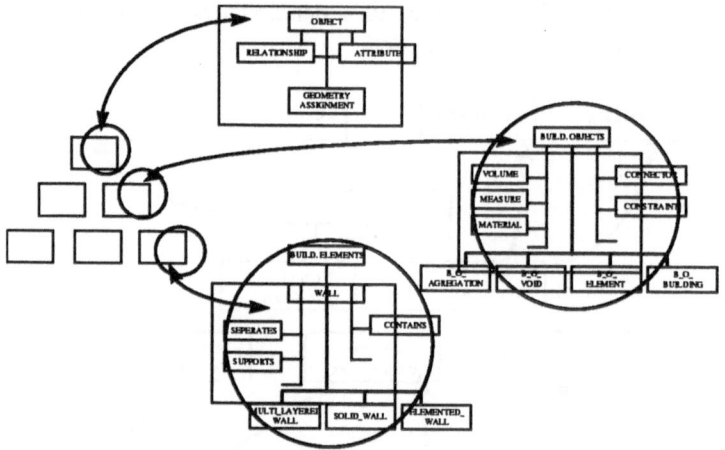

Figure 13. Components of the VEGA Kernel

subtype. The fact that walls are to be defined on domain model level only results in what can be called horizontal mapping or better translations between the different 'walls', e.g. the 'architects wall' and the 'structural engineers wall'. These mappings are much more pain making tasks compared with the horizontal mappings at the transitions between one model level to the next lower or higher one. The goal of the 'common concepts' is to avoid these horizontal translations and replace them with vertical mappings, in the case of wall from either 'architects wall' or 'structural engineers wall' to the common concept of wall.

Bjoerk, Bo-Christer [1995]
Requirements and information structures for building product data models
VTT Publications no245, Technical Research Centre of Finland, Espoo

Eastman, C., [1978]
The representation of design problems and maintenance of their structure.
Latcombe (ed.), Application of AI and PR to CAD, Elsevier, North-Holland

Gielingh, W., [1988]
General AEC Reference Model, ISO TC 184/SC4/WG1 DOC N.3.2.2.1

IMPACT [1991]
IMPACT Reference Model. Deliverable, ESPRIT Project Impact

COMBI, [1994]
Junge, R. Ammermann, E. Katranuschkow, Scherer, R.
ESPRIT Project 6909: deliverable A2, A3, B3

Junge, R. [1991]
Integration by Communication.
1st International Symposium on Buillding System's Automation - Integration
Conference Proceedings, Univ. of Wisconsin, Madison

Junge, R. and Storer, G., [1993]
B/C Application Protocol Planning Project. ISO TC 184/SC4/WG3/T12 AEC Working paper

Junge, R, [1994]
Bauproduktmodelle: Eine Einführung
Arbeitspapier, OFD Berlin, Bundesliegenschaftsverwaltung

Junge, R. and Liebich, Th., [1994]
Produktmodellierung im Architekturbereich.
Abstracts of IKM, HAB Weimar

Luiten, B., Luijten, B., Willems, P., Kuiper, P. and Tolman, F.P., [1991]
Development and Implementation of Multilayered Project Models.
Mangin, J-C., Kohler, N. and Brau, J. (edts), Computer Building Representation for Integration, Pre-Proceedings
of the second international workshop, Ecole de polytechnique federale de Lausanne, Lausanne,

Luiten, G., Froese, T., Bjoerk, B-C., Cooper, G., Junge, R. Karstila, K. and Oxman, R. 1993
IRMA, An information reference model for architecture, engineering and construction.
Mathur, K., Betts, M. and Tham, K. (eds.) Management of Information Technology for Construction., World
Scientific & Global Publication Services, Singapore 1993

Landsdown, J. [1983]
Expert Systems: Their impact on the construction industry
Report to RIBA Conference fund, London 1983

NEUTRABAS, [1991]
ESPRIT 2010 project: Outfitting Information Model, Deliverable 4.2.2

Richens, P. [1978]
The Oxsys system for the design of buildings.
3rd International conference on Computer in Engineering and Building Design CAD78., Ipc science and
technology press, UK

Tarandi; V. [1993]
Oblect oriented communication with NICC, neutral intelligent CAD communication.
Mathur, K., Betts, M. and Tham, K. (eds.) Managementof Information Technology for Construction., World Scientific & Global Publication Services, Singapore 1993

Tolman, F.P. and Wix, J., [1995]
ISO TC 184/SC4/WG3/T12 AEC, Part 106 Working paper

Turner, J., [1990]
Building Systems Model. ISO TC 184/SC4/WG1 Working paper

van Nederveen, G.A. and Tolman, F.P., [1992]
Modeling multiple views on buildings. Automation in Constuction, vol 1, 1992 nr. 3

Wager, D. [1984]
Expert systems and the construction industry.
CICA, Construction Industry Computing Association, Cambridge UK

THE VEGA PLATFORM
IT for the Virtual Enterprise

RICHARD JUNGE
Faculty for Architecture, Professorship for CAAD,
Technical University Munich

MANFRED KÖTHE
KARSTEN SCHULZ
European Applied Research Center
Digital Equipment, Karlsruhe

ALAIN ZARLI
Centre Scientifique et Technique du Batiment
Sophia Antipolis

WIM BAKKEREN
TNO Building and Construction Research
Delft

Abstract

One of todays manny buzzwords is 'virtual enterprise'. The objectives of the EPRIT project VEGA are the developement of an IT platform enabling such enterprises. Virtual enterprise means a number of people or smaller companies grouped together for a destinct contract, which none of them alone could or would able to get and to undertake. Modern decentralized, distributed IT solutions typically could support such virtual enterprises in their competituion against those who are big or strong enough to to carry out such contracts with their internal resources alone. VEGA gathers together the necessecary components as technically available and extends their cababilities as needed for a platform enabling collaboration in an flexible and distributed environment.

Introduction

The domain of architectural and engineering design and erection of a building is surely one of the most typical the term 'virtual enterprises' applies to. Always this has not been the domain if single centralized large companies as e.g. in car industry. Rather

591

R. Junge (ed.), CAAD Futures 1997, 591-616.
© 1997 *Kluwer Academic Publishers.*

592

these the task in the process of design and erection of a building are undertaken by relatively to really small entities forming teams (to use this more commonly used term instead of 'virtual enterprise') mostly on a project to project basis.

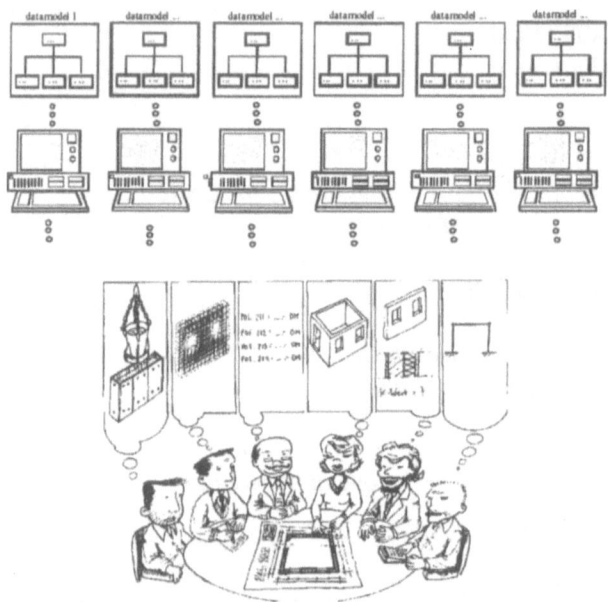

Figure 1 "Collaboration today and tomorrow

There are at least three scenarios encountered for the collaborative work in future design offices and teams:
• Parallel work of architects within a single office on a larger project
• Collaboration and parallel work of architects across scattered places
• Collaboration between architects and various design teams

Architects and/or engineers within a design team usually work in parallel on their specific part of the building project. This team specific part, mostly identical with one specific view type, should be represented by one common data model within the chosen AEC software application. This common data model covers all project objects, the team is working on. Each team member should have direct access to a common data model and be enabled to modify his or her part of it. All other team members should be informed, when one of these changes will affect their part of responsibility. This common data model, by no means, is a central model.

Current technology gives only little, or no support at all, to the parallel work even within one AEC application. In particular, the file locking mechanisms and in CAD the layer structures provided by current technology, forces designers to split their work topologically, e.g., into building sections or storeys. This mostly is in contradiction with the division of work that is more task oriented. An example is e.g. that one designer is responsible for the structural part, the other for the space layout and preparing the finish, and the third for the curtain wall. These tasks can not be separated topologically. In consequence, work has to be done separately and in sequencially.

The result of todays technology is, that the software applications do not provide enough support to ensure consistency between the different parts of the data model. Other current technologies, such as referencing other drawings (often called xref-technology) as background layers have their shortcomings as well, since conflicts have to be detected by the users only on a purely visual basis.

Figure 1 shows how different users/clients of an application could work in parallel within a networked solution of applications that are supported by a chosen distributed data management. Then there is no further need to structure the projects necessarily according to the requirements of data technology. The team members can now work in parallel with one common data model, and the software application is responsible for the automatic update of the data model and its forwarding to the application.

A second end user need concerns the whole process and interaction between the different design teams. Since each design team works on its own design task and with its own view on design, the communication of informtion or data sharing has to incorporate a way that allows for the translation between the different views. On the other hand the various design teams also work independently for a certain period of time (off line), usually until a next team meeting or deadline. Within this time they work on local versions, that might contradict with local versions of other teams or team members. These local versions are based on the last agreed or approved versions of the building project, but the work can proceed in different directions within the various teams.

It is, indeed, the task of collaborative work to plan for the entire interactions of the actors of the teams. This again incorporates advantages since each team can search for an optimal solution by leaving side effects out of scope for a moment and disadvantages since there is a risk, that some of the inconsistencies might not be detected at the right time. Aiming at the totally integrated and consistent working environment for the whole project would therefore mean a restriction on the freedom of design teams and facing them with unwanted efforts to check consistency for every design action.

On the other hand intelligent workflow management together with a model based exchange of partial information would prevent the high risk of inconsistency, as it would allow to update and check the projects' data base at certain times during the project. Especially during the co-ordination phase the incremental exchange of data by collaborative action needs IT support. The scenario of a second vision therefore includes that the designers are enabled to exchange distinct bits of data they are concerned about. The support of collaborative work between teams will include the update of the project data models on demand. It will help to support the flow of information needed for collaboration and guide the collaborative actions, such as approval or documentation according to the understanding of the building process.

The vision is, that whenever a design member needs actual data from another team to check its current version, the new technology allows him or her to retrieve this particular data set. He or she can also update the database of the partners after negotiation in order to provide the newest version at their disposal. At certain times, given by the workflow management, support is given to the update of the whole projects' data base, which includes that all relevant teams are being informed and that the results can be approved and documented as a certain reached design step.

The VEGA project aims at specifying and demonstrating a software architecture suited for such virtual enterprises. One of the key problems of the virtual enterprise is the sharing of information between the different companies. This problem includes the need for neutral specification of information, distribution of information, control of information flow, and security of information. To address these aspects of information sharing in the virtual enterprise, VEGA makes use of three different technologies:
- product-data technology for the specification of meaningful project information;
- middleware technology for the distribution of project information;
- workflow management technology for the control of the flow of information and work in the virtual enterprise.

Because the virtual enterprise requires an open solution that allows any company to use its own favourite applications, VEGA builds on existing open standards for the three technologies mentioned:
- STEP and especially EXPRESS for the neutral specification of product model data;
- OMG CORBA for the distribution of objects over networks and for information security;
- WfMC interface specifications for the common definition of workflows, the interaction between workflow management system components and for workflow system interoperability.

In addition to these standards VEGA also makes use of standards like SGML, HTML and VRML for specific areas such as exchange of administrative messages, document handling and presentation of information.

Fundamentals: Standards in Information Technologies

The VEGA project structure can be depicted as in figure 2. The envisioned VEGA Platform stands on for columns.
1. Product Data Modeling (PDM)
2. Technologies enabling communication between applications in an distributed product model environment (COAST)
3. Distributed information services (DIS)
4. Work- and information management and control (WFL)

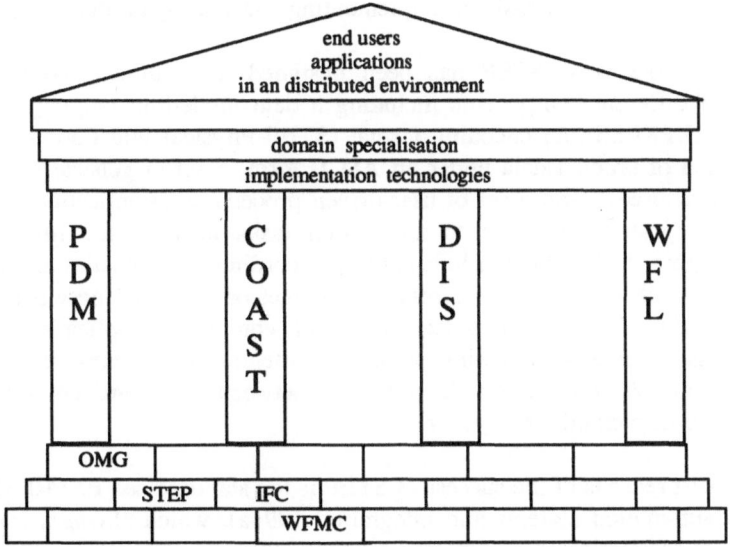

Figure 2 The structure of the VEGA project

These collumns can also be regarded as four fundamental standards: STEP, OMG and CORBA, Internet technology, Work Flow Management Coalition's Interfaces.

MODELING METHODOLOGY

The need of breaking down the islands of information in several industries - from aerospace to the car industry - has lead to dramatic research efforts to achieve effective product data exchanges. These efforts have become necessary due to the implication in the industrial processes of an increasing number of different actors on different sites using different systems and software. Consequently, wide enterprises have lead to the standardisation of methodologies, languages and technologies, especially in the context of STEP, the goal of which is a to allow uniform and complete representation of industrial product data, with general mechanisms for the exchange, the archiving of the data and the integration into any kind of application.

STEP is an emerging International Standard for the representation and exchange of product data, which is developed in ISO TC 184/SC 4 (Industrial data and global manufacturing programming languages). For a more comprehensive description, the reader could refer to [Fowler, J. 1995]. The main objective of STEP is to provide information sharing through neutral mechanisms, independent from any particular system or software, and allowing the description of product data throughout the life-cycle of a product. The nature of this description makes it suitable not only for neutral file exchange, but also as a basis for implementing and sharing product databases.

To reach its objectives, STEP has been designed as a quite modular standard, composed of different components including a data modelling language for product modelling (EXPRESS), an encoding format (STEP Physical File Format - SPFF) for the expression of product data under an ASCII form, a set of generic models (called Integrated Resources), and a set of user-driven procedures (Application protocols) to adapt the Integrated Resources to application tasks most of the time in specific industrial sectors. A STEP data file is built up according a set of rules and procedures, and on the ground of specific standardised languages or formats. These data files could then be read directly by STEP processors, STEP compatible applications or through STEP access interfaces, allowing a better integration of applications through standardisation of information interchange between different enterprises using different kinds of methods and systems.

One of the best reasons of the success of STEP is the standardised EXPRESS language [Industrial automation systems and integration 1994a], which allows a complete and unambiguous description of information related to the different conceptual components of a product. EXPRESS allows the designer to describe the data structures of the objects of his application, especially through the concept of entity describing a class of objects ; the corresponding instances are described in the SPF Format, which is a STEP standard for a neutral format of all the actual data in an application [Industrial automation systems and integration 1994b]. EXPRESS and SPFF are a first medium for the exchange of data, through ASCII files. STEP also offers the specification of an interface to STEP databases, the SDAI - "Standard Data Access Interface" - [Industrial automation systems and integration1995a], which is a functional Application Programming Interface (API). The SDAI allows the access and the manipulation of information contained in STEP databases in a quite transparent manner, i.e. independently of the proprietary archives methods of these databases, this separation being of course a fundamental key-point for industry needs.

Of course, all the STEP standards, including EXPRESS, SPFF and the SDAI, are defined in a complete independent way and with no direct connection with a given software language and implementation. But, to allow the use on a computer, some compilers for EXPRESS and SPFF are needed, and the SDAI must be "translated" in an implementation language : the specification of such an implementation is called a "Binding". Several bindings have yet been defined, especially for C [Industrial automation systems and integration 1995b] and C++ [Industrial automation systems and integration 1996] programming languages, which enables SDAI compliant calls thanks to functions written in C and C++ programs.

Besides STEP, another major effort is currently undertaken with the IAI (International Alliance for Interoperability), which is a non-profit alliance of the building industry including: architects and engineers, building project manufacturers, software vendors, research laboratories, and so on, divided in several national chapters (North America, Germany, United Kingdom, France, Singapore, Nordic Countries, and Japan). The IAI objective is to integrate the AEC/FM industry by specifying the IFC (Industry Foundation Classes) as a universal language to be a basis for collaborative work in the building industry and consequently to improve communication, productivity, delivery time, cost, and quality throughout the design, construction, operation and maintenance life cycle of buildings.

STEP and IAI/IFC share the same goals, which are applications interoperability, data exchange and actors co-operation. Nevertheless, they differ in their respective processes: the IAI promote a bottom-up approach, with an iterative and incremental development for quick bring in play, STEP is rather a long-term project, with a top-down approach and is concerned with broad standardisation. The IAI has partially adopted the STEP technology, mainly through an EXPRESS version of the IFC. In the future, an integration of the IFC within STEP is planned.

THE WEB INFRASTRUCTURE AN DISTRIBUTED ENVIRONMENT

The growth of world-wide data networks have generated a growth of methodologies, technologies, tools and interfaces for Intranet/Internet distributed solutions. The most well-known infrastructure integrating such tools and interfaces is the Internet/WEB. The Internet basically offers networks interoperability, and a multimedia-based exchange and access level. This is a different concept of the one of «exchange» or interoperability between applications, as it essentially concerns interoperability through networks connections. Technologies like CORBA are dedicated to distributed systems and environments, clearly allowing applications to «inter-work» and interact each other, invoking remote application resources in a dynamic way, and so on. The level of Wan, and Internet/WEB, is rather dedicated to information infrastructure which implies of course to allow the access to all the components of the infrastructure, but not real interoperability between applications: in fact, in that context, applications interoperability is a possible consequence of WAN/Internet/Intranet/WEB set of technologies as they provide network hardware and software support for that, on the other hand applications interoperability is the main objective of CORBA technologies. Even if the current developments on JAVA methodologies concern more integration and co-operation between applications around the WEB, they are at the moment far from the already huge set of specifications of the OMG with respect to distribution and interoperability of applications.

The Internet can be considered as a network of networks and offers a physical interconnection between computers spread all over the world: any end-user, should it be any public administration, private company or individual end-user, can connect to it through any kind of computer (PC, Apple Macintosh, Unix mainframe, and so on). This idea relies on the fact that the connection is done through a set of de facto standards, the main of them being TCP/IP (Transmission Control Protocol/Internet Protocol), for the different kinds of information in order to provide a better interoperability and opening to a large set of different platforms and systems. Various communications are ensured through a set of standard tools: the electronic mail (using SMTP: Single Mail Transport Protocol), the NEWS servers (using NNTP: Network News Transport Protocol), remote transfer of any types of file usinf FTP (File Transfer Protocol), connection to remote computers through Telnet mechanisms, etc. Thus Internet allows to access any kind of information, provided that this information is available under one of the standardized formats as defined within Internet.

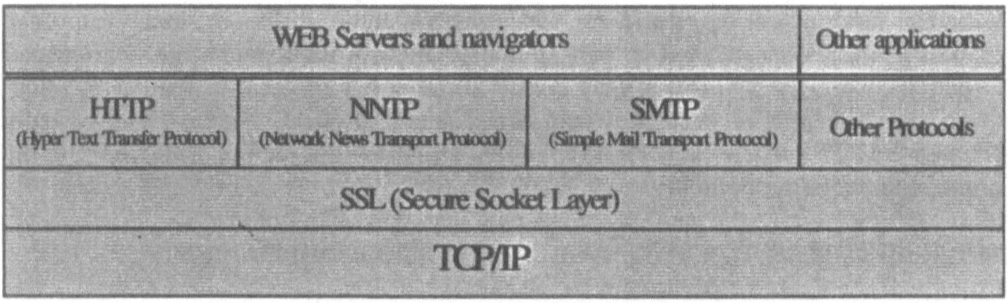

WEB Servers and navigators			Other applications
HTTP (Hyper Text Transfer Protocol)	NNTP (Network News Transport Protocol)	SMTP (Simple Mail Transport Protocol)	Other Protocols
SSL (Secure Socket Layer)			
TCP/IP			

Figure 3 Components of the WEB

The WEB is a powerful interface for Internet users, introducing two new protocols. Indeed, it associates two technologies which are Internet and the hypertext, this second one consisting in defining electronic links between the various documents of Internet, so that it is no more necessary to know the exact site where stand all the information as any information is located according to other information. It offers facilities for the organisation and the access to any type of manufacturing information: this is achieved by means of an encapsulation and distribution of the information through a generic transfer protocol which is http (HyperText Transfer Protocol), and a standardized representation for hypertext documents which is html (HyperText Mark-up Language).
All the Internet standard protocols are managed by all the WEB navigator tools (the two most famous ones are Netscape *Navigator* and Microsoft *Internet Explorer*), and this is quite transparent to the end user connected to the network, as shown in figure 3. Thus, the WEB nowadays federates almost all the Internet standards through a common friendly user interface. Even if it offers at the moment mainly electronic

management of hypermedia documents, the Internet/WEB is a world-wide support for communication, as anyone can connect to the network and access to any tool or service through http (whatever the hardware and software are).

The WEB can nowadays be considered as a basic support for multimedia communication as well as group-communication protocols, and one of the interests of the WEB is linked to the concept of groupware. Groupware is more than distribution or client/server computing: it is a collection of technologies that allow to represent and manage complex processes in the context of collaborative human activities. Groupware is built on several foundation technologies like remote actors and distributed processes, multimedia document management, workflow, electronic mail and (audio or video) conferencing, and so on. It is not the role of client/server or distributed systems technologies, for instance, to give a global answer to all the features involved in groupware, whilst the Internet/WEB offers effective fulfilment for some of the basic services of groupware like electronic mail, multimedia or the access to global wide public WEB databanks.

On the other hand, the Internet/WEB essentially collects highly unstructured data, most of the time grouped in WEB documents including text and images, under http-based formats. At the moment, the basic unit of exchange under the WEB is the document, which can be stored, viewed and even retrieved and replicated, and the WEB can be mainly considered as a multimedia document management architecture. Moreover, due to their intrinsic nature, the WEB protocols are too generic to vehicle the deep semantics of the different applications, and these protocols are not to be compared with powerful modeling languages as found in PDT. Another weakness of the WEB is the administration and mainly the access to all the local corporate information: no precise control is effective on what is available to other network users and what is kept private, and no access control or security considerations are really managed. Indeed, as the Internet/WEB offers a quite open and free world, this means many risks for private companies and their internal knowledge and databases, and also the need for powerful protection mechanisms in the context of commercial processes. Finally, it can be mentionned that the WEB essentially implements a pure client/server approach: the relationships between clients and servers are not bi-directional, while technologies like CORBA (see next section) specify a more general framework where clients may be servers too.

Thus, the integration of modeling methodologies, distributed technologies and the Internet/WEB suggests a new strategy for integrating legacy environments, providing more powerful mechanisms and services and offering a common interface for the access to any type of information, and the effective exploitation of an infrastructure based on PDT and distribution to deliver Value-added Web-based solutions. All these technologies should complement each other for the development of large scale information systems.

DISTRIBUTED ARCHITECTURES: CORBA

Standards like STEP or IFCs allows applications interoperability at a conceptual level, in fact through communication based on data model integration. They allow the exchange of information on the base of the same understanding and representation of the information semantics. But in practice and in a real computer-aided industrial and business processes context, there is a added need for operational exchange between applications, and applications software have to physically interoperate. This means that each application must be able to invoke, most of the time remotely, resources of the other applications. This can be achieved through powerful computer based mechanisms like client/server architectures or more recently object-oriented distributed systems. Among the various recently emerging technologies in this area, the OMG CORBA standard is one of the most important current developments and is becoming one of the most popular for applications interoperability. The following paragraphes hereafter shortly introduce to the major concepts of client/server and distributed systems, and then give an overview of the CORBA specification.

A standard for distribution as CORBA offers powerful means to make applications inter-operate each other on the base of a networked environment, but no specific prerequisite is given with respect to the network support, to all the various types of managed information or to the different accessible services. Thus, the applications may for instance work together all linked through some private net work or Local Area Network (LAN). However, in the context of the today increasing global competition, corporations have more and more to envisage partnerships over a large geographic and organisational spread, and there is a need for a general information infrastructure (information highways) able to support world-wide communication and interactions.

Client/server and distributed architectures

A distributed system can be roughly defined as two or more pieces of software sharing information with each other, and relying on:
- a set of (heterogeneous) hardware, software, and data components, connected by a network,
- providing a uniform set of basic services as naming, user registration and access control, remote process execution, management, security, and so on.

The first point mentions that a distributed system is first a networked system, with multiple heterogeneous and independent computers having their own CPUs, their own software and their own databases. The second point introduces the fundamental concept of services. These are components of the distributed environment providing generic features or behaviours quite independently of the various applications connected to the network and allow applications to really inter-operate. These services must be globally and uniformly accessible. They ensure the coherence of the distributed system as a whole rather than as a simple aggregate of interconnected computers and tools, thus contributing to provide a standard way to manage all the relevant generic-purpose information to ensure the smooth running of the distributed system.

The nowadays most popular distributed systems are client/server constructions. With this model, there are two major types of software: client software requesting the information to servers, and server software which accomplishes the requests. Despite some advantages, like the common location within the server of some basic services (access rights, transactions management, etc.), this model does not consider all the applications as similar independent entities: clients must be aware of distribution handling to find the location of the server on the network, clients and servers are not symmetric as clients are pro-active and servers reactive.

Distributed architectures as defined in CORBA are a step further than client/server system in their ability to allow in a seamless way the interoperability of applications and the sharing of information and resources between geographically spread corporations. They allow the sharing of resources, the use of any type of hardware and software on the base of a broker-service, and the parcelling of computing workload among many different machines. Distributed Systems authorise a great deal of autonomy through separated components which moreover can keep their own specific data close to their processing (and access other information on a remote server in the network only when necessary). They are naturally extensible and thus can grow in small increments over a large range of sizes, and maybe the most important property is the fact that in case one component crashes it does not necessarily imply that all the (remote) components cannot keep on running.

The CORBA specification

CORBA [Mowbray, T. J., Zahavi, R.][Otte, R., Patrick, P. Sr., Roy, M. 1995][Siegel, J., et al.1996] is the central component of the Object Management Architecture (OMA) defined by the OMG. In the context of a client/server architecture, its main goal is to manage requests from a client for specific operations on objects to the most appropriate object implementation for execution, on the base of the specification of a set of interfaces to objects. CORBA objects are effective means for hiding the characteristics of the underlying hardware and software implementations behind a high-level portable interface. They can invoke operations on each other even when implemented in different programming languages and when running on incompatible operating systems.

CORBA allows a flexible client/server relationship. In contrast to traditional client/server implementations, CORBA allows a piece of software that acts like a client for one request to act as a server for the next request. This flexibility is the result of CORBA's object-oriented approach defining an object model. Thus CORBA views all the applications in the CORBA-System as nothing else but a set of objects and operations associated to them. The location of the underlying application on the network and the operating system is transparent to the client. The CORBA specification is mainly composed of the following parts:

- The Interface Definition Language (IDL), a common neutral language to specify the object boundaries and its interfaces with potential clients. It is the way promoted in CORBA to separate interface from implementation and thus to provide language-neutral data types that make it possible to access the object with no matter of language the object is written in and system supporting it.
- The Object Request Broker (ORB), a software bus that establishes the client-server relationships between objects and seamlessly interconnects multiple object systems.
- The CORBA object services: a common collection of system-level distributed middleware services, complementing the functionality of the ORB. These services are packaged as components with IDL-specified interfaces: because IDL provides operating system and programming language independent interfaces to all the services and components that reside on a CORBA bus, this is true for the specified CORBA services as well as for specific business client applications.
- The inter-operation architecture, defined for the communication between different ORB implementations, which consequently can be interconnected using CORBA 2.0 inter-ORB services.

The ORB (see figure 4) plays the role of a broker between a client and a server. It is the middleware spinal chord, allowing to access objects independently of their locations on the network. Thus, when a client invokes a service attached to a given object, he does not need to know who (and where) is the server on the network implementing this service, and similarly, the server has no idea of the client invoking him. Everything is managed by the ORB acting as a link between the client and the server, essentially through message (request and reply) passing. Such a notion is a true conceptual improvement with respect to the underlying notions of classical client/server architectures. As the ORB is in charge of finding the appropriate server for a client request, it manages a specific repository in order to realise the task: the Implementation repository. The Implementation repository is a runtime repository of information that allows the ORB core to locate and activate implementations of objects. It is also the common place to store additional information associated with implementations of objects. Finally, the ORB also must have knowledge of the services offered by servers through the objects interfaces, with the help of the Interface repository. It is a runtime database containing persistent objects that represent the IDL information in a form available at execution, i.e. a description of all the IDL interfaces under the form of metadata usable by the ORB along with an API allowing components to dynamically access, store and update metadata information.

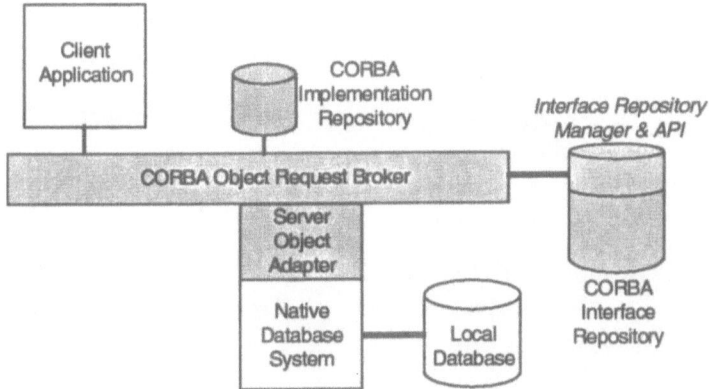

Figure 4 CORBA System Architecture

On the other side, the CORBA services [Object Management Group, 1995] are an added interesting feature with respect to classical client/server architecture: application developers can elaborate their components without any concern for middleware services, and then can mix the components with any combination of the CORBA services through IDL specifications (customising the application objects via sub-classing and multiple inheritance).

Introduction to Workflow Management

What workflow management comprises is probably best explained with the following quotation from Coleman and Khanna [Coleman, D. and R. Khanna. 1995]: 'Workflow is often explained with the analogy of the factory floor. In America, manufacturing made great strides in productivity during the late '80s and early '90s, most due to automation. Now, visionaries want to take the automated processes of the factory floor and apply them to the office.' Initially, there were some exciting success stories of applying workflow tools and techniques. Automating production-oriented processes like processing claims forms or loan applications has yielded dramatically lower cycle times for processing this paperwork and improved customer service. This has translated into a competitive advantage for organizations using these tools. Additionally, competitors are frantically trying to catch up and surpass early workflow tools users.

The strategic issue of workflow systems is to bridge the gap between the world of processes and the world of data modelling, as Gawlick et. al. [Gawlick, D., M. Hsu, and R. Obermarck. 1994] state. Figure 5 shows this. While business process engineering focuses on process management, computer support for operations focuses on data modelling.

604

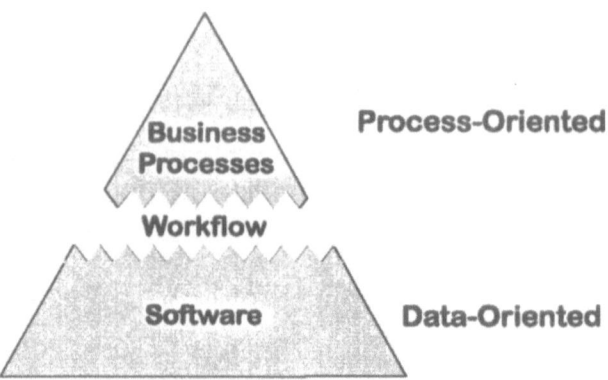

Figure 5   Workflow management bridges the gap between business processes and available software support.

The Workflow Management Coalition (WfMC), founded in 1993, is a non-profit, international organisation of workflow vendors, users and analysts. The WfMC's mission is to promote the use of workflow through the establishment of standards for software terminology, interoperability and connectivity between workflow products. Up to now, the WfMC has specified a Workflow Reference Model [WfMC. 1994], a glossary [WfMC. 1996 a] for workflow management, the interfaces 1, 2, 4 and 5 of the WfMC's reference model as shown in Figure 6 and a common Workflow Process Definition Language (WPDL) [10 WfMC. 1996 b]. Furthermore, the WfMC has shown demonstrations for the interoperability of Workflow Management Systems (WfMS).

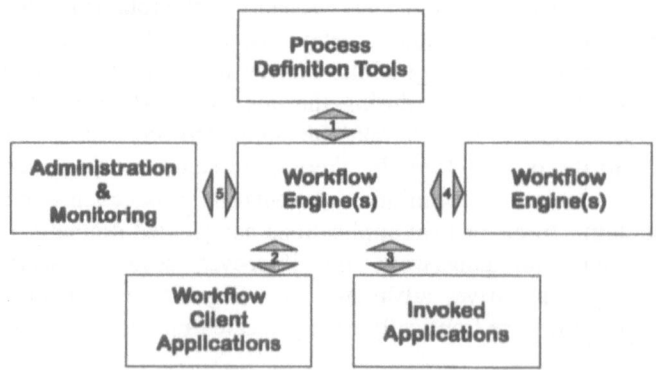

Figure 6   The Workflow Management Coalition Reference Model

The functionality of a WfMS is achieved by the interaction of several parts:

- process definition tools for the definition (i.e., the modelling) of workflows;
- workflow engines as the proactive computer systems that execute and control the processes;
- invoked applications, which are mostly legacy applications, e.g. word processors;
- workflow client applications, which act as the interface for an agent (i.e., a human being or another computer) with the WfMS;
- tools for administration and monitoring of the workflow process.

COAST: A global infrastructur for EXPRESS models

This section describes a more technical presentation of the COAST infrastructure [OMG. 1996], first by showing how COAST is embedded into CORBA, followed by an Introduction of its various components, and then giving a functional description on the basis of these components. The COAST is specified to offer powerful dynamic mechanisms as it is intended in a late binding approach.

ARCHITECTURAL CONTEXT

As the name suggests, the baseline technology for the COAST system is CORBA. A set of services is grouped around this middleware platform, which implement the special services of COAST. These services are partially standard OMG services, and partially native COAST services. The classification of COAST can be best described in the context of OMG's Object Management Architecture:

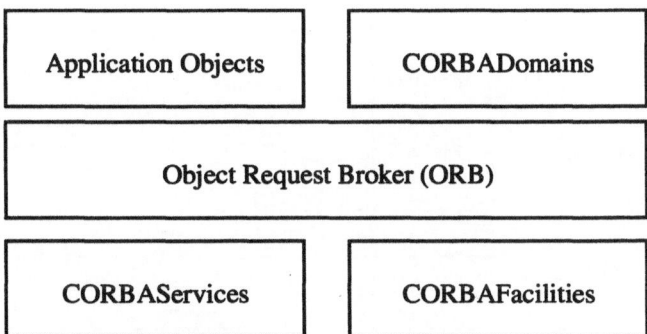

Figure 7 Object Management Architecture

The Object Management Architecture Guide (OMAG) describes OMG's technical objectives and terminology and provides the conceptual infrastructure upon which supporting specifications are based. The guide includes the OMG Object Model, which defines common semantics for specifying the externally visible characteristics of objects in a standard implementation-independent way, and the OMA Reference Model. [OMG. 1996]

The Reference Model identifies and characterizes the components, interfaces, and protocols that compose the OMA. This includes the Object Request Broker (ORB) component that enables clients and objects to communicate in a distributed environment, and four categories of object interfaces:

- CORBA Services are interfaces for general services that are likely to be used in any program based on distributed objects.
- CORBA Facilities are interfaces for horizontal end-user-oriented facilities applicable to most application domains.
- CORBA Domains are application domain-specific interfaces.
- Application Interfaces are non-standardized application-specific interfaces.

COAST can be considered as a new vertical service which is settled within the CORBA Domains section. To accomplish its work, COAST uses the Object Request Broker, CORBA Services and CORBA Facilities. In addition, COAST supplies additional services and entities as described below.

SYSTEM OVERVIEW

COAST is a true object-oriented access method, which supports by default distributed and heterogeneous environments. However, following the object-oriented paradigm, all details about distribution, heterogeneity and storage schema details are completely hidden from the application using COAST. All activities carried out by the COAST are transactional, thus supporting save information sharing.

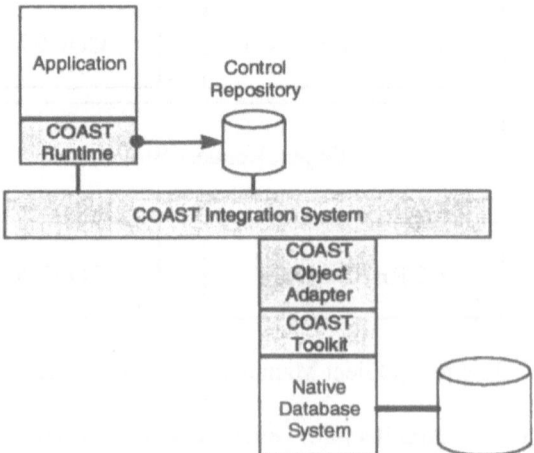

Figure 8: COAST System Architecture

Another classification can be done into client components, server components, and core components, complemented by some independent management components. The following list categorizes the various components and gives a brief description of each of them.

COAST Runtime System [client] This component implements the semantical functionality of the client part of the COAST system. It provides to the application an interface with late binding information access semantics to data models described by an EXPRESS schema. These models may be located at any storage server known to the COAST system. The distribution and explicit location of these servers are hidden to the application. Usually, the runtime system will be called by the Language Binding Interfaces, which provide the COAST API to the application conforming to a specific programming language.

COAST Control Repository [core] This repository stores information about available information models, available schemata, existing storage servers, etc. Furthermore, the control repository holds required information for the Object Relationship Service, the Object Transaction Service, and other CORBA services. The control repository is automatically self-replicating within a COAST domain.

COAST Object Adapter [core] It is the CORBA conform connection facility, linking the server methods with the CORBA core system. This specialized object adapter cooperates with the control repository to resolve invocations on distributed and multi-branch models linked by the Object Relationship Service. It also provides an extended object identification facility.

COAST Server Toolkit [server] This is a toolkit implementing the storage system independent server functionality. It will also provide assisting methods to bridge the gap to the native storage system functionality.

The *COAST Integration System* is the middleware level, relying on an ORB implementation and including an implementation of some of the OMG CORBA Services like the Object Relationship Service (ORS) to link distributed and multi-branch models and schemata together, the Object Transaction Service (OTS) for transactional behavior, and the Object Persistence Service (OPS) for objects long-term existence and management of object storage through a single common interface. It also connects the following components (see figure 8):

COAST Metadata Management System [core] A special object-oriented repository maintaining an internal representation of any schema following EXPRESS semantics. This component will intensively cooperate with the COAST Control Repository, and play a role similar to the Interface Repository in standard CORBA implementations, as a dynamic Metadata Repository for the COAST ORB.

COAST EXPRESS Compiler [management] This is a specialized EXPRESS compiler, which translates an EXPRESS schema into an object network conforming to the internal representation in the COAST Metadata Management System.

COAST Schema Loader Toolkit [management] A collection of routines to build a storage system specific schema loader as backed to the Metadata Management System.

COAST Management Tool [management] An interactive tool to fulfill all kinds of system management activities related to COAST. It allows the user to monitor and influence the operation of the Integration Services during runtime, assists application programmers, and provides fault isolation tools.

APPLICATION SUPPORT

Applications are not burdened with the need to know the actual distribution situation of stored information, nor is there any need to navigate to a particular storage server. Any request issued by an application is entered into the COAST system via a Language Binding routine which calls the appropriate Runtime System functions which submit the request to the COAST ORB Core. By using the Language Binding routine, the application only needs to specify the name of the model it intends to access, and the COAST runtime system locates the correct storage server and establishes the access path to the requested model. This is possible due to the additional context information from the session object and the navigational assistance from the control repository.

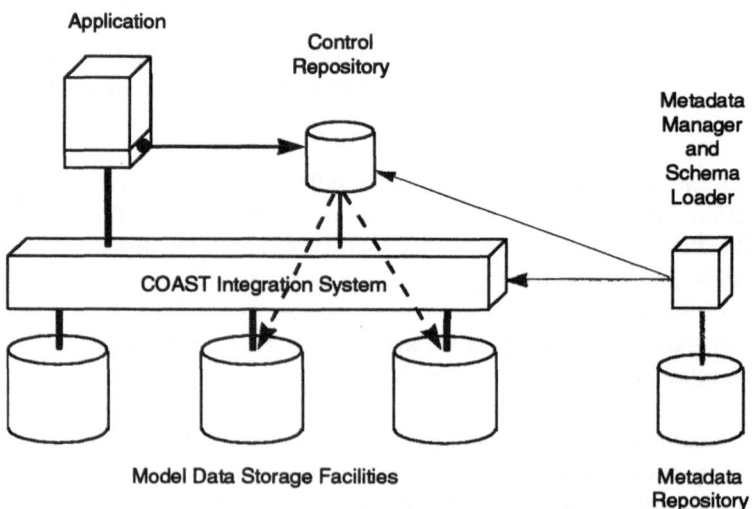

Figure 9: COAST Functional Overview

After an access path to a particular model has been established, the COAST runtime system will continue using this path until it is explicitly closed or an unrecoverable failure occurs.

If the request is a schema information request, the Control Repository implementation will directly satisfy the request with assistance from the Metadata Management System which holds the information of available schemata. If it is a data access request, the object adapter of the located storage server will activate the corresponding method routine. This routine will fulfill the request by accessing the storage system through storage system-specific access code based on the Server Toolkit. Any exception is handled finally by the top level method routine.

STORAGE SYSTEM AND SCHEMA SETUP

Any new application schema to be handled by the COAST system has first to be loaded into the Metadata Management System. This is done with the COAST EXPRESS Compiler. From the internal form, a database-specific schema has to be generated with an appropriate schema loader backend. The schema can then be loaded into the storage system using the normal CORBA transport mechanism (provided the storage system allows run-time schema loading). The top-level set-up of the storage structure is done with the interactive management tool. To ensure maximum information integrity, the COAST architecture relies on transactional technology. No access to stored information is permitted outside the context of a valid transaction. Therefore „access" and „open" operations are distinguished in several places. The „access" operation only establishes the communication path to the requested object and registers the access intend if applicable. However, to actually reach the stored information, a transaction has to be started and the object must be opened for the intended access mode.

RUNNING APPLICATIONS WITH THE COAST

As mentioned previously, two approaches can be envisaged (and possibly mixed) in CORBA, a static one based on IDL stubs and skeletons, and a more dynamic one based on ORB requests construction «on the fly». STEP-based applications having to inter-operate in a LSE project have to be plugged to the COAST either through IDL interfaces or using a COAST Application Programming Interface (API). This allows applications to either use a complete and natural integration or the COAST-Access on a rather wrapper-like behavior. The last approach allows the use of COAST without rebuilding a legacy application but assumes, the application has some proprietary interfaces which could be wrapped.

Workflow Management

The application of workflow management has to date been limited primarily to the control of administrative, document-based processes in individual organisations. The aim of the VEGA project is to apply workflow management to both the business and engineering activities across a virtual enterprise. This introduces several new requirements for workflow management in two different categories:
- requirements from managing processes in the LSE industry;
- requirements from managing processes across company boundaries.

The processes in the LSE industry include not only administrative but also technical activities, such as design and engineering, fabrication of elements, assembly and erection of the product, maintenance and demolition. Moreover, LSE projects are characterised by the large number of participants who are collaboratively and concurrently developing the LSE product. Consequently, there are many parallel processes that need to share information and need to be co-ordinated on a regular basis. In other words, unlike administrative processes in the traditional application domain of workflow management, LSE processes are complex processes that require flexible workflow management systems that can deal with frequent changes in activity sequences, parallel processes, co-ordination of work and information. In addition, because of the inherent complexity of LSE processes, it should be possible to control the workflow process on different levels of detail, ranging from the level of project milestones to the level of individual activities.

Another requirement, in addition to the support for complex processes, follows from the VEGA vision. The vision is a software architecture that enables project partners from different companies to work together in a virtual enterprise and share and access information without any problem. One of the three components of the VEGA solution is the use of product-data technology to describe and store meaningful product information. The VEGA vision requires an integration of product-data technology and workflow management. Document-based workflow management systems, as they can often be found in the traditional application domain of workflow management systems, are therefore not suited. The VEGA vision requires a tight coupling between the models describing the workflow process and product data models containing the information processed in the various activities in the workflow.
To summarise, workflow management for virtual LSE enterprises has to be able to deal with the complexity of collaborative and concurrent engineering activities and has to be able to link the relevant product data models to the workflow process.

CROSSING COMPANY BONDARIES

Workflow management in a virtual enterprise requires interoperability of the different WfMSs in the companies participating in the virtual enterprise. Because the communication between the different organisation will use a public network, firewalls are required to ensure security of project information, organisation specific information and organisation specific working procedures (see also [Amar, V. et al. 1997]). The workflow management systems (or rather the workflow enactment services in WfMC terminology) must be able to interoperate across the firewalls between the organisations, as shown in Figure.

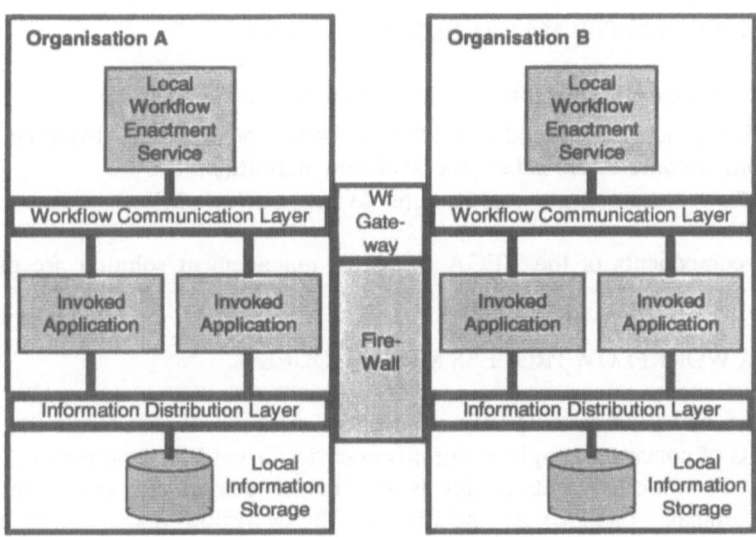

Figure 10   The WfMSs must be able to interoperate across the firewalls between organisations.

The interface between the workflow enactment services (interface 4) as described in the Workflow Reference Model of the WfMC (Figure 10) is of special interest here. As (sub)workflows are executed by different WfMSs, these systems need to interoperate and exchange information to achieve a systemwide workflow. A systemwide workflow is the sum of the distinct workflows in the respective participants of the virtual enterprise. The WfMC has demonstrated the interoperability of WfMSs based upon the interface 4 definition of the workflow reference model several times. The first demonstration of this kind took place only last year, in June 1996. Further demonstrations have been undertaken. Currently, the exchange of information is based upon email. This is a simple solution with many deficits. The VEGA platform will offer a more reliable solution.

As the different WfMSs have only a restricted view of the (sub)workflows they are involved in, a global monitoring service needs to be introduced to achieve monitoring of system wide workflows. Monitoring workflows means to analyse their runtime information. In addition, the objects which are created or used during this life-cycle are of interest. The global monitoring service also requires a solution that can cross the firewalls between the companies in the virtual enterprise.

To summarise, workflow management across company boundaries requires WfMSs that interoperate across firewalls and a global monitoring service that is able to monitor across firewalls.

THE VEGA WORKFLOW MANAGEMENT SUPPORT

VEGA will meet the requirements described above by developing:

- a workflow process meta-model to define workflow processes in virtual enterprises and to link product model data to the workflow definition.
- a workflow management architecture to manage workflow across company boundaries.

These two components of the VEGA workflow management solution are discussed below.

THE VEGA WORKFLOW PROCESS META-MODEL

The objective of VEGA is to specify a workflow meta-model that meets the specific requirements of concurrent engineering processes in virtual LSE enterprises. The core of the VEGA workflow meta-model is the model behind the workflow process definition language (WPDL) as specified by the workflow management coalition (WfMC) [WfMC, *Interface*. 1996]. The WfMC aims at the specification of international and open standards for the interoperability in workflow management. This is the main reason why this language has been chosen as the core of the VEGA model. Currently, the WfMC specifications are up for international ISO standard. Moreover, the WfMC WPDL is a generic and extendible meta-model of workflow processes. Taking this specification as a baseline for the VEGA meta-model increases the chance that useful extensions can and will be adopted by the WfMC and vendors of workflow management systems. This conforms to the objective of the VEGA project: to specify and realise an integrated software architecture based on open standards.

The main entities in the WPDL are the workflow process, the workflow activity, the transition, the workflow participant, the workflow application and the workflow relevant data, as shown by the EXPRESS-G diagram of Figure 11. The WPDL specifies the attributes for these entities and the relationships between them.

The VEGA workflow meta-model extends the WfMC WPDL in several places. Modelling constructs missing in the WfMC-WPDL schema are:

- a way to model the participant responsible for an activity
- a way to assign multiple participants to an activity instead of only one participant
- the ability to model the position of a participant in an organisational unit (It is only possible to model the role of a participant in a workflow process. VEGA deliverable

D501a specifies the requirement for modelling positions, or organisational roles, too)

- a way to model arbitrary collections of participants as a group
- a way to model the relation between a workflow activity and the relevant product model schema, e.g., an EXPRESS schema.

The VEGA workflow meta-model adds several constructs to address these shortcomings.

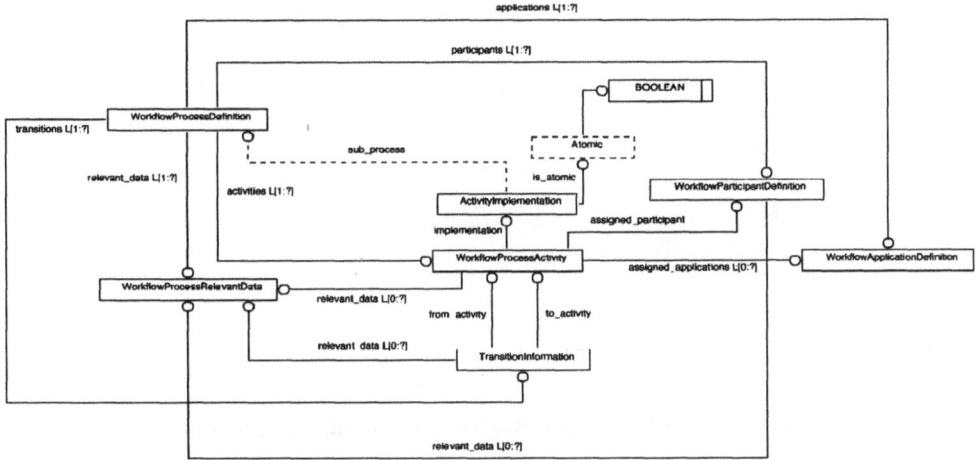

Figure 11 EXPRESS-G diagram of the minimal meta-model of the WfMC workflow process definition language.

THE VEGA ARCHITECTURE FOR WORKFLOW MANAGEMENT

The VEGA workflow architecture must enable WfMS interoperability and workflow monitoring across company boundaries.

Interoperability will be realised through a CORBA-binding as already demonstrated by Digital Equipment Corporation in the DeTeBerkom-project VorTel [VorTel. 1996] This is a more reliable solution than using Email. To achieve the interoperability across organisation boundaries, VEGA introduces workflow gateways, as shown in Figure. These are intended to connect the workflow engines of the respective WfMSs and to work on top of the organisation's firewalls to assure that workflow-related information can cross company borders. The workflow gateways assure, that only that workflow information is exchanged that is intended for partner companies. Any other workflow-information will be kept secure within an organisation's network.

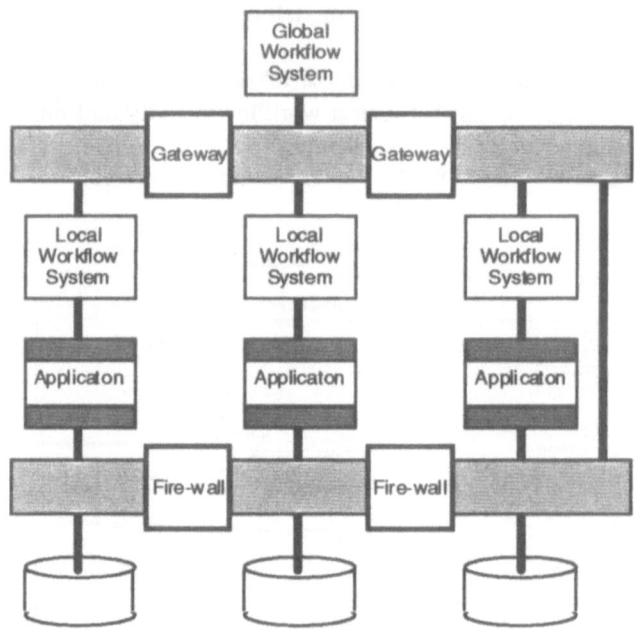

Figure 12   Workflow gateways to cross company boundaries.

For global workflow monitoring [Schulz, K. 1996]each local WfMS puts its information that is public within the virtual enterprise in a (logically) centralised database where all runtime-information of workflows is collected. An additional tool, a global workflow monitor, can access the database to show and analyse the information. Thus a workflow monitor is able to supply the following services:

- Global State Query: current state of workflows and their activities, which can be "Ready", "Blocked", "Started", "Paused" or "Finished".
- Improvement of capacities by redefining critical workflow-parts, i.e., activities or sub-workflows.
- Class-Improvement to apply workflows closer to reality.
- Object-Tracking for information about versions, actors, duration, etceteras of used objects within the workflow.

The WfMSs which interoperate in a virtual enterprise form a distributed workflow enactment service as described in Figure . To access workflow-related information like workflow process descriptions they access the CORBA Access to STEP information storage Architecture (COAST) to be developed in the VEGA-project. This architecture defines a new, third access method to stored information described by an EXPRESS schema. COAST is defined as a service layered on top of the Common Object Request Broker Architecture (CORBA), which has been defined by the Object Management Group (OMG). As a CORBA service, COAST conforms also to the OMG Object Management Architecture (OMA) [OMG, 1995][OMG, 1996].

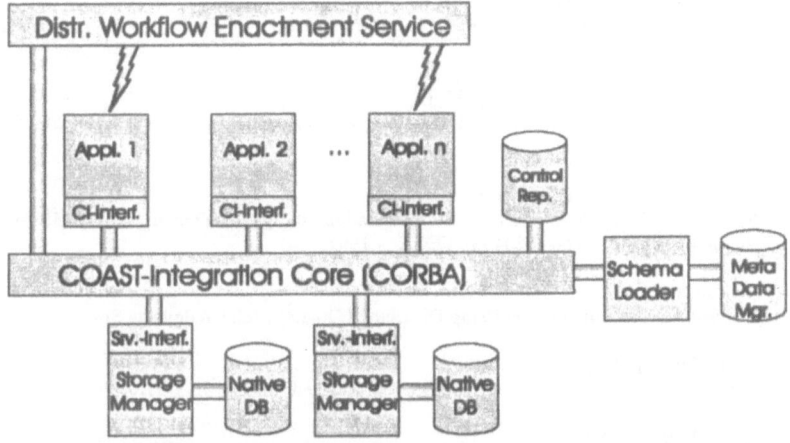

Figure 13 Workflow-COAST interaction within VEGA

COAST is intended to work independent from the WfMSs. The WfMSs form an additional COAST-Service. They access applications which participate as invoked application in a workflow via interface 3 of the WfMC Reference Model.

References

Amar, V., et al. [1997]
An open STEP-based distributed infrastructure: the COAST platform. in Concurrent Engineering in Construction. 1997. London, UK.

Coleman, D. and R. Khanna,[1995]
Groupware: Technologies and Applications. 1995: Prentice Hall.

Fowler J. [1995]
STEP for Data Management, Exchange and Sharing. Technology Appraisals 1995.

Gawlick, D., M. Hsu, and R. Obermarck,[1994]
Strategic Issues in Workflow Systems, ., Digital Equipment Corporation: Palo Alto, California, USA. 1994

Industrial automation systems and integration [1994a
Product data representation and exchange Part 11. Description methods: the EXPRESS language reference manual, 1994

Industrial automation systems and integration [1994b]
Product data representation and exchange Part 21 Implementation methods: Clear text encoding of the exchange structure, 1994

Industrial automation systems and integration [1995a].
Product data representation and exchange Part 22. Standard Data Access Interface,1995

616

Industrial automation systems and integration[1995b]
Product data representation and exchange Part 3. C Programming Language Binding to the Standard Data
Access Interface Specification 1995.

Industrial automation systems and integration [1996]
Product data representation and exchange Part 23. C++ Programming Language Binding to the Standard Data
Access Interface Specification 1996.

Köthe M., Schulz K., Amar V., Zarli A. [1997]
COAST Architecture: The CORBA Access to STEP Information Storage Architecture and Specification,
Deliverable D301 in the ESPRIT-Project 20408 „VEGA" - 1997.

Mowbray T. J. & Zahavi R.:
The Essential CORBA - System Integration Using Distributed Objects , John Wiley and Sons.

Object Management Group [1995]
CORBA Services: Common Objects Services Specification, Revised Edition, 95-3-31. 1995.

Object Management Group, [1995]
The Common Object Request Broker: Architecture and Specification Revision 2.0, . 1995, Object Management
Group.

Object Management Group [1996]
CORBA services: Common Object Services Specification, . 1996,

Object Management Group [1997]
Product Data Management Enablers; Request For Proposal 1997

Object Management Group [1996]
Document: mfg/96-08-01, 1996

Otte R., Patrick P. Sr., Roy M. [1995]
Understanding CORBA The Common Object Request Broker Architecture 1995.

Schulz, Karsten, [1996]
Monitoring of Systemwide Workflows. Fachhochschule Furtwangen, Germany 1996

Siegel J.& Al [1996]
CORBA Fundamentals and Programming, John Wiley and Sons, 1996

WfMC, [1995]
Coalition Overview, Workflow Management Coalition: Brussels, Belgium1995

WfMC, [1994]
The Workflow Reference Model,Workflow Management Coalition: Brussels, Belgium1994

WfMC, [1996]
Terminology & Glossary,Workflow Management Coalition: Brussels, Belgium1996

WfMC, [1996]
Interface 1: Process Definition Interchange, ,W.G. 1/B, Editor., Workflow Management Coalition1996

Workflow-Teleservices. [1996]
Joint project, sponsored by the DeTeBercom GmbH, Berlin. Finalised June 1996. Participants: Digital Equipment
; FHG Dortmund, GMD Darmstadt, IBM, TU Dresden.

A DYNAMIC PRODUCT MODEL
A base for distributed applications

RICHARD JUNGE
Faculty for Architecture, Professorship for CAAD
Technical University Munich

RASSO STEINMANN
Faculty for Civil Engineering, Professorship for Bauinformatik
Fachhochschule Munich

KLAUS BEETZ
Dipl. Math.
Nemetschek Programmsystem GmbH, Munich

Abstract

The project work described in this paper is a part of the ESPRIT VEGA Project. It is related to two companion papers issued in this conference proceedings. 'Product Data Model for Interoperability in an Distributed Environment' (Junge and Liebich. 1997, this volume) and 'The VEGA Platform' (Junge, Koethe, Schulz, Zarli, Bakkeren. 1997, this volume) are describing the technological basis for an application modeled to capture and convert the working environment of architects and building engineers, in short: the building design team, to an computer environment. The ESPRIT projects are increasingly forced into 'public and private risk funding and sharing policy. This part of VEGA is explicitly directed to exploitation of the EU funded project. This can be reached by a stepwise (small steps) transition from research to commercial implementation.

This paper is demonstrating the path VEGA project partners are heading. A major orientation point is an implementation based on the following technologies:
- use of object oriented design and implementation
- use of product data modeling
- dynamic schema evolution
- use of related standards as STEP, EXPRESS, IAI's IFC, OMG's CORBA, WFMC's Interfaces
- use of latest software engineering technologies as OO, COM/DCOM
- use of latest communication technologies based on Internet, Intranet, etc.

R. Junge (ed.), CAAD Futures 1997, 617-634.
© *1997 Kluwer Academic Publishers.*

1. Introduction

Several research initiatives have covered the application of product modeling for the building industry. These efforts have proven the applicability and usefulness of product modeling theoretically. However, only little or no experience have been made with introducing it into commercial implementations. Furthermore, the use of information technology in the architectural and engineering domain is characterized by a large number of applications for different realms in a scattered industry. Interoperability between these applications is not supported, it is reduced to data exchange by file exchange using whatever format.

Product modeling is an acknowledged method of describing the specific definitions of very complex products, such as ships, airplanes or in -our domain- buildings. Buildings are, although of normally very simple geometry, extremely complex structures. This is not the limited view of a 'building freak'. Buildings even when very simple are one-of-a-kind but consist of a multiplicity of components. These components are planned and constructed with the aid of many different applications. Each of these applications has its own data structure and generally the applications are incompatible with each other. Moreover, a common product model can not be derived from the specific data structure of all these applications

Today's modern communication technologies (Internet, Intranet, ISDN etc.) offer many possibilities for computer-based concurrent engineering. In order to use these technologies effectively there has be a holistic and unified product model providing the underlying semantic. The definition of product models is the goal of different initiatives, like STEP, IAI etc. is a critical opinion is that most members of these initiatives consider product modeling merely as an established concept with the sole, though important, purpose of semantic based data exchange. But why not go one step further and use product modeling also for application internal, rather than just inter-applicational requirements (Junge and Liebich 1995)

The vision of this project is an application environment which supports a building project during a whole life-cycle as a project server. This project server is able to store all relevant data of a building during its life time. It can be accessed by other applications using standards like AP225, IFC or others to come. It is using the Internet or some other network as transport medium. Furthermore, access through runtime using CORBA or COM/DCOM is a base functionality, thus enabling different client applications to behave as if their own internal data-structure were based on the same product model as the server's. This means that every application supporting such a standard can work with the project server and can retrieve stored data or add new data. The project server acts as a turntable for all relevant data, taking data security and data integrity issues into account.

The key is in information sharing in an distributed environment and by no means in data exchange between applications. This seems to be a fundamental difference to many existing product modeling approaches, where data exchange is the standard goal.

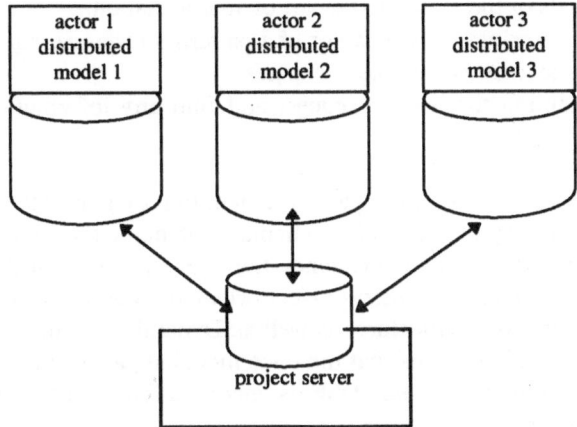

Figure 1: Data sharing using a project server

Another vision is that of applications based on an integration platform for applications tailored for the building industry linked together at the same time distributed through the VEGA Platform under development in the ESPRIT project VEGA.

2. The base technology

Although the project described in this paper in general is clearly positioned in the area of architecture, building engineering and construction it's first phase has been mostly occupied with the development of base technologies needed for the envisaged software solutions. The project does not claim to have invented the concepts of these underlying technologies. The achievements rather are on one hand bringing this number of concepts together to form an integral complete set and on the other hand to further develop this set of concepts to a very mature and stable state, which finally will allow to build commercial software solutions.

2.1 OVERVIEW

The core technology for the software application part of the project is pure product modeling technology as such. The advances made lie in a solution to overcome the normally rather static then flexible structures of product models. This kind of product model has named DPM Kernel (Dynamic Product Model Kernel). This name should indicate that while the application is directed to fulfill requirements for its use in the

620

building domain it generally is a generic solution from an IT point of view. The DPM Kernel consists of the following four basic components:

- Instances — are the concrete entities as stored in the data base.
- Patterns — are describing the properties of the instances. The properties may be attributes, methods or the ability to be part of an association.
- Methods — are describing the behavior of a pattern's instance, e.g. when the value of an attribute or an associated object changes.
- Filters —are enabling the user to request and find any information stored in the model.

In order to realize how these four parts are helping to gain the high flexibility, openness and extensibility of the loaded schema, that mark the main benefits of the approach, it is important to understand the generic base technology of the implementation. This technology allows object oriented modeling of the knowledge of a specific domain, such as architecture, as well as Dynamic Schema Evolution (DSE). Dynamic Schema Evolution means that the once modeled part of the world is not kept in a static and determined schema. DSE is modification, creation and deletion of schema information at runtime.

The DPM is a meta model and because of that the components of the DPM Kernel are not 'hardcoded' classes of the entities of a specific domain. There is for example no class 'wall' or a class 'room' in the DPM. The traditional 'hardcoded' implementation strategy would result in the implementation of a class together with its attributes as data members and additionally its methods for every entity of a domain. As a consequence, the entities could only be changed in a new version of the application (Figure 2). Thus the whole model is static and can not be expanded or modified after loading it to the DPM Kernel (see Section 2.6).

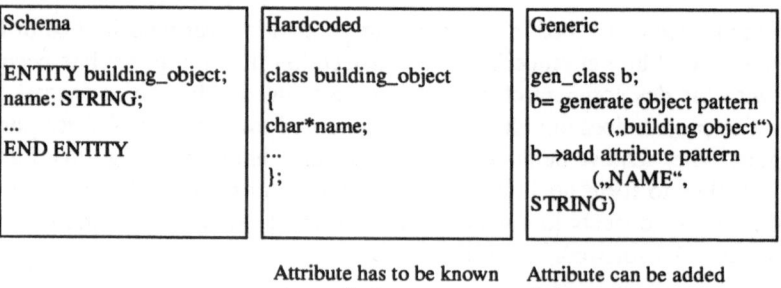

Schema	Hardcoded	Generic

ENTITY building_object;
name: STRING;
...
END ENTITY
	class building_object
	{
	char*name;
	...
	};
		gen_class b;
		b= generate object pattern
		("building object")
		b→add attribute pattern
		("NAME",
		STRING)

	Attribute has to be known at compile time	Attribute can be added during runtime

Figure 2: Generic vs. hardcoded approach

But the fact that the project didn't stick to the traditional 'hardcoded' concepts doesn't mean the paradigm is a non object-oriented one. In contrary, all important principles of the object oriented paradigm (Stoutrup 1992), such as abstraction, inheritance, information hiding or polymorphism are also fully available in the DPM Kernel (see Section 2.3).

The main constructs of the DPM Kernel are the patterns. Patterns are equivalent to classes in the traditional 'hardcoded' concept. They are providing the functionality to have attributes, methods and to be part of associations. Such a pattern gets its specific meaning of being, e.g. a wall, by a process called Dynamic Data Typing (DDT). Dynamic Data Typing is a functionality provided by the DPM-Schema-Loader (see Section 2.6).

2.2 DPM INSTANCES

DPM Instances are similar to C++ objects in the traditional hardcoded approach. The big difference to the hardcoded approach is that particularly, their internal structure, described by their attributes can be modified at runtime according to their pattern information. In the DPM Kernel four kinds of instances are available:

- objects — represent any concrete entity of a specific domain e.g. walls, windows, rooms in the architecture domain.
- attributes — store values (can be of various types, string, integer etc.) describing the internal state of an instance.
- associations — represent relationships among objects and store information about the relationships.
- taxonomies — manage a collection of subcategories as well as a collection of objects or attributes (targets of the arrangement) belonging directly to this category.

2.3 DPM PATTERN

Patterns are similar to C++ classes objects in the traditional hardcoded approach. They are dynamic class representations describing the properties and the behavior of their instances. The properties may be attributes, methods or the ability to be part of an association.

But in contrary to the 'hardcoded approach in the DPM approach an attribute or a method can be added to or deleted from a pattern at runtime, even if instances of this pattern exist. A pattern may be derived from another one, in the sense of the object oriented paradigm [2], and thus they may form hierarchies. According to the inheritance principle of the object oriented paradigm, attribute inheriting and overwriting and method inheriting and overwriting is supported.

Every pattern has its name which is unique throughout the model and its attributes and methods have names unique in the name space of the pattern.

Like the instances, the DPM Kernel distinguishes four kinds of patterns:

622

- object patterns — describe any entity of a specific domain e.g. walls, windows, rooms in the AEC domain
- attribute patterns — describe the attributes of an object pattern by specifying e.g. the data type, default value, unit etc.
- association patterns — serve for modeling relationships among objects
- taxonomy patterns — enabling instances to be structured in any hierarchical order, e.g. the topology ordering of a building in floors or in any other functional decomposition

2.3.1 The DPM Object Pattern

DPM object patterns are described by generic attribute patterns and methods. Moreover they can have relationships to other objects, so called associations. One can distinguish between one_to_one, many_to_one, one_to_many and many_to_many associations. In the example in Figure 3 the object pattern "Wall" is described by a set of attributes. It has a self referenced association describing the connection of two or more walls.

As shown in Figure 3 the geometry is an attribute among others just like the material. This is a major difference to traditional CAD systems, where an object can only exist by virtue of one favored geometrical representation. All other descriptions, especially all non-geometric attributes are regarded as peripheral ones, having only a supplementary nature.

In the DPM approach it is possible to have more than one geometrical attribute, e.g. a set of parameters, describing the geometry, a polyhedra as result of the geometrical parameter set, a polyhedral, including the recesses and openings, thus taking the information from associations into account.

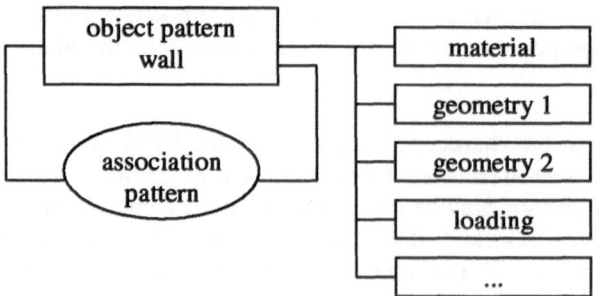

Figure 3: Example of the DPM object pattern wall

2.3.2 The DPM attribute pattern

In the DPM-Kernel two kinds of attribute patterns are distinguished, simple attributes and complex attributes. A simple attribute pattern has a common data type like string, real, date, currency, etc. and the corresponding DPM-attribute instance holds the actual value. Whereas a complex attribute pattern has reference to other attribute pattern as data type. These referenced patterns can be simple and complex attribute patterns. The corresponding attribute instance has a reference to the other attribute instance as a value. Thus, attributes are built that include a whole tree structure with simple attributes at the leaves of the tree. In the example in Figure 4 the complex attribute "Material" consists of two simple attributes, "Name" and "Description" and another complex attribute, called "Physical properties", which itself contains the two simple attributes, "Moment of inertia" and "Thermal conductivity".

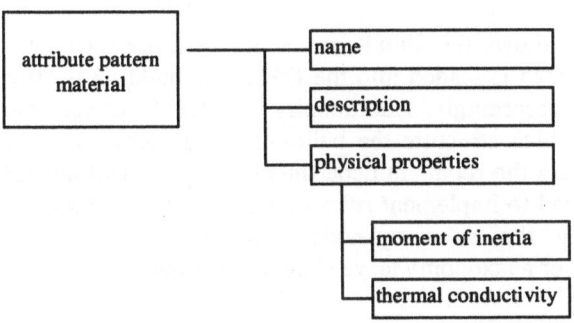

Figure 4: Example of a complex attribute pattern

2.3.3 The DPM association pattern

The association pattern has the same abilities as a common DPM pattern itself. This means the associations can be provided with attributes and methods and can be part of another association, as well. This ability may be used to describe the relationship of two objects and to specify the behavior of the objects, whenever an associated object is changed. For example, the "connected_with" association in Figure 3 has some attributes and rules, describing the contact of two walls. Moreover, it has some methods calculating the resulting polyhedra of the two walls, according to the attributes and rules that describe the contact behavior.

The ability to be part of another association can be used to model dependencies between associations. For example, in an early stage of the design the association "connect_with_rooms" determines that two rooms should have a connection. The association "connect_rooms" describes the concrete physical connection of the rooms in a later stage of the design. These two associations might be associated with another association which thereby expresses that the first association requires an association of the second kind.

2.3.4 The DPM taxonomy pattern

What applies to the associations holds for taxonomy patterns too. They are common DPM patterns and can be provided with attributes and methods. A taxonomy pattern's purpose is to describe the hierarchical arrangement of a set of DPM instances by categorizing them into a tree. A taxonomy pattern manages a collection of subcategories as well as a collection of DPM instances (targets of the arrangement) belonging immediately to this category. A taxonomy forms a tree which consists of taxonomy nodes and leaves. A taxonomy node refers to a collection of other nodes, to a leave node. A taxonomy leave contains only a collection of DPM instances.

The DPM Kernel allows the simultaneous use of more then one taxonomy. If for example STEP AP225 is loaded into the DPM, the building is structured in building sections or levels accordingly. Additionally, the DPM Kernel offers the use of any other taxonomy, which structure the building in any other way, e.g. in a functional hierarchy structuring the rooms in departments or functional units. Consequently, the methods can be used to implement rules that determine the behavior of the taxonomy node or leave, e.g. the taxonomy node 'building' only refers to nodes of a type 'building_section' or a taxonomy leave should only contain rooms.

2.4 THE DPM METHODS

The methods of a DPM pattern have to be implemented as a class in the sense of a name space. All the classes with the implemented methods of all DPM patterns are available for the DPM Schema Loader in dynamic link libraries.

A common DPM pattern provides several hooks at which the methods can be attached as callbacks. These hooks define the execution time of the methods, e.g. on creation, on deletion, when an attribute is updated.

An attribute may be provided with a domain rule. This domain rule is called by every assignment of a value. This domain rule can contain any functionality, e.g. it can prove that the value of the attribute "length" is always greater than or equal to zero.

Besides these predefined methods, every DPM pattern can contain other methods, which may be called within the hooks methods or by application code.

2.5 THE DPM FILTERS

Filters are similar to database queries. They allow a set of DPM Kernel objects to be selected that satisfy one or more conditions.

The DPM Kernel provides three classes of filters:
• object filters query a set of objects specified by a condition on the attribute set

- association filters query a set of objects associated with a specific object by an association and which satisfy some condition of the attribute set of the object or of the association.
- taxonomy filters query a set of objects classified under some taxonomy node and which satisfy some condition of the attributes of the object.

Furthermore these three can be regarded as 'atomic' filters which can be combined to compound filters using set theory operators (union, intersection, difference and complement). To speed up performance, objects and attributes can be supplied with indices that improve the execution time of filters, sometimes by several magnitudes.

2.6 THE DPM SCHEMA LOADER

The DPM Schema Loader fills the abstract data structures of the DPM-Kernel with the knowledge of a specific domain. Hereby the knowledge of a specific domain is formulated as an EXPRESS schema. This schema is loaded in our application and the methods that implement the functionality of the schema have to be available in a dynamic link library.

The DPM-Schema Loader first parsers the EXPRESS schema, then a process is started, called Dynamic Data Typing (DDT). This process of dynamic data typing transforms a common DPM pattern into a specific pattern according to the EXPRESS schema, e.g. transforms the EXPRESS entity "Wall" with all its attributes to a wall pattern and adds the functionality to the wall pattern. The same occurs with associations which are described in the EXPRESS schema, they were also converted to DPM associations with all its attributes and the methods were added to the pattern.
After the schema loading instances to the patterns could be created or loaded from an physical file.

2.7 CONCLUSIONS

With these four basic components in place the semantics of any specific domain can be described and stored. Moreover, description of the domain is not only a static one. It is dynamic in two respects.

First, methods are used to describe the behavior of the instances and associations are used to propagate the changes of an instance to the associated instances. Secondly, with the generic approach the model can be extended even at runtime.

The base technology described above is an abstract technology for modeling the knowledge of any specific domain. The problem is reduced to a appropriate formalized description method for the knowledge of the domain. This is the place for the product modeling technology itself.

A specific schema of the building domain is loaded in the application as it currently is, which includes the construction and the semantics of a building. But, it has to be emphasized that in principle any schema formulated in EXPRESS can be loaded.

To get real instances for the model, standards from STEP and IAI are used. In the current version the AP225 protocol is used to read in a whole building from any CAD system, which supports the AP225. An IFC interface is also existing as a first prototype.

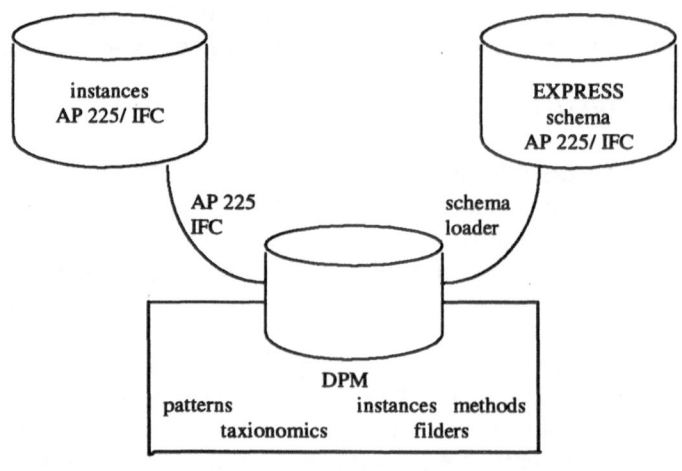

Figure 5: The base technology

The next step will be to distribute the information stored in and managed by the DPM Kernel over wide area networks, so that architects and engineers can work concurrently on the building model and have access to it regardless where they are working. We see the forthcoming COAST Platform from the VEGA project to be the appropriate way to do this. Therefore the DPM Kernel provides also an API, through which it can communicate with the COAST-platform.

3. The application

One can discuss the value, the potential extensions of functionality that DSE could have to end user' applications against the dangers of creating chaotic situations in an end user environment and the dangers DSE could have in information exchange. Without any doubts, however, are the advantages it opens in application development. The DPM constitutes a powerful and flexible basis for application development on top of a product data models. Together with foundations described in 'Product Data Model for Interoperability in an Distributed Environment' (Junge and Liebich. this volume) and 'The VEGA Platform' (Junge, Koethe, Schulz, Zarli, Bakkeren. this volume) it provides possibilities for new application software in the architectural domain. Only a few of those possibilities can be touched in the following.

The DPM itself is a neutral implementation method of any Express schema totally independent of any specific domain. What is needed to make it's use specific and useful to the architecture and building engineering domain is a core model of the kind described in 'Product Data Model for Interoperability in an Distributed Environment'. It follows a strategy to provide a core of semantics to allow domain models to be ' plugged in' to it. This core, in the project called 'BPM Kernel' (Building Product Model Kernel) provides the necessary semantic for communication between these domain models. Without such a level of commonly agreed semantics a technique like DPM would be useless in an implemented environment. The dynamically under runtime created extensions of a product model schema would not make any meaning to anybody besides the creator himself.

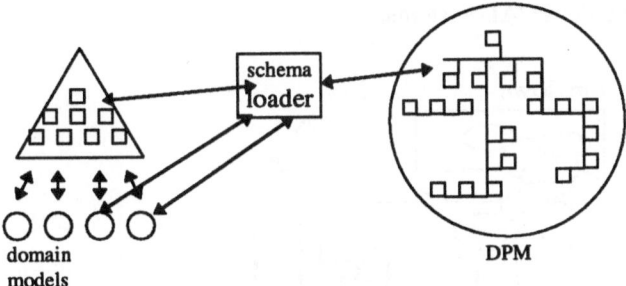

Figure 6 BPM Kernel and DPM

The applications using the DPM, as well as the VEGA platform together with the described product data model, currently under investigation are pointing in different directions, which are:
- CAAD kernel
- Coordinator Workstation
- A tool set (the current prototype implementation)

3.1. A CAAD KERNEL

Current CAD systems are 'geometric' systems not object oriented systems. The centerpiece of all these systems is the ability to create and manipulate geometrical objects. The only way to provide some meaning to these objects is to simply label these geometrical objects with names and in some cases also with attributes. A label, as for instance, wall or beam does not make more difference to the computer as that these two strings differ from each other. The objects themselves still remain to be pure geometrical objects. An architect or engineer however, when designing does not think in geometrical objects but in objects of his design task.

628

A computer could become a design assistant for architects and engineers only when he would know not only about drafting objects but rather about design objects and their meaning in the architects professional world. Traditional CAD Systems are the assistants of the draftsman. A new CAAD generation should become the assistents of the architects. For such a new CAAD environment it is the first prerequisite that the system would behave like a design assistant and not as a drafting tool. In (Junge 1995) and (Junge 1997) the author has described some functionality's to be fulfilled by such a new CAAD system:

- Continuous refinement.
  The design process is often called to go from the coarse to the fine or from vagueness to precision. This is a process of continuos refinement starting from the first conceptual ideas, where only vague representations of the designed objects are possible to be made to more and more precise definitions of the design. This clearly is a top down approach, but exactly this is in contradiction to the bottom up approach of todays CAD systems.

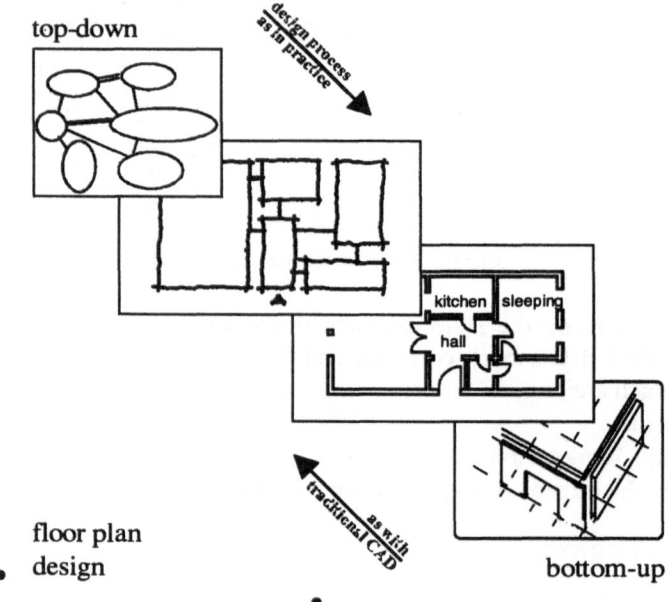

Figure 7 Aspect of continuous refinement

- Extendibility.
  The world of building elements seems to be infinite. There are so many that no one could bring them into one product model schema. In a schema only definition of classes down to a certain level can and should be defined. This however makes mechanisms necessary to extend definitions provided by a model. Design objects clearly have the tendency to be multifunctional and to be used for purposes they

originally where not meant for. This is happening during design but more often the
'use phase', in facility management. Mechanism that allow such changes in the
schema are necessary. See also the many papers Eastman (Eastman, C. 1995) is
describing his motivation for the EDM development.

- Multiple geometric repesentations
  During the process of a gradual definition of the design, the representation and
  presentation of it's form is having many expressions. For example it starts with
  simple symbols in early phases and goes in a stepwise manner finally to very
  precisely defined ones in shop drawings in late phases. The representation and
  presentation are not only dependent from phases of the design process but also from
  those engineering disciplines for which they are meant to be used. Consequently
  there are many valid representations of a design object coexisting at the same phase
  or time. For a certain task in structural analysis, for example, it may be appropriate
  to have only the center line of gravity and not the maybe the 3D volume definition
  of a girder. The representation of design objects is dependent from phases and
  disciplines, from the context in which they are used.

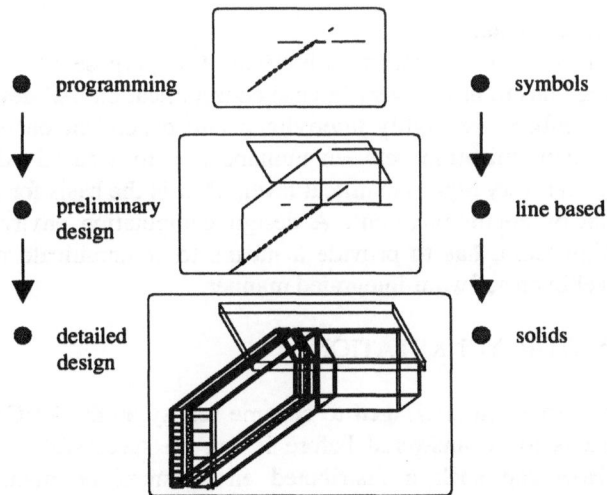

programming symbols

preliminary design line based

detailed design solids

Figure 8 Aspect of multiple geometric representations

- Continous specification.
  An architects thinking, especially in earlier phases of the design, is oriented to form
  and function, which seems to be merely geometry oriented, but in reality there is a
  second, parallel stream in the design work. This stream is dealing with all the 'non
  Gestallt', the non geometric, properties of the design object. These properties are,
  although often regarded less important and less sophisticated, compared with the
  Gestallt properties, almost the majority. Not only the 'Gestallt' but also these non
  geometric properties are defined in a stepwise manner during the design process.

All these specifications are made during the drawing process but they are not stored in the paper drawing neither they are in the CAD data base. They are almost lost, unretrievable. This information that has to be reconstructed later for the BoQ for example and that is done by people who almost where not involved in the design itself.

- Dynamic shift of focus.
During the design task the designer's interest is constantly shifting between various aspects of the design objects specification. So for instance at one moment he might be working one the functions to be fulfilled in a specific room and the relating spatial needs and adjacencies. In the next moment he might shift to the architectural appearance of the spatial arrangement, the surfaces, or finishes of the space boundaries, their structure color, etc. ;as an interior design aspect. The next focus might shift from the bounding wall as a spatial aspect, to the wall core as a constructive element itself. A question is what are the implications if these four aspects being objects of their own, existing in separate systems as they in today's CAD.

- Distribution and communication.
Design is done by a design team, which for the purpose of a coordinated or integrated design has to have a very intense communication. All design tasks of the design team members are highly interrelated and dependent on each other. The basis for this communication between humans is a to a certain degree common understanding on a very high semantical level. That is the basis for to use a slogan, for 'integration by communication'. A design computation environment, able to assist the design team, has to provide a means to communicate on such a high semantical level in an software-integrated manner.

3.2. A COORDINATOR WORKSTATION

The distributed environment envisaged to become reality by the VEGA Platform lets arise some questions to be answered before it can be successfully implemented in design offices. How can such a distributed environment be managed, be made transparent for the user? What are the functions needed? It seems to be an approach appropriate to be followed again is a recollection of processes and functions as in today's design offices. In today's practice a very important role in project work is on coordination. In larger projects a specific role is that of coordinating all participants, may it be in the home office itself, or with all engineering offices participating in the project. The tasks that such a coordinator has on his agenda have to be translated to this new computer tool. Candidate functions of the Coordinator Workstation are:
- Distribution and collection of models or model parts .
- Control of communication, network and workflow.
- Coordination of design tasks.
- Conflict detection and management.

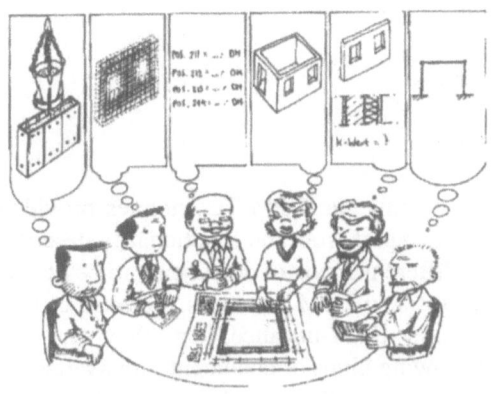

Figure 9 Coordination today

These functions are having two diverse aspects. Today both are fulfilled more or less perfect in an environment depicted in figure 9 First is the aspect of specific engineering discipline ones. These are among others:
- Technical coordination of design tasks.
- Tracking and backtracking of design tasks and results and their state.
- Management of human resources assigned to the project tasks.

The second aspect is on information technology issues. This is a wide area spanning from tools for:
- visualization of the technical processes
    - schema/model browser
    - model viewer
    - network viewer
    - conflict detector

- work flow management
    - work flow modeler
    - work flow controller

- communication flow management
    - translation, conversion
    - dividing and assigning of selected schema and model parts
        - distributor
        - delta storage

3.3. PROTOTYPE IMPLEMENTATION

The prototype applications which are implemented today upon the BPM/ DPM-Kernel contain, among other features, three main parts. These parts are giving access to the stored building for the user:

632

- The Project Explorer
- The 3D-Window
- The Spreadsheet Window

3.3.1 The Project Explorer

The Project Explorer is making the DPM-Taxonomies provided by the BPM/ DPM-Kernel visible for the user. Apart from the functionality described in section 2.3.4 the Project Explorer allows different kinds of documents to be attached to every taxonomy node. These documents are also DPM-instances. Because of this, they have the same functionality as a common DPM-instance, e.g. they can have attributes. In the prototype version the following kinds of documents are available:
- 3D-View documents include a 3D view of the whole building or a part of it, with a specific camera position.
- Report documents include the result sets of filters in a spreadsheet, which may be designed by the user.

The User can browse through the whole building taxonomy as well as view and modify the attributes of the building elements at the click of a button. Moreover, it is possible to import the same building in different versions in the same project. Executing the same reports on the different versions will show the consequences of the changes done in the CAD application.

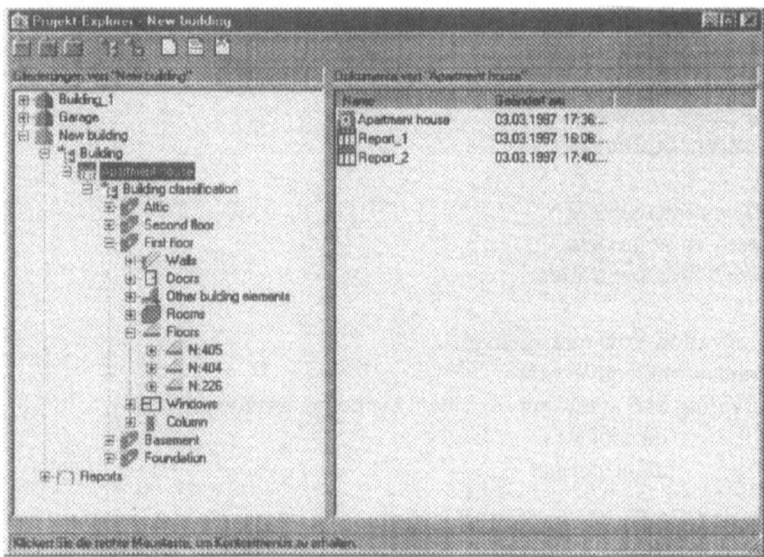

Figure 6: Example of the Project Explorer with a building taxonomy

633

3.3.2. The 3D-Graphic-Window

The 3D-Graphic-Window visualizes the geometric attributes. This means that the client gets a 3D-image of the whole building or of a part of it. It allows an animated navigation through the building. Identifying and picking of one or more elements is possible. However, the attributes of identified elements can be viewed and modified. Furthermore, it is possible to highlight all elements that are the results of a filter execution or that are identified in the project explorer. The attributes of these highlighted elements can also be viewed and modified.

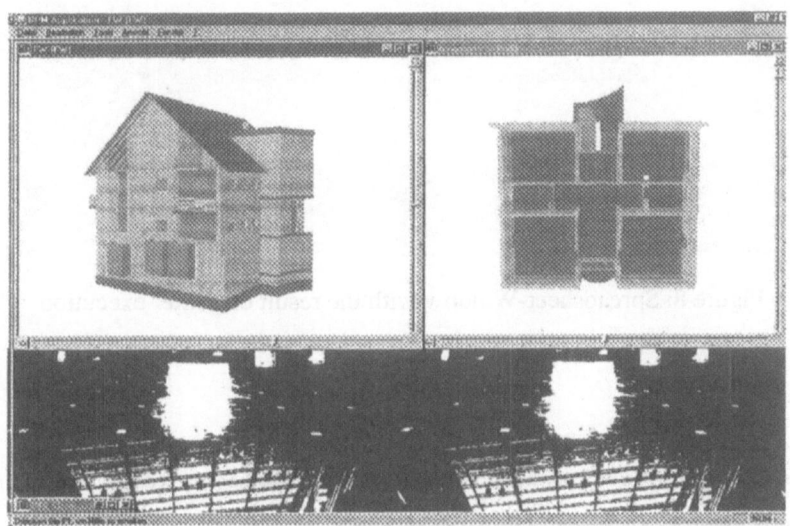

Figure 7: 3D-Window with the whole building(perspective view)
and 3D-Window with one floor (bottom view)

3.3.3. The Spreadsheet-Window

The Spreadsheet-Window shows the results of filters in a very comfortable way. The spreadsheet is fully EXCEL-compatible and offers all abilities like formatting, calculation etc., that are known from EXCEL. In the design window the user can define the whole layout of a report specifying which attributes it should include, defining how attribute values are sorted and so on. All defined filters can be stored in a catalog categorizing them under specific groups e.g. "All room filters". So a previously defined filter can easily be used in another project.

634

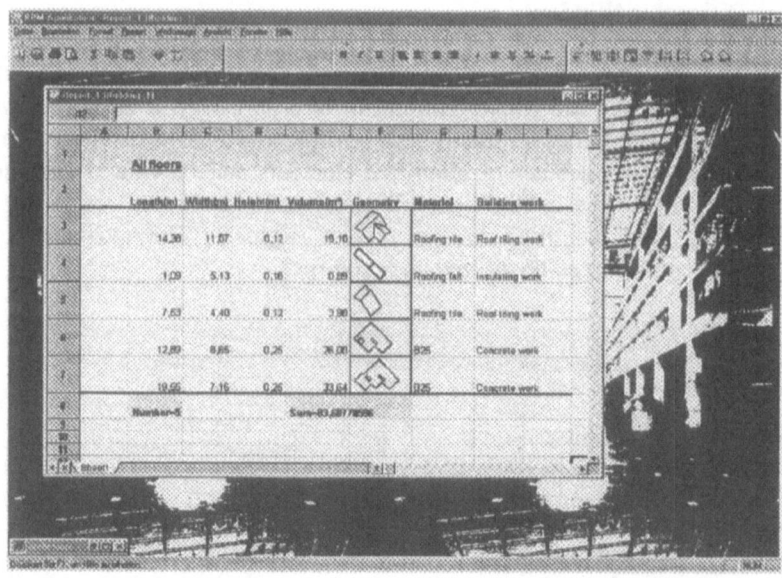

Figure 8: Spreadsheet-Window with the result of a filter execution

References

Eastman,C. (1995)
Structure of a Product Database supporting Model Evolution. CIB Workshop proceedings, Stanford University.

Junge, R. and Liebich, T., (1995)
Product modelling for applications: Model for next generation CAAD, Computing in Civil and Building Engineering, Pahl & Werner (eds), Balkema, Rotterdam

Junge, R. (1991)
Integration by Communication, 1st International Symposium on Building System's Automation-Integration. Conference proceedings, Univ, of Wisconsin, Madison

Junge, R. (1995)
Aspects of New CAAD Environments. CIB Workshop proceedings, Stanford University.

Junge, R (1997)
Building Product Model for Architectural Design Computing,, Kluwer. in print

Stroustrup, B.. (1992)
The C++ programming language, Addison-Wesley Publishing Company, Inc.

NOT JUST ANOTHER PRETTY FACE:

Images and Arguments in an Anthropology Web Site

JEN S. LEWIN, MARK EHRHARDT, and MARK D. GROSS

> *Sundance Laboratory for Computing in Design and Planning*
> *College of Architecture and Planning*
> *University of Colorado,*
> *Boulder, CO, USA 80309-314*
> *{ lewin@ucsub, ehrhardt@ucsu, mdg@cs } .colorado.edu*

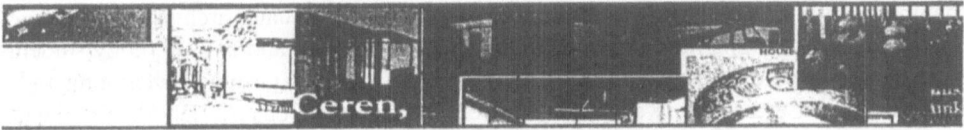

Abstract:

We are developing a web site with photorealistic animations and virtual reality walk throughs of architecture and artifacts at an archaeological site in El Salvador. The goal of the site is to support research and teaching about household anthropology in sixth-century Meso-America. To counter the false sense of realism and truth these experiences often convey we have developed Image Arguments, a scheme for integrating with images the arguments and data that they are based on. We provide this contextual information using a server side database and client side Java applets, enabling viewers to examine the assumptions and the data behind the images.

1. Introduction

Computer generated imagery has provided useful and fascinating reconstructions of ancient sites for archaeology and anthropology. However, the richness of information, the levels of discussion, the complexity of understanding and the ambiguity of decisions behind renderings are lost when images are viewed as isolated singular objects.
In constructing an anthropology web site for teaching and learning [http://wallstreet.colorado.edu/ceren/ceren.html] about village life at Ceren, El Salvador 1400 years ago, we are developing interactive computer-based imagery that links images directly to the information used to generate them. Seeing images in this context, viewers can formulate informed opinions about a rendering, perhaps developing

R. Junge (ed.), CAAD Futures 1997, 635-654.
© 1997 *Kluwer Academic Publishers.*

636

their own arguments. A rendering is no longer viewed merely as a pretty picture; instead it serves as a visible argument.

1.1 MODELING CEREN

In Fall 1995 we were invited to construct computer models for the archaeological site of Ceren, a pre-Columbian village in El Salvador buried in ash over 1400 years ago (Sheets 1992). Discovered in 1978 by Payson D. Sheets and in excavation since, Ceren is a fascinating site. Due to the sudden nature of a volcanic eruption villagers fled from ancient Ceren leaving everything as it had been used in daily life. Volcanic ash deposited on Ceren has prevented the decay of almost all objects. Thus Ceren offers exciting potential to modern anthropology by providing a glimpse of everyday Meso-American life.

Computer modeling offers an interesting view of the site and environment as it once appeared. Ceren's impeccable preservation and the large number of household artifacts found there promised rich, colorful and visually informative computer images. The project was especially interesting because it brought together students and faculty members with diverse skills and backgrounds: undergraduate architecture students versed in computer graphics and three dimensional modeling, and anthropology students studying household living patterns at the village of Ceren. Ground-penetrating radar suggests that Ceren had twenty-five household groups (Conyers 1995). We began production of computer models for one of the household clusters in excavation by Payson Sheets and his team.

The richness of Ceren, the vast quantity of information available and the abundance of artifacts makes Ceren a difficult site to understand without graphical representation. Computer models created for Ceren provided new views of the site and artifacts. Except for a few artist's sketches, the images we produced were the first real visualization anthropologists had of the site in its built form, the way it might have existed. Research at Ceren had previously consisted of analyzing numbers, charts and simple two-dimensional sketches and the computer renderings led to several new observations that were not apparent in the raw numerical data (Lewin and Gross 1996).

Our original goal was to graphically describe the site, creating full three dimensional computer models and renderings for all excavated households. Whether the computer renderings would go beyond the function of display was initially unclear. We believed computer images of Ceren as it looked 1400 years ago would be informative and we did not doubt that computer modeling would provide a highly flexible and effective way to construct images, capable of displaying multiple views, variation in lighting effects, materiality and environmental effects. We did not, however, understand the complexity of creating highly accurate architectural computer images from the archaeology site data, nor did we reckon on the potential for the images we made to evoke powerful viewer response and insight into village life at Ceren 1400 years ago.

1.2 THE ILLUSION OF REALITY

Computer renderings carry an illusion of reality. Miller's article "the Good the Bad, and the Down Right Misleading: Archaeological Adoption of Computer Visualization" (Miller and Richards 1994), and Kiernan's "Lies, Damned Lies and Slick Graphics" (Kiernan 1994) both emphasize that the apparent realism of computer generated images implies a degree of certainty that is not necessarily true. A photorealistic rendering evokes awe and wonder and can seduce the viewer to believe an image without question. While realistic computer imagery can be used to effectively display an architectural model of a no-longer-existing building, images also foster the illusion that the building once existed as it has been displayed. For example, models of Palladio's villas as they were originally designed suggest a reality that in some instances never existed (Mitchell 1992, p. 170) [http://andrea.gsd.harvard.edu/workshop/].

Interestingly, the renderings we made of Ceren did not create the illusion of existence for researchers familiar with the site. Although they were delighted to obtain realistic looking visualizations of their field data (they previously had relied solely on artist's interpretations), they treated the images we made with a degree of skepticism—questioning their interpretations of data. Each image we produced is the result of a multi-step decision process. Anthropologists worked closely with the modelers, resolving inconsistencies and making hypotheses about physical form that could not be resolved directly from the site data. Because creating each computer image required close examination and interpretation of site data, the process of rendering often entailed lengthy argument.

As we made the images, debates began about their content. For example, in one structure (# 11) the placement of artifacts was unclear from the archaeological evidence. To make the model, an opinion had to be formed and decisions taken regarding where artifacts might have been located in the original household. Often, after seeing an image, a researcher would disagree with the artifact placement it showed and request a new image with different artifact placements. The computer generated imagery of Ceren is not fixed; rather we continually produce images to explore new ideas that arise as the anthropologists examine and question the existing views.

However, as Kiernan argues (Keirnan 1994), our computer images play a different role for 'outside' viewers who are unfamiliar with the Ceren site. Without access to the site data and not having participated in the debate behind the renderings, outside viewers tend to be misled and see images as final and unquestionable views. To overcome this implicit acceptance our approach makes available along with the rendering the information and process behind its generation. By allowing outside viewers to see the process engaged by the Ceren research team we extend the function of computer generated images beyond mere display, and enable them to play a more integrated role in the underlying intellectual argument.

We are using Apple's QuickTime Virtual Reality (VR), the programming language Java, and database server software called Tango with a back end File Maker Pro data base to construct a web site structured around what we call 'Image Arguments'. An Image Argument allows a viewer to browse a rendering, triggering information displays that explain the components of the viewed structure and relevant issues, and to delve more deeply into the raw data and associated text about what the image represents. Thus an image is never viewed as a single isolated unit; rather, each image is seen in context of the information from which it was created.

The rest of the paper is organized as follows. We begin with an introduction to the use of computer images in archaeology. We then discuss false reliability in computer imagery and why this was not a problem for members of the Ceren research team. We introduce Image Arguments as a way to make visible to uninformed viewers the data and their interpretation that led to an image. We describe the methods and techniques we use to implement Image Arguments. Finally we conclude with a summary and discussion and outline the directions we are pursuing in the next phases of the project.

2. Computer Imagery in Archaeology

Computer graphics can be powerful and informative tools in archaeological research where physical models and artist's renderings have traditionally been the only visual aid available. The use of computer imagery in archaeology has grown increasingly popular as a way to display a site as it may have appeared in real life. Such imagery is the closest we can get to apprehending–through primary visual experience–structures that no longer exist. A number of recent projects have produced compelling views of various archaeological sites.

For example, a reconstruction of the temple precinct of the Roman town of Bath (UK) (Wallis et al. 1990) revealed information that could only be made apparent through such a visual experience. When the computer model was examined from several standpoints it became clear that views from one entrance of the precinct towards the temple were more impressive in contrast with views from the top of the steps of the temple towards the entrance (Reilly 1992). Such renderings provided the opportunity to see a site from the perspective of a user and make inferences that might not have been possible using other representations.

Other models such as the 1995 Lancaster University Archaeological Unit [http://www.lancs.ac.uk/users/archaeo/unit/luau.htm] reconstruction of Furness Abbey (deLooze and Wood 1990) and the 1989 model of Langcliffe Lime Kiln (Wood and Chapman 1992) provided a means to check reconstruction ideas, simulate structural systems and communicate results through images.

Brian Hayden and colleagues at Simon Fraser University constructed computer models of pithouse dwellings excavated at the Keatley Creek site in British Columbia, Canada, (Peterson et al. 1995). Their renderings show the distribution of found artifacts through real time interactive animation, and allow a user to view artifact placement with respect to surface slope and daylighting in three dimensions. This enabled them to "more readily identify potential relationships between artifact data distribution and the pithouse structure" (p. 54).

Other projects include experiments with visualization technology such as the medieval church of St. Giovanni in Sardinia (Nuero 1989), the Acropolis (Eiteljorg 1988) and reconstructions of the city of Messina prior to the 1908 earthquake (de Cola 1990).

Computer modeling is especially valuable for sites and structures that no longer exist and therefore cannot be seen. Computer modeling efforts in anthropology and archaeology are essentially an attempt to compile and present ideas about a once-built form; computer models are being built for many major archaeological sites. In short, modeling has become a popular and useful tool in archaeology and anthropology.

3. The Reliability of Computer Imagery

No doubt renderings can be useful, but there remains a question about their reliability. Unlike a hand drawn sketch, whose intrinsic graphic character reveals clearly that it is a subjective interpretation, computer images are treacherously seductive in their appearance of completion and correctness. Rendered images cannot display uncertainty, ambiguity and fuzziness, yet these characteristics are all prevalent in archaeology data (Miller and Richards 1994). Computer images suggest certainty, but in most cases they are at best theoretical interpretations of what may have been.

This danger of computer generated images is not limited to archaeology, but is problematic in other realms of computer visualization as well. Kiernan, (Kiernan 1994). noted that computer forecast air pollution images had an extraordinary influence over policy makers whereas dry tables of numbers and charts were relatively ignored. Commenting, Miller points out, "Worryingly, there is little, if any, quality control for computer graphics which are not subject to the same intense peer review as scientific papers" (Miller and Richards 1994). Seeing is believing: images tend to sway viewers more easily than numeric data or textual information.

The computer images we made of Ceren appear to display a complete and refined reality and they leave a viewer with the sensation of looking at photographs of the site as it once appeared. It is difficult to question the accuracy of these images. For example, looking at figures 1 and 2 an uninformed viewer is more likely to be awed than skeptical.

640

Although each figure represents a subjective view of Ceren, questions of ambiguity or the possibility of debate are not apparent.

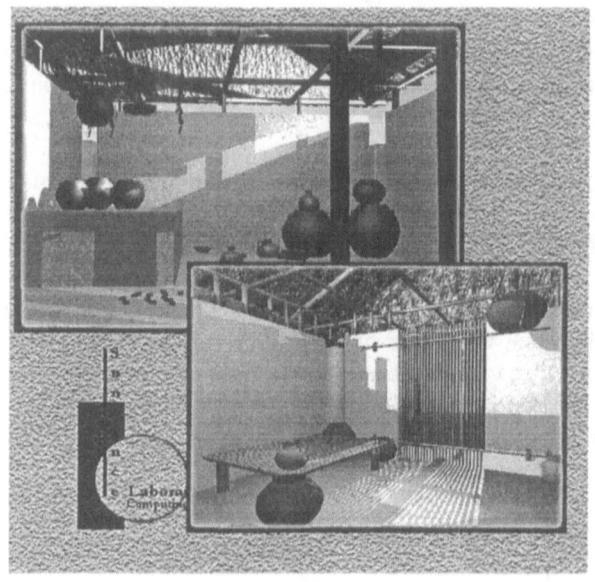

Figure 1. Renderings of the insides of Structure 1 and 7. Vertical placements of elevated artifacts, although determinate in these images, are actually the result of inference and supposition.

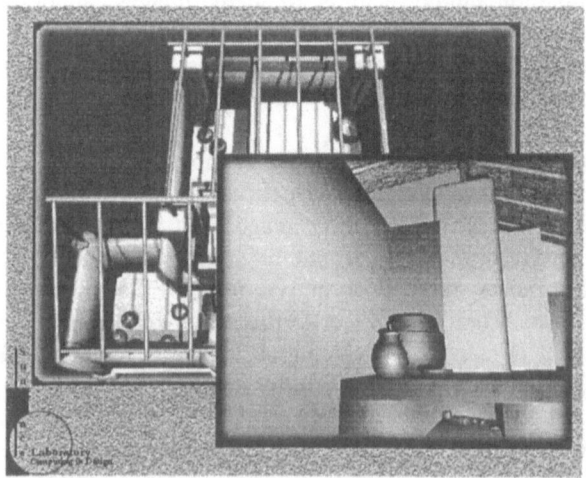

Figure 2. The inside of Structure 12 as we modeled it. Information found in excavation indicates a large painted mural on one exposed wall. The graphic content of this mural had not been fully determined, so it was not included in our renderings.

In many instances, renderings (either static views taken from the model, or walk-through animations) are the only experience that outside viewers are provided of the computer model, hiding the complex process of making the model and evaluating the site data. The image creation process is never revealed. For example, an award winning animation of the baroque Frauenkirche of Dresden destroyed in World War II [http://www.ibm.com/sfasp/mason.html] (Collins et al. 1994) seems more designed to evoke awe and wonder than to inform a viewer about the building (Reilly 1993) . Although the animation contains a great deal of information and does provide a view of how the structure once appeared, it blurs the line between animation for research and entertainment. Regrettably, the process of research and discovery experienced by researchers who produced the animation is largely lost in the final product.

A photorealistic image is based on a large amount of descriptive information about a site, and though this information is displayed visually in the final graphic presentation, it cannot be retrieved. Grant, in modeling the old city of Glasgow using his ISSUE program (Interactive Software System for the Urban Environment) (Grant 1993) , describes the frustration of working with the computer model's inability to respond to queries. Although the models and renderings he made successfully display spatial forms, they were inherently unable to respond to simple questions about their underlying data. According to Grant, due to a basic lack in the model's intelligence "answers remained locked inside the data structure" (p 558).

In summary, computer images are a dangerous medium. They hide uncertainty and ambiguity. They argue implicitly for the information that created them by displaying and describing a point of view. Yet viewers cannot examine this hidden information and therefore cannot truly understand the context an image represents.

4. Images as arguments in modeling Ceren

Perhaps predictably, anthropologists who were involved in the modeling process were not fooled by the apparent reliability of the images. Members of the Ceren research team did not treat the images as unquestionable facts. Rather, the images acted as catalysts for research and argument. The production of computer graphics for Ceren was a highly developed process requiring a sequence of specific decisions. Each image portrayed, collected, discussed, and evaluated information from the Ceren dig site logs and archives. Often an image served as an argument in support of a researcher's hypothesis.

For example, renderings for Structure 6 suggested an argument for new findings at Ceren. The adobe walls for Structure 6 were being rebuilt at the time of eruption with only the structural wood posts left standing. In collecting information from the

642

archaeological evidence and participating in the production of the image (figure 3), anthropologists observed an interesting fact. Although the structure contained a diverse collection of artifacts, most were of a larger size than the artifacts found in many other structures. Why? Several views suggested a theory: Without the adobe walls it would have been easy for a passing person to reach smaller artifacts through spaces between the wood posts. Larger artifacts, however, that would not fit easily through the spaces would be more difficult to steal. Perhaps smaller artifacts had been removed from the structure for safe keeping. Although this inference might have been made directly from the excavation data, in fact the renderings produced for Structure 6 acted as a catalyst for this discussion. The images became——in a sense—arguments for this idea.

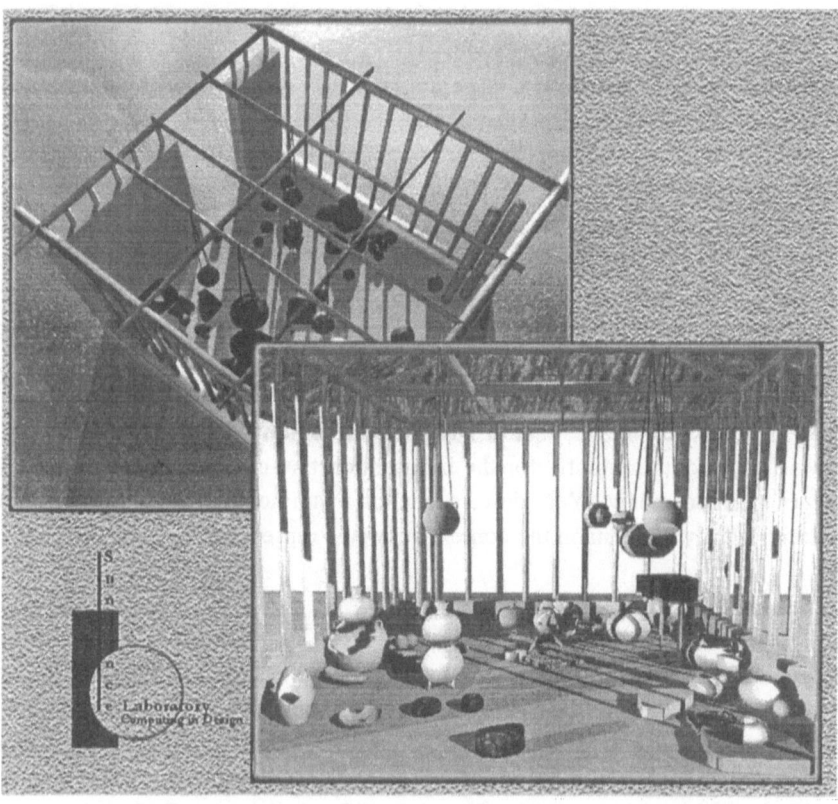

Figure 3. Interior renderings of Structure 6 showing the exposed posts and artifact distribution suggested a theory explaining why this structure contained mostly larger artifacts.

A second example involves the placement of artifacts in Structure 11. Figure 4 shows a plan drawing of Structure 11 and relevant excavation information. Descriptions of a higher shelf were available in published articles about the structure (Sheets and McKee

1989) but the shelf had never before been drawn. Due to the lack of data its placement in the original dwelling was unclear. We made renderings for Structure 11 that display an elevated shelf (see figure 5). These renderings were produced collaboratively with (then Ph.D. student) Scott Simmons. Simmons had to study the site data to infer a placement for the shelf (Simmons 1996). Thus the rendering reflects Scott Simmons's opinions about the placement of artifacts in Structure 11, and it is possible, even likely, that another expert could come to a different conclusion.

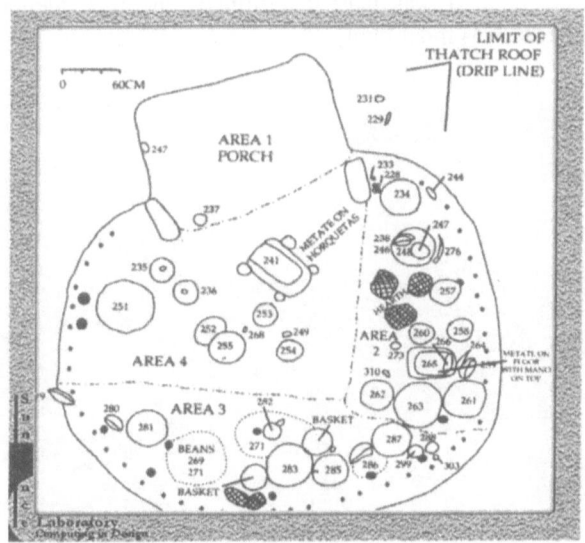

Figure 4. Plan sketch of Structure 11 showing artifact distribution.

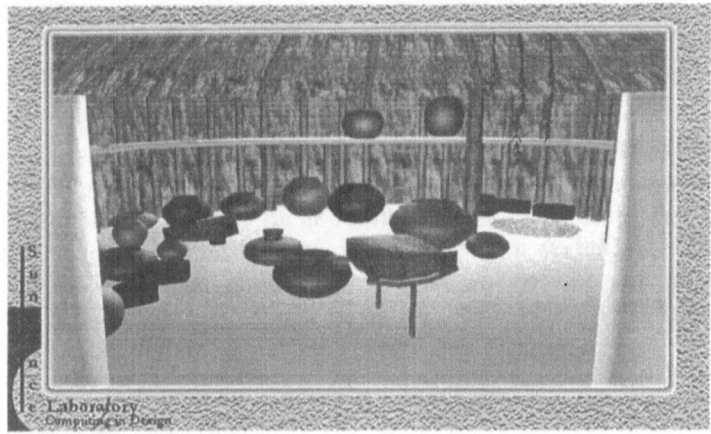

Figure 5. The height of the shelf in Structure 11 is based on the informed conjecture of Scott Simmons.

644

Renderings were often debated, sometimes provoking changes in the model and requiring the production of new renderings. Often an image did not satisfy the anthropologists' perceptions of the site and led to changes based on their arguments. For example, Figure 6 shows two views for two models of Structure 12 that were made in a sequence, as the images were changed and reevaluated. Each image describes a different interpretation of the data regarding roof structures. The two views of a hip roof structure in the top row, and the two views of a shed roof structure below are both plausible interpretations of site data. It was later determined, due to evidence of support pole deposits, that a shed roof was most likely.

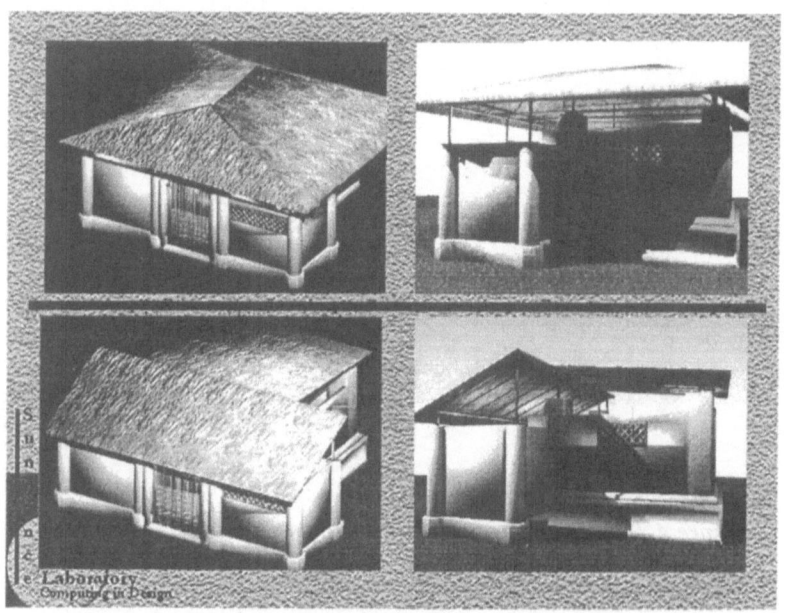

Figure 6. Alternate roofing schemes for Structure 12.

Computer images of Ceren, when viewed by researchers familiar with the site, serve in two ways in arguments about the interpretation of previously collected excavation information. First, each image is the result of an argument, a position that represents a set of hypotheses about how to interpret the archaeological record. Second, the image is a representation that aids certain inferences, enabling one to see things that might otherwise be difficult to understand.

5. Image Arguments

The difference between Ceren researchers and outside viewers is their knowledge of information about the Ceren site and the ability to formulate an opinion about the image. We are attempting to reduce this difference by using Web technology to develop a database of issues associated with the production of each image and to make these issues available to a broad audience of students and professional anthropologists. By linking the image with the information behind it, we aim to make the process of argumentation employed by the Ceren research team available to outside viewers.

Our system uses Apple's QuickTime VR tool kit to enable a Web user to view a fully interactive 360 degree rendering of all the currently excavated Ceren structures. Moving around the structure, the viewer can query areas of interest by clicking on an object or by typing a search command into a frame below.

Figure 7 shows an example of a QuickTime VR panel with a corresponding dialog box displaying the raw excavation data as the user examines specific objects. Excavation data about each object is stored in a server-side database, called up from the client side by CGI (Common Gateway Interface) requests and delivered to the user. The top left image is a QuickTime VR of the interior of Structure 12, with a triggered hot spot (the bottom center of the image, concentric circles). Directly to the right is a dialog box describing the object, a grinding stone, or 'metate', triggered by the QuickTime VR. Descriptions include basic dimensional data, as well as excavation comments. Below is a text area displaying general excavation remarks for Structure 12. Two applets are also displayed. The rightmost includes excavation comments describing other objects with important relationships to the metate. The applet window (bottom right) displays slides and text about continued use of the metate today.

Although it may be informative for an outside user to view the excavation data in a rendering as seen above, this will not necessarily leave the user with a complete understanding. Ceren researchers have a more general view of issues at Ceren and often formulate their ideas around these issues to motivate research about specific data. For example, there is anthropological evidence that Household Two has a higher status of wealth than Household One. Theories about this difference can be formulated and judged by studying images and detailed site information. An outside viewer trained in anthropology may frame a new issue by becoming familiar with the data, but issues provided by the Ceren anthropologists can also guide less experienced students in exploring the site.

Each structure at Ceren triggers a series of issues. When the user browses a Quick Time VR scene of Structure 12 the lower dialog box displays issues (phrased as questions) that are relevant to the structure (figure 8). The opening graphic for Structure 12 displays a rendering of the outside of Structure 12. To its right is a text area displaying pertinent

646

excavation information found in Structure 12. Areas of text are linked to their subjects. Below, a 'hint' about the structure is displayed providing the user with guidance if necessary. A menu applet (bottom left), and an image display window (bottom right) are shown.

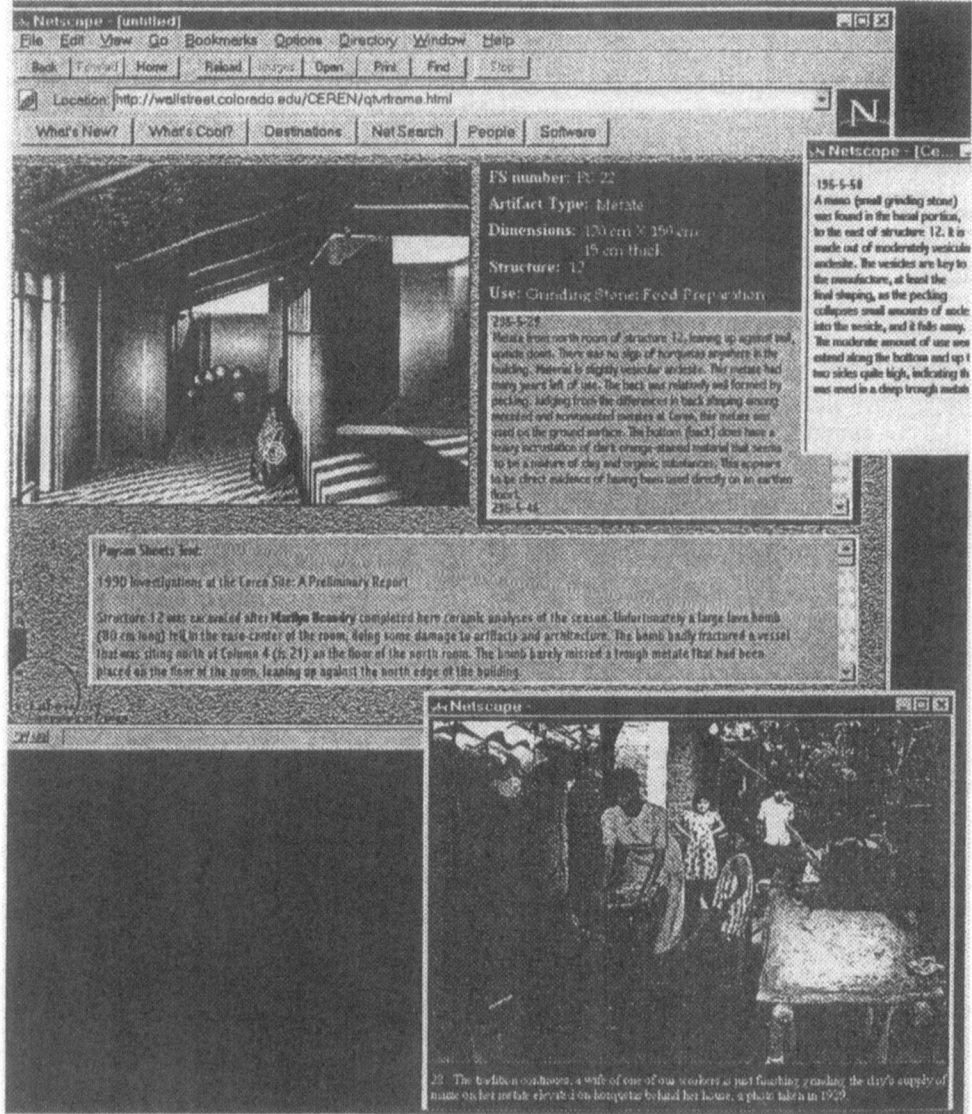

Figure 7. QuickTime VR image and related excavation records. Hot spots in the QuickTime VR image link to field specimen data and excavation notes.

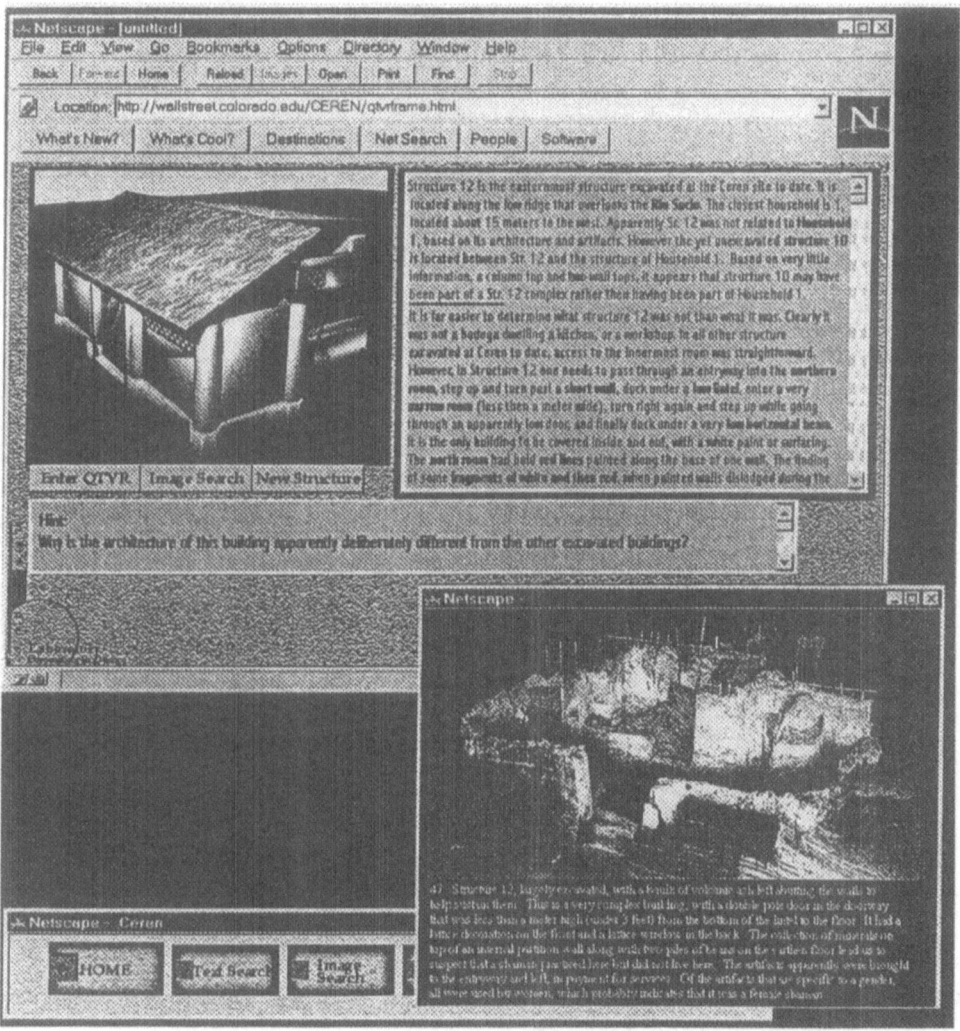

Figure 8. Outside view of Structure 12 links to the interior QuickTime VR shown in figure 9. The dialog below the rendering displays a more general issue for Structure 12.

Further exploration reveals more specific issues. As the user browses the Quick Time VR of the interior of Structure 12 (figure 9), a new issue is displayed. The previous site issue remains visible for the viewer to refer to. To the right of the QuickTime VR is a more general listing of the artifacts in Structure 12. Each artifact is linked to its specific information. The user can search for an artifact by clicking on this list, or by clicking on the artifact within the QuickTime VR.

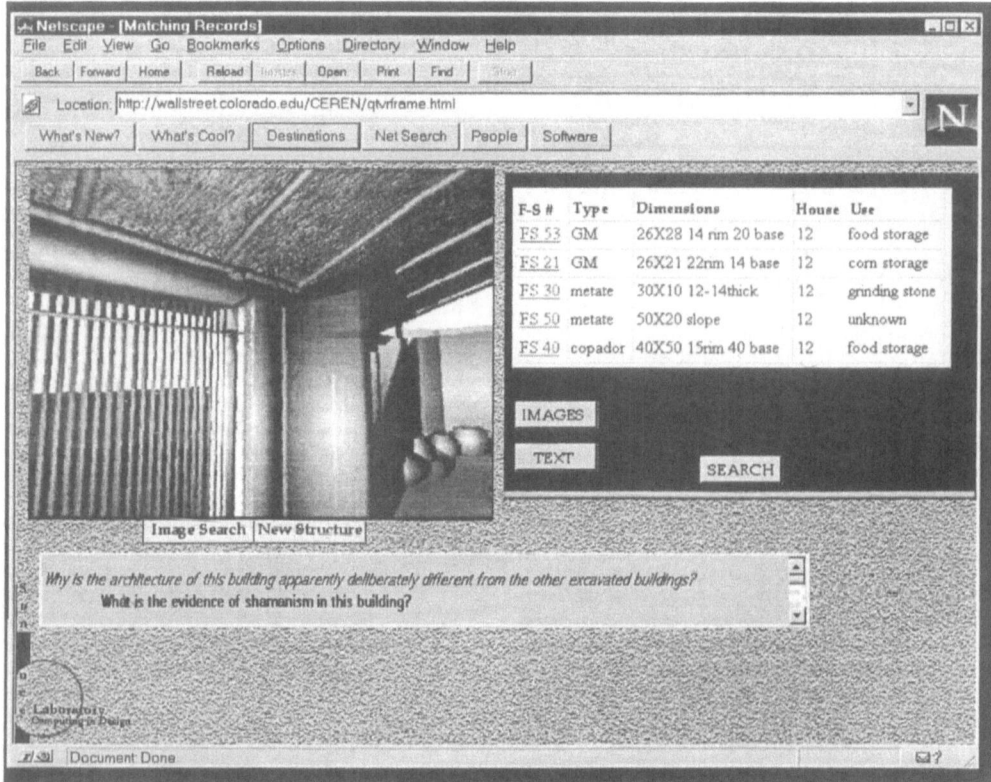

Figure 9. Internal Quick Time VR of Structure 12. As the user browses the QuickTime VR scene, relevant issues are displayed.

In summary, Image Arguments extend the capacity of images to convey needed context and they create more dynamic and information rich renderings. Viewers are no longer isolated from the process of image generation and they can participate in the questions, issues, and types of data required to understand the site. Viewers not only can see the image, but they can begin to explore the layers of information used in anthropology research. The result is a tool that provides a greater potential for understanding both a computer rendering and its context.

6. Implementation of Image Arguments

Constructing an Image Argument requires organizing both the graphic and the data set in a dynamic, aesthetic, and integrated fashion. The image acts as a gateway to the data as well as to the overall organizing argument. The data set is created by re-tracing the steps made in generating the image, including the data collection, arguments and discourse that occurred as part this process. We kept detailed records of the information and events in constructing each rendering. The record for each image includes raw data, charts, measurements, sketches, excavation text, excavation images, published text, references to journal articles, conversations, notes, corrections, and a history of changes. Because we needed this information to complete the models and renderings, compiling the rendering record was a relatively simple additional step.

After constructing a set of still rendering images from our initial models, we chose to use a more dynamic form of imagery and began producing QuickTime Virtual Reality (QuickTime VR), 360 degree images. QuickTime VR (Apple Computer 1992-1995) offers a degree of detail not yet available in other systems, as well as the ability to manage extremely large and complex models. We explored using Virtual Reality Modeling Language (VRML) based on Open Inventor (Wernecke 1994), but found that it was difficult to obtain real time performance with high graphical detail of the models we need for Ceren, which typically contain more than twenty thousand polygons. The drawback with QuickTime VR is that the representation of the modeled objects is lost in the final rendering, which is simply a color pixel map representation. VRML overcomes this limitation, and as it becomes faster we expect future implementations to use this technology.

The QuickTime VR solution has enabled us to rapidly deliver interesting images over the web. Users can browse different portions of each structure and select and inquire about artifacts that interest them. Because QuickTime VR uses image format files, areas of ambiguity can be shaded and changed through time within the animation to highlight problems or questions. QuickTime VR also makes it easy to specify hot spots in a rendering with URL links .

Figure 10 shows a block diagram of the Image Argument system, as well as the relationships between the image and its information. This system architecture provides a great deal of flexibility. We can quickly change the issues and information in the back end database and text searches are easy. Text stored in the server side database about Ceren, raw data, and images can be searched and viewed in conjunction with the computer rendering. For our server side data base we chose a combination of two commercial products: Tango (EveryWare 1997) and FileMaker Pro (Claris 1984-1985). Information from Ceren was compiled in a File Maker Pro database and made viewable on the web through Tango calls. Tango provides a link between a server-side CGI and a database, fielding database requests from Web clients and dynamically delivering this

650

information. Quick Time VR's hot spots trigger requests to the Ceren data base that Tango serves.

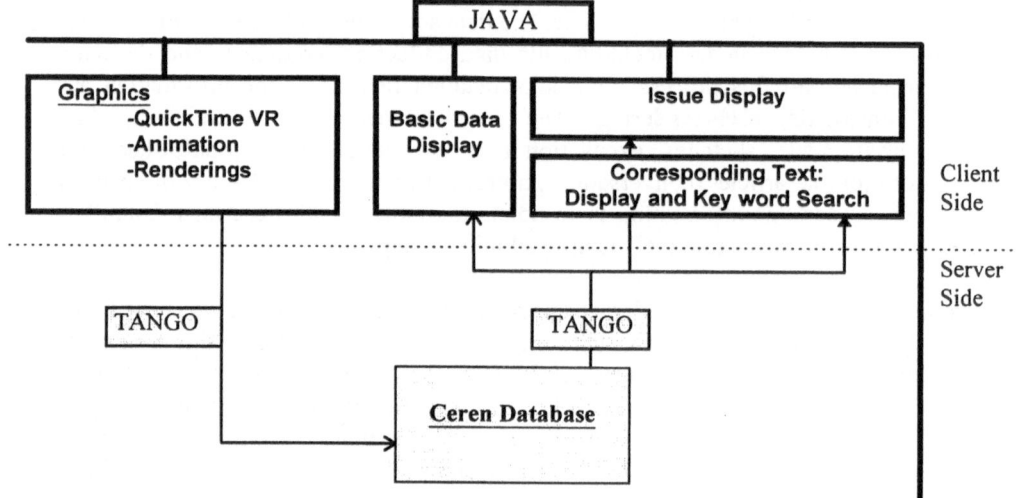

Figure 10. Graphical representation of the Image Argument system.

The client side uses Java applets (Sun 1996) to control the placement and viewing of the images and information and to guide the user throughout the Web site. Pop-up applets provide menu bars that enable users to navigate the Web site using a single consistent user-interface. We are developing additional user dialogs including internal maps that allow a user to log parts of the web site already visited and information viewed as well as a means to annotate and discuss areas of interest with other users.

7. Discussion and Future Work

7.1 SUMMARY

Computer renderings can be valuable visual aids to describe built form, but alone they have clear limitations in both their reliability and their function as research tools. Renderings create an illusion of existence; yet they reflect only a small portion of much larger bodies of knowledge that lie behind them.

Although animations and other computer graphics technologies have created exciting imagery, typically images are viewed separately from the information that led to their creation. In constructing computer models, renderings, and interactive animations for anthropologists studying Ceren's images served as arguments about a large and complex body of information. However, this richness of information, the discussion and the complexity of arguments about Ceren are lost when images are considered as solitary isolated objects. Although the Ceren anthropologists can connect images with the information behind their creation, outside viewers cannot.

We have described Image Arguments, an approach to extending the function of computer generated renderings to provide more informative tools for exploring—in this case—an anthropology/archaeology data set. The creation of Image Arguments, dynamic images, linked to the information used to construct them as well as new issues they raise can extend the function of computer imagery. Computer renderings viewed in conjunction with the rich data set behind them become not only more informative, but also more impressive. They no longer stand alone as a single portrait, but instead, like the cover of a book, provide an entrance into a larger story. They no longer ignore the context from which they have emerged.

7.2 USE IN INSTRUCTION

We have slated the Image Argument system for use in an undergraduate level anthropology course in the academic year 1997/98, and we expect that this will enable us to observe the effectiveness of our system. We aim to challenge students to become anthropologists, exploring the Ceren site and the arguments that have been made about it, to answer questions posed by the instructor. Students will be asked to explore the Ceren data set to answer anthropology questions, such as

It was a surprise to us to find that household #1 had more than 70 ceramic vessels. Do you know of any households, in your experience, with that many containers? What are the range of uses to which ceramic vessels can be put?

Questions like this require a student first to browse the Ceren data (in this case, to determine if there are any other households that contain more than 70 ceramic vessels), but then to look also at the interpretations and arguments of anthropologists who have written about the site, in this example, to find the range of uses to which the vessels can be put. Of course, some uses are a matter of common sense and do not require exploring the texts further. But other uses are particular to Meso-American agricultural life, and it will require looking through the sources of the site to discover them. We believe that this process, which requires students to do more work than if they merely looked up the answers in their textbook, will engage them more seriously, even at an introductory level, in the questions of anthropology.

7.3 APPLICATION TO ARCHITECTURE

The Ceren modeling project fortuitously offered us the opportunity to engage in a fascinating set of questions and to develop a set of tools and techniques that we plan to apply in our home domain of architectural design. Many of the same issues arise with respect to computer generated images of buildings. We have intentionally developed Image Arguments as a 'shell-infil' system in which the anthropology/archaeology content is distinct from the technology used to provide the links between interactive animations and back end data and arguments. Architectural computer renderings in this format can provide interesting tools for education, helping students understand the decisions and development behind a building's design.

7.4 FUTURE WORK

We have argued that any computer generated image reflects a combination of data, opinions, and arguments from which it was constructed. Our Image Argument system aims to link an image with the data and the arguments behind it. In our current system, the arguments serve essentially as explanations, but the user cannot explore the consequences of other resolutions of the arguments behind the images. "How, for example, would Structure 11 appear if ...?" We envision a more fully interactive system that makes accessible to the end user the process of producing a rendered model from the data, a process that now requires the assistance of skilled computer graphics modelers. That way, users could argue about how to interpret the site data and generate images themselves, interactively, to serve their arguments. An obstacle is the interface to the modeling programs; even simple editing requires the user to be an expert operator. A speech or text interface to the models combined with a pointer into a three-dimensional rendering would enable a user to edit by simply saying, for example, "move the shelf up a foot. Place the ceramic vessel --Field Specimen 415 -- on the shelf."

7.5 CONCLUSION

Computer renderings can extend their potential to communicate. Images no longer need to act as stand alone displays, but can themselves become innovative tools to display and describe vast quantities of information. By itself computer imagery is little more than entertaining, but combined with information, it becomes exceedingly more successful as a communication tool. As Vicki Goldberg remarks in an article The Power of Photography, "bearing witness is what photographs do best" (Goldberg 1991). In this spirit, computer imagery should bear witness not only to its subject, but to the process that creates it.

Acknowledgments

The Ceren research team included anthropologists Payson Sheets, Linda Brown, and Scott Simmons and undergraduate modelers Natat Poomviset, Ian Page, Matthew Bayless, Justin Call, Ethan DeFrees, and Steve Perce. The work has been supported by student stipends from the University of Colorado Undergraduate Research Oportunities Program (UROP) and a 1997 University of Colorado President's Grant 'Changing the Learning Paradigm Through Technology'.

References

Apple Computer Inc. (1992-1995). Quick Time VR Authoring Tools Suite.

Collins, B. et al. (1993). From Ruins to Reality: The Dresden Frauenkirche. IEEE Computer Graphics and Applications Vol 13: pg. 13-15.

Conyers, L. (1995). The Use of Ground-penetrating Radar to Map the Buried Structures and Landscape of the Ceren Site, El Salvador. Geoarchaeology Vol. 10: pp. 275-299.

Delooze, K. and Wood J. (1990) Furness Abbey Survey Project: The Application of Computer Graphics and Data Visualization to Reconstruction Modeling of an Historic Monument. Computer and Quantitative Methods in Archaeology 1990. K. Lockyear and S. Rahtz. British Archaeological Reports (BAR) International Series, Oxford.

Claris Corporation, (1984-1985). Filemaker Pro 3.0v1.

de Cola, S., B. de Cola, et al. (1990). Messina 1908: The Invisible City. The Electronic Design Studio, Cambridge 1990, MIT Press.

Eiteljorg, H. (1988). Computing Assisted Drafting and Design: new technologies for old problems, Center for the study of architecture, Bryn Mawr, Pennsylvania.

EveryWare Corp. (1997). Tango Editor.

Goldberg, V. (1991). The Power of Photography. New York, Abbeville.

Grant, M. (1993). ISSUE Interactive Software System for the Urban Environment. CAAD Futures '93, Pittsburgh, PA, New York.

IBM Corporation (1995). Freedom Rises From the Rubble; http://www.ibm.com/sfasp/mason.html, IBM.

Kiernan (1994). Lies Damned lies and slick graphics. New Scientist.

Lancaster University Archaeological Unit (1996) http://www.lancs.ac.uk/users/archaeo/unit/luau.html.

Lewin, J. and M. D. Gross (1996). Modeling Archaeological Site Data: The Case of Ceren. Proceedings of the Association For Computer Aided Design in Architecture, ACADIA 1996. Tucson AZ.

Lewin, J., et al. (1997). The Ceren Site; http://wallstreet.colorado.edu/ceren/ceren.html, Sundance Laboratory, University of Colorado.

Miller, P. and J. Richards (1994). The good, the bad and the downright misleading: archaeological adoption of computer visualization. Computer Applications in Archaeology, University of Glasgow, Tempvs Reparatvm.

Mitchell, W. J. (1992). The Reconfigured Eye: Visual Truth in The Post-Photographic Era. Cambridge, Massachusetts, The MIT Press.

M.I.T. Course 4.183 Architectural Design Workshop -- Software Design (1997) http://andrea.gsd.harvard.edu/workshop/

Peterson, P., F. D. Fracchia, et al. (1995). Integrating Spatial Display with Virtual Reconstruction. IEEE Computer Graphics and Applications pg. 40-46.

Reilly, P. Rahtz, S. (1992). Three-dimensional modeling and primary archaeological data. Archaeology and the information age, London: Routledge.

Reilly, P. (1993). Access to insights: simulating archeological visualization in the 1990's. in A. Suhadja, K. Rio (eds.) The Future of our Past. Hungarian National Museusm. Budapest.

Sheets, P. D. (1992). The Ceren Site: A Prehistoric Village Buried By Volcanic Ash in Central America. Fort Worth, Harcourt Brace College Publishers.

Sheets, P. D. and B. R. McKee (1989). 1989 Investigations at the Ceren Site, El Salvador: A Preliminary Report. Manuscript, Department of Anthropology, University of Colorado, Boulder.

654

Simmons, S. E. (1996). The Households of Ceren: Form and Function in Middle Classic Period El Salvador. Unpublished Ph.D. dissertation. Department of Anthropology, University of Colorado, Boulder.

Soprintendenza Archeologia Per Le Provincie Di Sassari E Nuoro (1989). Sipia-Progetto SITAG Archaeologia del Territorio. Territorio dell' Archeologia:immagini di un' esperienza di catalogazione informatica dei beni culturali dell Gallura. Chiarella-Sassari. Tempio Pausania.

Sun Microsystems Inc. (1996). Java Programing Language. Mountain View, CA.

Wallis, D., A. Bowyer, et al. (1990). Solid modeling of Roman Bath. In precirculated papers for Information Technology themes at World Archaeological Congress 2, Winchester: IBM UK Scientific Centre.

Wernecke, J. (1994). The Inventor Mentor. Silicon Graphics, Addison-Wesley.

Wood, J. and G. Chapman, Eds. (1992). Three-dimensional computer visualization of historic buildings-with particular reference to reconstruction modelling. Archaeology and The Information Age: a Global Perspective. London, Routledge.

The Palladio Web Museum

A Heterogeneous Database of Architecture and History

DANIEL E. TSAI, *Research Associate, Harvard University*
Visiting Lecturer, Massachusetts Institute of Technology

This paper presents the overall information system architecture and the approaches used for creating the Palladio Virtual Museum - a heterogeneous database of history and architecture. Creating a virtual museum is treated as an information system engineering task[Martin90]. The World Wide Web (the Web) is used as the open access platform for both presentation and input. Client-server database transaction technology [Date95] is used to provide a concurrent real-time system for consumers (visitors) and producers of information. The system is a test bed for structuring, searching, and presenting historical, architectural, spatial information.

1. Introduction

The Palladio Museum is a virtual museum dedicated to the works and life of Andrea Palladio (1508-1580). This museum of history and architecture is by necessity, virtual. The artifacts - buildings - are not physically transportable or containable. Furthermore, collections of physical fragments are decontextualizing. Electronic media is uniquely suited for creating flexible, dynamic, interactive, context-sensitive representations of historical, architectural space and time.

Although a virtual exploration does not replace the corporeal experience of traveling to Italy in person to visit Palladio's villas and palaces, it does provide a wide audience with an information-rich alternative. In particular, many works are not in existence. Some were designed but never built. Others were only partially executed. Still others are in existence but differ from the documented designs, as Palladio laid out in the Quattro Libri dell'Architecttura [Palladio]. A virtual museum of architecture can facilitate the piecing together of both artifacts and information to form a coherent view. Analytical reconstructions of unbuilt, destroyed or incomplete works can act as surrogate representations where none would otherwise exist [Mitchell93]. The historian's efforts of piecing together facts, and making sense of drawings and physical fragments can be made manifest in 3-dimensional computer models. Although they are analytical and interpretative works, such virtual representations of (lost) architecture can serve vital educational ends.

R. Junge (ed.), CAAD Futures 1997, 655-662.
© 1997 *Kluwer Academic Publishers.*

There are a few areas in which virtual museums in general, and the Palladio museum specifically, can extend and redefine what a museum is. First, a virtual museum can present a more complete experience of times and places past, things destroyed or never built. Second, a virtual museum can be an online center of activity and information on its area(s) of specialty. The Palladio Museum project is a venture by its sponsoring academic institutions [1] to create a resource suitable for both public and research use. The public face will appear as the Palladio Museum on the World Wide Web, and as dedicated interactive sites. The research face will also appear on the web but as an optional and possibly regulated resource. Both will draw from a common collection of material. The museum will strive to present an experience of architecture and history in virtual medium - Palladio's built and unbuilt works, 16th century Veneto, the people, events and things that influenced his unique influential designs. The research center will strive to provide an electronic locus of information on the study of Palladio and a virtual community where information can be shared. Instead of crafted presentations, the ability to pose specific questions and find novel inter-relations between people, places, and events will be sought.

In between these two reference points - the research center and the public museum, lie many possibilities. The resource can become a virtual community of scholars and students studying Palladio's architecture. Exhibit-making can become a distributed enterprise in which shared and private materials and tools are used to craft a view of Palladio's works. The electronic carnation of the museum can be an ever changing, evolving, electronic focal point on Palladio. Electronic artifacts do not have to be precious items kept inside glass cases, but instead be resources to be reformed and used anew. Serious "visitors" to the virtual museum can go beyond passive spectatorship, to engage the material experimentally.

The creation of a virtual museum of architecture and history can be approached in many ways. The major areas this project has addressed include:

2. Web based retrieval

In order to provide access to the widest possible audience, all information within the system is retrievable via the Web. This presents particular challenges when dealing with visually intensive material. Browsing material from the Web not only involves being able to display the multimedia content in a browser, but also organizing the material in a logical manner, providing access paths and tools to explore relationships. The Palladio Museum's approach has been to present information without a high degree of crafted narrative. Places should be explored via interactive maps that allow the user to select regions to explore. Buildings should be explored in terms of places, images, models, written materials, and so on. The internal information 'web' is spun around an underlying relational and object oriented database. This database holds facts about people, places, events, buildings and other things [Booch94] [Ross87] and digital representations [Mitchell95] such as searchable texts, images, HTML pages, and VRML models [Ames96].

Example Web browser clippings[2] (http://andrea.gsd.harvard.edu/palladio/):

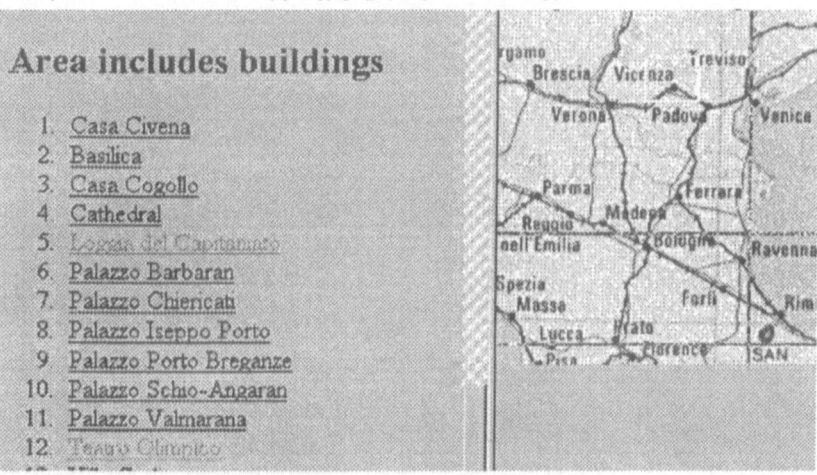

Area includes buildings

1. Casa Civena
2. Basilica
3. Casa Cogollo
4. Cathedral
5. Loggia del Capitaniato
6. Palazzo Barbaran
7. Palazzo Chiericati
8. Palazzo Iseppo Porto
9. Palazzo Porto Breganze
10. Palazzo Schio-Angaran
11. Palazzo Valmarana
12. Teatro Olimpico

Figure 1 : Map applet searches for places and buildings

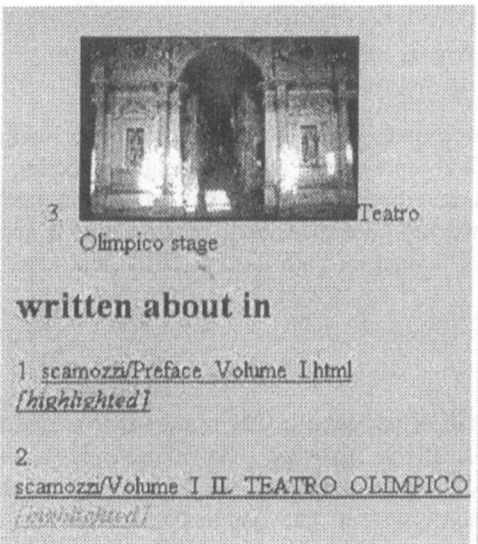

3.   Teatro
Olimpico stage

written about in

1 scamozzi/Preface_Volume_I.html
[highlighted]

2.
scamozzi/Volume_I_IL_TEATRO_OLIMPICO
[highlighted]

THE BUILDINGS AND THE DESIGNS OF
ANDREA PALLADIO

Collected and Illustrated by Ottavio Bertotti
Scamozzi

Translated into English by Howard Burns

IL TEATRO OLIMPICO

THE TEATRO OLIMPICO

The most noble Academy of the Olimpici ,
which was founded in the year 555 , and which
still flourishes with rare decorum and splendour
in the field of Vicentine letters , is indebted for
its foundation to the inclination to literary
studies of some learned and prominent

Figure 2 : Digital text search results.
(e.g. on "The Teatro Olimpico")

Figure 3 : Images and text documents found

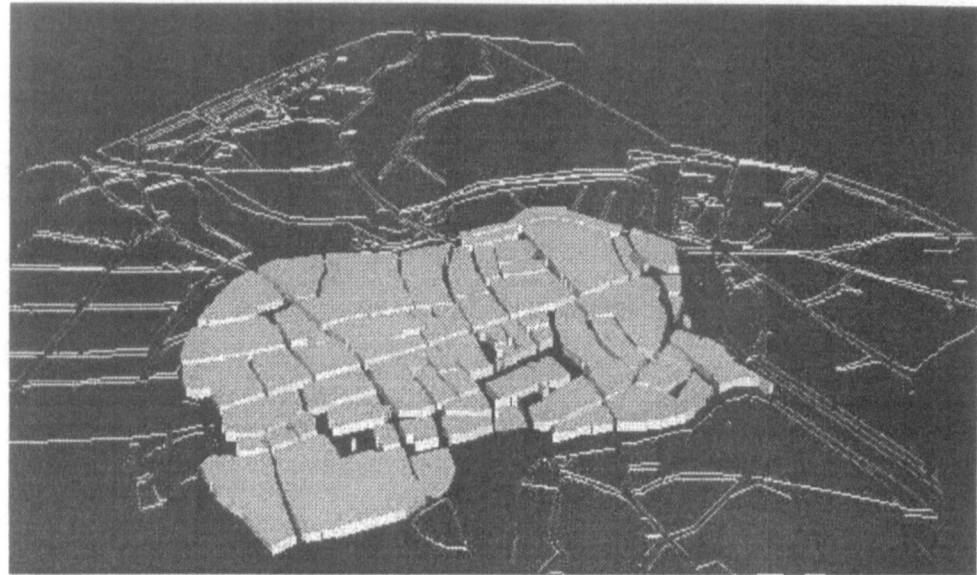

Figure 4 : VRML model of Vicenza. Interactivity is coordinated with 2D browser material.

The above clips show examples of an interrelated web of information and media. Selecting a region on a map causes a search for places and buildings. Selecting a building shows images and documents on that building. 2-dimensional media (text, photographs, drawings, etc.) can be combined with 3-dimensional representations such as VRML models, to provide a coordinated browsing of both spatial, visual and textual information.

3. Web based data entry

Browsing the museum on the Web is only half of the picture. The system is intended to be contributed to via the Web. A scholar working in a library, with internet access and authorization, should be able to contribute to the information within the system directly. This requires an underlying transaction database system. The displayable content has to be dynamically updated to reflect the current state of the system. Web pages therefore have to combine static content with dynamic content. The combination of Web entry, concurrent transaction based updating, and dynamic content serving over the Web makes for a fully open Web database.

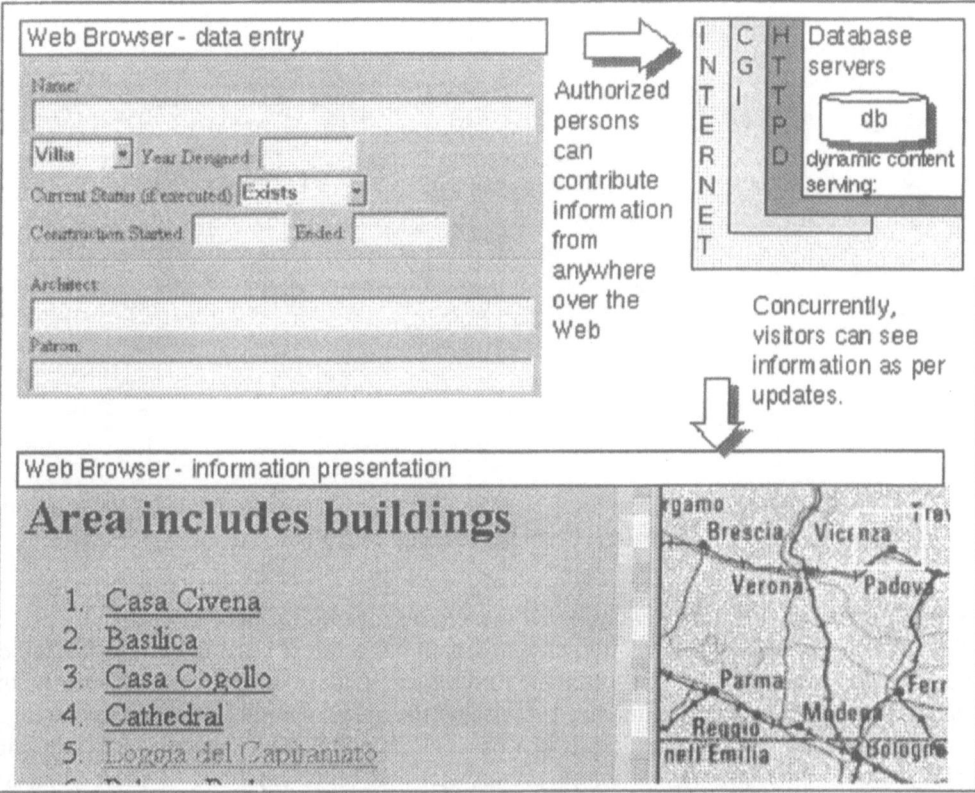

Figure 5 : Contribution and retrieval to the system over the Web.

Web sites that are primarily based on static pages require updating documents. The present system trades some degree of custom authorship for custom dynamic on-demand publishing. A common set of facts and media are stored in a database. Changing the underlying data or meta data reconfigures the served pages automatically.

4. Digital media procurement

An information system is only as good as the information it holds. One great impedance to achieving a functional system is the procurement of content in suitable digital format. Images have to be made digital. Texts have to be converted into machine readable form. Facts have to be organized and stored in logical structures. Representations have to be identified and made searchable. This process is both a logistical and technological one. Some of the pathways from source materials to digital representations are shown below. Increasing the ease of data assimilation into the system is critical. Direct to digital recording - such as with digital still and video cameras - is one step in this direction. Automated / smart cataloging methods are also critical.

660

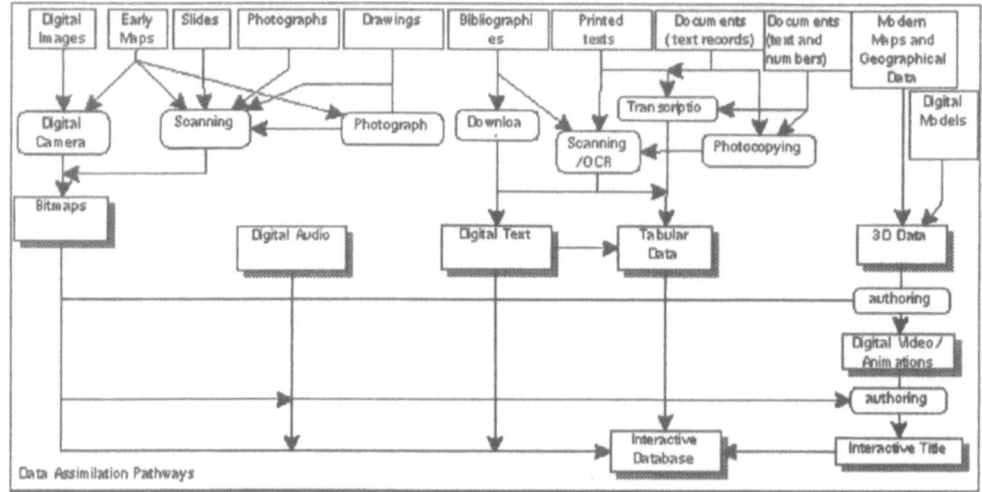

Figure 6 : Data assimilation pathways.

5. Architectural and historical information representation

The explicitness of how information is represented has a direct bearing on how it can be manipulated, searched for, and ultimately presented. A page of text may contain facts that are comprehensible by a human, but without the aid of natural language processing, it is only searchable by a computer as word strings. The Palladio Museum stores explicit facts about people, places, buildings, events, etc. and representations of these objects. Information models define what attributes exist and how they relate to each other. Representations are digital media that have meta-information about how they can be displayed.

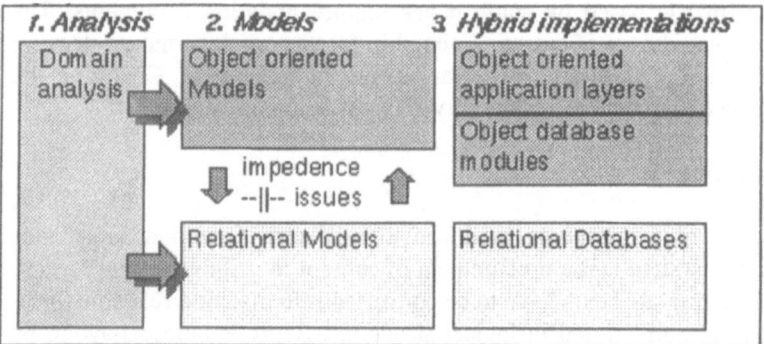

Figure 7 : Analysis, Models, Implementations.

Current trends in information analysis and modeling are object-centric. However, traditional entity-relationship modeling techniques are still important because of the high cost and complexity of object databases and because relational searching offers advantages. Both methodologies have been utilized in the analysis of the historical architectural domain at hand. Impedance problems between object and relational models have been approached via hybrid OO/Relational solutions in which distinct functional layers are realized in one technology.

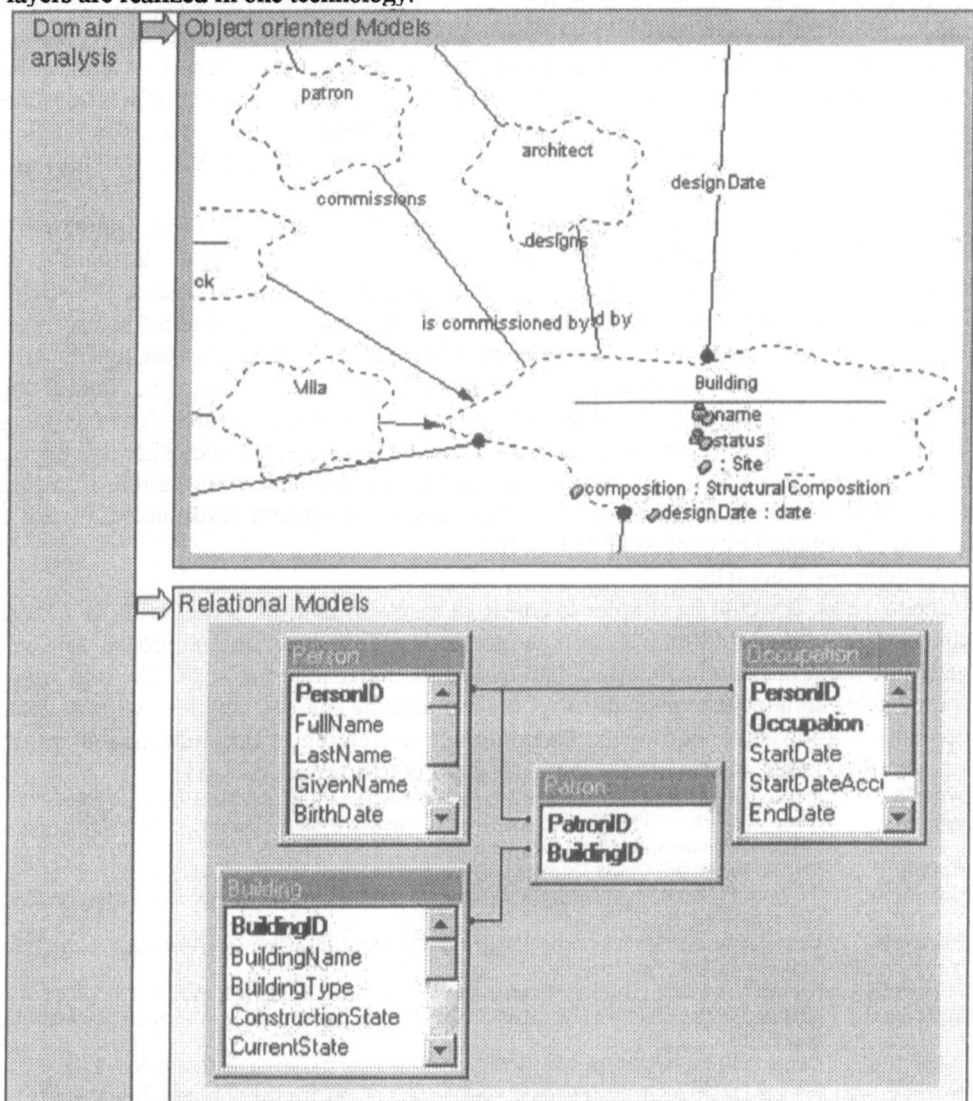

Figure 8 : Object and relational designs (fragments shown).

The creation of an underlying information framework (whether Object oriented or entity relational) codifies a domain of knowledge in a particular (computable) manner. This model becomes an evolving working representation that is tested on 2 fronts - from the contributor who has information to put into the system, and from the viewer who explores and asks questions.

6. Search techniques

A narrative provides an authored perspective on a subject. A research database needs to go beyond prepared narrative to provide content based on logical searches. Different media require different search techniques. Texts such as a fully indexed version of Palladio's Four Books of Architecture can be searched by word phrases. A real world object (such as a building) can be searched for as an object, attribute or relation. For example, a search for the "Barbaro" may find it as a person's last name, as in Daniele Barbaro and Marcantonio Barbaro; as a building's name, as in the Villa Barbaro, and Tempietto Barbaro.

The current approach to searching is a direct extension of the underlying object/relational framework. A search can be initiated on a word (e.g. "Barbaro"), an object type (e.g. Person ("Barbaro") vs. Building("Barbaro")). In all cases, the search hits each underlying database - indexed text base, relational tables, and an object meta-base. The system is now using the Relational Navigation Engine [3] that dynamically creates Web pages by searching and extracting information from the underlying relational database based on an object-role schema. This system creates dynamic pages that are self-propagating and implicit. New links are generated on each successive page based on the object schema being traversed and on the available information. Complex form based searching which requires a high degree of upfront information to ask a question is replaced by contextual search-links.

7. Concluding Remarks

In many ways, it would be simpler to construct a system that, as is the norm, is a large collection of crafted multimedia documents/exhibits. This traditional approach does not make the material searchable other than by standard syntactic methods. The making of computable architecture and history, the structuring of information within these domains of knowledge, and the representing and presenting of this information within an open Web based system are fundamental goals of this research.

[Ames96] Andrew Ames, David Nadeau, John Murphy. (1996) The VRML Source Book, John Wiley, NY.
[Booch94] Grady Booch. (1994) Object-oriented Analysis and Design with Applications, Benjamin/Cummins, Redwood City, Calif.
[Cox86] Brad J. Cox. (1986) Object-Oriented Programming, Addison-Wesley, Reading, MA.
[Date95] C.J. Date. (1995) An Introduction to Database Systems. Sixth Edition, Addison-Wesley, Reading, Massachusetts.
[Martin90] James Martin. (1990) Information Engineering. Book II, Planning and Analysis, Prentice Hall, Englewood Cliffs, New Jersey.
[Mitchell93] William J. Mitchell. (Dec 1993) Virtual architecture, Architecture, v82, n12, p39 (3).
[Mitchell94] William J. Mitchell. (1995) Electronic Design Media. Second Edition, Van Nostrand Reinhold, New York.
[Palladio] Andrea Palladio. (1570) The Four Books of Architecture. Dover edition (1865), New York.
[Ross87] Ross, R. (1987) Entity Modeling: Techniques and Application, Database Research Group. Boston, MA.
[1] Acknowledgment of sponsored research : The Palladio Museum project is sponsored jointly by CISA AP, MIT and Harvard University. Project headed by Professor Howard Burns (University of Venice) and Dean William Mitchell (MIT). Principal investigator : Daniel E. Tsai.
[2] Visit the Palladio Museum on the World Wide Web at http://andrea.gsd.harvard.edu/palladio
[3] Relational Navigation Engine, to be detailed in: "Abstract Schema-based Navigation of Relational Data on the Web :A Model for Automated Relational Web Publishing." Ownership, creation and rights, D. Tsai ©1996, separate from the Palladio Museum Project. Used here by permission of the author.

MAKING SENSE OF THE CITY

VASSILIS BOURDAKIS[*]
Centre for Advanced Studies in Architecture (CASA)
University of Bath, UK

Large-scale, three dimensional, interactive computer models of cities are becoming feasible making it possible to test their suitability as a visualisation tool for the design and planning process, for data visualisation where socio-economic and physical data can be mapped on to the 3D form of the city and as an urban information repository. The CASA developed models of the City of Bath and London's West End in VRML format, are used as examples to illustrate the problems arising. The aim of this paper is to reflect on key issues related to interaction within urban models, data mapping techniques and appropriate metaphors for presenting information.

Keywords: urban modelling, virtual environments, navigation, data mapping, VRML.

1. Introduction

With the recent developments in computers, in terms of raw power and graphics hardware performance as well as new software technologies, we experience a switch in architectural visualisations from still images and pre-rendered animations to interactive three dimensional (3D) models and Virtual Reality (VR). VR is developing on a fast pace and, as more and more people are acquainted with the new generations of computer games and their capabilities, expectations are growing. VR developments over the last 4-5 years in many ways resembles the CAAD anticipated development over the 60ies and 70ies only the latter took a few decades to materialise—and to date has still not satisfied all visionaries. We have now a situation where the technology is available, has been applied successfully in other fields (aeronautical, automotive, chemical engineering etc.) and it is mainly a matter of finding ways to use it, as opposed to arguing what the technology can /should do and then trying to materialise it (Negroponte, 1971). However, applying 3D modelling techniques in order to create interactive computer models of cities is still a rare phenomenon mostly due to resources limitations and, to a lesser extend, doubts on its utility.

---

[*] E-mail: V.Bourdakis@bath.ac.uk URL: http://fos.bath.ac.uk/

R. Junge (ed.), CAAD Futures 1997, 663-678.
© *1997 Kluwer Academic Publishers.*

Two years ago in the last CAAD Futures conference, Day et al (1995), reporting on the Centre for Advanced Studies in Architecture (CASA) work at the time, explained that the urban models developed are a means to exploring issues on urban design, a test-bed for new theories and practices. At that time, technology was the main limiting factor; rendering an image of the Bath city model would take at least twelve hours on a single computer. The author is currently using an interactive version of the Bath city model where an image is rendered in a fraction of a second—the possibilities for exploring the city are immense. This enables the development and testing of new theories as well as the creation of urban models not only as interactive urban planning and design tools but as front end to a series of underlying databases. Furthermore, developments on the Internet, enable fast communications and distribution of digital urban models, facilitating tele-working and teleconferencing for the evaluation process as well as training and education.

As the problems governing this new medium are quite different to classic CAD ones, we should grasp them the sooner possible if we are to create successful, compelling VE that can be useful to both research and practice. In this paper, issues related to interaction within urban models are considered and data mapping techniques are discussed together with some appropriate metaphors for presenting information.

2. Background

CASA has been involved in 3D computer modelling for the last six years. In 1991 CASA received a grant to construct a 3D computer model of Bath. The model was constructed in order to assist in making the planning and development control process more democratic by providing a means by which proposals could be visualised and alternative schemes for a site compared. In addition to its use in development control, the model has also been used to widen the public debate on how the city should develop in the future. The model was created from aerial photographs of the city in 1:1250 scale using a stereo digitiser. It is accurate to 0.5 metre and covers the whole historic city centre, an approximate area of 2.5x3.0 km along with the landform of the Avon valley covering a total area of 10km x 10km. During 1995 and using similar techniques, the core of London's West End covering 1.0X0.5km was modelled in similar accuracy but on a lower level of detail. It was initially used for transmitters' signal propagation experiments by British Telecom (BT); CASA has been using it as a test-bed to navigation and mapping information research. Throughout the models' construction, commercially available software has been used wherever possible with in-house applications being limited to program customisation (Bourdakis et al, 1997).

Two years ago, the London's West End 3D model was tested against the then new Virtual Reality Modelling Language (VRML) and an online version of it was created—the Map of the Future (Bourdakis, 1996). Following, the much larger Bath city model was imported into VRML. One of the initial concerns was how VR technologies and VRML in particular would scale and adopt to the visualisation of a whole city. So far, many small VR models exist, but not any at this scale. The Bath city model

665

database is, to our knowledge, the largest and most detailed one produced as yet; the UCLA Dept. of Architecture and Urban Design (AUD) is currently building a model of the entire Los Angeles basin covering an area in excess of 10000 square miles as part of "The Virtual World Data Server Project" which is still under construction. Cybercity's model of Berlin is another large application of VR focusing on urban planning. However it seems that most VR urban models currently available are commercially rather than research driven (Virtual Soma, Virtual Derby, Bigbook etc.). As a consequence, the level of detail is quite low, since the focus is not on architectural features and accuracy but in simply recreating a recognisable and commercially viable image of the city. In this paper, the commercial feasibility of the issues discussed and analysed is not considered; after all, commercially viable solutions will eventually find their way into products.

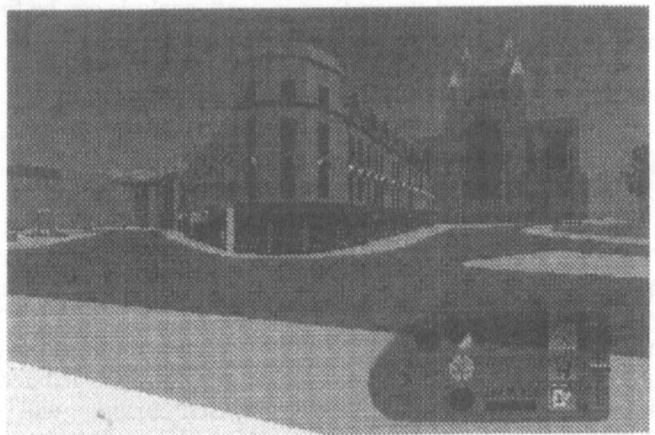

Figure 1. Bath City VRML Model

3. Characteristics of VR urban models

Digital models can be accurate on small scale 3D objects but do not easily scale up to the size of an urban model. There are indeed many different ways of structuring and creating a 3D computer model (be it a chair, the interior of a room, a building, a hotel complex or a whole city) however, using the engineering approach to layering (BS-1192) or Constructive Solid Geometry (CSG) modelling techniques are definitely not the right ones. Most problems are caused by the fact that VR databases are conceptually completely different to the CAD ones that engineers are used to. In many ways, VR databases are closer to the way architects work by reducing levels of uncertainty and improving understanding of the project in the design stage (Brawne, 1992) but far away from the way CAD programs usually force us to work. CAD databases can be extremely helpful in designing and constructing an architectural project, but they are quite problematic when used unmodified in a VR system.

As Negroponte (1995) explains, the whole world around us is analogue, nothing alternates from black to white or on to off and as such changes are continuous. Simulating such physical, continuous phenomena on a computer generated and maintained digital environment is difficult and in certain cases impossible. Therefore, designers and researchers have to resort to various tricks/techniques in order to create/imitate this continuity of the real world. One of the most important techniques is the use of levels of detail (LOD) to represent the same object with increasing detail to match the viewer to object distance. The overall breakdown of the urban model into entities (usually based on physical urban blocks) and the use of photographs to enhance the images created is another technique employed. Billboards are used to replace high polygon count representations of trees, humans and inanimate street furniture. Finally, instances are applied to reduce the file sizes and thus transmission times across networks. Since the structure of VR urban models is not the focus of this paper, the above techniques are not analysed any further; in depth analysis can be found elsewhere (Bourdakis, 1996).

4. Interaction within VR Urban Models

We are still a long way from Gibsonian states of hardware where a pair of glasses is a complete VR experience carrying a database of a whole city (Gibson, 1994) and even longer away from hardware devices interacting directly with the human brain turning us into cyberorgs (as pictured on the 1983 Trumbull film "Brainstorm"). However, a properly constructed digital urban database will be suitable for such devices when they become available. With the advents in nano-technology (as demonstrated both in sci-fi, cyberpunk culture and real world) an Infinite Reality SGI computer will probably be the size of a portable CD player in a few years time (although undoubtedly not as affordable as a CD player), expanding the availability and usability of urban VR systems. In a few years time, engineers could well be working on site seeing what is proposed (super-imposing it on VR glasses) and even editing and feeding back their comments to the designers in the office.

In the following section, the issues of navigation, modes of interaction (immersive, non-immersive) and interface design are discussed.

4.1 NAVIGATION

Early in the process of creating urban VR models in CASA, the problem of navigation was identified not only with occasional users of the models but with the creators as well which made the whole issue alarming. The main problem in exploring and navigating within digital urban models is orienting oneself. This means being able to identify areas, streets, buildings. In other words making the virtual environment (VE) recognisable. There is a great difference between aiming for a recognisable VE and pursuing realism

—the former does not focus on imitating reality (Carr et al, 1995). Creating realistic VE of such scale is still not feasible and in many cases both pointless and inappropriate.

In real life, when a person becomes disoriented in an urban environment, the natural action taken is to scan the surroundings for *information*. As Lynch (1960) explains, there is a consistent use and organisation of definite *sensory cues* from the external environment into a mental map. It seems that in virtual environments many problems occur due to users not understanding the rules of movement. As a result the provision of a simple, easy-to-learn navigation solution (Hughes, 1995) is exceedingly important for the user.

Computer games development is a field that one can review various design decisions; a very efficient, competitive environment enforcing quick development cycles. Tomb Raider by EIDOS Interactive, seems to epitomise the developments in navigation techniques employed by the games industry. Here, the player's avatar is fully included in the VE. The player is not a spaceship dashboard or a weapon carrying hand anymore but a fully articulated avatar giving better sense of scale and participation (Gibson, 1986). Real time rendered texture mapped animated characters are used in collision detection and gravity aware environments. The most interesting feature is the *Director's Camera Movement* where the camera zooms and moves around the action (that is the player's avatar) creating a feeling of a well directed action movie rather than a simple game playing. Wandering in the rooms becomes a worthwhile experience itself! The scale issue addressed with the full avatar presence is the most notable improvement along with the use of perspective and camera movement to follow the action.

VR software employ a series of navigation metaphors; the three main ones are *walk, fly* and *examine*. (Wernecke, 1994). The examine mode enables the spinning of objects in order to study them from all possible viewing positions. Within an urban model this is hardly a suitable navigation metaphor; for example, there is a very limited need for underground camera positions (mapping the city's infrastructure, services channels, etc.). The walk metaphor is the ideal navigation metaphor, being the closest one to the way people interact with the environment in real life. However, due to the lack of sufficient detail at street level, the "identity" and "structure" of a digital urban image is much "stronger" from an elevated position—when more distant cues are visible. This especially applies to the aerial photograph based CASA models. Therefore, the work carried out in CASA's focuses to a great extend on the flying mode of navigation.

Among the problems linked to the *fly* navigation mode is that it effectively removes any degree of immersion by switching the person to map reading mode. It can be even argued that it defies the reason of having a VE in the first place. According to Ihde (1993) there is evidence that there are important connections between the bodily-sensory perception, cultural perceptions and the use of maps as a navigational aid. He relates the bird's eye view of flying over a VE to the God's eye map projection identified in the literary cultures. The fly mode of interaction has definite advantages although it also introduces an often unknown perspective of the city. This perspective seems more comprehensible to engineers and architects who are used to working with scale models of sites which inherently introduce the bird's eye view.

4.2 IMMERSIVE VERSUS DESKTOP VR SYSTEMS

According to Slater et al (1995a), an immersive system is one that is capable of delivering varying degrees of *"...inclusive, extensive, surrounding and vivid illusion of reality to the senses of a human participant"*. A higher level of inclusive and surrounding illusion can be created using head mounted displays (HMD) which "block" the real world and data gloves that enable greater flexibility and input range. Vividness is affected by the actual quality of the image projected and as such is not clearly related to the hardware interface mechanisms but more to the underlying software. Hardware tracking head movement and even iris movement, enhance immersion by closely matching participant's proprioceptive feedback with body movements. Probably the ultimate immersive system is the one demonstrated in the 1995 cyberpunk film by Cameron, "Strange Days"; a hardware device that clamps over the skull, interfacing with the brain directly, without the need for headsets, loudspeakers, tactile feedback mechanisms etc.

Projection VR is another immersive system, improving the illusion created by the surrounding environment. Users do not wear helmets, but lightweight stereo polarised glasses while they move inside a large cube made out of display screens (CAVE)[†]. Again, body tracking mechanisms are used for navigation within the environment.

Desktop VR on the other hand employs a computer screen, a preferably high degree of freedom input device and if possible spatialised sound. Due to CASA's architectural 3D modelling background and the focus of the research on planning evaluation, it was decided to adopt a desktop VR approach. It is also believed that since we are still in the early stages of development, immersiveness is not crucial and furthermore can be achieved in the future without having to re-structure the digital urban databases currently under development.

Among the problems identified in early observations in CASA, was the lack of the concept of time and distance. Walking on the streets of a virtual city is an effortless exercise in contrast to the real experience. It is possible to fly over or walk across a whole city model in a matter of seconds and that can be instrumental on loosing orientation. It is therefore interesting to note a few recent attempts in immersive VE which introduce body interfaces to navigation. Shaw (1994) used a bicycle as the metaphor for navigating in his installation *The Legible City*, whereas Slater et al (1995b) used a "walking on the spot" metaphor. Finally, Davies (1996) has created a synthetic environment called "Osmose" exploring the relationship between exterior nature and interior self where the immersant's body movements are triggering events that position their body within the VE.

The very first technique employed in CASA in order to enhance sensory cues from the VE was to increase the field of view. This technique has been successful in architectural

---

[†] Heim (1995) gives a detailed explanation of the differences between HMD VR and CAVE VR or as he calls them, Tunnel VR and Spiral VR.

photography, where the author rarely uses anything but a 20mm lens on a 35mm SLR camera, giving a field of view close to 90 degrees. However, the results of a small case study with students of architecture and others where discouraging. Among the users tested, architects identified the problem as being that of "wrong perspective" compressing the depth of the image and generally creating a "false" image of the city. Others could not easily identify the problem but found the environment confusing nevertheless. Consequently, it was decided to use only the "normal" lens on all VR projects although it does limit the perceived field of view which in real life is much higher than 45 degrees. Experiments on level of detail degradation in the periphery of HMDs (Watson, 1995) demonstrate that more efficient VR interfaces can be achieved (compared to non-immersive ones) without necessarily hitting on the main VR problem; CPU capabilities.

Another problem faced in non-immersive VR is that of the direction of movement versus direction of sight. Due to the two dimensionality of the majority of input devices used in non-immersive VR, it is assumed that the user looks at the exact direction of the movement. This is true in many cases but while investigating a new environment, movement and direction of viewing should be dealt as two individual variables. Immersive VR headsets with position and orientation tracking mechanisms are again the easiest and more intuitive solution to this problem only matched by high quality joysticks and some home-brew solutions (Hollands, 1996).

4.3 INTERFACE CONTROL DESIGN

Navigation is by no means the only activity catered for in urban VE. Interacting with elements of the VE, editing buildings, inserting objects etc. are all essential features of an interactive planning tool. Having decided to take the desktop approach to VR, the interface design follows similar concepts. Typical human computer interface (HCI) design often implies that the fastest way to achieve a particular task (assuming you are proficient with the tools available) is a cryptic, complex, highly personalised set of commands (Negroponte, 1971). Users are forced to tune the way they think and work according to the tools available, whereas it should really be the other way round. It was therefore decided that CASA's work would follow Hughes (1995) recommendation and "keep it simple".

Figure 2. Bath Model Control Panel

A dashboard (Figure 2) made out of 3D elements (buttons, scroll bars, etc.) and 2D ones (compass, text, images) was developed and placed on the lower right corner of the VE display screen. This method is best suited to desktop VR systems but should not cause problems in an immersive setup. The elements within the dashboard can be classified in 2 distinct groups; the *informative or feedback* ones and the *actions*. The compass on the far left and the X, Y and Z co-ordinates of the avatar's position together with the height and scale fields are informative. Buttons simulating seasons and lighting, toggling alternative schemes for a site, inserting trees and other elements and dynamically altering properties of existing urban elements are actions. Some of the actions are exclusively carried out in the dashboard, however, inserting objects and moving them within the VE is done by directly manipulating the VE. Links to urban related information are achieved by direct interaction with the VE; moving the cursor over a building reports its address point while selecting a building triggers links to internal as well as external databases. Finally reports with positions of new elements inserted and modifications on existing ones can be created as a record of the editing session and as a feedback to the designer of a particular scheme.

5. Data Mapping and Visualisation Metaphors

Over the years, theoreticians and researchers claimed that digital models could be used amongst others for information visualisation: *"Since computers have been the most important instrument in creating a new system of cartography for weather, in studying DNA, in mapping atomic surfaces and subatomic particles, and especially in enabling the visual maps that explore chaos theory, then why not hope this juxtaposition of masses of visual information and high-tech appliances will produce a new map for the city and for architecture, a map that will describe non random order suddenly appearing in the midst of seeming disorder?"* (Boyer, 1996). However, using an 3D digital urban model for mapping and visualising information is an entirely different process to urban planning and evaluation implying a different approach to structuring, organising and visualising digital urban models. The points of what can be mapped on an urban model, how it can be mapped, how accessible is the information and whether it is worth the effort are addressed in the following sections. The whole issue closely relates to the field of Geographic Information Systems (GIS) where software solutions, based on 2D and 3D maps, have been developed since the late 60ies.

5.1 WHAT CAN BE MAPPED?

Data mapped on a digital urban model can be generally classified as either *static* or *dynamic*. Among the former are all census data, land use and value, crime rate etc. Engineering services and facility management are also included in this category together with business directories and services similar to A-Z directories. Finally, historic and tourist data sum up the static data that can be mapped on an urban model. Dynamic data imply a real time (or at least near real time) stream of data from various sensors placed in the physical environment. Traffic and pollution together with temperature and energy consumption can be considered dynamic (although in most cases they will be treated as static). Information related to safety and security management are the most typical dynamic data forms.

Following the above classification, there is a real question as to whether it is important or indeed useful to map information dynamically on a digital 3D model. It has been argued that such a technique gives a better overview of the data, but the same can be true on a pre-rendered animation of similar data. *Dynamic* implies that the visualisation will be used for decision making and action taking in real time, else it defies the reason of having dynamic data mapped in the first place. Out of the data types listed above, traffic, safety and security are the only ones that would genuinely benefit from a dynamic organisation. A VR traffic model could be used and linked to the computers controlling traffic lights within the city or integrated to a security model utilising Global Positioning Systems (GPS) to monitor the position of emergency or other selected vehicles within the city. Following, closed circuit television (CCTV) remotely operated from within the VR model could feedback live video of the real city to the control room and facilitate decision making and navigation.

5.2 HOW CAN DATA BE MAPPED?

Tomas (1991, p.35) claims that *"The 'abstract representation of the relationships between data systems' in cyberspace is, however, highly plastic, and can take any form ranging from pure geometric color-coded copyrighted shapes or architectural representations signifying corporate ownership to 'photo-realistic' illusions. Such sites are the essence of a postindustrial society-pure information duplicated in metasocial form: a global information economy articulated as a metropolis of bright data constructs, whose plasticity is governed by a Euclidean model based on a given problematic of visualising data, a problematic subordinated, in Gibsonian cyberspace, to the dictates of a transnational computer-based economy"*.

The two main issues related to mapping information on a digital urban model are the model's *physical entities* on which data will be mapped and *data granularity*. Regarding the former, the geometric properties of the data mapped and their appearance are considered. Starting with, the easiest way to map information is by using colours or patterns on the existing urban forms. This can be done on either a 2D or 3D basis. The next possible way, is to map information onto the third dimension—height—altering it according to the data values. Mapping more than one variable in the same urban representation can be achieved by combining colour or patterns with 3D height mapping. Finally, constructing new geometrical shapes, either deriving from the existing urban structure or abstract as in Space Syntax maps (Hillier, 1996) is another way to map data onto a digital urban model. It should be noted that there are data types better suited to colouring (such as categories) whereas height is more suited to ranked and continuous numeric data.

Data granularity determines the breakdown of the digital model into data primitives. Assuming an almost infinite amount of information, *unit sized cubes* can be used—cube size dictated by data volume and detail. Alternatively, the *building borders* can be used as the geometrical entities for data mapping. In certain cases, *property borders* can be considered instead. When the amount of data available is small, whole building rows can be used instead. Finally, the urban block contour and the streets can be utilised for census data mapping (Figure 3). The whole issue is a function of data breakdown and quantity versus urban LOD.

Using the Z axis—height—to map data within the urban model *alters* the image of the city upsetting the mental image of the city that the visitor/viewer possesses. In order to reduce the effects of this phenomenon, an abstract city form representation is often more appropriate where certain elements of the city may be better omitted. It should be generally attempted to use the 3D digital model as a 2D map (no ground undulations, building heights etc) where data are mapped on the 3rd axis (Z) (Wen, 1995) without attempting to imitate the actual city image. However by doing so, landmarks may not be as prominent as in real life, may be obstructed by disproportionally extruded buildings or may even be completely removed, hindering orientation and creating navigation problems similar to the ones described earlier. Subsequently, name tagging and textual information of addresses, street names etc. can be employed to improve navigation. Alternating screens between the abstract data urban model and the "real" urban VR model, running the two models in parallel, or employing aerial oblique view video of the area as a support tool (Shiffer, 1995) may also help in way finding and interacting with the data model.

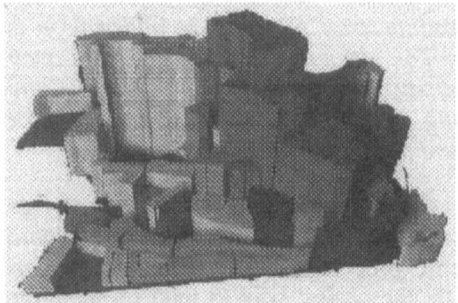

Figure 3. Land use and value in the Bath model

A flexible solution to the data mapping problem is to use a series of *unit cubes* arrayed so that they cover the whole city (Figure 4). Each cube is named and addressable with the two main variables of colour and height accessible to external databases. Such databases can manipulate the two main variables and thus create various representations of available data. Furthermore, it is abstract enough to produce urban models of different granularity (LOD) by combining units to match the type of data mapped. Cube size in plan may vary according to the level of detail needed, the resources available and the size, resolution and flexibility of the underlying database—typically ranging from one metre to five or even ten. However this solution is not without weaknesses. The image of the city created is distorted and may end up as being incomprehensible especially if units are grouped together. The whole approach implies a GIS like structured approach where data are stored in a database and according to a set of rules, different representations are created, limiting the portability of the developed digital model.

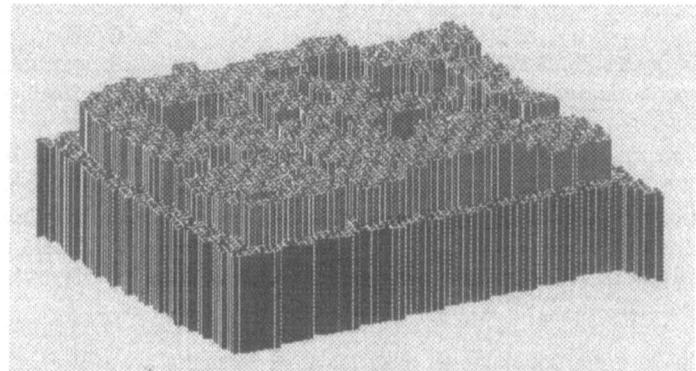

Figure 4. Unit mapping on part of the Bath model

Another shortcoming of the unit cube approach is that the resulting model is orders of magnitude larger than the original VR model. A way around this problem is to increase the level of abstraction in the digital model structure, leading to a lightweight version of the above concept where each property is modelled as an individual entity irrespective of the size, proportions and positioning of the actual building within it making it also suitable for a greater range of data.

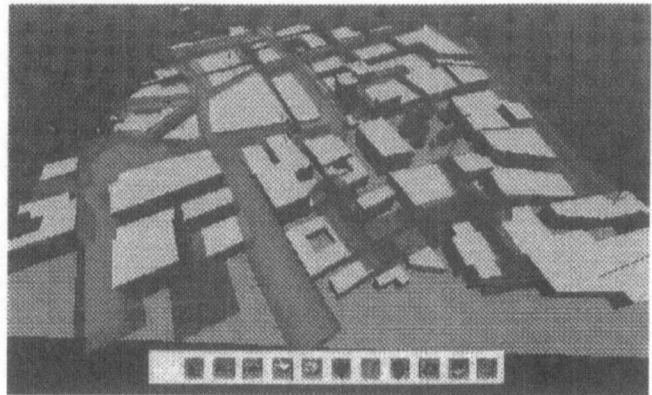

Figure 5. Pollution on the Map of the Future

As Calvino (1996) demonstrates in the *Invisible Cities*, events can be reduced to abstract patterns facilitating the procedures of logical operations (in Boyer, 1995). This can be expressed by a completely new abstract way of representing the city as a series of "links" or wires where information flows and data can be mapped. Alternatively and more conventionally, the same can be modelled as data tubes of varying diameters, colour and translucency where each of the aforementioned properties relates to individual data sources (Figure 5).

5.3 HOW ACCESSIBLE IS THE DATA

"The technology that delivers immense bundles of data does not simultaneously deliver a reason for accumulating so much information, nor a way for the user to order and make sense of it... The pressing challenge of multimedia design is to transform information into usable and useful knowledge" (Friedlander, 1995). As Boyer (1996) explains, arrays of information and knowledge is not the same, *"Information is merely data, devoid of an abstract processing framework that can make comparisons, draw connections, recognize exemplars, and set and accomplish goals"*. There is however a general belief that a VR approach to the subject is beneficial mainly by extrapolating from other applications of VR technology.

It is still early to draw conclusions, after all most attempts are still experimental and fairly primitive. However, employing an Internet based, hypertexted visualisation system, such as the one used in the Map of the Future (Figure 6), data are compatible across platforms and computer operating systems making it a true global solution. Furthermore, such a system is easy to use without introducing more interface/interaction metaphors. In the Map of the Future, an HTML/VRML interface, utilises a textual based pane at the left hand side, where all the queries from either the control panel or the VR model itself are directed. This way, all visualisation takes place on the top right pane with navigation on the right hand side panes (both the VR and the control ones).

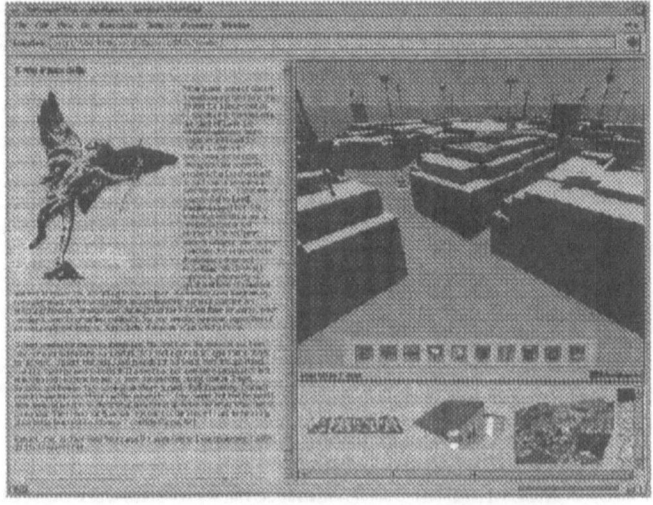

Figure 6. Multi-windowed display of the Map of the Future

Using the Internet, databases can be distributed and used around the globe without the need for expensive software solutions. This can be a serious consideration in large collaborative projects. On the other hand the limited security provided by the Internet (although improved lately) has to be considered very carefully. Concluding, using digital 3D urban models data are presented in ways that are more comprehensible than the typical spreadsheet or 2D coloured map. Furthermore, multiple data sources may be mapped more intuitively in a digital 3D model.

5.4 IS IT WORTH THE EFFORT?

We should not fall in the trap of technological development by using new technology just for the sake of using it. There are quite a few drawbacks in using a VR system for data visualisation. In this section, the limitations of conventional systems (be it GIS or simple 2D maps) are presented and compared to the ones of VR systems described in the previous sections.

5.4.1 Limitations of conventional systems

Among the main limitations of using 2D maps is that relationships and comparisons between data is difficult to both model and visualise leading to substantial replication and repetition in the effort involved. Furthermore, the data mapped and presented are mainly static, the only exception been in computer animated live feeds of data that may modify colours. There is also a important semantic differentiation between the data and visualisation since the database and visual representation are two distinct linked functions whereas in VR environments (and VRML in particular) the database *is* the visualisation. Finally, it is difficult to relate to other information—a diagram visualising crime rate will neither let you pinpoint the address of a property (unless it is a GIS database) nor compare crime rate with land value or use.

5.4.2 Pros and Cons of VR systems

The *freedom of movement within the data* is a major feature of a VR based system. It is possible to experience the data from different viewpoints, camera angles, heights creating a more inclusive map of the information involved. The drawbacks of the fly navigation mode turn into advantages when large amounts of abstract data are visualised. The *lack of scale limitations* of the data mapped is also extremely beneficial. One can zoom into an area, get a higher resolution representation and still move back and have an overview of the whole area using familiar metaphors. It is possible to assess and model relationships between seemingly unrelated data sets making it a more suitable research and teaching tool. Finally, it is easier to *dynamically alter* the data feed and thus have a more real time experience presenting live information on the city making it highly appropriate for certain types of data—temperature, pollution, traffic, shadow calculations mapping etc.

The disadvantages of employing VR systems for data mapping and visualisation are limited to that of decision making on the appropriate data and suitable data resolution. Potential problems due to abstract 3D representations of cities must be also considered.

6. Conclusions

Large scale 3D computer urban models highlight a series of shortcomings in existing technology and most important in theory and research. Navigation is the topic better covered in past research and it seems that a combination of fly mode with suitable interaction metaphors is providing a satisfactory solution. Controlling events within the VE can be dealt with using onboard control panels. However, data mapping and visualisation is still an area in need of further research.

It is essential to identify the types of data suitable for 3D VR representation. Dynamic datasets are suited to VR urban models much more than conventional static ones. Finally, one must be careful and critical in deciding on abstraction levels to match the data types mapped. Failing to achieve the right balance, leads to digital urban models that are either too detailed for the information mapped (i.e. resources wasted in modelling and visualisation) or too coarse to map accurately the available data.

Overall, digital versions of our cities are extremely useful and are starting to enable seamless integration in the design and evaluation stages, the planning and finally the education of researchers and the public in general.

7. Biographical sketch

Vassilis Bourdakis is an architect, completed a PhD at Bath University and is currently working on virtual reality systems for urban design.

8. References

Bourdakis, V. and Day, A. (1997) *The VRML Model of the City of Bath*, Proceedings of the Sixth International EuropIA Conference, europia Productions.

Bourdakis, V. (1996) *From CAAD to VRML: London Case Study*, The 3rd UK VRSIG Conference; Full Paper Proceedings, De Montfort University.

Brawne, M. (1992) *From Idea to Building ; Issues in Architecture*. Butterworth Architecture.

Boyer, C.M. (1996) *Cyber cities; Visual Perception in the Age of Electronic Communication* Princeton Architectural Press.

Calvino, I. (1996) *Invisible Cities* Translated from Italian by W. Weaver (First published in Italian, 1972) Secker & Warburg. London.

Carr, K. and England, R. (1995) *Simulated and Virtual Realities: Elements of Perception.* Taylor and Francis, London.

Charitos, D. (1996) *Defining Existential Space in Virtual Environments* In: Virtual Reality World96 Proceedings.

Charitos, D. and Rutherford, P. (1997) *Ways of aiding navigation in VRML worlds* Proceedings of the Sixth International EuropIA Conference, europia Productions.

Davies, C. and Harrison, J. (1996) *Osmose: Towards Broadening the Aesthetics of Virtual Reality,* ACM Computer Graphics: Virtual Reality (Volume 30, Number 4).

Day, A., Bourdakis, V. and Robson, J. (1996) *Living with a Virtual City* In Architectural Research Quarterly, Vol2, pp.84-91.

Day, A. and Radford, A. (1995) *Imaging Change: The Computer City Model as a Laboratory for Urban Design Research* In: CAAD Futures 95, Papers received Volume 2. National University of Singapore.

Friedlander, L. (1995) *Space of Experience: On designing Multimedia Applications.* In: Contextual Media: multimedia and interpretation E. Barrett and M. Redmond (eds). MIT Press

Gibson, J.J. (1986) *The Ecological Approach to Visual Perception.* London

Gibson, W. (1994) *Virtual Light* Penguin Books

Graham, C. *Data Visualisation and VRML* http://www.best.com/~cyber23/virarch/article.html

Heim, M. (1995) *The Design of Virtual Reality* In: Cyberspace/Cyberbodies/Cyberpunk; Cultures of Technological Embodiment M. Featherstone & R. Burrows (eds). Sage, London

Hillier, B. (1996) *Space is the machine: a configurational theory of architecture* Cambridge University Press.

Hollands, R. (1996) *The Virtual Reality Homebrewer's Handbook* John Wiley & Sons.

Hughes, K. (1995) *From Webspace to Cyberspace* Enterprise Integration Technologies (EIT) Version 1.1 http://www.eit.com/~kevinh/writings/cspace/

Ihde, D (1993) *Postphenomenology,* North Western University Press, Minnesota.

Lynch, K. (1960) *The image of the city* MIT Press, Cambridge, Mass.

Mitchell, W.J. (1996) *City of Bits.* MIT Press.

Negroponte, N. (1971) *The architecture machine* MIT Press.

Negroponte, N. (1995) *Being Digital.* Hodder and Stoughton, London.

Shaw, J. (1994) *Keeping Fit,* @Home Conference, Doors of Perception 2, Netherlands Design Institute, Amsterdam.

Shiffer, M.J. (1995) *Multimedia Representational Aids in Urban Planning Support Systems* In: Understanding Images; Finding Meaning in Digital Imagery, F.T. Marchese (ed) TELOS.

Slater, M. and Wilbur, S. (1995a) *Through the Looking Glass World of Presence: A Framework for Immersive Virtual Environments* In Framework for Immersive Virtual Environments FIVE'95 Esprit Working Group 9122, QMW University London.

Slater, M., Usoh, M. and Steed, A. (1995b) *Taking Steps: The Influence of a Walking Metaphor on Presence in Virtual Reality,* ACM Transactions on Computer-Human-Interaction (TOCHI) Vol.2, No3.

Tomas, D. (1991) *Old Rituals for New Space: Rites de Passage and William Gibson's Cultural Model of Cyberspace* In: Cyberspace: First Steps M. Benedikt (ed). MIT Press

Wen, J. (1995) *Exploiting Orthogonality in Three Dimensional Graphics for Visualizing Abstract Data.* Department of Computer Science Report CS-95-20, Brown University.

Watson, B., Walker, N. and Hodges, L.F. (1995) *A User Study Evaluating Level of Detail Degradation in the Periphery of Head-Mounted Displays* In Framework for Immersive Virtual Environments FIVE'95 Esprit Working Group 9122, QMW University London.

Wernecke, J. (1994) *The Inventor Mentor: programming Object-oriented 3D graphics with Open Inventor, release 2* Addison Wesley

World Wide Web resources referenced in the text:

CASA   http://www.bath.ac.uk/Centres/CASA/

VRML2 Spec	http://vag.vrml.org/VRML97/
Berlin	http://www.artcom.de/projects/stpl/WWWpaper/CyberCity.html
BigBook	http://www.bigbook.com/
Virtual Derby	http://www.virtual-derby.com/
Virtual LA	http://www.gsaup.ucla.edu/bill/LA.html
Virtual Soma	http://www.hyperion.com/planet9/vrsoma.htm

ISSUES OF ABSTRACTION, ACCURACY AND REALISM IN LARGE SCALE COMPUTER URBAN MODELS

A RADFORD[1], R WOODBURY[1,2], G BRAITHWAITE[1], S KIRKBY[2], R SWEETING[3] AND E HUANG[1].
[1]School of Architecture, Landscape Architecture and Urban Design, The University of Adelaide, Australia
[2]Key Centre for the Social Applications of Geographic Information Systems, The University of Adelaide, Australia
[3]Corporation of the City of Adelaide, Australia

Abstract
The availability of large scale computer urban models promises to radically improve the effectiveness of urban design policy-making and development control. A key question in the implementation of such models is how the balance between abstraction, accuracy and realism influences the effectiveness of their use. This paper discusses and illustrates the issues involved, with a computer model of the City of Adelaide as example.

1. Urban Design and the Development of Cities

Urban design encompasses "the spatial and territorial organisation of human realms: namely, the study of the relations between indoor and outdoor spaces and private and public territories on scales or levels ranging from single dwellings and dwelling clusters to service centres and towns and cities. Furthermore, appropriate urban design aims to provide, facilitate, and rationalise the creation of pleasant, livable, familiar, controllable, communicative, unambiguous and sustainable urban and living environments" (Binno, 1995). Urban design has become a major concern in recent years, spurred by widespread professional and public dissatisfaction with the cumulative effect on Australian and overseas cities of a multitude of individual development proposals initiated in the 1970's and 80's and recognition of the interrelation of design, economic, environmental and social well-being. The city of Adelaide, Australia, provides a prime example of this dissatisfaction, with a widespread view that 'better' urban design would have enabled the creation of a City Centre that was more attractive, and more supportive of economic and other objectives, than we now have. Urban design guidelines, providing advice for future developments, and urban design panels, providing expert critiques of development proposals, have both proved useful but limited by the difficulty in visualising and comprehending the implications of the development of any object as integrated and complex as a city.

Public authorities charged with the responsibility for development control have adopted public consultation as a mandatory stage in the approval process at least for 'non-complying' or 'subject to consent' forms of development. The process for this public

R. Junge (ed.), CAAD Futures 1997, 679-690.
© 1997 Kluwer Academic Publishers.

consultation is now quite well established and the authorities normally rely on architects and planners to supply varying media to describe the development and its impact on the surroundings. Traditionally, the written documents (plans, elevations, perspectives, photomontages, etc) have been static two dimensional representations of proposals, subject to abstract representation and 'artistic license' and unable to fully convey the interaction of the development with the surrounding urban fabric. Through work that the City of Adelaide undertook with the University of Adelaide it became evident that the majority of the public that come into contact with the development control process can not read plans and other two dimensional drawings.

2. Public and Council Reaction to Computer Models of Cities and Developments

An obvious approach to avoiding these problems is to simulate the present and possible future appearance of urban environments through computer modelling. Users of such models have emphasised their value in discussions about the future of the modelled areas (Day, 1993; Day and Radford, 1995). In Adelaide, a large-scale 3D computer urban model was established early in 1996 and trials of smaller-scale computer modelling of the neighbourhoods of development proposals were held with the City Council, using retrospective case studies and current applications. Working with the public in these trials, it was fascinating to see how individuals reacted to animated 3D models compared with 2D plans. The majority, if not all, accepted the technology and were keen to interrogate the models to satisfy their needs in terms of impacts from their perspective. This 'interrogation' aspect offered the public a new dimension not before offered as part of the process. Once people had gotten over the 'razzle dazzle' of the models, they often became quite demanding, asking for views into or out of the development sites, overshadowing predictions, etc. The end result of the trials was certainly that the public were better informed and armed with the information they needed to make a decision on their representations, and the Council's Development Assessment Committee was left with no illusions about the concerns of the public about the modelled developments. If, to quote the old saying, 'a picture is worth a thousand words', how many words is a fully textured and animated urban model worth?

3. Abstraction, Accuracy and Realism

How far to go in seeking 'realism' was a question which constantly arose. The two main concerns from the City Council's point of view were time and cost. In an orthodox modelling of a building and the presentation of that model there are several stages. The following table shows experience in making models of two 1996 development proposals and their immediate neighbourhoods in Adelaide, using the two CAD systems ArchiCAD and AutoCAD. Turning block models into 'realistic' models is a time-consuming process.

In general, high levels of *abstraction* (for example, representing a building by a block projection of its ground plan) facilitates both the making and imaging of computer models. There is an enormous increase in computational and human effort in *ensuring accuracy* and adding *accurate detail* through increased degrees of measurement of the 'real' world and precision in its representation in the computer model, and increased levels of detail. There is an important distinction between *accuracy* and *realism*, where the latter is the impression of a 'real scene' that is given by an image of the model. Realism, by borrowing or generating images of texture, sky, and detail, can be readily achieved in computer modelling, but the result may have little real bearing on the 'reality' of the situation. For example, images of nature generated using fractal geometry look 'real', but do not represent any actual object or scene. There are complex interrelations between accuracy, precision, realism, and distortion in the presentation of such images.

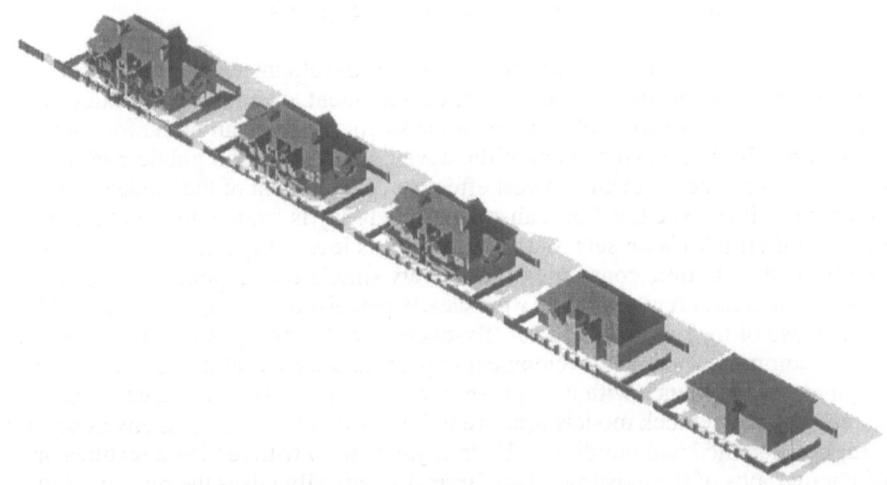

Figure 1. Design proposal for an Adelaide house modelled with varying levels of detail with shadows at mid-day in mid-winter. For many purposes the more abstract models are adequate.

Project stage	Project 1 (Modelled in ArchiCAD 4.5) % of whole time	Project 2 (Modelled in AutoCAD 13) % of whole time	Average % of whole time
Preparation (Review application documents and set up files)	26	10	18
Create 'Block' Model	33	19	26
Create and render 'Realistic' Model	30	59	44
Assemble Presentation	11	12	12
Totals	100	100	100

Table 1. Percentages of total times for different stages of a development proposal modelling and presentation process.

3.1 ABSTRACTION

Issues in abstraction include:

• The relative importance of roof (skyline), facade, and massing detail in the simulation of large-scale urban form.
• The role and importance of colour.
• The role and importance of shadows and reflections.
• The role and importance of ground detail and texture.
• The role and importance of 'activity indicators' such as traffic, people, etc.
• The role and importance of modelling trees and other plants.

The Adelaide City Council's aim was that models of development proposals be required from applicants as a part of submissions for development approval in the future, with such models to be 'dropped into' a model of the surrounding urban fabric for context. Strict timelines from the submission of the developments through public consultation to approval necessitate efficient and cost effective development of the models to make the process feasible. The level of realism possible today is evident to the public when they turn on their television sets, but to emulate this level of quality graphics is still extremely costly and time consuming. Relatively simple block models with a medium degree of architectural representation were clearly perceived as being superior to 2D plans. The use of the block models for 'fly-overs', 'walk-throughs', overshadowing, etc., better communicated the development impacts and were useful in demonstrating the developments' compliance with urban planning guidelines. Although crude abstractions of the real world, the block models used were dimensionally accurate as envelopes of the more highly modelled real buildings. Their augmentation with rendered textures or actual photographs of the existing urban fabric dramatically raises the realism of the models and the public more readily identifies with the locality. However, the time, effort and cost of this augmentation is significant and research is needed to demonstrate the cost/benefit of going this further step.

3.2 REALISM

Issues in 'realism' include the effectiveness in imparting realism of techniques such as:

• Mixing 2D images from a model with digital photographs or video from the site situation, and images 'borrowed' from other locations, to provide the sense of visual detail, colour, texture and 'grain' familiar in photographs.
• Mapping of 2D elevational images onto 3D block forms
• Using generative rule-based computer systems to simulate urban detail.

In most visualisation systems so-called 'material textures' are represented as precaptured images, procedurally generated images, and through changing the reflection and transparency properties of surfaces and objects through combinations of the above. The first of these, precaptured images, provides a way to bring images taken from a real environment into a visualisation system. In essence, a digital photograph of a building facade is used as the basis for a texture, which is assigned to the surface of an idealised building form. Such images are then placed in perspective by the visualisation program, and visually approximate the appearance of the building. Facade images need not be taken perpendicular to a facade to be used in this technique. 2D image manipulation programs provide perspective transformations by which a bit map of a building facade in perspective can be differentially stretched (or shrunk) to the shape of the actual building facade being modelled.

The advantage of using precaptured images is clearly speed; it is generally faster to photograph and paste than to model. Such images provide good context in a visualisation. Their disadvantage is revealed, literally, on close examination, for they are of finite resolution and are adamantly two-dimensional. When a pixel in the precaptured image begins to occupy several pixels in a visualisation, the illusion begins to dissolve.

Many neighbourhoods, especially the ones we recognise as having 'good character', comprise buildings with common design, construction and siting characteristics. As shown by Flemming (Flemming 1985, Flemming 1987), plausible buildings can be generated in such neighbourhoods by a spatial grammar developed from a sample of exemplary buildings from the neighbourhood. Adaptations of the grammar can produce buildings with different functional characteristics (for example, attached garages) while retaining many formal characteristics from the neighbourhood. In urban modelling, we are concerned both with the representation of existing building form and with the prediction of possible future forms.

With respect to existing form, a spatial grammar system can capture some aspects of buildings typically covered by urban design guidelines. Buildings generated by such a grammar could be arranged (again following grammatical rules) into neighbourhoods. Changes to the underlying grammar would then model changes to the urban design guidelines and the example neighbourhoods generated by the grammar would stand as examples of likely futures given by a set of guidelines. To our knowledge, no such application of grammars has yet been made, although Rabie (1991) also proposes the generation of urban 'texture' in exploring the future form of cities and Radford and Day (1996) use the implicit rules of urban design in Georgian Bath in the generation of hypothetical extensions to that city.

3.3 ACCURACY

Issues of accuracy include:

• Accuracy of location, geometry and form
• Accuracy of colour and texture
• Accuracy of lighting
• Accuracy in the representation and transparency of vegetation (for example, how much one can see 'through' trees obscuring a building)
• Legal liability

Accuracy is particularly important with issues such as overlooking and view maintenance, where residents affected by new development are likely to be very unhappy if they feel that the model has mislead them. Sometimes it is necessary to be 'inaccurate', as in showing more transparency in tree vegetation than is present in reality in order to simulate the impression of a view through trees of a building facade that is obtained by a person moving view angles slightly to get a series of partial images between leaves and branches.

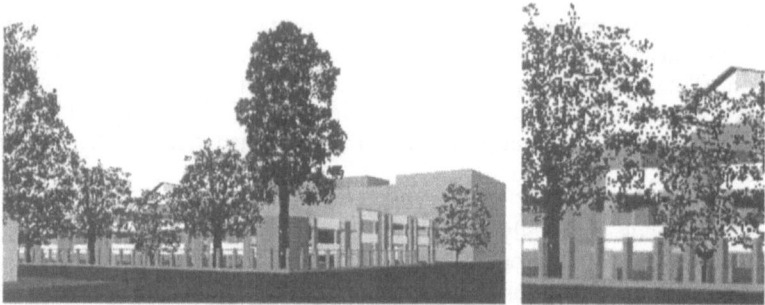

Figure 2. Street trees, using 2D images of trees which are represented on two vertical planes which intersect on the tree trunk. The tree image has been severely 'pruned' to allow the building to show behind the tree. Overall view (left) and detail (right).

One aspect of the task of ensuring accuracy presents considerable technical difficulty. When using digital photographs, it is essential that the location from which the photograph was taken and the direction (altitude and azimuth) of the camera be known to sub-metre accuracy. Three options were considered. Conventional surveying techniques were thought to be too labour intensive. Global positioning technology with this degree of accuracy appears to be too expensive and unreliable in an urban context. The preferred technique, at this stage, appears to be use of a developed version of the Adelaide City Model that provides sufficient site features for accurate relative location of a camera. This technique would require use of a computer on site, a requirement which would already exist for downloading the digital images on site to verify them.

Such accuracy is not always important. For the University of Adelaide New Science Building project, computer modelling simulations were used to establish a general impression of how the view from the adjoining street to the portico of the University's Library would appear if a major obstructing building was demolished, as a part of the debate about redevelopment strategies for that part of the campus. The camera was necessarily positioned on the library side of the building to be demolished, and the view constructed using a model viewpoint some 30m further away from the library. The photograph image was then distorted to fit the appropriate surfaces of the model.

There is a mix in this single image of different degrees of abstraction, accuracy and realism. The two flanking buildings are abstract; they were not yet designed, and are indicative only of general massing. The basic geometry of the urban space is accurate, taken from the University's 2D CAD database of its grounds. The rendering of the library portico is realistic (from a photograph) but inaccurate (the photograph was not taken from the position of the image viewpoint). The foreground trees are from a photograph taken from the correct view position, but digitally edited to allow more of the view to appear. Overall, although inaccurate, the impression is sufficient to assist in the decision making process.

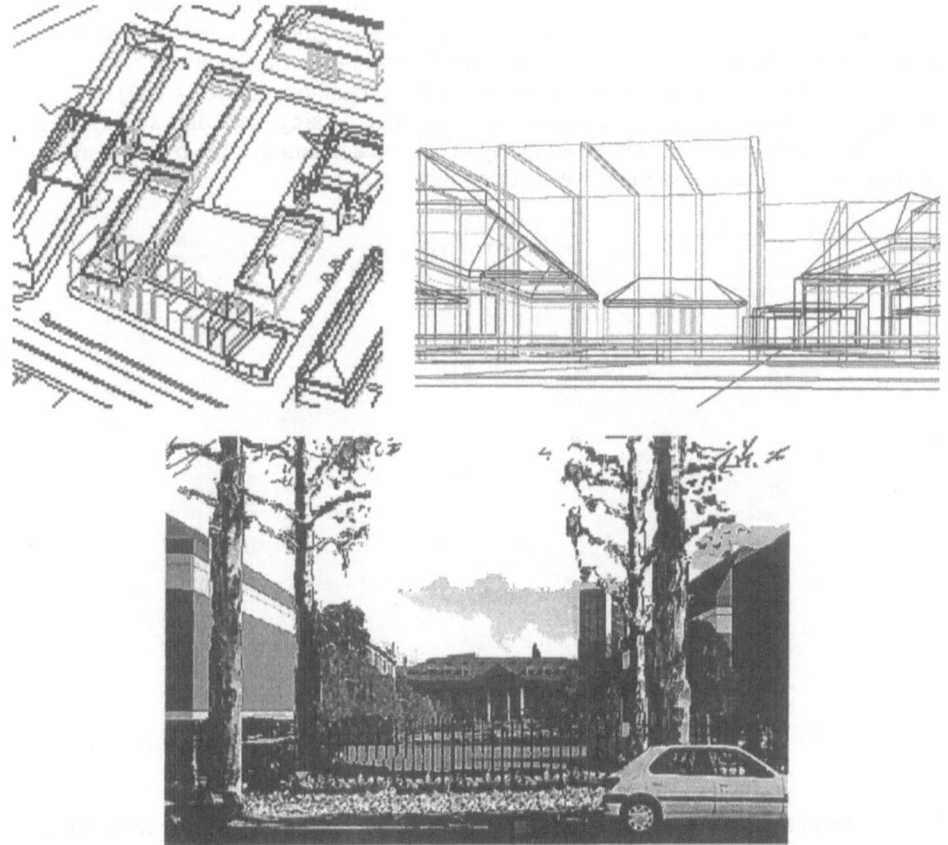

Figure 3. Part of 3D model of the Adelaide University campus (top left) with view from street 'through' existing building (top right) and 'pasted' parts of photographs from actual viewpoint and viewpoint beyond existing building.

Another interesting issue is accuracy in modelling lighting, particularly night-time lighting. With the complexities of luminaire fixtures, it is difficult to accurately represent illumination on a model. Radiosity models can produce effective photometric simulations of diffuse lighting with soft shadows given accuracy of colour rendering and distribution data, but the effort required is considerable.

4. Abstraction, Accuracy and Realism in the City of Adelaide Model

Adelaide University's National Key Centre for the Social Applications of Geographic Information Systems (GIS), in association with the City of Adelaide and Maptek (a software developer/provider), has created a 3D block model of the City which was demonstrated at the UN Habitat Conference in Istanbul, June 1996. It was also used extensively in promotion of proposals for revitalising the City Centre under a joint Federal, State and City government under the title of 'Adelaide 21'.

With the development of 2D GIS in the early 1960's, and their general acceptance as a spatial data storage and analysis tool during the 1980's and the early 1990's, a data collection revolution commenced. At least five different government institutions (local/state levels) collected spatial data for the central business district (CBD) of Adelaide. The initial 3D Adelaide Model was constructed using these data sources, with no new data being acquired. The availability of this data meant that the base 3D model could be constructed in less than two working days.

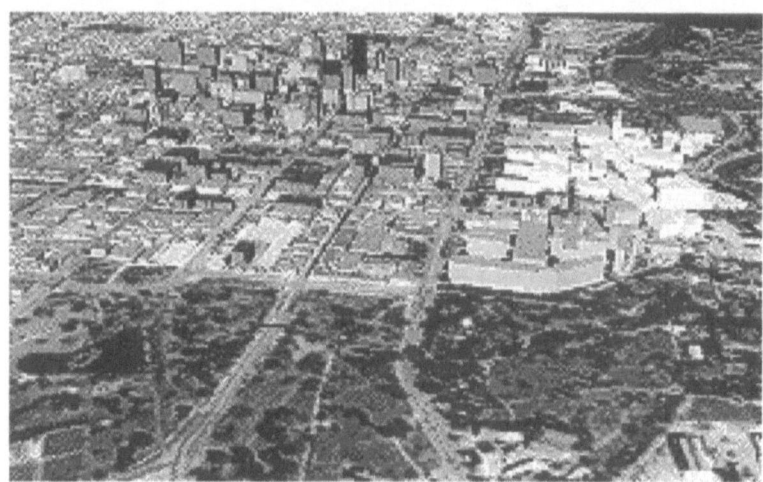

Figure 4. The Adelaide City Model, essentially a 'block' model mixed with digital photographs.

The data collected for the CBD of Adelaide was road centre lines, building footprints, sewer pipes water pipes, electrical cables, electrical ducts, precinct boundaries, zoning boundaries, and a digital elevation model. The geometric accuracy for the spatial data is unknown as the root means square errors for registration of the data and other meta-knowledge regarding the spatial data transformation were not correctly recorded. Yet in defence of each of the five participating agencies, it must be noted that when their data sets were fused together the small number of overlap errors between the datasets was very encouraging. Transposition of the 2D GIS data into the 3D environment entailed attributing Z values to each spatial feature. The first step in this process was the registering of each data layer to the digital elevation model. This process attributed a Z value to every X Y node/vector within the data model. Extrapolation of the building footprint polygons to their 3D heights entailed allocating the specific height as listed in the database to each vector. Once allocated, each polygon was then triangulated using software options into a block model. Colour attributed to each building represented the building height as a function of the number of floors. Sub-surface feature Z values were not listed in the database, thus expert heuristic's were used to allocate them to their appropriate depth. The database values that were attributed to the graphic primitives were the diameters of the pipes and the material from which they were constructed. The visual representation of this information provided some interesting insights into the sub-surface networks, for example pipe diameters varied in an interesting pattern throughout the city.

The block model is combined with aerial photographs of the surrounding park lands, suburbs and street network, and of the major squares in the City. The result is surprisingly realistic when the views are those of the city as a whole, or of large segments of the city. It also works well in animated sequences where the lack of modelling detail is less apparent when images change quickly.

The data in the initial model provides a base and context for local areas modelled at much greater detail. Interestingly, small areas of detail (such as well-known 'heritage' buildings) add much to the overall sense of realism in the model.

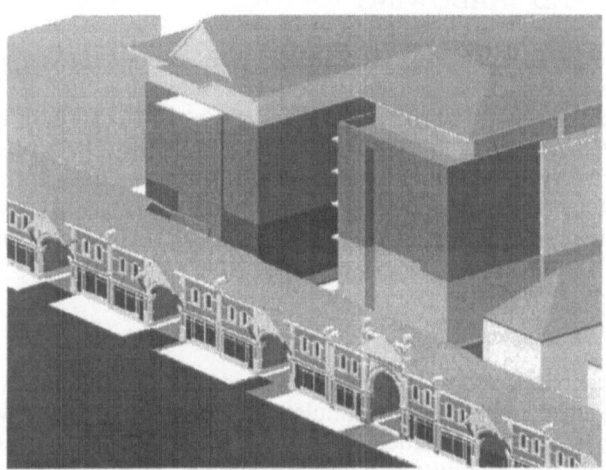

Figure 5. Model of part of the East End of the City of Adelaide, with detailed facade model for the retained 'heritage' edge of the East End Markets.

The Adelaide City Model is intended as a tool for strategic planning in many fields, including infrastructure and emergency service provision. Our main use of the data so far has been in the provision of context information for models of a development proposal. The use and gradual enhancement of the Adelaide City Model provides an efficient means of providing this information.

5. Modeling Purpose and Modeling Criteria

The appropriate level of detail depends on what the issues are for a specific development proposal.

5.1 STREETSCAPE

Streetscape contribution probably depends more on what is happening at ground level, what is perceived by the passer-by, than on the actual building. Example issues are how transitions are made from inside to outside, vertical to horizontal, and artificial to natural. Issues of massing, street presence, use of materials and detailing of the building can be evaluated in its future context from virtually any viewpoint. In addition, because the building model and its future context can be created in real world units of measure, all surfaces can be modelled to represent the proposed building to whatever level of detail is required to understand how it will look when finished.

Normally very little detail beyond windows, doors, balconies, verandahs, and roofs is necessary to grasp the designer's concept, but the latitude exists for more and is not limited by issues of scale. The appropriate level of "realism" necessary to provide contextual information (such as footpaths, vegetation, cars and people) remains, for us, a matter of some debate.

5.2 SUN SHADE AND SHADOWING

Our experience suggests that a basic block model with accurate roof representation is sufficient to generate the sun shade and shadowing information. For clarity, neighbouring buildings could be modelled to a slightly simpler form or in another colour to connote their role as "background". However, at too abstract a level, the uninitiated viewer may lose the relation between the blocks on the screen and the buildings which they represent. Solar access issues can be explored with a simple massing model which includes adjacent and pertinent surrounding buildings to the same level of detail.

5.3 OVERLOOK

Overlook issues require the representation of openings in the building envelope of both the proposed and adjacent buildings to simulate views from those windows and into the adjacent buildings windows, or garden. Windows and doors help the viewer to understand the scale and proximity of adjacent buildings. Our experience suggests that general overlook issues are flagged with the basic block modelling of the proposed development and its closest neighbours. However, the addition of windows, doors, and balconies to all buildings modelled would be preferable for more in-depth study. With this simple articulation, simulated views out of the proposed building and into adjacent properties and buildings can be created. Such elements also add a sense of scale and interpretability to the model, so would be important for viewers with little or no experience reading 2D or 3D building representations.

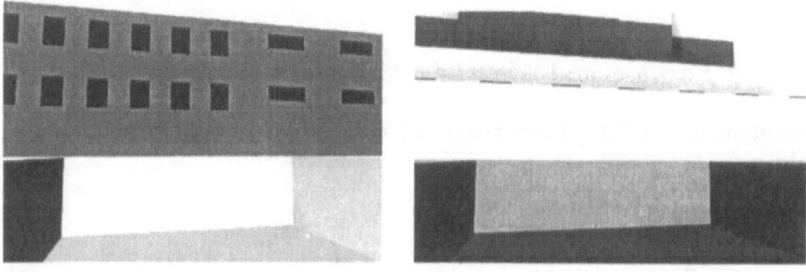

Figure 6. Comparing overlook with existing (left) and proposed (right) building developments on adjacent land. Realism is unimportant for the purpose of the image.

5.4 COMPATIBILITY

Compatibility issues may only require the simplest massing model if scale or distance to boundaries is the problem, but may require a model of relatively high detail if appropriateness in the streetscape from the perspective of a passerby is the problem. It may be necessary to provide greater detail where lay people need to understand and relate to the resulting images than where they will be viewed and understood by professional staff used to abstraction in representations.

The compatibility of the basic massing of a proposed building is best evaluated in the larger context of its surroundings, not necessarily limited to the adjacent properties sharing a boundary with the site in question. Beyond this general appropriateness, varying levels of detail are required to assess its contribution to the streetscape and neighbourhood overall. Use of an city model comprising only geometric block forms (as in the extant Adelaide City Model shown in Figure 1) as context for controversial proposed developments within the city square mile is reasonable and efficient. The only draw back to this large database of 3D information is the lack of facade and building detail information required for proper interpretation of the buildings on close inspection. Adding facades and surface details to the buildings in the Adelaide City Model is currently being explored through the use of digital imaging. This is an important extension of the City Model, and will increase its usefulness to the city to a large degree. We have found that digital imaging, used as both foreground and background, may also be a fast and reliable way of adding information about the existing surroundings to our representation of the development application being modelled.

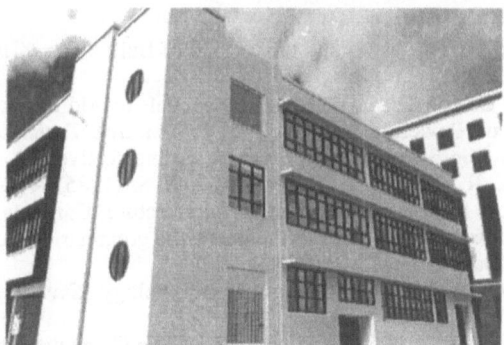

Figure 7. View of new car park as backdrop to existing building, seen from street level. Considerable detail is needed for such close comparisons. (Model by Adrian Price)

5.5 MARKETING

For marketing and public relations, there is generally opportunities to mix images from high and low detail parts of the model and to mix 'real world' video with model animation. The Adelaide block model has been used very effectively in this way in promoting the City. The detail in the model needs to be related to the video story board, and in our experience it is important to decide the story board while still constructing the model.

6. Conclusion

Any attempt to provide a realistic representation may be counterproductive if the model makers' view of reality does not match the final product. There are often advantages in visible abstraction; several CAD vendors offer paint routines to reduce the realism of their models. With detailed aspects such as matching materials and colours for heritage developments we need to use care or clearly state the limitations of the models. Environmental aspects such as lighting levels, sun light, glare, shadow and reflection can dramatically influence the viewers impressions of a model and a good deal more work is needed to establish guidelines on the appropriate integration of these aspects.

At the time of writing, the City of Adelaide is continuing to develop ways and means to have, at least major, developments modelled in 3D at the time of their development application and is continuing to work with the University towards this end. Both players have their own goals. The City is concerned *inter alia* with the improvement of both its processes and the physical fabric of which it is comprised. The City is also cautious of the public's acceptance of these models, believing that once they are introduced the public will expect them to be a normal part of the development approval process and will become more demanding. The University sees opportunities for a virtual urban laboratory for students, for community service in development of a comprehensive model, and for research opportunities in the social and professional applications of technology. These goals appear to share objectives through the joint development of urban models and urban modelling expertise.

References

Binno, R. (1995) Urban Design Issues and the Design of Built Form, *Australian Planner*, Vol.32, No.3, pp172-174.

Day, Alan. (1993) The Use of Urban Visualisation Models to Aid Public Participation in the Planning Process, in: Powell, J.A. & Day, R., *Information Technologies for Construction, Civil Engineering and Transport*, Brunel University, pp.141- 149.

Flemming, U., Gindroz, R., Coyne, R., and Pithavadian, S. (1985) A pattern book for Shadyside. Technical Report, Department of Architecture, Carnegie-Mellon University.

Flemming, U. (1987) More than the sum of the parts: the grammar of Queen Anne houses. *Planning and Design*, 14, pp323-350.

Rabie, J. (1991) Towards the simulation of urban morphology, *Environment and Planning B: Planning and Design*, vol.18, pp 57-70.

Day, A.K., and Radford, A.D. (1995) Imaging Change: The Computer Urban Model as a Laboratory for Urban Design Research', *The Global Design Studio*, ed. M Tan and R Teh, Centre for Advanced Studies in Architecture, National University of Singapore, pp495-506.

Radford, A.D., and Day, A.K. (1996) Growing Georgian Bath, *Environment and Planning B: Planning and Design*, in press.

A Network-based Kit-of-parts Virtual Building System

A. SCOTT HOWE, ARCHITECT
Kajima Corporation and The University of Michigan
School of Architecture and Urban Planning
email: ash@ipc.kajima.co.jp

Abstract

This paper describes an experimental browser / modeler which will allow the user to collect and assemble virtual kit-of-parts components from "component libraries" located on the Internet (such as manufacturer's databases) and assemble them into a virtual representation of a building. The fully assembled virtual building will provide a basis for ordering and manufacturing actual components and preparing for construction. The browser will allow the designer to affect a limited degree of remote fabrication at real manufacturing facilities, and facilitate eventual interface with built in sensors and actuators. The browser will manipulate and display interactive three dimensional objects using Virtual Reality Modeling Language (VRML). Upon assembly, actual components will have sensors built into them for providing data about the real building, which could be viewed during a walkthru of the virtual building by clicking on parts of the model. The virtual building will work as a remote facility management tool for monitoring or controlling various architectural devices attached to the real building (such as electrically driven louvers, HVAC systems, appliances, etcetera).

1. A Research Programme

Increasingly, powerful computer-aided design tools have enjoyed greater roles in the design process. In parallel, the Internet with its World Wide Web has proved to be a revolution in information dissemination, providing real-time access to sources located around the world. Since information is the ultimate substance from which designs are conceived, a logical question could be: how can a design process be enhanced by direct links to information sources? Considering the increased proliferation of information-based automated manufacturing processes, a second question would logically follow the first: how would direct instantaneous access to manufacturing processes affect the design process? Finally, instantly gleaning performance data from the constructed object itself, how would an information feedback loop affect current use and future design improvements to future objects of a similar type? While these questions will probably remain unanswered for many years, the development of an experimental environment which sets the stage for linking design, manufacturing, and use can be facilitated.

R. Junge (ed.), CAAD Futures 1997, 691-706.
© 1997 *Kluwer Academic Publishers.*

As a research programme, it is proposed that a computer tool be developed which can affect such an experimental environment. The computer tool has been conceived in the form of a plug-in to a common Internet browser. Some functions of the plug-in will eventually be made available to a selected number of designers for further research purposes.

2. VBuild Browser

2.1 VBUILD CONCEPT

The browser plug-in (hereafter called VBuild) makes maximum use of two powerful concepts: object-oriented programming and kit-of-parts philosophy. In a way, the two concepts work hand in hand.

2.1.1 Object Oriented Programming

VBuild was programmed using an object-oriented structure in the C++ language. Object-oriented languages utilize data and code in discrete structures called classes. Once instantiated, the class defines types of data called objects. The data itself is protected and can only be manipulated via pre-defined methods. The methods are interfaces with the rest of the program. Once the interfaces are defined, the actual coding for implementing the interfaces can take on any form as long as it supports the interface methods. In this way specific elements in a model can be defined independently of other code according to function or behavior. Methods and attributes specific to that element's behavior can be added as needed to give the coding completely expandable capabilities.

In an object-oriented program, once a class has been debugged it becomes a reliable building block in the entire coding of the program. Object-oriented design becomes a clean way of defining functionality without having to deal with loose ends associated with the complexities of unstructured raw coding.

2.1.2 Kit-of-parts Philosophy

The active use of kit-of-parts philosophy was an important element in the conception of VBuild. A kit-of-parts is a collection of discrete building components that are pre-engineered and designed to be assembled in a variety of ways, much similar to an erector set or toy Lego blocks. When assembled, the entire kit-of-parts can define a finished building or artifact. The components fit together according to rigorously designed interfaces which provide for flexible configurations. Components are sized for convenient handling or according to shipping constraints. Since a well designed component can be used over and over again, fabrication processes can be worked out in advance for real-time manufacturing at time of need.

Kit-of-parts components can be thought of as objects in an object-oriented programming environment. With well-defined interfaces which are rigorously followed, the component itself can assume any form. Interfaces can include mounting points, rules for the transfer of loads, specifications for thermal performance, and maximum cost constraints. In short, a kit-of-parts approach lends itself to cheaper and more efficient manufacturing, and is a clean way of demonstrating a network-based virtual building system without having to deal with loose ends associated with the complexities of unorganized raw materials.

2.2 VBUILD CLASSES

The programming classes or objects which have been devised for the browser mostly fall into three main categories: Geometry classes, Assembly classes, and Construct classes. The Geometry classes consist of classes which define geometrical representations of objects, and include 0D, 1D, 2D, and 3D geometry. The Assembly classes define specifications, function, and fabrication processes associated with individual components in a kit-of-parts. An assembly would be a discrete component which is manufactured using a combination of different fabrication processes, and follows interface rules for connection to other components (in this paper, "assembly" and "component" may be used interchangeably and refer to the same thing). Construct classes define ways of organizing the assemblies.

GEOMETRY CLASSES

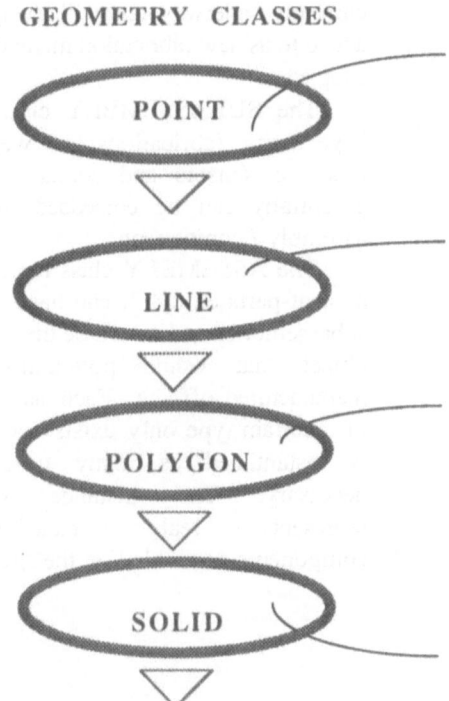

The POINT class represents the most basic form of geometry or 0D geometry. One point object would contain XYZ coordinates as data, and include various methods for manipulating the coordinates.

The LINE class represents 1D geometry and includes both lines and segments. A line will have two point objects as data and include methods for manipulating itself.

The POLYGON class represents 2D geometry of enclosed shapes. Eventually it will also represent compound 2D elements such as splines, curves, and open shapes as well, but for now only closed shapes constructed of a series of connected vertices is implemented.

The SOLID class represents 3D geometry. Eventually it will also represent curved surfaces as well, but for now only faceted solids are implemented.

ASSEMBLY CLASSES

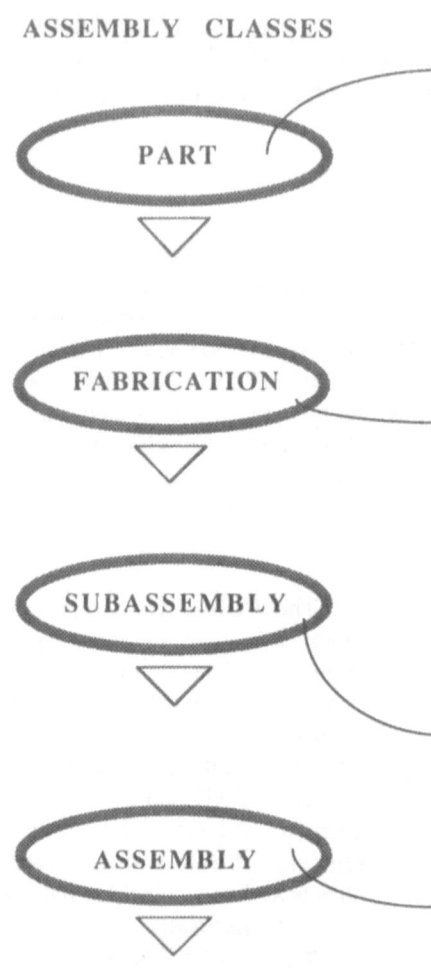

The PART class potentially contains one solid or one polygon or both, as well as a material definition. A part is meant to represent a raw material with limited definition of a potential shape. The solid would be representative of a cast object or folded sheet fabrication, where the polygon would represent a section for extruded objects or the shape of a cutout from sheet stock.

The FABRICATION class has exactly one part as well as paths for extrusion, and includes methods for creating G-code for actually machine fabricating the part (G-code defines paths for manufacturing tooling). Implementation of these methods is currently underway, and will always be added to as new fabrication methods are adopted.

The SUBASSEMBLY class can have many fabrications, as well as links to sensors and actuators that potentially can be embedded in the assembly / component.

The ASSEMBLY class represents a kit-of-parts entity. It can have many subassemblies and would be the largest object that would potentially be manufactured off-site. Each assembly of a certain type only exists once and is instantiated as many times as necessary. The instance would represent real manufactured components assembled on the site.

CONSTRUCT CLASSES

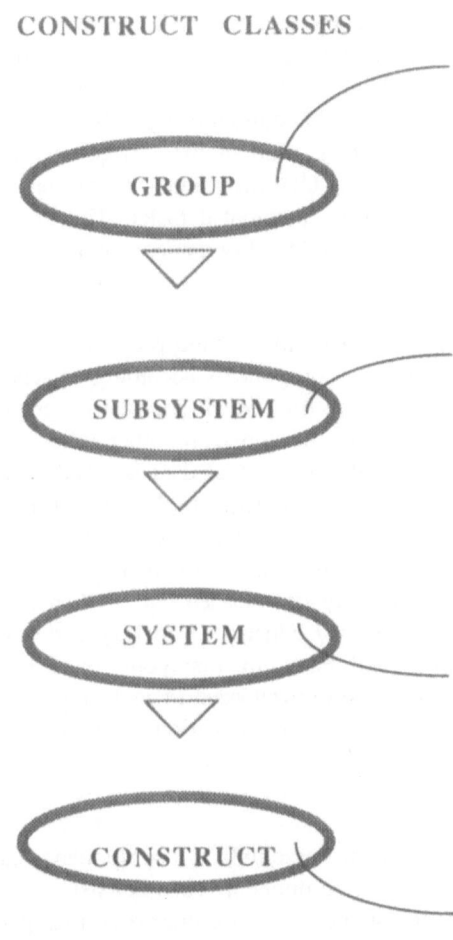

The GROUP class is a special collection of instances of assemblies which have some common purpose or function, such as all the column assemblies of a certain type that belong to a certain floor of a building. A group can have many instances. Methods in the group class would manipulate groups.

The SUBSYSTEM class is a collection of groups which have a common purpose or function, such as all the columns of a building. Each floor may have its own group of columns, which added together would constitute the column subsystem. The subsystem would contain methods which would be for the express purpose of manipulating subsystems.

The SYSTEM class is a collection of subsystems, such as the structural system of a building. The structural system would include column subsystem, floor framing subsystem, and foundation subsystem. System class methods would manipulate, view, or analyze entire systems.

The CONSTRUCT class is the finished artifact, which includes all the systems. Construct methods would define behavior of the construct, and include mechanisms for viewing it in various ways.

In addition to the three main groups of classes, VBuild also has attribute classes and utility classes for the viewing and general manipulation of the various data types. In each step of the hierarchy, methods specific to that data type are implemented. More methods can be added later as the need arises.

2.3 PROPOSED VBUILD CONFIGURATION

VBuild will actually consist of several programs functioning in unison in various locations on the Internet. The browser Plug-in portion is merely the local tool which helps the user view and manipulate the data. Locally the user will create a construct which can be saved as a file locally. Assemblies / components will be returned on request from remote kit-of-parts virtual librarians. Fabrication of real parts will be coordinated by a virtual contractor, and remote control and monitoring will be facilitated by a virtual facility manager. Except for the plug-in, each of the virtual servers will consist of simple Common Gateway Interface (CGI) programs, which are small programs linked to web pages that perform simple automated tasks. The process is delineated in the three steps of design, manufacture, and facility management.

2.3.1 Design

When the user wants to insert another component, the VBuild plug-in will contact remote virtual librarians which will return requested components according to type. The assemblies / components for the most part are high-level parametric primitives consisting of collections of cuboids, cylinders, cones, frustums, and other solids that can be defined with a limited amount of data. Since the amount of data is small, Internet transfer can occur very quickly. The local plug-in then fills in the rest of the data according to a predetermined formula based on the type of primitive and creates an instance of the component. Each time the same type of assembly is requested by the user, another instance is created from the previously downloaded data. When the user saves the construct, the actual data of each assembly / construct is deleted and only the instance references and their locations are preserved. Each time that model is opened, the real data associated with each instance is once again downloaded in real time from the source. Using filters, the user will be able to view the data in various ways which may include sections and plans.

2.3.2 Manufacture

Once the building is virtually assembled, the assemblies / components can be manufactured and delivered. Each assembly / component can consist of many fabrications. Since the fabrications contain the data necessary for their own manufacture, the VBuild plug-in will contact a virtual contractor and request an estimate based on fabrication type. The contractor will select a manufacturer based on price and wait list and return an estimate. The ideal setup would have a simple list on the manufacturer's server which would hold current setup and processing rates, with another list on the material supplier's server which would hold current material costs. A queue could also be maintained on the manufacturer's side which would hint at possible wait lists. When the manufacturer finishes processing the fabrication, it will be shipped directly to the designer or another designated address (such as the site). In the case of multiple fabrication types in a single component, a system of bar codes will facilitate the forwarding of one finished fabrication to the next manufacturer, who will build upon the previous work until the entire component is built and sent to the site.

2.3.3 Facility Management

Once the building is actually assembled, a Hypertext Transfer Protocol (HTTP) or remote access server can be paired with a virtual facility manager CGI on a local computer installed in the finished building. An HTTP server would be used to facilitate an Internet connection, as opposed to a remote access server which supports private phone connections. Each subassembly has the capacity to link to a CGI program which can handle bitwise communication with a LonWorks, CEBus, or X-10 interface device connected to the computer's serial port. LonWorks, CEBus, and X-10 hardware are brand name pseudo-standards which facilitate plug and play monitoring and control of other devices on a powerline network or dedicated bus. These devices can plug into a standard 110 volt outlet, and send and receive signals along the power wire to and from other similar devices. The user would be able to fly-through the VRML model of the building and click on various parts of the model to affect monitoring and control. Upon clicking on the link, special CGI programs would bring up a Java control panel especially prepared for that device.

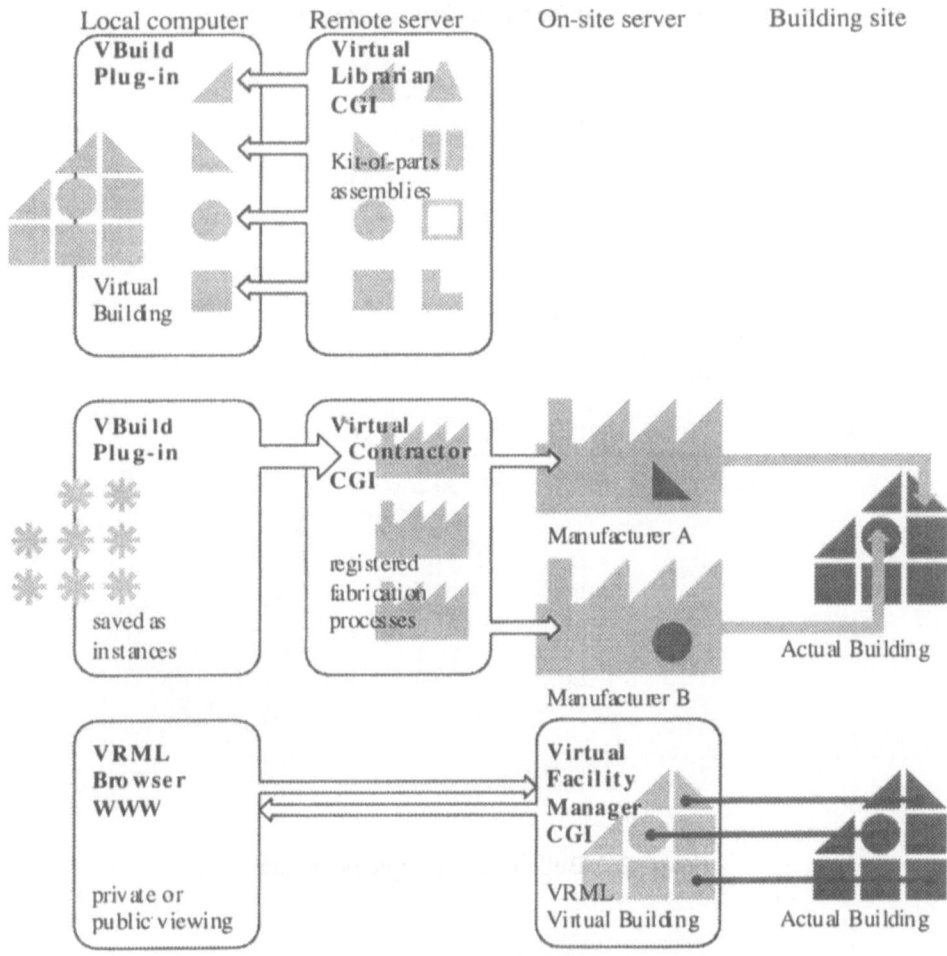

Figure 1: VBuild Proposed Implementation

Figure 1 describes the processes associated with the three phases of design, fabrication, and facility management. In the design stage a designer on a local computer would install the VBuild plug-in in a World Wide Web (WWW) Internet browser such as Netscape. VBuild would allow the designer to collect kit-of-parts assemblies and assemble them together in an intuitive way much the same way a child would construct an object out of Lego blocks. The assembled virtual building could then actually be manufactured component by component through the brokerage of the virtual contractor which would have various manufacturer's fabrication processes on register. Once the actual building is built, a VRML version of the virtual building would be loaded into a computer installed in the actual building, and would have access to CGI's which provide an interface to the LonWorks, CEBus, or X-10 hardware (symbolized in Figure 1 by lines connecting the virtual building and actual building). Full Internet access to the VRML virtual building could be facilitated for public access, or limited phone connection only could be facilitated for remote private use.

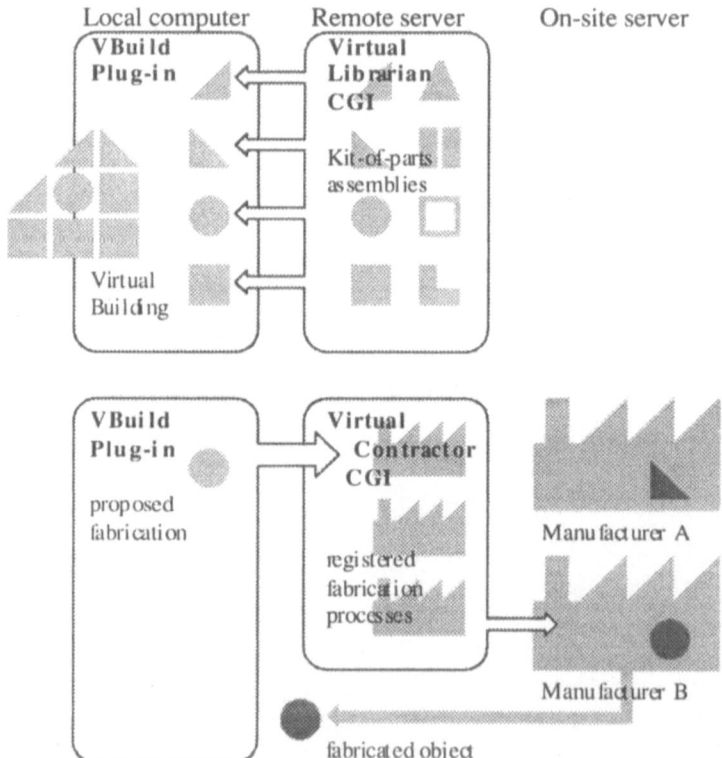

Figure 2: VBuild Current Implementation

2.4 CURRENT VBUILD IMPLEMENTATION

Currently the VBuild plug-in has skeletal appendages of most of the proposed implementation. All geometry classes are complete and operational. The assembly classes and construct classes are present in a skeletal form with limited implementation. Data structures, including linked lists, arrays, and overall file formats are fully implemented, with several experimental parametric primitive assemblies prepared for insertion into a sample kit-of-parts library. Conversion of GEDIT PH3-SET's into full blown VRML models (complete with dummy lighting and camera nodes) is implemented.

The code for actually selecting a component from the web and inserting it into the model has yet to be written, but a demonstration application is being prepared. None of the final CGI programs have been prepared as of yet, but simple implementations have been written and tested. A quick implementation of a link to a manufacturer has been developed and tested.

3. Related Research

The VBuild project is supported on a triple foundation of extensive research in the areas of architectural information visualization, automated construction, and remote facility management.

3.1 INFORMATION VISUALIZATION

During the course of information visualization research, a simple experimental network-based kit-of-parts library was established using the VRML standard. The experiment consisted of three simple virtual building components located on a server in Japan, with a building model located on a computer in Michigan. Using a VRML web browser the Michigan model could be opened, whereupon the components were automatically loaded from Japan in real time and placed in their proper locations and orientations (see Figure 3). Clicking on a component downloaded its specification page. During the course of the visualization research, valuable insight was gained in understanding VRML structures and overall Internet File Transfer Protocol (FTP), Hypertext Markup Language (HTML), and hypertext linking and anchoring concepts. Past experience with design and computer modeling proved to be a valuable resource in the research.

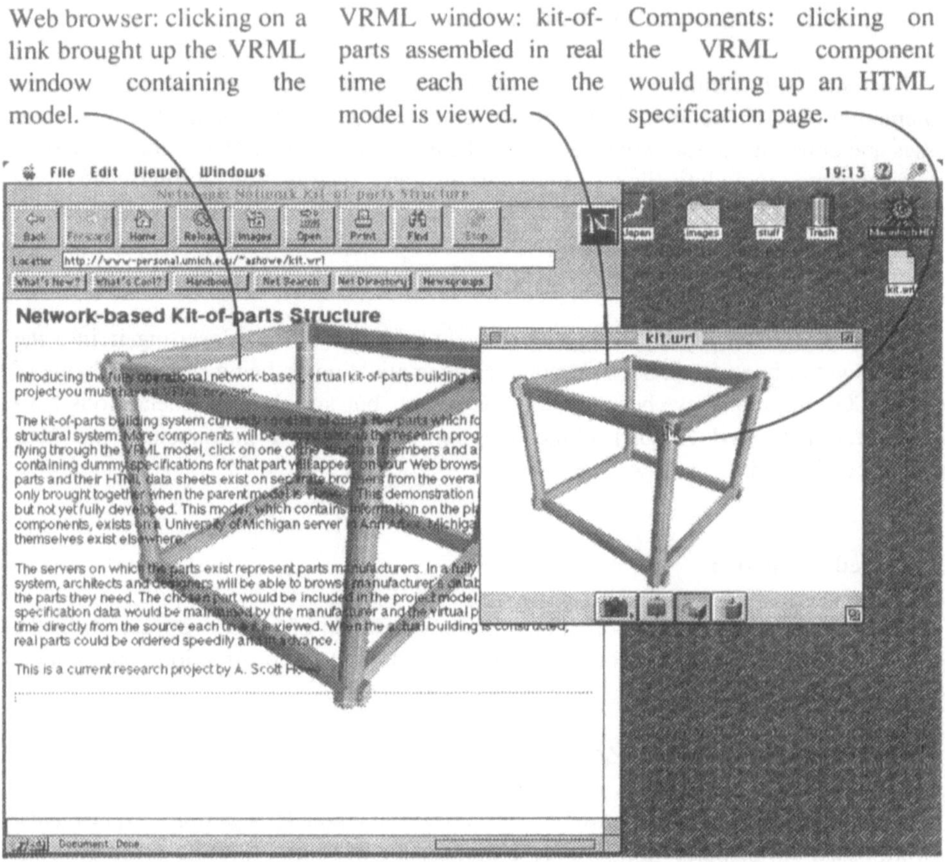

700

Web browser: clicking on a link brought up the VRML window containing the model.

VRML window: kit-of-parts assembled in real time each time the model is viewed.

Components: clicking on the VRML component would bring up an HTML specification page.

Figure 3: Architectural Information Visualization

3.2 AUTOMATED CONSTRUCTION

An exhaustive study of the state of the art of automated construction was conducted, followed by controlled simulations using industrial robots. A model kit-of-parts building system was devised for the purpose of deriving principles of design for automated construction. The simulations involved the automated assembly and disassembly of the model kit-of-parts from remote locations over the Internet. The components incorporated mechanisms which could interlock with other components that were actuated for deployment by the robot's end affector (see Figure 4). In addition to numerous successful simulations conducted from various computers located on the University of Michigan's local network, control of assembly and disassembly of the Michigan-based model was satisfactorily affected from both Denmark and Japan. Control of the robot was facilitated by the use of a laptop computer connected by modem to the local telephone system using Compuserve telnet. The simulations and construction of the model and components resulted in the derivation of a set of design principles which

701

can be used to design building components with characteristics that easily lend themselves toward automated construction processes. The research also contemplated the development of shape grammars that would not only act as a basis for the design of the building components, but would guide the design of robotic construction machines as well. A simple shape grammar was devised based on the derived design principles and an example conceptual kit-of-parts building system presented, including concepts for robotic systems that could be used to assemble the components on site. The example system illustrated possible applications of the automated construction design principles. Past experience with robotic work cell simulation and conceptual robotic construction system design contributed to the study. The automated construction research provided valuable insight into general principles of design for kit-of-parts systems which are manufactured and constructed using automated means.

Components incorporated mechanisms which could be actuated by the robot to facilitate connection to other parts

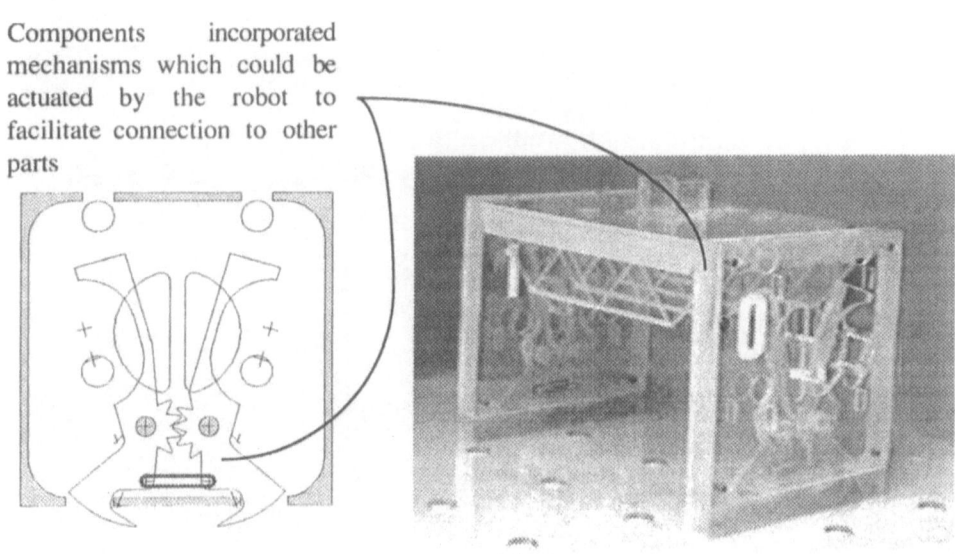

Figure 4: Model kit-of-parts building

3.3 FACILITY MANAGEMENT

The remote facilities management research involved the study of techniques used for remote device monitoring and control. As an experimental project, a facility web page was devised which acted as a home page for an experimental environmental controls facility at the University of Michigan which was wired with various sensors. The web page contained a VRML model of the facility complete with representations of virtual sensors (see Figure 5). Clicking on a virtual sensor would display a recent history of the data downloaded from the actual sensor, using tables and Java graphs. The excercise involved writing a CGI program which linked to the web page. When viewers of the web site initiated a request, the CGI would search through data dumped from the sensors and compile a web page displaying the requested parameters. The remote facilities management research contributed to an understanding of CGI programming structure, Java applets, and a general knowledge of "plug and play" device control and data collection systems such as LonWorks, CEBus, and X-10.

Web browser

Sensor descriptions

Facility description

VRML model (the numbers
represent sensor locations)

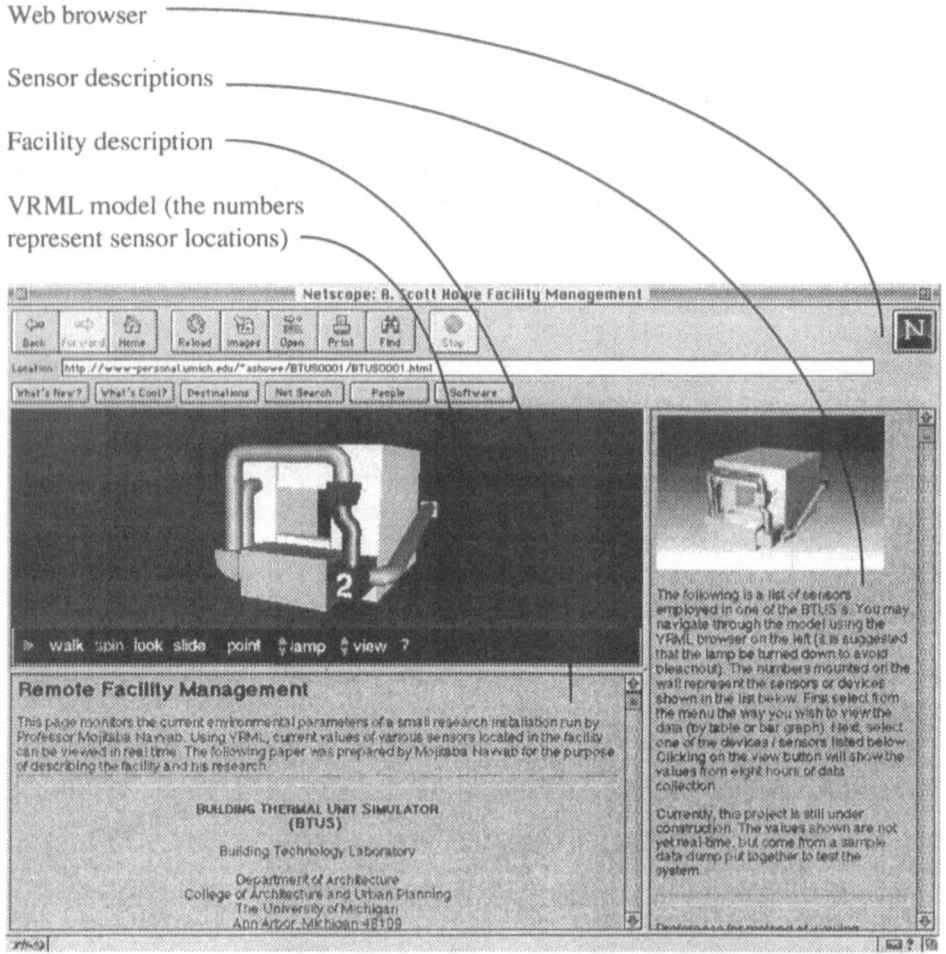

Figure 5: Remote facility management web page

The VBuild project began with a series of C programs which were designed to demonstrate simple network-based modeling functions. A CGI program was written which allowed Internet users to create and view various solid primitives by interactively specifying their parameters on an HTML form. The solids were created by the CGI and returned to the web page for viewing. This excercise proved to be a vital link between the earlier network kit-of-parts experiment completed in the architectural information visualization research and the CGI programming done in conjuction with the remote facility management studies. With the simple web-based solid modeler in place, a study of C++ and the GEDIT modeling classes commenced. The C++ studies included a general familiarization of object-oriented principles and the analysis of advanced data structures such as trees and linked lists. The switchover from C to C++ proved to be extremely valuable in the overall object-oriented / kit-of-parts concept conceived from the beginning. With the C++ foundation, the VBuild classes could be implemented one at a time in order of complexity.

Starting with the POINT class which uses only built-in data types for XYZ coordinates, a series of methods were devised for manipulating points. The methods included creation of point, deletion of point, calculation of distance between points, and the transformation of points. A GEDIT utility class, 3D MATRIX, was adopted early on and facilitated transformation functions. The LINE class built upon the point class, using points as data types representing end points. Methods included the transformation of lines, calculation of segment length, and the equal division of segments. The POLYGON class adopted GEDIT's PG3-SET data structure and borrowed some of its algorithms, such as area and normal calculation. Many other algorithms were added from scratch, such as perimeter calculation and the line extraction function. The SOLID class also adopted GEDIT's PH3-SET data structure and fuctions such as the calculation of volume, surface area, and centroid. Other algorithms were added from scratch, including polygon extraction and spread centroid calculation. In the same way, each of the classes were defined and tested using special programs designed to load data and use each of the methods.

The object-oriented structure of C++ proved to be a tremendously powerful programming foundation. After each class was implemented and tested, the objects they defined could be completely relied upon. In the hierarchical structure of VBuild, the dependence upon lower level classes made rock-solid reliability a necessary requirement. In addition, the opportunity to build upon the tremendous amount of work put into the GEDIT modeling functions by staff and students of the University of Michigan Architecture and Planning Research Laboratory contributed tremedously to the functionality of VBuild.

4. Future Research

Proposed work will initially continue with the implementation of various unfinished functions. A robust interactive web interface will be implemented which will allow the designer to intuitively add, move, position, and delete assembly instances from the model. DXF interpretation will be facilitated, and all CGI programs will be put in place. Once the program is in a usable state, the plug-in will be made available to selected designers.

As a test for the experimental design environment, a small model kit-of-parts library will be prepared with complete fabrication processes incorporated in each component. In addition, an X-10 device will be included in each component's subassemblies. With data prepared, the full sequence of a design / build / monitor simulation scenario will be initiated and documented, using real manufacturers and designers.

Other possible extensions to the experimental software could include collaboration with other researchers and organizations. The development of shape grammars that embody typical configurations and connections could function as generic components much similar to the way ASCII text relates to different font styles. The shape grammar would define the interfaces between components, and the component itself could be designed to a certain style. A family of such components would form a library "font style". Designers would quickly be able to switch between styles due to the fact that each of the libraries are designed according to the same shape grammar. Another possible avenue for joint research could be the development of web-based analysis software and expert systems for evaluating the performance of virtual buildings and constructs. Expert systems could take information based on the underlying shape grammar and inform the designer of inadequacies or violations of pre-engineered parameters.

5. Conclusion

For simplicity's sake kit-of-parts philosophy is utilized in this work. When individual designers begin to use the system they may want to be able to define their own components rather than use those already designed by someone else. Along with the ability to harness various fabricatioin processes for the purpose of facilitating the manufacture of a pre-defined kit-of-parts, real time design and manufacture could be a next step. Regardless of whether the design consists of extemporaneously thought-out elements or pre-designed components, the ability to create a virtual artifact and link it back to the real one should for the purpose of design, manufacture, and facility management prove to be a powerful tool.

The eventual goal would be to have a real building that adapts to the users needs through user-initiated instructions affected by the manipulation of its virtual building counterpart. Initial construction and renovative construction could eventually utilize robotic systems which are either brought in from elsewhere or are actually incorporated into the components of the building itself. Redesign and renovation would be facilitated by changing the virtual building to meet the new needs, and executing the automated construction and assembly features to bring about the changes in the actual building. Though this research does not directly address the use of automated construction, it contemplates its eventual use and attempts to forsee possible preparations for future seamless integration.

NOTES & REFERENCES

A. Scott Howe, architect, is a designer for Kajima Corporation of Tokyo, Japan and a PhD candidate at The University of Michigan School of Architecture and Urban Planning. For more information see Home Page URL: http://www-personals.umich.edu/~ashowe

VRML REFERENCES

Mark Pesce (1995) *VRML: Browsing & Building Cyberspace*, New Riders Publishing, Indianapolis, Indiana.
John R. Vacca (1996) *VRML: Bringing Virtual Reality to the Internet*, Academic Press, London.

OBJECT-ORIENTED PROGRAMMING REFERENCES

Gregory Satir, Doug Brown (1995) *C++ The Core Language*, O' Reilly & Associates, Inc., Sebastopol, California.
Mark Allen Weiss (1996) *Algorithms, Data Structures, and Problem Solving with C++*, Addison-Wesley Publishing Company, Menlo Park, California.
John December (1995) *Presenting Java*, Sams.net Publishing, Indianapolis, Indiana.
Suleiman Lalani, Kris Jamsa (1996) *Java Programmer's Library*, Jamsa Press, Las Vegas, Nevada.
Thomas Boutell (1996) *CGI Programming in C & Perl*, Addison-Wesley Publishing Company, Menlo Park, California.
Apple Computer, Inc. (1994) *Inside Macintosh: Devices*, Addison-Wesley Publishing Company, Menlo Park, California.
James Turner, et.al., *GEDIT solid modeler*, The Architecture and Planning Research Laboratory, University of Michigan, Ann Arbor, Michigan.

KIT-OF-PARTS REFERENCES

Robert W. Marks (1960) *The Dymaxion World of Buckminster Fuller*, Reinhold Publishing Corporation, New York, New York, pp88-91.
Stephen C. A. Paraskevopoulos, Harold Borkin, et.al. (1968) *Research on Potential of Advanced technology for Housing*, Architectural Research Laboratory, The University of Michigan, Ann Arbor, Michigan.
Udo Kultermann (1970) *Kenzo Tange: Architecture and Urban design 1946-1969*, Praeger Publishers, New York, NY.
Moshe Safdie (1970) *Beyond Habitat by 20 Years*, Tundra Books, Montreal, Quebec.
Emilio Ambasz (April 1972) *Italy: The New Domestic Landscape, Achievements and Problems of Italian Design*, The Museum of Modern Art, New York, NY.
Kisho Kurokawa (1977) *Metabolism in Architecture*, Westview Press, Inc., Boulder, Colorado.
Frank Russell (1985) *Richard Rogers + Architects*, St. Martin's Press, New York, NY.
Toshio Nakamura (May 1988) "Norman Foster 1964-1987", *A + U Architecture and Urbanism*, Extra Edition.
Colin Davies (1988) *High Tech Architecture*, Thames and Hudson, London, Great Britain, pp42-55, pp68-85.
A. Scott Howe (1989) "Intelligent Living" and "Intelligent Office" in: *Kawasaki Safety Intelligent Plaza*, University of Utah Master of Architecture Thesis, Salt Lake City, Utah: University of Utah.
Wim J. Van Heuvel (1992) *Structuralism in Dutch Architecture*, Uitgeverij Publishers, Rotterdam, The Netherlands.
Martin Pawley (1993) *Future Systems: The Story of Tomorrow*, Phaidon Press Limited, London, England.
Vittorio Magnago Lampugnani (1995) *Renzo Piano: Progetti e architetture 1987-1994*, Electa, Milano, Italy. Book in italian.

LONWORKS, CEBUS & X-10 REFERENCES

V. Elaine Gilmore (September 1990) "U.S., Japan, Europe: The World's Smartest Houses", *Popular Science*, pp56-65, 102.
Denny Radford of Intellon Corp. (November 1996) "Spread-spectrum data leap through ac power wiring", *IEEE Spectrum*, Vol33 No11, pp48-53.

VIRTUAL BUILDING REFERENCES

James Turner, et.al. (August 1990) "AEC Building Systems Model", *ISO TC/184/SC4/WG1, Document 3.2.2.4*, The University of Michigan, Architecture and Planning Research Laboratory, Ann Arbor, Michigan.
James Turner (May 1991) "Guide to Reading NIAM Diagrams", The University of Michigan, Architecture and Planning Research Laboratory, Ann Arbor, Michigan.
A. Scott Howe, Hirata Yasutoshi (July 1991) "A Feasibility Study for a New Architectural Design Approach Using 3D Solid Modeling CAD Systems", *Fourth International Conference on Computing in Civil and Building Engineering*, Tokyo, Japan.
Edward F. Smith (Summer 1992)"Virtual Buildings: Knowledge Based CAD Models for Design, Analysis, Evaluation and Construction", *Computer Solutions*, pp30-32.
A. Scott Howe, Hirata Yasutoshi (1993) "Architecture, Urban Planning Example 1: Design Using 3D CAD", *IBM CATIA: State of the Art 3D CAD / CAM, Technical Papers '92 / '93*, Tokyo, Japan pp6-9, 36-40.

INFORMATION VISUALIZATION REFERENCES

George W. Furnas (April 1986) "Generalized Fisheye Views", *Human Factors in Computing Systems CHI '86 Conference Proceedings*, Boston, pp16-23.
Benjamin B Bederson, Larry Stead, James D. Hollan, "Pad++: Advances in Multiscale Interfaces", SIGCHI '94 short paper.
Sougata Mukherjea, James D. Foley (April 1995) "Visualizing the World Wide Web with the Navigational View Builder", *Proceedings of the 3rd International World-Wide Web Conference*, Darmstadt, Germany, Computer Networks and ISDN Systems; v 27 n 6, pp1075-1087.
A. Scott Howe (May 1996) "Internet-based Architectural Visualization", presented at the *ACSA European Conference*, Copenhagen, Denmark.

AUTOMATED CONSTRUCTION REFERENCES

Roozbeh Kangari, and Daniel W. Halpin (1983) "Potential Robotics Utilization in Construction" , NSF research report under grant no. CEE-8319498, National Science Foundation, Washington, D.C.
Leonhard E. Bernhold, Dulcy M. Abraham, and Davis B. Reinhart (April 1990) "FMS Approach to Construction Automation" in *Journal of Aerospace Engineering*, Volume 3, No.2, pp.108-121.
Taisei Corporation (2 November 1990) "Announcing the Age of Construction Automation, Now!" in: *Nihon Keizai Shinbun* newspaper article.
Miroslaw J. Skibniewski, and Stephen C. Wooldridge (1992) "Robotic materials handling for automated building construction technology" in: *Automation in Construction volume 1*, Amsterdam, Elsevier, pp.251-266.
Akinaga Makoto (7 June 1993) "Four Firms Start Construction in the Search for a New Architectural Manufacturing Paradigm" in: *Nikkei Architecture*, pp.160-165. Article in Japanese.
H. Kurita, T. Tezuka, and H. Takada (May 1993) "Robot Oriented Modular Construction System - part II: Design and Logistics" in *Automation and Robotics in Construction X*, proceedings of the 10th International Symposium on Automation and Robotics in Construction (ISARC), Houston, Texas, Amsterdam, Elsevier, pp.309-316.
Colin Bridgewater (1993) "Principles of Design for Automation applied to construction tasks" in: *Automation in Construction volume 2*, Amsterdam, Elsevier, pp.57-64.
Kajima Corporation (1994)"AMURAD Grow-up System" in project report, also conceptual robotic system by A. Scott Howe, Tokyo: Kajima Corporation. Report in Japanese.
A. Scott Howe (1994) "A Genesis System" in: *Special Research Report of the Sensitivity Engineering Product Development Research Group: A Sensitively Designed City*, Tokyo: Japanese Ministry of International Trade and Industry, pp.73-92. Report in Japanese.
Japanese Ministry of Construction (1994) "Development of Automated Systems for the Construction of Reinforced Concrete Structures" in: *Ministry of Construction Technology Development Project Collection: New Construction Technology Development for the Construction Industry*, volume 2, Tokyo, Japan Ministry of Construction, pp.14-95. Report in Japanese.
Javier Ibanez-Guzman (1995) "Modeling of on-site work cells for the simulation of automated and semi-automated construction" in: *Construction Management and Economics*, pp.427-434.
Intelligent Manufacturing System (IMS) Collaborative (1995) "Automated Materials Delivery System for High-rise Buildings" in product pamphlet, Tokyo, IMS Collaborative. Pamphlet in Japanese.
A. Scott Howe (April 1997) "Designing for Automated Construction", presented at the Second Conference on Computer Aided Architectural Design Research in Asia, Hsinchu, Taiwan.

REMOTE FACILITY MANAGEMENT REFERENCES

Jeffrey M. Hamer (1988) *Facility Management Systems*, Van Nostrand Reinhold Company, New York, pp79-84.
Ken Sakamura and Richard Sprague (April 1989) "The TRON Project", *Byte*, Vol14 No4, pp292-301.
H. Michael Newman (1994) *Direct Digital Control of Building Systems: Theory and Practice*, John Wiley & Sons, Inc., New York, New York.

SHAPE GRAMMAR REFERENCES

William Mitchell (1990) *The Logic of Architecture: Design, Computation, and Cognition*, MIT Press, Cambridge, Massachusetts.
Shuenn-Ren Liou (1992) "A Computer-based Framework for Analyzing and Deriving the Morphological Structure of Architecture", doctoral dissertation, The University of Michigan.

DESIGN SPEECH ACTS. "HOW TO DO THINGS WITH WORDS" IN VIRTUAL COMMUNITIES

Anna Cicognani and Mary Lou Maher
Key Centre of Design Computing
Department of Architectural and Design Science
University of Sydney
NSW2006 Australia
anna,mary@arch.usyd.edu.au

Abstract
Cyberspace is language based (Cicognani, 1996; Cicognani, 1997; Winograd and Flores, 1986), and so are Virtual Communities (VCs). We propose that VCs are ideal places to experience and enhance a language for design. Design in a VC can actually be performed using speech acts that in-real-life wouldn't perform any design. We call these acts 'design speech acts'. We present, as a starting point, a list of verbs which can be used in a VC for design and the implications of using these verbs to design cyberspace. We present a methodology for structuring and defining design speech acts, so that a language for design in a VC can be subsequently developed. We are developing a specific environment for a virtual community in which designers can articulate their needs and produce text-based design objects.

1. Introduction

Text-based realities are becoming a subject of study by a large variety of researchers. From studies of Computer Mediated Communication (CMC) to sociology of cyberspace, to cyberpolitics, to AI, many researchers share the view that Virtual Communities (VCs), and more generally Collaborative Virtual Environments (CVEs), have found legitimacy for applied research. Virtual Communities have a sense of place as well as being when they exist in cyberspace "cities" or "buildings" (eg LambdaMOO). The metaphor of buildings and rooms provides a strong basis for understanding, navigating, and communicating in cyberspace. We often think of the result of architectural design as the specification of drawings, schedules, and symbols that provide instructions for a builder (often not the designer) to construct the structure. We also think of computer-aided architectural design as facilitating the development of such specifications. A major difference in the design of cyberspace is that the specifications ARE the design, the designer is the builder. This difference means that a language for designing cyberspace needs to be expressive enough to carry the feeling as well as the structure of the places. We claim that computer-aided architectural design of cyberspace, being language-based, needs a language of design to complement the existing languages for communication and navigation.

Speech Acts (SA) are utterances which contain information needed to assert and perform actions, or, according to Austin "things that people do with words". Speech

R. Junge (ed.), CAAD Futures 1997, 707-717.
© 1997 *Kluwer Academic Publishers.*

Act Verbs are verbs used in speech acts utterances, to perform actions. Several studies on speech acts have been conducted since the first apparition of Austin's book "How to do things with words" (Austin, 1962). In particular, Searle (Searle, 1969; Searle, 1971; Searle, 1976; Searle, 1991; Searle, Kiefer, and Bierwisch, 1980) has developed specific explanations on how speech acts and speech act verbs work and perform in natural language. Speech acts are used in everyday life in the form of, for instance, promising, wishing, booking, complaining, forgiving, and so on (Verschueren, 1977; Verschueren, 1980).

In this paper we apply the theory of Speech Acts to text-based Virtual Communities, such as MOOs (MUDs Object Oriented). This follows the research conducted in two different fields, linguistics and design, and joins the two with the vision of drafting a first language for design for and of the electronic space.

Firstly, we will introduce a brief framework to understand how this theory is presented by its first authors (Austin and Searle), and why and how we intend to apply it to VCs. We intend to push the theory beyond the limits of computer languages and abstractions, and beyond some already seen applications of design of programming languages which use speech act theory and natural language interpreters.

Secondly, we will present some examples of Design Speech Acts, within a Dictionary of Design Speech Acts, which we give to students to solve design tasks (and, in a near future, professional designers). We also sustain and develop the idea that "how to do things with words" in a VC can be formulated into "how to design with words."

Finally, we discuss the considerations on the subject of improving and implementing research in this field of text-based virtual realities and design. We are particularly interested to explore possibilities of designing electronic space using text-based linguistic interfaces.

It is intuitive to think about programming language as a performative one. A sequence of commands can make computers "do things with words", literally. The programming language stands between the user and the machine, in such a way that they both can react and interact using the common language. Part of this language can be considered the "structure," or what that language allows to be recorded; and another the "content," or what has been recorded. A set of macros (groups of instructions which facilitate the execution of a command) can help designers to model and reproduce their ideas in a CAD system (content). This set of macros is called metalanguage (Wierzbicka, 1987). The structure of metalanguage in Virtual Communities underlies and supports the possibilities of performing actions. This language is, on the one hand, a programming language, with which the software is able to run and handle the interface; on the other hand, it constitutes the content and the substance of the interaction among users and the interface.

Our goal in this paper is to show:

- how and why speech act theory can be applied to text-based VCs;
- why the electronic space (also called cyberspace, in this paper) can be considered language based;
- how and why a language for design can be created in a VC;
- how such an approach can combine the function and the appearance of the designed objects;
- our choice of Design Speech Acts to be applied to a particular VC (MOO).

For this purpose we introduce a terminology and a point of view about design which may have more in common with linguistics and philosophical theories than GUI or CAD system design approaches. In this paper we do not consider CAD systems interfaces as relevant in the definition of design speech acts in a text-based VC. We recognise that recent developments in CAD and GUI interfaces include the use of natural language in a computer environment. We also recognise that CAD systems focus on the specification of the appearance of the design and lack an intrinsic means of creating a correspondence between the appearance and the functionality of the design.

In other words, the CAD tools create objects which remain graphical metaphors of real-life objects--they need to acquire physical form before being used--whereas design speech acts create at the same time the object appearance and its functionality. Objects designed using CAD systems belong to the metaphor of the interface. In a text-based virtual reality, such as a MOO, the correspondence between the appearance of the object designed and its functionality is complete.

It is beyond the scope of this paper to give an exhaustive analytical review of the debate around speech acts, on a philosophical point of view, and their applications in computer programming. These are, however, of a primary interest both for the us and other researchers in the field.

2. Framework of Linguistic Theories

Linguistic theories address questions about the nature and use of language, its construction and syntax, its reference to 'real facts.' Among these questions, there have been some which interest us such as, "What kind of actions do we perform in language?" and "How does an utterance have meaning?" (Winograd and Flores, 1986) These questions and the resulting theories are relevant here because the use of a design language in a Virtual Community environment has the effect of performing actions that change the environment, and the "meaning" of the words in the language determine directly the effect on the environment. We will focus on Speech Act Theory as the basis for a linguistic theory.

John Austin (Austin, 1962) in his book "How to do things with words" is the first to introduce the idea of Speech Acts (SA), analysing the relationships between utterances and performance. Speech Acts usually appear in the first person, and use the simple present tense, indicative (I promise I'll come tomorrow). Speech Acts are not descriptive; instead they are pronounced to affect an actual situation; they usually do not refer to past events.

A speech act is the action performed by language to modify the state of the object on which the action is performed. It represents an action effectively fulfilled by a sentence:

- I name this ship the Queen Elizabeth
- I pronounce you husband and wife

Austin presents two kinds of utterances: *constative* and *performative*.

"The *constative* utterance, under the name of statement, has the property of being true or false. The *performance* utterance, by contrast, can never be either: it has its own special job, it is used to perform an action. ...

To issue a *performative* utterance is to perform the action: I name this ship *Libertè*, I welcome you, I apologize. ...

The performative must be issued in a situation appropriate in all respects for the act in question: if the speaker is not in the conditions required for its performance (and there are many such conditions), then his utterance will be, as we call it in general, 'unhappy'." (pp.13-14)

The conditions of 'happiness' of performative utterances are important to state how and when utterances are valid, in a real situation. A performative utterance can be void, unhappy, in two ways: 1) the conditions in which the utterance is performed do not satisfy the requirements for the utterance to be successful (I baptize penguins); and 2) the utterance is issued *insincerely* (such as, I am not in the position to utter a certain sentence, but I do. "I fire you" without being in a position which allows me to do so). Also, it is possible that after the utterance is issued, there may be a "breach of commitment", where the speaker doesn't operate under the performance of the utterance.

Austin does not present his theory in a strict logical-symbolic way - such as, for instance, Wittgenstein does, with a definitive scheme to analyse each sentence. Instead, he works with examples of everyday language to show how performances can happen. He admits that the distinction between various kinds of utterances is not always clear, but he claims five general classes of performative verbs:

1. Verdictives, which give a finding or verdict
2. Exercitives, the exercise of a power or right
3. Commissives, which commit you to an action
4. Behabitives, expressing attitudes about social behavior
5. Expositives, which fit utterances into conversations

Searle (1965; 1979) criticises Austin's taxonomy of performative verbs and proposes an alternative taxonomy with the following categories:

1. Assertive: to commit the speaker to something's being the case, to the truth of the expressed proposition (eg I warn you that the bull is about to charge; It's cold here)
2. Directive: attempt by the speaker to get the hearer to do something (I warn you to stay away from my house!; Please give that to me)
3. Commissive: to commit the speaker to some future course of action (eg I promise that I'll come tomorrow; I promise to return)
4. Expressive: to express the psychological state specified in the sincerity condition about a state of affairs specified in the propositional content (eg. I thank you for paying me the money, I congratulate you on winning the race, I am sorry to hear that)
5. Declarations: successful performance guarantees that the propositional content and reality correspond to the world (if I successfully perform the act of appointing you the chairman, then you are the chairman; or "I declare: your employment is (hereby) terminated")

Other researchers have studied, criticised and applied Speech Act theory to both the analysis of natural spoken and written language, and formalisation into logical speech systems. (Auramaki, Lehtinen, and Lyytinen, 1988; Button, 1995; Croft, 1994;

Fishman, 1971; Holdcroft, 1978; Hymes, 1962; Kearns, 1984; Longacre, 1976; Norrick, 1978; Rodden, 1993; Sadock, 1988; Stubbs, 1983; Tsohatzidis, 1994; Vanderveken, 1990; Verschueren, 1977; Verschueren, 1980; Wastell and White, 1993; Wierzbicka, 1985; Wierzbicka, 1987)

The Theory of Speech Acts has been applied, deeply analysed and criticised mainly within the boundaries of linguistic theories. The theory can be applied to natural language, to examine text and discover how the syntax places a Speech Act (SA) in one category. We propose the application of SA Theory to design tasks in text-based Virtual Communities, so that identifiable categories of performance would correspond to specific design acts, or as we call them, Design Speech Acts. We introduce an hypothesis for developing some classes of Design Speech Acts, based on the assumption that the correspondence between a speech performance that produces a design, and the resulting design represent at the same time its appearance and its function. We believe that the renewed strength of this application, against many other unsuccessful applications, has to be found in this correspondence (Button, 1995).

3. Speech Acts Verbs for Text-Based Virtual Communities

Text-Based Virtual Communities are online computer mediated communities in which users (also called players, or characters) interact using various commands and natural language. In this paper, when we refer to VCs, we intend these text-based VCs.

Communication in VCs has been (and is being) broadly studied by both linguists and researchers of Computer Mediated Communication (cf. Journal of Computer Mediated Communication), and by sociologists interested in finding relationships between the change of linguistic register and the cultural formations of the community (cf. Cherny, 1995). The CMC researchers have reported some interesting results on how and why online communities, and electronic communication, have developed a particular set of expressions (a register) characteristic of that medium. Others have demonstrated how the electronic space can be considered as language based (cf. Cicognani, 1997; Winograd and Flores, 1986). However, research on how SAs can produce design and, perhaps, architecture in an online community is still to be developed. We believe that a similar register for design can be developed using SAs. We deal here with a particular kind of Virtual Communities called MOOs. This acronym comes from MUDs (Multi User Dungeons) which are virtual environments in which players chat, build objects, leave traces of their presence. MOOs are MUDs Object Oriented, from the characteristic of their programming language. This choice does not restrict the application of speech act theory to MUDs more in general.

Speech Acts are performed by speech act verbs. As seen above, in English some verbs can be considered performatives (eg. promise, permit, declare, inform), whereas others cannot (eg. walk, laugh, dress, open). We may say that the latter are simply 'descriptive' of a situation: 'I walk home,' for instance, describes the condition of me moving on my feet to go home. The performance is such only when the utterance produces a change of a real situation, when it affects reality in some forms. In-real-life if I say: "Open the door," unless I am giving order to someone (in which case the sentence means: "I order you to open the door"), the door remains closed. In a VC, I may be able to open the door simply typing the command "Open door." We believe that beyond the communicative aspect of SAs in a Virtual Community (such as 'I cheer you' or 'I promise), there is a further aspect of performance in verbs like open, close, walk, take, drop, go, and so on, which affect the virtual reality as much as Speech Acts - intended in Austin's way as communicative acts - affect reality. On the contrary, we think the SAs as known lose (partially or totally) their force and significance when uttered in a VC. For instance, 'I promise I'll be in the MOO tomorrow' is a performative SA which has more to do with an in-real-life situation rather than a MOO one. This distinction is relevant when we try to define categories of Speech Acts in VCs.

We can group verbs in Virtual Communities which may not be considered performatives in spoken language, but they are in virtual environments. Verbs such as open, close, lift, move, are performatives in a text-based VC. We propose here categories of classification for Speech Acts in a text-based Virtual Community. The following categories group some of the basic commands of a MOO. These verbs are part of the database of an experimental MOO (StudioMOO) that the authors are using as support for research activities and education, which derives from the LambdaCore Database. Some categories of classification for commands which issued in a MOO perform actions are:

- communication (say, whisper, emote, page, think, etc.)
- navigation (go, teleport, move, etc.)
- manipulation (open, close, move, give, take, drop, lock, etc.)
- design (create, dig, recycle)

These categories identify four different types of actions in a VC. The communication acts are developed to provide flexibility and expressiveness in text-based communication that mimics the gestures and body language that are used in speech-based communication. The navigation acts provide alternative ways and modes of moving around the VC environment. The manipulation acts allow the user to do things with (and on) the objects in the VC. The design acts are less developed than the other three categories, since so far the emphasis has been on effective interaction with other objects/people in the VC rather than in the design of the VC.

The speech acts in the communication category (say, whisper, etc.) do not take into consideration the performative SA verbs as considered by Austin (such as promise, book, order, and so on). For example, a player issues the following command:

say Hello! I am here, or

say I call this room my Office

In this case 'I am here' and 'I call this room my Office' are the content of the message and not the SA itself. These may not be considered Speech Acts in a VC for the VC's sake, but they may be SAs with reference to the 'outside' (or real life). This distinction is carried through all the categories of speech acts in a VC. The navigation verbs effectively move the user around the space, where the equivalent statement in real life may only be a description of an intention. The manipulation verbs in a VC actually make changes to the object they refer to, again in real life they may only describe an action. The design verbs make changes to the environment, effectively allowing the use of language to include both the specification and the "making" of the design.

4. Classification of Design Speech Act Verbs in VCs

Now we look more carefully at the category of Design Speech Acts in a VC. We first present the verbs that are currently available in VCs that fit in the category of design speech acts, then we look at what is possible. We propose a set of verbs and their classification as categories within the design speech act category as a means of improving the expressiveness of a designer in a VC.

The current LambdaCore Database (the basis from which we started our examination) has very few design verbs. Strictly speaking, there are only two: @create and @dig (the '@' is a syntax required by the software). The @create verb clones any specified 'fertile' object in the MOO. The @dig verb is a specialisation of the @create verb and it is used to create rooms with exits and entrances. The manual assignment of properties (using @set for example) and the programming of new verbs extend the ability to design using the current database 'Designing' means 'changing properties' in a MOO, similar in a way to real life situations where designing is planning the changes of matter, from one status to the other. To design a room using the current database, the specification is:

@dig in|out to anna's office

The room can only be specified as having a name for the room object and names for the entrance and exit. Further elaboration on the design of the room requires the use of the @describe verb.

We have selected some verbs from an English dictionary which are suitable for a designer to perform Design Speech Acts in a text-based Virtual Community. The selected verbs describe several aspects of design, not just the 'creative/programming' part, but also the description, use, organisation, and analysis of the design. The initial dictionary counted more than 17,000 words (not only verbs); we selected about 700 verbs that we can group into the following:

manipulative verbs
- model; verbs which assign shape properties (fold, append, paint, cut)
- create; verbs which literally clone objects and place them in the environment (create, build, construct)
- destroy; verbs which literally destroy objects previously created (blow, explode, crash, recycle)
- change; verbs which modify directly characteristics and properties (elaborate, stretch, scale)
- move; verbs for positioning objects or parts of them in the environment (lift, suspend, raise)
- activate; verbs dealing with the use of the object which change their properties (start, stop, switch)
- copy; duplicating verbs (clone, imitate)
- clear; verbs to assign default properties to objects (wipe, clean)

descriptive verbs
- specify; verbs which assign special characteristics and properties (compare, denote)
- describe; textual description of the object (comment, label, state)
- set or assign; to set attributes

analytic verbs
- organise; verbs which are intended for the organisation (such as the position or the locks) of objects in the virtual environment (plan, combine)
- calculate; evaluating verbs, of help while designing (measure, count)
- associate; verbs which lock objects with other objects (arrange, attract)
- select and search; verbs which help the designers during the creation and modification of properties

The verbs proposed affect one or more properties of a particular object. The above groups are what we understand being relevant for a designer to work in a VC. These verbs are not yet covering all the aspects of design in VCs, neither they are exclusive, and some of them may be considered matching equal functions (eg. change and specify, organise and associate). Each of these groups includes verbs which are of a similar nature, or synonyms. This reflects the nature of MOOs, which makes coincident the function (and functionality of an object) and its appearance.

We can imagine, for instance:

@scale table to <value>

or

@calculate area of kitchen

In the first example, the <value> may also be a verbal description, such as "big" or "bigger" so that a relative dimension is specified. The second example is about an evaluation of the object kitchen which could be useful to the designer.

In real life, performative Speech Acts only need to be pronounced to perform. In a MOO, SA verbs need to be programmed to affect other cells of data (properties), reiteratively calling other verbs, in order to modify the environment. It is like saying that in-real-life performatives change some properties when uttered. For instance, pronouncing someone man and wife, means changing their civil status. Similarly, SAs in a MOO will change specific properties.

It is relevant to note that, as appearance and functionality are coincident in the MOO environment, it is possible for design speech acts to create and describe, with a single action, MOO objects. We believe that a further step in the application of SA Theory can be taken: from the interpretation of the natural language content - as for CMC, to the application of that content for the creation/design of a virtual environment. The analysis of the content, in the case of design SAs, is no more a strict question of linguistic logic, which has demonstrated weaknesses and discrepancies. (Suchman, 1994; Winograd, 1994)

5. Discussion

Speech Act theory can be applied to text-based virtual environments so that design can be facilitated. The background of this application has to be found in the linguistic nature of electronic space. The performance of a command issued in such environments affects specific properties of the "virtual objects." Beyond the communicative content of the commands issued, a design dictionary of Design Speech Act verbs can be defined. We have presented some categories of how these design verbs can be applied in the construction of a design language in a MOO.

In this paper, we have deliberately left out two fundamental aspects of Speech Act Theory, as developed by Searle: meaning and intentionality. (Searle, 1989) According to Searle, the manifestation of the intention of performing an act, is enough for that act to be performed. Moreover, he proposes intentionality as a characteristic to group some SA verbs. SA verbs alone, even though categorised as performatives, may not reach the performance attributed if intentionality by the speaker is missing. Meaning is related to the situation in which the utterance is issued, and to the interpretation by the hearer(s). Design speech acts as proposed do not take in consideration that, for instance, '@create wall' has other meaning a part from creating a MOO object called wall, with special characteristic. We argue that it is probably at this stage that SA theory finds its strength and validity when applied to text-based VCs. We have not entered the debate about interpretation issues, but we understand these as starting points for a discussion.

Another problem encountered is that there are commands (MOO acts) which we cannot classify into any of the MOO categories (communication, manipulation, navigation, design) or the proposed design ones (manipulative, descriptive, analytic), such as '@describe.' In section 4, we have classified the verb 'describe' as a design verb, but it might as well considered a communication one. When a player examines an object (@examine) the textual description of that object is displayed. This is, indeed, part of the communicative aspect. We are aware of the complexity of the classification, in-real-life, of Speech Acts, and we consider Design Speech Acts at least of the same complexity.

Finally, it is relevant to note that, as it is somehow impossible to separate the design from the communication aspect, the interface which supports design SAs in a MOO has a double-faced direction: toward the designer's communicative skills, and his planning ones. The reduction of design skills to linguistic ones might create solutions which only reflect the complexity of a language-based interface.

References.

Auramaki, E., Lehtinen, E. and Lyytinen, K. (1988) "A speech-act based office modelling approach." *ACM Transactions on Office Information Systems* **6** pp.126-152.

Austin, John Langshaw. (1962) *How to do things with words.* Boston: Harvard University Press.

Button, Graham. (1995) "What's wrong with speech-act theory." *CSCW* **3** (1), pp.39-42.

Cherny, Lynn. (1995) "The Modal Complexity of Speech Events in a Social MUD." *Electronic Journal of Communication* **5** (4),

Cicognani, Anna. (1996) "Which language for Cyberspace?" In *Collaborative Virtual Environments 1996*, Conference Proceedings, Nottingham, University of Nottingham, 19-20 September 1996.

Cicognani, Anna. (1997) "On the linguistic nature of Cyberspace and Virtual Communities." In *CVE, Special Journal Issue.* Edited by Dave Snowdon, Nottingham: Submitted.

Croft, William. (1994) "Speech Act Classification, language typology and cognition." In *Foundations of Speech Act Theory. Philosophical and linguistic perspectives.* pp.460-477. Edited by Savas L. Tsohatzidis, London: Routledge.

Fishman, J. A. (1971) *Sociolinguistics: a brief introduction.* Rowley, MA: Newbury House.

Holdcroft, David. (1978) *Words and Deeds. Problems in the Theory of Speech Acts.* Oxford: Clarendon Press.

Hymes, Dell. (1962) "The ethnography of speaking." In *Readings in the Sociology of Language.* pp.99-138. Edited by J. Fishman, The Hague: Mouton.

Kearns, John T. (1984) *Using language: the structure of speech acts.* Albany: University of New York Press.

Longacre, R. (1976) *An anatomy of speech notions.* Lisse: de Ridder.

Norrick, Neal R. (1978) "Expressive illocutionary acts." *Journal of Pragmatics* **2** pp.277-291.

Rodden, T. (1993) "Technological Support for Cooperation." In *CSCW in Practice: An Introduction and Case Studies.* pp.1-22. Edited by Dan Diaper and Colston Sanger, London: Springer-Verlag.

Sadock, J. (1988) "Speech act distinctions in grammar." In *Linguistics: the Cambridge survey.* pp.183-197. Edited by F. Newmeyer, Vol. 2. Cambridge: Cambridge University Press.

Searle, John R. (1965) "What is a Speech Act?" In *The Philosophy of Language.* pp.130-140. Edited by A. P. Martinich, Oxford: Oxford University Press.

Searle, John R. (1969) *Speech acts: an essay in the philosophy of language.* London: Cambridge University Press.

Searle, John Rogers,(1971) edited by. *The philosophy of language.* London: Oxford University Press.

717

Searle, John Rogers. (1976) "A Classification of illocutionary Acts." *Language in Society* 5 pp.1-23.

Searle, John R. (1979) "A Taxonomy of Illocutionary Acts." In *The Philosophy of Language.* pp.141-155. Edited by A. P. Martinich, Oxford: Oxford University Press.

Searle, John Rogers. (1989) "How performatives work." *Linguistics and Philosophy* 12 pp.535-558.

Searle, John R. (1991) *(On) Searle on Conversation.* Amsterdam/Philadelphia: John Benjamin.

Searle, John R., Kiefer, Ferenc and Bierwisch, Manfred,(1980) edited by. *Speech act theory and pragmatics.* Synthese language library. Dordrecht, Holland: D. Reidel.

Stubbs, Michael. (1983) *Discourse Analysis.* #4. Language in Society, Oxford: Basil Blackwell.

Suchman, Lucy. (1994) "Do Categories Have Politics?
The language/action perspective reconsidered." *CSCW* 2 (3), pp.177-190.

Tsohatzidis, Savas L.,(1994) edited by. *Foundations of Speech Act Theory. Philosophical and linguistic perspectives.* London: Routledge.

Vanderveken, Daniel. (1990) *Meaning and Speech Acts.* Cambridge: Cambridge University Press.

Verschueren, Jef. (1977) *The analysis of speech act verbs: theoretical preliminaries.* Bloomington, Indiana: Indiana University Linguistic Club.

Verschueren, Jef. (1980) *On speech act verbs.* Amsterdam: J. Benjamins.

Wastell, D.G. and White, P. (1993) "Using Process Technology to Support Cooperative work: Prospects and Design Issues." In *CSCW in Practice: An Introduction and Case Studies.* pp.105-126. Edited by Dan Diaper and Colston Sanger, London: Springer-Verlag.

Wierzbicka, Anna. (1985) "A semantic metalanguage for a cross-cultural comparison of speech acts and speech genres." *Language in Society* 14 pp.145-178.

Wierzbicka, Anna. (1987) *English Speech Act Verbs.* Sydney: Academic Press.

Winograd, Terry. (1994) "Categories, Disciplines, and Social Coordination." *Computer Supported Cooperative Work* 2 (3), pp.191-196.

Winograd, Terry and Flores, Fernando. (1986) *Understanding computers and cognition: a new foundation for design.* Norwood, NJ: Ablex.

THE ARCHITECTURAL DESIGN OF VIRTUAL ENVIRONMENTS

ALAN BRIDGES and DIMITRIOS CHARITOS
University of Strathclyde

Abstract

The paper discusses the use of precedents from architecture, urban design and film to propose guidelines for the improvement of navigation and wayfinding in virtual environments.

1. Virtual Environments

The predominant use, so far, of virtual reality technology in relation to architecture, has been as a means of visually simulating architectural designs. In this respect, virtual reality may be considered as the ultimate medium for producing representations of architectural designs, as it is the only technology capable of simulating the experience of being and moving within a designed environment, prior to its construction. This paper, however, is concerned with how architectural design may contribute to the design of virtual environments themselves. More specifically, it refers to the problem of designing virtual environments from the point of view of enhancing the users' spatial awareness and, consequently, aiding the task of navigation and wayfinding within virtual environments. The need for addressing this issue is evident in the number of virtual environments which are not very efficient in orientating their operators (and are usually aesthetically displeasing as well).

This study has been carried out in relation to a particular class of virtual environment. World Design Inc., in a report on the design of virtual environments (1993, p.12), classified virtual environments in terms of the represented level of realism. They defined:

- *Hyper-realities*, which aim at representing the material world and reflecting its complexity and in which the creative design element is often limited to modeling interaction with the user.
- *Selective realities*, which are simplified representations of the material world and where some aspects of the environment are distorted or transformed and others are accurately represented.
- *Abstractions*, where the virtual environment represents a very complex material world information or information that has no physical representation. In this case, the designer must conceive an effective

719

R. Junge (ed.), CAAD Futures 1997, 719-732.
© 1997 *Kluwer Academic Publishers.*

abstraction of the complex information (which implies a transformation of meaning of what is being represented) and make this understood and responsive for the operator.

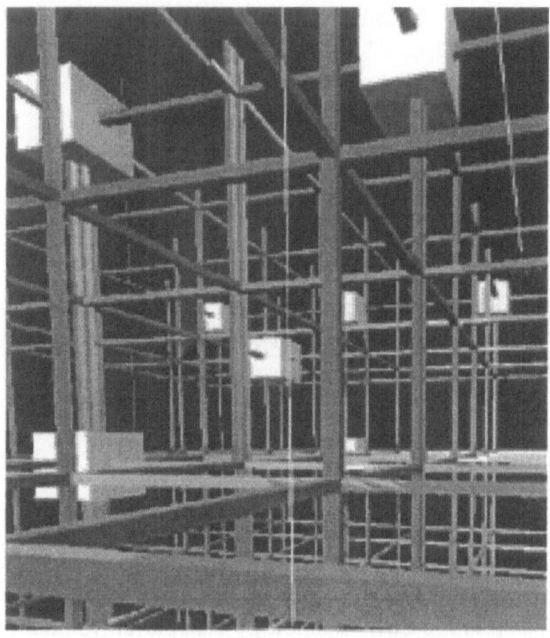

Figure 1. Abstract visualisation of data sets

This paper refers to the third type of virtual environment, "Abstraction". Such a virtual environment may consist of several spatial entities and events which have no real-world counterparts and which accommodate human activities such as navigation, interaction or communication within their three-dimensional, spatial, representational context. To this extent, the design of such a virtual environment is an architectural problem, and this paper considers architecture and urban design as appropriate precedents for informing virtual environment design. Virtual environments, however, have more aspects to their design than purely spatial elements. Time and movement are important and the precedents we draw on here are from cinema.

Metaphor is often employed as a means of abstractly representing complex information sets. Indeed, Bryson (1994) suggests that the employment of a metaphor is essential for the design of such a virtual environment. He refers to the different levels of metaphor that might be considered in a virtual environment:
1. Overall environment metaphor
2. Information presentation metaphor
3. Interaction metaphor

Metaphors, however, may be limiting in that they carry with them associations which might be irrelevant or inconsistent with what is being represented and therefore distracting for the application task.

This paper suggests that employing metaphors is not the only way of facing this problem. Although architecture and urban design are employed as appropriate precedents, an attempt is made to identify generic ways in which these disciplines may inform the design of virtual environments. Due to the lack of real world constraints, elements of space in a virtual environment do not need to resemble any kind of particular real-world spatial elements. However, in the generation of new methods of composing form in order to define space in virtual environments, we believe designers should build on what is known about space in the real world. This paper discusses the composition of space in such virtual environments, addressing the specific problem of wayfinding and navigation in the virtual environment.

2. Navigating Within Real Environments

In considering navigation in real environments, Passini (1992, p.82) suggests that three forms of environmental information (sensory, memory and inferential) are used in order to decide where to go next and how to get there. This environmental information is the essential criterion in determining the wayfinding solution. This paper restricts itself to information of a spatial nature, since it is mainly concerned with wayfinding as a result of the spatial arrangement within the virtual environment.

Passini (1992, p.90) suggests that environmental information may be obtained from various sources, either directly (by means of information booths, signs, maps, etc.) or indirectly (the architectural and spatial characteristics of a setting). This paper mainly focuses on this latter, indirect, information implied by the arrangement of objects in the environment in such a way that the sense of space, conveyed by this arrangement, helps operators anticipate forthcoming events and directs them towards spaces, which are significant for the fulfillment of the application task. Signs, being direct sources of environmental information, are considered, in the context of this paper, more as a part of the objective and static set of environmental objects than as dynamic tools which subjectively refer to each operator. Such specific tools and other direct ways for aiding navigation in virtual environments have been described in Darken and Sibert (1993) and Charitos and Rutherford (1996, 1997). Other aspects of virtual environment design in relation to their characteristics and sensory limitations have been discussed elsewhere (Bridges and Charitos, 1996).

3. Direct Spatial Meaning in the Virtual Environment

The arrangement of objects in a virtual environment provides the operator not only with a purely plastic, experience but with a certain meaning, as well. This meaning may range from a philosophical to a purely practical level, which is informative for her orientation and wayfinding within the virtual environment.

Meaning in the virtual environment is also directly conveyed to us by means of signs and symbols; signs are indicating the past, present or future existence of a thing, event or condition and symbols are vehicles for the conception of things (Thiel, 1961, pp. 45-46). These signs and symbols may be inherent or implicit in the configuration or relationship of the space in the virtual environment or they may be parts of the objects themselves. They may also be visual and/or auditory. It is essential that a correspondence between the outer form and the inner meaning exists, as without this isomorphism the intended message may be misunderstood.

Signs according to Passini (1992, pp.90-92) communicate environmental information needed to make wayfinding decisions; they tell the viewer what is where and they also specify when and how an event is likely to occur.

Passini classifies signs as:
- Directional signs, which designate direction towards a place, an object or an event in form of a name, a symbol or a pictograph and an arrow.
- Identification signs, which are the most elementary state of description of a location, usually perceived when the destination is reached; they identify an object, a place or a character in a virtual environment.
- Reassurance signs, which act as checkpoints after a wayfinding decision is made, to reassure the subject that they are on the right track.

A sequence of directional signs may be employed for the purpose of aiding operators in finding their way towards a particular destination within an unfamiliar virtual environment. Several key issues involved in the use of signs in urban environments are identified by Passini (1992. pp.92-107), many of which are of relevance to virtual environments:
- familiarity with form and design of the sign. Brief, clear and visually structured message and indication of the decision needed to be taken by the operator, are all factors which enhance the effectiveness of a sign,
- signs with a similar message, or those which are a part of the same directional system, should be consistent in their graphic identity and also in the location that they are placed within the virtual environment,
- the continuation of directional signs should not be interrupted or discontinued as this will result in certain disorientation,
- the complexity and intensity of sensory stimulation provided by the surroundings may reduce the reception of information from the operator and the effectiveness of the sign system.

4. The Elements of Space in a Virtual Environment

Virtual environments, as a medium for creating synthetic, interactive experiences, are still in their infancy. When one tries to comprehend the intrinsic nature of this medium and in particular the characteristics of space and time in a virtual environment, other established time-based media such as film (or video) may prove useful precedents. Here, the work of the French philosopher Deleuze on image and movement in film, is used as a starting point for describing the structure and components of the spatiotemporal experience in a virtual environment.

The elements of cinema relate to Virtual Worlds insofar as they operate within a closed system. Following Deleuze (1983), we view the virtual environment as a closed system. Within this system framing is the art of choosing the elements which become part of a set (a relatively closed system) which includes everything which is present in the image. These elements consist of characters and props.

The closed system determined by the frame can be considered in relation to the data that it communicates to the viewer: it is "informatic". Considered in itself and in the nature of its parts, it is geometrical or dynamic-physical. Considered in relation to the point of view it is an optical system, simply related to the angle of framing. Given that there is a frame, there is inevitably an "out of frame". The closed system determines this out-of-frame, sometimes in the form of a larger set which extends it, sometimes in the form of a whole into which it is integrated. In film the out-of-frame implies that a character has not yet arrived, or more pertinately here, that the character is momentarily in a "zone of emptiness" and is invisible. Similarly, a virtual environment may be designed so that the framed part of the set includes cues that imply elements of the out-of-frame part of the set, which may be essential for an application task or to aid navigation.

Cutting is the determination of the shot, and the shot the determination of the movement which is established in the closed system, between elements or parts of the set. Thus movement has two facets: on the one hand, it modifies the respective positions of the parts of the set, which are like its sections, each one immobile in itself; on the other it is itself the mobile section of a whole whose change it expresses. From one point of view it is relative, from the other, absolute. Similarly, movement in virtual environments is experienced as change of position of the set's parts or of the viewpoint (virtual camera) in relation to the whole virtual environment. However, virtual environments differ significantly in that movement within them is largely determined by interaction of the operator with the system, whereas in film it is determined by the shot and consequently by cutting. Movement in a virtual environment is also partly determined by the virtual environment designer, who defines the constraints for navigation and the appropriate input/output devices for interaction.

Nature is framed in a different way to people or things. One critical feature in nature is the horizon. With no horizon as a reference point, everything becomes relative and subject to varying interpretations. Individuals are not framed in the same way as crowds; and sub-elements of scene may appear in various sub-frames. Doors, windows, mirrors, etc. are all frames in frames. In this dovetailing of frames the various parts of the set are not only separated, but may also be seen to converge.

This multiplicity of sub-frames means that the set itself is no longer the object of geometric divisions, but of physical gradations. Ultimately, the film is projected onto a screen which becomes the ultimate frame of frames. In desktop virtual environments the computer screen is the ultimate frame and in immersive virtual environments as well, the field of view of a head mounted display is such that there always exist a certain frame around the display.

The frame is related to the angle of framing. This is because the closed set is itself an optical system which refers to a point of view on the set of parts. These points of view may be extraordinary in real life but, in the film, must appear to be normal and regular - either from the point of view of a more comprehensive set which includes the first, or from the point of view of an initially unseen, not given, element of the first set.

Space is no longer a particular, determined space, it has become any-space-whatsoever. Any-space-whatsoever is not an abstract universal, in all times, in all places. It is a singular space which has merely lost its homogeneity. That is, it has no principal of metric relations or connections between its parts. Linkages may be made in an infinite number of ways. It is a space of virtual conjunction, grasped as pure locus of the possible. What in fact manifests the instability, the heterogeneity, the absence of fixed links in such a space, is the richness in potentials or singularities which are, as it were, prior conditions of all actualisations, all determination.

5. The Components of an Image in Virtual Environments

Continuing this argument, we suggest that the design of a virtual environment may be seen as the act of framing or in other words the determination of a closed system which includes all elements which are present in the image. The "props" consist of the "solid" objects which define the spatial elements in a virtual environment. The "sets" are comprised of the "void" spatial elements in the virtual environment and the characters are the dynamic, animated objects inhabiting the virtual environment.

5.1 PROPS

As is the case in real environments, we survive by orienting ourselves to objects which are distributed in space and which allow for the spatial experience. Similarly a virtual environment consists of visual and auditory objects which bind and subsequently define space. Props are these inanimate objects which define space in a virtual environment, by means of their arrangement. Props may be "landmark" objects (see figure 2), which mainly function as a point of reference and cannot be entered, or "binding" objects, which generally define space (as distinct from void) by binding it in some way.

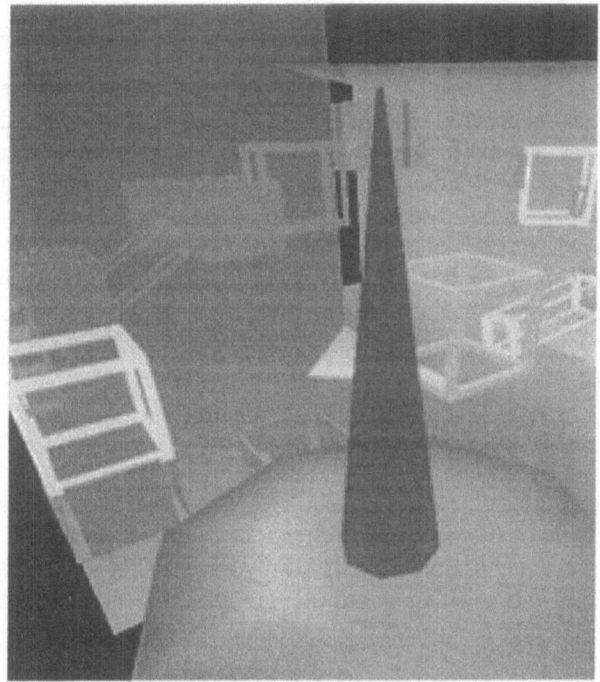

Figure 2. Landmark object

Lynch (1960, p.48) has defined landmarks as point-references within the urban environment which may vary widely in scale. Their use involves the singling out of one element from a host of alternate possibilities. They are frequently used clues of identity and even of structure. Some landmarks are distant and can be used as radial references whilst others symbolize a constant direction. To a certain extent, they may even be mobile if their motion is sufficiently slow and regular, but in this case they may be included in the characters category of elements.

When designing landmarks within urban environments, it is important to know the principle criteria which affect their significance within the context and scale of the environment. When designing a navigable virtual environment, these criteria must be adapted to the intrinsic characteristics of the virtual environment. Following Lynch (1960) and Appleyard (1969), this paper suggests that the following aspects of landmarks contribute to their significance within the context of a virtual environment:

- *Form*. If an object has prominent physical features (sharp contours, clear form, bright surfaces) or definite size disparities between itself and the environment then it more easily influences memory. This is due to its ability to be singled out from a multitude of other possibilities by virtue of its formal qualities.

- *Function* and *Association*. If an object has a high usage or performs a symbolic function, or if for example there is an activity, sign or history associated with it, it may signify a landmark.
- *Location*. If the object is located at an intersection (where path decisions are made), especially in a smaller scale context, then we would associate it as a clear landmark. Smyth et al. (1994, p.312) elaborated on this by concluding that landmarks become positioned in space when they are crossed by many routes, therefore becoming a major part of the organizational framework for the map of the environment.

Essentially, the object should be visible from many locations and also contrast with its background. This "figure-background" contrast does not necessarily refer to the physical characteristics of the landmark, but to any of its characteristics (age, style, etc.), therefore overlapping with the functional and formal aspects.

5.2 SETS

An attempt is made here to suggest a taxonomy of the generic elements of space in virtual environments, which are considered and analysed in terms of their spatial qualities and significance and of the effect they may have on the operator's wayfinding behaviour. These elements together constitute the sets that the virtual environment consists of. They may either be: places, paths, domains or thresholds.

- A *place* is a space where particular activities are carried out. More specifically, a place implies static action taking place within its boundaries. When its boundaries are well defined and the relation between "inside" and "outside" is also clearly defined then the operator may feel safe and secure to engage into an activity in the specific place. A place may be a goal, or a focus for an event or a point of departure towards the rest of the environment. (Norberg-Schulz, 1971, p.20). (See Figure 3).
- A *path* is a kind of space which implies movement, that is dynamic action, and within which directions are always evident, due to the formal qualities of its spatial arrangement. A path consists of a starting point, a direction to be followed through a sequence of places and events and finally a destination. (See Figure 4).
- A *domain* is a subset of the whole environment which consists of a system of paths and places. Domains are mainly thought of rather than experienced, since we cannot directly experience them as a whole. It is necessary to structure an environment by means of domains, in order to be able to conceptualise it as a whole. In real environments, domains (or *districts* as Lynch (1960) has named them) expand in two dimensions, due to physical constraints. In virtual environments, which are devoid of such constraints, domains may expand in three dimensions.

- *A threshold is an* intermediate space which functions as interface or intersection between the other spatial elements. The nature of this interface may range from a state where one space flows into the other to a case where one space clearly ends and the other begins.

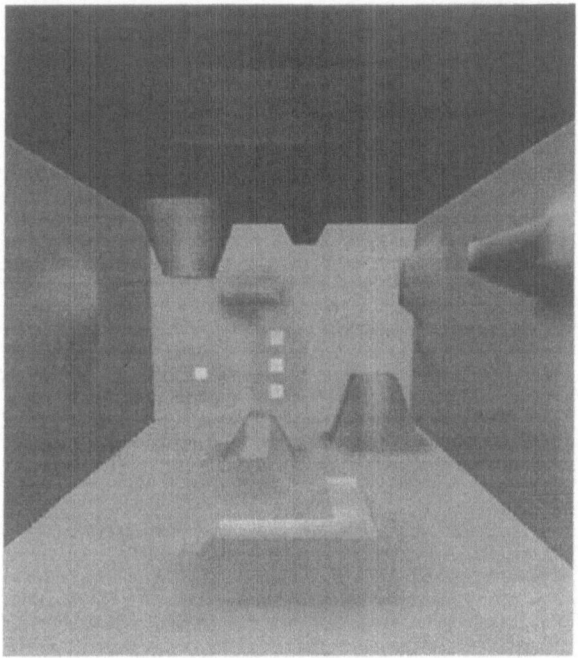

Figure 3. A Place

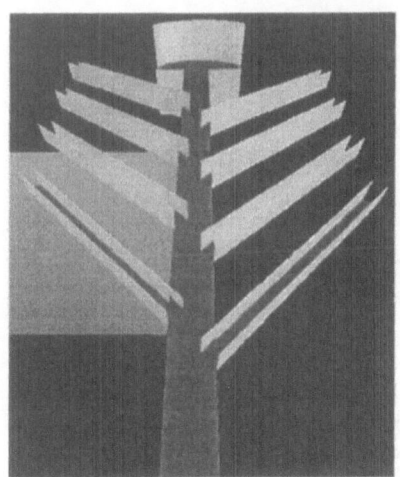 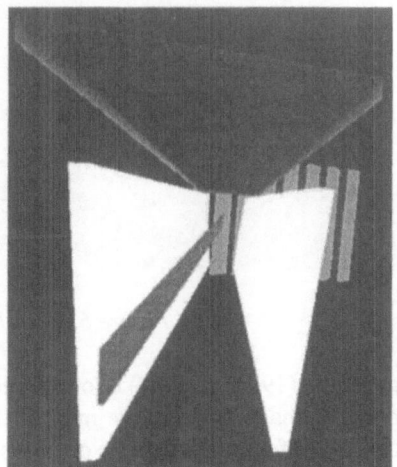

Figure 4. Two examples of Paths

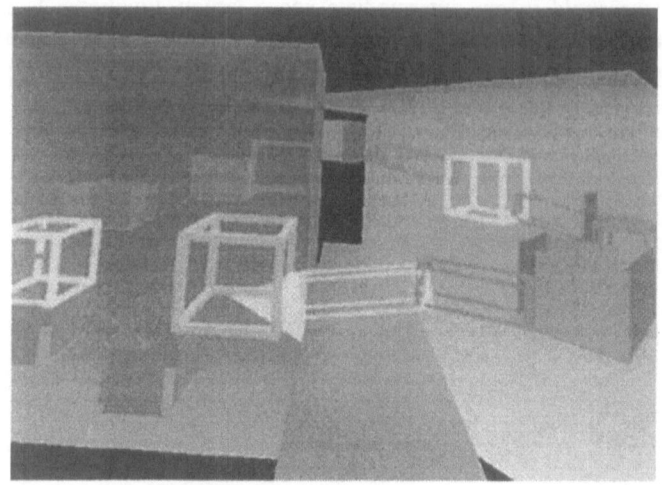

Figure 5. A Domain

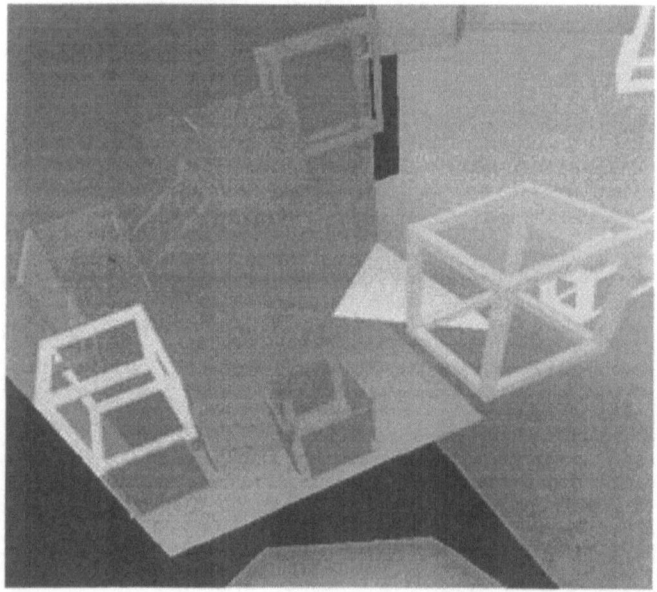

Figure 6. Thresholds

Norberg-Schulz (1971, p.29-30) quotes Lynch (1960) on the fact that the generation of an informative image or cognitive map of the environment is facilitated by the design of distinct and unforgettable places or nodes, paths with a clear end and districts with a particular character.

Thresholds, although void spatial entities, may also establish space in that they partly bind and subsequently define spatial elements, such as paths or places: they designate a gate from a space and an exit to another space. They are the intersections of the possible routes of an operator within the environment. Thresholds afford views to more than one space and involve decision-making (where to go next). The combination of the actual or potential navigation choices and the greater complexity of the scene, coupled with the decision-making process (Gale et al., 1990, p.21) and the action of passing through a threshold make them the foci of route knowledge. When an operator is acquiring spatial knowledge through navigation, more information is being registered in thresholds, than in spaces between them. Thresholds are significant for the action of navigating in a virtual environment, since important decisions for wayfinding are taken there. Therefore, they have to be carefully designed so as to aid the operator's overall spatial awareness.

5.3 CHARACTERS

Characters are dynamic, animated objects. They may exhibit a certain behaviour, or may move within the virtual environment, or respond to an action of the user. They may also have intelligence.

6. The Form of Spatial Elements and its Significance for the Operator

Spaces may be defined implicitly or explicitly. Thiel (1961) classifies spaces by their degree of spatial definition as
- vagues - spaces vaguely defined by objects in a random or statistical distribution. Vagues are of an indefinite and ambiguous form.
- spaces - areas of intermediate degree of explicitness, more or less implicitly suggested and of a fluctuating quality.
- volumes - explicit, completely defined spaces resulting from the use of complete and contiguous surfaces in all positions.

The quantity of space-establishing elements is not necessarily an indication of the explicitness with which a space is defined; the form of the elements and the manner in which they are arranged in space is generally more important. Spaces and Volumes have a perceptible form and according to this form they imply a particular response from the operator.

730

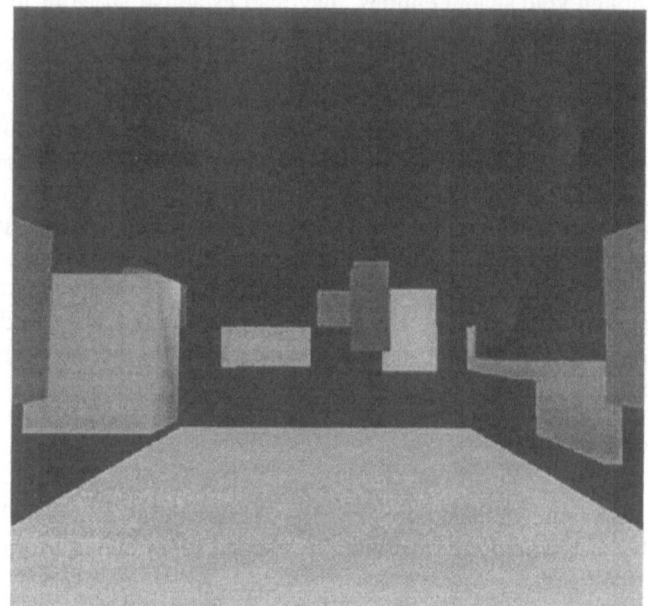

Figure 7. A Vague

The form of a place, of course, affects the way that we experience it. "A place that is being experienced as an 'inside' should generate a spatial sense of proximity, centralisation and closure." (Norberg-Schulz, 1971, p.20). The properties of the enclosure (dimensions, shape, configuration, surfaces, edges, openings) determine the qualities of the place (form, proportion, scale, definition, colour, texture) (Ching, 1979, p.175) and accordingly how we experience it. A centralised form means concentration and results in limiting the size of a place.

A space or volume which has any one overall dimension two or more times greater than any other dimension is called a run. This type of space implies dynamic action, that is movement, and is likely to be a path of some sort (horizontal or vertical). The rest of the spaces are named by Thiel as areas, and following the above taxonomy are likely to be places, if they are clearly defined by means of binding objects. Such spaces imply static action generally relating to a centre or the periphery of their spatial configuration.

Referring to this distinction of form quality, Keppes (as quoted in Thiel, 1961, p.41) suggests that our environments mainly consist of configurations which may be seen as either patterns of a static character, with order, closure, a tendency towards a centre, cohesion and balance or, alternatively, patterns of a dynamic character, exhibiting mobility, freedom, change and opening. These characteristics may be recognised in almost every pattern emanating from real-world visual sensory stimuli and we become aware of them as a result of the effect they have on our motivations, feelings and states of mind.

Similarly, Norberg-Schulz (1971, p.26) mentions this distinction as the basic dichotomy between the concepts of place and path which is experienced as the tension between centralisation and longitudinality in our environments: "Whereas centralisation symbolises the need for belonging to a place, the longitudinal movement expresses a certain openness to the world, a dynamism which may be physical as well as spiritual." Quite often the loci of tension between centralisation and longitudinality are thresholds.

7. Conclusion

The guidelines outlined in this paper are presented as hypotheses. The next stage of this project is the evaluation of the effectiveness of the guidelines. This will be done by practical experimentation.

A virtual environment functioning as a three-dimensional interface to a repository of images and sounds has been designed following the guidelines suggested by this paper. User feedback will indicate whether the way that space in this virtual environment has been designed does inform operators sufficiently for them to navigate within this representation of data sets. Aspects of preliminary models for this virtual environment are presented here to give a visual indication of the generic elements of space in such an environment.

8. References

Appleyard, D. (1969) Why Buildings Are Known, Environment and Behaviour, 1, 131-159.

Bridges, A.H. and Charitos, D. (1996) On Architectural Design in Virtual Environments, Proceedings of the 2nd Creativity and Cognition International Symposium, Candy, L. And Edmonds, E. (eds), Loughborough University, April 96, 184-192.

Bryson, S. (1994) Approaches to the Successful Design and Implementation of VR Applications, Proceedings of the Virtual Reality Applications Conference, June 94, British Computer Society, Leeds.

Charitos, D. and Rutherford, P.(1996) Guidelines for the Design and Exploration of Virtual Environments, Proceedings of the 3rd UK Virtual Reality Special Interest Group Conference, July 96, De Montfort University, Leicester, 93-111.

Charitos D. and Rutherford, P. (1997) Ways of Aiding Navigation in VRML Worlds, Proceedings of the 6th International EuropIA Conference, Coyne, R., Ramscar, M.,Lee J. and Zreik, K. (eds), April 97, Edinburgh, 119-132.

Ching, F.D.K. (1979) Architecture: Form, Space and Order, Van Nostrand Rheinhold, London.

Darken, R.P. and Sibert, J.L. (1993) A Toolset for Navigation in Virtual Environments, Proceedings of ACM User Interface Software and Technology 1993, 157-165.

Deleuze, G. (1983) Cinema 1, L 'Image-Mouvement, Les Editions de Minuit, Paris, translated by Hugh Tomlinson and Barbara Habberjam as 'Cinema 1:The Movement Image', Athlone Press, London, 1986.

Gale, N., Golledge, R., Pellegrino, J.W. and Doherty, S. (1990) The Acquisition and Integration of Route Knowledge in an Unfamiliar Neighborhood, Journal of Environmental Psychology, 10, 3-25.

Lynch, K. (1960) The image of the city, MIT Press, Cambridge, Mass.

Norberg-Schulz, C. (1971) Existence, Space and Architecture, Praeger Ltd. New York.

Passini, R. (1992) Wayfinding in Architecture, Van Nostrand Reinhold, New York.

Smyth, M.M., Collins, A.F., Morris, P.E. and Levy, P. (1994) Cognition in Action, Lawrence Earlbaum Associates, Hove.

Thiel, P. (1961) A sequence-experience notation, Town Planning Review, vol.32, April.

World Design Inc. (1993) Designing Virtual Worlds, presented in Meckler VR 93 Conference, San Jose, June 93, adapted from WorldDesign's 'A Report to the Evans and Sutherland Computer Corporation: A Virtual Worlds Guidebook' published in April 93.

THE LEGACY OF SURREALISM IN THE ELECTRONIC DESIGN STUDIO

RICHARD COYNE and FIONA MCLACHLAN
Department of Architecture
University of Edinburgh

We examine how Surrealist themes are evident in the world of information technology, and in the electronic design studio. We show that much of the current popular appeal of the computer in design schools is attributable to the computer's apparent surrealistic possibilities rather than the potentialities traditionally put forward by exponents of formalism, design methods and systems theory. We discuss developments on Surrealism, including the application of Freud's concept of the uncanny and Lacan's understanding of the image, before concluding that Surrealism and its developments support a "hermeneutics of suspicion," which is one way of interpreting what occurs in the design studio.

Surrealism began as an art movement in art and literature in the 1920s. The movement was formally disbanded in the 1960s with the death of its founder, Andre Breton. But Surrealism has left a strong legacy in three areas pertinent to design. First is the design studio tradition, particularly where pedagogy incorporates abstract exercises and explorations, the appropriation of found objects, strange juxtapositions of elements, objects and materials, montage, collage, and the language of shock, reversal, dialectic and deconstruction. The second area is popular culture. Although initially a movement of the avant garde, many of the tenets of Surrealism have found expression in the mass media, particularly through the early "commercialisation" of Surrealism by Dalí, the films of Alfred Hitchcock, the comedy of *Monty Python's Flying Circus*, certain genres of television commercial and music videos. Surrealism also informed the psychedelic cults of the 1970s, and it is easy to find links between the Beetles in the 1970s and the avant garde. Third, the tenets of Surrealism seem to pervade aspects of what is loosely termed "IT culture" or the "discourse of cyberspace," particularly in the way some cultural critics present the mass media, computing, and the medium of the Internet and the World Wide Web.

Surrealism does not feature prominently in architectural discourse, but the Surrealist legacy is advanced through the reflections and critiques of cultural theorists influenced by the Frankfurt School, such as Baudrillard, and poststructuralists, particularly those influenced by Derrida. Some commentators argue that the influence of Surrealism, and its developments in the late modern age, has eclipsed, or perhaps subsumed, those of other art movements, such as abstract expressionism, cubism and functionalism.

Surrealism in turn draws on several traditions, including the culture of the absurd and the carnivalesque. According to Esslin's (1961) study of the absurd in theatre the behaviour of the clown "arises from his inability to understand the simplest logical relations." (p.235) In such plays the unities of time and place were violated, and the

R. Junge (ed.), CAAD Futures 1997, 733-748.
© 1997 *Kluwer Academic Publishers.*

workings of dreams, hallucinations, and paradox prevailed. Then there is also the tradition of "verbal nonsense" as exemplified in the writing of Lewis Carroll, which, according to Esslin, aims at the destruction of language by naming things in an arbitrary way, which in turn expresses a mystical yearning for unity with the universe. (p.248) There is also the influence of the theatre of the absurd, which called on sources as diverse as vaudeville and Nietzsche.

The cyberspace discourse inherits the legacy of Surrealism in popular culture, but also amplifies it. People seem to be rediscovering and reinventing Surrealism in a new medium. We outline some of the Surrealist themes here, and show how they are realised in the IT world. A cursory browse of the Web reveals a plenitude of overtly Surrealist imagery. We give examples of how these concepts interact with the work of architectural designers using computers. Work shown here is by students involved in the design of a museum. The projects encouraged students to address the changing notions of exhibition, participation, the mass media, and the body. Students inevitably used the WWW in gathering information and exploring precedents, and used image processing, 3d modelling, rendering and multimedia tools, as well as manual media. The Surrealist theme did not occur to us until after the projects were completed, and the concept was never raised with the students. The Surrealist influence pertains to the presentation as well as the design of buildings. Clearly, the computer is caught up in a field of influences that includes Surrealism, and this field interacts with the program, the emphasis of the school, and the architectural culture in which the designers participate.

1. Surrealist Themes

(i) Some IT commentators claim that when immersed in the Web we are barraged with data from various sources juxtaposed in apparently random ways. The Web is formed much as the Surrealists considered montage or collage—as the juxtaposition of objects out of their normal contexts. For the Surrealists these included the juxtaposition of the torso of a woman, a tuba and a chair depicted as clouds floating above the ocean (a painting by René Magritte), or a gramophone with legs protruding from the horn in Domínguez's *Never*. Surrealism shares with structuralism the Hegelian appropriation of difference, realised through the juxtaposition of image against image. According to Chénieux-Gendron (1990, p.65), Surrealists eschewed the literal expression of comparisons (the use of "like," "such as," "just as"), preferring the enigmatic, "grammatical indeterminacy" of phrases such as "the hand holds the night by a thread," or "the ruby of champagne." As invented and explored by Max Ernst, collage functions in much the same way, involving "irrational" juxtapositions of ready-made elements. Such theories resonate with the tenets of structuralism, and of metaphors as creators of meaning. The concept of juxtaposition is a celebrated feature of studio teaching and practice, which finds ready amplification through computer imaging tools (Figures 1 and 2). Studio commentators such as Novak (1995) claim that "morphing" has replaced collage as the operative mechanism in the electronic design studio: "True to the technologies of their respective times, collage is mechanical whereas morphing is alchemical." (p.46) Both concepts belong within the Surrealist repertoire, including the concept of alchemy, particularly the transformative, alchemical function of words and images.

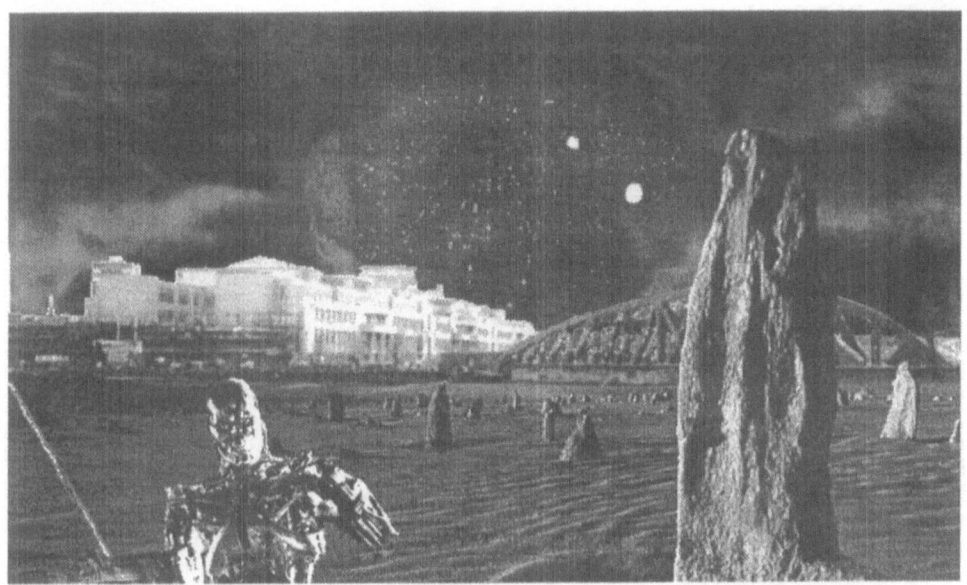

Figure 1. A "site study" exploiting collage, by Adrian Shilliday

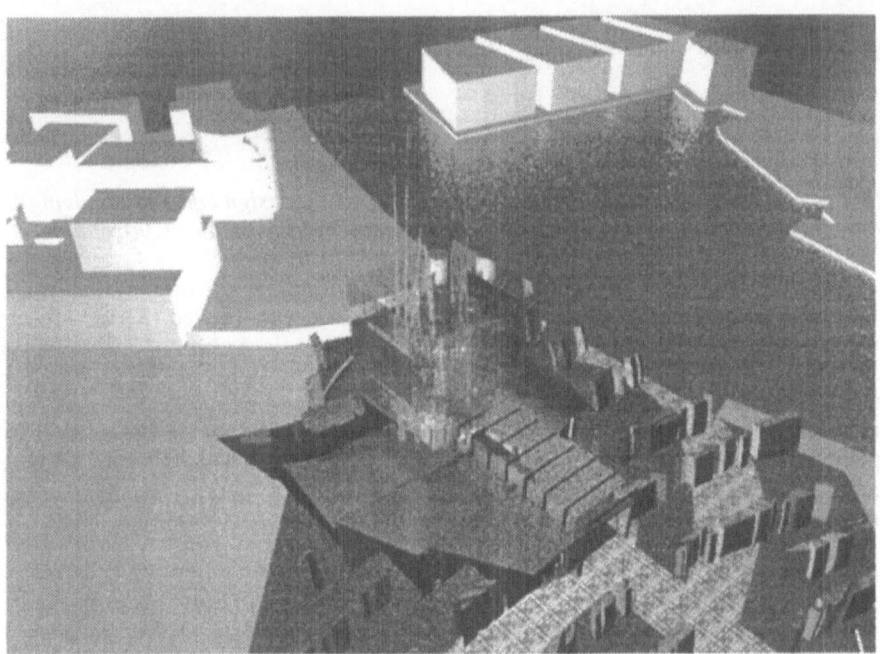

Figure 2. The appropriation and juxtaposition of elements from around the site, including cranes and other hardware of the waterfront, by Nico Warr.

(ii) The Web involves a juxtaposition of media: text, sound, images, and movies. Surrealism was concerned with art in all its manifestations: poetry, painting, sculpture, photography and film, and Surrealist exhibitions would include all of these, combined in diverse ways. The concept of networked multimedia further animates this theme (Figure 3).

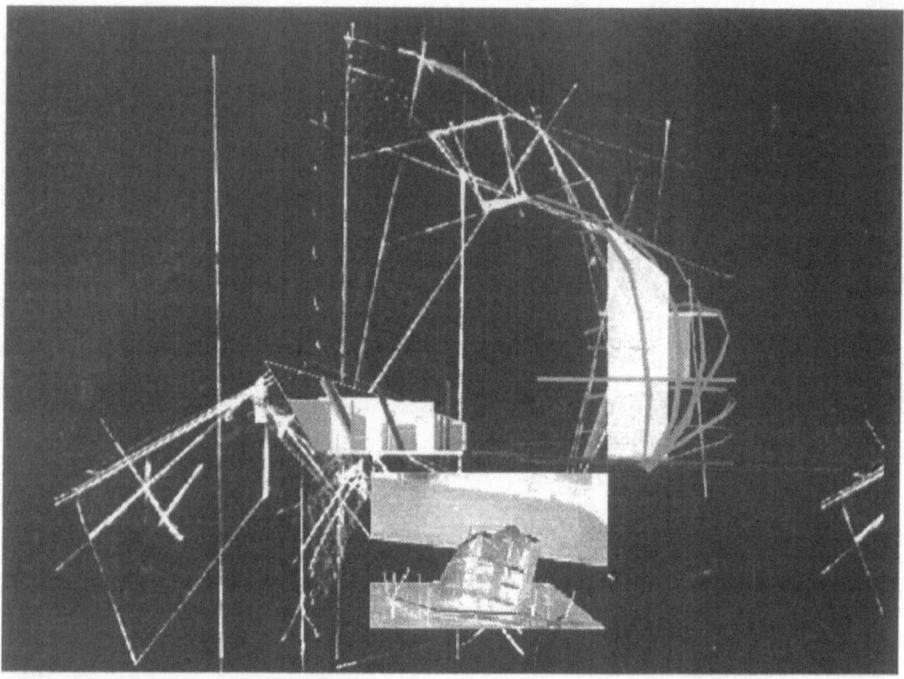

Figure 3. Mixed media presentation and exploration of a design concept involving line drawing, 3d modelling, physical model making and digital video, by Claire Robertson.

(iii) To search the web is to explore a vast "city" within which one stumbles across strange objects and encounters surprise. The Surrealists were excited by this aspect of vibrant cities such as Paris, as evident in Walter Benjamin's *Arcades Project* and Andre Breton's novel *Nadja*. The artist is to be in a state of active expectation. This theme of discovery pervades computer games, such as *Myst*, though one of the ultimate sources of dissatisfaction with such media is the finitude of discoveries, all of which have to be pre-planned. The Web offers no such limitations. The Surrealist legacy already informs concepts of the architectural manipulation of space and the invocation of a sense of place, and even alienation. Working with computers seems to bring these issues to the fore in design (Figure 4).

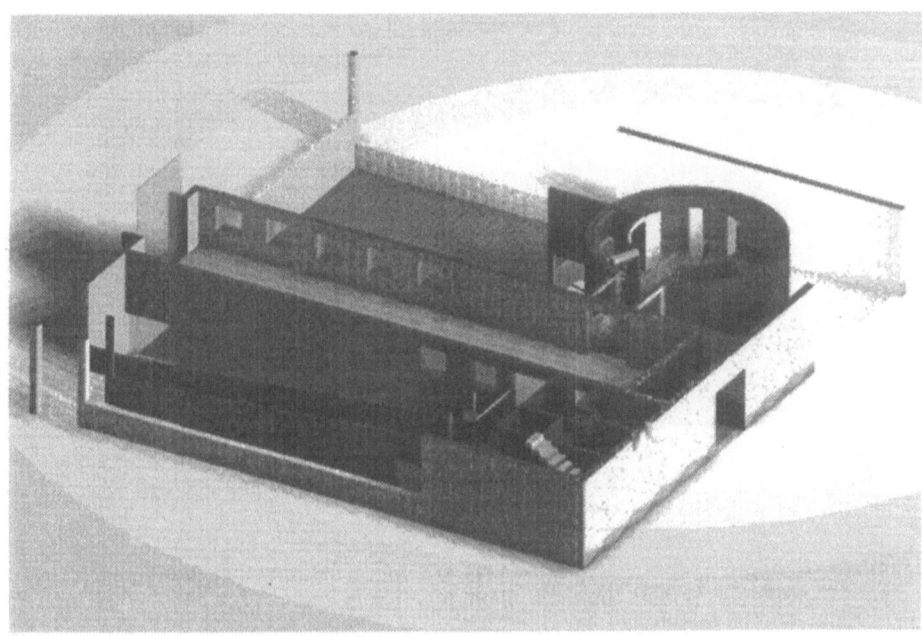

Figure 4. Design for a museum in which concepts of expectation, discovery and exploration were at the fore, by Manuel Figueroa.

(iv) In accord with the Surrealists' interest in the world of imagination and dream, the computer (the Web, computer games, and electronic role playing) seems to provide opportunities to celebrate the marvellous, dreams, fantasy, and the labyrinthine. The Surrealists saw correspondences between the production of art and the dream state, valorised the imagination, fairy tales, and the state of childhood. Surrealist art was inspired by the symbolism of dreams, the latent content of which could be revealed through psychoanalysis. But according to Alexandrian (1970), in all this: "Surrealism cannot accurately be described as fantasy, but as a superior reality, in which all the contradictions which afflict humanity are resolved as in a dream." (p.49) The task of the Surrealist writer, painter and sculptor was that of "calling existence into question." (p.50) In his *Manifestos*, and quoting Baudelaire, Breton (1972) claims that Surrealism "acts on the mind very much as drugs do. ... It is true of Surrealist images, as it is of opium images, that man does not evoke them; rather they 'come to him spontaneously, despotically.'" (p.36) In a similar vein, some IT commentators report that in virtual reality: "There is both a fluidity and speed of movement that are more akin to dreams than waking life." (Franck, 1995, p.20) Computer imagery provides opportunities for the exploration of transparency and ephemera, which can also feed into the design process. The designs in Figures 5 and 6 were animated by notions of narrative and personal journey, and resonate with the Surrealists' reverence for dream.

738

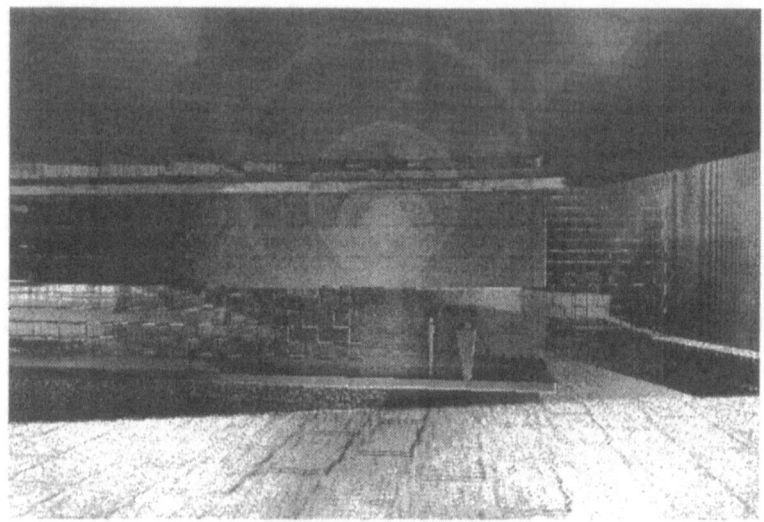

Figure 5. A building that places store on unreal glowing surfaces and the effects of light, by Dulcimer Taylor.

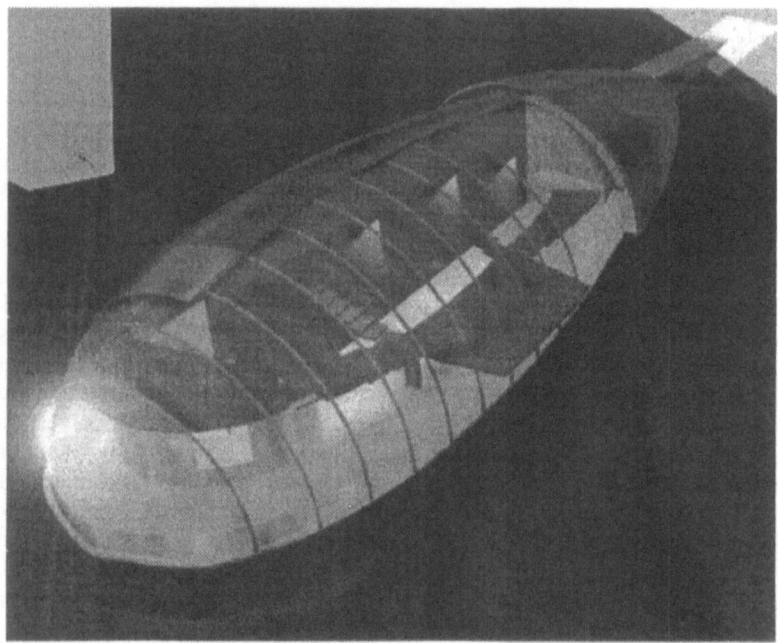

Figure 6. A design around the theme of a journey of discovery in the first person, by Asrul Sani Abdul Razak.

(v) Designers can use ray tracing and radiosity software for creating synthetic photorealistic imagery, and image manipulation software is used for presenting and experimenting with the photographic image. Computer imagery brings the tension between the real and the imaginary into sharp relief in ways that would have delighted (or horrified) Salvador Dalí (Figures 7 and 8).

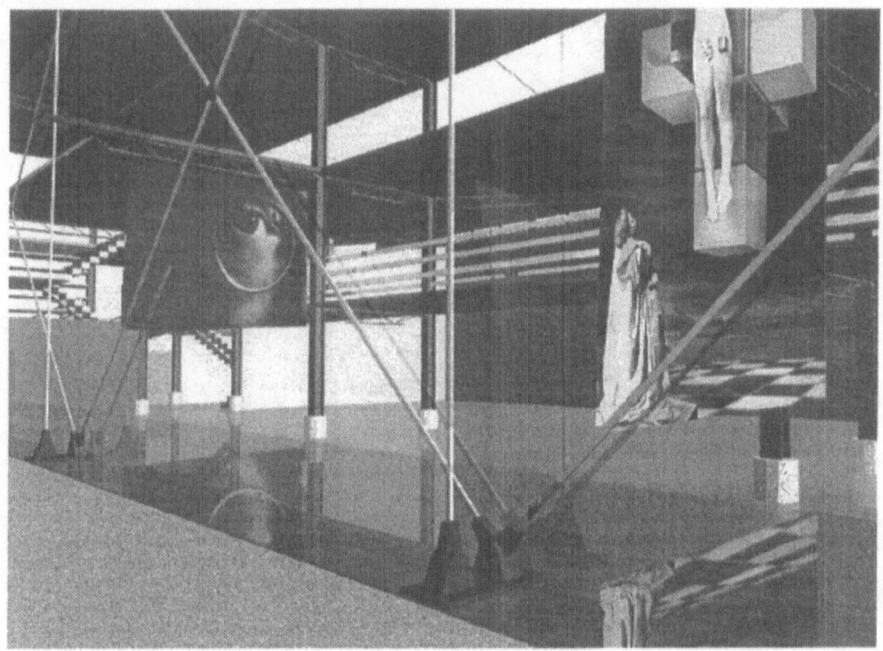

Figure 7. A photorealist/surrealist interior by Simon Hamilton.

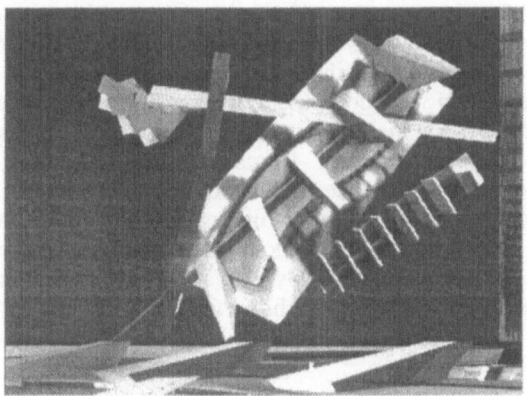

Figure 8. A familiar plan view in an unfamiliar orientation and in photorealistic mode, by Max De Rosee.

(vi) Aspects of IT culture, cyberpunk and the Web, seem to trade in the carnivalesque (Figure 9) and "unreason." At the very least, the Web treats information and imagery in a fluid and non-deterministic way.

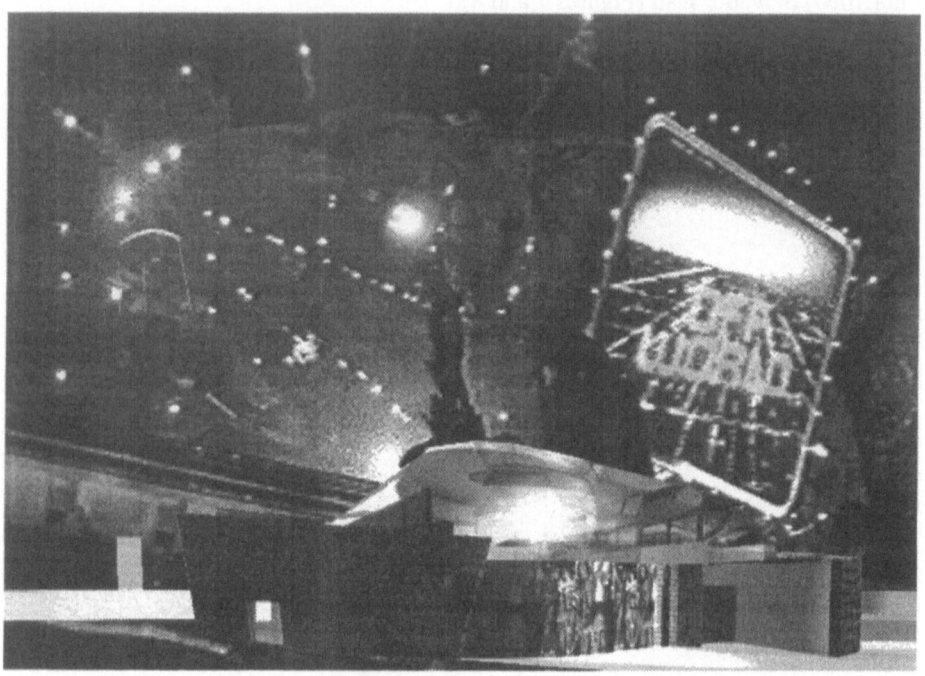

Figure 9. Collage, appropriation and carnival, by James Barnett.

The Surrealists were explicitly interested in the emergence of creativity from random behaviour, as exemplified in early experiments in automatism (automatic writing): the generation of random streams of words to gain access to the subconscious. The Surrealists were also interested in the machine as a means of generating chance occurrences. Designers who use computers commonly report that they have been constrained or enabled by what the computer allows them to do, and some, as in Figure 10, exploit the chance occurrence of forms and configurations presented by the idiosyncrasies of the computer interface, data structures and so on. Of course, the use of the computer to generate form has a long history, though the formalist and systems approaches generally deny the value of randomness. Recent generative experiments that may be construed as surreal are reported by Novak (1995) and Frazer (1995), and we show a design student's generative approach in Figure 11.

741

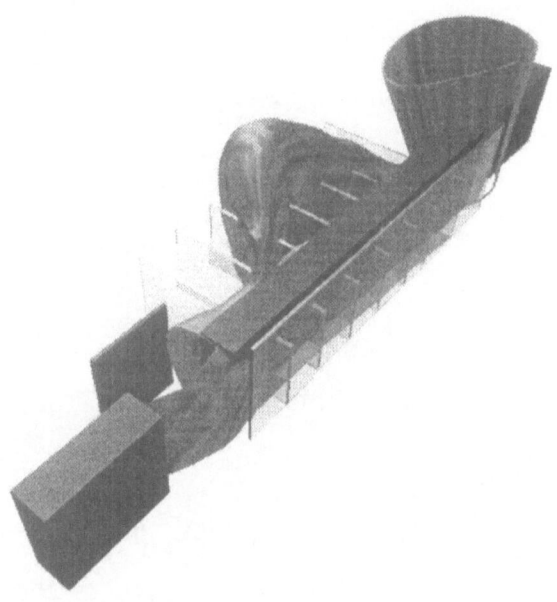

Figure 10. Forms suggested by the capabilities of the modelling software, in a concept design by Magnus Weightman.

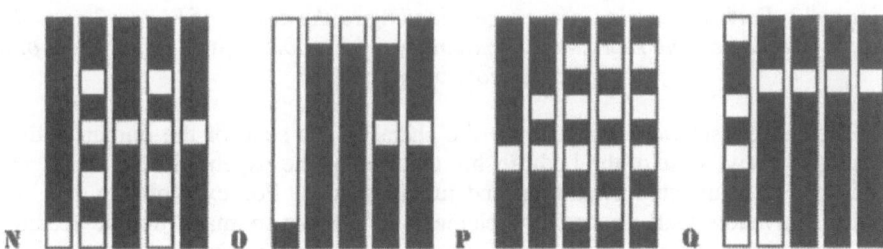

Figure 11. The use of a generative schema for exploring layouts by Rodrigo O'Malley Diez.

(vii) According to many IT commentators, and certainly the advocates of cyberpunk, IT is thought to challenge various empiricist "conventions," including our sense of identity, subjectivity, and the nature of reality. According to Frazer (1995, p.76), "Virtual reality has caused us to reassess reality ... the transcendence of physicality in the virtual world allows us to extend our mode of operation in the physical world," a theme illustrated in Figure 12. Surrealism was founded on similar radical ambitions.

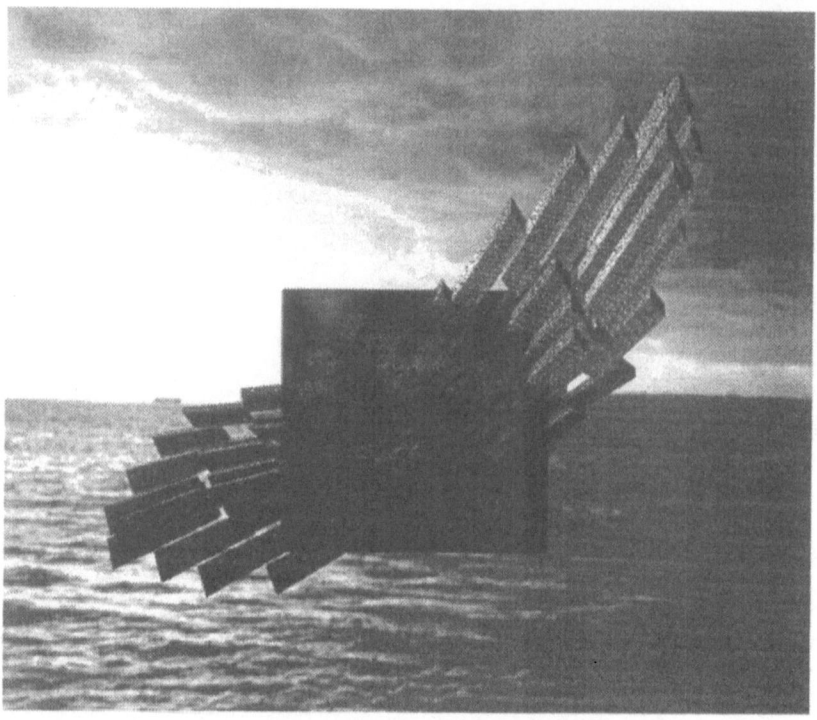

Figure 12. Early expression of a concept leading to the design of Figure 2, in which there is full play on the real and the imaginary, with priority given to the flight of the imagination, by Nico Warr.

(viii) The Surrealists were intrigued by the character and state of the human body, and appropriated Freud's use of the body in his theories of the psyche. In Surrealist art the body is placed in unusual situations and juxtapositions. For example, in one of his paintings Salvadore Dalí arranged the elements of a room to make up the features of Mae West's face. There is Vitor Brauner's strange use of the eye in *The last journey*, featuring a man sitting on a giant eye, and René Magritte's *Philosophy in the boudoir*, featuring a night dress with breasts attached and a pair of high healed shoes with fleshy toes. For Surrealism, the body provokes, and it is highly sexual. Similarly, the human body regaled in a data suit has informed contemporary cultural commentary. Harraway's (1991) concept of the "cyborg" is a provocation for radical action, and spawned the critical genre known as "cyberfeminism."

Giving little thought for surrealism at the time, we organised one of the design projects around the theme of a "Corporium," a museum for the exhibition of body technologies. The theme of the body seemed to fit very comfortably with the emerging surrealist themes of the design work. The design of Figure 13 treats part of the building as a prosthetic arm that reaches over the water and mirrors the "prosthetic" nature of the port of Leith, the site of the project. Figure 14 hints at the ambiguity of the body as the site of the visual sense and the body as object of display.

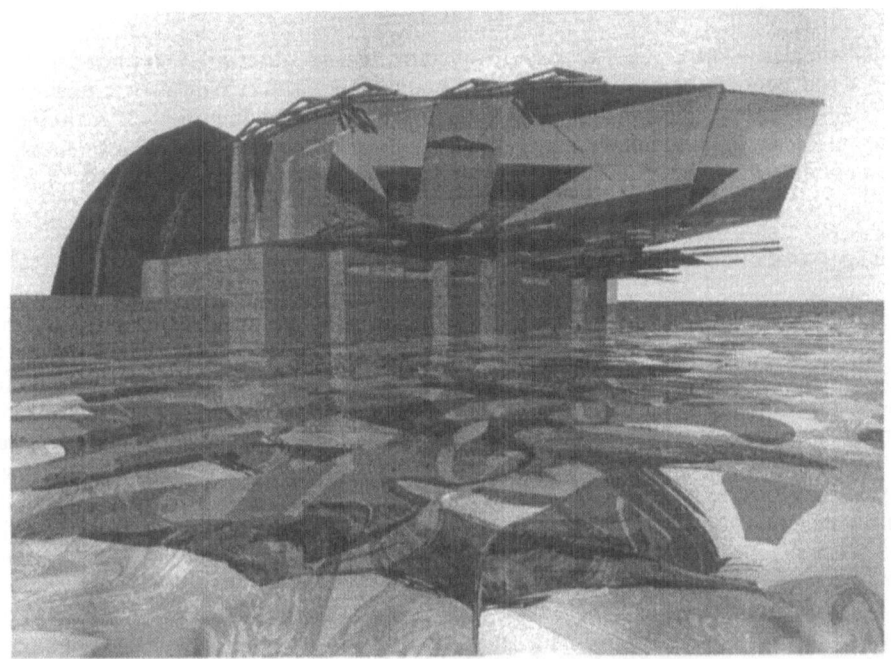

Figure 13. Building as prosthesis, by James White.

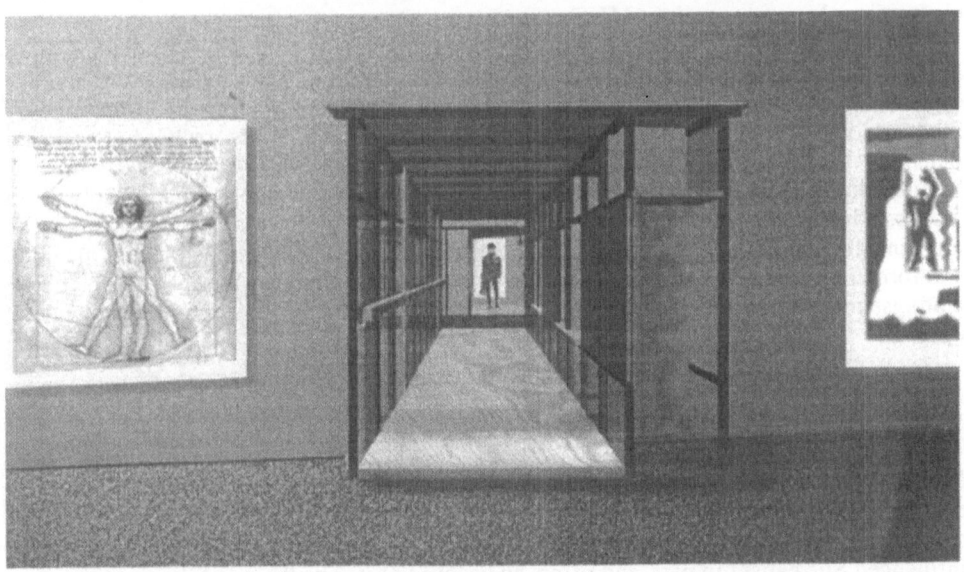

Figure 14. Body, gaze, reflection, image, by Eleanor Egan.

Surrealism also adapted the methods of mediums and spiritualists, and played with the Neoplatonic concept of ecstasis, the out-of-body experience, which also resonates with the cyborg literature. The design in Figure 9 was informed by this theme, involving the metaphorical progression of the museum visitor from a state of sensory awareness to a state in which they were "out of their bodies" in a virtual reality environment.

2. Computers and Romanticism

Surrealism, of its own admission carries the trappings of the romantic movement. There is a contempt for the strictly logical, an emphasis on individual genius, a focus on the subject, the power of the imagination, and so on. It broke away from, and outlived the anti-art movement known as Dada, which sought simply to shock and scandalise. Surrealism had a political edge. According to Alexandrian (1970), Surrealism declared the rights of fantasy against "a world racked by war, with boring dogmas, with conventional sentiments, with pedantry, and the art which did nothing but reflect this limited universe." (p.29) It was a highly politicised, inflammatory movement which had a radical concept of freedom. It aimed to liberate the resources of the unconscious mind through art.

In spite of its claims to radicality, aspects of IT culture seem to be as firmly grounded in the romantic movement as Surrealism. The WWW is currently populated with home pages in which individuals assert themselves as dreamers, free spirits, and fountainheads of original ideas that can at last be published to the world. People also claim that they can assume and experiment with different identities in elaborate electronic role games, such as MUDs.

Engagement with computers as a Surrealist enterprise is one of many factors at play as designers use computers, which is also informed by the design methods movement, systems theory, formal geometry, empiricism—in architectural terms the trappings of functionalism and modernism. But the Surrealist influence presents itself as at variance with this "rationalism." In 1995, the journal *Architectural Design* published a special feature edition entitled *Architects in Cyberspace*. This publication has probably done more to advance the enthusiasm for computers amongst training architects than any of the scholarly books and articles published on CAD, 3d modelling, computer methods and formal systems. The surrealist rhetoric of cyberspace seems to have greater purchase amongst many designers than the systems approach, though it is no doubt parasitic upon it.

3. Cyberspace and the Uncanny

Much of the "philosophy" of Surrealism has been eclipsed by its progeny, including neo-Freudianism and poststructuralism. Freud was no admirer of Surrealism, but his account of the uncanny resonates with Surrealist themes. His concept of the uncanny influenced Surrealism in part, and offers an account of it. The theme of the uncanny also contributes further to the Surrealist understanding of design on the computer.

The uncanny (the unhomely) is the feeling of not being at home, an impression easily conveyed through computer imagery (Figure 16) and Surrealist art. As with many aspects of mind, Freud was able to link the sense of the uncanny with the Oedipus complex and early childhood development. He removed the uncanny from the notion of dolls and other machines that may behave as if alive. According to Freud, since childhood we are readily able to attribute life to inanimate objects through the power of the imagination. So we need to look for how computer narratives reveal the Oedipus myth rather than to the computer as autonomous or intelligent entity if we are to appreciate how it presents to us as uncanny.

Freud, posits several sources of the uncanny. He identifies an encounter with one's "double" as uncanny, as when you see someone who looks just like you, or even see your own picture. An image can instil a sense of longevity, as in the case of the Ancient Egyptians making images of the dead in lasting materials. For Freud this reverence for one's image reflects an early stage in childhood development or of primitive society, namely that of self love, narcissism. But when we surmount the narcissistic phase, in early childhood, the significance of the "double" takes on a reversal. Whereas it was initially an assurance of immortality, it later becomes "the uncanny harbinger of death." (Freud, 1990, p.357) At an earlier stage the double wore a more friendly face, but later, the "'double' has become a thing of terror, just as, after the collapse of their religion, the gods turned into demons." (p.358) In this light, the fact that a designer can place himself and his colleagues in the picture, as in the right of Figure 11, presents to us as uncanny, as is any exercise in which we see ourselves in the machine.

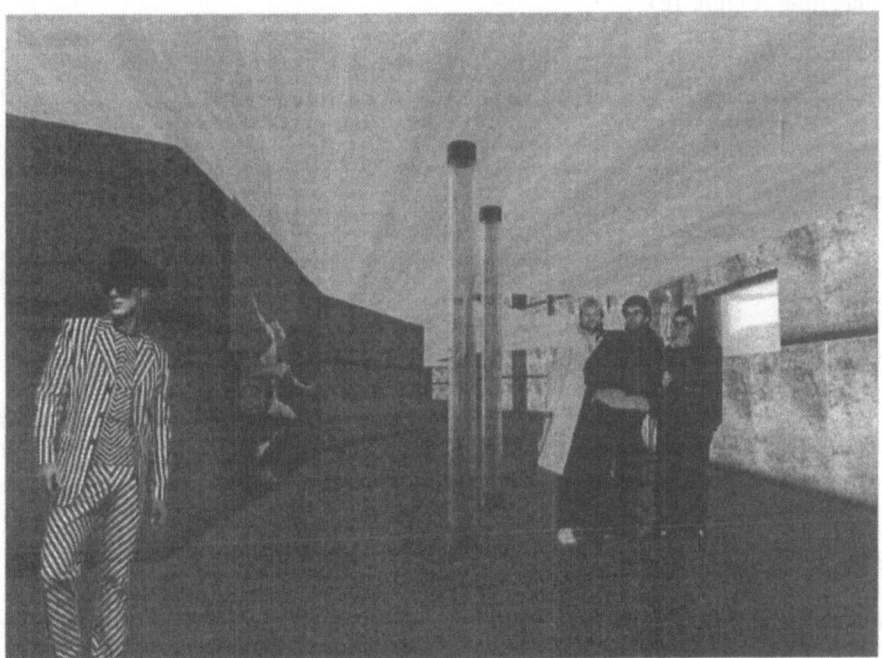

Figure 15. Museum interior by Peter Maxwell, inhabited by friends wary of the avant garde.

746

A second instance of the uncanny is where seemingly random events exhibit a pattern, as when the number on a theatre ticket turns out to be the same as the seat number on the train. For Freud, repetition itself is uncanny, as in the child's apparently obsessive repetition of simple games. According to Freud, "whatever reminds us of this inner 'compulsion to repeat' is perceived as uncanny." (p.361) The computer trades in obsessional repetition, in its internal functions and in the modes of practice that develop around its use. This could help explain the strength of opinion that gathers around computers amongst designers. It is not the threat to creativity that intrigues us but the obsessive modes of repetition it demands.

A third case of the uncanny is where we take what is imaginary for reality, as when old, discarded beliefs in ghosts, death wishes, animism, and so on present themselves as confirmed after all: "As soon as something *actually happens* in our lives which seems to confirm the old, discarded beliefs we get a feeling of the uncanny." (p.371) For Freud, an uncanny effect is often produced when the distinction between imagination and reality seems to be removed: "as when something that we have hitherto regarded as imaginary appears before us in reality ..." (p.367) For Freud, we encounter the uncanny "when infantile complexes which have been repressed are once more revived by some impression, or when primitive beliefs which have been surmounted seem once more to be confirmed." (p.372) By this reading the play between the real and the imaginary made possible through computer imagery presents to us as uncanny, and therefore disconcerting (Figure 16).

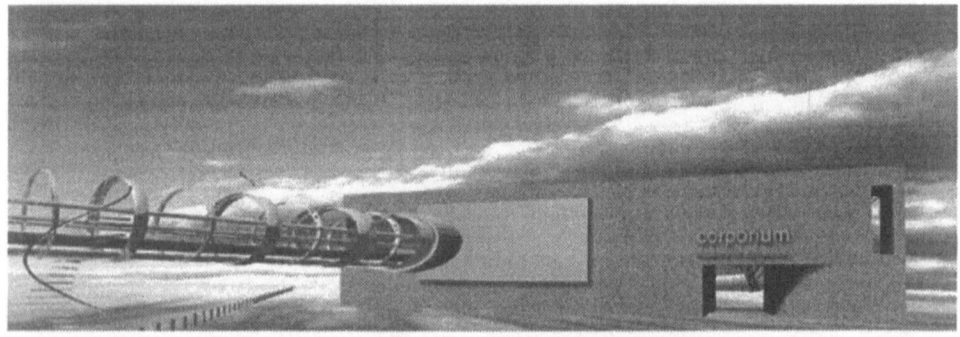

Figure 16. The presentation of a museum design that deliberately denies its context, demonstrating the ability of computer imagery to evoke the uncanny. No one is at home. The work is by Adrian Shilliday.

4. The Refracted Image

Surrealism played on the theme of the image, including the interreferentiality of images, that one image refers to another, that images are reflections of each other, and the ambiguity of image and reality. The mirror provides a potent metaphor of the Surrealist concept of the image. Mirrors feature prominently in absurd (*Alice Through the Looking Glass*) and Surrealist iconography: the eye, the mirror, the look, the gaze, aperture, window, frame, mimicry, perspective. Such metaphors also resonate with the IT world, which purports to present openings into worlds, presenting windows and hyperlinks that can return to themselves, and that suggest the interreflections of a chamber of mirrors. For IT commentators such as Chaplin (1995), Alice's looking glass is a "precursor of cyberspace" (p.32) in which the fundamental laws of physics, logic and language are inverted. It functions as a "substitute reality." (p.33) For others the mirror also suggests a divided self, subject to surveillance. According to Tabor (1995): "Each form filled, card swiped, key stroked and barcode scanned, replicates us in dataspace—as multiple shadows or shattered reflections." (p.17) Computer imagery also presents copies of other images, modified and distorted, and renders the concept of an original uncertain or ambiguous.

The image, particularly the mirror image, features prominently in the theories of Lacan, who was a leading psychoanalyst and controversial interpreter of Freud. As for Freud, Lacan's ideas now provide a potent resource for cultural critique (Zizek, 1992). Lacan's interpretation of Freud is informed by structuralist language theory and Surrealism. Surrealism influenced Lacan's view of language but also his use of it, as he exploited puns and word games, and used language to provoke, shock and even confuse. According to Bowie (1991), Lacan steps forward in the company of the Surrealists "as a writer and demonstrates his prowess in a self-conscious parade of puns, pleasantries, conceits, learned allusions and whimsical etymologies." (p.67)

For Lacan, biology is imbued with concepts of mimicry, which is even more basic than concepts of fitness and survival. Lacan was impressed by Caillois account of mimicry in nature, particularly the form assumed by the praying mantis, which exemplifies a creature being captivated by the image, that is, assuming the form of a twig. As Sarup (1992) summarises Lacan's position: "the human being, like the praying mantis, is captivated by the image. ... we are dominated by a structure of images and that this has a toxic, poisonous effect on the human subject." (p.25) For Lacan (1979) this concern with the image eventually expresses itself in terms of consciousness, which trades in the "illusion of *seeing itself seeing itself*," (p.82) which is based on the "inside-out structure of the gaze." (p.82) Lacan constructs a theory of human development at variance with Freud's and that builds on the concept of a child's first encounter with a mirror, the first moment when the child realises that she is other than her parents and the world around her. Lacan implicates language in this encounter, attributing the concept of identity with the separation between signifier and signified, which is also the separation between self and world, as encountered in the mirror. For Lacan the essence of self resides in rift and division. In this light the seduction of computer imagery and the production of designs in the context of computer graphics touches at the core of selfhood and identity, even unsettling it.

5. Surrealism and Suspicion

We can regard the Surrealists, Freud and Lacan, as practitioners of the "hermeneutics of suspicion." They purport to probe beneath the surface, which is a process of "a tearing off of masks, an interpretation that reduces disguises," according to Ricoeur (1970, p.30). It even ends with suspicion about the very process of reflection, as there are things about ourselves hidden from introspection. They dwell in the unconscious, and we do not really know ourselves. The tenets of Surrealism encourage us to "psychoanalyse" ourselves, and our designs, or at least the narratives we construct.

748

An alternative mode of interpretation engages a "hermeneutics of trust," which is at variance with Surrealism. This is a discourse that preserves the notion of truth and meaning, and recognise the role of community, context and history. It is a discourse that engages with the issue of metaphor, and the social situations in which we construct them. It presents design and design critique as richly dialogical activities, informed by the technologies we use, and the practices we are engaged in, or that engage us. It also adopts Gidden's (1992) criticism of Lacan and deconstruction: "Meaning is defined through difference, certainly; not in an endless play of signifiers, but in pragmatic contexts of use. There is absolutely no reason why, on the level of logic, acknowledgement of the context-dependent nature of language dissolves continuity of identity." (p.114) However, the "trusting" kind of interpretation acknowledges the role of discourses such as Surrealism and the language of provocation. The Surrealist artist does not have privileged access to creativity or radicality, but is caught up in particular modes and communities of practice, sometimes parasitic on the concept of the norm. If this is so of art then it is certainly true of design. From this point one can embark on further readings of computers in the design studio, that focus on metaphor, technology, narrative, and practice (Coyne, 1995).

References

Alexandrian, Sarane. (1970) *Surrealist Art.* trans. Gordon Clough, London: Thames and Hudson.

Bowie, Malcolm, (1991). *Lacan*, London: Fontana.

Breton, André, (1972) *Manifestoes of Surrealism*, trans. Richard Seaver and Helen R. Lane, Ann Arbor: University of Michigan Press. (First published 1929-1953.)

Chaplin, Sarah, (1995). Cyberspace: lingering on the threshold (architecture, post-modernism and difference), *Architectural Design Profile No. 118: Architects in Cyberspace*, 32-35, London: Academy Edition.

Chénieux-Gendron, Jacqueline, (1990) *Surrealism*, trans. Vivian Folkenflik, New York: Columbia University Press.

Coyne, Richard D. (1995). *Designing Information Technology in the Postmodern Age: From Method to Metaphor*, Cambridge, Mass.: MIT Press.

Esslin, Martin, (1961) *The Theatre of the Absurd*, London: Eyre and Spottiswood.

Franck, Karen, A. (1995) When I enter virtual reality, what body will I leave behind? *Architectural Design Profile No. 118: Architects in Cyberspace*, 20-23, London: Academy Edition.

Frazer, John H., (1995) The architectural relevance of cyberspace, *Architectural Design Profile No. 118: Architects in Cyberspace*, 76-77, London: Academy Edition.

Freud, Sigmund, (1990) The 'uncanny,' in *The Penguin Freud Library, Volume 14: Art and Literature*, ed. Albert Dickson, 335-376, Harmondsworth, Middlesex: Penguin. (First published in German in 1919.)

Giddens, Anthony, (1992) *The Transformation of Intimacy: Sexuality, Love and Eroticism in Modern Societies*, Cambridge, UK: Polity Press.

Haraway, Donna J., (1991) *Simians, Cyborgs, and Women: The Reinvention of Nature*, London: FAb.

Lacan, Jacques, (1979). *The Four Fundamental Concepts of Psychoanalysis*, trans. Alan Sheridan, London: Penguin.

Novak, Markus, (1995) Transmitting architecture: transTerraFirma/TidsvagNoll v2.0, *Architectural Design Profile No. 118: Architects in Cyberspace*, 43-47, London: Academy Edition.

Ricoeur, Paul, (1970) *Freud and Philosophy: An Essay in Interpretation*, trans. Denis Savage, New Haven: Yale University Press.

Sarup, Madan, (1992) *Jacques Lacan*. New York, Harvester Wheatsheaf.

Tabor, Philip, (1995). I am a videocam: the glamour of surveillance, *Architectural Design Profile No. 118: Architects in Cyberspace*, 15-19, London: Academy Edition.

Zizek, Slavoj, (1992) *Enjoy Your Symptom! Jacques Lacan in Hollywood and Out*, New York: Routledge.

BABYLON S M L XL

The missing Language of Cyberspace

FLORIAN WENZ
Chair for CAAD and Architecture
Swiss Federal Institute of Technology

We first discuss the future role of the CITY as a main generator of cultural fiction and suggest a superimposition of the PHYSICAL city and the DIGITAL city. We then draw parallels between the original intentions behind the World Wide Web and Hyper Text Markup Language and its expected follow up CYBERSPACE and Virtual Reality Markup Language. The development of three-dimensional SEMANTIC CODES for interactive environments is identified as one main task of the future. Within this framework, Babylon S M L XL, a series of research experiments conducted at the Architectural Space Laboratory at the professorship is investigating concepts and methods. The images display some scenes from this work in chronological order, while the captions provide content descriptions and METACODE abstractions.

1. The Crisis of the City

Cities, as the main processor units of civilisation seem to be on the decline. After most of the flows of humans, information or goods have been focused into urban hot spots for centuries, some of these streams are now being diverted by other forces. One of the main reason for the urban concentration process, the simplification of complex system flows through short transmission distances is becoming increasingly less dominating.

This, of course, has to do with the accelerated separation of the communication body from the physical body in networked communities. The architect Toyo Ito describes the Japanese society as "a society permeated by information and penetrated by communications systems. A society in which each individual has two bodies: a 'real'

749

R. Junge (ed.), CAAD Futures 1997, 749-756.
© 1997 *Kluwer Academic Publishers.*

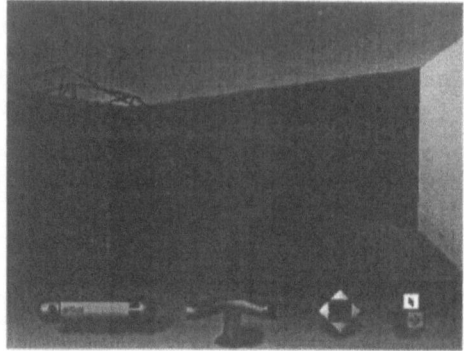

{{anim}{belief}}{true_space}

Babylon_S, "True Spaces and Beliefs"
September 1995
An early experiment, that investigated the combination of animated objects and static objects and consists of a hyperlinked web of scenes, which are to be read as personal signals of the author, forming a cinematic 3D story-space. The sequences use simple rooms (True Spaces) and juxtapose them with "translated" truisms (Beliefs) by the well known communication artist Jenny Holzer. i.e. "Action causes more trouble than thought".

body consisting of its physical presence, and a 'fictional' body, shaped by the information directed at or received by it." While face-to-face communication as a basic emotional need remains untouched, a wired person can perform functional communication from anywhere to anyone. As a consequence, Ito sees the elimination of the very concept of the city: "A kind of de-socialisation will take place within the city which will then be perceived as a 'fictional' structure, its spaces no longer needed to serve the needs of a 'real' population. At this time the non-city will emerge as the 'real' answer." (1)

Really? What can be observed at present are major global shifts of economic resources in online workforce markets, that can transform cities in very short time spans. Public access to the Internet was opened in India on August 15, 1995. Only 18 months later, the city of Bangalore, the capital of Karnataka state, is known as the Silicon Valley of India, being the home of some of the biggest and most profitable software companies in India, with a highly educated networked workforce and an extremely attractive market base to potential foreign investors. As part of a global urban infrastructure, a metacity of connected markets, cities are apparently still in high demand.

A very different image of the city was on display at the exhibition "The Archaeology of the Future City" (Museum of Modern Art, Tokyo, Japan, 1996), which interpreted the city as a materialised memory map of cultural history: "The City is an accumulation of spaces cobbled together in time. The City in time reflects the powers, rules, and economic systems of each age, succeeding as an actual space linked to that particular age's technology and styles. Personal aspirations can be found swirling in every nook and cranny of this city space, leaving traces of their collisions with public systems. The actual existing city is thus a complicated object, an amalgam of the traces of diverse 'times' and 'spaces', and the reality of the city is a compound made up of logic and contradiction." (2)

With a minor shift of internal perspective, the author could also be talking about the Internet in its present state. All that needs to be adjusted is the mental configuration CITY, the cognitive map of imagery, that every one of us makes up to create his own personal navigation maps. PLACES, STREETS, CROSSINGS are still there, but they are now being rapidly superimposed by traces of interactions with new associative object-systems like NODES, LINKS and NETS. In her essay "Associative Assemblages" Christine Boyers investigates this gap, that exists between the city that we can visualise and the invisible city that is constituted in and through its fields of information circulation: "These spaces or systems, which combined discourses and architectures, programs and mechanisms, also seem to be dislocated from space, deeply hidden within the electronic matrices of a global computer network that connects all points in space and directs our lives from some ethereal 'other' location." (3)

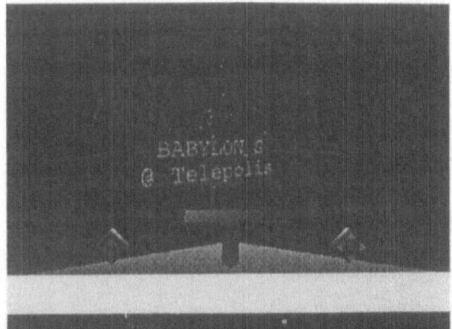

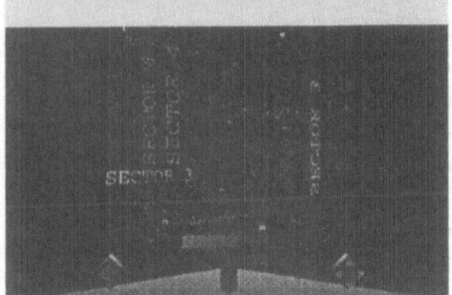

{{LOD}{url{symbol}}{url{shape_ext}}}

Babylon_S, "@Telepolis - The City"
November 1995
This was the centrepiece of our installation at the exhibit to the international conference "Telepolis - the Networked City". (Luxembourg 1996). It served as the entrance gateway to the 3D website of Telepolis and provided an interface to all the other VRML models in the exhibit.
All the extensions to VRML are written in Open Inventor, which was at that time the only way to program interactively in 1.0. Here are some of the concepts, we explored:
- Dynamic loading of "sectors" from different locations on the Internet, into on consistently perceivable environment
- hierarchical structured model with "city code" in its top level, controlling position, transformation and lighting while authors only control the geometric shape of the sectors.
- Multiple formal representations, depending on the user's relative position:
from far away - a symbol
from just outside - spacial outlines

2. The Crisis of the Net

The strategy, that is responsible for the enormous success of the World Wide Web from its original conception in 1991 to its present omnipresence is clearly revealed in an early proposal by its inventor Tim Berners-Lee, now Director of the World Wide Web Consortium [W3C]: "The WWW project merges the techniques of networked information and hypertext to make an easy but powerful global information system. The project represents any information accessible over the network as part of a seamless hypertext information space." (4)

This original vision was written before anybody knew that the WWW will actually work and well before the first webbrowser. Just like the engineers of packet switching must have had a vision of what the consequences of this technology will be, we can, in retrospect, read between the lines some of the implications of these design guidelines: "easy but powerful" points to Hyper Text Markup Language, a very simple encoding syntax for flow-layout documents and the Uniform Resource Locator, a universal locator mechanism for data sets, "hypertext information space" is the non-hierarchically evolving rhizomatic content mass, now known as The Web which was first made "seamless" by the Mosaic Webbrowser, probably the single most important piece of software ever written. What followed is already history, but is is safe to say, that despite all extensions of functionality, these ingredients are still the substance of the Web.

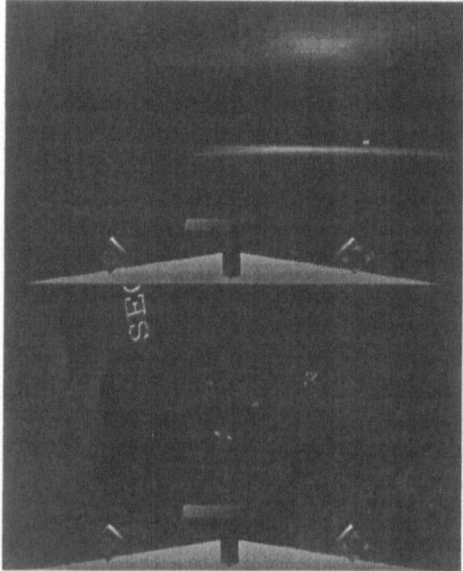

Babylon_S, "@Telepolis - The City"
November 1995

Top: Different perceptual resolutions for environment (exterior) and spaces (interior). On entering this sector, it becomes hyperreal, as it is being lit by physical lighting mode and begins to simulate material presence.

Bottom: Abstracted form or spacial drawing, the opposite of hyperreality. From the inside, this sector dissolves into a line drawing of itself, and transforms from a closed space to the perspective idea space.

{{LOD} {abstract} {physical}}

During the First International Conference on the World Wide Web, in Geneva, Switzerland, 25-27 May 1994, Mark Pesce, Peter Kennard, and Toni Parisi presented a paper entitled "Cyberspace", describing a visualisation tool for WWW, "Labyrinth", which uses WWW and a proposed protocol, Cyberspace Protocol (CP) to visualize and maintain a uniform definition of three-dimensional objects, scene arrangement and spatio-locations. As the next logic step beyond HTML, they proposed VRML (Virtual Reality Markup Language) an unsophisticated, but effective description language for objects, scenes and binding of external objects similar to URLs. It is important to note here, that "Labyrinth" was never meant to mirror or represent the existing physical world, but to superimpose a continuously smooth layer of abstraction (3D.SPACE) on top of the existing networked reality: "At its fundamental level, cyberspace is a map that is maintained between a regular spatial topology and an irregular network topology. The continuity of cyberspace implies nothing about the internetwork upon which it exists. Cyberspace is complete abstraction, divorced at every point from concrete representation." (5)

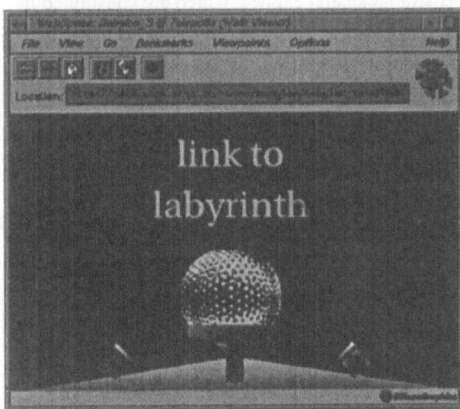

Babylon_S, "@Telepolis - The City"
November 1995

Macroworlds with enclosed Microworlds.
Animated 3D logos form symbolic anchors and are entry and exit points to other linked worlds by other authors.
Text in space adds another level of abstraction and gives the User a sense of what to expect after the anchor is activated.

{{anchor world_URL}{symbol} }

The proposal was received with great enthusiasm by the Web community and supported by a large number of individuals, institutions and companies. A mailing list was created, which consciously utilised the collective intelligence of a large number of experts and within a very short period of time, a language draft for VRML 1.0 was issued, followed by the first browser software from Silicon Graphics Industries. In its present state, VRML 2.0, which extends 1.0 by interactive scripting functions is among the core technologies of the Internet and large efforts are being made to establish a commercial market for VRML worlds.

In those research fields, where results can be communicated effectively through three-dimensional scientific visualisation, i.e. molecular chemistry, VRML is already established as an Internet standard, as VRML scene descriptions also offer excellent content compression, as opposed to images or animations. Companies like "Worlds Inc." or "Black Sun Interactive" market three-dimensional chat world applications based on VRML, where users, represented through interactive models called AVATARS can move around abstract WORLDS and communicate with each other through text or audio signals, thus trying to establish new forms of social interaction. "Community ... The Next Generation from Black Sun: Communicate in real time with people from around the world, explore dynamic 3D worlds, lead tours of your site, host moderated events and play interactive games - this is the Internet as you always imagined it." (6)

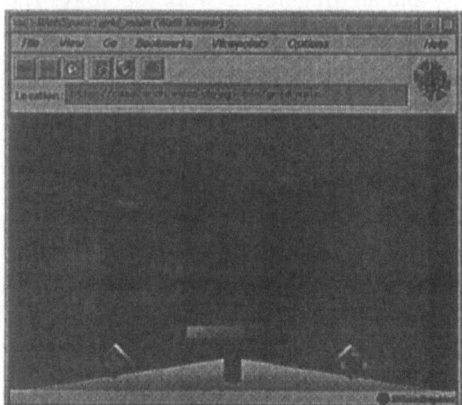

Babylon_S, "Matrix - The Space"
November 1995

Matrix consists of a theoretically endless grid of self-similar spatial entities, which are constituted in real time by a remotely-connected machine and surround the user with a coded architectural syntax during interactive exploration. Matrix creates spaces from a system of lighting, colour and aperture typologies, which can be read as private interior rooms and can be personalised within the constraints of the coded architectural syntax.

$$\{x*\{\ \{y*\{node\_proc\}\}$$

3. The missing Language

It is obvious that the Internet community is waiting for the promises of William Gibson's "Neuromancer" or Neil Stephenson's "Snow Crash" to finally materialise. But for a number of reasons, the evolution of Cyberspace is not following the footprints of the World Wide Web as predicted. For one, browsers are still too slow to allow for high resolution, complex worlds being displayed in real time on average machines, a problem that will be solved by time and the advances of computer graphics subsystems. Further and more seriously, the inventors of cyberspace have overlooked the fact, that there is no common language for content description in interactive 3D scenes, as there is for web pages. The web is still essentially based on the syntax of the Gutenberg galaxy, which was developed over the last 2000 years: text, image and diagram.

The usual claim that 3D must be better than 2D, simply because it offers an extra dimension, does not hold up under critical inspection. On the contrary, one can argue that 2D media have been developed to inject meaningful CONTENT into abstract SHAPE and 3D is actually one step backwards. With its physical presence stripped from its shape, an object in Cyberspace is literally asking of the user "What does it look like?", while at the same time, what should be communicated is "What does it mean?", which is precisely, what most existing VRML scenes fail to answer. These questions of semantics are especially important in communication networks, because most files on the Net are consciously created as communication signals by somebody or somebody's software and thus only makes sense in terms of their relations to other signals by other authors.

To establish these new codes of interaction, the semantics of Cyberspace, will be a major task for the CAAD community in the near future, as it demands a thorough understanding of spacial systems, interface design and the mechanics of code. Until now, most of the research work in this area has been done by game designers, where the genre of interactive first-person 3D game has already established a new code of spacial interaction. Millions of juvenile gamers, playing endless hours of "Wolfenstein 3D", "Doom" or "Descent" and now "Quake" have already evolved a semantic system of spacial environment interaction, tuned to the needs of the gaming industry, whose evolution is comparable to the progress from line input terminals to the present graphical user interface with windows and icons.

In his abstract to "Aquamicans", one of the advanced works in VRML done at the Architectural Space Laboratory at the chair in 1996, its author etoy writes: "One has to realise that precisely now, when the simulation of the material world in 3D browsers seems to be most obvious, the development of new archetypes of form symbolic, that are closely related to Internet culture will generate a new form of sense in Cyberspace. This must be a privilege of architecture, if one sees its main task in discovering relations and improving the legibility of the world, even if this means that we have to redefine our formal understanding of architecture."(7)

756

4. Babylon S M L XL

To develop this missing language of cyberspace and to generate concepts and principles for networked immaterial architecture, a series of experiments was initiated at the Architectural Space Laboratory in fall 1995. The term "Babylon" refers to the myth of Babel and the ancient City of BAB-ILU , which dates back almost 2000 years before AD. Almost 2000 years after AD the western culture is again facing similar issues of identity in a massively connected and complex environment. The Babylon Series "S M L XL" points to a research mode, that has to follow the highly accelerated networked evolution of this platform in rapid prototypes from small and simple (S) to large and complex (XL). It is also important to note, that the substantial body of work that is connected to the Babylon Series was created by a large number of students and junior faculty in the spirit of the Internet, where collective intelligence is at the basis of individual achievement. The author wishes to acknowledge the substantial contributions of the following individuals and multiples: etoy, Patrick Sibenaler, Tristan Kobler, Tom Sperlich, Urs Hirschberg, Christian Waldvogel, Claudia Weinmann, Urs Kuehni, Peter Mackes and all the participants of !hello_world? and "Real Fiction - Virtual Realities".

References

(1) Toyo Ito. 'Experimental Architecture'. World Architecture No 34. London, March 1995.
(2) Invitation to 'The Archaeology of the Future City', Museum of Contemporary Art, Tokyo, 24 July - 16 September, Hiroshima Museum of Art, 22 September - 4 November, Gifu Prefectural Museum, 12 November - 22 December, Tokyo, 1996.
(3) M. Christine Boyer. 'CyberCities'. Princeton Architectural Press, New York 1996.
(4) Tim Berners-Lee. 'An Introduction to WWW'. CERN, 1991.
(5) Mark D, Pesce et al. 'Final Amputation: Pathogenic Ontology in Cyberspace ', CERN, Third International Conference on Cyberspace, Texas 1993.
(6) Black Sun Interactive, 'The Black Sun Community 2.0', http://www.blacksun.de/launch/index2.html, 1997
 (7) etoy. 'Die Architektur und Cyberspace', CAAD and Architecture, ETHZ, http://caad.arch.ethz.ch/projects/aquamicans/html/page4.html, 1995.

Online

Babylon S M L XL: http://caad.arch.ethz.ch/~wenz/babylon
TRACE: http://caad.arch.ethz.ch/trace
Aquamicans: http://caad.arch.ethz.ch/projects/aquamicans
!hello_world?: http://caad.arch.ethz.ch/hello_world

THE APPLICATION OF CONJOINT MEASUREMENT AS A DYNAMIC DECISION MAKING TOOL IN A VIRTUAL REALITY ENVIRONMENT

J. DIJKSTRA[a], H. J. P. TIMMERMANS[b]

Eindhoven University of Technology,
Faculty of Architecture, Building and Planning
P.O. Box 513, 5600 MB Eindhoven, The Netherlands
[a] *email*: j.dijkstra@bwk.tue.nl
[b] *email*: h.j.p.timmermans@bwk.tue.nl

Abstract

This paper describes an innovative aspect of an ongoing research project to develop a virtual reality based conjoint analysis system. Conjoint analysis involves the use of designed hypothetical choice situations to measure subjects' preferences and predict their choice in new situations. Conjoint experiments involve the design and analysis of hypothetical decision tasks. Hypothetical alternatives, called product profiles, are generated and presented to subjects. A virtual reality presentation format has been used to represent these profiles. A profile consists of a virtual environment model and dynamic virtual objects representing the attributes with their respective levels. Conventional conjoint choice models are traditionally based on preference or choice data, not on dynamic decision making aspects. The status of this new approach will be described.

1. Introduction

Conjoint analysis or experimental choice analysis represents a widely applied methodology for measuring and analyzing consumer preferences (Carrol and Green, 1995). Conjoint analysis is a generic term coined by Green and Srinivasan (1978) to refer to a number of paradigms in psychology, economics and marketing that are concerned with the quantitative description of consumer preferences or value trade-offs (Timmermans, 1984; Louviere, 1988). Conjoint analysis sometimes referred to as stated preference modelling involves the use of hypothetical choice situations generated according to the principles underlying the design of statistical experiments to measure subjects' preferences, examine consumer behaviour and/or predict their choice in new situations (Oppewal, 1995). In a conjoint study, a researcher (1) selects the characteristics (attributes) that are assumed to influence the choice behaviour of interest, (2) classifies these attributes into numerical or categorical levels, and (3) combines these attribute levels into profiles according to some statistical design.

757

R. Junge (ed.), CAAD Futures 1997, 757-770.

Implicitly, it is assumed that choice alternatives can be viewed as a set of attributes. Subjects are assumed to trade-off the attributes of interest according to some algebraic rule to arrive at an overall preference for each profile. In order to estimate the preference function, a set of profiles designed according to some experimental design is presented to subjects who are requested to express their overall preference. While a verbal description might be a valid means of describing profiles in many contexts, one could argue that some attributes are better presented visually.

A framework for a virtual reality based system of conjoint analysis has been outlined in Dijkstra *et al* (1996). Such a system is of particular interest when subjects have to experience the context of choice and/or the attributes describing the choice alternatives. To take this argument one step further and to explore the possibilities of using this framework as a decision-making tool for virtual wayfinding environments is described in Dijkstra *et al* (1997). This paper explores the essentials of this methodology to utilize the application of conjoint measurement as a dynamic decision making tool in a VR environment.

The paper is organized as follows. First, we will discuss some basic principles of conjoint measurement and virtual reality and their integration, and also their integration. Then, in section 3, we will look back at some simple illustrations of conjoint experiments. This is followed by a section about decision making. Finally, we draw some conclusions.

2. Conjoint Measurement & Virtual Environment

1.1. CONJOINT MEASUREMENT

Conjoint analysis is a family of related for measuring consumer preferences or choice behavior. It helps to understand why consumers prefer or choose certain products (or services or new conditions).

1.1.1. Traditional conjoint analysis
The application of conjoint analysis technique implies the study of the joint effects of multiple product attributes on product preferences or product choice. The researched products (or services or new conditions) are described in terms of product profiles. Each profile is a combination of attribute levels for the selected attributes. Conjoint analysis has two major objectives: (1) to determine the contributions of predictor variables (attribute levels) to consumer overall preferences, and (2) to establish a valid model of consumer judgments useful in predicting the consumer acceptance of any combination of attributes, even those not originally evaluated by consumers (Hair *et al*, 1995).

In order to achieve these objectives, coefficients called 'utilities' (or part-worths) are estimated for the various attribute levels making upon the alternatives of interest by decomposing measured preferences for product profiles into these part-worth utilities according to some a priori defined combination rule which specifies how subjects are assumed to integrate those separate part-worth utilities to arrive at an overall preference or choice. The profile utility is an overall utility or 'worth' of a profile calculated by summing all utilities of attribute levels defined in that profile.

$$U_j = \sum_{i=1}^{N} x_i$$

, U_j = overall utility of the profile alternative j.

x_i = 'part worth' of the i-th level of the x-th attribute

N = number of attributes

Figure 1. Preference measurement in conjoint analysis

The estimation of these part-worth utilities is based on experimental designs.

1.1.2. Experimental designs

Profile construction in conjoint analysis involves determining which attributes to present to subjects, and how to present these attributes. In a conjoint experiment, first the key dimensions (attributes) of products or services are defined. Next, the specific levels of each attribute are specified. The chosen attributes and their levels should be realistic and relevant to the problem. Also, the ultimate definition of attributes and their levels will be influenced by the possibilities of constructing a suitable experimental design. That is, the design should satisfy the necessary and sufficient conditions, required to estimate the assumed preference or choice model that describes the way in which subjects are assumed to arrive at some choice or preference. Traditionally , in axiomatic conjoint measurement, the focus is on testing the structure of preference functions, i.e., whether preference functions are additive, multiplicative or combined additive-multiplicative. More recently, preference functions are estimated from experimental design data using an appropriate multivariate statistical technique. Conjoint experiments thus require subjects to express their preference for various experimentally designed, hypothetical alternatives in terms of their most relevant attributes. Two or more fixed levels are defined for each attribute and these are combined to create different profiles. Subjects are invited to express their preference for the experimentally varied profiles by rating or ranking these in terms of overall preferences. Alternatively, subjects may be asked to choose the profile they like best. Preference functions are estimated from this data. Obviously, the number of possible combinations increases exponentially with an increasing numbers of attributes and/or levels. Fortunately, the data collection can be greatly reduced by using fractional factorial design techniques (Montgomery, 1991). In analysis-of-variance terms, this often means that only main effects are estimated. Therefore, an experimental design is defined by an optimal subset of profiles of a fractional factorial design, which can be presented to a subject without negatively influencing responses in terms of boredom or fatigue.

1.1.3. Design and analysis of conjoint experiments

The current discussion, strictly speaking, relates to a preference model. A subject's preferences are decomposed into part-worth utilities, which represent the contribution to the individual's overall preference or utility of the attribute levels that were used to generate the profiles. If one wishes to construct a choice model, the attribute profiles have to be placed into choice sets. Subjects are then asked to choose one alternative from each choice set, or alternatively, to allocate some fixed budget among the choice alternatives (Oppewal and Timmermans, 1991).

We will illustrate some of the above issues by presenting a simple example.

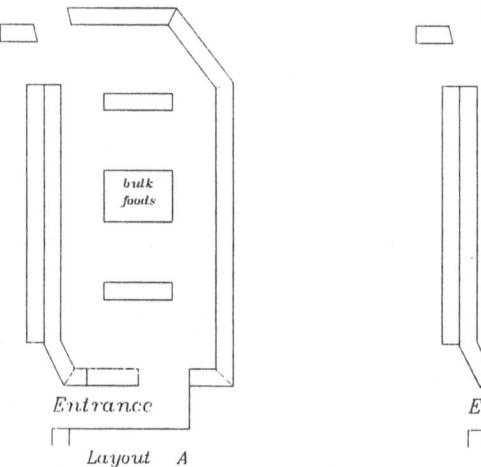

Let us assume that a retailer has to plan the first aisle after passing the entrance in a supermarket. In making his decision, the retailer has to consider a number of factors. These factors will be presented by attributes and their levels. Let us assume that the following attributes and their levels are important:

- *Merchandise indication (MI)*
  do you want that there's a merchandise category indication near at a merchandise area (for instance 'fresh meat', 'cheese chop'), which is good visible at a distance?
  No = no indication, Yes = indication
- *Bargain offers near at the entrance (BOE)*
  do you want the bargain offers are exposed near at the entrance?
  No = not desirable, Yes = desirable
- *Layout (LO)*
  which layout do you prefer?
  A = layout A, B = layout B

Summarized, we have the following attributes with their levels:

Attributes	Levels
MI	Yes, No
BOE	Yes, No
LO	A, B

This scenario results in 8 (=2^3) possible profiles.

Profile 1 .... Profile 8
- MI : No
- BOE: No
- LO : A

- MI : Yes
- BOE: Yes
- LO : B

The question then is the choice of the respons format

In case of ratings data, subjects are requested to rate these profiles on some psychological scale, this format provides information about both order and degree of preference.
For example, a question could be: What is your opinion about this arrangement?
Profile 1
- MI : No
- BOE: No
- LO : A

                    X
1   2   3   4   5   6   7
very   average   very
poor   good

In case of choice data profiles are placed into choice sets and subjects are asked to choose among two or more profiles.
For example, which profile do you prefer?
Profile 1
- MI : No
- BOE: No
- LO : A

Profile 8
- MI : Yes
- BOE: Yes
- LO : B

Finally, in case of budget allocation, budget-points will be allocated among a set of profiles. For example, a budget allocation with profiles 2 and 7 as hypothetical alternatives, could be:

Allocate 20 budget-points	Basic Profile .... points	Profile 2 .... points	Profile 7 .... points

'Design Descriptions'
In experimental designs, attributes are termed 'factors'. The goal is to structure the data collection process in such a way that the identification possibilities for the utility function are maximized. In a full profile approach, we distinct a full factorial design and a fractional factorial design. A Full Factorial (FF) design contains descriptions of all possible combinations of attribute levels. It enables one to independently estimate all main effects and all interaction effects of each attribute. On the other hand, a Fractional Factorial design contains a fraction of a FF design. It assumes that certain interaction effects among the attributes are not statistically significant.

The following simple example will illustrate this. Suppose we have 3 attributes with 2 levels each, with level indications 0 and 1. A 'Full Factorial' (FF) design exists of all possible combinations.

Combination/Profile	Levels
1	0 0 0
2	0 1 0
......	
8	1 1 1

A 'Fractional Factorial' design exits of a fraction of a FF design, for instance:

Fraction 1		Fraction 2	
Profile	levels	Profile	levels
1	0 0 0	2	0 1 0
4	0 1 1	3	0 0 1
6	1 1 0	5	1 0 0
7	1 0 1	8	1 1 1

1.2. VIRTUAL ENVIRONMENT

In architectural and real estate simulations, it is interesting to get as realistic as possible impressions of a designed model by means of virtual reality. Consideration can be given to modeling autonomous objects and to the simulation of operations on objects. The visionary design studio embedded in a VR environment is an example of a vision of a new design environment, equipped with a modeling tool that allows intuitive and interactive modification of intelligent objects (Engeli and Kurmann, 1996).

What distinguishes VR is the crucial role played by the user. That is, the user has an active involvement and is not a passive observer. The user becomes an essential participant in the virtual environment with unlimited freedom to explore, control and change it. The only limits are those set by the designers of the virtual environment. VR techniques can be used to create an interface that allows modeling in an intuitive way. Through the imitation of behavioral aspects of the real world, the interface gets predictable and recognizable characteristics. In fact, simulation based design technologies within VR technologies enhances the capabilities of the design to manufacturing process. The product can be visualized, design changes can be made, and new concepts, without the traditional expense of prototyping, can be tested.

This last aspect is subject of another area of research and application area for VR. Advances in VR techniques now enable consumers to be immersed in new environments and experience new products or services. This aspect is of interest to get a better insight into consumer behavior and support product testing in its most general meaning.

1.3. INTEGRATION OF CONJOINT MEASUREMENT AND VIRTUAL ENVIRONMENT

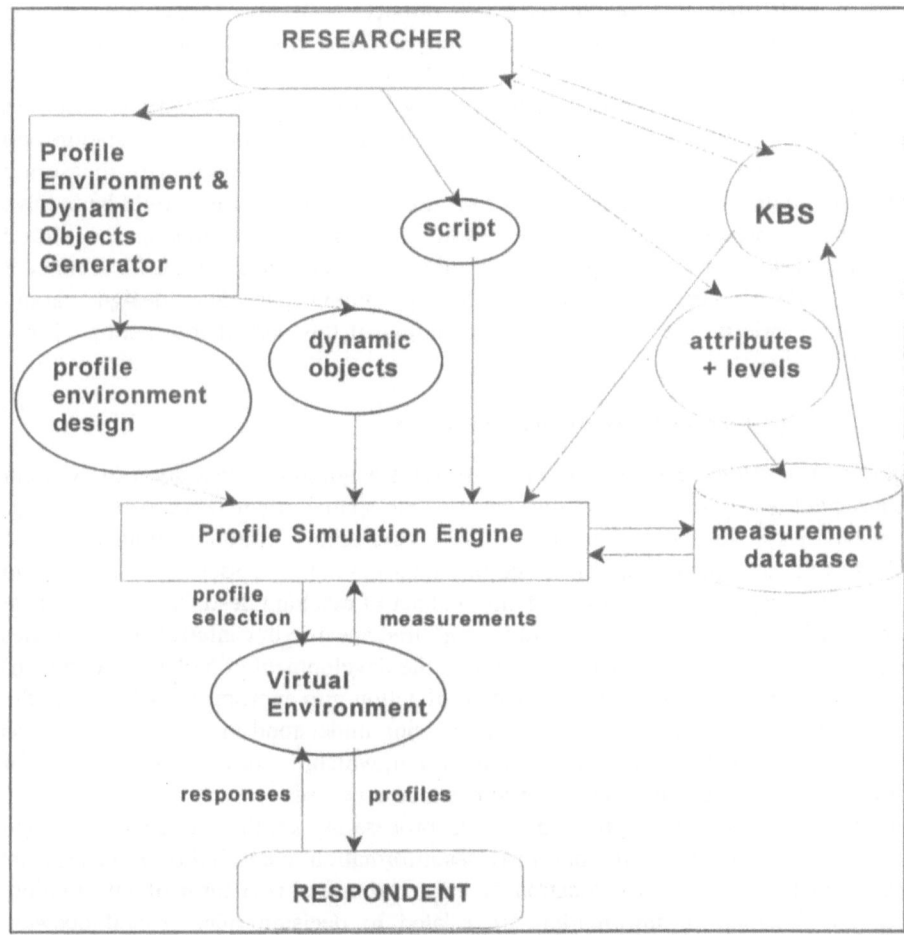

Figure 2. CA & VR system approach

Most studies of conjoint analysis involve verbal descriptions of product profiles, although some studies have used a pictorial presentation of product profiles. Klabbers *et al* (1996) propose a multimedia engine for stated choice and preference experiments, which enables researchers to use varying presentation formats thereby measuring the influence of the presentation format. In this research project, profiles presented by a virtual reality presentation format, are for conjoint measurement. A profile consists of a virtual environment model and dynamic virtual objects representing the attributes with their respective levels. Each attribute level is a different state of the concerned virtual object. In the case of a virtual walk through a building design, the system can be viewed as a visual simulation of the environment. In this research project, the focus is on profiles presented by a virtual reality presentation format, thereby emphasizing yet other aspects of decision-making compared to multimedia engines.

The ultimate objective of the research project is to investigate and develop possibilities of a VR-DIS (which is an acronym for Virtual Reality - Distributed Interactive Simulations system) - environment for those cases where choice behavior and decision-making processes of consumers may be of importance. This is based on the idea that a VR-presentation of a design is not only a matter of simulation and visualization, but virtual reality research could be a key to a mechanism for measuring responses particular design characteristics. One means of measurement could be to have subjects choose between profiles by exploring the building features in the virtual environment. Taking prospective users of a building into account, the CA&VR concept could be a decision-making tool in choosing among particular design variant characteristics. Briefly, the scheme at the beginning of this part of the section, shows the CA&VR concept.

3. Simple Illustrations of Conjoint Measurements

In Dijkstra *et al* (1996) and Dijkstra *et al* (1997) a wayfinding illustration of conjoint mmeasurement was given. Wayfinding concerns the spatial organization of the setting, the circulation system and architectural as well as graphic communication. It can be described as all perceptual, cognitive, and decision-making processes necessary to find one's way. That is, it is as a mental and physical act of reaching destinations and is best defined as spatial problem solving, comprising three specific but interrelated processes (Arthur and Passini, 1992): (i) decision making, the development of a plan of action, (ii) decision execution, which transforms the plan of action into appropriate behavior at the right place in space, and (iii) information processing understood in its generic sense as comprising environmental perception and cognition, which, in turn, are responsable for the information of the two previous decision-related processes.

In addition to the spatial problem solving process aspect, there is also the design aspect of wayfinding. The design part provides information, identified by three aspects: (i) content of information, (ii) location of information and (iii) form of information. Content and location of information are related to decision making and decision execution. The form of information comprises environmental perception and cognition and is related to information processing.

In the design part of wayfinding, the emphasis is on the graphic components. In the experimental designs of the illustrations only the design part of wayfinding, especially graphic communication as part of the environmental communication would be considered. In the first illustration, information about the settings as well as information directing to the location and information identifying the destination would be emphasized.

Attribute	Level	Description
◊ Orientation	⇒ Floor plan ⇒ Directory	Ability to perceive an overview of a given environment
◊ Directional signage	⇒ Text besides arrow sign ⇒ Text inside arrow form	Guides people along a designated route to a destination
◊ Identification	⇒ Name ⇒ Sign with name	Information provided at the destination

Identification: Name   Identification: Sign with name

For instance: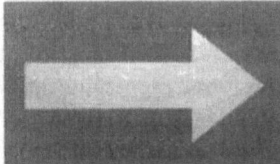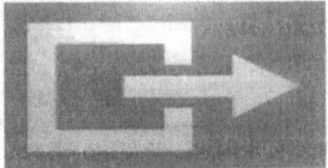

Figure 3. Design specifications Illustration 1

In the next illustration, we will focus on information directing people to the way out. This would be realized by graphic communication on exit-signs. How to test exit-signs for their suitability, is the underlying idea. Besides the aesthetic aspect we could measure, we would also have a mechanism to measure the effectiveness of exit-signs.

Attribute	Level	Description
◊ Directional Exit-sign Location	⇒ Fixed at wall, column ⇒ Fixed at ceiling	Guides people along exit-route to exit
◊ Directional Exit-sign Type	⇒ Exit sign I ⇒ Exit sign II	Guides people along exit-route to exit
◊ Exit Identification	⇒ Exit Font I ⇒ Exit Font II	Information provided at exit

Exit sign Type I   Exit sign Type II

For instance:

Figure 4. Design specifications Illustration II

Especially, exit-signs could be tested in a fire drill simulation. In that case, we have an example of virtual wayfinding.

4. The Decision Making Aspect

4.1. DYNAMIC DECISION MAKING

Decision making is the heart of human behavior (Kerstholt, 1996). The main goal of behavioral decision making is to explore how subjects make their choices. Many real-world decision tasks are dynamic in nature, which often means that the decision context changes over time. Dynamic task situations are rather difficult to model, which can be traced back to the exponential growth over possible outcomes over time: as the decisions influence each other, each combination of decisions made over time has its own expected value. The main reason for modeling the task is to have a normative solution to which subjects' choices can be compared. We can summarize some characteristics of dynamic tasks:

- The environment in which the decision is set may be changing, either as a function of the sequence of decisions, or independently of them, or both.
- Decisions need to be made in real-time. This factor added an extra dimension to dynamic decision making, as the decision maker has to consider the dimension of time explicitly. It is not enough to know what should be done but when it should be done. Thus, in dynamic tasks subjects typically control a changing system, they receive feedback about the state of the system and they need to make a sequence of decisions.
- Dynamic decisions change over time. The main effect of the changing nature of dynamic tasks is that the time dimension has to be taken into account explicitly.

The given illustration of virtual wayfinding is really a dynamic decision making process. Decisions will be made real-time and also the time dimension had been considered. Indications like signs presented as different virtual objects could be tested for their suitability. After a fire-alarm and during a smoke production an individual should find his way to the exit within a certain period. Thus, the time-pressure aspect had been involved. Maybe, we could get answers on the questions how subjects deal with dynamic tasks and about how adaptive their behavior is. The perception of virtual objects in the virtual environment gives the necessary feedback. The illustration of virtual wayfinding is a matter of functional simulation. As a consequence, a decision about the actual signage can be proposed. It is possible that a potential difference between this actual signage and the preferred signage is the consequence.

Another aspect of decision making is the distinction between action-oriented decision making and judgment-oriented decision making. Due to the time pressure, the decision making is more action-oriented such as during a smoke production after a fire-alarm. Judgments about characteristics of architectural design features represented as virtual objects is a judgment-oriented decision making process, just like the exploration of virtual objects. On the other hand, the perception of graphic components of signage in the first illustration of wayfinding is more judgment-oriented.

4.2. METHODOLOGY

Within the context of VR-DIS, the aspect of dynamic decision making is an unconventional part of ongoing research. This aspect will be realized in the development of a conjoint and a virtual reality system. Besides a profile generation part, a profile simulation part and a data-analyzing part, the system essentially consists of a virtual environment part for a subject's virtual walk-through of selected profile alternatives. Such an application is likely to better incorporate the idiosyncrasies of virtual reality systems, but does not satisfy the typical assumptions underlying conjoint models.

Conjoint analysis originally was developed to measure users' utilities for multi-attribute choice alternatives. The design of the experiments should allow one to reflect the assumed preference function or choice model. Consequently, to measure preferences or utilities, subjects are requested to rank or rate the experimentally varied attribute profiles on some preference scale. The observed values can then be decomposed into the utility contributions of the attribute levels. Similarly, if the focus is on testing an assumed choice model, then the profiles are placed into choice sets and subjects are requested to choose from each choice set the alternative they like best.

These response formats can also be applied to the problem of wayfinding if one is primarily interested in understanding the contribution of attributes of the guidance system on users' preferences. However, if may well be that the researcher's or designer's interest goes beyond this problem. For example, the question of interest might be whether users were successful in finding an exit, or in the time it took them to find an exit. In these situations, the dependent variable of the model shifts from a ranking, rating or choice to a dichotomous yes/no variable or a time variable respectively. The multivariate statistical analyses that are commonly applied in conjoint analysis are no longer appropriate for these situations and hence alternative techniques need to be chosen.

If the focus of interest concerns the question whether users are successful in finding an exit, then the binary logit model is a candidate for analysis. It allows one to estimate the contribution of attribute level on the probability that users find an exit. Hence, the result of the analysis may be used to identify the most critical attributes of the guidance system. Alternatively, if the focus of interest concerns the issue of how long it takes to find an exit, the dependent variable of the model consists of positive numbers only. Poisson regression analysis is an appropriate multivariate analysis tool to analyze such data. In this case, the coefficients of the regression equation indicate the effect of the attribute levels on the time it takes to find an exit.

5. Remarks on Choice Models and Exploration of the Virtual Environment

We will make some briefly comments on developing a conjoint analysis based virtual reality system.

5.1. CHOICE MODELS

Choice models are suitable for the use of experimental designs to construct hypothetical products or services, and observe subjects' choices. Hereby, the researcher has control over the attributes and their correlation. The basic principle is:
- Ask subjects to choose from Profile A and Profile B.
- Choice of A vs. B indicates preference of the preferred Profile attribute levels.
- Systematically varying attribute levels allows building models of such trade-offs.

Subjects are assumed to attach an overall value to a choice alternative. This utility is arrived at by combining evaluations of attributes. Originally, choice sets will be formed and subjects decide which profile alternative from the choice set is chosen. In a virtual reality based conjoint measurement experiment, it is not appropriate to choose from more than one choice set without negatively influencing responses in terms of boredom or fatigue. For subjects, only a small number of reduced choice sets could be dealt with. If we want to perform an experiment on this manner, many mores subjects are needed. Adaptation of accepted utility functions needs to be developed.

5.2. EXPLORATION OF THE VIRTUAL ENVIRONMENT

A virtual environment is very convenient to explore virtual objects. By clicking a virtual objet, a specific attribute level will be displayed. Each attribute will be presented by a virtual object and each virtual object has a number of appearances. Each appearance is a level of the attribute. By exploring the attributes in this way, the subjects' preferred profile can be generated. Due to the conjoint measurement in a virtual environment, it is also feasible to record a subjects' walk through a virtual environment. Data about actions, decisions, judgments, perception and time-span can be collected and evaluated. Obviously, a virtual environment is a particular environment for applying conjoint measurement experiments. The conjoint measurement can be affected by:
- Experience. That is, experience about virtual reality, experience about the model, and experience about virtual reality techniques. It is important that the virtual environment is familiar to subjects participating in the conjoint measurement experiment.
- Effects of attributes that are of interest may be confounded with the effects of other factors in the virtual environment.
- Also the order of displayed attributes as virtual objects can act on the measurement.

6. Conclusion

Recent development in virtual reality systems allow the creation of interactive environments that can be used to observe user reactions and decision making in not yet existing environments. It will be evident that such systems offer the potential of an evaluation of building performance in advance. The power of such systems can be enhanced if a modeling approach underlies the observations of user reactions. Such an approach offers an opportunity for generalizing the findings beyond the actual environments that were incorporated in the virtual environments.

In the present paper, we have discussed such a framework for interactive virtual environments and illustrated its potential use in the context of wayfinding in buildings. We proposed to link the technology of virtual reality systems to conjoint analysis and discussed the implications of such an endeavor. We argued that the methodology for conventional conjoint analysis is well developed, but that for some more advanced problems, additional methodology needs to be developed.

The ultimate test of the relevancy of such systems, however, depends on empirical testing. One critical issue relates to the question whether user behavior in virtual reality is systematically related to user behavior in real-world situations. We hope to report on such research in the near future.

References

Anderson, N.H.: 1981, Foundations of Information Integration Theory. Academic Press, New York.

Arthur, P. and Passini, R.: 1992, Wayfinding, People, Signs and Architecture. McGraw-Hill Ryerson, Toronto.

Carrol, J.D. and Green, P.E.: 1995, Psychometric Methods in Marketing Research: Part I, Conjoint Analysis, Journal of marketing Research, Vol. XXXII , pp. 385-391.

Dijkstra, J., Roelen, W.A.H. and Timmermans, H.J.P.: 1996, Conjoint Measurement in Virtual Environments: a Framework, in H.J.P. Timmermans (ed.), 3rd Design and Decision Support Systems in Architecture and Urban Planning Conference, Vol. 1: Architecture Proceedings , pp. 59-71.

Dijkstra, J. and Timmermans, H.J.P.: 1997, Exploring the Possibilities of Conjoint Measurement as a Decision-making Tool for Virtual Wayfinding Environments, Paper presented at CAADRIA'97.

Engeli, M. and Kurmann, D.: 1996, A Virtual Reality Design Environment with Intelligent Objects and Autonomous Agents, in H.J.P. Timmermans (ed.), 3rd Design and Decision Support Systems in Architecture and Urban Planning Conference, Vol. 1: Architecture Proceedings , pp. 132-142.

Green, P. E. and Srinivasan, V.: 1978, Conjoint analysis in consumer research: issues and outlook, Journal of Consumer Research, 5, pp.103-152.

Hair, J. F., Anderson, R. E., Tatham, R. L. and Black, W. C.: 1995, Conjoint Analysis. in Multivariate Data Analysis, Prentice Hall, Englewood Cliffs NJ, pp. 556-599.

Kerstholt, J.H.: 1996, Dynamic Decision making, Ph.D. Thesis, TNO Human Factors Research Institute, Soesterberg.

Klabbers, M. D., Oppewal, H. and Timmermans, H. J. P.: 1996, ESCAPE: (Multimedia) Engine for Stated Choice and Preference Experiments, Working paper 3rd Design and Decision Support Systems in Architecture and Urban Planning Conference.

Louviere, J. J.: 1988, Analyzing Decision Making: Metric Conjoint Analysis, Sage University Paper Series on Quantitative Applications in the Social Sciences, series no.07-067. Beverly Hills: Sage Publications. Montgomery, D. C.: 1991, Design and Analysis of Experiments, John Wiley, Chichester.

Oppewal, H.: 1995, Conjoint Experiments and Retail Planning: Modelling Consumer Choice of Shopping Centre and Retailer Reactive Behaviour, Ph.D thesis, Bouwstenen 32, Eindhoven University of Technology.

Timmermans, H. J. P., Van der Heijden, R. E. C. M. and Westerveld, H.: 1984, Decision-Making between Multi-Attribute Choice Alternatives: a model of spatial shopping behaviour using conjoint measurements, Environment and Planning A, **16**, pp. 377-387.

MANAGING INFORMATION WITH FUZZY REASONING SYSTEM IN DESIGN REASONING AND ISSUE-BASED ARGUMENTATION

Quinsan Cao
Department of Architecture
Virginia Tech.
Blacksburg, VA 24061

Jean-Pierre Protzen
Department of Architecture
University of California at Berkeley
Berkeley, CA 94704

Design by argumentation is a natural character of design process with social participation. Issue-Based Information System (IBIS) is an information representation system based on a structured database. It provides a hierarchically linked database structure to manage design information and facilitate design by argumentation. In this paper, we explore the enhancement of IBIS with FRS (Fuzzy Reasoning System) technology. The FRS adds computationally implemented dynamic links to the database of IBIS. Such dynamic links can represent logic relations and reasoning operations among related issues which allows further clarification of relations among issues in IBS. The enhanced system provides a general framework to manage design information and to assist design reasoning, which in turn will contribute to machine assisted design. The final goal is to formulate a system that can represent design knowledge and assist reasoning in design analysis. The system can help designers in clarifying and understanding design related issues, requirements and evaluating potential design alternatives. To demonstrate the system and its potential use, we reexamine a design experiment presented by Schon and represent the design knowledge and reasoning rules of the architects with our system, FRS-IBIS.

1. Introduction

As the scale of today's design and construction projects expands, and their technological sophistication grows, so does the quantity and complexity of the information upon which the projects depend. The high levels of investments, the scope of committed resources, and the magnitudes of the impact of modern projects demand ever broader participation in architectural design and planning processes, and involve more elaborate discussions, extensive collaboration, and rigorous analyses. The tasks of generating, preserving, organizing, presenting and processing information become momentous and difficult.

It is widely accepted that today's architectural design process is a social activity with expanding participation of many parties (Cuff, 1991). The involved parties are no longer just architects, clients and engineers, but a variety of scientific and technological experts (e.g., health authorities, technical experts, special scientists, etc.), bankers, lawyers, community representatives, environmentalists, government agencies, media and general public.... The list may go on and on. With this wide social participation, designs, especially early phases of designs, are made through meetings of discussion, deliberation

R. Junge (ed.), CAAD Futures 1997, 771-786.
© *1997 Kluwer Academic Publishers.*

and debate. Such process is formalized as design by argumentation. During such process, effective representation of arguments and communication of information are among the most crucial factors for design progress. To meet the challenges, the industrial and academic communities are striving to develop more effective methods and tools.

One approach, which we call the instrumental approach, strives to provide modern computer, communication and information processing technologies to assist the architectural design process. Thus, Computer Aided Design (CAD) and Information Technology (IT), like image data generating, editing, storing, organizing, distributing, processing and representation, are now widely used in architectural design practices. Another approach is pursuing Artificial Intelligence (AI) technology applications for design professionals.

The instrumental approach, despite some impressive progress made, has not overcome all difficulties and leaves many frustrations. CAD, since its introduction into design practice some decades ago, has become a mainstay in most architectural offices. However useful and cost-effective CAD has become, it still leaves many design tasks, particularly in the early stages of a project, unassisted. CAD does not support, for example, the generation, exploration, development, and evaluation of initial design alternatives.

The Artificial Intelligence community strives to fill the gap left by the instrumental approach by attempting to capture the designers' knowledge and their reasoning patterns. What is known as expert system represents one such attempt. Expert systems are most successful in solving problems for which there is a definable goal, a clear set of initial conditions, and the outline of a procedure or a set of rules by which the goal can be reached. These conditions, as Rittel has shown, do not apply to design problems (Rittel, 1972, 1973). The expert system paradigm relies on a universal description that is applicable to all situations in practice, yet design problems "are essentially unique," and extremely context dependent. Design problems do not have "a definitive formulation," that is, what the goal is becomes clear only once the problem is solved. Design problems do not have a specific set of rules by which they get solved.

To overcome the difficulty of establishing and using rigid rules and procedures in design, many explorers turned to a new approach called Case-Based Reasoning (CBR). CBR attempts to directly use historical examples in forming new design rather than extracting rules. Much effort has been aimed to the development of record storing, editing, and distributing tools in order to establish case libraries. However, lacking of effective abstraction, (such as classification and recognition), reconstruction and evaluation schemes, CBR developments essentially remain as building case libraries or archives of historical examples. The case libraries provide little direct help in design for lack of mechanism to identify the relevant essence of the cases and associate it to a design goal, although they might help designers in seeking inspiration and reusable components from the history.

A new design, especially an exceptional one, is not an ensemble of old components made by following a set of rigid rules. For instance, to design a new type of "cash-less bank" that completely relies on electronic transactions, there is little can be learned from traditional bank designs. No case library could have examples of the new type. Moreover, there is little chance to build an expert system to help design, since nobody has ever seen one before not to mention establishing design procedures. Different people may have different ideas and expectations about the new type of bank. It is a great challenge for designers to systematically sort out existing information and knowledge related to the new design and innovatively produce a new design. However, some architects could handle the situation better than others. What is their skill? What is the process? Can it be understood? Similarly, many fundamental questions about design are yet to be answered, such as: What is design? Is design knowledge manageable and communicable? How do knowledge and intelligence contribute in design process?

In seeking answers to these questions, some explorers concentrated on observing designers' design process and analyzing their cognitive patterns. This approach is more directly aimed to address the fundamental questions about design. We call it analytical approach. While the instrumental approach is making admirable progress in providing convenience to designers, the analytical approach is seeking for long range and fundamental breakthrough. According to many explorers' observations of actual design process, such as Schon (1988) and Akin (1993), logic reasoning based on knowledge represented by rules (explicit and implicit) seems to be a fundamental element in the process of design.

1.1 A MODEL OF DESIGN PROCESS – COMPOSITION AND EVALUATION

Design is "a kind of dialectic between the designer's prestructuring of the world and the world as it is seen to be when examined in these terms" (Hillier). Obviously, "prestructuring" and "examine" are two different types of activities. It is like an argumentation involving: suppose I do such and such, then this and that would happen, and then is this acceptable? This line of self argumentation could repeat again and again with different initial assumptions until an acceptable solution is reached. Therefore, we can consider design an iterative process with two different phases alternately playing the dominant role. The two phases are speculative composition (prestructuring) and predictive evaluation (examining), respectively. Although the two phases are not always distinct, the differences and iterative progress of the two types of activities are evident.

In terms of speculative composition, the intelligence propelling such activities is definitely one of the most mysterious part of human brain's aptitude. It is responsible for the most innovative concepts and greatest discoveries. Although little is understood about such process, clearly such process is led by a goal and eventually constructs a form to accommodate the goal. Many attempts have been made to artificially reproduce or substitute such process, such as formulating design problems as finding a solution that fits all constrains or an optimal solution search with given criteria. However, the success of such attempts is severely limited by its linear and deductive approach, while inductive and nonlinear nature is essential in speculation.

Recently becoming more and more widely used evolution programs (such as evolution strategy and genetic algorithm) offer a significant breakthrough from the linear and deductive nature of earlier optimization schemes. It emulates the natural evolution process with selection based on the principle of best fits survive. The speculation (offspring) generating is essentially random. The evolution driven by random mutation over generations is directed by evaluative selection. Such method offers a "soft" connection between the speculations and the goal. Compared to other known methods, this method is more similar to human designers' behavior, a trial-and-error process guided by the evaluation regarding the goal. This method has been successfully used in electronic device layout design where the requirement and thus the evaluation is objective and relatively simple. It is difficult to predict how much such methods would succeed in general design, such as architectural design and planing. One of the most distinctive difference between electronic device layout and architectural design is in the design criteria. In architectural design, design criteria are much more complicated, often subjective and constantly evolving.

In general design process, design goals are often quite complex. It is often a set of many different and even conflict desires and concerns. For architectural design and planning, the situation is more severe due to broad social participation. It is often very difficulty to clarify and understand concerns of different parties, not to mention aggregating them into a single decisive opinion.

To address broad participation in design and planning practice as well as the complexity of design related issues, concerns and knowledge about them, Rittel developed the Issue-based Information System (IBIS) in late sixties with the objective of facilitating argumentative approach of design (Kuntz, 1970). Rittel pointed out that design process with social participation is conducted by argumentation. The system provides a framework to document, clarify and represent design and planning processes. Its hierarchical structure of linked topics, issues, positions, arguments and references provides a natural and convenient form to capture design related information and knowledge.

Due to the complexity of design goals and criteria, design evaluation for selection is also complicated. Rittel studied hierarchical structure of building performance evaluation and developed the concept of Deliberated Performance Evaluation (DPE) (Musso, 1967). The DPE method was further enhanced by Cao and Protzen for general applications and addressing uncertainties (Cao, 1992, Cao and Protzen, 1994). The DPE offers a framework to aggregates complicated design criteria into a single measurement index. However, the tree structure hierarchy and symmetric aggregation limited the use of PDE as a general form to describe design criteria and to represent design analysis and reasoning. To address such limitations, a general hierarchical mapping system based on fuzzy logic theory, Fuzzy Reasoning System (FRS), was developed (Cao, 1994). The FRS offers a more general format to describe design criteria.

1.2 THE NATURE OF DESIGN --- REASONING

As we have seen from above discussion, design process is really an argumentation process: a self dialectic arguing process within a single designer or a group arguing process among participants. Such arguing processes are reasoning processes, self reasoning and group reasoning. The composition and evaluation iterative process can be viewed as a iterative reasoning process of composing an hypothesis (an hypothetical premise) and then derive a conclusion. Schon (1988) observed that architects' design activities are nothing but reasoning with knowledge of "types" and "rules". In such reasoning process, "their patterns of inference were entirely familiar and conventional" and "the logic by which they passed from premises to conclusions was indistinguishable from the logic of everyday discourse." However, Schon also warned us about context relevance of design reasoning and implicit nature of design rules. Realizing the essential nature, argumentation or reasoning, of design process, solving the mysterious puzzle of design problems seems to become hopeful.

Obviously, the value of traditional wisdom about the model of reasoning, logic, is indisputable. There is no doubt that logic is very helpful in sorting and clarifying premises and to draw conclusions without conflicts. Tweed (1994) has demonstrated an example of using IBIS together with traditionally understood general reasoning pattern to design regulatory codes and standards.

However, traditional principle of inference, or logic, is strict and linear. Although it is mathematically perfect in theory, it's application in real life is limited. The "logic of everyday discourse" is not quite the same as the logic in theory. "The logic is linear and the world is nonlinear." In all our lives we are engaged in making guesses and conjectures, and finding evidences to support (not necessarily prove) them. We take reasonable judgments and actions based on reasonably well supported conjectures without worrying about a solid proof.

1.3 PRACTICAL REASONING --- PLAUSIBLE, SHADED AND FUZZY

The non strict logic in "casual" reasoning has been recognized and studied for long time. Some scholar argued that such studies can be traced back to the thoughts of Buddhism and Taoism thousands of years ago (Kosko, 1993). Black (1937) studied the vagueness in logical analysis mathematically with a multivalent logic model, which might be considered the beginning of formal study of such unconventional logic.

Polya (1954) recognized the importance of such practical reasoning in the course of scientific discoveries and named it plausible reasoning, in contrast to the traditional logic reasoning that he called demonstrative reasoning. He pointed out, "We secure our mathematical knowledge by demonstrative reasoning, but we support our conjectures by plausible reasoning ... Anything new that we learn about the world involves plausible reasoning, which is the only kind of reasoning for which we care in everyday affairs." Instead of having rigid rules and standards, plausible reasoning has fluid standards. In the effort of uncover and describe such standards, Polya used the concept of shaded inductive inference and, further, used probability theory to describe the inference patterns mathematically. The goal was to describe patterns of inference in which the premises and conclusions are neither absolutely true nor absolutely false, and the rules governing such inference are neither absolutely reliable nor completely unreasonable. In such theory, truth is relative (not absolute) and quantitatively measurable. Although, as Polya admitted himself, the detailed numeric formula of probability theory may not hold strictly and even show conflicts, the key concept that the plausible inference lies between the extreme limits set by demonstrative inferences (traditional logic) is well and sound.

Zadeh developed fuzzy logic and fuzzy set theory in early sixties that can be considered as another mathematical form of describing shaded logic and plausible inference. Compared to probability theory, it seems to have more freedom and flexibility in describing heuristic inferences, since the theory does not have to be associated with random experiments that are essential in probability theory. Recent development on fuzzy reasoning techniques based on fuzzy set and fuzzy logic theories offers a powerful tool to represent heuristic (mostly plausible) inference patterns. This paradigm is powerful in dealing with concepts and rules with uncertainty and vagueness, especially in real life situations where absolute precision has little relevance while a robust representation of relative trend is more valuable.

With the encouragement of Zadeh and Protzen, Cao (1994, 1996) developed Fuzzy Reasoning System (FRS) to describe reasoning process in design analysis. In FRS, design knowledge is represented as fuzzy rules (or functions) describing relations among various issues and their status. Fuzzy logic technique in such framework allows the representation to be more flexible and compact.

In this paper, we explore the enhancement of IBIS with FRS technology. The final goal is to formulate a system that can represent design knowledge and assist reasoning in design analysis. The system is expected to help designers in clarifying and understanding design related issues, requirements and evaluating potential designs. To demonstrate the system and its potential use, we reexamine a design experiment presented by Schon (1988) and represent the design knowledge and reasoning rules of the architects with our system.

2. Issue-based Information System (IBIS)

IBIS offers a natural framework to record information as argumentation (Grant, 1992). It consists of a hierarchical network of essentially four types of nodes: topic, issue, position and argument. The nodes are connected by various types of links. Figure 1 shows the basic structure of IBIS network with all major types of nodes and links.

A topic serves as a starting point and defines the domain of the discussion in terms of its relevance. Issues related to the topic are raised by the participants during the discussion. Positions are taken by participants regarding their opinion about the issues. To justify their opinions, participants often support or object certain positions with various arguments. On the other hand, issues often trigger or introduce other issues in various occasions. To record and represent the information generated in such process, the framework of IBIS shown in Figure 1 is obviously quite natural and convenient. There have been many applications and developments of IBIS, such as IBIS with hypertext programming (Hashim, 1990).

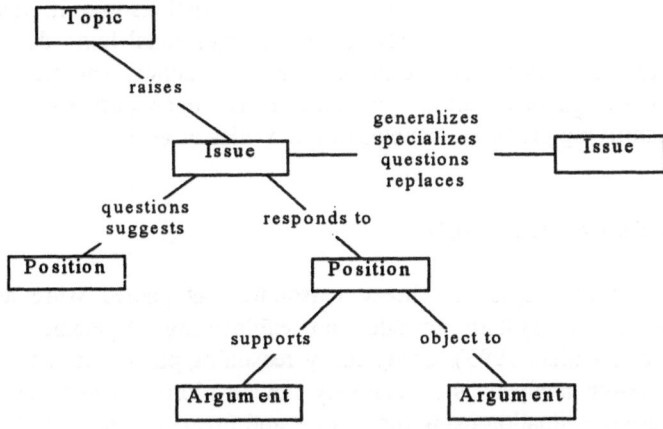

Figure 1. Framework of IBIS with four types of nodes and various types of links for recording argumentation.

Traditional wisdom about general patterns of reasoning has also been usedto enhance IBIS to in resolving design and planning problems. Tweed (1994) used the general framework of reasoning (Toulmin, 1984) to further detail IBIS when representing knowledge to design regulations and standards. The general framework of reasoning is illustrated in Figure 2.

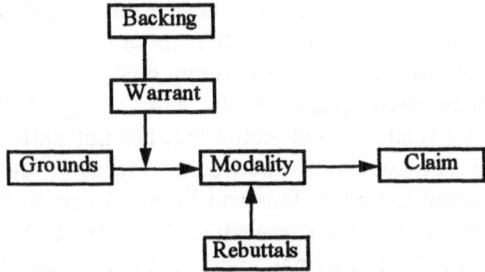

Figure 2. A general framework of reasoning, which offers a micro-structure to clarify the link between the position and argument nodes in IBIS.

This framework can be used to specify the micro-structure between the position and argument nodes in IBIS, where the "claim" represents a position, "grounds" represents the arguments, and "rebuttals" support and modify the applicability of the argument. Obviously, with such a micro-structure, the IBIS representation of knowledge can describe argumentation and reasoning with more clear definition of relations.

On one hand, the detailed micro-structure allows more precise description of the knowledge. On the other hand, the micro-structure also makes the system more strict and rigid. Such rigidity makes the use of the system cumbersome in many practical situations. As we discussed in the introduction, the traditional logic does not support shaded or plausible inferences that are more useful in practice. The paradigm of fuzzy logic and fuzzy reasoning offers an improved mechanism to support shaded and plausible reasoning by quantifying premises, conclusions and rules of inference.

3. Fuzzy Reasoning System (FRS)

The paradigm of fuzzy logic and fuzzy reasoning has gained wide application in information technology, system science and engineering, especially in intelligent systems (Yager and Zadeh, 1993). Using fuzzy reasoning paradigm to represent design knowledge for design analysis is a relatively new exploration, such as treatment of uncertainty in design evaluation using fuzzy logic and fuzzy numbers and functions (Cao and Protzen, 1994), and design analysis system using FRS (Cao, 1996). The main function of such system is to provide a general framework to describe "soft" (fuzzy, shaded or plausible) inference rules, premises and conclusions. One of the most important properties of such fuzzy reasoning is that the limiting extremes of the inferences must agree with traditional logic, while intermediate situations fall in-between the extremes.

The essential concept of the FRS is illustrated in Figure 3 and Figure 4. The FRS consists of Fuzzy Reasoning Charts (FRC) and Fuzzy Reasoning Equations (FRE) (Cao, 1994). The FRC and FRE are two different forms of representation of a fuzzy reasoning network with nodes and links. Each node corresponds to a fuzzy variable that represents an interested issue and its status. The links represent different types of relations among the issues. With the FREs, the links in the FRCs not only depict relations among issues but also quantitatively describe the relations among their status. With such a system, the reasoning can be implemented quantitatively with a digital computer. It has the capability of representing relations and status that are not quite crisply clear (neither "true" nor "false" but somewhere in between). Figure 3 demonstrates the FRCs describing four fundamental logic relations and Figure 4 demonstrates those describing two typical combinations of them for aggregation. It is obvious that the FREs presented in the figures agree with the traditional logic at the extreme limits, absolutely true (represented with value 1) or false (represented with value 0).

Compared to IBIS, FRS emphasizes on the positions taken on various issues and the interactive relations among these positions. With such relations, changing of the position on one issue is propagated through the network which properly represents (predicts) new positions on other related issues. FRS offers a dynamic format to implement the essence of IBIS.

It is straight forward to see, from Figure 3 and Figure 4 in comparison with Figure 1, that the FRS can be conveniently used to implement and describe IBIS including the micro-structure described in Figure 2. In FRS, topics, issues and arguments are represented by nodes in the network of FRC and corresponding variables in the FRE. The positions taken on each issue are simply represented by the status measurement of the corresponding node or the value of the corresponding fuzzy variable. The links between issues are represented by fuzzy logic functions or fuzzy rule based mappings that specify the relation between the position of linked issues. Compared to the original IBIS, FRS presents the links in more detail with clearer definitions.

For example, the relation between an issue and the issues specializing the original issue in IBIS are defined more clearly in FRS as aspect aggregation or alternative aggregation, shown in Figure 4. Different aggregation functions represent different ways for the positions on the specialized issues to contribute to the position on the general issue. The aspect aggregation, for instance, requires a positive position on every specialized issue (numerically represented by 1.0 or near 1.0) for a positive position on the general issue;a negative position (0.0 or near 0.0) on any specialized issue leads to a negative position on the general issue.

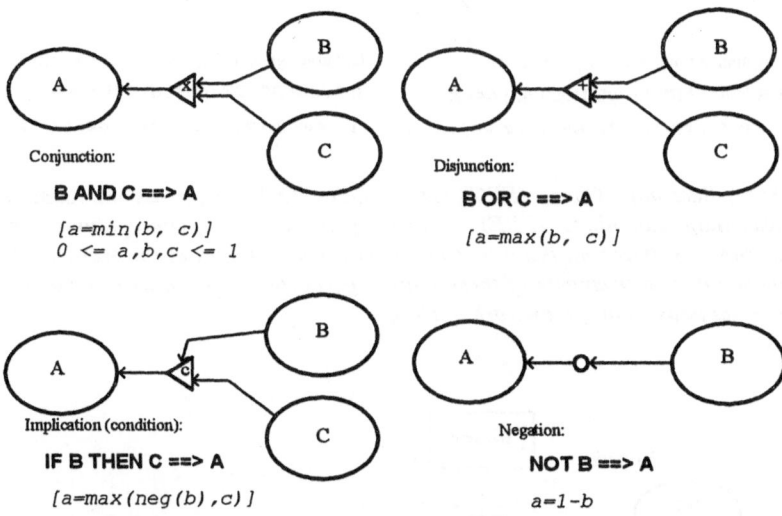

Figure 3. *Fuzzy Reasoning Charts (FRC) of four basic types of logic relation, with a name of the relation, corresponding logic statement, and associated Fuzzy Reasoning Equations (FRE). The FRE operates on the variables whose values represent the status measurement of their corresponding nodes. An issue in IBIS corresponds to a node (issue) in FRS, and a position suggesting an issue in IBIS corresponds to the status measurement (value) of the node (variable) in FRS. An argument supporting or objecting a position in IBIS corresponds to a node in FRS whose status measurement contributes (directly or through a function, conjunction, disjunction, etc.) to the status measurement of the original issue to which the position regards.*

We have shown in Figure 2 the general reasoning framework depicting an argumentation supporting a position. Such framework can also be conveniently represented by a combination of the four fundamental relations shown in Figure 3. The FRS corresponding to the reasoning pattern of Figure 2 is shown in Figure 5.

780

In Figure 5, the "backing" that supports the "warrant" to the "claim" is given as an inactive comment that does not affect the network's operation actively. The modality is represented by the function (warrant) AND NOT (rebuttals). Although the construction of FRS corresponding to a reasoning pattern may not be unique, there is little room for ambiguity once an FRS representation is established. In addition to the basic first order logic relations, the FRS network also describes the strength measurements of the nodes, such as grounds, rebuttals and claims. In other words, FRS can describe not only how but also how much.

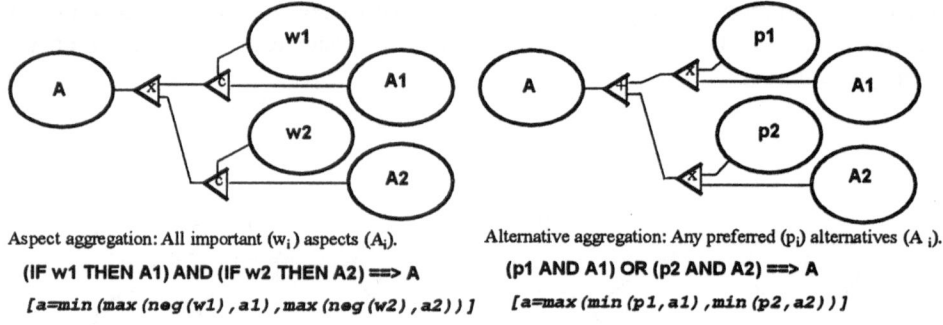

Aspect aggregation: All important (w_i) aspects (A_i).

(IF w1 THEN A1) AND (IF w2 THEN A2) ==> A

$[a=\min(\max(\text{neg}(w1),a1),\max(\text{neg}(w2),a2))]$

Alternative aggregation: Any preferred (p_i) alternatives (A_i).

(p1 AND A1) OR (p2 AND A2) ==> A

$[a=\max(\min(p1,a1),\min(p2,a2))]$

Figure 4. Fuzzy Reasoning Charts (FRC), corresponding logic statements and associated Fuzzy Reasoning Equations (FRE) of two typical types of aggregation, aspects and alternatives. When an issue specializes into several issues in the framework of IBIS, the status measurements of these issues aggregate to the status measurement of the original issue in the framework of FRS.

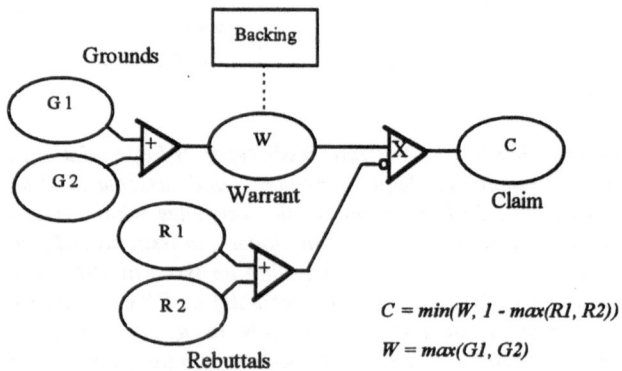

$C = \min(W, 1 - \max(R1, R2))$

$W = \max(G1, G2)$

Figure 5. FRS representation of general reasoning pattern with FRC and corresponding FREs. All elements are represented by active nodes except the "backing" that is given as an inactive comment.

With the new paradigm of FRS-IBIS, different designers' design knowledge can be represented separately or collectively. Representation of subjective preference and context dependence can be accommodated either by separated networks or collective network with conditioned connections. The fuzzy variables, fuzzy rules and fuzzy logic functions allow complex relations to be described in a compact form that is suitable for computer processing. In addition to the fuzzy reasoning patterns shown in Figure 3, 4 and 5, fuzzy rule based general mappings can be used to describe more general and complicated relations if necessary. Above all, the FRS-IBIS is not only a compact, graphical and digitally implementable representation of design information, argumentation and knowledge, it is also a very flexible and thus dynamic media. Since the design problems is a dynamically varying problem with its definition being evolved and modified during the whole problem solving and resolving process, a flexible and dynamic media for information management during such process is critically important. The net work of FRS-IBIS system is conveniently modifiable during such dynamic evolving process to properly reflect the changes of problem definition, argumentation knowledge, and solution.

4. Representation of Architects' Reasoning Patterns

As an example to demonstrate the use of the FRS-IBIS system to represent design knowledge, in this section we reexamine an experiment of design exercise reported by Schon (1988). In the experiment, invited designers were presented with a drawing shown in Figure 6. They were told that the drawing represents the "footprint" of a generic design of suburban branch library. Their task was to analyze the entrance locations and provide guidelines for entrance location selection. Each participant was asked to independently describe his opinion and the reasoning supporting it.

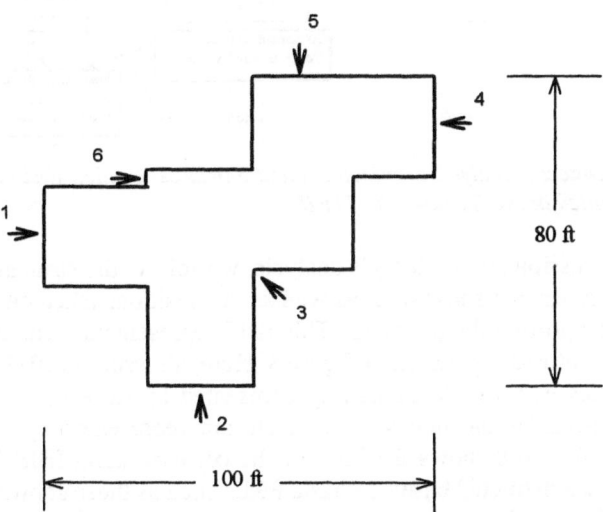

Figure 6. "Footprint" of a building with six generic alternatives of entrance positions.

A practicing architect's, Henry's, description can be represented in the framework of IBIS shown in Figure 7. Regarding the topic of good entrance position, he raises two issues, workability and interesting (features). Each issue further specializes into more issues. The issue of workability specializes into three issues: (1) spatial order of reading rooms and stacks, (2) clarity of entrance from street, and (3) efficient traffic control of materials with narrow end entrances. The positions and supporting arguments are: narrow end entrances offer series internal order and clarity of street entrance as well as efficient traffic control and therefore works fine. The issue of easy traffic control is conditional, depending on the location of the library. It is only important if the library is in a city center, otherwise the issue is negligible. The other issue, beside workability, interesting (features) can be replaced by the issue of poetic attraction. The position and arguments are entrances coming from middle of the form are rather poetic. However, as Henry said, "Poetry is only for poets."

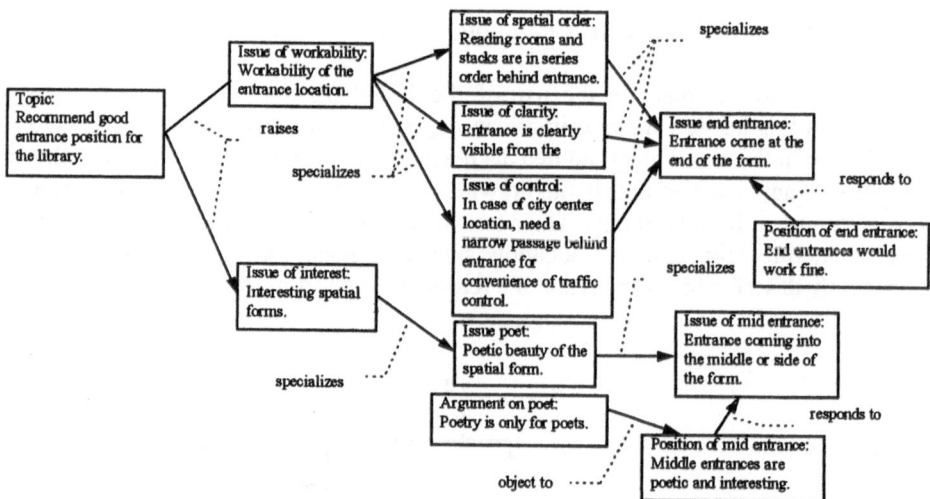

Figure 7. An architect's analytic arguments about entrance location design of a county library represented in the framework of IBIS.

In the above discussion about Henry's analysis, we follow the structure of IBIS starting from a topic to issues and sub-issues, as well as the positions taken on various issues and the arguments supporting the positions. This IBIS representation can be further clarified in the framework of FRS as shown in Figure 8. Here, all issues in IBIS are represented by nodes or variables in FRS. Positions are represented by the values of the variables or status measurements of the nodes. Arguments are represented by functions that link nodes to nodes. Figure 8 shows the FRC of the corresponding IBIS in Figure 7, while FREs that can be constructed based on basic FREs such as those shown in Figure 3, 4 and 5 are omitted for brevity. Compared to the IBIS, FRS carries more detail of design reasoning information. It is more definitive and, therefore, requires more clear specification for the designers. In the reinterpretation of this particular example, we took the freedom of assuming that Henry implied all specialized issues' positive positionsare necessary for a positive position on the issue of workability and thus form a component aggregation relation.

Henry grouped the entrance alternatives shown in Figure 6 into two groups. The first group consists end entrances, number 1, 2, 4 and 5, that"come in at the end." The other group consists of middle or side entrances, number 3 and 6, that"coming in the middle of the form." Supported by the given reasoning and arguments, he concludes, as one would find by checking each entrance alternative with the FRS shown in Figure 8, that the entrance number 3 and 6 are more difficult, although poetic. He recommends the other alternatives, number 1, 2, 4 and 5.

To further clarify Henry's reasoning, Henry could specify his criteria regarding the classification of end, middle and side entrances. The specified criteria can be described in FRS. If the criteria can be clarified down to objectively measurable parameters, then the evaluation can be conducted automatically with the FRS in a computing and measuring machine.

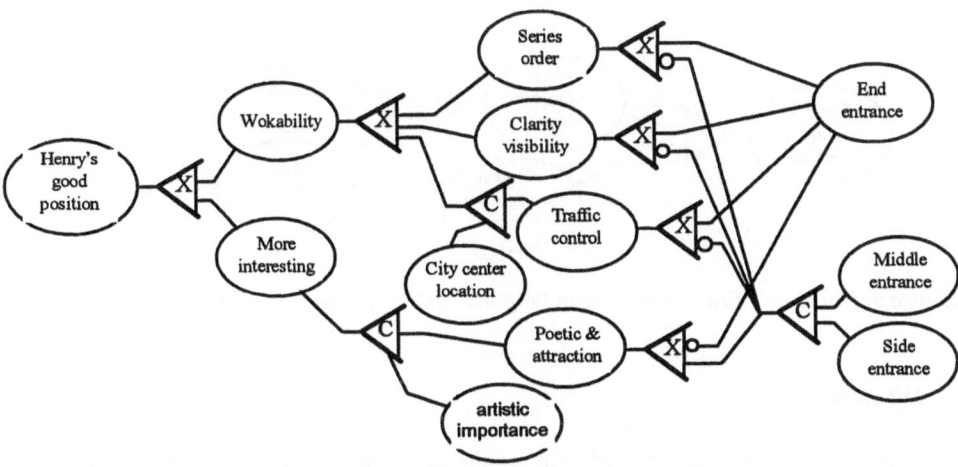

Figure 8. FRC representation of Henry's reasoning and analysis for good choices of entrance position. Examining each alternative of entrance positions through the network, one would find the alternatives that near the end of the building (gives high truth measure as "end entrance") agree with Henry's recommendation.

Quite different from Henry, Benny, a design instructor, argued that middle entrances (#3 and 6) are better than end entrances. His main argument is that end entrances make large area of the construction into traffic area for people to walk-by instead of useful area for people to study. Yet, he also realizes that such argument depends on the size of the library. Furthermore, he pointed out that if the library is on the side of a street, an end entrance near the street side is convenient.

784

Another participant, Franz, is a practicing architect and also a design teacher. His conclusion is similar to that of Benny's, although based on slightly different argument. His major point is that entrances should be close to geometric center except for certain exceptional uses. Benny and Franz's reasoning can be expressed in FRCs shown in Figure 9.

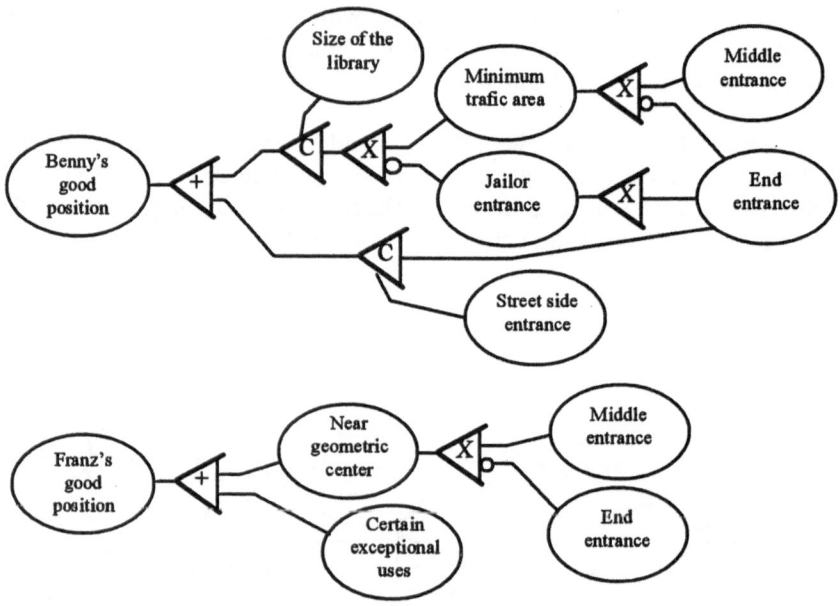

Figure 9. Benny and Franz's reasoning for good choice of entrance position.

5. Discussions

FRS-IBIS system not only records the architects' opinion but also carries the architects reasoning system and capable of conducting analysis accordingly.

Instead of trying to exhaust all possibilities (every possible entrance position in the example) with countless number of rule or cases, the FRS uses a mapping function based on reasoning that links certain characteristic measurements (such as distance to the geometric center in the example) of the possibilities to the conclusion. Therefore, the system is capable of dealing with new situations on its own without additional specific instructions from the architect regarding each and every case. In fact, this is the way human keep and use their knowledge. We do not and can not learn and remember all possible cases one by one. We use our general knowledge and reasoning skill to deal with new situations and new possibilities all the time, especially in the process of creative design.

The paradigm introduced in this article is different from expert system's approach where a universal applicable and complete knowledge system is expected. The approach we take is to allow an architect to express his/her opinion, knowledge, reasoning pattern and analysis criteria, at a certain condition under a particular context. The system is constructed for a particular design domain and even a particular design problem. For a different design problem a different system with different nodes and links are constructed. Furthermore, the system is being continuously modified and reconstructed during the process of problem resolving, for design criteria are evolving throughout the process.

6. Conclusions

In this paper we introduced Fuzzy Reasoning System (FRS) as an enhancement to the Issue-based Information System (IBIS). The enhanced FRS-IBIS not only qualitatively but also quantitatively describes the relations among various issues involved in a design problem. This system forms a general tool to facilitate design by argumentation with objectified evaluation criteria and evaluation process.

With the Fuzzy Reasoning Charts (FRC), the FRS-IBIS describes the relations among various issues in a not only more detail but also more clear form. With the Fuzzy Equations (FRE), the FRS-IBIS allows the design reasoning process be communicated not only between human designers but also between human and machines. In FRS-IBIS, the design reasoning and argumentation process are quantified through fuzzy logic techniques. With such quantification, design reasoning and argumentation process can be dictated to and, in turn, performed by a machine, a digital computer for instance. This automated reasoning capability provides an essential element, evaluating system, for an automated design assisting machine, for evaluation and composition are the two most fundamental components in a design process.

An example of hypothetical design problem given in this article, and an example of a real design analysis (Cao, 1994), demonstrated the potential of using FRS-IBIS in practices. Applications of the system in design of practical scales need to be further studied. As the FRS-IBIS offers a potential sub-component in a design assisting machine, the possibilities of combining the FRS-IBIS into an automatic/semi-automatic design assisting machine also calls for further investigation. A potential candidate for such design assisting system is based on evolution programs, where the FRS-IBIS can be used for evaluation and selection and the final design will be evolved through such selection and random mutation. Such a automated design assistant system is calling for further investigation.

7. References

Akin, O. (1993) "Architects' reasoning with structures and functions," *Environment and planning B: planning and design*, Vol. 20, 273-294.

Black, M. (1937) "Vagueness: an exercise in logical analysis," *Philosophy of Science*, Vol. 4, pp. 427-455.

786

Cao, Q. (1992) "Deliberated performance evaluation and application in comfort assessment," Ph.D. dissertation, University of California, Berkeley, California.

Cao, Q. and Protzen, J-P. (1994) "Aggregation and uncertainties in deliberated performance evaluation," in *Models and Experiments in Risk and Rationality*, Kluwer Academic Publisher.

Cao, Q. (1994) "Design Analysis Based on Fuzzy Reasoning," *Artificial Intelligence in Design '94*, pp. 679-696, Kluwer Academic Publishers.

Cao, Q. (1996) "A Knowledge-Based Design Analysis System," *Environment and Planning B: Planning and Design*, 1996.

Cuff, D. (1991) *Architecture: The Story of Practice*, MIT Press, Cambridge, pp. 72-84.

Grant, D. (1992) "Argumentative information and decision systems," *Design Methods: theories, research, education and practice*, 26, 1524-1588.

Hashim, S. H. (1990) *Exploring Hypertext Programming: writing knowledge representation and problem-solving programs*, Windcrest Books, Blue Ridge Summit, PA.

Hillier, W. and Leaman, A., "How Is Design Possible? A Sketch for a Theory," DMG/DRS Journal, Vol. 8, No. 1.

Kosko, B (1993) *Fuzzy Thinking: The New Science of Fuzzy Logic*, Hyperion, New York.

Kuntz, W. and Rittel, H. W. J. (1970) "Issues as Elements of Information System," Working Paper 131, Center for Planning and Development Research, University of California, Berkeley, California.

McCall, R. (1991) "PHI: a conceptual foundation for design hypermedia," *Design Studies*, Vol. 12, No 1, January.

Musso, A. And Rittel, H. R. (1967) "Measuring the performance of buildings," report about a pilot study, Washington University, St. Louis, Missouri, September.

Polya, G. (1954) *Induction and Analogy in Mathematics*, Volume I & II of mathematics and plausible reasoning, Princeton University Press, Princeton, New Jersey.

Rittel, H. (1972) "Interview with D. P. Grant and J. P. Protzen", in D. Grant (edt.), *Design methods group: 5th anniversary report*, DMG OP1, Dept. of Architecture, University of California, Berkeley.

Rittel, H. (1973) "Dilemmas in a general theory of planning", *Policy Sciences*, 4, 155-67.

Toulmin, S., Rieke, R. and Janik, A. (1984) *An Introduction to Reasoning*, 2nd eds., Macmillan Publishing Company, Inc., New York.

Schon, D. (1988) "Designing: Rules, types and worlds," *Design Studies*, Vol. 9, No 3, July, pp. 181-190.

Tweed, C. (1994) "Intelligent Authoring and Information System for Regulatory Codes and Standards," The International Journal of Construction Technology, Vol. 2, No 2, pp. 53-63.

Yager, R. and Zadeh, L. A., (eds.) (1993) An Introduction to Fuzzy Logic Applications in Intelligent Systems, Kluwer Academic Publishers.

ANALOGICAL REASONING AND CASE ADAPTATION IN ARCHITECTURAL DESIGN: COMPUTERS VS. HUMAN DESIGNERS

MAO-LIN CHIU
Department of Architecture
National Cheng Kung University
No.1, University Road, Tainan, Taiwan, R.O.C.

SHEN-GUAN SHIH
Department of Architecture
National Taiwan Institute of Technology
No. 43, Section 4, Keelung Road, Taipei, Taiwan, R.O.C.

This paper depicts the studies of the differences between human designers and computers in analogical reasoning and case adaptation. Four design experiments are undertaken to examine how designers conduct case-based design, apply dimensional and topological adaptation. The paper also examines the differences of case adaptation by novice and experienced designers, and between human judgement in case adaptation and the evaluation mechanism by providing similarity assessment. In conclusion, this study provides the comparative analysis from the above observation and implications on the development of case-based reasoning systems for designers.

Keywords: Case-based reasoning, analogical reasoning, case adaptation, computer-aided architectural design

1. Introduction

Design cases were considered as the design solution or condensed knowledge of previous design experience (Dave et. al. 1992, Rosenman et. al. 1991). Cases are described as design stories, scripts, frames, etc. (Leake 1996, Oxman 1994). Kolodner (1993) defined "a case, which generally represents a concrete situation, integrates a multitude of complex information in a very concrete way." Reasoning with cases involves case retrieval, adaptation, and justification.

Analogical reasoning consists of (1) transferring knowledge from past problem solving episodes to a new problem that shares significant features with corresponding past experience, and (2) applying the transferred knowledge to construct solutions to new problems (Chen 1991). In the analogical reasoning process, case adaptation is the

R. Junge (ed.), CAAD Futures 1997, 787-800.
© 1997 *Kluwer Academic Publishers.*

788

fundamental task for solving the problem. Case-based reasoning (CBR) is a research paradigm that also uses the adaptation machanism for solving a new problem from past experience. Analogical reasoning can be case-based reasoning, but is more general in the problem-solving process.

Furthermore, in the case-based reasoning process as shown in *Figure 1*, dimensional, topological, and other factors may affect the adaptation of design solution (Maher 1995, Hua et. al. 1992). To make case adaptation useful, justification need to be evaluated through similarity assessment.

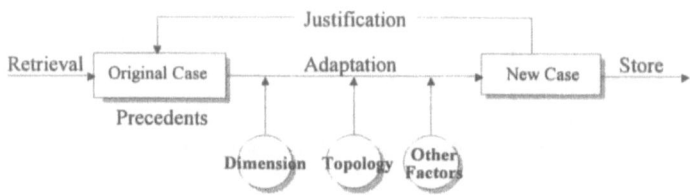

Figure 1. The Case-based Reasoning Process and Adaptation

In the following sections, ways of performing case-based design, making case adaptation, and a computational approach for case adaptation are explored.

2. Case-based Design

Design is a situated activity. The design process is considered as a learning process in which designers are learning from analogy by deduction and induction (Chen 1991). Two situtations may be occurred in adaptation as shown in *Figure 2*. In the situation A, one case may generate many new cases by analogical reasoning by deduction. In situation B, many cases may be retrieved and developed into to one new case by analogical reasoning by induction. In the adaptation process, dimensional or topological adaptation can be used.

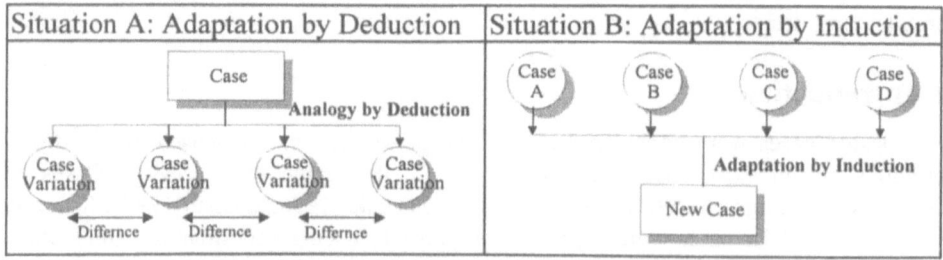

Figure 2. Two Situations of Case Adaptation

In the past three years, four case-related design experiments undertaken in NCKU are examined as shown in *Table 1*. These experiments provide the fundatation for the following discussion, i.e., the comparison between human designers and computers in case adaptation.

Table 1. Summary of Design Experiments

	Design Experiments	Contents	Objectives
1.	Elementary School Layout	Case-based Design	1.The role of case studies in design 2.Adaptation mechanism
2.	Color Museum	Case-based Design	Case adapation
3.	Urban Housing Design	Analogy by Deduction	Prototypes, generation, and variations
4.	Street Facade Renovation	Analogy by Induction	Transformation and substitution

2.1 The Role of Case Studies in Design

In the first experiment, the elementary school layout, the role of case studies play a critical factor for design. Cases are often used as "short-cut" for conceputalization and transformation of knowledge. For example, the spatial relationship of original design was reconfigured by adaptation from case studies as shown in *Figure 3*.

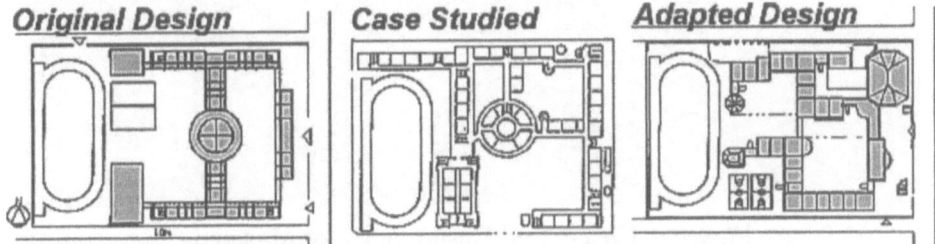

Figure 3. Case Adaptation in Elementary School Layout

2.2 The Use of Cases in Design

The second experiment, the color museum project, adopts case-based design approach which designers retrieve cases for initiating conceptual design. Cases selected are driven by the concept, the structure, and the form. Detailed descriptions can be found in (Chiu 1997). For example, Frank Gehry's Vitra Design Museum was chosen because of the form. While the site is different in size, new design was generated by abstraction of initial case, and dimensional and proportional change, followed by topological changes, as shown in *Figure 4*.

The above two experiment provides why and how cases can be used in design. More importantly, adaptation are applied by analogical reasoning. To make the demonstration explicit, we conduct two other design experiments based on two situations as mentioned. Each experiment is assigned to two testing groups. Group A consists of novice designers who have no previous design experience and Group B consists of experienced designers.

790

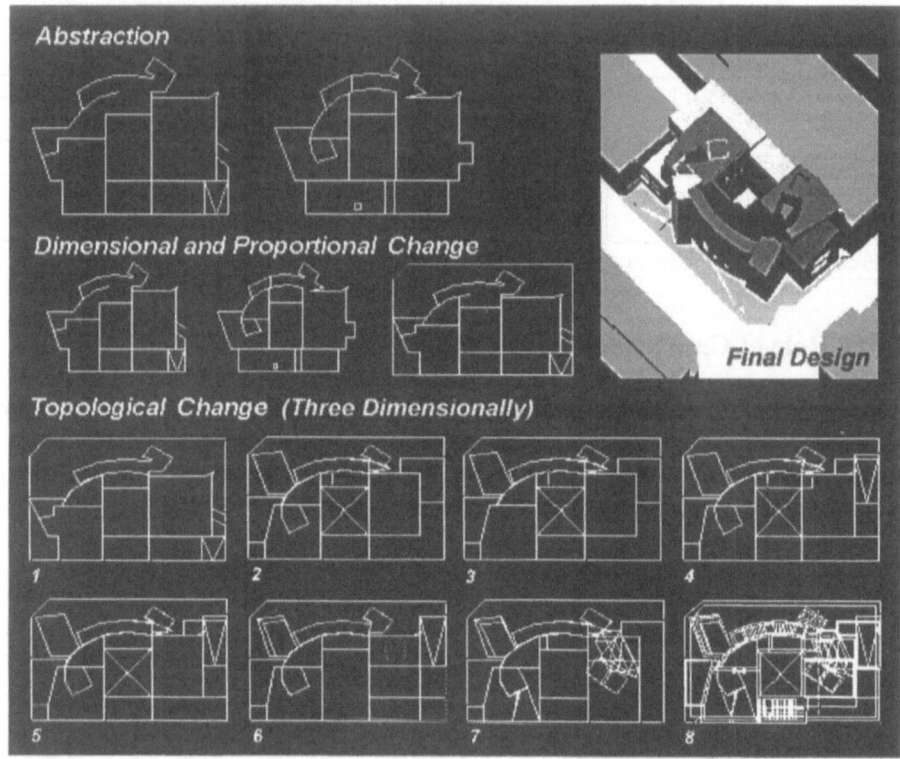

Figure 4. Case-based Design in Color Museum

Therefore, four steps are undertaken: (1) first examines how designers apply dimensional adaptation and topological adaptation; (2) continues the previous experiments and explores the differences of case adaptation by novice and experienced designers; and (3) finally examines the differences between human judgement in case adaptation and the evaluation mechanism by providing similarity assessment.

2.3 Dimensional and Topological Adaptation

The third design experiment of a housing project was given to Group A for exploring the dimensional and topological adaptation. Details are given in (Chiu 1996). Prototypes are derived from design knowledge and developed as a conceptual model (Gero 1991). Two prototypes of each housing unit are given to designers. Based on their preference and judgement, designers can only choose one prototype and try to fit into different sites of incremented width of 1.5 meter from 6 meter to 10.5 meter, as shown in *Figure 5*.

The design result indicates that one third of designers basically maintain the original structure, one third maintains the structure with minor changes of width, and one third change the structure greatly. Dimensional changes, including change of proportion and scales, work easily within the 1.5 meter incremental range, and topological changes are occurred beyond the range. Designers tend to give up further development beyond the 3 meter range. Only few designers are able to change dimension and topology simulateously. The most difficult task for designers is to maintain the characteristics of the prototype.

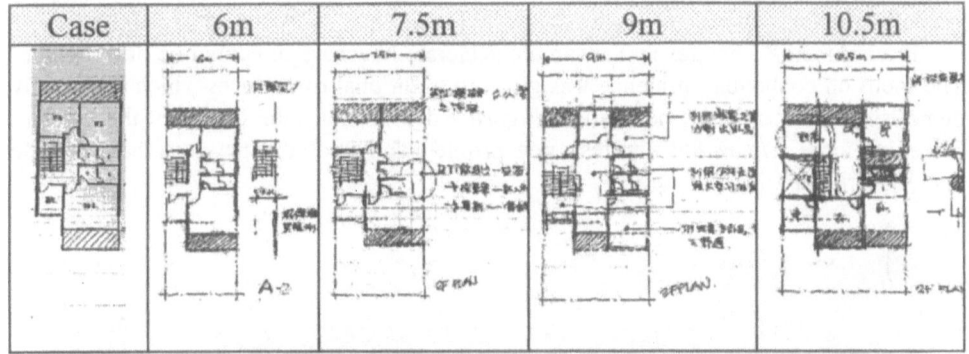

Case	6m	7.5m	9m	10.5m

Figure 5. Changes of The A-Type Unit Plan in Various Widths

Parameter adjustment is a technique for interpolating values in a new solution based on those from an old one. The above adaptation process can be modeled as the parameter adjustment of the housing unit. The SAR (Stichting Architecten Research) theory was applied for defining the structure (Wiewel 1976). The supports (the structure and the partitions) and the infills (space units) are shown in Figure 6. As accepted and feasible solutions, most of the developed schemes by designers can be simulated by parameter adjustment. Therefore, new cases can be developed by dimemsional and topological adaptation. However, the limitation of dimensional and topological adaptation is clear as shown in the experiment.

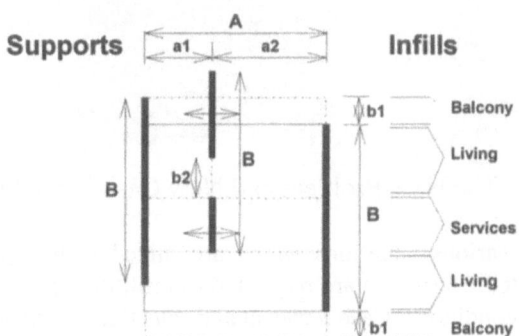

Figure 6. Parameter Adjustment of A Housing Unit

2.4 Substitution and Transformation

The fourth experiment is also given to Group A for exploring the substitution and transformation in case adaptation. Reinstantiation is used to instantiate an old solution with new objects for suggesting substututions. Transformation is a process of adapting an old solution for a new problem by structurally deleting, inserting, substituting, or adjusting parts of the solution.

Environmental changes in Taiwan are acclerated because of economic development. The focus on contextual influence was given to traditional row houses which are rapidly demolished and needed to be infilled. *Figure 7* demonstrates the facade of the Di-Hwa Street in Taipei. *Figure 8* illustrates a new problem and a possible new case based on the analytic structure.

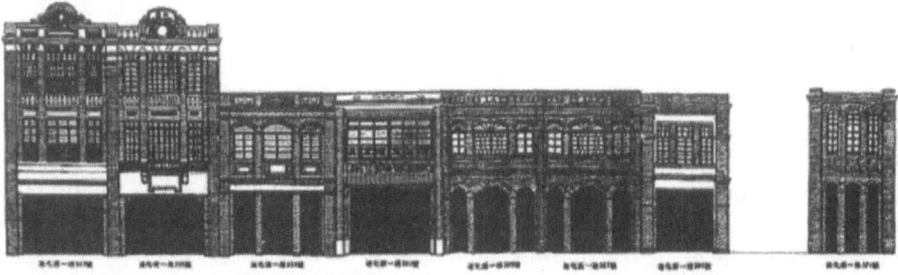

Figure 7. The Original Facade of the Di-Hwa Street in Taipei

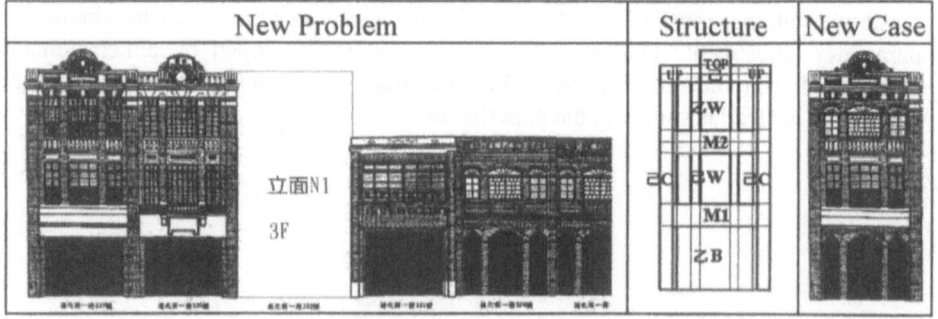

Figure 8. The New Problem and New Case in the Di-Hwa Street

Furthermore, various situations which are mixed with two-story and three-story houses are simulated. Designers are requested to infill the empty block in each situation based on existing conditions. Case retrieval and selection is critical to adaptation. *Figure 9* demonstrates the adapation of facade C derived from neighbour A and B, i.e. components of A and B are retrieved. In the case of substitution, facade A is selected, partial substitution is implemented by using componts of facade B. In the case of transformation, a two-story building can be tranformed into three-story building by inserting the middle section.

Reintantiation of an old solution may be used when the structure of the solution to the problem and an old case are the same, but roles in the problem solution may be filled differently than roles in the old case. Local reinstantation may be applied when only parts of an old soultion need to be reinstantiated. As shown in *Figure 9*, the supports of the street facade is connected to certain contexts. Based on the present of certain features, each building facade as precedent can be organized and retrieved. Coyne and Yokozawa (1992) indicate that a connectionist approach can be used for automating the classification and designing with precedents.

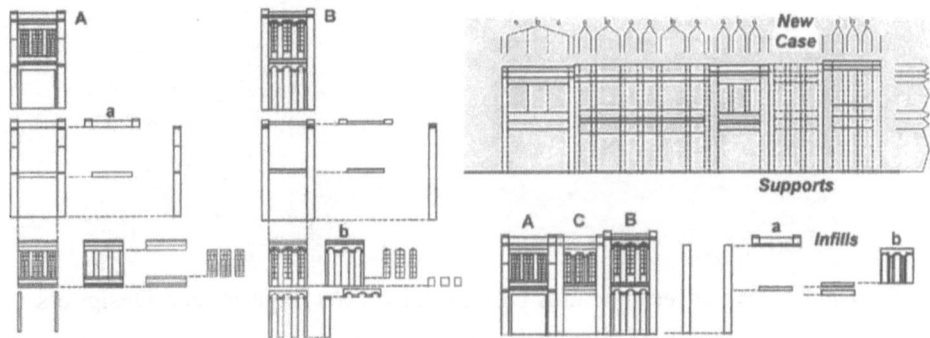

Figure 9. Case Adaptation by Substitution and Transformation

The result of 17 designers from Group A shows that most of designers conduct analogical reasoning based on the direct neighbours (right and left sides) of the empty blocks. Few designers are based on all facades in the street. This phenomena can consider that designers tend to use partial substitution of the all facade. Meanwhile, all designers are based on supports and infills to construct the basic frame, while few rules are implemented and constraints are imposed.

3. Case Adaptation by Novice vs. Experienced Designers

3.1 Preliminary Comparison

The above design experiments were also given to the second group of designers, i.e. experienced designers. In the third experiments, the results show that experienced designers are more capable to resolve topological adaptation, and topological adjustment generally follows dimensional changes. The reasoning time of experienced designers is usually shorter than the novice designers. Meanwhile, novice designers occcassionally misuse cases. Generally, in Group A, the ability of topological adaptation is weaker than the dimensional adaptation. One the other hand, in Group B, experienced designers are more capable to conduct dimensional and topological adaptation simultaneously.

In the fourth experiements, experienced designers typically have individual intentions and consideration based on semantic relations. More importantly, attitudes toward historical and environmental context vary, and could be harmony, neutral, or constrast to the existing situations. *Figure 10* demonstrates two examples by experienced designers in the street facade experiment.

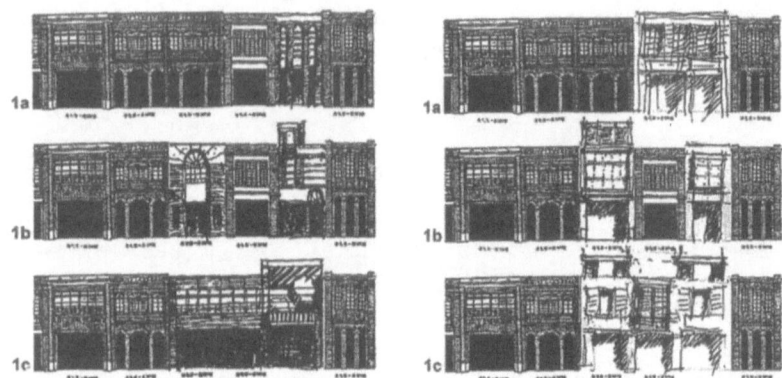

Figure 10. Two Street Facades Designed by Two Experienced Designers

Can CBR systems provide innovative or creative design become an interesting issue? Innovative design may arise during routine design when a new requirement is introduced that takes the design away from routine, requires new components and techniques. Since some design knowledge remains constant, designers do not have to re-develop a new design schema, parts of the original building are transformed while some new parts are built. Often designers modify other architects' schemata because their design requirements are almost identical.

3.2 Case Similarity Assessment

Furthermore, similarity is addressed by most designers in the design experiments. Therefore, similarity assessment is a major concern in adaptation. Apparently, the selected street facade is built in the same period. The building styles can be said as similar or uniform. Designers tend to retrieve adjacent cases for developing the new cases. It would be interested to examine in the semi-uniform or hetrogeneuous situation of the street facade. However, no designers want the new cases to be identical to the old ones. Variation are generated in the substitution and transformation process. When rules or contraints are applied, the level of variation can be manipulated.

Case-based research assumes that cases whose solutions are most similar to a new solution will most likely be useful in designing it (Dzeng 1995). Generally, a new case is developed and retrieved by key features (the structure or components). The similarity assessment by human judgement may be inconsistent due to the complexity of contexts and designers' intentions.

Kolodner (1993) argued that some classes of matches, "easy-adapted" matches, should be referred over "hard-to-adapted" matches during retrieval. This study uses nearest neighbour retrieval in case retrieval instead of using the index-based method which have been studied in most CBR research. Typically, the query case (Q) and a case (C) in the case-base S(Q,C) is the weighted sum of similiarity of each attribute: $\Sigma Wi *$ s(Qi,Ci) / SWi, where Wi is the weight of the attribute, s(Qi,Ci) is a similarity between the value of the i-th attribute of Q and C. Traditional implementations would compute the similarity value for all cases, and sort cases based on their similarity. However, this is a time consuming task as computing time increase linearly as the number of cases in the case-base and as the number defined attributes.

If the user can specify and assign similarity value, then the differences between human judgement in case adaptation and the evaluation mechanism by computers is quite narrow. Therefore, for each assessment, a similarity value will be assigned using similarity between values of attributes specified by the user and values of case created in the previous stage. Calculation is similar to weighted nearest neighbour, except that not all attributes are involved. Then computation time will be saved. While design computation will be beneficial from converting heuristics into mechanism, problem-solving requires the transformation of non-routine problems into routine problems (Maher et. al. 1995).

4. A Computational Approach

Learning from the findings of design experiments, we have implemented a case match system as shown in *Figure 11* for examining case adaptation mechanism.

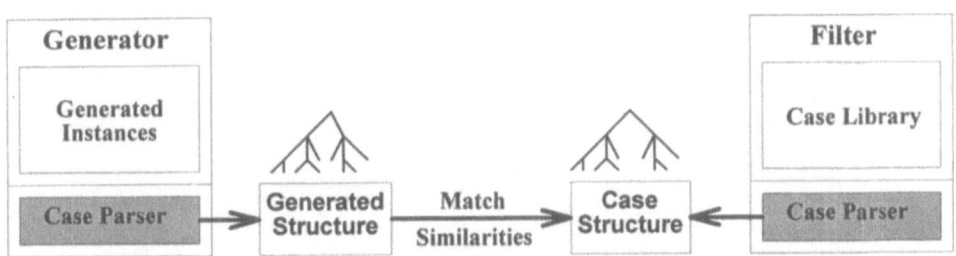

Figure 11. A Case Match System

4.1 Adaptation Algorithms

In most situations, solutions from retrieved cases needed to be adapted so they can solve a new problem. Generally, case adaptations are considered as a constraint satisfaction problem in design computation. Kolodner's (1993) case-based reasoning textbook describes ten methods by which adaptation can be done. Among these methods, adaptation algorithms fall into four categories: (1) substitution, (2) transformation, (3) special-purpose adaptation and repair, and (4) derivational replay.

More than one of these algorithms may be used in the CBR system. Different systems implement their algorithms of each category differently. This paper focuses the first two categories, i.e., substitution and transformation methods. Substitution methods substitute values appropriate for the new situation for values in the old solution. Reinstantiation and parameter adjustment are frequently used by CBR systems. Transformation methods are used to transform an old solution into one that work in a new situation.

4.2 Search Mechanism

The elementary school layout project is used to do experiments on case adaptation. The SAR design method is also used to define problem spaces that can be systematically enumerated using various search strategies. In the design process using SAR method, a layout problem is structured into a system of supports, which divides the layout site into zones and sectors by analyzing environmental factors; and infills, which are the components to be placed in the layout. Various layout alternatives can be systematically generated and tested according to the relationships between infills and supports.

The system of supports and infills as shown in *Figure 12* defines a problem space that can be enumerated in various ways. Each way of enumeration imposes a structure to the problem space, on which the positions and relations of all derived layout alternatives can be defined. Based on the notion of positions and relations defined in this manner, major issues in CBR such as case similarity and adaptation can be formally discussed and experimented. *Figure 13* demonstrates alternatives generated by the forward search process.

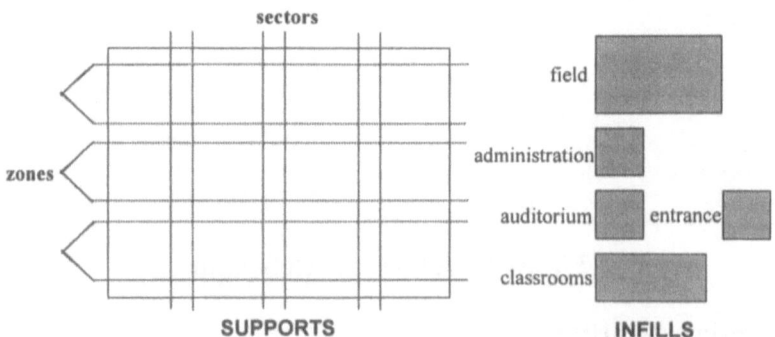

Figure 12. A System of Supports and Infills.

Two types of enumerative processes are used to study CBR in the domain of school layout planning using SAR design method. The two types of enumerative processes can be distinguished as first, generation by insertion and second, generation by replacement. In the process of generation by insertion, a sequential order is introduced to the layout components so that the entire problem space can be enumerated by traversing all possible insertion of layout components following that order. Each different order of layout components defines a lattice structure upon the problem space. The similarity of two different layouts can be measured by calculating the distance of the shortest path through their closest common ancestor in the lattice. Case adaptation can be executed by systematically examining layout variations that are most similar to the original case. In the process of generation by replacement, a layout alternative can be transformed by insertion, deletion or replacement of layout components. The least operations that are required to transform one layout to another can be used as a measurement to the distance between cases. Case adaptation is then carried out by searching through the closest neighbors of the original case. Based on these two types of operations, variations of search methods can be derived.

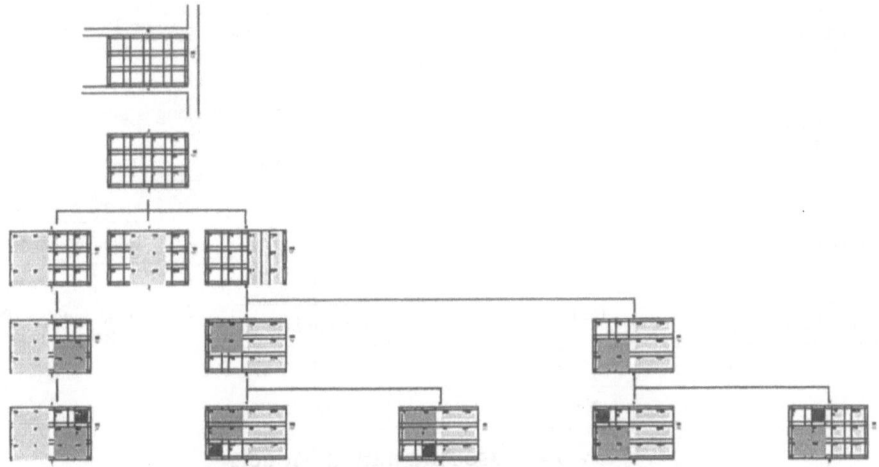

Figure 13. Forward Search of School Layout

The diagrams in *Figure 14* illustrate three methods of case adaptation. The first method, is basically a process of back-tracking, according to an order imposed to the layout components. The component that is most critical to the solving of the new problem is reallocated first, and the least critical component is reallocated the latest in the search hierarchy. The second method does not distinguish different component in separated levels of search hierarchy, although weights may be imposed to calculate case similarity. The third method is a combination of the first and the second methods, it switches strategies according to the result of evaluations. Upon the problems defined for our experiments, all of the three methods are capable of deriving layout alternatives that solve the given layout problem if solutions do exist.

As shown in *Figure 15*, generation of school layout by the forward search and backward search method increases the alternatives quickly. Detailed descriptions can be found in (Hsieh 1997). While alternatives generated from the system facilitate the design development, further comparison of these methods of case adaptation are yet to be investigated.

798

(1) Backward and Forward search

New Problem

↓

Case retrieval

↓

Specification Comparison

↓

Delete Unwanted Objects

↓

Insert Objects

↓

Checking Rules — No

↓ Yes

Forward Search

↓

New Case

(2) Replacement

New Problem

↓

Case retrieval

↓

Specification Comparison

↓

Delete Unwanted Objects

↓

Insert Objects

↓

Replacement ←

↓

Checking Rules — No

↓ Yes

Replacement

↓

New Case

(3) Hybrid (Combination of 1 & 2)

New Problem

↓

Case retrieval

↓

Specification Comparison

↓

Delete Unwanted Objects

↓

Insert Objects

↓

Checking Rules

Yes / No

Replacement ←

Checking Rules — No

Yes

Forward Search

↓

New Case

Figure 14. Case Adaptation Models

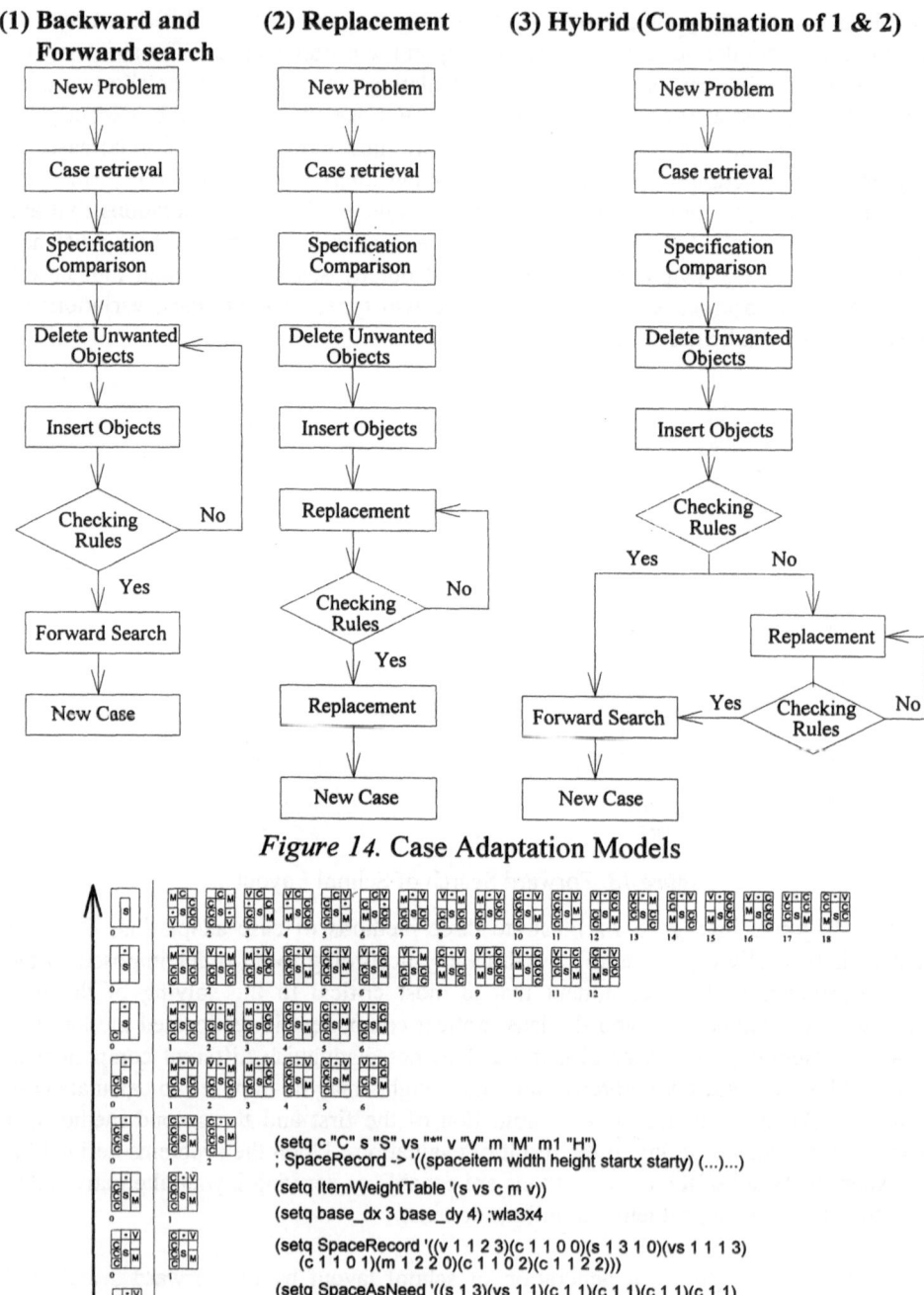

```
(setq c "C" s "S" vs "*" v "V" m "M" m1 "H")
; SpaceRecord -> '((spaceitem width height startx starty) (...)...)

(setq ItemWeightTable '(s vs c m v))

(setq base_dx 3 base_dy 4) ;wla3x4

(setq SpaceRecord '((v 1 1 2 3)(c 1 1 0 0)(s 1 3 1 0)(vs 1 1 1 3)
    (c 1 1 0 1)(m 1 2 2 0)(c 1 1 0 2)(c 1 1 2 2)))

(setq SpaceAsNeed '((s 1 3)(vs 1 1)(c 1 1)(c 1 1)(c 1 1)(c 1 1)
    (m 1 2)(v 1 1)))
```

Figure 15. Generation of The Elementary School Layout by Backtracking

5. Conclusion

This paper provides a basic understanding of the human design and computers in case adapation by observation from design experiments. The findings indicate the limitation of dimensional adaptation and the complexity of topological adaptation. Case adaptation by novice and experienced designers are different in approaches of dimensional adaptation and topological adaptation. While design computation will be beneficial from converting heuristics into mechanism, problem-solving requires the transformation of non-routine problems into routine problems. If the user can specify and assign similarity value, then the differences between human judgement in case adaptation and the evaluation mechanism by computers is quite narrow. Without an understanding of how these above conditions are met, further study of what computational tools are needed for case-based design and reasoning cannot be reached.

In general, the complexity of design case adaptation has been underestimated, with a tendency for it to be modeled as the manipulation of simple algorithms and heuristics. From the preliminary observations, existing CBR systems are constructive tools which produce correct answers but not necessarily good ones because:
(1) They do not consider characteristics of an individual user or the problem. The outcomes are standardized regardless of problem constraints and the context of the domain.
(2) They do not address semantics and design justification. Architectural plans are usually constrainted not only by their dependency relationships through spatial reasoning, but also by the designer's experience and heuristic rules.

In conclusion, design is a significant human activity. Computers open up new realms of possibilities for design assistance. However, a computer is not a designer. It is best understood as useful tools which provides means of storing information and carrying out potentially useful computations (Sun 1993). The research indicates that case-based design in architectural design can proceed effectively through the cognitive science of views for proceeding future research.

Acknowledgments

The work presented here has been supported by the Taiwan National Science Council, under grant NSC 86-2211-E-006-054 and 055.

References

Chen, C.C. (1991) Analogical and Inductive Reasoning in Architectural Design Computation, Ph. D. Dissertation, ETH Zurich

Chiu, M.L. (1997) Analogical Reasoning in Architectural Design: Comparison of Computers vs. Human Designers in Case Adaptation, In the Proceedings of the Second International Conference of CAADRIA '97, Taipei, Taiwan

Chiu, M.L. (1996) Prototypes, Variation, and Composition: A Formal Design Apporach in Urban Housing Design with Computer Assistance, in the proceedings of the First International Conference on CAADRIA'96, Hong-Kong, ISBN 9627-75-703-9, p.287-298

Coyne, R.D. and Yokazawa, M. (1992) Computer Assistance in Designing from Precedent, Environment and Planning B, vol. 19, p.143-171

Dave, B., Schmitt, G., and Faltings, B., Smith, I. (1994) Case Based Design in Architecture, in J.S. Gero and F. Sudweeks (eds.), Artificial Intelligence in Design '94, p.145-162, Kluwer Academic Publishers

Dzeng, R.J. (1995) CasePlan: A Case-based Planer and Scheduler for Construction Using Product Modeling, Doctoral Dissertation, U. of Michigan, USA

Gero, J. (1991) Design Prototypes: A Knowledge Representation Schema for Design, AI Magazine, 1991 Spring, p.25-36

Hsieh, C. (1997) A Case-based reasoning system for spatial layout problems, Technical report for National Science Council, R.O.C., Dept. of Architecture, National Taiwan Institute of Technology

Hua, K., Smith, I., Faltings, B., Shih, S. and Schmitt, G. (1992) Adaptation of Spatial Design Cases, in J.S. Gero (ed.), Artificial Intelligence in Design '92, p.559-575, Kluwer Academic Publishers

Kolodner, J. (1993) Case-Based Reasoning, Morgan Kaufmann

Leake, D.B. (ed.) (1996) Case-Based Reasoning: Experiences, Lessons, and Future Directions, The MIT Press

Maher, M. L., Balachandran, M.B., Zhang, D.M. (1995) Case-based Reasoning in Design, Lawrence Erlbaum Associates, Inc.

Oxman, R.E. (1994) Precedents in design: a computational model for the organization of precedent knowledge, Design Studies, Vol. 15, No.2, p.141-157

Rosenman, M.A., Gero, J., and Oxman, R.E. (1991) What's in a case. The use of case bases, knowledge base and databases in design, in G.N. Schmitt (ed.), CAAD Future'91, Zurich: ETH, p.263-277

Sun, D. (1993) Memory, design, and the role of computers, Environment and Planning B, vol. 20, p.125-143

Wiewel, W. (1976) Variation - The Systematic Design of Supports, English Edition, translation from the Dutch edition, by Boekholt, J.T., Thijssen, A.P., Dinjens, P.J.M., Habraken, N. J., 1961

FORMALISING SITUATED LEARNING IN COMPUTER-AIDED DESIGN

JOHN S. GERO AND GOURABMOY NATH
Key Centre of Design Computing
University of Sydney

In this paper, we propose and begin to formalise an approach to machine learning in design called situated learning with the purpose of providing a foundation to developing better design tools in an agent-based framework. Situated learning theory postulates that the situations that an expert is exposed to forms the developmental conditions of expertise. We extend and adapt that theory for computer-aided design with the primary objective of learning the use of existing knowledge, rather than simply the knowledge itself. The idea behind situated learning is to learn situations and associate them with some knowledge with the intention of using the knowledge in similar situations.

1. Motivation

Machine learning for design has largely been concerned with acquiring design knowledge from examples. An implicit assumption behind such work has been that the knowledge is always applicable. However, the situation within which the knowledge is learned plays a role in its future applicability. Thus, in addition to learning design knowledge it is important also to learn the situation so that a basis exists for the reuse of that knowledge. One way to achieve this is to represent situations and contexts explicitly. However, current machine learning design tools do not represent situations and contexts and therefore are situation and context free. We propose a situated learning paradigm applicable in designing. In this approach, we borrow and adapt a theory from educational instruction called "situated learning" which states that situations which one experience as a novice form the developmental conditions of an expert (Ingold 1995, Brown et al. 1989). Our primary concern is with the use of knowledge rather than the knowledge itself. Our first aim is therefore to make a design tool learn situations which existed when some knowledge was learnt, with the intention that when similar situations exist in the future it could apply the knowledge. A secondary goal of this approach is to emerge high level structure-behaviour relationships that serve as a means of improving on existing knowledge, and consequently improve the performance of the design tool especially in the conceptual design stages (Gero 1996). In this paper we discuss the foundations of such a design tool based on agents and begin to provide a formalisation of the key concepts of situated learning in an agent-based framework.

R. Junge (ed.), CAAD Futures 1997, 801-808.
© 1997 *Kluwer Academic Publishers.*

2. Agent-Based Design

Design is essentially a collaborative and distributed activity requiring heterogeneous and multiple knowledge sources and often different kinds of problem solving (Rosenman and Gero, 1996). This often conveniently and intuitively maps onto specialists in design and computationally to computer programs that are assigned to perform various functional tasks or coordinating various subdisciplines in design. For each of these specialist agents it is neither possible nor needed to map the world in its full complexity and richness. Each agent thus has its own useful view of the world that allows it to make suitable problem representations. For example a structural engineer conceives a building in terms of structural components, while an architect sees the same building in terms of spaces. Researchers in machine learning in design (Reich 1996) also acknowledge the need for addressing the "perspectives" problem, which we believe can be addressed by the agent-based framework. In our system, each agent maps the world in its own useful way which we will call the "relevance field" of the agent.

3. What are Situations in Design

Many authors who refer to situations in computational design as well as in artificial intelligence implicitly use it as a synonym for a state as in AI and problem solving or equivalently as a snapshot of the world. Some researchers like Oki and Lloyd-Smith (1991) and Muller and Pasman (1996) state the importance of situations in terms of applicability of knowledge. In our view, a situation is a partial state composed of a set of a facts that is relative to an agent. Thus two agents may at the same time be sensing the same world but the situations sensed will be different. What is the situation and what is not the situation is determined by the focus of the agent. This focus of the agent analogically maps onto the figure-ground hypothesis in Gestalt psychology, Figure 1, which state that focus forms a foreground while the rest is the background. This focus also results in interchangeability of the figure and the ground with the restriction that both cannot be focused at the same time. Situations, analogically, are also that part of the relevance field of the agent that is not in focus (background). A situation is also like a pattern that is a mechanism for indexing other knowledge or defining the applicability condition of the foreground.

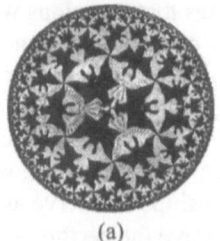
(a)

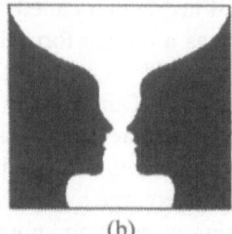
(b)

Figure 1. Figure-ground hypothesis of Gestalt psychology: (a) M. C. Escher's "Circle Limit IV", and (b) E. Rubin's Vase and facial profile (Bruce and Green 1990)

4. What, When and How does our System Learn?

Since the focus in this paper is on the applicability of knowledge, we aim to learn something like "If situation S then apply knowledge K with some variable bindings" and finally arrive at a generalisation-specialisation hierarchy of S. We also want to build heuristics from detailed design analysis procedures that extend the capability of the design tool to early design phases. A learning algorithm will be executed only if there exists "interesting" facts in the world with respect to an agent. What is interesting, depends on which of the facts in the world can be used to build new concepts utilising existing highly believed concepts.

5. Beliefs as a Mechanism in Learning

We define belief as a numerical utility measure of a situation-knowledge pair. This should not be confused with belief as in classical philosophy[1] . Beliefs can be generated and propagated as per existing theories of probabilistic or inexact reasoning as in artificial intelligence. Beliefs serve the purpose of assigning credits or blames to situation-knowledge pairs. At first a high value of belief would be assigned to all learnt situation-knowledge pairs. If that pair is utilised in the subsequent design iterations, then credits will be assigned that will reinforce the belief based on a normalised value of the number of times that that situation-knowledge pair has been used. If it is not used, the belief will slowly decay according to a decay function and thereafter "die".

6. Formalisation

In this section we present the formalisation of the concepts for situated agent-based learning in computer-aided design.

6.1. DESIGN STRUCTURE STATE

A subset of this set is sensed by agent Z and is called the sensed design structure state and is denoted by $D_t{}^Z$. Note that this set is internal to the agent while D_t is outside the agent .

$$D_t =< s_1, s_2, ..., s_n > \quad n \in I^+ \tag{1}$$

where D_t = design structure state defined at t, Figure 2

I^+ = set of positive integers, typically very large

n = the maximum bound of an index

s_i = structure variable such that $s_i =< O_j > < A_k > < V_k >$: Object, attribute, value triple. i, j, k are appropriate indices.

---

[1] Although refutations exist, the most commonly accepted relation between belief and knowledge in classical philosophy is that " knowledge is justified true belief" (Ackermann, 1972).

6.2. DESIGN BEHAVIOUR STATE

A subset of the design behaviour state posted onto the world by agent Z is called the actual design behaviour state and is denoted by $B_t{}^Z$.

$$B_t = < b_1, b_2, ..., b_n > n \in I^+ \qquad (2)$$

where $B_t =$ design behaviour state defined at iteration t, Figure 2

$b_i =$ behaviour variable such that $b_i = <O_j> <A_k> <V_k>$: An object, attribute, value triple i, j, k are appropriate indices.

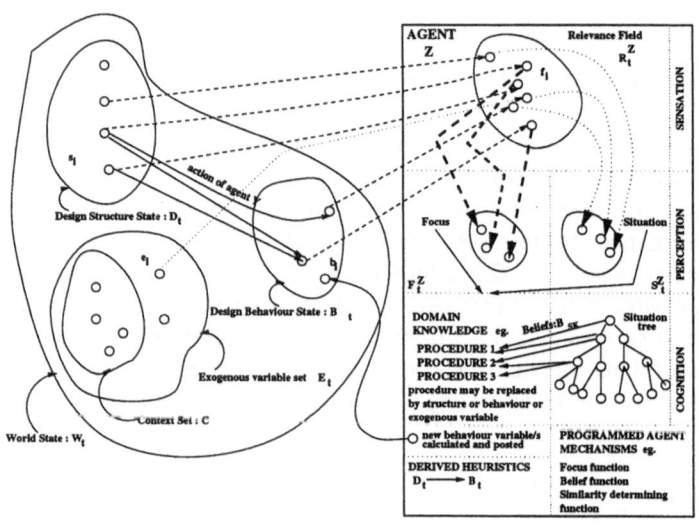

Figure 2. A high level description of an agent-based framework for situated learning

6.3. EXOGENOUS VARIABLE SET

The subset of this set, sensed by agent Z, is called the sensed exogenous set and is denoted by $E_t{}^Z$. Note that this set is internal to the agent.

$$E_t = < e_1, e_2, ..., e_n > \quad n \in I^+ \qquad (3)$$

where $e_i =$ exogenous variable such that $e_i = <A_k> <V_k>$: An attribute value pair, j, k are appropriate indices

6.4. CONTEXT SET

The set of exogenous variables in the world has a subset called context where the variables and values do not change during the course of designing. That subset, C, is called the context set. The context set is defined as:

$$\forall t, \ \exists C : C_t \subset E_t \ and \ C_t \equiv C_{t+1} \qquad (4)$$

6.5. WORLD STATE

The world state is the union of all the states in the system.

$$W_t =< \varphi_1, \varphi_2, ..., \varphi_n > n \in I^+ \tag{5}$$

where $\varphi = b \mid s \mid e$

W_t = world state at iteration t defined as a n-tuple of facts, Figure 2.

6.6. RELATIONS BETWEEN DESIGN STRUCTURE STATE, EXOGENOUS VARIABLE SET AND BEHAVIOUR STATES

A relationship between these three states is :

$$W_t = D_t \cup B_t \cup E_t \tag{7}$$

The sets in W_t are also mutually exclusive which means that

$$D_t \cap B_t = \phi, \quad B_t \cap E_t = \phi, \quad D_t \cap E_t = \phi \tag{8}$$

Thus it follows that $D_t \cap B_t \cap E_t = \phi$

6.7. RELATIONS BETWEEN THE ELEMENTS SENSED

The mapping from a design structure state to a design behaviour state is called analysis in design. Exogenous variables are used in this process. Thus, for any agent Z we have the following:

$$\mathcal{A}^Z \{D_t^Z, E_t^Z\} : (D_t^Z, E_t^Z) \longrightarrow B_t^Z \tag{6}$$

where $\mathcal{A}^Z$ = analysis function specific to agent Z.

6.8. RELEVANCE FIELD AND RELEVANCE FUNCTION OF THE AGENT

The relevance field of the agent typically comprises of some structure, behaviour and exogenous variables which form a subset of the world state, Figure 2. Thus, we have:

$$R_t^Z =< \varphi_1, \varphi_2, ..., \varphi_m > \tag{9}$$

where R_t^Z = Relevance field of the agent Z such that typically $m < n$

m = a positive integer denoting the maximum bound of an index.

Also $R_t^Z \subset W_t$. Typically, the relevance fields of two agents Z_i and Z_j cannot be the same, ie. $R_t^{Z_i} \neq R_t^{Z_j}$.

The relevance field at any point of time completely describes the portion of the world the agent has access to. The function that generates the relevance field of the agent Z is defined as follows :

$$f_R^Z\{S^Z \xrightarrow{B_{SK}} K^Z\} : S^Z, B_{SK}, K^Z \longrightarrow R_t^Z \tag{10}$$

where f_R^Z = Relevance function of agent Z

$S^Z \xrightarrow{B_{SK}} K^Z$: A list of situations S^Z in agent Z connected to a list of knowledge K^Z with a belief B_{SK}.

6.10. FOCUS FIELD AND FOCUS FUNCTION OF THE AGENT

The focus field is a subset of the relevance field. It is the foreground or the focus of attention of the agent. The focus field is derived by a function called the focus function. Hence, we have the focus function as:

$$f_F^Z\{S^Z \xrightarrow{B_{SK}} K^Z, R_t^Z\} : S^Z, B_{SK}, K^Z, R_t^Z \longrightarrow F_t^Z \tag{11}$$

where f_F^Z = focus function of agent Z
F_t^Z = focus field of agent Z

$$F_t^Z \subset R_t^Z \text{ and. } F_t^Z = <\varphi_1, \varphi_2, ..., \varphi_l> \quad l < m \tag{12}$$

where l = a positive integer denoting the maximum bound of an index.

Once, the focus field is determined the situation is defined .

6.11. SITUATION

The situation is defined as the difference between the relevance field and the focus field for an agent:

$$S_t^Z = R_t^Z - F_t^Z \tag{13}$$

ie, S_t^Z is defined as:

$$S_t^Z =< \varphi_1, \varphi_2, ..., \varphi_p> \quad p < m \tag{14}$$

where p = a positive integer denoting the maximum bound of an index.
It is obvious that $R_t^Z = S_t^Z \cup F_t^Z$. Also at any instant of time:

15

$$S_t^Z \cap F_t^Z = \phi. \tag{15}$$

6.12. FOCUS-SITUATION DUALITY

A situation and focus or vice-versa has a undirectional duality iff:

$$(S \bullet F)_{t,t+i}: \ \exists S, F, i: \ (S_t \equiv F_{t+i}) \tag{16a}$$

or

$$(F \bullet S)_{t,t+i}: \ \exists F, S, i: \ (F_t \equiv S_{t+i}) \tag{16b}$$

There may be a bidrectional duality $(F \oplus S)_{t,t+i}$ between situation and focus iff $(F \bullet S)_{t,t+i} \wedge (S \bullet F)_{t,t+i}$ is true. In this case $R_t^Z \equiv R_{t+i}^Z$.

7. Discussion

The concept of machine learning is design is extended through the introduction of the world within which learning occurs. This world is bifurcated into situation and focus, where focus maps onto potential knowledge to be learned. This extension allows for learned knowledge to be situated. The effect of this additional situation knowledge is to contextualise the learned knowledge and have a system be capable of determining when the learned knowledge could be applicable. This begins to address one of the difficult questions in machine learning: when is the learned knowledge useful.

This paper has presented an outline of the basic building blocks of the concepts involved and provided a formal representation of them.

Acknowledgments

This research is supported by a scholarship from the Australian International Development Assistance Bureau. We wish to thank Simeon Simoff, of the Key Centre of Design Computing, for his comments.

References

Ackermann, R. (1972) Belief and Knowledge, Anchor Books, New York.

Brown, J., Collins, A. and Duguid, S. (1989) Situated cognition and the culture of learning, Educational Researcher 18(1): 32-42.

Bruce, V. and Green, P. (1990) Visual Perception: Physiology, Psychology and Ecology, 2nd edition, Lawrence Erlbaum, London.

Gero, J. S. (1996) Design tools that learn: a possible CAD future, in B.Kumar (ed.), Information Processing in Civil and Structural Design, Civil Comp, Edinburgh, pp. 17-22.

Ingold, T. (1995) Lecture at workshop on situated learning within post secondary education, http://www.dur.ac.uk/~dps8zz2/Lave/TimIngold.html.

Oki, A. and Lloyd Smith, D. (1991) Metaknowledge reasoning in civil engineering expert systems, Computers and Structures, 40(1): 7-10.

Muller, W, and Pasman, G. (1996) Typology and organisation of design knowledge, Design Studies 17(2), 111-130.

808

Reich, Y. (1996) Modelling engineering information with machine learning, Artificial Intelligence in Engineering Design, Analysis and Manufacturing **10**(2): 171-174.

Rosenman, M. and Gero, J. S. (1996) Modelling multiple views of design objects in a collaborative CAD environment, Computer Aided Design **28**(3): 193-205.

REAL-TIME MODELING WITH ARCHITECTURAL SPACE

DAVID KURMANN, NATHANEA ELTE AND MAIA ENGELI
Architecture and CAAD
Swiss Federal Institute of Technology - ETH Zurich, Switzerland
[kurmann, elte, engeli]@arch.ethz.ch

Space as an architectural theme has been explored in many ways over many centuries; designing the architectural space is a major issue in both architectural education and in the design process. Based on these observations, it follows that computer tools should be available that help architects manipulate and explore space and spatial configurations directly and interactively.

Therefore, we have created and extended the computer tool Sculptor. This tool enables the architect to design interactively with the computer, directly in real-time and in three dimensions. We developed the concept of 'space as an element' and integrated it into Sculptor. These combinations of solid and void elements - positive and negative volumes - enable the architect to use the computer already in an early design stage for conceptual design and spatial studies. Similar to solids modeling but much simpler, more intuitive and in real-time this allows the creation of complex spatial compositions in 3D space.

Additionally, several concepts, operations and functions are defined inherently. Windows and doors for example are negative volumes that connect other voids inside positive ones. Based on buildings composed with these spaces we developed agents to calculate sound atmosphere and estimate cost, and creatures to test building for fire escape reasons etc.

The paper will look at the way to design with space from both an architect's point of view and a computer scientist's. Techniques, possibilities and consequences of this direct void modeling will be explained. It will elaborate on the principle of human-machine interaction brought up by our research and used in Sculptor. It will present the possibility to create VRML models directly for the web and show some of the designs done by students using the tool in our CAAD courses.

R. Junge (ed.), CAAD Futures 1997, 809-819.
© 1997 *Kluwer Academic Publishers.*

1. Space and Architecture

1.1.   THE REPRESENTATION OF SPACE IN ARCHITECTURE

The following section deals with the issue of architectural space representation. A central topic of architectural design is the space, for example in [Tschumi 90]. Every built intervention, even that of a single wall, creates architectural space. The design of a building involves the conceptualisation of complex, interdependent and changing spatial groupings. For a designer, it is important to have a variety of tools available that facilitate modelling, presentation, testing and further development of the spaces and spatial compositions that they conceive.

A few of the traditional methods used for spatial design [Joedicke 85, Van de Ven 87] will be described in order to illustrate the applicability of Sculptor for this aspect of the design task. Sculptor represents a logical extension of these methods and can be used effectively in design.

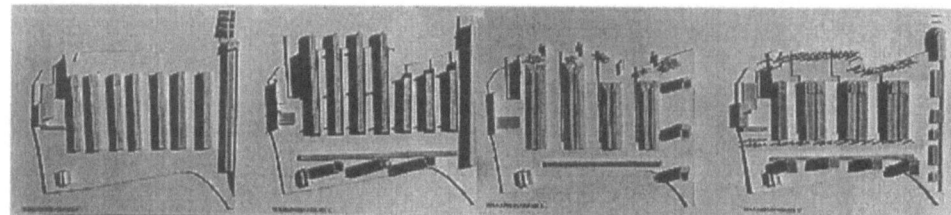

Figure 1. Massing models for volumetric studies

1.2.   VOLUMETRIC REPRESENTATION

In the first phase of a design project - be it the planning of a new section of a city or the design of single object - massing models are often used to create volumetric studies. Such studies involve the placement of blocks on a site plan or in imagined space. These constellations of blocks represent complex arrangements of spatial elements. When the elements of a massing model are placed during a design process, the exterior space of a project are simultaneously delineated. Exterior spaces can also be considered architectural spaces, as they are defined by the existence of architectural elements. By shifting, exchanging and regrouping the elements of a composition, students can observe the effects of these decisions on their design as a whole. Students can learn to recognise the spatial components of their designs by carefully studying such models, photographs and video tapes; which ideally enables them to design more consciously.

These conceptual studies are used to master a given design problem and prepare for the refinement of the design and the accompanying shift in dimension and detail. To be able to develop a design effectively, students must be able to imagine the spatial consequences that their ideas might have. A physical model is often used to efficiently test and express an idea at this stage. Naturally, perspective drawings can also be used for the same purpose. Although it has been seen that perspectives are more often used for presentation purposes and are rarely an effective part of the design process. It is very difficult to work with a drawn perspective. However, the possibility to directly and constantly test the spatial consequences of design decisions remains crucial. This capability can strengthen a persons ability to imagine a proposed space, particularly in the case of students who do not yet have actual building experience. They can learn to judge more effectively whether or not the spaces that they have proposed actually fit the conceptual goals of their design.

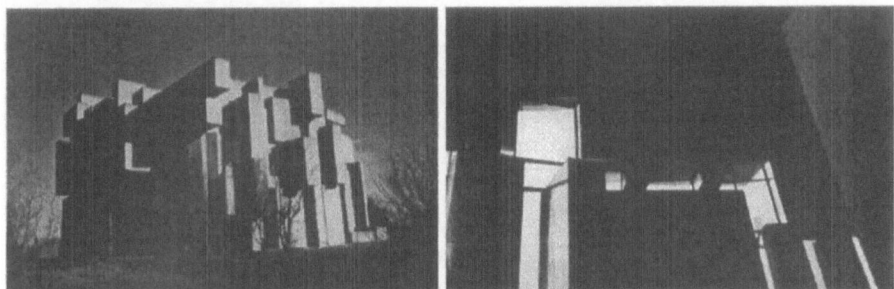

Figure 2. The church 'Zur Heiligsten Dreifaltigkeit' by Fritz Wotruba, complex spaces seen from inside and outside.

1.3. INTERIOR AND EXTERIOR

The next step of the refinement process addresses the formal design of individual spaces. A church in Vienna, 'Zur Heiligsten Dreifaltigkeit', by Fritz Wotruba, illustrates how complex spaces can develop and how important the relationships between interior and exterior spaces are. It is difficult to represent the spatial quality of such a church using only plans, sections and a few perspective drawings. Sometimes, students design spatial configurations that are also difficult to visualise, test and present via the more traditional media. However, especially in such cases, this is very important. Additionally, it must be possible to make and visualise changes in a project during the design process. The ability to directly change a design model and to immediately be able to view those changes help the designer develop and judge the design.

There are also spaces that do not involve such expressive massing, but that are nevertheless, very complex. Frank Lloyd Wright's 'Falling Water' is one such example. In it, various spaces and spatial zones - both interior and exterior - blend and overlap throughout the design. Differentiated ceiling design serves to create spatial zoning as well. Similarly, differences in floor levels are also used to express shifts in spatial order. Imagining the spatial aspects of such a project using only the available drawings is not easy. Such spaces can actually only be truly understood when one is able to move about in them; a type of spatial perception through movement. In built projects, one can enhance perception by walking through and around it. For un-built designs, it is valuable to at least be able to simulate such a tour.

A further design concept that has interesting spatial consequences is the development of a facade from the inside out. This approach to facades is based on the idea that a facade not only contribute to and effect the streetscape and spatial aspects of the site and its surroundings, but even more strongly the perceptions of the people who use the building. As shown in the example of Wotruba's church, interior and exterior aspects of a design often effect one another very directly. The selected representation method must, if it is truly intended to be used as an instrument of design, clearly represent this direct connection.

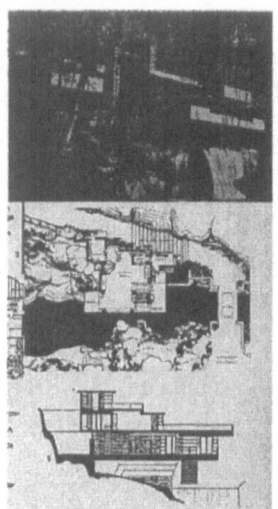 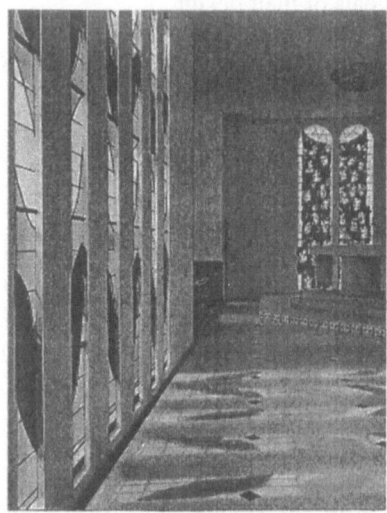

Figure 3. 'Falling Waters' by Frank Lloyd Wright.
Figure 4. 'Chapelle du Rosaire' Henri Matisse

1.4.   SPACE DEFINING ELEMENTS

The nature of space defining elements, not just the form of a space, effect the perception of a space decisively. One can collect and display material samples, but this, depending on the material, can be quite costly. Being able to simulate the effects of various materials on the perception of a designed space is therefore very desirable and potentially a powerful aid in the design decision making process. The effects and meaning of light in Henri Matisse's 'Chapelle du Rosaire' can, for example, be clearly shown. Being able to simulate such aspects of a design, allows the student to design more exactly and consequently. The implications of light and material on design change constantly during the course of a day. Completely different impressions of a space can occur depending on the time of day. These changes are nearly impossible to convey with drawings, especially when one is attempting to combine it with the total spatial effect of a project.

2. Modelling the Space

2.1.   SPACE AS AN ELEMENT

This importance of space should be reflected in computer design tool for architecture. Even though this idea is not new - tools that use solid modelling are well known [Mäntylä 88] - we want to propose a slightly different approach. Let's assume that there is a 'space element' ('void element' or 'negative volume'). An element that consists of no material and carves out space when it intersects with a solid element. This would have several interesting and important consequences. The operation with these elements would be different from a solid modeler. Instead of using volumes and boolean operations like subtraction, union, or difference, two *types* of volumes, positive (solid) and negative (space) volumes are defined. A negative volume always creates a space inside a solid. Where it intersects with a positive volume this becomes visible, but there is no effect where it intersects with another negative volume. This requires also a different data structure and a new description of objects needs to be implemented for CAD. In the following we will show implementation details and results achieved by students using this approach in Sculptor.

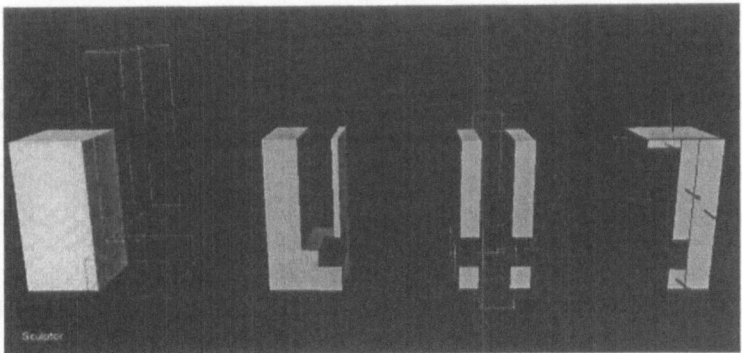

Figure 5. Positive and negative volumes: A sequence of interactively moving two voids (wire frame) into one solid. One special case solved: Entirely separated face of one positive volume. The face normals of the resulting object are indicated on one object.

In our implementation the calculation of objects that are composed of positive and negative volumes is based on four steps: 1) finding out which faces get intersected, 2) cutting each face according to the intersection lines, 3) deciding, whether the trimmed face is shown or not, and 4) finding out, in which direction the face is oriented. Technically, the principle used is a surface modelling technique, used by other authors in a similar way, for example in [Yessios 87].

There are some principles that make this concept easier to realise. A possible face of any composed object will always be a part of an original face of the composing objects. Therefore it does not matter what shape an object has or how it is oriented in space, as long it is described by a closed surface. The intersection of an object with another object's face will give the points and lines that define the shape of the resulting face. This trimming process is the most time consuming part of the calculation. The normals of the resulting faces are defined easily: a solid volume will always create faces looking outside, faces looking inside are produced by a void. The fact, that the void defines the corresponding faces makes it possible, that the attribute of a void defines colour and texture of the faces it cuts out.

Two representations, the way an object is composed and the final composition, are stored. This dual representation is needed for two reasons: The final composition is kept because the face calculations should be done as few times as possible. To gain performance, a re-calculation is done only when one of the objects changes. The description of how an object is composed is necessary to do this re-calculation.

This representation of objects is implemented for an orthogonal world with cubes as the only possible voids. The algorithm is generally applicable for every closed object freely oriented in space, but is much faster if it can be reduced to orthogonal cases. The orthogonal world makes interesting and complex objects possible and allows to work with spaces in an intuitive and direct way. With the currently available computing speed, direct manipulation would not be possible in the general case if there are a larger number of volumes in the scene.

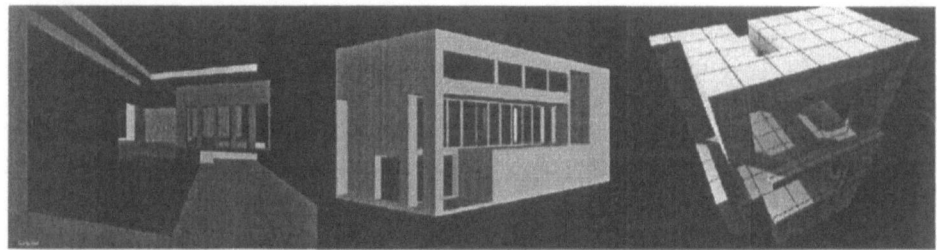

Figure 6. Different architectural models built by using voids and solids in Sculptor: a cave, a rough reconstruction of the 'Casa Cavagli' by Luigi Snozzi and a light simulation rendered with Radiance.

2.2.   ROOMS, WINDOWS, DOORS, AND SECTIONS

With this concept of negative volumes, the relevant objects and operations for an architect in an abstract phase are defined. Windows and doors are negative volumes that connect other voids inside a positive one. A room can be seen as a composition of a void within a solid. This group then can be intersected with other room groups interactively, a correct composition is always assured. While resizing a room group (a pair of a solid and a void) the thickness of a wall stays constant. Using the negative volume concept, sections through buildings are possible by using a large void and intersecting it with the building.

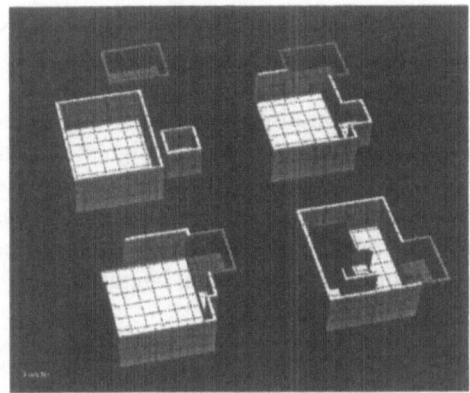

Figure 7. A sequence of how to compose a room (= group of a void and a solid).
The rule also allows hierarchical 'room in room' buildings.

Besides the ease of manipulation, the introduction of rooms offers new possibilities in
the analysis of buildings. By defining rooms that compose a building (instead of the
walls that create the rooms), it is easy to retrieve data for additional support such as:
navigation in a building, spatial sound effects, cost estimations, or energy consumption
calculations. This will be emphasised later in this article in the description of some
intelligent agents.

3. The Design Tool Sculptor - Space Interactive

This approach of using 'void element' has been implemented to prove and demonstrate
that it works. We wanted to show that interactively creating and manipulating spaces
using our approach is more intuitive. Therefore we added this approach to Sculptor, our
computer tool for virtual design that has been developed over the last four years
[Kurmann 95]. It combines these novel spatial features and functions of the void
elements with general concepts of modelling in 3D and focuses on the early, conceptual
stage [Van Vries 90] of the architectural design.
A key factor in the program is the human machine interaction. In contrast to many
existing CAD tool, Sculptor allows very direct, intuitive and immersive access to three
dimensional design models. Through interactive modelling in a virtual space and this
introduction of positive and negative volumes, an easy way of generating and
manipulating architectural models is made possible. Interactive parameter specification
of objects, and models with attributes like form, geometry, colour, or material are
supported. Objects can be grouped together hierarchically. Objects, groups and virtual
worlds can be changed in real time by scaling, resizing, rotating, reshaping and moving
them in space. [Laseau 89]
Different points of view can be chosen as well as functions invoked for walking and
flying through 3D space. All manipulations happen immediately by moving the mouse

or one of the possible 3D input devices and the scenes are changed and rendered in real time. Multiple windows, complex text input and sliders or buttons are avoided. The interface is almost widget-less.

A direct approach also enables inspiration and creativity when computers are used in design. A direct visual control of what is being designed is critical. It facilitates the development of new ideas, allows a designer to inspect design ideas from any point of view and to discover new solutions to a design problem. It also enables modeling while being inside this architectural virtual space. Sculptor offers the possibility to visualise design, but does not dictate a specific architectural language.

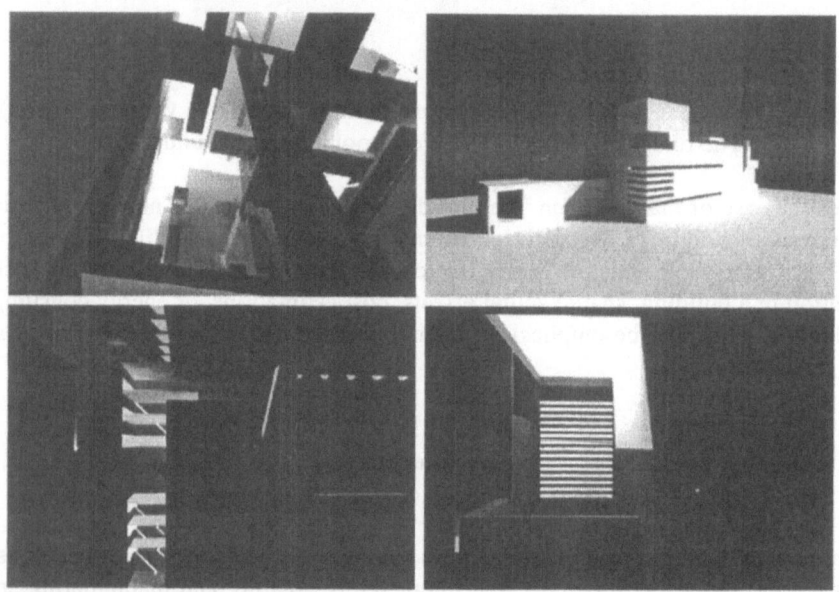

Figure 8. A collection of models that were created by our students in the CAAD courses showing very different conceptions of space and architecture.
Images by Habegger, Chladek, Papanikolaou and Nielsen.

Besides assisting the users when creating the three-dimensional geometry of objects, Sculptor also supports models of behaviour based on principles of mechanics and dynamics [Barzel 92]. One is collision detection while objects are changed in size or moved in the scene. Another is the concept of gravity: an object will fall down if it is not supported by another object or standing on the ground. Using these constraints, the interaction with objects in the virtual worlds is enhanced since users experience in a direct way the act of moving objects to valid positions or combining them by following physical principles. Different sorts of feedback are an important feature of virtual reality tools to understand complex scenes in space.

Sculptor designs can be exported to CAD construction tools like AutoCAD for further refinement as well as to VRML files for the World Wide Web. Light simulations with Radiance are possible as shown in figure 8.

4. Agents and Space

We were implementing two basic kinds of agents using these space modeling approach: Design assisting agents, that help the user by providing information and executing background tasks, and design generating agents, that interfere with the design and autonomously come up with new solutions. We started working on design assisting agents first and implemented three prototypical interface agents [Kozierok et al 93]: The Navigator, the Sound Agent and the Cost Agent. They are meant to be personal assistants, trained by each user to adapt to one's individual preferences. All of these agents are based on and take advantage of this data structure which describes the openings and the different kinds of rooms as negative volumes with attributes like room type, size, colour, etc.

The **Navigator** acts like a guide in the virtual world. It can follow different kinds of instructions like: moving to a specified place, room or building, moving in a specific direction, or composing a tour. He understands commands like: go, show, go to, jump to, that describe the action to be taken, followed by a description what to show (this floor, the building), a place to go to (the kitchen, the entrance, the living room), or the direction to be taken (left, right, up, down, forward, backward). The navigator is built directly into Sculptor. From the information of the spatial composition the Navigator builds a graph. According to the command by the user the Navigator can then search the shortest path in this graph and walks through the building without colliding with walls and furniture.(Fig. 10).

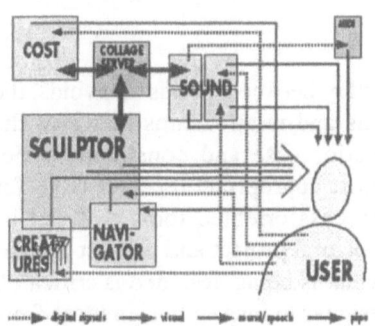
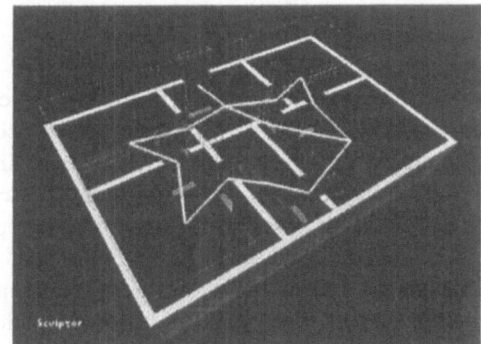

Figure 9. The components and interaction channels of Sculptor & Agents
Figure 10. The graph built by the Navigator.

The **Sound Agent** is a companion of the Navigator. It will try to enhance the perception of a space by adding an auditory component to the visual impression. It can play "sound labels", generate the sound of footsteps, change the spatial sound effects, and play a sound track, in accordance to the purpose, shape and colour situation of different rooms. The sound agent is a program that communicates with Sculptor using the NCSA Data Transfer Mechanism (DTM) [NCSA 92]. Sound labels and footsteps can easily be

818

generated from the information available from Sculptor. It describes the room the Navigator is in (the purpose, size, and colour situation) and the current position within this room. Which spatial sound effect to use and which sound track to play has to be learned first. [Maes 93]

The **Cost Agent** estimates the costs of the project and displays the result graphically. The vertical bar represents the costs and turns from green to red when the costs are getting too high. This small interface can be used to define a cost limit, to indicate the desired level of luxury, to access a database of projects and to look at a graph that illustrates the calculation. Like the sound agent the cost agent is a separate program that communicates with Sculptor using the NCSA DTM. Kind and size of each room are considered for the calculations. Using these formulas windows and doors can be treated like rooms. This corresponds to the way the data is represented in Sculptor.

To test the design in different situations, **Creatures** have been implemented. Creatures are sets of simple agents which can interact with each other and recognise some characteristics of Sculptor models. Their behaviour can be programmed for simulating specific situations, like escaping from a fire, traffic jams or the queue at the cloakroom in a theatre. [Resnik 94]

For the next step we plan to implement agents that connect to information sources that are outside of the actual design studio, like Controlling Agents, Information Agent, HVAC agent or Illumination Agents. Later design agents, that generate and suggest design solutions, will be added. They will be based on a more knowledge based and holistic approach than can be built into Sculptor's intelligent objects.

Conclusions

Architectural design consists primarily of a complex play between: solids and voids, the development of, movement through, the connections and relationships between, the refinement of space defining elements, light, function, use and construction. For architects and design, it is important to have the most appropriate tool available for every phase of the design process. Particularly, process oriented tools that can be directly used during the designing phases of an architectural project and not just used as a means for presentation. A direct visual control of what is being designed is critical. It facilitates the development of new ideas, allows a designer to inspect design ideas from any point of view and to discover new solutions to a design problem. Sculptor offers the possibility to visualise design, but does not dictate a specific architectural language. We try to enable modelling while being inside this architectural virtual space.

Acknowledgements

This research is supported by the Swiss National Science Foundation - Special Program Computer Science.

References

Barzel, R. (1992) Physical-Based Modelling for Computer Graphics, Academic Press

Joedicke, J. (1985) Space and Form in Architecture - A Circumspect Approach to the Past, Krämer Verlag, Stuttgart

Kurmann, D. (1995) Sculptor - A Tool for Intuitive Architectural Design, in: CAAD Futures '95 - The Global Design Studio, M. Tan and R. Teh (Eds.), University of Singapore, P. 323-330

Kozierok, R., P. Maes (1993) Learning Interface Agents, in Proc. of the 11th National Conference on Artificial Intelligence, AAAI

Maes, P. (1993) Behaviour-Based Artificial Intelligence, in: Proc. of the 2nd Conference on Adaptive Behaviour, MIT Press

Laseau, P., (1989) Graphic Thinking for Architects and Designers, Van Nostrand Reinhold, New York

Mäntylä, M.(1988) An Introduction to Solid Modeling, Principles of Computer Science Pronciples; 13, Computer Science Press, Maryland

NCSA (1992) Data Transfer Mechanism Programming Manual, ftp://ftp.ncsa.uiuc.edu/DTM/dtm.manual.ps, National Center for Supercomputing Applications

Resnik, M. (1994) Turtles, Termites and Traffic Jams, Complex Adaptation Systems Series, MIT Press, Cambridge

Tschumi, B. (1990) Questions of space - Lectures on Architecture, Architectural Association, London

Van de Ven, C. (1987) Space in Architecture: The Evolution of a New Idea in the Theory and History of the Modern Movements, Van Gorcum, Assen/Maastricht, The Netherlands

Van Vries, M., and H. Wagter, (1990) A CAAD Model for Use in Early Design Phases, The Electronic Design Studio - Architecural Knowledge and Media in the Computer Era. McCullough, Mitchell, Purcell (Eds.), MIT Press, P. 215-228

Yessios, C.I. (1987) The Computability of Void Architectural Modelling, in: Computability of Design - Principles of Computer-Aided Design, Y.E. Kalay, ed., Wiley-Interscience, New York, 141-172.

COMPUTABLE FEATURE-BASED QUALITATIVE MODELING OF SHAPE

JOHN S. GERO AND SOO-HOON PARK
Key Centre of Design Computing
University of Sydney

This paper introduces and describes a qualitative approach to the modeling of shapes applicable at the early stage of designing. The approach is based on using qualitative codes at landmarks to describe shapes. These strings of codes can be analysed to determine patterns which map onto features. An analogy with language is drawn to assist in articulating the modeling ideas. An example is presented which demonstrates the utility of the approach.

1. Introduction

In modeling architectural designs, it is usual to consider shape and space as two fundamental primitives. Designers as well as researchers have an increasing concern about how to handle these primitives better through the development of better design descriptions that are more appropriate to the design situation than current description techniques. There has been considerable success in modeling architectural designs using quantitative approaches. For example, CAD package which are nowadays often synonymous with drafting programs, are based on quantitative descriptions of shapes, ie., those descriptions requiring all the necessary detailed measurements sufficient to display the designed object on the computer screen. Unfortunately, those detailed measurements are only available at the final stages of designing. Therefore, modeling of designed objects is not readily available during the early or conceptual stage of designing because of the unavailability of the numerical data. In order to fully utilise the power of computational support tools, it should be possible to have design aids from the early stages of the design process regardless of the accuracy of the available design information. This leads us away from numerical descriptions to symbolic descriptions which represent qualities rather than quantities. It is proposed that shapes can be represented through a range of qualities which are available at the early stages of designing. The ability to represent shape qualities provides opportunities to link such shape qualities to space qualities since it is the space which is being designed with the shapes a consequence.

Qualitative approaches based on symbolic modeling provide different viewpoints in the sense that no accurate measurement of shape or space is needed to model the design primitives. There have been a number of symbolic schemes developed to handle shape and space. One of the most common approaches to handling shapes is based on contour

R. Junge (ed.), CAAD Futures 1997, 821-830.

lines of the shape which are de-segmented using directional vectors (Freeman 1961, Weinberg et al 1992, Jungert 1993). Most symbol systems for modeling space are concerned with capturing spatial relationships such as topological relations (Clarke 1981, Egenhofer and Al-Taha 1992, Randell et al 1992) or orientation relations (Allen 1983, Chang et al 1987) among spatial objects. Nevertheless, these symbolic modeling schemes display restrictions and limitations when applied to architectural shape and space since little effort is given to understanding the correspondence among shape characteristics, design semantics, and spatial characteristics of designed objects. Therefore, a more suitable and flexible symbolic scheme is needed in order to overcome the limitations of currently available symbol systems.

In this paper we demonstrate a qualitative modeling scheme for handling shape based on descriptions of the shape features. We will leave for another paper the issue of modeling space qualitatively and how to link the modeling of shape with that of space.

2. Methods

Here we use concepts derived from feature-based modeling to capture design knowledge related to the qualitative character of shapes. This will be presented in three steps. The first step is to develop an appropriate representation scheme: this is based on previous work on syntactic pattern and contour representation methods (Freeman 1961, Fu 1976), with which we can describe distinctive shape characteristics of drawings. The second step is to turn those qualitative descriptions of shape into a meaningful shape semantics. The third step is to discover meaningful shape, and then later design, semantics by using syntactic pattern matching of features from the given description of shape. Although the syntactic approach to the representation is not new, our syntactic representation scheme for qualitative characteristics of shape extends current generic methods and specialises them for design-related tasks which have previously either been unavailable or have been difficult.

2.1. Q-CODE REPRESENTATION OF SHAPE

Humans seem to recognise and identify complex forms by registering their characteristic features and their peculiar configurations (Treisman and Gelade 1980). Drawing, which is a fundamental tool for designers to express and communicate design ideas to others, is concerned with forms and figures which are entities made up of shapes. A shape is taken to be a finite arrangement of lines (straight, curved, open or closed) in the plane drawn in a finite area in a finite amount of time (Stiny 1978). The shapes we are concerned with are made of closed and connected lines, which define a boundary contour of a spatial object.

The qualitative representation of drawings looks for general features with which we could distinguish one drawing from another. The term "feature" refers to any geometric and topological entity (Shah 1991), or just a named entity with attributes of both form and function (Stiny 1989). The basic features that capture the physicality of a shape are mostly "the shape attributes" with characteristic variable names and values. We therefore use shape features to encapsulate design significances and to associate them with their geometry, functionality and design semantics.

2.1.1. *Encodings formalism*
Setting qualitative values to shape attributes follows a strict formalism suggested by researchers from the qualitative reasoning community (deKleer and Brown 1984). Qualitative values are set to variables by mapping their numeric values (where they exist) to a finite and discrete set of symbolic values. Where the numeric information does not exist, a lexigraphic ordering of the concepts can be used. The simplest way of setting qualitative values from the real number range of values is to use a "landmark set" with qualitative values in the range $\{-\infty, 0, \infty\}$. Then the set of intervals becomes $\{(-\infty, 0), [0, 0], (0, \infty)\}$, which corresponds to the qualitative set Q with the sign values $\{+, 0, -\}$ (Wertner 1994). This simple yet effective process of symbolic mapping is useful in modeling design variables and attributes, transforming possible numeric value ranges into small sets of discrete and finite symbol values. This formalism of setting qualitative values can be applied to most design variables since most values can be measured in terms of "polarity" and "granularity".

2.1.2. *Shape-attributes and Q-codes*
Q-codes are formulated by combining symbols with sign values. Four types of shape attributes are considered to be fundamental in describing shapes in this qualitative way. These are:

(i)   angle measured at a node, A-code;
(ii) relative length of line segments, L-code;
(iii) angle measured at a node for two tangents, C-code;
(iv) relative curvature of a line segment, K-code

Q-codes in (i) and (ii) are basic shape attributes necessary to describe arbitrary polygonal shapes, while Q-codes in (iii) and (iv) are needed to describe arbitrary curvilinear shapes. Symbols stand for categories of shape attributes and sign values stand for their qualitative values.

The A-code describes the qualitative measure for the inner angle at a node between two contiguous line segments in a shape. The C-code is a generalisation of the A-code to curvilinear line segments at a node. L-codes describe the comparisons of the lengths of the two adjacent line segments with values of "bigger", "equal", and "smaller". The K-code describes the curvature of a curvilinear line segment with values of "convex", "straight", and "concave".

Q-code descriptions of shapes result in sequences of Q-code strings. The general characteristics of Q-codes are:

(i)   direction of scanning is counterclockwise;
(ii) encoding can start from any code (normally from a node); and
(iii) components of Q-codes are symbol(s) plus sign value(s).

Table 1 shows how qualitative values are set to shape attributes. The granularity of Q-codes can be changed using a sectioning method, which is a way of segmenting the intervals. Sign values are thus extended, for example, from {+} to {++, +0, +-} as a result.

Table 1. Qualitative value assignments to shape attributes

	A-code / C-code	L-code / K-code
Numeric value range	$0 \leq \theta < 2\pi$	$-\infty < l, (k) < \infty$
Landmark set	$\{0, \pi\}$	$\{-\infty, o, +\infty\}$
Interval set	$\{[0,0], (0,\pi), [\pi,\pi], (\pi,0)\}$	$\{(-\infty,0), [0,0], (0,+\infty)\}$
Q-code set	$\{A_{nil}, A_+, A_0\ A_-\},$	$\{L_-, L_0, L_+\},$
	$\{C_{nil}, C_+, C_0\ C_-\}$	$\{K_-, K_0, K_+\}$

2.1.3. *Syntax of Q-code encodings and syntactic operators*

In a qualitative representation process, a shape is converted into a sequence of Q-codes resulting in a line of strings called its "primitive code". Some syntactic regularities can be discovered from the structure of Q-codes. Coding theory has been developed for handling this kind of pattern analysis task (Leeuwenberg and Buffart 1983) with three syntactic operations identified as "iteration", "symmetry", and "alternation" (Helm and Leeuwenberg 1986, Martinoli et al 1988) and converts a Q-code to a structured pattern of simpler codes called "end codes".

Since the major task of the Q-code representation of drawings is to handle shape features, the three types of syntactic regularities of Q-codes become the most generic characteristics of shape description. These syntax can be applied not only to the simple atomic Q-codes but also to bigger chunks of Q-codes with significant design meanings associated with them.

2.2. SYNTACTIC STRUCTURE FOR SHAPE FEATURES

Shape features are recognisable, definable, structural patterns, ie., shape features are named entities with distinctive structural patterns from which the physical characteristics of design objects are understood and explained. These shape features are found in the drawings describing designed objects, but when they are encoded, it is often not clear how to distinguish one from another. It is possible to draw an analogy with the structure of natural language in order to explain how the syntactic structures in Q-codes are related hierarchically to each other.

Table 2 shows levels of shape features described in linguistic terms, referring to different aspects of shapes. A word, as a minimum and discrete unit of information, can contain a basic shape feature. The words are then aggregated to construct more complicated expressions in the form of a phrase. A Q-phrase displays a certain syntactic structure of one or more Q-words by explicitly describing the structure with a set of syntactic operations such as "iteration", "alternation", and "symmetry". These smaller shape features are aggregated to form a complete and closed shape termed a Q-sentence as another level of shape features.

Table 2. Various levels of shape features with their linguistic analogy

Levels of shape feature	Reference to the shape
Q-code	A simplest symbol which refers to an atomic component of a shape attribute.
Q-word	A sequence of Q-codes which refers to a shape pattern with distinctive design significance – a shape feature.
Q-phrase	A sequence of Q-codes in which one or more Q-words show a distinctive pattern of structural arrangements.
Q-sentence	An aggregation of Q-codes, Q-words, and Q-phrases so that it refers to a closed and complete contour of a shape.
Q-paragraph	A group of Q-sentences where necessary spatial relationships are described with specific connectives.

2.3. SHAPE FEATURES FOR DISCOVERING DESIGN SIGNIFICANCES

Recognising patterns that match the description of an object is a method widely used in pattern matching and pattern recognition. The pattern matching algorithm identifies some interesting patterns in the given descriptions of shapes. A primitive description of a shape is nothing but a sequence of symbols which in itself does not reveal any design significances or design meanings. No display of syntactic structure in the primitive encoding of a shape leads to recognition of possible functions those structures might perform. Identification of shape features from the given primitive encodings will eventually fill the gap of correlating structure descriptions with function descriptions. Identification of Q-words and Q-phrases from a given Q-sentence therefore becomes one of the major tasks in the qualitative modeling of shape.

2.3.1. *Methods for syntactic shape feature recognition*
The following steps outline the analysis of the qualitative representation of shapes using syntactic shape feature recognition.
- Given shapes are encoded with a set of Q-codes to form Q-sentences. Both A-codes and L-codes are used at the coarsest level of granularity.
- All the possible Q-words are systematically generated by using words of increasing length and then the codings are searched to determine if the Q-sentence contains the matching Q-word. The number of appearances as well as encodings of the Q-words are counted.
- The matching results are plotted in a graph with the length of the Q-word and the number of appearances as the axes. The significance of the Q-word increases as either the length of the Q-word or the number of appearances increases.
- Significant Q-words are analysed in terms of Q-phrases. Hence the qualitative design characteristics of the shape are analysed in terms of some meaningful design semantics.
- To determine symmetry a Q-phrase is firstly searched by looking for the inverse pattern of the Q-word. Then the reflective symmetry pattern (palindrome pattern) is checked for.
- Finally, the results are interpreted.

826

2.3.2. *Three examples*

Figure 1 and Table 3 show three shapes with their Q-code encodings. These three shapes are chosen from a group of church plans (Schnell 1974) which illustrate shape patterns for church spaces such as "altar", "bench" and "entrance".

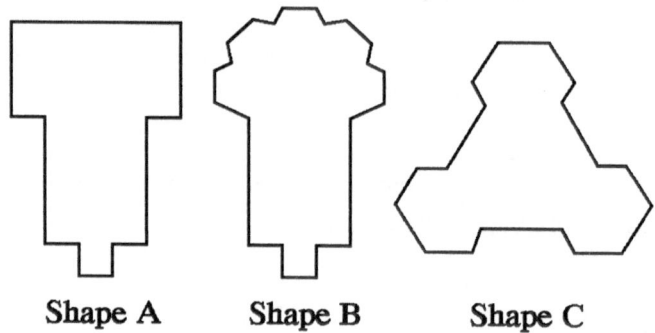

Shape A   **Shape B**   **Shape C**

Figure 1. Three shapes chosen for shape feature recognition

Table 3. Q-code encodings of the three shapes

Q-code	Shape A	Shape B	Shape C
A-code	$(A_+\ A_+\ A_+\ A_+\ A_-\ A_+$ $A_-\ A_+\ A_+\ A_-\ A_+\ A_-)$	$(A_-\ A_+\ A_+\ A_-\ A_+\ A_+\ A_-\ A_+$ $A_+\ A_-\ A_+\ A_+\ A_-\ A_+\ A_+\ A_-$ $A_+\ A_-\ A_+\ A_+\ A_-\ A_+)$	$(A_+\ A_+\ A_+\ A_+\ A_-\ A_-\ A_+$ $A_+\ A_+\ A_+\ A_-\ A_-\ A_+\ A_+\ A_+$ $A_+\ A_-\ A_-)$
L-code	$(L_+\ L_-\ L_+\ L_+\ L_-\ L_-\ L_+$ $L_-\ L_0\ L_0\ L_0\ L_0)$	$(L_+\ L_-\ L_0\ L_-\ L_0\ L_+\ L_-\ L_0\ L_+$ $L_-\ L_0\ L_+\ L_-\ L_0\ L_+\ L_0\ L_+\ L_-$ $L_0\ L_0\ L_0\ L_0)$	$(L_+\ L_-\ L_+\ L_0\ L_0\ L_-\ L_+\ L_-$ $L_+\ L_0\ L_0\ L_-\ L_+\ L_-\ L_+\ L_0$ $L_0\ L_-)$

2.3.3. *Q-word recognition*

Possible shape features (Q-words) are searched for from the given shape descriptions using a generate-and-test method. Firstly, all the possible Q-words are generated. At the coarsest level of granularity, A-codes and L-codes are composed of two and/or three basic Q-codes respectively, namely $\{A_+, A_-\}$ and $\{L_+, L_0, L_-\}$. Possible numbers of Q-words for a length n are, therefore, 2^n and 3^n for each A-code and L-code case. Secondly, the appearances of each Q-word are checked in terms of "iteration (alternation)" and "symmetry" operations. As for iteration, even overlapping Q-words are counted respectively. For example, when the Q-word "$(A_+\ A_-\ A_+)$" is searched for, its iterative appearance in the Q-sentence $(A_+\ A_-\ A_+\ A_-\ A_+\ A_-)$, it is counted as "3". For symmetry, Q-words (Qw) are counted if it is possible to describe a Q-phrase using symmetric operations as S[Qw1 Qw2]. Figure 2 shows the result of Q-word recognition in terms of the iteration (alternation) operation from the given shapes A, B, and C.

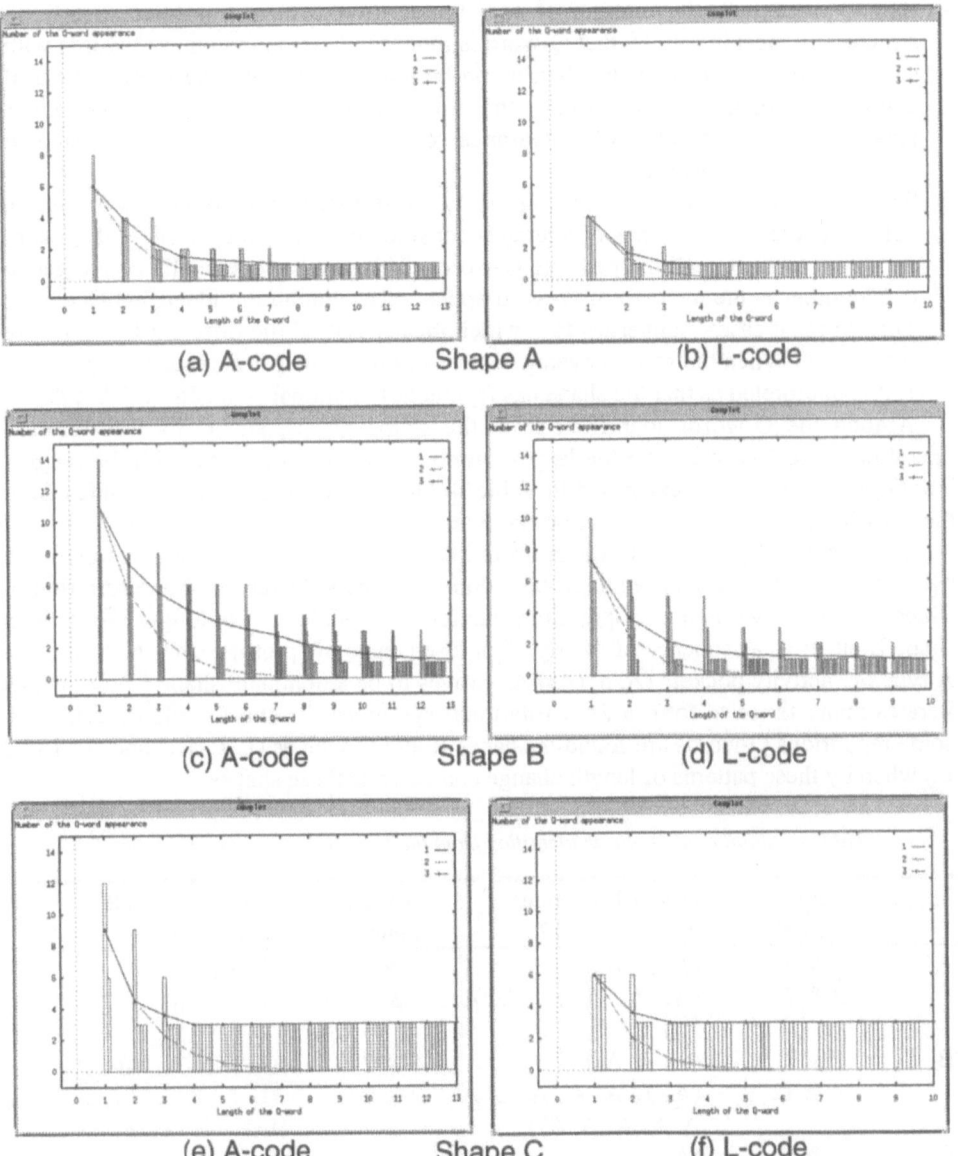

Figure 2. Identified shape features from the three shapes in Figure 1, derived by counting the occurrence of Q-words in the Q-code representations

3. Results

The following can be observed from these results.

- When the total length of a Q-sentence is n, there are n Q-words, ie., therefore, including all the overlapping Q-words, the sum of all the appearances of each Q-word of the same length equals n.

- Any Q-word with the number of appearance greater than 1 implies something structurally meaningful. If that Q-word is longer in length than other Q-words with same number of appearances, then it can be interpreted as having more structural significance than the other. Hence the most meaningful Q-word(s), with longer length and larger number of appearances can be interpreted as being related to design concept of that shape.
- As the length increases, the average of Q-word appearance converges to certain values. There are two types of averages considered. One is the average of Q-word appearances among all the possible Q-words. This number eventually converges to 0. The other is the average of Q-word appearances among the identified Q-words. This, in most cases, converges to 1 which means that all the different Q-words are recognised. When the second average converges to an integer number i greater than 1, it is interpreted as that the shape has (syntactic) rotational symmetry of "$2\pi/i$"

Among the Q-words identified from the shapes A, B, and C by the iteration operation those Q-words with the largest number of appearances and with the longest lengths are chosen to be examined in Table 4. Q-words of length 1 are excluded from the table because of their lack of information as a "word".

Table 4 displays some findings from the shape-feature matching results. The "wedge shape" $(A_+\ A_.\ A_+)$ seems to be a dominant shape feature in the description of shapes A and B, while a multiple interpretation is possible for the longer Q-words in shape B either as a sequence of "wedges" or a sequence of "protrusions" $(A_.\ A_+\ A_+\ A_.)$. A type of "activity pocket" $(A_.\ n^*(A_+)\ A_.)$ proves to be a dominant shape feature which iterates many times to form a $2\pi/3$ rotational symmetry. In the L-code descriptions, some repetitious Q-words are found in shapes B and C such as $(L_.\ L_0\ L_+)$ and $(L_+\ L_0\ L_0\ L_.)$ whereby these patterns of length change characterise those shapes.

Table 4. Significant Q-words identified from the shapes A, B, and C in Figure 1

Shape	Appear. & Length	Q-word (A-code) up to length 12	Appear. & Length	Q-word (L-code) up to length 9
A	4 & 3	$(A_+\ A_.\ A_+)$	3 & 2	$(L_0\ L_0)$
	2 & 7	$(A_+\ A_+\ A_.\ A_+\ A_.\ A_+\ A_+)$	2 & 3	$(L_0\ L_0\ L_0)$
B	8 & 3	$(A_+\ A_.\ A_+)$	6 & 2	$(L_0\ L_+),\ (L_.\ L_0)$
	3 & 12	$(A_+\ A_.\ A_+\ A_+\ A_.\ A_+\ A_+\ A_.\ A_+\ A_+\ A_.\ A_+)$	2 & 9	$(L_.\ L_0\ L_+\ L_.\ L_0\ L_+\ L_.\ L_0\ L_+)$
C	9 & 2	$(A_+\ A_+)$	6 & 2	$(L_.\ L_+)$
	3 & 12	$(A_.\ A_+\ A_+\ A_+\ A_+\ A_.\ A_.\ A_+\ A_+\ A_+\ A_+\ A_.)$	3 & 9	$(L_+\ L_0\ L_0\ L_.\ L_+\ L_.\ L_.\ L_+\ L_0\ L_0)$

Another interesting aspect of shape feature matching is the variation in the numbers of significant Q-words, ie, those Q-words that appear more than once. There are two patterns of either "increase-decrease" or "increase-steady". Shape features $\{(A_+\ A_.\ A_+)\}$, $\{(A_+\ A_+\ A_.\ A_+\ A_+\ A_.\ A_+),\ (A_+\ A_.\ A_+\ A_+\ A_.\ A_+\ A_+),\ (A_.\ A_+\ A_+\ A_.\ A_+\ A_+\ A_.)\}$, $\{(A_+\ A_+\ A_+\ A_+),\ (A_+\ A_+\ A_+\ A_.),\ (A_+\ A_+\ A_.\ A_.),\ (A_+\ A_.\ A_.\ A_+),\ (A_.\ A_+\ A_+\ A_+),\ (A_.\ A_.\ A_+\ A_+)\}$ are found at word lengths 3, 4, and 7 for shapes A, B, and C respectively. The dominant shape features seem to emerge around these points such as $(A_+\ A_-\ A_+)$ for shape A, $(A_-\ A_+\ A_+\ A_-)$ for shape B, and $(A_-\ A_+\ A_+\ A_+\ A_+\ A_-)$ for shape C.

4. Discussion

Qualitative modeling of shape distinguishes itself from quantitative modeling in a number of significant ways. It describes with classes of shapes as opposed to quantitative modeling which describes individual shapes. This distinction is significant because qualitative modeling does not require the specificity of values which are only available later in the design process. The utility of qualitative modeling is partly founded on this distinction. The effects of the availability of qualitative modeling of the kind introduced in this paper can be grouped into the following categories:

- (i) tools to support conceptual designing;
- (ii) reasoning about shape; and
- (iii) relating shape to space.

It now becomes possible to construct computational tools to support conceptual designing since there is no requirement that drawings be precise as is the case with most other tools. The designer does not need to take numerical decisions until later and can concentrate on shape as a class descriptor rather than shape specified through geometry.

Since shape features are now modelable it becomes possible to reason about shapes through their features unrelated to their specific geometries. This has the potential to open up new areas of computational support for designers.

Acknowledgments

This research is supported by an Overseas Postgraduate Research Award. Computing resources are provided by the Key Centre of Design Computing.

References

Allen, J. F. (1983) Maintaining knowledge about temporal intervals, CACM 26(11): 832-843.

Chang, S.-K., Shi, Q. Y., and Yan, C. W. (1987) Iconic indexing by 2-d strings, IEEE Transactions on Pattern Analysis and Machine Intelligence PAMI-9(3): 413-427.

Clarke, B. L. (1981) A calculus of individuals based on 'connection', Notre Dame Journal of Formal Logic 22(3): 204-218.

de Kleer, J., and Brown, J. (1984) A qualitative physics based on confluences, Artificial Intelligence 24: 7-83.

Egenhofer, M. and Al-Taha, K. (1992) Reasoning about gradual changes of topological relations, in Frank, A. U, Campari, I., and Formentini, U., (eds) Theories and Methods of Spatio-Temporal Reasoning in Geographic Space, Springer-Verlag, Berlin, pp. 196-219.

Freeman, H. (1961) On the encoding of arbitrary geometric configurations, IRE Trans. on Electronic Computers EC-10: 260-268.

830

Fu, K. S. (1976) Syntactic (linguistic) pattern recognition, in Fu, K. S. (ed.) Digital Pattern Recognition, Springer-Verlag, Berlin, pp. 95-134.

Helm, P. , and Leeuwenberg, E. (1986) Avoiding explosive search in automatic selection of simplest pattern codes, Pattern Recognition 19(2): 181-191.

Jungert, E. (1993) Symbolic spatial reasoning on object shapes for qualitative matching, in Frank A. U, and Campari I. (eds), Spatial Information Theory (COSTI'93), Springer-Verlag, Berlin, pp. 444-462.

Leeuwenberg, E,., and Buffart, H. (1983) An outline of coding theory: Summary of some related experiments, in Geissler, H.-G. (eds), Modern Issues in Perception, North-Holland, Amsterdam, pp. 25-47.

Martinoli, O., Masulli, F., and Riani, M. (1988) Algorithmic information of images, in Cantoni, V, Gesu, V. D., and Levialdi, S. (eds), Image Analysis and Processing II, Plenum Press, New York, pp. 287-293.

Randell, D. A, Cui, Z., and Cohn, A. (1992) A spatial logic based on regions and connection, Proceedings Third International Conference on Knowledge Representation and Reasoning, Morgan Kaufmann, San Mateo, pp. 165-176.

Schnell, H. (1974) Twentieth Century Church Architecture in Germany, Verlag-Schnell & Steiner, Munich.

Shah, J. (1991) Assessment of feature technology, CAD 23(5): 331-343.

Stiny, G. (1978) Generating and measuring aesthetic forms, Handbook of Perception Vol X, Academic Press, New York, pp. 133-152.

Stiny, G. (1989) Formal devices for design, in Newsome, S., Spillers, W. R. and Finger, S. (eds) Design Theory '88, Springer-Verlag, New York, pp. 173-188.

Treisman A. M, and Gelade, G. (1980) A Feature-Integration Theory of Attention, Cognitive Psychology 14: 97-136.

Weinberg, J., Uckun, S., Biswas, G., and Manganaris, S. (1992) Qualitative vector algebra, in Faltings, B. and Struss, P. (eds), Recent Advances in Qualitative Physics, MIT Press, London, pp. 193-207.

Wertner, H. (1994) Qualitative Reasoning: Modeling and the Generation of Behavior, Springer-Verlag, Vienna.

LEVEL-OF-DETAIL VISUALIZATION OF ARCHITECTURAL MODELS

S. BELBLIDIA, J.P. PERRIN
CRAI (Research Center in Architecture and Engineering),
School of Architecture, Nancy, France

Abstract

The work presented in this paper aims to use level-of-detail representation in realizing interactive walkthroughs or ignoring useless details in large architectural models. In order to choose the right representation of a model, we have to evaluate the error comitted when using a simplified version instead of the full description of an object. This error depends on the object deformation during the simplification process but also on the importance of this object in the current viewing conditions. This "visible" error is used with different visualization strategies to find the model representation which satisfies either a quality criterion or a cost condition.

1. Introduction

When visualizing large architectural databases, even powerful graphic workstations can not ensure interactive walkthroughs. Furthermore, the great accuracy of some models are useless in many applications or viewing contexts. The level-of-detail representation allows to render a model with different accuracies which depend on the application requirements.

Before the visualization step, a level-of-detail model must be generated following these two steps :
- organizing the database in a hierarchical structure which contains simple and composite objects.
- creating several versions of these objects using an automatic simplification algorithm.

In a previous paper, we presented the improvements we introduced [1] to an existing simplification algorithm [2] . We focus here on the visualization stage where the main problem is to choose the adequate model representation. We present two rendering algorithms which allow to find the representation that matches either a quality criterion (section 4.1) or a rendering cost (section 4.2). To make it possible, we need to evaluate the approximation error when using a simplified version (section 2) and the rendering cost of this version in time units(section 3). The main apport of these algorithms in

R. Junge (ed.), CAAD Futures 1997, 831-836.
© 1997 *Kluwer Academic Publishers.*

regard to other published works [3][4] is the use of an approximation error deducted from the simplification stage.

We are now experimenting the software on a complex architectural model composed of thousands of polygons (section 5).

2. Error evaluation

2.1 GEOMETRIC ERROR

When we simplify an object using the automatic simplification algorithm, we evaluate the difference between the original shape and the simplified version by storing the maximum vertex displacement [1]. This purely geometric factor is independent from the viewing context and the object importance.

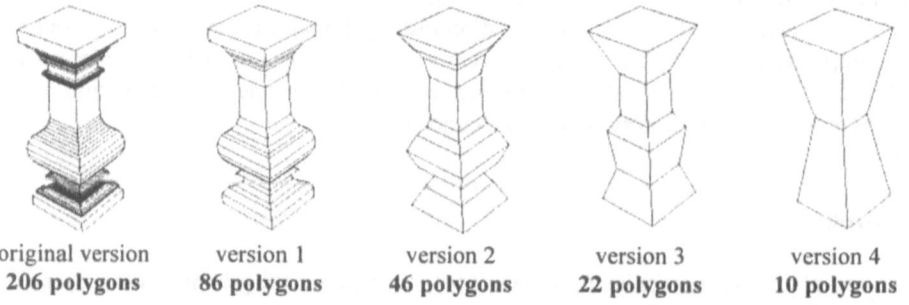

original version	version 1	version 2	version 3	version 4
206 polygons	**86 polygons**	**46 polygons**	**22 polygons**	**10 polygons**

Figure 1: Simplified versions automatically generated.

2.2 IMPORTANCE CRITERIA

The geometric factor computed above does not include any information about the object position or importance in regard to other objects of the model. What we need to evaluate is the "visible" error on the screen which is computed by weighting the geometric error with several factors depending on the following criteria.

2.2.1 Focal distance

The visible error is direcly proportional to the focal distance. The approximation error is more visible with a 200 mm camera than a 35 mm camera.

2.2.2 Object-camera Distance

The visible error is in inverse proportion to the object-camera distance. The error is more perceptible with close views.

2.2.3 Object Position on the Screen

In some visualization contexts, objects which are close to the center of the screen are more important. The relationship between this importance and the distance to the center of the screen is a gaussian function which aspect is user-controlled.

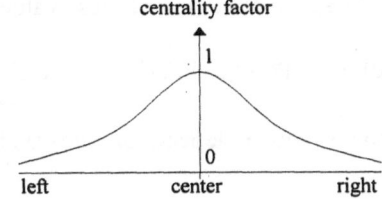

Figure 2: Variation of the centrality factor

2.2.4 Object semantics

Some objects could be more important in a complex architectural model. For these objects, a semantical importance factor is fixed by the user in order to major the visible error.

2.3 VISIBLE ERROR

The visible error for a given triplet (object, version, camera) is computed by weighting the geometric error with the importance factors.

$$\text{VisibleError}(o,v,c) = \underbrace{\text{GeometricError}(o,v) \cdot \frac{\text{FocalDistance}(c)}{\text{Distance}(o,c)}}_{projection} \cdot \text{Centrality}(o) \cdot \text{Semantics}(o)$$

The two first terms of this expression represent the projection of the geometric error on the screen and are always used. The two last ones are optional factors.

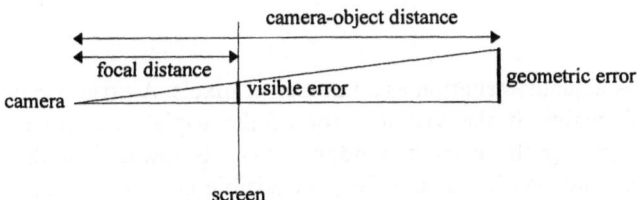

Figure 3: Visible error before weighting with centrality and semantics factors

3. Cost evaluation

The rendering cost of an object version depends on two values : the number of polygons it contains and the number of pixels it covers on the screen. The cost of a given triplet (object, version, camera) is a linear combination of these values :

$$\text{Cost}(o, v, c) = C_1 \cdot \text{Polygons}(o, v) + C_2 \cdot \text{Pixels}(o, v, c)$$

where C_1 and C_1 are constants which depend on the rendering algorithm and the accuracy options used.

4. Rendering strategies

When a complex architectural model is composed of hundreds of simple objects, each one with several versions, the number of possible representations is very large. The more accurate and the less detailed are two particular ones.

The visualization algorithm must be able to find the adequate representation according to user-specified criteria. We have implemented two strategies which handle either a quality criterion or a cost condition.

The algorithms presented in the sections 4.1 and 4.2 operate only on simple objects. They also accept composite objects in extended versions which use recursive calls.

For both strategies, the different versions of an object O are sorted in a quality decreasing order . The full version is indexed 0 and the less detailed is indexed n-1.

4.1 THE QUALITY STRATEGY

4.1.1 Principle
The user specifies a quality criterion as a number of pixels. A given version of an object has the required quality if the visible error of the triplet (object,version,camera) – converted in pixels, for the current window size – is lower than the user-specified threshold. If this condition is not satisfied, the visible error is evaluated with a better version.

With the quality strategy, the algorithm determines independently for each object the version to use for given viewing conditions.

4.1.2 *Algorithm*

```
objectVersion (obj : object, cam : camera, qualityThreshold : real) : integer
  boolean stop ;
  integer ver ;

  stop ← FALSE ;
  ver ← numVersions - 1 ;
  while (stop = FALSE and ver > 0) do
    if (visibleError (obj, ver, cam) < qualityThreshold) then
      stop ← TRUE ;
    else
      ver ← ver-1 ;
    endif
  endwhile
  return version
```

4.2 THE COST STRATEGY

4.2.1 *Principle*

The user specifies a target rendering time as a number of seconds. The rendering cost for the whole model must be lower than the user-specified threshold.

4.2.2 *Algorithm*

```
modelRepresentation( initList : quadrupletList, c : camera, costThreshold : real) : quadrupletList
  quadrupletList list ;
  quadruplet quad ;
  real modelCost ;
  object obj ;
  integer ver ;

  list ← initList ;
  modelCost ← cost (list) ;
  quad ← head (list) ;
  while (modelCost < costThreshold and not EndOf (list)) do
    obj ← object (quad) ;
    ver ← version (quad) ;
    if (cost (obj, ver+1) - cost (obj,ver) <= costThreshold - modelCost) then
      removeQuadruplet (list, quad) ;
      insertQuadruplet (list, obj, ver+1, visibleError (obj, ver+1, cam), cost (obj, ver+1, cam)) ;
      modelCost ←modelCost + cost (obj, ver+1, cam) - cost (obj, ver, cam) ;
      quad ← head (list) ;
    else
      quad ← next (quad) ;
    endif
  endwhile
  return list
```

The data structure we use is a list of quadruplets (object, version, visibleError, cost). In its initial state, this list contains all the objects in their less detailed version. If the initial cost of the model is greater than the user-specified threshold, some objects can not be displayed. For each camera position, this initial list is sorted with a visible error decreasing order. The refinement procedure begin with the first objects of the list since they are those with a greater visible error. This algorithm ensures :

- the rendering time is optimally used. If the head of list can not be refined, the algorithm attempts on the next object in the list.
- the rendering time is equally dispatched on the objects. The algorithm avoids to render some objects with a great accuracy when others are not refined at all.

5. Experimentation

All these features have been developed on Silicon Graphics workstations using Open Inventor C++ graphics library which provides interface and rendering facilities. We are experimenting the software on an urban area : *Place Stanislas* in *Nancy* (France). The place is surrounded of seven classical buildings (City Hall, Opera, Grand Hotel, Museum, ...) with ornemental details. The level-of-detail approach matches very well this kind of built environments.

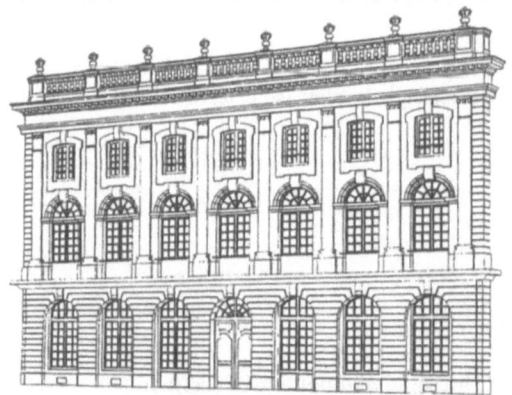

Figure 4 : Opera Building, Place Stanislas, Nancy

6. References

1. Belblidia, S., Perrin, J.P., Paul, J.C. (1995) Multi-Resolution Rendering of Architectural Models, CAAD Futures '95 Conference.
2. Rossignac, J., Borrel, P., (1993) Multi-Resolution 3D Approximations for Rendering Complex Scenes, Conference on geometric Modeling in Computer Graphics, 453-465.
3. Funkhouser, T.A., Séquin, C.H (1993) Adaptative Display Algorithm for Interactive Frame Rates during Visualization of Complex Virtual Environments, Siggraph '93 Conference, 247-254.
4. Maciel, P.W.C., Shirley, P. (1995) Visual Navigation of Large Environments using Textured Clusters, Symposium '95 on Interactive 3D Graphics, 95-102.

AN EXPERIMENT ON HYBRID ARCHITECTURAL FORM-MAKING

SHUENN-REN LIOU, EMMANUEL-GEORGE VAKALO,
and KUO-CHIN CHANG

Liou and Chang:
Department of Architecture
Tunghai University
Taichung, Taiwan
E-Mail:
shuenn@s867.thu.edu.tw

Vakalo:
College of Architecture and Urban
Planning
The University of Michigan
Ann Arbor, MI 48109-2069 USA
E-Mail: egvakalo@umich.edu

This paper illustrates an approach to hybrid architectural form-making. A hypothetical project - the Des Moines Art Center 3rd Addition - is employed as a design experiment. The computer is used as a form-searching medium in the form-making process. Suggesting an addition to the existing center designed by Saarinen, Pei, and Meier, the designer is confronting the problem of how to respond to the three distinct architectural styles. The proposed solution to this problem is to create a hybrid building which inherits architectural properties from those precedents. Potentials of the use of the computer for such task are discussed.

1. Hybrid Architectural Form-Making

The purpose of hybrid architectural form-making is to create an architectural work which embodies significant formal attributes of two or more architectural styles. One way to achieve this is based on the concept of design rules. A hybrid architecture is derived through the application of the design rules of two or more architectural styles. (Liou, 1992:91) The design rules can be precisely defined, as shape rule schemata in shape grammars, or appear diagrammatically, as patterns in pattern languages. (Chen and Liou, 1993:427-434) Further development of this approach has encountered difficulties for two important reasons. One has to do with the difficulty for implementing design rules on computers. The other is that the rule-based approach, however correctly it may carry the knowledge of form, is an alien mode of reasoning for designers. (Archer, 1984:348)

This paper illustrates an alternative approach to hybrid architectural form-making. A hypothetical project - the Des Moines Art Center 3rd Addition - is employed as a design experiment. The computer is used as a form-searching medium in the form-making process.

2. The Experiment - the Des Moines Art Center 3rd Addition

The original design of the Des Moines Art Center is the product of work by three well-known American architects. Specifically, Eliel Saarinen first established a U-shaped

R. Junge (ed.), CAAD Futures 1997, 837-842.
© 1997 *Kluwer Academic Publishers.*

exhibit building in 1948. Following that, I. M. Pei provided an addition to the south to close the U shape and formed a sculpture court in 1965. Finally, Richard Meier attached a number of fragmented volumes to the northwest of Saarinen's building in 1984. (See Figure 1.) It is assumed that a third addition is requested to accommodate an architectural exhibit and study center. Suggesting an addition to the existing center, the designer is confronting the problem of how to respond to the three distinct architectural styles. The proposed solution to this problem is to create a hybrid building which inherits architectural properties from the existing precedents.

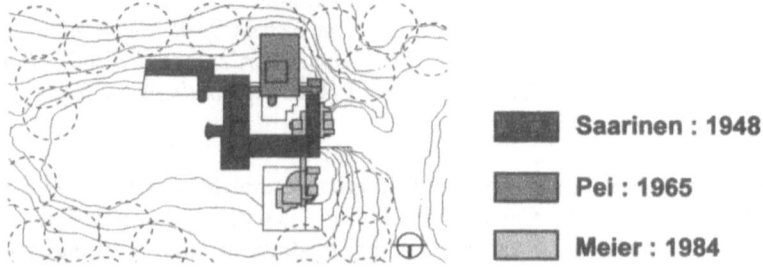

Figure 1. The Original Design of the Des Moines Art Center

As shown in Figure 2, the procedure of the form-making experiment is made up of three major stages. They are (1) the derivation of prototype models, (2) the derivation of a preliminary hybrid model, and (3) the derivation of a final architectural model. In the first stage, a review of the three architects' work is conducted. It constitutes the basis for the designer to develop two prototype models, i.e. a solid form-model and a void space-model, for each of the architects. A total of three form-models and three space-models are thus established. (See Figure 3.)

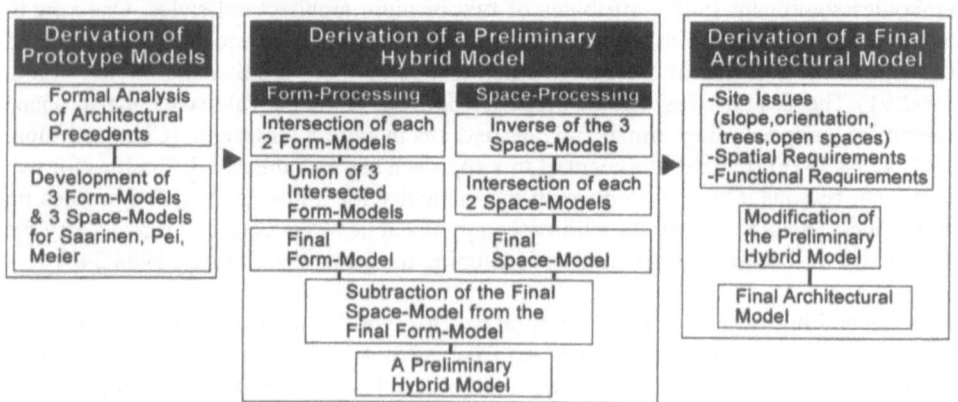

Figure 2. The Procedure of the Hybrid Architectural Form-Making

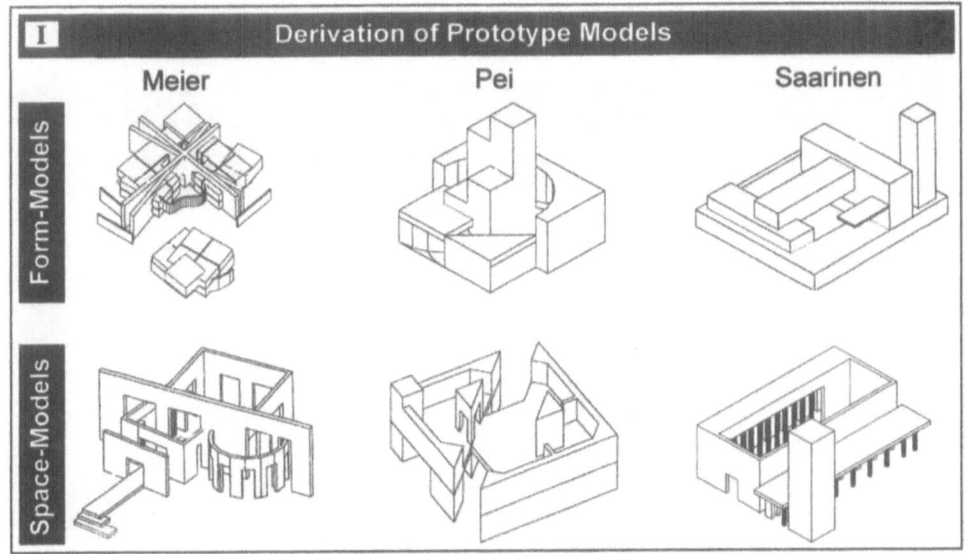

Figure 3. Six Prototype Models

The second stage consists of a number of significant manipulations on the aforementioned solid and void models. Its goal is to derive a hybrid solid model and a hybrid void model. Conceptually, the former can be viewed as the blending of the formal attributes and the latter as the blending of the spatial attributes of the three distinct styles. Moreover, the solid model subtracts the void model would result in a hybrid model filled with spaces. Therefore, as shown in Figure 4, the manipulations can be classified as form-processing and space-processing. Methodically, the concept of union and intersection in set theory is applied. Each two form-models are "synthesized" under the Boolean operation. The synthesis is carried out through the application of "intersect" of AME in AutoCAD. Then, three intersected models are taken together to derive a final form-model. This is carried out through the application of "union." Similarly, a final space-model is derived. Note that in order to manipulate the space-models as "solids," each of them has to be inverted in advance. At last, a preliminary hybrid model is derived through the subtraction of the final space-model from the final form-model.

In the third stage, the hybrid model is placed onto the site. This model is further modified taking into account the site issues such as slope, orientation, trees, and open spaces, as well as the spatial and functional requirements of the program. (See Figure 5.) For example, the preliminary model is located on the east side of the original center to create an entrance court. As well, the whole model is "reflected" to utilize the volumes dissociated from the corner to signal out the new addition. Furthermore, detailed manipulations are taken to resolve the architectural problems such as internal spatial organization and users' needs. Most of the manipulations of form and space are carried out on computers.[1]

840

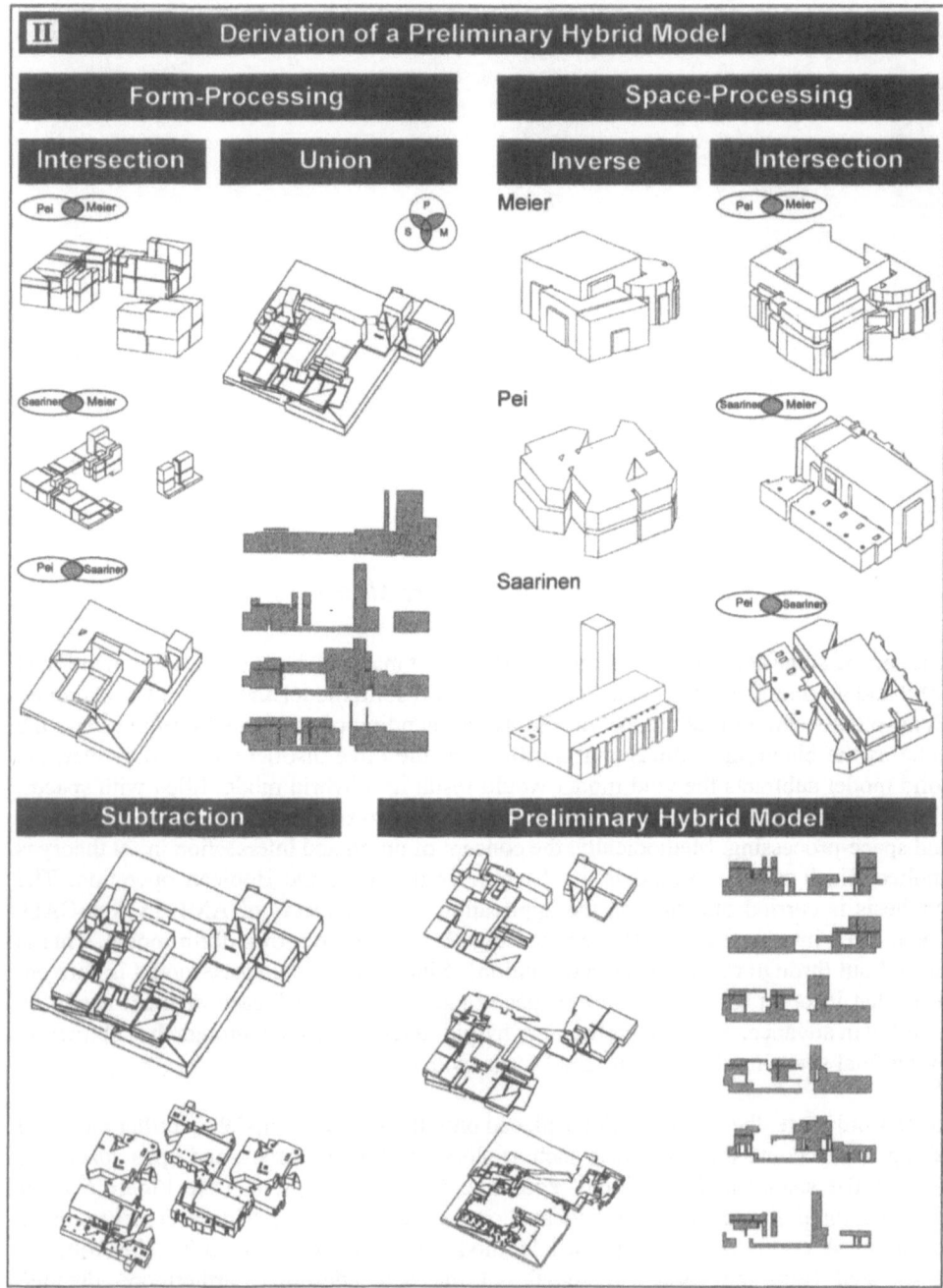

Figure 4.. The Derivation of a Preliminary Hybrid Model

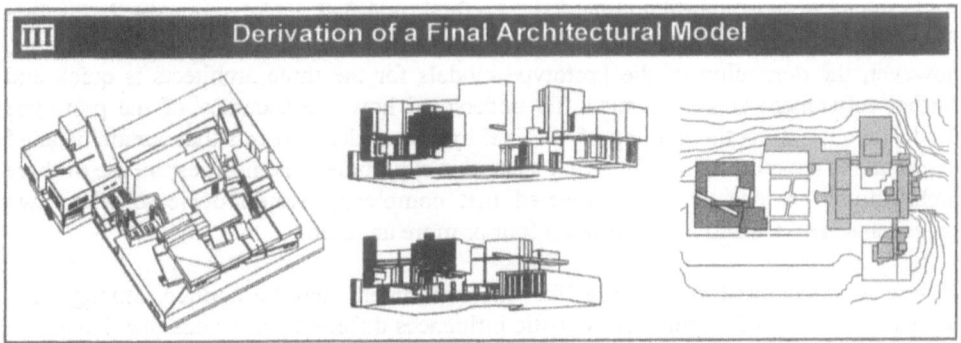

Figure 5.. The Final Architectural Model

3. Discussions

The discussion of the experiment illustrated above may proceed along two different, but interrelated lines. The first is the computer-aided architectural design and the second hybrid architecture. In current use of the computer to aid architectural designs, most emphases are placed on the work of drafting (e.g. in producing work drawings) and final presentation (e.g. 3D animations). It is argued in Liou (1996) that, in addition to "servant" there are various roles such as "assistant," "partner," and "expert" that the computer can play. To facilitate the use of the digital environment in the creation of architectural designs, Liou suggests that the designer (1) establish a new relation (other than master to servant) with the computer, (2) be more aware of her/his own design procedure, and (3) solicit critiques on design process instead of product.

In this experiment the computer is used not as a drafting or presentation tool. Rather, it is demonstrated that the computer, when incorporated properly, can be a powerful form-searching assistant in a design process. It is extremely exciting to derive many interesting and unpredictable 2D and 3D compositions. The spatial composition of the sections is particularly noteworthy. (See Figure 4.) In contrast to the vague image normally preoccupying the designer taking the traditional design approach, the unpredictable form solutions derived through computation constitute an important basis for decision-making in the design process. Moreover, the new digital work environment provides a great opportunity for 3D thinking and manipulation, which were once difficult, if not impossible tasks.

It should also be noted that the design procedure of the experiment is flexible and in many episodes designer-oriented. For example, the intersection of models may not results in satisfactory forms and the designer may has to go back to adjust the locations of the models. This and other similar actions allow the designer's creation to be exercised within the procedure. As a matter of fact, it is not surprised that the ability to derive a good solution is dependent strongly on the designer's discipline of architectural design.

842

As to the purpose of hybrid architecture, the final modified model expresses, in a subtle form, the stylistic characters of the three architects. Due to the limitation of time, however, the derivation of the prototype models for the three architects is quick and further refinement is not pursued. To understand how the accuracy of the prototype models would affect the making of the final hybrid model, it seems that certain detailed experiments are required. Last but not least, this experiment exemplifies the use of three architectural precedents. It is expected that complexity would decrease when two precedents are used and increase when four or more are used.

Hybrid architectural form-making like this experiment may be applied, among other things to discern and capture the stylistic influences different precedents may have on a specific building design. As well the concept of hybrid architecture may be particularly useful in situations where a designer is asked to propose a design for a site whose physical context comprises buildings that are distinctly different stylistically. Different from the rule-based approach, this experiment offers a greater flexibility for the designer to interpret architectural precedents, and the power of the computer is easier to be incorporated into the design process. As such it may also be seen as a digital design game that can be used to hone one's form-making ability.

Note:

1.   The softwares employed includes AutoCAD R12 for MS Windows 3.1, 3DStudio R4, Animator Pro, Photoshop 2.51, Lotus Ami Pro 3.0, MediaStudio 1.0/Album/ScreenCapture/VideoEditoe /ImageEditor, AuthorWare Professional 2.0 for MS Windows 3.1

References:

1.   Archer, L. (1984) Whatever Became of Design Methodology? Developments in Design Methodology, John Wiley & Sons, New York, 347-349.

2.   Chen, L.-F. and Liou, S.-R. (1993) A Study on Hybrid Architectural Design. Proceedings of the 6th Annual Conference of Architectural Institute of the Republic of China, Taiwan, 427-434.

3.   Liou, S.-R. (1996) Computers Applied to the Creation of Architectural Designs. Proceedings of the 9th Annual Conference of Architectural Institute of the Republic of China, Taiwan, 211-216.

4.   Liou, S.-R. (1992) A Computer-Based Framework for Analyzing and Deriving the Morphological Structure of Architectural Designs. Doctoral Dissertation of the University of Michigan, Ann Arbor.

HYPERMEDIA STRUCTURING OF THE TECHNICAL DOCUMENTATION FOR THE ARCHITECTURAL AIDED DESIGN

BIGNON J.C.[ab], HALIN G.[ac], HUMBERT P[a].
[a]CRAI (Research Center in Architecture and Engineering)
Ecole d'Architecture de Nancy. France.
[b]Ecole d'Architecture de Strasbourg. France.
[c]University of Metz. France.

Abstract

The definition of an universal structuring model of the technical documentation is arduous, indeed utopian considering the great number of products and the diversity of relative information. To answer this situation we are trying to develop a general approach of the documentation. The document is the base entity of documentation structuring and it represents a coherent informative unit. We propose a model of document hypermedia structuring. This model allows the definition, the presentation, the navigation and the retrieval of general information on building products by a document manipulation. It is associated with a hypermedia design method adapted to document management. This method proposes, after the identification of the user, three phases of hypermedia definition : data definition, navigation definition and user interface definition. The model of a hypermedia structuring of the technical documentation proposed in this article is at once independent of avalaible information on products, open, and makes easier the addition of new navigational functions.

1. Introduction

The design activity and more widely the building activity generate and are generated by a complex system of information exchanges. The exchanged data are numerous and structured according to many and different points of view. The complexity of such a system has created the definition of a great variety of exchange formats expressing the large diversity of exchanged information usages. Some usages, which are strongly codified by practices, such as the building site management have rapidly obtained standards [1].

Until now, other usages have been the subject of only few information structuring agreements. That is why the technical documentation domain which have been the subject of works as the French GT5/EDICONSTRUCT or the SFB [2], does not have obtained any federative and operating results yet. The definition of an universal structuring model of the technical documentation is arduous, indeed utopian considering the great number of products and the diversity of relative information. To answer to this

R. Junge (ed.), CAAD Futures 1997, 843-848.

situation we are trying to develop a general approach of the documentation. It is based on the fact that eighty percent of realized, exchanged and consulted information with computer tools is on a document form.

We propose a model of document hypermedia structuring. This model allows the definition, the presentation, the navigation and the retrieval of general information on building products by a document manipulation. It is associated with a hypermedia design method adapted to document management. This method proposes, after the identification of the user, three phases of hypermedia definition : data definition, navigation definition and user interface definition.

This model and method have been used to realize the DOMITEC application which is described in this article.

2. Hypermedia & Technical Documentation

The advantages and drawbacks of hypermedia structuring used to represent a set of information are exposed in many books and articles [3].

The technical documentation and the product cataloguing are domains where hypermedia can propose many functions : intuitive navigation, precise information retrieval, animated presentation (video), or commented (sound), product selection, search for product components aided with a graphical presentation of nomenclature.

These services are essential considering the important volume and the multidimensional feature that the information on the building products and on those who manufacture them have :

- multi-media : a product can be described by an image, a video, a 3D representation, a text, a sound ...
- multi-lingual : the documentation can be consulted by architects or prescriptors speaking different languages,
- multi-structure : information avalaible on products has to respect different presentation standards in order to make easier its integration into many design aided tools,
- multi-view : products can be presented according to different points of view : description, execution (implementation) , prescription, norm ...
- multi-culture : information on products can be presented through different manners according to the cultural origin of the users.

Considering this situation, the information must be organized with a coherent structure associated with many functions of access, manipulation and exchange. An hypermedia structuring can answer this objective if the representation of navigations and accesses are based on a strong organization of data. In order to build, to manage and to communicate this hypermedia structuring, we have defined an adapted method of hypermedia design.

3. Documentation Hypermedia Structuring

Our hypermedia structuring method, as similar methods [4], contains three definition phases : data definition, navigation definition and user interface definition.

In the data definition phase, all the avalaible information on a technical documentation are described with a specific data model where the based description unit is the document. This step reveals the existing links between pieces of information and manages the coherence.

The navigation definition allows the description of browsings and accesses the hypermedia will propose. This phase generates a navigational schema composed of a nodes graph. Each node represents a manipulation function of the information described in the data schema of the previous phase. We have defined three sorts of information manipulation : information consultation, information retrieval and information exportation.

Finally in the user interface definition phase, the visual aspect of each node and the interactions synchronization are determined.

3.1 INFORMATION DEFINITION

The data model used is inspired by the one of the hypermedia MORE system [5]. Its principal characteristic is to exhibit, in the data schema, the potential navigations and the types of media used in the future hypermedia (figure 2).

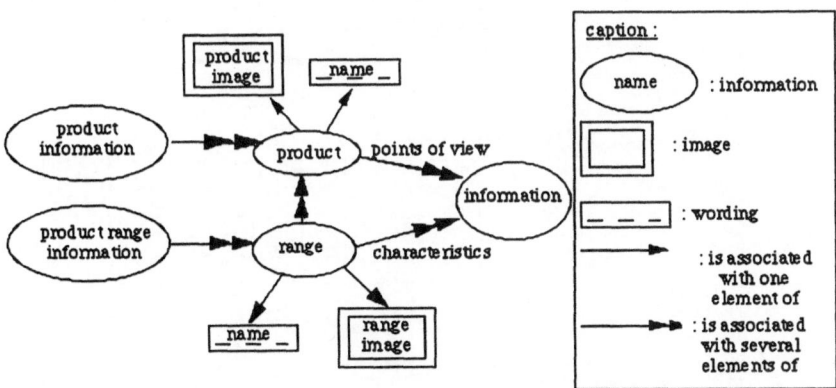

Figure 1: information on products and ranges.

The main information which is manipulated is a company representing a manufacturer of products. The building culture being different from a country to another, the information linked with a company is also dependent of a country. This information includes the description about the company and also the one about its ranges and its products. The information description is done in the language of concerned country.

The multimedia information is present in every descriptive facets of a company. For example, the information about the ranges or products is structured as a set of information illustrating the main properties of one range or the product viewpoints (figure 1).

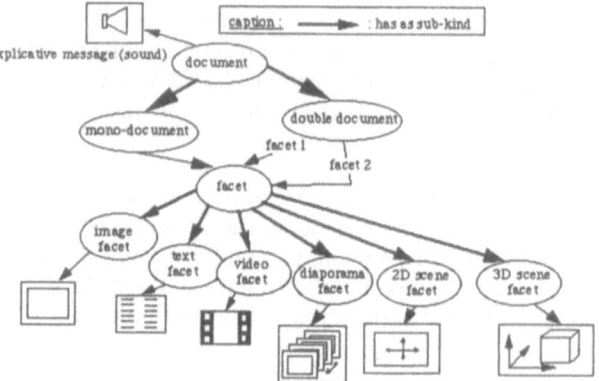

Figure 2 : Document structure.

An information is a hierarchical organization which is defined in the following manner : a piece of information could be either documented - it holds a set of documents - or structured - it holds a set of information where each piece of information could be either documented, or structured .

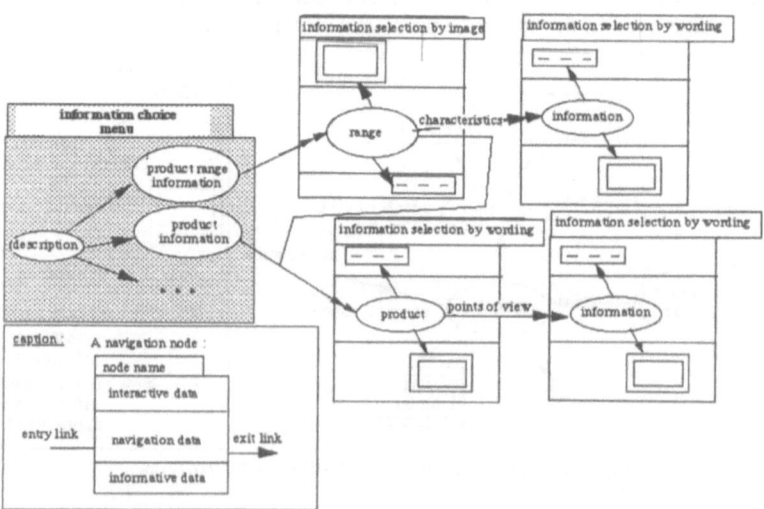

Figure 3 : navigation schema of the products and ranges

Although every range or every product has an image that illustrates it, the multimedia information is located essentially on the level of the document held into a documented information. The document represents the elementary information which can be consulted, but also the exchange unit between tools participating in the process of architectural conception. A document can be associated with a sound and have one or two constituent named facets. A facet is linked to a multimedia information : movie, picture, text, schema, 2D scene (QuickTime VR), 3D scene (QuickDraw 3D, V.R.M.L.) (figure 2).

Once the structuring of data obtained , the schema is going to be used as a holder to the construction of the navigation schema.

3.2 NAVIGATION DEFINITION

The navigations that we defined lean on an arrangement of typed nodes. A node allows the user to handle some information through a predefined function. For example, the browsing of a set of documents is going to be depicted with a node whose function is the browsing, and whose information is a set of documents. The set of nodes and their links also define the graph of the states of the hypermedia.

The navigation schema can be built by tracing the chosen nodes on the data schema. The figure 3 illustrates the possible navigation on the ranges and products of a company.

The identified navigation are those suggested by the data structure. Other forms of navigations can be proposed in order to improve the access to pertinent information : product retrieval by image, search by navigation through 2D scenes of buildings, component search by 3D product scenes spliting, multi-criteria search ...

4. the Domitec Application (User interface Definition)

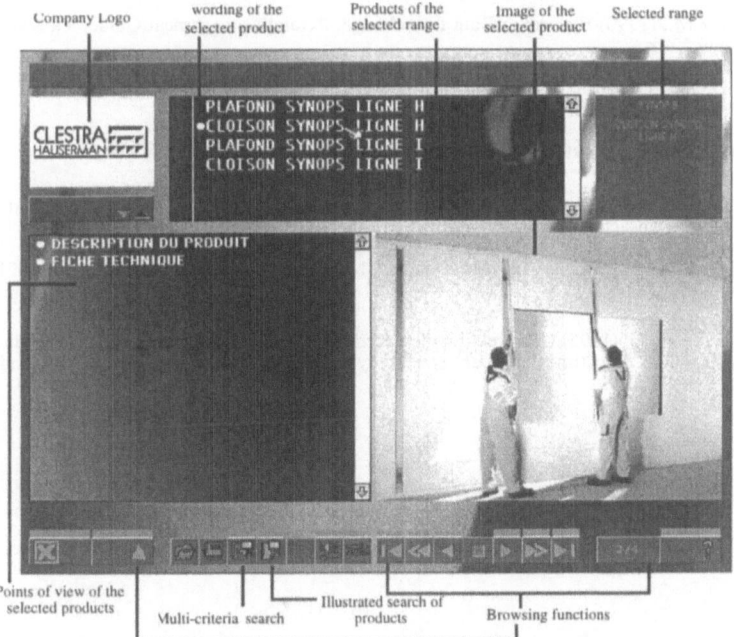

Figure 4 : printed screen of the DOMITEC application.

The DOMITEC application [6], developped in Macromedia Authorware, has been the experiment field of this hypermedia structuring. It contains a part of the functions presented previously. The printed screen of figure 4 shows the user interface definition corresponding to the nodes of the navigation schema presented in figure 3. We can see the selected product by the "big dot" before its name and the display of its image and of its list of points of view.

5. Conclusion

The classical methods of cataloguing are based on the use of DBMS technics. These tools are useful for a logical and deductive search but they are not adapted to an intuitive discovery. An hypermedia structuring seems to be able to achieve this kind of information access and navigation.

The hypermedia structuring of the technical documentation and the associated method proposed in this article are at once : independent of avalaible information on products, open because they allow a fine information structuring according to standards in the process of development, evolutive because they make easier the addition of new navigational functions.

These qualities are essential for the data exchange and for the document integration into a design process.

References

1. Emmelhainz M., (1990) Electronic Data Interchange. A total Management Guide, Van Nostrand Reinholld,New York.

2. Havenne D.(1991) La structuration de l'information via le BB/SFB, Note de synthèse, Louvain la Neuve, Unité d'architecture-cellule SFB belge.

3. Nielsen J. (1990) Hypertext and Hypermedia, Academic Press edtition, San Diego.

4. Garzotto F., Paolini P., Schwabe D. (1995) HDM - A Model-Based Approach to Hypertext Application Design. In ACM Transactions on Informations Systems, Vol 11, n°1, January 1993, pp 1-26.

5. Lucarella D. & al., (1993) MORE : Multimedia Object Retrieval Model.. In Proceeding of Hypertext 93, Washington, pp 39-50.

6. Bignon J.C., Halin G. (1995) Construction d'Hypermedias "ouverts". Application à la documentation technique des produits du bâtiment. Hypertextes et Hypermedias, réalisations, outils, et méthodes. Edition Hermes, Paris, pp 251-261.

HYPERSKETCHING:

DESIGN AS CREATING A GRAPHICAL HYPERDOCUMENT

RAYMOND MCCALL, ERIK JOHNSON, AND MARK SMITH
Sundance Laboratory for Computing in Planning and Design
College of Architecture and Planning
University of Colorado

Abstract

There are empirical and theoretical reasons for believing that current CAD does not adequately support the early, conceptual stages of design. Hand-done design drawing has a several advantages over current, CAD-based approaches to generating form in these stages. One advantage is the indeterminacy of hand drawing--i.e., its abstractness and ambiguity. Another is a non-destructive drawing process, where new drawings are created without modifying old ones. A third is designers' creation of large collections of inter-related drawings--i.e., graphical hyperdocuments. A fourth is the unobtrusive character of conventional drawing tools. We have created two prototypes that incorporate these features into a new type of CAD based on sketching with electronic pens on LCD tablets. The first prototype, called HyperSketch, is a stand-alone system that simulates tracing paper. It creates a hypermedia network in which the nodes are sketches and the links are primarily *traced-from* relationships recorded automatically by the system. The second prototype adds the HyperSketching functionality to our existing PHIDIAS HyperCAD system. This aids design by using the sketches to index and retrieve multimedia information that is useful for a variety of design tasks.

1. Problem

Computer-aided design (CAD) is increasingly used by architects for design. It is are finding greatest use in the later stages of design, especially *design development*. It is, however, finding far less acceptance in the *conceptual* and *early schematic* stages. For these early stages most architects continue to devise solution form primarily by sketching with pencil and pen on paper.

Evidence for this state of affairs was produced in a recent survey by two students from the University of Colorado (Greg Eddy and Scott Saia). They surveyed 30 architects in the Boulder-Denver area to determine in what ways--if any--they are using computers in design. Only two of the architects surveyed made no use of computers for design. Of the 28 remaining, 25 use CAD for *design development* and 17 use it for *schematic design*. But, only 7--i.e., only one quarter of those using CAD extensively for design--said that

R. Junge (ed.), CAAD Futures 1997, 849-854.
© 1997 *Kluwer Academic Publishers.*

they *ever* use CAD for the *conceptual design*. In-depth interviews with a number of the surveyed designers suggests that the use of CAD in conceptual stages is often a minor part of those processes. Most conceptual design is still done by hand-based sketching.

2. Problem Analysis

2.1. MORE THAN A MATTER OF TIME

CAD researchers have long been aware of reluctance by architects to use CAD for design. It has often been said, however, that the solution to this problem is merely a matter of time. Some believe it is "a generational thing," and that when today's computer-literate children grow up they will naturally prefer the computer as a design medium. Others are waiting for Moore's Law to deliver inexpensive, high-powered graphics hardware.

We argue that the problem is more fundamental--at least in the early stages of design--and thus requires a more fundamental software solution. We perceive a mismatch between the graphical processes architects use in early design and the processes current CAD supports. Thus, CAD is unlikely to become the preferred design medium even for computer enthusiasts.

Ultimately, however, the acceptance or non-acceptance of CAD by architects is not the central issue. Far more important is the notion that the form-making processes that current CAD supports for conceptual design are inferior in crucial respects to those supported by pencil and paper. If this is so, then we would hope that CAD--at least in its current form--is *not* accepted by architects for early design. We are therefore seeking to understand how hand-done drawing supports early design and to use this understanding to devise features for future CAD systems.

2.2. DIFFERENCES THAT MAKE A DIFFERENCE

There are a number of obvious differences between hand-done design drawing and current CAD modeling. These have largely been overlooked--or dismissed as inconsequential--by the CAD community. We take as our starting point the possibility that these differences might represent important advantages of hand drawing for early design. Among these differences are the following:

Hand-done design drawing is indeterminate in varying degrees, but current CAD models are highly determinate. By this we mean that hand-done design drawing is typically approximate, abstract, vague or ambiguous to some degree. It uses thick, wobbly or multiple lines; shapes are often rounded, lines only approximately parallel. CAD models, by contrast, use hard-line drawing and are typically based on precise dimensions and angles. That the indeterminacy of hand-done drawing is deliberate is indicated by the fact that its degree varies systematically during design, generally becoming less indeterminate as design progresses. By being indeterminate, hand-done design focuses on larger issues of design while ignoring temporarily the many detailed issues that arise in determinate drawing. Indeterminate drawing thus enables designers to use a divide-and-conquer strategy for attacking the complexity of architectural design.

Hand-done design is based on non-destructive drawing, but all current CAD is based on destructive editing. Devising new solution states with CAD is accomplished by editing one model to create a new one. This is a destructive process in that old states of the solution are destroyed--i.e., modified--to create new ones. Hand-done design, however, generally does not involve destroying old drawings to create new ones. Instead, designers create new drawings--often by tracing over previous ones. Destructive drawing, i.e., by erasure, generally has a small role in early design. Use of non-destructive drawing preserves an *episodic history* of the project, the episodes corresponding to individual drawings. This facilitates backtracking. It also enables evaluation of the current solution state by examining the solution history.

Hand-done design creates a large collection of inter-related drawings--i.e., a graphical hyperdocument--but CAD creates only a single model. Non-destructive drawing, by definition, produces multiple drawings. In fact, a large number of drawings is typical. In one case, we saw more than a thousand sketches created in a three-week period. Often, less than three minutes is spent on a drawing, and rates can exceed 20 drawings an hour.

Our empirical studies show that design drawings are highly heterogeneous. They are at many different levels of detail and deal with many aspects of a building--function, structure, appearance, circulation, lighting, user interactions, views, relation to site and surrounding context. They show and compare alternatives at many different levels of aggregation and abstraction. Many drawings, if not most, are not recognizable as being--even conceptually--edited versions of other drawings. Paper and pencil thus function as more than a "poor man's graphical editor," as has often been supposed.

Current CAD, of course, is designed to support repeated editing of a single model. CAD can, of course, be made to support more than one model. But current systems do not have the database capabilities to manage thousands of drawings. Furthermore, few systems support creation of drawings in under three minutes.

In hand-done design, there are typically important relationships between drawings. Tracing is one source and indicator of such relationships. Some drawings represent alternative solution possibilities to other drawings. Some drawings are done to resolve design problems that arise in doing other drawings; often the former are done in smaller scale on the side of the latter drawings. All of these represent crucial relationships between drawings. In addition, design drawings are commonly annotated with text, arrows, and numerical information. Thus--in addition to small "study" sketches--diagrams, notes, tables and calculations are often found on the sides of larger drawings.

Current CAD systems have not been designed with the hypermedia capabilities needed to represent the relationships between drawings, much less to manage extensive networks of drawings. And, of course, CAD systems do not generally support extensive annotation.

The tools for hand-done design--pencil, pen and paper--are effectively invisible, direct and unobtrusive, but current CAD software is visible, indirect and obtrusive in design.

During traditional, hand-done design, designers effectively devote full attention to thinking about the project--the problem and its solution--while drawing. They do not focus significant attention on the pencil, pen and paper. With current CAD, far more attention must be devoted to the tools for design--the software and how to use it. Even when the CAD software has been mastered, considerable time and attention are devoted to multi-step processes for manipulating form. CAD seldom achieves the "transparency" or directness of pencil or pen. Typically, the computer interface is the computer "in your face." Consequently, interacting with the computer disrupts design thinking.

3. System Prototypes

We have created two prototypes that are modeled on the way architects design with pencil, pen and paper. With both systems, designers create form by sketching with electronic pens on LCD tablets. Both systems are based on the idea of designing as creating a graphical hyperdocument--a collection of linked sketches--a fundamental alternative to the CAD paradigm of designing by destructive editing.

3.1. FIRST PROTOTYPE: A STAND-ALONE HYPERSKETCH SYSTEM

Our first prototype was called HyperSketch. Running on PCs with pen-sensitive LCD tablets and using the metaphor of *sketching on tracing paper,* HyperSketch enables designers to draw directly on the screen. As they create a set of sketches, *HyperSketch automatically links the sketches--e.g., using "traced-from" and other links--to create a graphical hyperdocument having individual sketches as nodes.* Stacks of digital tracing paper can be created; and branching of stacks supports generation of solution alternatives. Sets of drawings are organized in *projects.* Individual drawings can be scaled between full-screen and postage-stamp size; they can also be shown as icons or hidden completely.

We videotaped three professional architects using HyperSketch. All had little experience with computers and a strong aversion to CAD. They were generally unhappy with the slow speed of our initial prototype and the feel of the electronic pen on the tablet. Nevertheless, at several points, *all three architects spontaneously began developing solution form for real-world projects they were then working on.* This happened, furthermore, without prompting from us and within the first hour of system use.

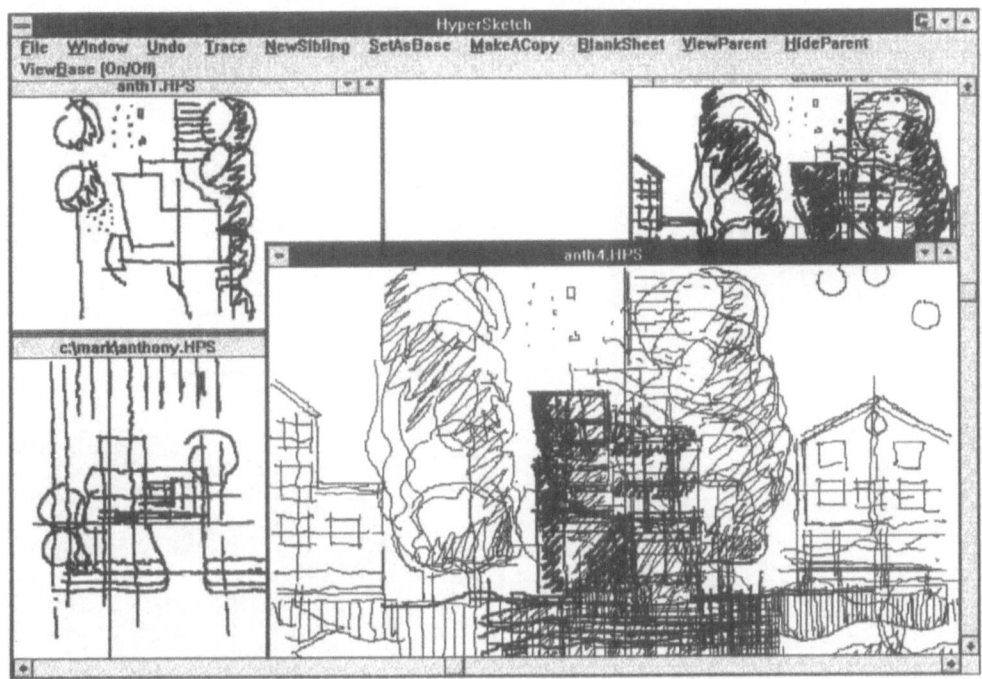

Figure 1. A screen image from HyperSketch. These drawings represent a house that the user--an architect--was designing for himself. The drawing in the lower right shows three layers of electronic "tracing paper," with the darker lines being on the top layer.

3.2. SECOND PROTOTYPE: PHIDIAS WITH INTEGRATED HYPERSKETCHING

We are creating a second prototype by integrating hyper-sketching functionality into our existing PHIDIAS HyperCAD system [McCall, Bennett, and Johnson 1994]--which, in fact, was our intention all along. PHIDIAS has vector-graphics, knowledge-based computation, hypermedia navigation and search; yet it implements all these using only hypermedia mechanisms. Complex vector-graphic objects are represented by composite hypermedia nodes. Knowledge-based computation is accomplished. by semantic networks represented in hypermedia networks. Search uses the node-link structure.

Figure 2 shows a screen image of PHIDIAS with integrated HyperSketching. Here a lunar habitat is being designed. The lines defining a table have been grouped and labeled as a "table." This creates an object (node) with an *is-a* relationship to the *table* class, thus associating the sketched object with all information on tables. The sketched table also inherits information from higher classes--such as *furniture*--inherited by the *table* class. Double-clicking on the sketched table brings up the associated information for design and placement of tables in the lunar habitat. This includes text, raster graphics, video, and links to external information in Web sites, spreadsheets, word processors, etc.

854

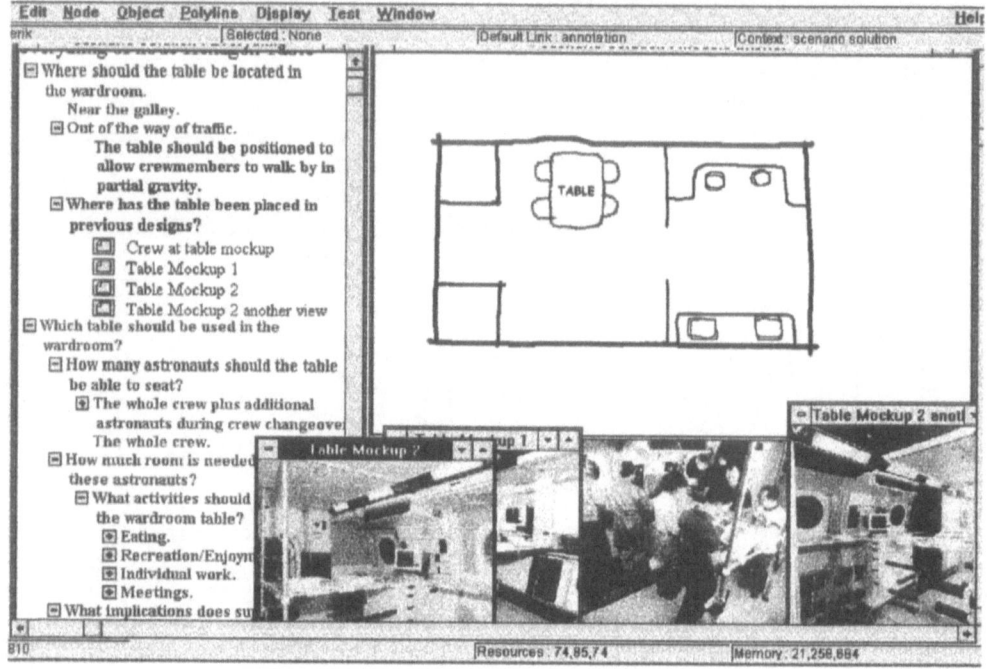

Figure 2. A screen image from the PHIDIAS prototype with integrated HyperSketching. The application is the design of a lunar habitat for NASA. The table in the sketch has been labeled and then used to access information for the design and placement of tables.

4. Conclusion

HyperSketching enables users to do conceptual design yet does little to enhance that process graphically in the way that CAD systems attempt to--e.g., by generating sweeps and other complex forms. In this sense it has little advantage over pencil and paper. But unlike pencil and paper--and conventional CAD---PHIDIAS provides multimedia information to aid design decision making. The information can include issues, solution ideas, and arguments, as well as cases of prior design projects represented with text, video, photos, and vector graphics. PHIDIAS with HyperSketching thus provides a fundamentally different approach to aiding design, an approach better suited to support of early design than the approach of current CAD.

References
McCall, R. ; Bennett, P.; and Johnson, E. An Overview of the PHIDIAS HyperCAD System. In *Reconnecting*, proceedings of the 1994 conference of the Association for Computer-Aided Design in Architecture (ACADIA'94), A. Harfmann and M. Fraser (eds.), Association for Computer-Aided Design in Architecture, 1994, pp. 63-76.

AN EVOLUTIONARY APPROACH TO GENERATING CONSTRAINT-BASED SPACE LAYOUT TOPOLOGIES

JOSÉ C. DAMSKI AND JOHN S. GERO
Key Centre of Design Computing
University of Sydney

This paper describes a system to produce space layout topologies for architectural plans using an evolutionary approach. The layout specification is defined as a set of topological and directional constraints, which are used as a fitness function in the evolutionary system. The halfplane representation is used to represent the genotypes in the evolutionary system, for both arrangements of halfplanes and the figures generated from those arrangements. As the halfplane representation proposed here does not distinguish between straight and non-straight boundaries, at the symbolic level the spaces and the layouts produced can also be bounded by straight or non-straight lines. The well known rectangular (polyomino) arrangements become a particular case only.

1. Introduction

Space layout planning problems have been addressed by many researchers (Buffa et al, 1964; Liggett, 1980; 1985; Steadman, 1983; Akin et al, 1992; Yoon and Coyne, 1992; Jo and Gero, 1997) among others. They have presented many different approaches, synthesizing layouts using generative grammars, constructive placements, genetic algorithms, etc., addressing topological, directional and geometrical issues. The most common representation is placement (or generation) of rectangular units on a plan in dimensionless form (Steadman 1983). In this type of representation a coarse granularity layout of a house is similar to that shown in Figure 1(a). Many interpretations can be derived from this type of representation, such as topological relations and symmetries. The other major research direction is concerned with determining, from the constraints on area, width and length of each space, the optimal dimensions according to some criteria. It is quite common to map the layout shown in Figure 1(a) into the graph shown in Figure 1(b). The graph represents the topological relationships between rooms, where the nodes represent the rooms and arcs represent the adjacency between them (Miller, 1971). Many interpretations can be based on these graphs, such as conditions for planarity, coloured and weighted graphs, etc.

R. Junge (ed.), CAAD Futures 1997, 855-864.
© 1997 *Kluwer Academic Publishers.*

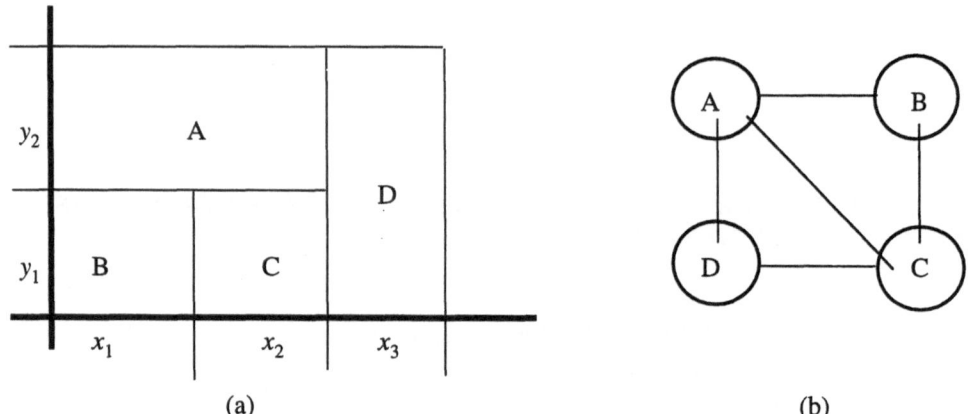

Figure 1. (a) A coarse granularity layout of a house in a dimensionless space, and (b) its equivalent graph of adjacency between rooms.

The dimensioning of floor plans have been tackled using various optimization techniques, including linear programming (Mitchell et al, 1976), and nonlinear and dynamic programming (Gero, 1977). Jo (1993) attempted to solve the topological and geometrical problems together by using a evolutionary approach, where a set of shape rules generates a space plan and the geometrical constraints are evaluated with a fitness function based on multiple criteria.

2. Background

The planning of space layouts has two levels of solutions: topological and geometrical. At the topological level there is interest in the relation among the spaces as invariant properties of the layout under any geometrical transformation. At the geometrical level the focus is on numerical values of each space, such as area, length, distance, etc. In this paper we proposed a system to tackle the first level: finding arrangements of lines that generate a set of spaces valid under the desired topological constraints. As an example of such constraints we may have the following statement of requirements which can be treated as a set of topological constraints:

> A house has 7 spaces: room1, room2, room3, living, kitchen, bathroom and a corridor. The space room1 must be adjacent to room2 and to the corridor. The space room3 must be adjacent to the bathroom and to the corridor. The living must be adjacent to the corridor but not adjacent to room1, room2 or room3. The kitchen should be adjacent to the corridor. The space room3 should be on the left of room1.

This set of constraints has the potential to produce a large set of possible solutions and is computationally complex to solve. To deal with such complexity we develop an evolutionary system that starts with some basic layouts and evolves them based on their suitability (fitness) compared to the topological constraints.

The basic representation for use in the genotype of the evolutionary system is the *halfplane*. In this paper it will be used to construct spaces.

3. The Representation

The basic representation used in the genotype of the evolutionary system is a halfplane. The halfplane representation has been used successfully in other applications (Damski 1996, Damski & Gero 1996, Gero et al. 1995) as the basis of a formal system of representing shapes founded on logic. Such a formal representation has been used to reason about spaces. Here it is used to construct spaces.

In the halfplane representation there is no line dividing a plane, but an abstract border. This abstract border divides the plane into two non-overlapping areas, as shown in Figure 2. The division of a plane into only two halfplanes has the advantage of reducing the complexity of the logical representation to two-value logic, such as propositional and predicate logic. In this way we can arbitrarily assign the truth value **true** to one side and **false** to other.

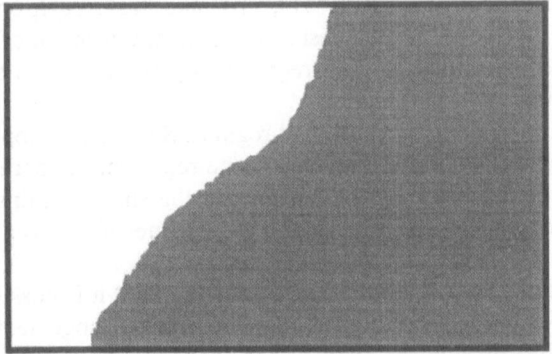

Figure 2. Two halfplanes – one shaded and the other unshaded

In order to represent a shape using the halfplane representation it is necessary to map each geometrical line in a figure into a halfplane, and then map the halfplanes into logic. Figure 3(a) shows a figure with 3 lines, which is re-represented as halfplanes in Figure 3(b). The shape S_1 has the logical expression hp(a) $\wedge$ hp(b) $\wedge$ ¬hp(c), which means the shape is on the **true** value side of halfplane a and b and on the **false** side of halfplane c. The logical expression that defines this arrangement is given by the formula ¬hp(a) $\wedge$ ¬hp(b) $\rightarrow$ ¬hp(c), because it is always true that the region defined by ¬hp(a) $\wedge$ ¬hp(b) is inside the halfplane ¬hp(c). With this information it is possible to determine all possible regions with the halfplanes and the topological information among these regions. It is interesting to note that the representation is the same regardless of whether the boundary is a straight line or not. With the halfplane representation it is possible to reason about the shapes at both the topological and directional levels (Damski, 1996).

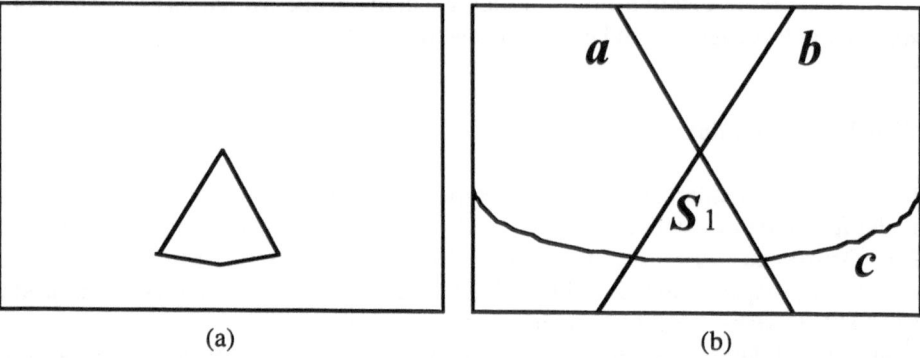

Figure 3. (a) Primary shape, and (b) the halfplanes a, b and c derived from the primary shape

We can label one halfplane as "0" and the other as "1" (the geometrical line which may be thought to exist at the border can reside in either sides without changing the representation). For any given two halfplanes, it is possible to divide a plane into four or fewer distinctive parts, labelled: "00", "01", "10" and "11". Each one of these parts, defined as a region, represents the smallest space in an arrangement of halfplanes. For a given arrangement of n halfplanes, each region is represented as an ordered bit-string $[00101...1]$ with n bits.

Shapes are formed by the composition of regions. Shapes have topological relations among them. In architectural examples each shape represents a space within a layout. The evaluation of the topological relations among the shapes against the constraints initially specified composes the fitness function value of a given arrangement of halfplanes.

Using the genetic algorithms approach (Goldberg, 1988) it is possible to represent a set of halfplanes in a genotype as a bit string. From this arrangement it is possible to generate a population of shapes. Each set of shapes can be evaluated against the topological and directional constraints. The shapes and their configurations are evolved until a desired level of fitness is reached. The best result is passed on to the initial arrangement of halfplanes. In this cycle it is possible to evolve halfplane arrangements at a non-numerical level and derive space layouts that fulfill the initial requirements (used in the fitness function). All these operations are performed using the logic representation.

4. The Evolutionary System

The evolutionary system operates at two hierarchical levels: arrangements of halfplanes and arrangements of figures. Each arrangement of a set of halfplanes produces a set of regions. A population of arrangements of halfplanes, initially randomly generated, is evolved according to the results given from the possible set of shapes generated from the regions resulting from each arrangement. The standard roulette wheel system of evolutionary systems (Goldberg 1988) selects some arrangements of halfplanes to be "parents" of the next generation using both crossover and mutation mechanisms.

For each particular arrangement of halfplanes a number of possible figures are generated. This population of figures is evaluated against the desirable constraints. The Pareto optimization technique is used as a selection criteria for the next generation of figures. Pareto optimization was chosen to handle the disparate criteria defined in the topological constraints. The fitness value of the best figure in a given population is passed upwards to the related arrangement of halfplanes.

The description of this system is:

- a set H of halfplanes is generated
- a population of arrangements of halfplanes is generated, where every member H_i is a subset of H
- for each H_i, generate a set of layouts (set of figures) L
- each layout in L is evaluated against the set of constraints
- the layouts are evolved according to the results of the fitness function
- each arrangement in H_i is evolved according to the evaluation of the population of layouts in that topology.
- arrangements are evolved in H.

This is a hierarchical evolutionary system. The first level evolves arrangements of halfplanes and the second level evolves layouts. At the second level it is necessary to use a multicriteria system in the fitness function, such as the Pareto optimization schema.

5. Implementation and Results

The evolutionary system was implemented using part of the system developed in Damski (1996). The evolutionary component was implemented in Prolog. The system initially generates a population of halfplanes at random. For simplicity we generated halfplanes with straight boundaries. In addition, because of the building layout application, all the boundaries of the halfplanes are set parallel or orthogonal to each other. Examples of those arrangements are shown in the Figure 4.

From the population of halfplanes, a population of arrangements of halfplanes is generated. For each arrangement we generate a population of figures and evolve them. This evolution is based on a set of criteria. In the example shown in this article we used only three criteria:

- Shape connection: this criterion checks if all parts of a shape are contiguous, because the selection at random can create non-contiguous shapes. In order to improve the algorithm we generate and evolve them contiguously.
- Shape overlapping: this criterion looks after shapes that overlap another shape. In the layout case shapes should not overlap each other.
- Shape adjacency: while the first two criteria were basic for any layout, this one sets how we want to relate, topologically, all shapes in the same layout. In our example we define a layout with three shapes (spaces) A, B and C, where A should be adjacent to B, B adjacent to C and A should not be adjacent to C.

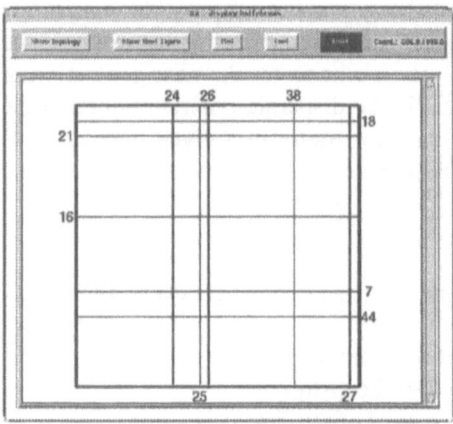 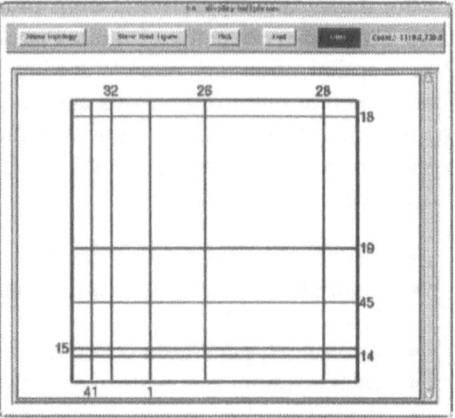

Figure 4. Examples of the halfplane arrangements

With these three criteria we use Pareto sets to calculate the best solutions and select randomly some of them to be evolved for the next generation. Figure 5(a) shows the result after 20 generations of best, average and worst cases in each generation. The value plotted is the "distance" of the gene. This distance is calculated as the square root of the sum of squares of each criterion. In Figure 5(b) the average value of each criterion is shown as the height of the bar chart. In this way it is possible to see the evolution of each criterion across the generations. In this case it simple to see that the shape connection criterion does not change much because we already generate shapes that are connected (they may be loosely connected after some generations. The criterion of shape overlapping is solved along with the top criterion, shape adjacency, across the generations. At the end of the evolution all layouts generated satisfied all the criteria.

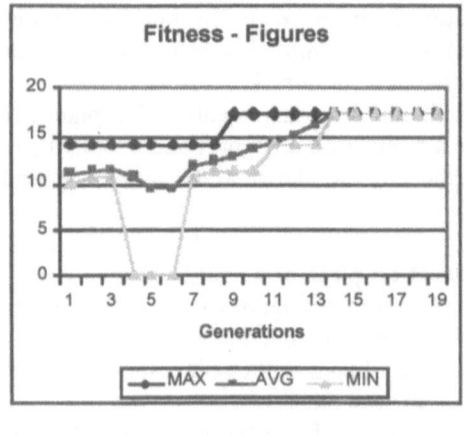 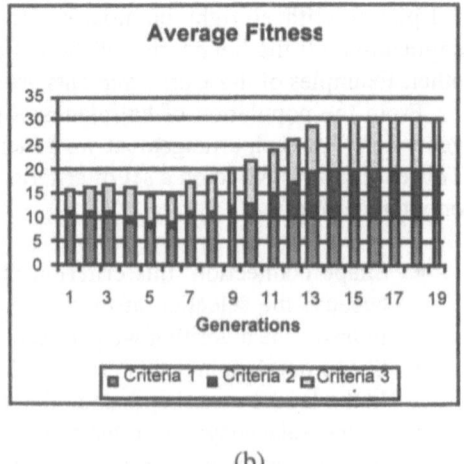

(a)   (b)

Figure 5. Evolution of figures (layouts) (a) the distance of the best, average and worst cases (b) the average value of each criterion.

The final result of the evolution of the layouts is passed upwards for the particular halfplane arrangement used for this evolution. Once all the populations of one arrangement have been calculate it is possible to start to evolve arrangements in the same way layouts were evolved. The best two arrangements are selected and the crossover operation applied on them. The two worst solutions are removed from the population. Figure 6(a) shows the "distance" of the best, average and worst arrangement in the population for each generation. Figure 6(b) the average value of each criterion is shown as the height of the bar chart.

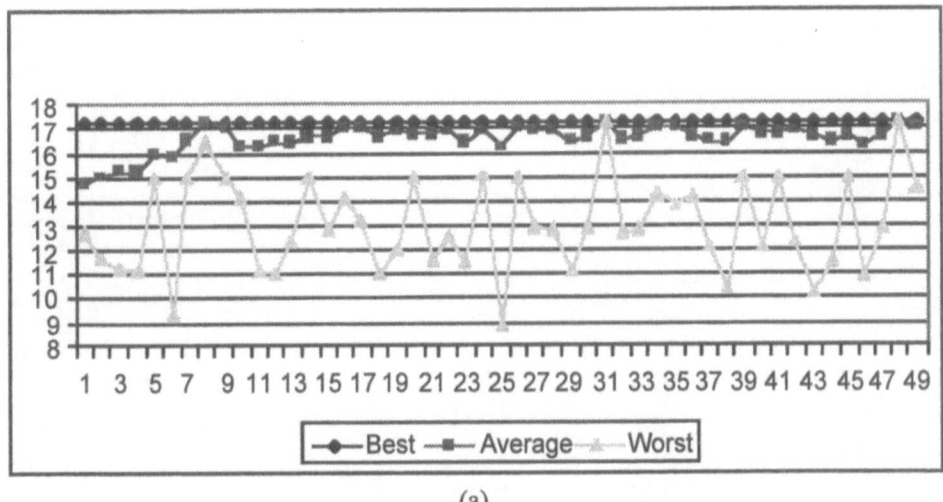

(a)

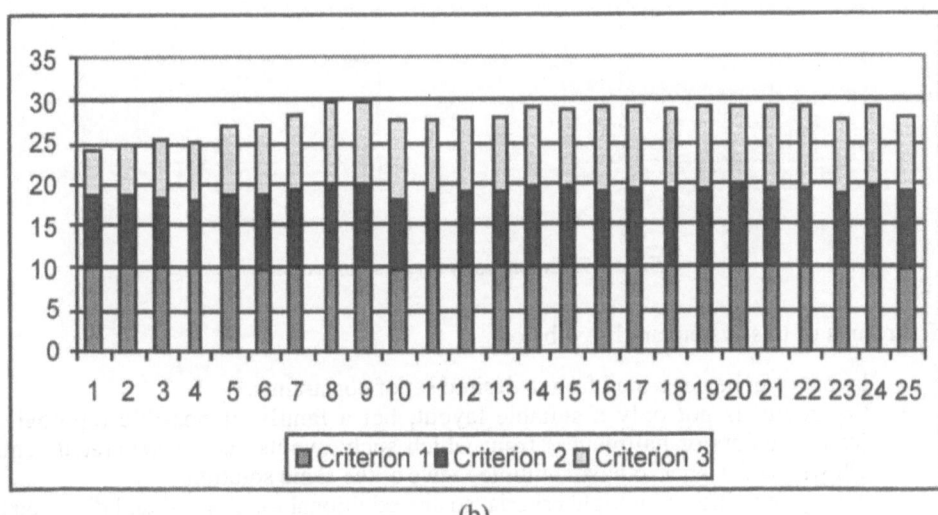

(b)

Figure 6: Evolution of the halfplane arrangements (a) the distance of the best, average and worst cases (b) the average value of each criterion

862

Examples of the layouts produced after the evolution of figures in each halfplane arrangement is shown in Figure 7.

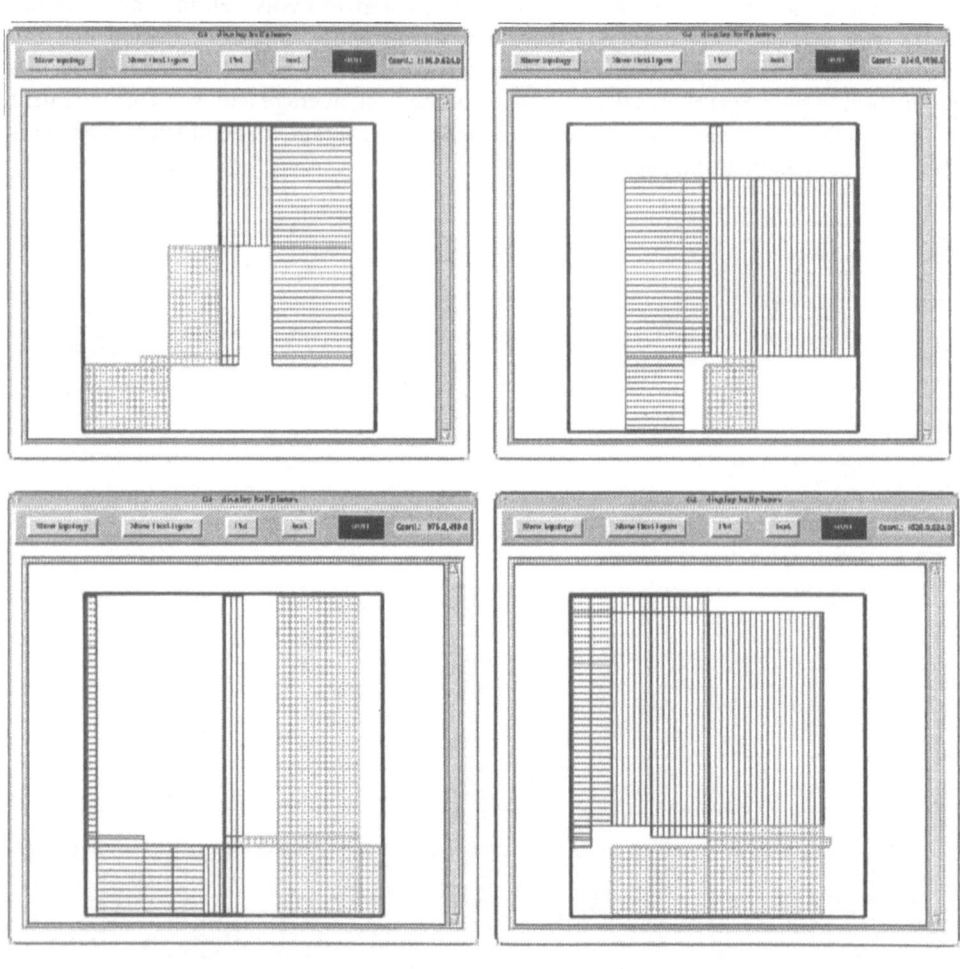

Figure 7. Examples of layouts generated

The results of this system are listed below.

- layouts can be generated from a desired set of constraints.
- the output is not only a suitable layout, but a family of possible topologies (arrangements of halfplanes) from which such layouts can be generated. This allows the designer to have multiple views of the same solution.
- The system allows multiple criteria, so any additional topological and directional requirements can be expressed in the system.

863

6. Conclusions

This article presented an evolutionary system to generate space layouts. The layout specification is defined as a set of topological and directional constraints. While the halfplane representation is not limited to any type of boundary, the example shown in this article have straight and orthogonal boundaries (lines). This system completes the generation of legal topologies is the first of two stages in space layout planning. The second stage, that of dimensioning topologies is a well-known optimization problem with a variety of well-established techniques available for its solution.

Acknowledgements

This work is supported by a grant from the Australian Research Council. Computational support is provided by the Key Centre of Design Computing.

References

Akin, O., Dave B. and Pithavadian, S. (1992) Heuristic generation of layouts (HeGel): based on a paradigm for problem structuring, Environment and Planning B **19**: 33-59.

Buffa, E. S., Armour, G. S. and Vollman, T. E. (1964) Allocating facilities with CRAFT, Harvard Business Review **42**(2): 136-140.

Damski, J. C. B. and Gero, J. S. (1993) Using logic to represent graphical shapes, in C. Rowles, H. Liu and N. Foo (eds), AI í93 - The Sixth Australian Joint Conference on Artificial Intelligence, World Scientific, Singapore, pp. 90-95.

Damski, J. C. B. (1996). Logic Representation of Shapes, PhD Thesis, The University of Sydney, Sydney, Australia.

Damski, J. C. B. and Gero, J. S. (1996). A logic-based framework for shape representation, Computer-Aided Design **28**(3): 169-181.

Gero, J. S., Damski, J. C. B. and Jong, H. J. (1995). Emergence in CAAD system, in M. Tan and R. Teh (eds), The Global Design Studio, Centre for Advanced Studies in Architecture - National University of Singapore, pp. 423-438.

Gero, J. S. (1977) Note on íSynthesis and optimization of small rectangular floor plansî of Mitchell, Steadman, and Liggett, Environment and Planning B **4**:81-88.

Goldberg, D. (1988) Genetic Algorithms in Search, Optimization and Machine Learning, Addison-Wesley, Reading, Massachusetts.

Jo, J. H. (1993) A Computational Design Process Model Using a Genetic Evolution Approach, Ph.D. Thesis, Department of Architectural and Design Science, University of Sydney, Sydney, Australia.

Jo, J. H. and Gero, J. S. (1997) Space layout planning using an evolutionary approach, Artificial Intelligence in Engineering (to appear).

Ligget, R. S. (1980) The quadratic assignment problem: an analysis of applications and solution strategies, Environment and Planning B **7**: 141-162.

Ligget, R. S. (1985) Optimal spatial arrangement as a quadratic assignment problem, in J.S. Gero (ed.), Design Optimization, Academic Press, New York, pp 1-40.

Miller, W. R. (1971) Computer-aided space planning, an introduction, DMG Newsletter **5**: 6-18.

Mitchell, W. J., Steadman, J. P. and Liggett, R. S. (1976) Synthesis and optimization of small rectangular floor plans, Environment and Planning B **3**: 37-70.

Steadman, J. P. (1983) Architectural Morphology - An Introduction to the Geometry of Building Plans, Pion, London.

Yoon, K. B. and Coyne, R. D. (1992) Reasoning about spatial constraints, Environment and Planning B **19**: 243-266.

EMERGENT SHAPE GENERATION IN DESIGN USING THE BOUNDARY CONTOUR SYSTEM

PHIL TOMLINSON AND JOHN S GERO
Key Centre of Design Computing
University of Sydney

This paper discusses the boundary contour system as the basis of a computational model of emergent recognition applicable in design. Details of this system which make it appealing as a computational approach for emergent recognition are introduced. The performance of a system implementation is covered and an extension to improve its performance is discussed.

1. Introduction

In the field of design computing, emergent recognition is defined as,

emergent recognition (ER):
   the process of seeing properties, features, functions not originally intended
   in a designed artefact.

The ability to recognize and complete without difficulty incomplete emergent shapes is an important aspect of human visual perception. There has been increasing interest in producing computational analogs of shape emergence. Part of the reason for this interest is based on the hypothesis that emergence can play a role in design in the visual domain and in creative designing in particular. Thus, a computational system capable of shape emergence could play a role in computer-supported creative designing.

1.1. COMPUTATIONAL MODELS OF EMERGENT RECOGNITION

From this interest in ER, various computational models of emergent perception have been investigated (Gero and Damski 1994, Gero and Jun 1997, Gero and Yan 1994, Liu 1994). These recent works have generally approached the problem in a symbolic, rule based, depth first search manner. The concepts involved (infinite maximal lines and shape correspondence) are largely derived from the vector representation of shapes while the ER was defined as the identification of isomorphisms with schemas after perturbing in characteristic ways the vector representation. Henceforth, this approach is referred to as the CS (conceptual schema) approach.

865

R. Junge (ed.), CAAD Futures 1997, 865-874.
© 1997 *Kluwer Academic Publishers.*

A different approach has been developed in the domain of real-time neural networks and has been labelled the Boundary Contour (BC) system by its co-developers Grossberg and Mingolla (1985). It is differentiated from the CS approaches in a number of significant ways.

1.1.1. *Non-Schema Based*

Previous computational models of shape emergence have largely been of two kinds: symbolic systems based on schemas and sub-symbolic systems also based on (teaching the system) schemas. The effect of these models is that not all emergent shapes can be found computationally because they need to be recognised a priori by the system developer.

The BC system does not rely on schemas. Instead it uses a small number of rules applied at all locations and orientations of a two and a half dimensional matrix which represents the scene. Application of the rules occurs in parallel and in real time (Grossberg 1987a, 1987b) such that Gestalt effects can result from the computation. It would seem that this approach may provide a richness not found in previous ER systems.

1.1.2. *Illusory Reality*

The BC system was developed as a cognitive theory used to analyse and explain a variety of perceptual grouping and segmentation phenomena. The theory was approached from the perspective that the perceptual system faces a major problem in the ambiguous nature of the input data available to perceive from. This ambiguity comes from several sources which include:

- imperfections in the human perceptual system, such as the veins and blindspots in the lens which tend to mask real input data and create illusory input data. These imperfections also result from the discreteness of the visual system and the problems associated with discrete edge detection.
- noise, a uniform level of activation across a visual field, which can be the result of differences in lighting conditions.
- and the Gestalt nature of perception where

> "it has been widely recognised that local feature of a scene, such as edge positions, disparities, lengths, orientations, and contrast, are perceptually ambiguous but that combinations of these features can be quickly grouped by a perceiver to generate a clear separation between figures and between figures and ground." (Grossberg and Mingolla 1985)

Starting from this condition, the theory offers the explanation that the processes required to perceive reality from illusory and ambiguous input data also can provide the perception of illusory shapes from real data. ER is not an anomaly of normal perception but an emergent effect itself. To design computing, this suggests that the BC system does not require precision vector based drawings as input. Instead, fuzzy hand drawn sketches are valid input. This positions ER as a potential tool at the early stage of designing.

1.1.3. *Preattentive*

The boundary contour system is not a recognition system but rather a stage of image preprocessing whose output is theoretically an input to the recognition system. The output is an activation pattern of the same computational representation as the input. It is not a representation which could be efficiently generalised from and as such is not directly considered recognition. The input-output transformation carries little knowledge, learns no knowledge of particular shapes and is recurrent. Without such knowledge the resulting performance of the system is unpredictable.

1.1.4. *Selective*

Unlike the CS systems, the BC system is selective. The selected output is a pattern which best satisfies a small number of competing constraints applied at each location in the input data matrix. Mathematically described in terms of differential equations among neurons of the matrix, the BC system is a dynamic non-linear system which approaches an attractor defined by both the input data and the constraints imposed by the input-output transformational rules.

2. BC System in Design Computing – The ER Interface

Beyond being simply an alternative computational approach, it is conceivable that the differences described above have value in the performance of an ER system. To consider this, it is first necessary to state the current vision of utility and operation of an ER system in practice. From the perspective of visual design computing, the BC system appeals, first and foremost, as a computational approach to emergent recognition. As such, it was envisaged that the process could be used as a background image processing algorithm which modifies a drawn/sketched scene. This process is expected to have value if it provides alternative shapes which are in some non-conventional way derived from the initially drawn shape. As an alternative to prior ER systems, the BC system would present a single recognition for the input scene that is consistent across the entire scene. Without knowledge of any particular shapes, it was expected that the BC system would be a good ER system for it would not rely on recognising the known.

3. The BC System Implementation – Providing the Interface

The operating principle of the BC system is simple. Input and output data exist as patterns of activation on a two and a half dimensional matrix of activations in a neural network. The activation patterns are transformed through sequential and recurrent sets of filters which implement simple rules designed to reduce the ambiguity in local information. An example of this input output transformation of a (non-emergent) shape is shown in Figure 1.

The overall transformation occurs through the individual transformations which make up the BC system. First the input scene is edge detected via a contrast sensitive mask. This identifies the edges of the scene but in doing so creates considerable ambiguity. An example of the edge detector output is shown in Figure 1. As can be seen the process has a side effect that line ends, edges and corners are fuzzified (cannot be determined by local evidence). It is just this type of input which begins to justify why alternative shapes can appear at all. Figure 2 shows the output scene from the BC system.

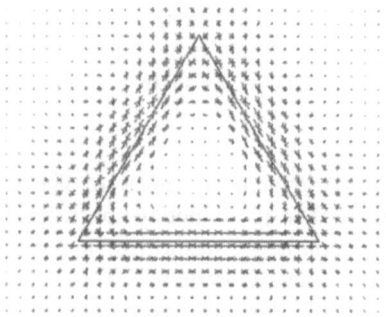

Figure 1. Input scene to the BC system and edge detection

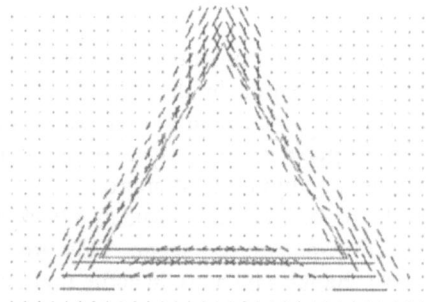

Figure 2. Output scene from the BC system

The remaining transformations seek to disambiguate the border of the scene. A number of rules are utilised whose effect is to combine non-local and local evidence to infer local evidence.

Heuristically these rules are:

Rule 1 Edge Location Rule: An edge cannot exist at one direction and orientation and at the same direction and a location orthogonally nearby.

Rule 2 Edge Direction Rule: A region edge cannot go in two directions at the same location.

Rule 3 Edges As Populations Of Edgelets: An edgelet cannot remain active without receiving long range feedback from other edgelets on the same edge.

That these rules can be specified in terms that are computable by a neuron and a small neighbourhood makes the BC system computationally plausible. They are expressed in the BC systems in terms of several On-Centre Off-Surround competitive processes and a cooperative feedback process (Grossberg 1987a, Grossberg 1987b, Carpenter and Grossberg 1991). Individually these processes can be linked together in a computational loop, giving the overall view of the image processing shown in Figure 3.

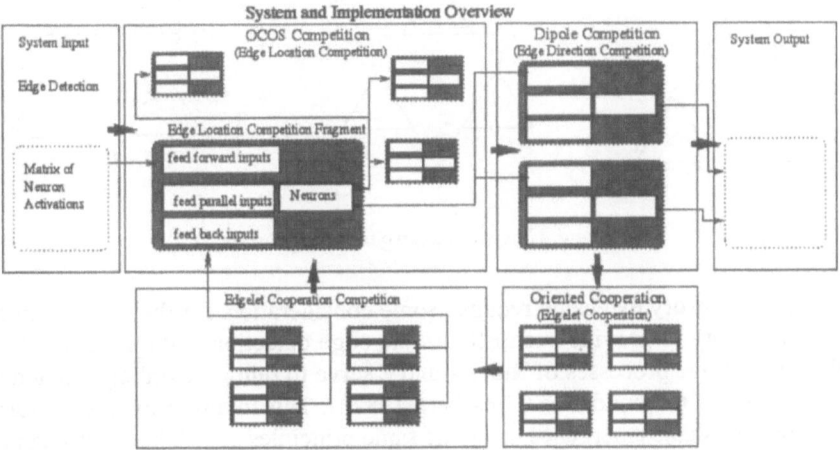

Figure 3. BC system overview

4. BC System Performance

The capability of the BC system to produce emergent recognitions has been verified through a reimplementation of the original system and testing examples provided in the original literature (Grossberg 1987b). Though important, this verification lead to an appreciation of the BC system as something of an incomplete system for the purpose of ER in design computing.

The BC system as an approach towards a design computing tool theoretically provides the capability to select both embedded shapes and to produce illusory shapes. In the former no new lines or boundaries are constructed, only new interpretations of existing boundaries are formed. In the latter the new shapes are bounded by some or even all boundaries that did not exist in the initial shapes. In illusory shape emergence we draw a further distinction between those emergent shapes which have boundaries which are linear extensions of existing boundaries, those which have boundaries which include intersections and those whose boundaries are not extensions of preexisting shapes, Figure 4.

Demonstrated in the literature and verified through our implementation is the capability of the BC system to produce recognitions derived from the linear continuous extensions of pre-existing boundaries. This capability to select embedded shapes has been provided by other computational approaches to emergent recognition through it is conceivable that the BC system approach would facilitate it. The BC system as originally designed was demonstrated not to be capable of corner extension illusory ER. As this is a limitation in its utility for designers, an extension of the BC system to provide corner extension illusory ER was investigated and is described below.

Continuous Illusory   **Corner Extension Illusory**   **End Cut Illusory**

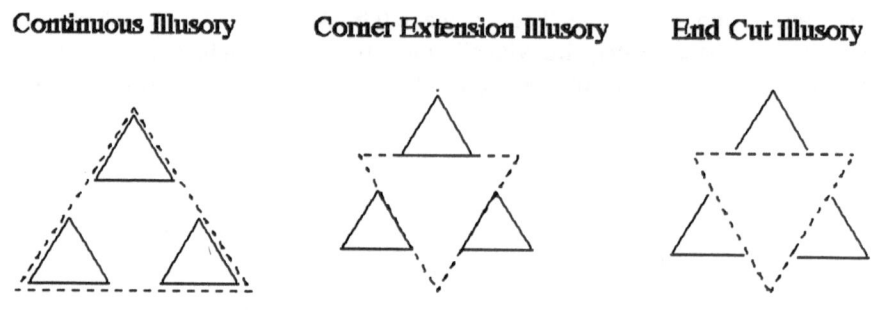

Figure 4. Classes of emergent recognitions

The end cut illusory extension requires some consideration, for the BC system does suggest a capability to provide such ER. As the edge detection ambiguates the end of lines, the competitive processes of Rules 1 and 2 serve to reinforce orthogonal and near orthogonal end cuts to boundaries (Grossberg 1987b). This in turn provides the starting point for illusory shape emergence using the same principles as will be described below to provide continuous extension ER. As the only computational approach theoretically capable of providing such a capability, the BC system deserves further consideration.

4.1. ORIENTED COOPERATIVE FIELDS

To appreciate the BC system and its capability to produce emergent recognitions, it is necessary to specifically consider one stage of the system: the oriented cooperative feedback process (Rule 3 above) of the BC system which both facilitates continuous illusory extension ER and prohibits corner extension illusory ER. Oriented cooperation takes the form of positive feedback from the activation of an intermediate neighbourhood of a neuron back to the single neuron. It is a top down reinforcing filter which was designed to enforce the rule that evidence for an edge should be reinforced by extensions of this edge (Rule 3). Feeding back the reinforcement requires the input from the edge extension and in the original BC system description such neighbourhoods were lobed fields as shown in Figure 5.

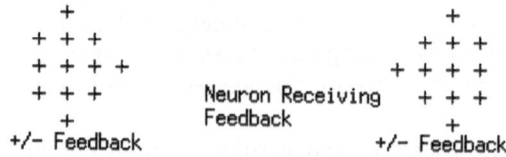

Figure 5. A cooperative field

Cooperative fields, as originally proposed, were two opposing lobed regions. Quite obviously these are limited to linear boundary completion. Grossberg and Mingolla (1985) hinted at curved boundary completion using a slightly curved cooperative field.

However, the logical extension of this idea to corners does not work. The BC system was run with angular cooperative fields of the form shown in Figure 6 in an attempt to achieve illusory corner extensions recognition.

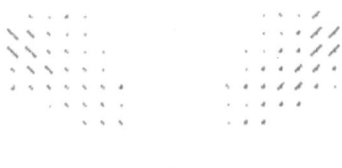

Figure 6. Angular cooperative field receiving positive feedback and no negative feedback

Simple use of such angular cooperative fields has a disastrous side effect in that the feedback from orthogonal cooperative fields is equal. This feedback is mutually annihilating in the dipole competitive stage of Rule 2 as can be seen in Figure 7.

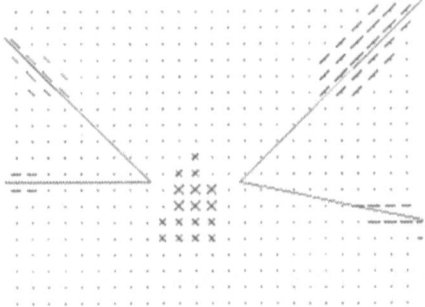

Figure 7. Feedback from angular cooperative field which is equal and orthogonal and hence eliminated by the edge direction rule (Rule 2)

4.2. LOCATION/ORIENTATION SPECIFIC ANGULAR COOPERATIVE FIELDS

To overcome this problem the definition and use of cooperative fields which distinguish their location relative to a corner was tried. These cooperative fields are consequently referred to as location/specific cooperative fields. Specifically, with standard fully lobed cooperative fields the field would be activated independently of its root orientation and root position relative to the corner, Figure 8. With the semi-lobed fields it is possible to choose configurations which feed back orientation only when the lobe root is on one side of a corner. Using these lobes, it became possible to achieve corner illusory extensions, Figures 9 and 10.

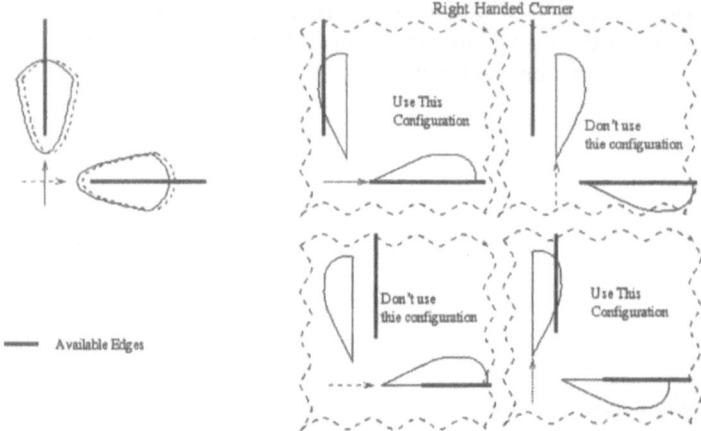

Figure 8. *Location/orientation specific angular cooperative field*

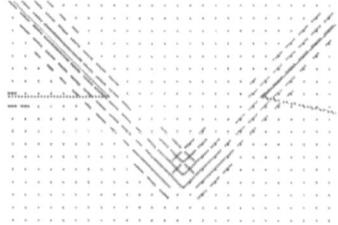

Figure 9. *Example of corner extension illusory recognition*

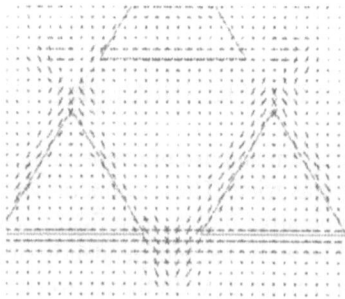

Figure 10. *Example of corner extension illusory recognition*

5. Conclusions

These investigations into the BC system as a computational model of emergent recognition for use in visual designing have yielded the following observations.

The performance of the BC system was disappointing in its robustness. The system is extremely sensitive to the correspondence of cooperative field parameters to the scale of possible boundary completions. To obtain the emergent results which are presented above, the system required extensive fine tuning. Essentially the problem here is that not enough computational power could be brought to bear in this exercise to simultaneously allow the system a variety of parameter settings. The BC system requires a lot of computational power for it is computing concurrently a large number of parallel constraints. To proceed towards the use of the BC system or a variant as a viable design support tool, much optimization or better, significant parallelization of the computation is required.

The BC system is also disappointing in that the output often remains locally quite ambiguous. By no means does the system approach a definite and precise edge definition. Often the competitive and cooperative processes are not able to produce a winner and maintain the ambiguity present in the edge detected data. As such it would seem the BC system is suited towards a human design aid rather than input to a following recognition system.

In spite of these disappointments, the appeal of the BC system or a variant of it as an alternate to previous computational models remains valid. Systems characterized by selective, non-schema based approaches operating from imprecise drawings do offer an alternative.

Acknowledgments

This work is supported by a grant from the Australian Research Council. Computational resources are provided by the Key Centre of Design Computing.

References

Carpenter, G. and Grossberg, S. (1991) Pattern Recognition by Self-organizing Neural Networks, MIT Press, Cambridge, MA.

Damski, J. and Gero, J. S. (1994) Object emergence in 3D using a data-driven approach, in Gero, J. S. and Sudweeks, F. (eds), Artificial Intelligence in Design, Kluwer, Dordrecht, 419-435.

Grossberg, S. (1987a) Competitive learning from interactive activation to adaptive resonance, Cognitive Science, 11: 23-63.

Grossberg, S. (1987b) The Adaptive Brain I, Elsevier, Amsterdam.

Grossberg, S. and Mingolla, E. (1985) Neural dynamics of perceptual grouping, Perception and Psychophysics 38: 141-171.

Jun, H. and Gero, J. S. (1997) Emergence of shape semantics of architectural shapes, submitted to Environment and Planning B.

874

Liu, Y-T. (1994) Encoding explicit and implicit emergent subshapes based on empirical findings about human vision, in Gero, J. S. and Sudweeks, F. (eds), Artificial Intelligence in Design, Kluwer, Dordrecht, 401-418.

Yan, M. and Gero, J. S. (1994) Shape emergence by symbolic reasoning, Environment and Planning B: Planning and Design, **21**: 191-212.

A GENETIC PROGRAMMING APPROACH TO THE SPACE LAYOUT PLANNING PROBLEM

ROMUALD JAGIELSKI[*] AND JOHN S. GERO
Key Centre of Design Computing
University of Sydney

The space layout planning problem belongs to the class of NP-hard problems with a wide range of practical applications. Many algorithms have been developed in the past, however recently evolutionary techniques have emerged as an alternative approach to their solution. In this paper, a genetic programming approach, one variation of evolutionary computation, is discussed. A representation of the space layout planning problem suitable for genetic programming is presented along with some implementation details and results.

1. Introduction

Algorithmic solutions of spatial allocation problems were first developed more than thirty years ago. In this paper we discuss space layout planning formulated as a quadratic assignment problem. This is a difficult combinatorial optimisation problem of great importance, reaching beyond formal architectural design. Lately, such spatial allocation problems have again attracted attention, yielding solutions which employed genetic algorithms (Jo and Gero 1997). An overview and history of the automated layout problem of the pre-genetic period can be found in Liggett (1985).

The quadratic assignment problem can be defined in the following way: m distinct objects $M = \{1,2,...,m\}$ are to be placed uniquely in n distinct sites $N = \{1,2,...,n\}$, where $m \leq n$. In other words, we want to find a one-to-one mapping of the set of facilities M into a set of locations N:

$$\rho : \{ M \rightarrow N \}, \quad j = \rho(i), \ i \in M, \ j \in N$$

The quadratic assignment problem (Tate 1995) can be stated as the task of finding the minimal total cost Cost(A) of allocation A, usually with some constraints to be satisfied as well:

---

[*]On leave from School of Computer Science and Software Engineering, Swinburne University of Technology.

R. Junge (ed.), CAAD Futures 1997, 875-884.
© 1997 *Kluwer Academic Publishers.*

$$\text{Cost}(A) = f_{i\,\rho(i)} + \sum_i \sum_j q_{ij}\, c_{i\,\rho(i)\,\rho(j)} \tag{1}$$

where

$f_{i\,\rho(i)}$ fixed cost of assigning element $i \in M$ to element $j \in N$,

q_{ij} intensity of interaction (traffic) between elements $i, j \in M$,

$c_{i\,\rho(i)\,\rho(j)}$ cost of interaction between elements $i, j \in N$, (often referred as the distance between i and j).

2. Introduction to Genetic Programming

Genetic (or evolutionary) computing is a powerful method of programming which relies on developing systems that demonstrate self-organization and adaptation in a similar, though simplified, manner to the way in which biological systems work. The best known technique are genetic algorithms(Goldberg 1989, Koza 1992, Michalewicz 1994). Koza has proposed a system which evolves Lisp computer programs. This method, *called genetic programming*, is extensively described in his book (Koza 1992). Genetic programming starts with an initial population of hundreds or thousands of randomly generated computer programs. Then using the Darwinian principle of survival and reproduction of the fittest, new offspring populations are created. The reproduction operation involves selecting from the current population programs (individuals) in proportion to their fitness and copying them to the next population. The best individuals survive, and finally optimal or near optimal programs are generated. Following is the pseudo-code for genetic programming:

> *Create an initial population of randomly generated programs*
> *REPEAT*
>   *Execute each program in the population and evaluate its fitness*
>   *Create a new population of programs by*
>   *- reproduction*
>   *- crossover and/or*
>   *- mutation*
> *UNTIL the termination condition is satisfied.*

The solution is the best program that appeared in any generation (called the *best so far*). The individuals of the populations originally were so called S-expressions (or Lisp-expressions), but it is possible to develop genetic programming in any other language, with individuals in the form of syntax trees. The program used in this investigation, called GENPRO, has been written in C++ and uses C-like syntax trees to represent programs (individuals) and finally those programs constitute the outcome of genetic programming.

The initial population consists of programs which are randomly generated syntax trees. These programs are produced from a set of non-terminals (or functions):

$$\mathbf{F} = \{f_1, f_2, \dots, f_n\}$$

and a set of terminals

$$T = \{t_1, t_2, \ldots, t_m\}$$

Each terminal and nonterminal can be represented by a node on the syntax tree. If we think about them as functions, then terminals have arity zero (they do not have arguments, and in the tree they are always leaves). In a program, terminals are variables or constants. Nonterminals have one or more arguments, so in the tree they are always nodes with children. includes arithmetic and Boolean operators, mathematical functions, conditional and iterative operators and any other primitive functions predefined by the user.

The initial population and subsequent operations involve random generation of symbols which, sooner or later, create a tree not having any conventional meaning or containing an illegal operation. Therefore the system must have a closure property. The closure property ensures that any combination of symbols and data types is always legal. That is why many of the operators have to be redefined. For example, the normal division will be replaced by a special division, which allows division by zero. The initial population of programs is converted to a new population by the genetic operators of reproduction, crossover and mutation. Reproduction is an asexual operation, ie, it requires only a single individual. For this operation a tree is selected from the current population according to a criterion based on fitness, and is copied into the new population. Typically fitness-proportionate reproduction is used (the biased roulette wheel). Recently the tournament selection has proven to be popular (a small group of individuals is selected randomly and then the fittest individual in the group is determined and reproduced). Crossover is a sexual operation and requires two individuals. Two programs (parents) are chosen with the probability proportional to their fitnesses. A crossover point at each tree is selected randomly, and the fragments of the trees which follow the crossover points are swapped. Mutation is a small random change in a program. For example one node of a tree can be replaced by a randomly generated different one. Usually mutation is used sparingly.

The process of transforming the current population by means of reproduction, crossover and mutation is repeated. As a termination condition, the number of generations can be chosen and the best program that appeared in any generation *(the best so far* individual) is the solution. Alternatively the program may be terminated when the fitness reaches a certain level (for example a sufficiently small error). These automatically produced programs are incomprehensible to humans, but are legitimate C functions and can be included without any modification to an ANSI C or C++ program.

3. An Office Layout Problem

We will apply the genetic programming approach the problem of an office layout planning (Liggett 1985). There are a number of office departments to be placed in a four-level terraced building. The building is divided into 17 zones, as it is shown in Figure 1 (letters A, B, C, ... denote zones). The whole area of the building is divided into smaller units (modules - each letter in Figure 1 represent one module), thus the size of each zone can be expressed as a number of modules (Table 1).

```
25              N  N           O  O           P  P                 Q  Q
24              N  N           O  O           P  P                 Q  Q
23              N  N           O  O           P  P                 Q  Q
22              N  N  N  N  O  O  O  O  O  P  P  P  P  Q  Q  Q  Q
21              N  N  N  N  O  O  O  O  O  P  P  P  P  Q  Q  Q  Q

20              J  J           K  K           L  L                 M  M
19              J  J           K  K           L  L                 M  M
18              J  J           K  K           L  L                 M  M
17              J  J           K  K           L  L                 M  M
16              J  J  J  J  K  K  K  K  K  L  L  L  L  M  M  M  M
15              J  J  J  J  K  K  K  K  K  L  L  L  L  M  M  M  M

14              F  F           G  G           H  H                 I  I
13              F  F           G  G           H  H                 I  I
12              F  F           G  G           H  H                 I  I
11              F  F           G  G           H  H                 I  I
10              F  F           G  G           H  H                 I  I
 9              F  F  F  F  G  G  G  G  G  H  H  H  H  I  I  I  I
 8              F  F  F  F  G  G  G  G  G  H  H  H  H  I  I  I  I

 7  A  A  A  A  A  A     B  B        C  C        D  D           E  E
 6  A  A  A  A  A  A     B  B        C  C        D  D           E  E
 5  A  A        A  A     B  B        C  C        D  D           E  E
 4  A  A        A  A  A  A  A  B  B  R  R  R  C  C        D  D  R  R  R  E  E
 3  A  A  A  A  A  A     B  B        C  C        D  D           E  E
 2  A  A  A  A  A  A     B  B        C  C        D  D           E  E
 1              B  B  B  B  C  C  C  C  C  D  D  D  D  E  E  E  E
 0              B  B  B  B  C  C  C  C  C  D  D  D  D  E  E  E  E

    0  1  2  3  4  5  6  7  8  9  0  1  2  3  4  5  6  7  8  9  0  1  2  3  4  5
```

Figure 1. Graphical representation of zone definition

Table 1. Cost of interaction (matrix D)

	0	1	2	3	4	5	6	7	8	9	10	11	12	13	14	15	16	17	18	19
0 =	0	0	0	0	0	0	0	0	0	0	0	0	0	0	0	0	0	0	0	0
1 =	0	0	4	9	21	27	9	10	23	28	10	11	24	29	11	12	25	30	6	24
2 =	0	4	0	5	17	23	5	6	19	24	6	7	20	25	7	8	21	26	2	20
3 =	0	9	5	0	13	18	6	5	18	23	7	6	19	24	8	7	20	25	2	15
4 =	0	21	17	13	0	5	23	18	5	6	24	19	6	7	25	20	7	8	15	2
5 =	0	27	23	18	5	0	24	19	6	5	25	20	7	6	26	21	8	7	20	2
6 =	0	9	5	6	23	24	0	5	18	23	5	6	19	24	6	7	20	25	3	21
7 =	0	10	6	5	18	19	5	0	13	18	6	5	18	23	7	6	19	24	3	16
8 =	0	23	19	18	5	6	18	13	0	9	23	18	5	6	24	19	6	7	16	3
9 =	0	28	24	23	6	5	23	18	9	0	24	19	6	5	25	20	7	6	21	3
10 =	0	10	6	7	24	25	5	6	23	24	0	5	18	23	5	6	19	24	4	22
11 =	0	11	7	6	19	20	6	5	18	19	5	0	13	18	6	5	18	23	4	17
12 =	0	24	20	19	6	7	19	18	5	6	18	13	0	5	23	18	5	6	22	4
13 =	0	29	25	24	7	4	24	23	6	5	23	18	5	0	24	19	6	5	17	4
14 =	0	11	7	6	23	24	5	6	24	25	5	6	23	24	0	5	18	23	5	23
15 =	0	12	8	7	20	21	7	6	19	20	6	5	18	19	5	0	13	18	5	18
16 =	0	25	21	20	7	8	20	19	6	7	19	18	5	6	18	13	0	5	18	5
17 =	0	30	26	25	8	7	25	24	7	6	24	23	6	5	23	18	5	0	23	5
18 =	0	6	2	2	15	20	3	3	16	21	4	4	22	27	5	5	18	23	0	18
19 =	0	24	20	15	2	2	21	16	3	3	22	17	4	4	23	18	5	5	18	0

Some zones have a fixed allocation (here zone 1- executive office, and zones 18 and 19 - the public access). We have in total 280 modules to which 17 departments are to be allocated. The departments require 277 modules in total. For practical reasons we will add three "dummy" departments of one module size each, and in this way the total required area equals the total available area (that is 280 modules). Now, having a fixed order of assigning (as in Figure 2), our space layout planning problem has been reduced to finding a sequence of departments d_1, d_2,...,d_{20}, which are allocated correspondingly to zones in that order. Since we don't consider any constraints here (for example, that each department should be located on one floor only), each permutation of 20 departments is a feasible solution. Our task, however, is to find such a sequence of allocations, which minimises the costs defined by formula (1). In this problem the fixed cost $f_{i\,p(i)}$ is ignored, the cost of interaction (travel) $c_{i\,p(i)\,p(j)}$ is specified in Table 1 and intensity of interaction between departments q_{ij} is given in Table 2. Since the departments can be split between different zones, the calculation of the total cost has to be performed as a sum of costs of all 280 modules.

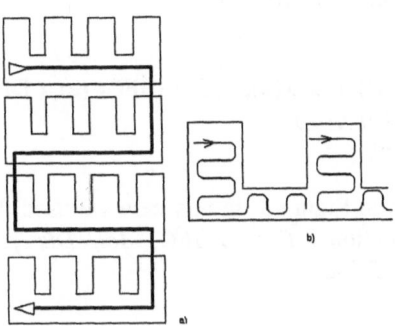

Figure 2. Order of the departments assignment (a) over floors of the building, (b) over a floor

Table 2. Intensity of interaction between departments (matrix I)

	0	1	2	3	4	5	6	7	8	9	10	11	12	13	14	15	16	17	18	19	20
0	0	0	0	0	0	0	0	0	0	0	0	0	0	0	0	0	0	0	0	0	0
1	0	0	10	10	5	5	5	0	0	10	0	0	0	0	0	0	5	0	0	0	10
2	0	10	0	10	5	5	5	0	0	10	0	0	0	0	0	0	5	0	0	0	0
3	0	10	10	0	10	5	5	0	0	10	0	0	0	0	0	0	5	0	0	0	0
4	0	5	5	10	0	10	5	0	0	10	0	0	0	0	0	0	5	0	0	0	0
5	0	5	5	5	10	0	5	0	0	10	0	0	0	0	0	0	5	0	0	0	0
6	0	5	5	5	5	5	0	0	0	10	0	0	0	0	0	0	5	0	0	0	10
7	0	0	0	0	0	0	0	0	10	0	0	0	0	0	10	0	5	0	0	0	0
8	0	0	0	0	0	0	0	10	0	0	0	0	0	0	10	0	5	0	0	0	0
9	0	10	10	10	10	10	10	0	0	0	0	0	0	0	0	0	5	0	0	0	0
10	0	0	0	0	0	0	0	0	0	0	0	0	0	0	0	0	5	0	0	0	10
11	0	0	0	0	0	0	0	0	0	0	0	0	0	0	0	10	5	0	0	0	0
12	0	0	0	0	0	0	0	0	0	0	0	0	0	0	0	10	5	0	0	0	0
13	0	0	0	0	0	0	0	0	0	0	0	0	0	0	0	10	5	0	0	0	0
14	0	0	0	0	0	0	0	10	10	0	0	0	0	0	0	0	5	0	0	0	10
15	0	0	0	0	0	0	0	0	0	10	10	10	0	0	5	0	0	0	0	0	0
16	0	5	5	5	5	5	5	5	5	5	5	5	5	5	5	0	0	0	0	0	10
17	0	0	0	0	0	0	0	0	0	0	0	0	0	0	0	0	0	0	0	0	0
18	0	0	0	0	0	0	0	0	0	0	0	0	0	0	0	0	0	0	0	0	0
19	0	0	0	0	0	0	0	0	0	0	0	0	0	0	0	0	0	0	0	0	0
20	0	10	0	0	0	0	10	0	0	0	10	0	0	0	10	0	10	0	0	0	0

4. Genetic Programming Solution to the Office Layout Problem

A solution to the space layout planning problem is a sequence of numbers (or a vector), representing allocations of the particular facilities, and in this form is very convenient for to genetic algorithms. A string of bits or numbers is a natural representation for genetic algorithms but is not the case for genetic programming. Genetic programming produces solutions in a form of programs, or more precisely, in a form of syntax trees. Therefore, applying genetic programming to the problem of space layout planning, we assume that the expected solution is a sequence of terminals read in order from a syntax tree during the evaluation process. The following code is a sample solution produced by genetic programming (an exact output from GENPRO):

```
GPvalue =
( IFG( (Act((d),((j)))), (i), ( Dis((k), ( IFG( ( IFG( (e), ( Dis((e), ( Act((j)
, (q)) )) ), (i)) ), (q), (h)) )) )) )
; // end of the GPvalue
```

An equivalent tree is shown in Figure 3. In this case we used the set of terminals: T = {a, b,...,s}, and the set of functions: F = { *IFG, Act, Dis* }. The terminals denote the departments (as specified in Table 3).

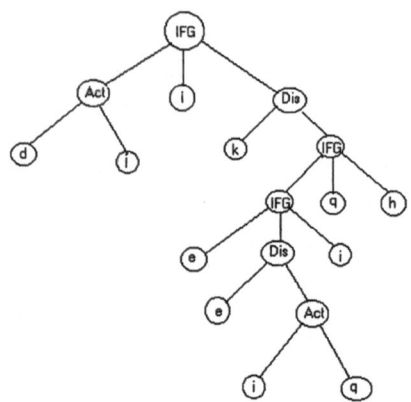

Figure 3. A sample parse tree

The functions *IFG, Dis, Act* have been chosen arbitrarily, with the underlaying aim to build parse trees rich in information. *IFG* is an *if-greater-then* function, which implements conditional branching on the trees. Functions *Dis* and *Act* are in fact Boolean functions, and are used as arguments for *IFG*. The meaning of these functions is given by the following pseudo-code:

Table 3. List of zones and departments of the office layout planning problem

List of zones and their sizes (in modules)

no	symbol	size	no	symbol	size	no	symbol	size
1	A	39	7	G	20	13	M	16
2	B	20	8	H	18	14	N	14
3	C	22	9	I	18	15	O	16
4	D	20	10	J	16	16	P	14
5	E	20	11	K	18	17	Q	14
6	F	18	12	L	16	18	R	6

List of departments and their sizes (in modules)

no	symbol	size	no	symbol	size	no	symbol	size
1	a	2	8	h	7	15	o	31
2	b	2	9	i	6	16	p	61
3	c	8	10	j	12	17	q	1
4	d	8	11	k	53	18	r	1
5	e	15	12	l	10	19	s	1
6	f	13	13	m	16	20	t	6
7	g	15	14	n	18			

IFG(z, y, z)   // *if-then-else* function
Evaluate z (z can be a terminal or a function *IFG*, *Dis* or *Act*)
Allocate (z)
IF $z > 0$
THEN
          Allocate(x)
          Return x
ELSE
          Allocate(y)
          Return y

Act(x, y): // data-dependent function (intensity of interaction in matrix I)
IF $I(x, y) > 0$
THEN
          Allocate(x)
          Allocate(y)
          Return x
ELSE
          Allocate(y)
          Allocate(x)
          Return 0

Dis(x, y): // data-dependent function (distance in matrix D)
IF $D(x, y) > 0$
THEN
          Allocate(x)
          Allocate(y)
          Return x
ELSE
          Allocate(y)

Allocate(x)
Return 0

The function Allocate(x) allocates x, if x is a department that hasn't been yet allocated, to the currently available space. A successful allocation of a department x returns the value x>0, an unsuccessful allocation returns 0.

These set of functions has been selected intuitively. The only way to confirm the validity is experimentally. During the process of evaluation any repetitions are ignored. The evaluation of the tree from Figure 3, for example, will result in sequence: A = { j, d, k, e, q}. For each sequence the total cost is calculated according to the formula (1) and this cost is used as a measure of fitness (with this assumption solutions with smaller fitness are better). Solutions such as the one above, which is an example of a sequence that doesn't allocate space for all the departments, remain in the population, however a very high cost is attached to them.

The best solution was obtain in generation 157 with the fitness 2185.5 and the sequence of allocations:

$$A = \{j, k, g, r, h, p, f, n, l, o, m, q, c, a, i, b, d, e, s\}$$

Figure 4 shows the solution graphically. The setting for the genetic program were as follows: population size 1200, tournament selection with a group size of 5, probability of mutation of the terminals 0.02, one point crossover. Figure 5 shows the best solutions developed in the process of evolution.

```
25        j j        k k        k k           k k
24        j j        k k        k k           k k
23        j j        k k        k k           k k
22        j j  j k k k k k k k k k k k k k k
21        j j  j k k k k k k k k k k k k k k

20        p p        p p        h h           k k
19        p p        p p        h r           k k
18        p p        p p        h g           k k
17        p p        p p        h g           k g
16        p p p p p p p p p h g g g g g g
15        p p p p p p p p p h g g g g g g

14        p p        f f        n n           o o
13        p p        f f        n n           1 o
12        p p        p f        n n           1 o
11        p p        p f        n n           1 o
10        p p        p f        n n           1 o
 9        p p p p p f f f n n n n 1 1 1 o
 8        p p p p p p f f f n n n n 1 1 1 o

 7        e s        i i        m m           o o
 6        e e        i i        m m           o o
 5        e e        i a        m m           o o
 4        e e t t t i a        m m t t o o
 3        e e        b c        m m           o o
 2        e e        b c        m m           o o
 1        e e d d d d c c c q m m o o o o o
 0        e e d d d d c c c m m o o o o o o

    0 1 2 3 4 5 6 7 8 9 0 1 2 3 4 5 6 7 8 9 0 1 2 3 4 5
```

Figure 4. The best solution obtained for the office layout

These results are better than those obtained previously using a classical approach where the lowest cost obtained was 2405.0 (Liggett 1985, and using genetic algorithms where the lowest cost evolved was 2254.5 (Jo and Gero 1997). These two solutions were re-evaluated by our system, to avoid differences due to small implementation dissimilarities. Direct comparison with Liggett's result has limited value, since we allowed splitting of a department between floors.

Solution	Cost	Generation
e h g o m r s p n j l k f d b i a c q	2346.5	39
e h g o m r s p n j l k q f d b i a c	2330.5	41
e h g o m r p n j l s k q f d b i a c	2322.5	42
e h g o l r s p n j q m k f d b i a c	2321.0	44
j k l r s p f m o n g h c b a i q d e	2302.0	50
j k l r h p f m o n s g c b a i d q e	2288.0	67
j k n r s p f m o l g h c b a i d q e	2280.0	68
j k g r h p f m o n l s c b a i q d e	2261.5	85
j k g r h p f m o n l s c b a i d e q	2249.5	86
j k m r h p f g o l s n c a i b d q e	2241.5	144
j k g r h p f n o l m q c a i b d e s	2192.5	145
j k g r h p f n l o m q c a i b d e s	2185.5	157

Figure 5. The final twelve best so far *solutions*

5. Discussion

Evolutionary computing can be successfully used for such NP-hard problems as space allocation planning where it shows a clear and demonstrable improvement path. In this paper we have demonstrated that genetic programming can be used to model the space layout planning problem and produce results equal to those obtained by genetic algorithms and gives consistently better results than the best known heuristics from previously published research. As with many such general solution methodologies there is still considerable skill required in formulating the space layout planning problem appropriately for solution using genetic programming. This is the primary disadvantage of these general solution methods.

The space layout planning problem is an example of problems which belong to the class of NP-hard problems in terms of their complexity. A problem of the size we have solved in this paper is already not trivial. The advantage of the genetic programming approach is that it is largely independent of the size of the search space unlike many other approaches. As the size of the problem gets larger the quality of the solution generated may drop but the robustness of the method is unchanged.

884

References

Goldberg, D. E. (1989) Genetic Algorithms in Search, Optimization, and Machine Learning, Addison-Wesley, Reading.

Jo, J. H. and Gero J. S. (1997) Space layout planning using an evolutionary approach, Artificial Intelligence in Engineering (to appear).

Koza, J. (1992) Genetic Programming, MIT Press, Cambridge, MA.

Liggett, R. S. (1985) Optimal spatial arrangement as a quadratic assignment problem, in J. S. Gero (ed), Design Optimization, Academic Press, New York, pp. 1-40.

Michalewicz, Z. (1994) Genetic Algorithms + Data Structures = Evolution Programs, Springer-Verlag, Berlin.

Tate, D. M. (1995) A genetic approach to the quadratic assignment problem, Computers Opns Res. **22**: 1, 73-83.

THE USE OF GENETIC PROGRAMMING IN EXPLORING 3D

DESIGN WORLDS

A REPORT OF TWO PROJECTS BY MSC STUDENTS AT CECA UEL

T.Broughton, A.Tan, P S Coates
Centre for Environment and Computing in Architecture,
University of East London.

Abstract

Genetic algorithms are used to evolve rule systems for a generative process, in one case a shape grammar,which uses the "Dawkins Biomorph" paradigm of user driven choices to perform artificial selection, in the other a CA/Lindenmeyer system using the Hausdorff dimension of the resultant configuration to drive natural selection.

1) Using Genetic Programming in an interactive 3d shape grammar (AmyTan &P.S..Coates) A report of a generative system combining genetic programming(GP) and 3D shape grammars. The reasoning that backs up the basis for this work depends on the interpretation of design as search In this system, a 3D form is a computer program made up of functions (transformations) & terminals (building blocks). Each program evaluates into a structure. Hence, in this instance a program is synonymous with form. Building blocks of form are platonic solids (box, cylinder....etc.). A Variety of combinations of the simple affine transformations of translation, scaling, rotation together with Boolean operations of union, subtraction and intersection performed on the building blocks generate different configurations of 3D forms. Using to the methodology of genetic programming, an initial population of such programs are randomly generated,subjected to a test for fitness (the eyeball test). Individual programs that have passed the test are selected to be parents for reproducing the next generation of programs via the process of recombination.

2) Using a GA to evolve rule sets to achieve a goal configuration(T.Broughton & P.Coates) . The aim of these experiments was to build a framework in which a structure's form could be defined by a set of instructions encoded into its genetic make-up. This was achieved by combining a generative rule system commonly used to model biological growth with a genetic algorithm simulating the evolutionary process of selection to evolve an adaptive rule system capable of replicating any preselected 3-D shape. The generative modelling technique used is a string rewriting Lindenmayer system the genes of the emergent structures are the production rules of the L-system, and the spatial representation of the structures uses the geometry of iso-spatial dense-packed spheres

R. Junge (ed.), CAAD Futures 1997, 885-915.
© 1997 *Kluwer Academic Publishers.*

886

1) Using Genetic Programming in an interactive 3d shape grammar

This paper attempts to show how evolving creative forms can be cast as a problem of induction. And more importantly, that there is a way to solve the problem of induction -- by genetic programming. The genetic programming paradigm described in the following provides a way to do program induction. The basic idea of program induction is the inductive discovery of a computer program from a space of possible computer programs. In our case the computer program is a sequence of operations or a composition of functions that evaluates to a geometric output.

Given that the computers are more effective in generating proposals according to rules and control strategies, the human critic is used to inspect and test them. There is no question that there are many analysis programs that will do the task of evaluating the performance of each proposal. As the main purpose of this paper is to explore and create imaginative and beautiful forms, the proposals must pass the 'eye' test. Criterion for high performance is that it must appeal to the critic enough for it to be selected to evolve further. The final structure is the form selected that will not be bred any further.

The structure of the program is equivalent to the form because it contains all the information that will be manifested in the geometric properties of the form. When reference is made to the computer program, it implies the structure.

Mapping of the code to a virtual model is simple and direct. Three dimensional geometrical shapes pose no problem to programs that use the constructive solid geometry paradigm, such as AUTOCAD and a host of others too numerous to mention. What better way is there to represent properties of geometrical objects than to use the in-built functions of CAD software. Because AUTOLISP, a version of LISP unique to AutoCad, has been employed to implement the genetic programming paradigm, the result or emergent computer program that has evolved can be easily translated into its visual geometric form using the EVAL function in AutoLISP. The process of genetic programming takes on a search for highly fit individual computer program in a space of computer programs. In particular, the search space is the space of all possible computer programs composed of functions and terminals appropriate to the problem domain. Breeding using reproduction of the fittest along with genetic recombination (crossover) operation appropriate for mating computer program applies. A computer program that solves, or approximately solves a given problem (or meets a target) may emerge from this combination of Darwinian natural selection and genetic operations.(Koza, Genetic programming - on the programming of computers by natural selection, MIT, 1992)

The set of possible structures in genetic programming is the set of all possible compositions of functions that can be composed recursively from the set of Nfunc functions from F = { f1, f2, ..., ffunc} and the set of Nterm terminals from T = { a1, a2, ..., afunc }. Each particular function fi in the function set F takes a specified number of z(fi) of arguments z(f1), z(f2), ..., z(fNfunc).

The function in the function set may include

* arithmetic operations (+, -, *, etc.),

* mathematical functions (such as cos, cos, exp, and log),

* Boolean operations (such as AND, OR, NOT),

* conditional operators (such as IF-THEN-ELSE),

* functions causing iteration (such as DO-UNTIL),

* function causing recursion, and

* any other domain-specific functions that may be defined.

We are interested in the last category of functions because the functions applicable to this paper are uniquely defined so as to give desirable results - generating interesting forms. Bearing in mind that genetic programming was developed to solve a wide range of problems, it is probably more widespread and easier to cast problems of a mathematical nature as a computer program. However, this important aspect of the versatility of genetic programming has been harnessed to bridge the gap between creativity in design on the one hand and the mathematical nature of geometrical forms on the other.

Consider the function set

F = { UNION, INTERSECT, SUBTRACT }

and the terminal set

T = { S0, S1}

where S0 and S1 are Boolean variable sets that serve as arguments for the functions.

A combined set of functions and terminals is as follows:

C = F U T = { UNION, INTERSECT, SUBTRACT, S0, S1 }

888

The terminals in the combined set C can be considered as functions that does not require any arguments.

Take for example the INTERSECT function with two arguments. A set that occurs both in the two arguments is returned if there is an overlap in arguments sets (i.e., S0 and S1) and a NIL if there is none. A typical Boolean function can be expressed in disjunctive normal form (DNF) by the following LISP S-expression;

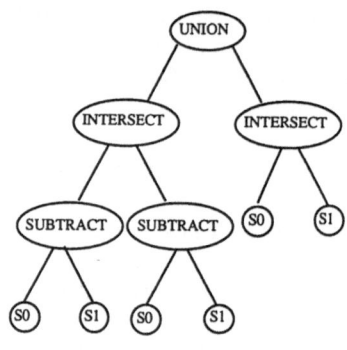

(UNION (INTERSECT (SUB-TRACT S0 S1) (SUBTRACT S0 S1)) (INTERSECT S0 S1))

Fig. 1: S-expression depicted as a rooted, point-labeled tree with ordered branches

The above LISP S-expressions can be graphically shown as a rooted, point labelled tree with branches. The leaves (external points) of the trees are labelled with terminals S0, S1, S0, S1, S0 and S1 respectively. The internal points are labelled with functions UNION, INTERSECT, SUBTRACT, SUBTRACT and INTERSECT. The root of the tree (the UNION) is the first function just inside the outermost left parenthesis of the LISP S-expression. This tree is equivalent to the parse tree which most compilers construct internally to represent a given computer program.

All possible trees of this description that can be generated from any possible recursive combinations of the available functions and terminals appropriate to the problem at hand is the search space to be explored by genetic programming. This is equivalent to searching through all LISP S-expressions consisting of the available functions and terminals.

This is the distinction between the conventional genetic algorithm and genetic programming. While one-dimensional strings, be they finite or variable length, are the structures that undergo adaptation in genetic algorithms, the structures that undergo adaptation in genetic programming are hierarchical structures (rooted point labelled trees with ordered branches).

Closure of the Function set and Terminal Set

Due to the hierarchical nature of the trees, one has to take care that the functions in the function set should be well defined and closed such that each function is able to accept any value or data type that may be returned by any function in the function set and any value and data type that may possibly be assumed by any terminals in the terminal set. This is especially relevant where ordinary computer programs contain arithmetic operations, conditional comparative operators, and conditional branching operators. Computation breaks down when the operators are given an undefined variable (eg. division by zero, logarithm of zero) or unacceptable data (e.g. NIL, square-root of a negative number).

Closure of the function set in such cases can be dealt with by defining a protected function that does not evaluate division by zero or returning an absolute value when a nonpositive argument is encountered or avoiding non-numerical logic operators.

Sufficiency of the Function Set and Terminal Set

In order to use genetic programming effectively, the set of terminals and the set of primitive functions selected should be capable of expressing a solution to the problem. It is up to the designer to identify the set of function and terminals that has sufficient explanatory power for the problem at hand. Depending on the problem, this may be obvious or may require considerable insight. No doubt, design knowledge and experience comes into play where a good choice is concerned. This step of identifying the right variable to solve a particular problem is common to virtually every problem in science. In the sphere of design, personal sets of strategies which designers adapt to particular design circumstances can be expressed in the set of functions and terminals. It is not unknown that many architects already have strategies that are often pronounced and consistent to the point where their individual projects are instantly recognisable.

This project also attempts to show that the designer has complete control over how he/she would like to express his/her design concepts. Designers are free to define their own functions set and terminal set. The results of the experiments that I have performed are solely based on the operations implied in the functions that I had written and the operations are performed on the terminals which I have selected from the domain of all possible terminals. Terminals can be any 3 dimensional shape. They can be simple primitive geometrical objects (eg. cube, wedge, etc.) or a composite of them or even any 3 dimensional form that can be mathematically generated by the computer.

The Initial Structures

The initial structures are made up of individual S-expressions that form the initial population in the first instant. This first population is completely generated at random. For each individual S-expression, the root of the tree is labelled by randomly selecting one function from the set of functions F . Selection of the root function is confined to functions in the function set. The purpose is to generate a program tree with branches extending profusely to form a hierarchical structure rather than creating a depraved tree with a single terminal.

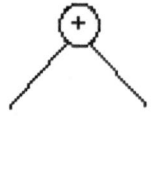

Fig. 2 Beginning of the creation of a random program tree, with the function + with two arguments chosen for the root of the tree

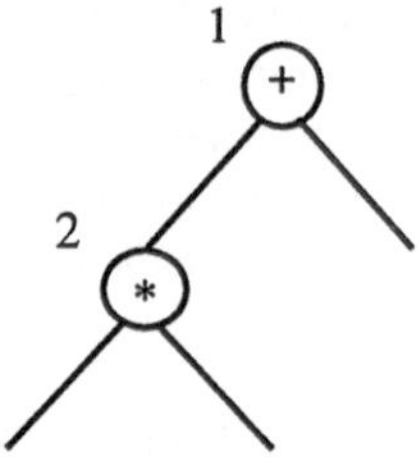

Fig. 3: Continuation of the creation of a random program tree with the function * with two arguments chosen for point 2.

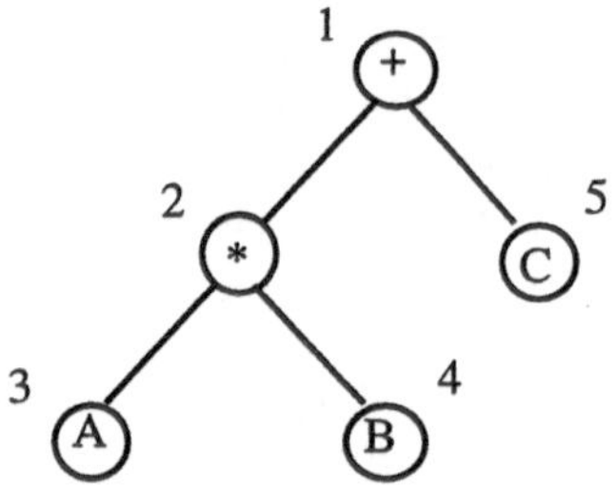

Fig. 4: Completion of the creation of a random program tree, with terminals A, B & C chosen

Fig 2 shows the first node of the random program tree. As the root of the tree, the function chosen (+) has two arguments. Two lines where two is the number of arguments of the function radiate out from the node. For each function, F with z(f) arguments, there are z(f) number of radiating lines. Then, for each such radiating line, an element from the combined set C = F U T of the functions and terminals is randomly selected to be the label for the endpoint of that radiating line. In the case where a function happens to be selected for the endpoint of a radiating line, then the process is repeated as described above. A recursive generative process continues if functions are repeatedly selected In Fig 2, the function * was selected for the second point (internal node) from the combined set C = F U T. The function * has two arguments which is represented by two lines radiating from node 2. Suppose then, that a terminal, A was chosen to be the label of an endpoint, for example at node 3. At this point, the generating process will terminate. Similarly, if terminals B and C were selected to be the labels of the other two radiating lines, the generating process is complete and a fully labelled tree is created, as shown in Fig 4.

Although the example shown has a tree of depth of 3, due to the random nature of selecting functions and terminals, resulting trees may be of variable shapes and sizes. The depth of the tree is defined as the length of the longest non backtracking path from the root to an endpoint. There are several ways to achieve some control over the size of a tree.

One method of generating the initial random population is to define a fixed maximum depth. The length of every non backtracking path between the root and endpoint is restricted to a specified maximum length. This requires that for points at less than the maximum depth, the random selection of labels is taken from the combined set C = F U T. However, once the depth has reached the maximum specified, selection of labels is restricted to the terminal set T. The number of functions in the function set F and the number of terminals in the terminal set T influence the expected size of the tree some- what. This method of generating initial trees has been termed as the "grow" method by Koza(1994).

Other methods of generating initial trees are the "full" method and "ramped half-and- half" method described by Koza (1994). In the "full" method, selection of labels for all internal nodes at less than specified maximum depth are taken from the function set and only from the terminal set T at maximum depth. The resulting trees tend to have the same size. The "ramped half-and-half" method is a combination of the "grow" and "full" methods. For example if the maximum specified depth is 6, 20% of the trees will have depth 2, 20% of the trees will have depth 3 and so forth up to depth 6. Then for each value of each depth, 50% of the trees are created via the full method and 50% of the trees are produced via the grow method. The "ramped half-and-half" method pro- duces a wide variety of trees of various shapes and sizes, but for simplicity sake, the "grow" method was chosen for this experiment.

For reasons of creating genetic diversity and to avoid wasting computational resources, I have taken steps to eliminate duplicate trees in the initial population of generated trees, although this is not necessary. Each newly created S-expression was subject to a check for uniqueness before it is inserted into the initial population, otherwise it is dis- carded and the process is repeated until a unique S-expression has been created. Variety of the population is maintained at 100% for the initial population. This may be not be the case in later generations but should be expected as an inherent part of the genetic processes.

Fitness

The rule of Darwininan natural selection piv- ots about a measure of fitness. In nature, fit- ness is the measure of a living thing to sur- vive an propagate itself . This is adapted to the sphere of computer algorithms in genetic algorithms and genetic programming, where it takes on the form of some control over the process of when and how artificial reproduc- tion is determined. In other words, the indi- viduals in a population is evaluated by some procedure and rank in order of performance according to the procedure. This procedure can be made explicit, as in many applications of a mathematical nature, or it may be implic- it. In my case, the fitness can be called the

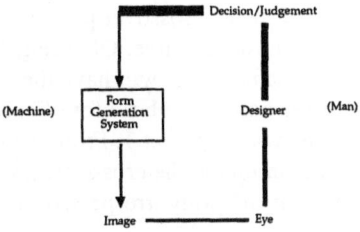

Fig. 5 : Interaction between man and machine, the 'see and decide' relationship

'eye test'. Individuals in a population will be scrutinised for its aesthetic appeal and the best looking one(s) selected to live on. The user is given the opportunity to make subjective judgements before moving on. The selected individual(s) are the chosen parents of the next generation.

Fitness can also, in this context, be understood as a steering mechanism. The act of choosing parents with desirable characteristics is to lead to a preferred direction in the evolutionary path. Since we do not have a million years to spare, an evolutionary process is accelerated by purely manoeuvring through the shortest path to arrive at a possible solution. Anything else not in favour is omitted.

Reproduction.

The individuals in each generation of the genetic programming undergo adaptation via the choice of three operations

- ° crossover (sexual recombination)
- ° asexual reproduction
- ° mutation

CROSSOVER

New offspring are produced from the parent(s) that has been selected from the population according to the 'eye test'. In the case where crossover (sexual recombination) is chosen to be the breeding process, two parents are selected. The crossover operations begins by randomly selecting a point or node in each parent. The nodes serve as crossover points for the parents. Offspring are produced like this: The fragment of one tree lying below the crossover point in one tree will be exchanged for the sub-tree fragment of the second tree. Offspring 1 keeps everything else above the crossover point of the parent 1 but will have the sub-tree fragments of parent 2 attached to it at the crossover point. Similarly for offspring 2, parts of parent 2 is combined with the subtree fragment of parent 1. From the point of LISP S-expressions, crossover swaps the sublists starting at the crossover point. This will always produces legal LISP S-expressions as offspring irrespective of parents or crossover points.

There are several cases of crossover worth discussing here due to the random nature of selecting crossover points.

If the crossover point of a parental tree is chosen to be the root, the crossover operation produces an offspring 1 by replacing the entire tree of the first parent with the sub-tree fragment of the second parent while offspring 2 will include parent 2 and the whole tree of parent 1 at the point of crossover point. In the case where the roots of

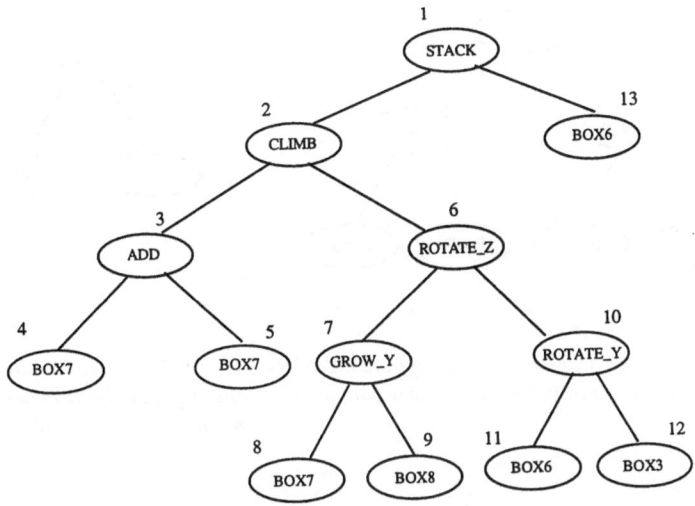

(STACK (CLIMB (ADD (BOX7 BOX7) (ROTATE_Z (GROW_Y BOX7 BOX8) (ROTATE_Y BOX6 BOX3))) BOX6)

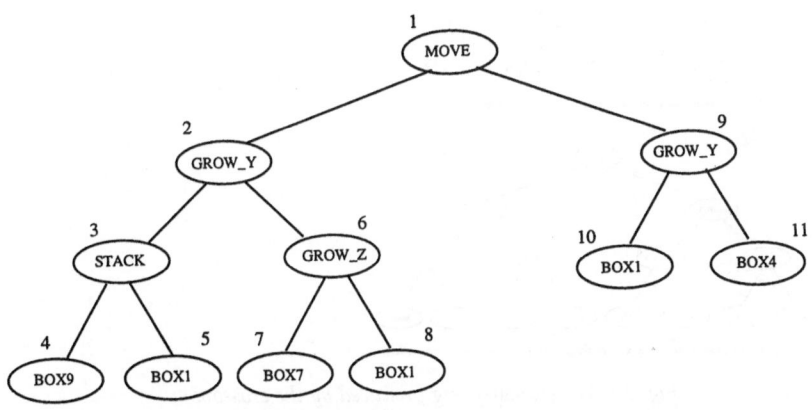

(MOVE (GROW_Y (STACK BOX9 BOX1) (GROW_Z BOX7 BOX1)(GROW_Y BOX1 BOX4))

Fig.s 6 & 7: Two parental computer programs. (S-expression shown below each parse tree)

both parents are chosen to be crossover points, the resulting offspring are repeats of their parents.

Crossover produces trees of considerable variety. Shapes and sizes of trees tend to be influenced by whether a function or a terminal occupies the node chosen as crossover point. In the event where the node of parent 1 selected is a terminal, then the terminal will be be inserted at the location of the subtree of the second parent while this subtree

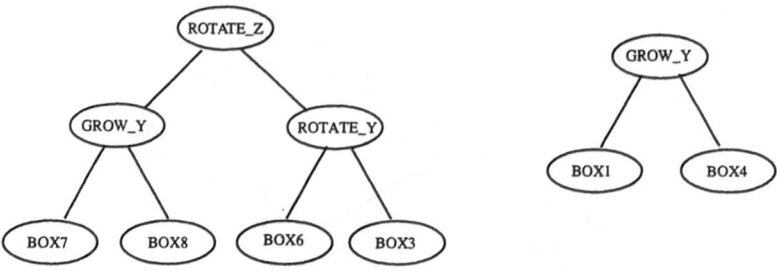

Fig. 8: The crossover fragments resulting from the selection of crossover point of point 6 of the first parent .

Fig. 9 : The crossover fragments
resulting from the selection of
crossover point of point 9 of the sec-

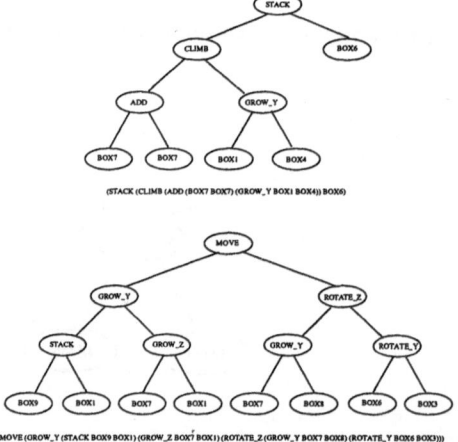

Fig. 10: The two offspring produced by the crossover

now occupies the position of the terminal. The first offspring has 'grown' and the second offspring is a shorter tree. If the crossover points are both terminals, the crossover operation has the effect of just swapping terminal from tree to tree.

ASEXUAL RECOMBINATION

This is the situation where one individual mates with itself. Practically, only one parent is chosen and the other is a copy. The operations are the same as for crossover except that now the two parents are identical. The two resulting offspring are likely to be different because the crossover points selected are going to be different anyhow. However, I am not ruling out that the choosing the same crossover points will not happen.

This is not the case with genetic algorithms. What is different between genetic algorithm and genetic programming is that in conventional genetic algorithms, only one crossover point is selected and applied to both parents. Two similar fixed length strings crossing over at the same point cannot produce dissimilar offspring. The implication of this is that premature convergence occurs in conventional genetic algorithms and the population is directed towards over production of similar offspring.

In 'The Selfish Gene' (Richard Dawkins, 1974) writes; " ...Darwinian adaptation precludes that selection cannot produce adaptations unless there is a hereditary difference among which to select." Darwinian selection has to have gene variation to work on.

MUTATION

The third method of reproduction, mutation introduces variation in the gene pool. The aim is to cause random changes in the structures in the population. The use of mutation most useful on two aspects in this project. The first is to do a quick random walk (which in principle is what mutation alone achieves) to search for a suitable form (or forms) and thereafter to perform crossover or asexual reproduction on the selected forms to hone in on desirable characteristics and features. Secondly, mutation brings about leaps from hill to hill. Occasionally, this might be beneficial when the need to add diversity to the population arises.

The operation of mutation is asexual and only one parental S-expression is involved. One offspring S-expression is produced in a mutation. Once again, a node is selected at random. The subtree below this point is discarded and a new randomly generated subtree is inserted in its place. A high rate of mutation is said to have taken place if the node is chosen is higher up the tree nearer the root and a low degree if chosen node is near the bottom of the tree. Obviously an offspring produced at a lower rate of mutation has near all the characteristics of the parent and more besides. But the degree of similarity depends very much on the new subtree. It may be possible that an offspring with a low rate of mutation resembles its parents in very few places, especially if the subtree is of huge size or contain features that overshadows the original.

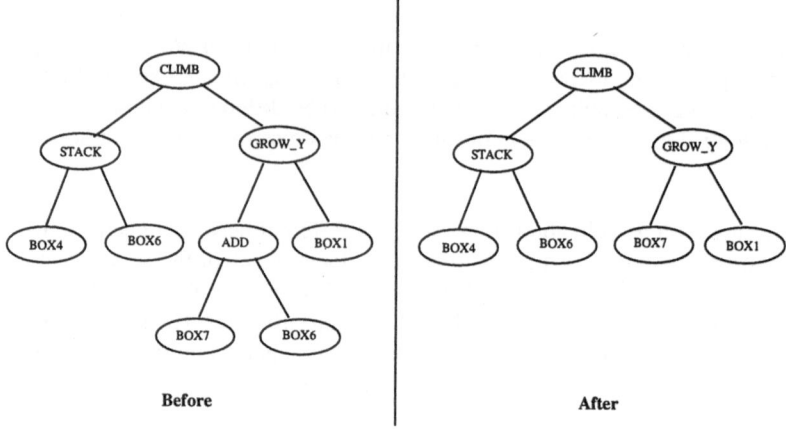

Fig. 11: A computer program before and after the mutation operation

The mutation operation in genetic programming is quite different from that of genetic algorithms. In conventional genetic algorithms, mutation helps to prevent against premature loss of potentially useful genetic material. 1's and 0's at particular locations may be lost occasionally. Mutation is the random alteration of the value of a string position, i.e. flipping a 1 to a 0 or vice versa. Mutation is generally considered a minor operation in genetic algorithms.

RESULT DESIGNATION AND THE TERMINATION

The genetic program paradigm adopted in this project aims to imitate nature and that foreordains a never-ending process. In principle, the evolutionary process continues on and on. Practically, it is usually terminated by either one of the following whichever comes first; a desirable structure(s) has been discovered and designated as the result of the experiment, or the processing power of the computer does not permit any further processing.

Using Genetic Programming

There are five major steps in preparing to use the genetic programming paradigm to solve a problem:

• determining the set of terminals

• determining the set of functions

• determining the fitness measure

• determining the parameters and variables for controlling the run, and

• determining the method of designating the result and the criterion for terminating a run.

Note that for each problem, solutions can be found on numerous runs. However, since the genetic programming paradigm is a probabilistic method, different runs almost never yield precisely the same S-expression. No one particular run and no one particular result is typical or representative of all the others.

The terminals can be viewed as the input to the computer program being sought by genetic programming. In turn, the output of the computer program consists of the value(s) returned by the program.

FORM GENERATION AND SEARCH

In my problem of form search, the information we want to process is the transformations that can be applied to solids and shapes to generate interesting forms. Thus the functions set for the problem consists of three dimensional transformation operators. They can be pure affine transformations or composites of them. The terminal set for this problem should then contain the objects or shapes on which the transformations are to be acted upon. Thus, the terminal set selected is :

T = { box, cone, cylinder, sphere, torus, wedge }

primitives of solid shapes available in AutoCad.

The function set F selected is:

F = { (grow_x), (grow_y), (grow_z), (move), (stack), (rotate_x), (rotate_y) (rotate_z), (climb), (add), (subtract), (intersect) }

Each of these two functions has an two arguments. For reasons of computer processing power, the function arguments have been limited to two. Theoretically, it can be any number of arguments. However, the resulting LISP tree would be too complicated for more than 3 arguments.

The function (grow_x) grows the second argument by a random factor and attaches itself to the first argument along the x-axis. (grow_y) and (grow_z) is similar to (grow_x) except that the attachments are along the y and z- axis receptively. Functions (move), (stack) and (climb) translates the second argument along the x, y, and z-axis by the extents of the first. Functions (rotate_x) (rotate_y), (rotate_z) rotates the second

898

argument with respect to the first. Functions (add), (intersect), (subtract) are standard Boolean operations performed on both arguments simultaneously.

The third major step in preparing to use genetic programming is to identify the fitness measure. This may be considered as the test mechanism at the local level, i.e., at the solution generating level. On a global level, the form will be subject to more formal analysis often associated with the function of the structure.

Visual feedback and interaction are an important part of this experiment. The key to whether an individual form survives depends on its appeal to the eye. The fitness can be known as aesthetic fitness. No number crunching is involved in the fitness measure. The best-of-run individual(s) in the population is the individual(s) chosen as the parent(s) of the next generation.

Breeding can take place using 3 methods, by crossover reproduction, by asexual reproduction, or by mutation. In crossover reproduction, two individual of the population are selected from whom 16 other offspring are bred. One parent is chosen for both asexual reproduction and mutation. The parents are not copied to the next generation. Every new generation consists of 16 new re-combinations of the chosen parent(s). This cuts down on having to re-evaluate the same parent(s) twice and since I am interested in exploring a wider expanse of search space, no effort should be spared to produce as many possible forms as possible in each generation. 16 forms per generation has been tested to be generally suitable.

At every generation, the 16 S-expressions for each individual is evaluated and its form presented to be tested for aesthetic fitness.

Any number of generations can be tested, there is no limit to the number of generations. Once a form has been identified as the final one with no further transformations required, the run can be terminated. The resultant satisfactory structure would be deemed as having emerged from the search of possible forms defined by the functions and terminals.

As the populations are bred from generation to generation, a history of the evolution is created. This is duly recorded and they form a gene bank. Each structure is a potential choice as a new form to be adapted. An interesting forms from the gene bank can be added to the set of terminals. The starting point of the evolution process using a terminal set consisting of forms taken from the gene bank has a head start many generations from the primitive shapes from which they were made up.

Results of the experiments illustrated in Appendix A. based on representation set, R1 where the set of objects is the set of nine possible pattern of wedges, terminal set T =

{wedge1 wedge2 wedge3 .... wedge9} and the operators, function set F = {grow_x grow_y grow_z rotate_x rotate_y rotate_z climb move stack add intersect subtract}.

Thus by varying the sets F and T, a different search space is created every time. It is up to the user to discover what to define in representation set in order to create his/her modelling world.

A constant seed was chosen while the reproduction methods were varied. Two different evolutionary paths taken for each type of reproduction demonstrates the power of selection in steering the evolution of form. Each different type of reproduction chosen produces it own family of structures.

Asexual and crossover reproduction breed trees that bears resemblance to each other only to the extent that no extra information is added to the first population that is initially generated. Outcomes are a recombination of information amongst the individuals of each population. For each generation, information is limited to that contained in one tree selected for asexual reproduction while crossover reproduction has twice the number.

Clearly by changing the seed, the consequence is a new set of initial representations, I. Starting from a different point in the search space and choosing different paths allows one to visit virtually every state in the search space. Due to the randomness of traversing paths one cannot rule out the chance of a fortuitous occurrence of the same structure and should therefore not be piqued by it. In most cases, similar looking structures can have entirely different S-expressions (hierarchical trees), one may be of a depth of 8 and the other a considerable depth of 17. It goes to prove that there are more ways than one to achieve a 'solution', but the paradigm does not take into account the 'economical efficiency' of arriving at it. This is also found in nature. Up till today, no one has been able to completely explain the extraneous human genetic material that apparently cannot be accrued to any feature . (Dawkins)

As the populations are bred from generation to generation, a history of the evolution is created. This is duly recorded and they form a gene bank. Each structure is a potential choice as a new form to be adapted. An interesting forms from the gene bank can be added to the set of terminals. The starting point of the evolution process using a terminal set consisting of forms taken from the gene bank has a head start many generations from the primitive shapes from which they were made up.

900

Fig. 12 traces the genealogy of an emergent form. .History of evolution of the emerging structures can be referred to in Appendix A.

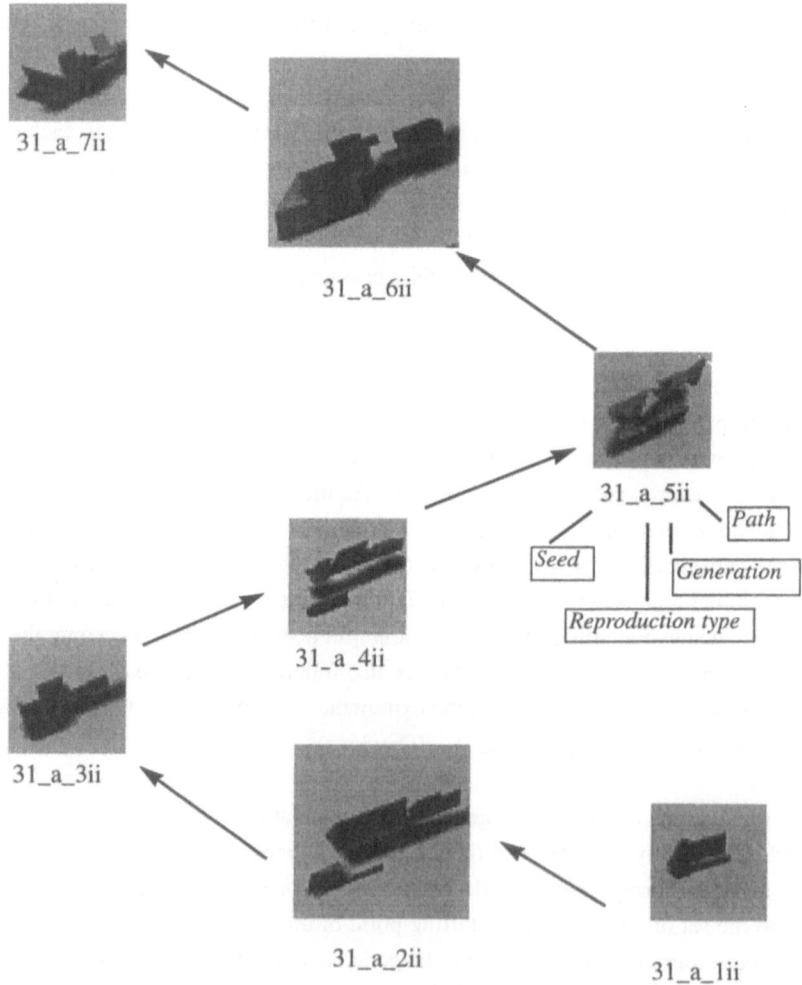

Fig. 12 : Evolving tree of forms, parent forms only

Conclusion

A device to execute design as search necessitates three subclasses of jobs. The first, the representation of the problem through structuring and restructuring the problem space, also known as the design representation. The control strategy (design concepts) that the architect chooses to produce a final solution has to be translated into the description of a three-dimensional space. The second, a solution generation mechanism applying the evolutionary concepts of Darwinian principle of natural selection and reproduction is set up. Structure is defined by its 'genetic' code scripted as S-expressions (hierarchical trees). Structure undergoes adaptation via the process of reproduction and crossover of individual structures, the vehicle for creation of new population structure. A test mechanism that evaluates the 'aesthetic fitness' of a structure is put to use. This drives the evolutionary path to generate interesting and appealing structure. Third, the designation of a candidate solution (or solutions).

The search paradigm stands in harmony with, and perhaps partially motivates, the present recurrence of compositional ideas in architecture. At its core, the model views search as the action of a set of operators on a representation all being guided by a search strategy. When compared to a current understanding of human problem solving, the model suggests that a complementary relationship exists between human designers and computer-based search systems.

The search paradigm owes intellectual debts to many sources outside architecture. These include formal language theory and computational linguistics for grammatical concepts, set theory for representations and proof procedures, cognitive psychology for models of human problem solving, and artificial intelligence for representation and search. However, the paradigm is more than an amalgam of these gleanings, it is enlivened by its own logic and firmly anchored in the field of design.

Design exploration through search will improve the ability to create in fundamental ways, the generation of visual alternatives in a short period of time allows for making selections and new associations possible. A fast feedback of results and the immediate visual control enhances imagination. For intelligent users, the proposed model provides more opportunities for creativity.

Design intelligence located in the generation mechanism assumes that an acceptable solution based on the control strategy will be quickly produced; the evaluation has nothing to do. This simulates the existence of God, a smart designer with no need for a critic. Designer intelligence located in the test mechanism indiscriminately produces alternatives and it is up to the evaluation mechanism to sort out the acceptable ones by bringing a knowledge to bear, drawing inferences and exploring entailments of alternatives. This is evolution; indiscrimate generation but deadly effective criticism.

2) Using a GA to evolve rule sets to achieve a goal configuration

PRIOR WORK

Prior work involving the use of the essential techniques for form generation and analysis in the experiments - L-systems, genetic algorithms and the Hausdorff distance equation.

Various computer-generated models of morphogenesis have been used to help understand the emergence of complex forms in living organisms since Turing proposed the reaction-diffusion process in 1952. Diffusion limited aggregation models have been used to simulate crystal formation in super-saturated solutions and J. Kaandorp in "Fractal Modeling:Growth and form in Biology uses an aggregation model to investigate the growth of corals. L-systems were developed by Astrid Lindenmayer in 1968 to model the morphology of organisms using string re-writing techniques. These techniques have been applied in a variety of studies to the production of abstract models of biological forms as an aid to interspecial comparison and classification see (deBoer, Fracchia and Prusinkiewicz 1992) . An L-system form generation engine is the main ingredient in the following experiments.

L-SYSTEMS

An L-system model starts with an initial axiom and one or more production rules. Axioms and production rules consist of symbol strings whereby individual symbols in the axiom are replaced by a string of symbols designated by the production rule. This process of character recognition and string substitution is carried out iteratively with each successive iteration producing a symbol string of greater complexity.

The resultant symbol string is interpreted as a series of drawing instructions which produce an abstract representation of the desired organism. The success of the L-system model lies in the self similarity of cell structure that many biological systems exhibit which is mirrored in the grammar based string re-writing process.

The L-system method of modeling developmental processes has been the subject of considerable research. Our approach is to use the L-system biological model in conjunction with an evolutionary algorithm and is an area which has seen relatively little investigation, recent work is reported in (Jacob 96)

A standard Genetic algorithm and the associated Genetic Programming strategy are models which utilise the processes involved in the evolution of biological organisms

The Hausdorff distance measures how close each point comprising a given shape is to a point in a second shape and vice versa and is used photographic image recognition .

Visualisation of the growth model was carried out in Autocad, a 3-D modelling appli-

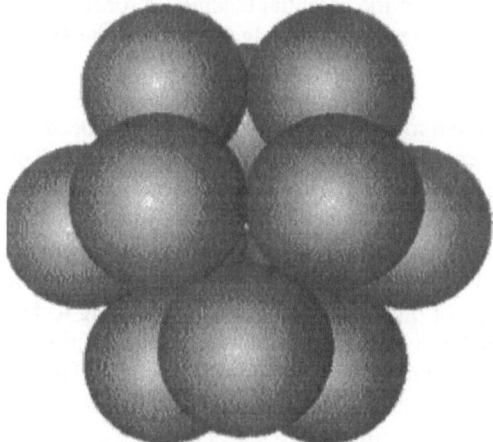

Fig 13 : the 12 spheres of the Iso-spatial dodecahedral array

cation using the Autolisp programming language.

Due to the constraints imposed by using Autocad we kept to the tri-axial Cartesian co-ordinate system rather than the alternative , more 'elegant' four axial system preferred by Frazer. The twelve neighbours of any given point-'*apoint*- are given by the autolisp function '*getneighbours*':

```
(defun getneighbours ( apoint / neighbours offsets p np )
    (setq neighbours '()
                offsets(list '(1 1 0)
                            '( 1 -1 0)
                            '( -1 1 0)
                            '(-1 -1 0)
                            '(1 0 1)
                            '(0 1 1)
                            '(-1 0 1)
                            '(0 -1 1)
                            '(1 0 -1)
                            '(0 1 -1)
                            '(-1 0 -1)
                            '(0 -1 -1)))
        (foreach p offsets
        (setq np (mapcar '+ a point p)
                neighbours(cons np neighbours))
        )
```

Fig 14: the offset list for con-structing the 12 neighbours of a point

The method of representing form will be through the insertion of spheres drawn and saved in separate drawing files. Different coloured spheres represent different levels of recursive branching. The geometry of the 3-D space into which the spheres are inserted is an iso-spatial grid where the co-ordinates of the grid are the vertices of close packed cuboctahedrons as used by J. Frazer in "Data Structures for Rule-based and Genetic Design" . Spheres can only be inserted at these vertices and each sphere has 12 equally spaced neighbours. A variation on this geometry is the one used by Carter Bays' in "Patterns for Simple Cellular Automata in a Universe of Dense-Packed Spheres" which has a four axial system where 12 neighbouring points are the vertices of a dodecahedron.

There are three primary genetic operators used in this experiment, initialisation, crossover and mutation. The first stage of the program is initialisation where a function is called to produce an initial gene pool. The genes are a randomly generated series of nested brackets containing three types of symbols;

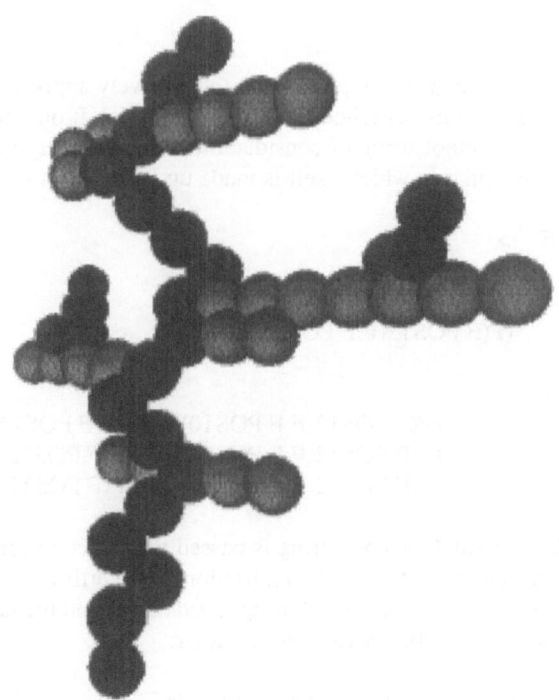

fig 15 : example of a 3D branching morphology using the Iso-spatial array & L system

INITIALISATION

°An instruction to insert a sphere;
  this is represented by the symbol **F** (for "forward")
°A positional variable which refers to which of the 12 neighbouring positions the next
  sphere inserted will occupy;
  these are given in the symbology as **POS1** .. **POS12** for each of the 12 possible
  orientations from any sphere
°A bracket
   an open one (indicates a branching point,
   a closed one) is an instruction to return to a position on a lower 'limb' where it
   last branched.

These symbol strings which make up the gene pool are the production rules of the L-system.

PRODUCTION RULE

The next stage of the program is to iteratively apply a production rule to the initial
axiom - in this experiment a simple 'f symbol. It only requires 3 or 4 iterations to generate a symbol string of considerable length as every instance of 'f is substituted by the
production rule which itself is made up largely of 'f's.

   F ->

   (F(F POS12(F F POS10)F)F) ->

      (F(F POS12(F F POS10)F)F)((F(F POS12(F F POS10)F)F) POS12
      (F(F POS12(F F POS10)F)F)(F(F POS12(F F POS10)F)F)POS10)
      (F(F POS12(F F POS10)F)F)(F(F POS12(F F POS10)F)F))

The re-written symbol string is passed to a function for interpretation of the symbols,
ie its genetic code, and the realisation of the artificial organism in the drawing database. Every symbol in the string is evaluated and the corresponding function is called
which carries the instruction for either;

(1) changing the positional variable or
(2) inserting a sphere at a point on the grid.

Before a sphere is inserted clash detection is carried out and if a collision is imminent the insertion command is ignored and the next symbol evaluated.

MUTATION

The mutation genetic operator works by counting the levels of recursion or nested brackets in an individual's genetic code and selects, at random, a self-contained balanced chunk of code to be replaced by a similar, randomly generated, chunk. The chunk of code selected always starts with a branching bracket symbol and ends with closed bracket. The symbol string is not of fixed length so mutated organisms can have varying lengths of genetic material.

CROSSOVER

Crossover works by choosing two organisms, selecting suitably balanced sections of code and swapping them. As in mutation the amount of genetic material an individual has can change, producing differences in generations ranging from slight to radical.

OBJECTIVE FUNCTION

After choosing a datastructure and method of representation and defining the type of genetic operators to be used, the third element of the algorithm is defining the objective function or 'fitness' function. This was carried out by designing a series of structured experiments which build on the results of each experiments precursor.

Initial experiments were carried out to test the crossover function. In the figures below, the left hand and centre configurations are two objects generated from a random production rule, the right hand object is generated from the resulting rule after crossover of the other two rule sets.

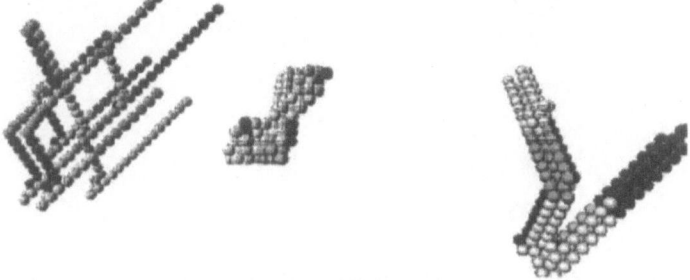

fig 16: Crossover test 1

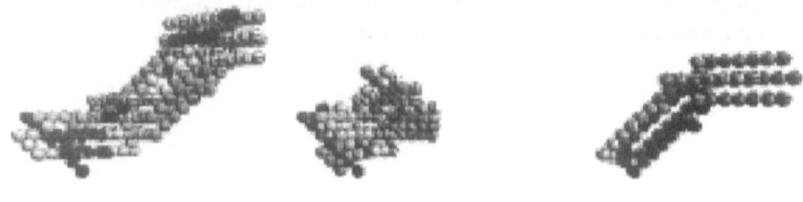

fig 17 Crossover test 2

Our immediate goal was to develop a model which would respond to the user's selection of the characteristics of two individuals which would survive and combine and become accentuated over successive generations - the 'eyeball' test as used in Dawkins' Biomorph program. Figs 18 -21 show screen shots of two experiments.

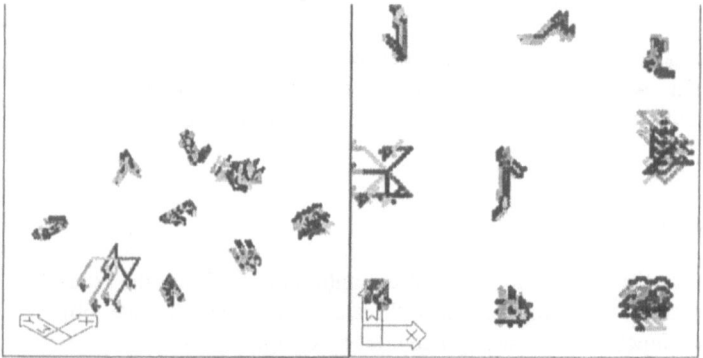

Fig 18 : Artificial selection trial 1- Starting random field of 9 objects

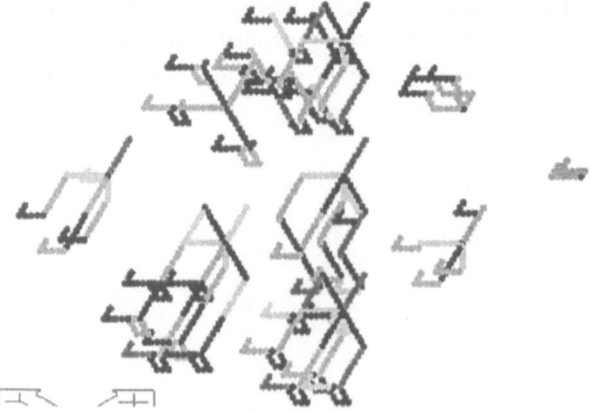

Fig 19 : Artificial selection trial 1-After 6 generations of artificially selecting for "spideryness" the field consists of predominantly spidery objects

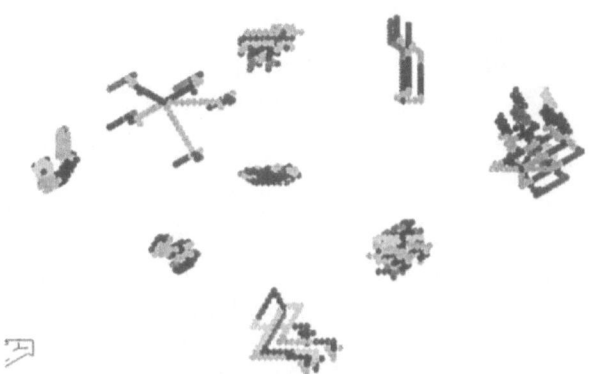

Fig 20 Artificial selection trial 2, the initial field of 9 random genotypes

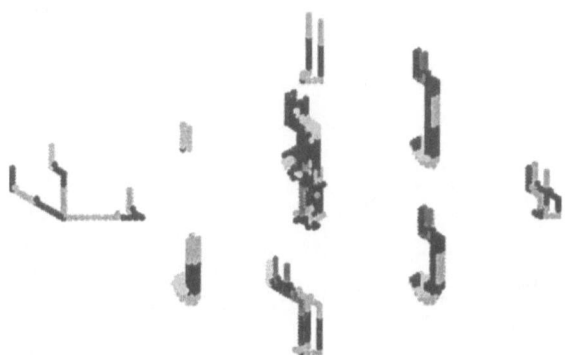

Fig 21 Artificial selection trial 2, after selecting for long thin genotypes

This was completed successfully. The mechanism used here is similar to that employed in section one of this paper, except that whereas there a range of breeding options were allowed (mutation, crossover and asexual reproduction) here the operation is restricted to crossover. At each generation 9 phenotypes are displayed and the user chooses two parents from which to breed the next generation. These two candidates' production rules (genetic material) are them crossed over, 9 times to ,create the 9 genotypes for the next generation, which are then developed into the 9 phenotypes for display.

910

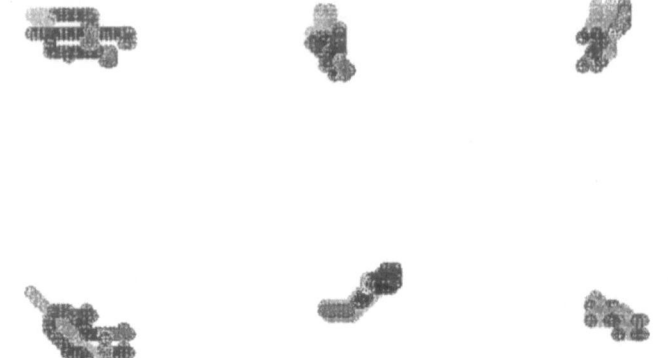

Fig 22 : starting phenotypes for the aspect ratio natural selection experiment

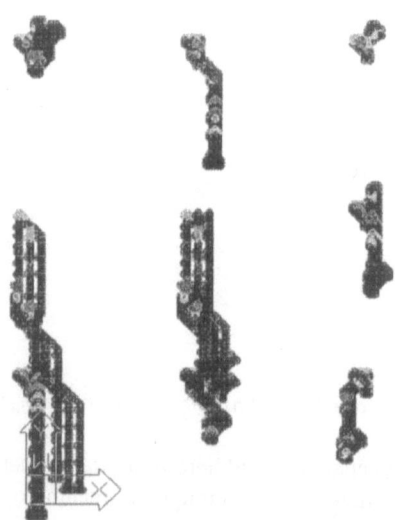

Fig 23 : result of setting a 1:10 X-Y aspect ratio after 20 generations

The next stage was to define an ideal form in terms of a desirable length over width ratio and run the algorithm until it produced forms matching the target within a given tolerance Figures 22 and 23 illustrate a typical run, showing convergence to the target (ie the initial random morphologies slowly adapt to assume more and more etiolated forms).

More complex forms can now be defined as a target. By using the Hausdorff distance equation we can analyse how close an individual has come to matching the pre-defined shape. Successful individuals are used to breed the next generation. These experiments are still continuing, and will be reported to the conference.

FUTURE WORK

The work reported here is at an early stage, and there are two main areas which will be developed next. Firstly the basic function set of **F** and **P** can be enriched to provide a wider range of behaviours; secondly the problem of adopting arbitrary fitness functions can be addressed by using the "arms race" paradigm where a competitive scenario can be introduced by growing two forms simultaneously where each form is its opponent's environment and the fitness function becomes pure survival.

References

Bays C.(1988): Classification of semi-totalistic cellular automata in 3-D, Complex Systems 2

Bays C (1987).Patterns for Simple Cellular Automata in a Universe of Dense-Packed Spheres, Complex Systems 2

DeBoer,Fracchia,Prusinkiewicz (1992), "Analysis and Simulation of the Development of Cellular Layers" Artificial Life II ed. C Langton

Das, S., Franguiadakis, T., Papka, M., DeFanti, T. A., Sandin, D. J.(1994):A genetic programming application in virtual reality. IEEE Computational Intelligence Evolutionary Computation Conference Proceedings, June .

D'arcy Wentworth, Thomson. (1917/61). On growth and form. Abridged ed. (Tyler Bonner, John ed.) Cambridge University Press.

Dawkins, Richard (1991). The Blind Watchmaker. London: Penguin.

Dawkins, Richard (1972). The Selfish Gene. OxfordUniversity Press.

Fraser, John (1995). An evolutionary architecture. London: Architectectural Association.

Frazer J Datastructures for rule-based & Genetic Design

Graves, Michael (1977). "The Necessity of Drawing: Tangible Speculation." Architectural Design, 47, no 6, pp. 384 - 394.

Gero, John S. & Schnier, Thorsten. Evolving representations of designcases and their use in creative design.

Goldberg, D. E. (1989). Genetic Algorithms in Search, Optimization, and Machine Learning. Addison-Wesley.

Horling B Implementation of a context-sensitive Lindenmayer-system modeler Dept Engineering and Computer Science Trinity College Hartford USA

912

D.P.Huttenlocher, G A Klanderman, W J Rucklidge. (1992): "Comparing images using the Hausdorff distance under translation" Computer Vision and Pattern Recognition, pages 654-656 Champaign-Urbana Ilinois,

Holland, John (1975). Adaptation in Natural and Artificial Systems. Cambridge, Massachusetts: MIT Press.

Jacob, C.,(1996) "Evolving Evolution Programs:Genetic Programs and L-Systems". Proceedings of first annual conference on genetic programming, Stanford USA MIT Press pp 107-115.

Jo, H. Jo & Gero, John (1995) Representation and use of Design Knowledge in evolutionary design. CAAD Futures '95. Singapore.

Kaandorp J.(1994)Fractal Modeling: Growth and Form in Biology, Springer Verlag,

Langton C. G. (ed.) (1990).Artificial Life II. Proceedings of the workshop on Artificial Life. Santa Fe. Feb Addison-Wesley.

Koza, John R. (1992). Genetic Programming, on the programming of computers by means of natural selection. Cambridge, Massachusetts: MIT Press.

Langton C. G. (Ed)Artificial Life I II & III Addison-Wesley Publishing Company

Lawson, Bryan (1990). How designers think? 2nd ed. Butterworth Architecture.

Lawson, Bryan (1994). Design in mind. Butterworth-Heinemann.

Lindenmayer A. and Prusinkiewicz P.The Algorithmic Beauty of Plants, Springer Verlag, 1988

Lionel March (ed.) (1976). The Architecture of Form. Cambridge: Cambridge University Press.

March, Lionel & Steadman, Philip (1971). The Geometry of Environment. London : RIBA
Mitchell, William J. (1990). The Logic of Architecture, design, computation & cognition. Cambridge, Massachusetts: MIT Press.

Neutra, Richard (1969). Survival Through Design. London : Oxford University Press.

Pearce, Peter (1978). Structure in Nature as a Design Strategy. Cambridge, Massachusetts: MIT Press.

Pask, Gordon (1972). An approach to cybernetics. London : Hutchinson & Co Ltd

Rowe, Peter G. (1987). Design Thinking. Cambridge, Massachusetts: MIT Press.

Schimitt, Gerhad (1988). Microcomputer aided design for architects and designers. New York : John Wiley & Sons, Inc.

Schnier, Thorston & Gero, John. "Learning genetic representations as alternative to hand-coded shape grammars."Artificial Intelligence in Design '96. pp. 35-57.

Schnier, Thorston & Gero, John.(1995) "Learning representations for evolutionary computation." 8th Australian Joint Conferenc on Artificial Intelligence. AI '95. pp387-394.

Steadman, Philip (1979). The Evolution of Designs. London: Cambridge University Press.

Stiny, G(1980). Introduction to shape and shape grammars. Environmental and Planning B 16, pp 253-287.

Todd, Stephen & Latham, William (1992). Evolutionary Art and Computers. London : Academic Press Ltd.

Tschumi, Bernard (1987). Cinegram Folie Le Parc De La Villette.Butterworth Architecture.

Walker, Miles (1993). Digital evolutions. Dissertation for MSc Computing and Design, University of East London.

Watt, Alan (1993). 3D Computer graphics. Addison-Wesley. pp 1-13

Appendix A

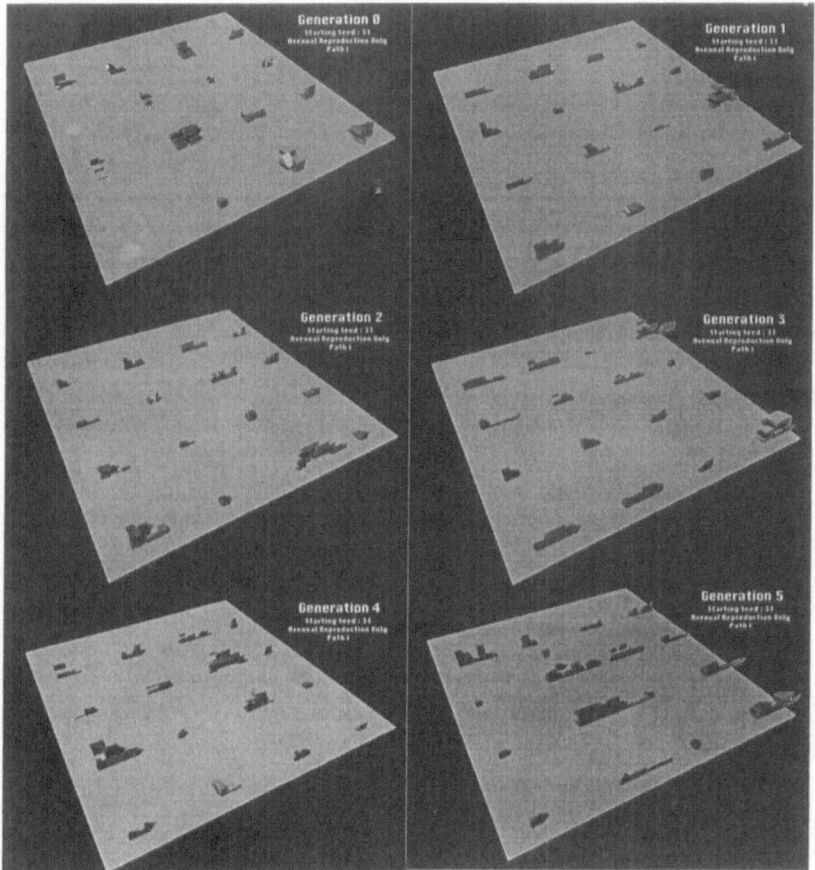

Fig 24 Asexual reproduction only

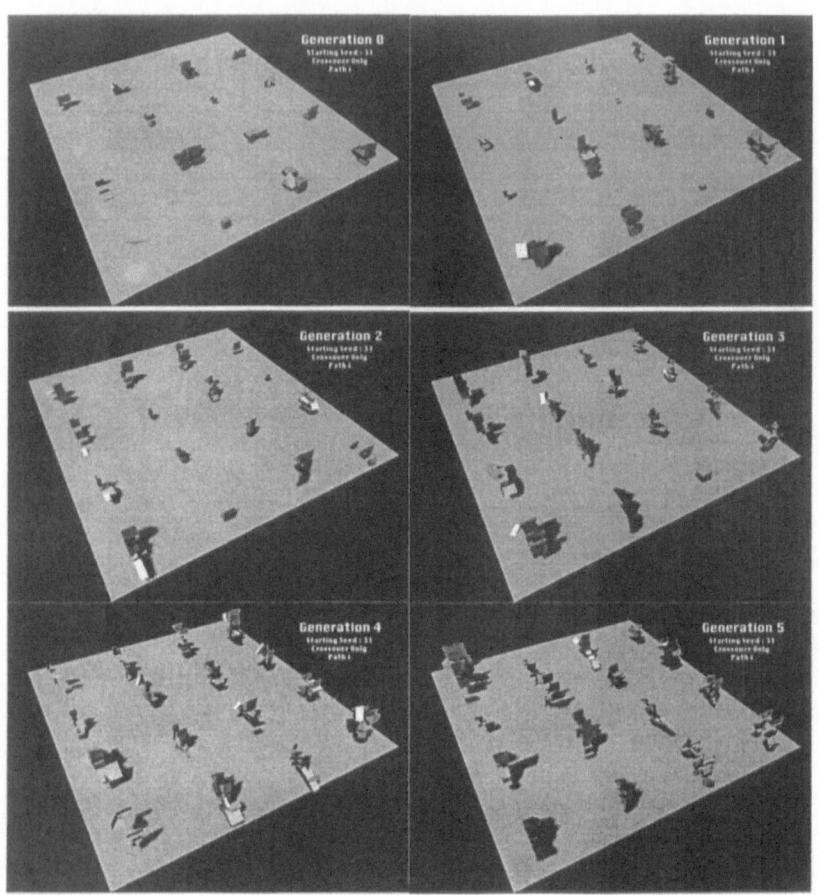

Fig 25Crossover only

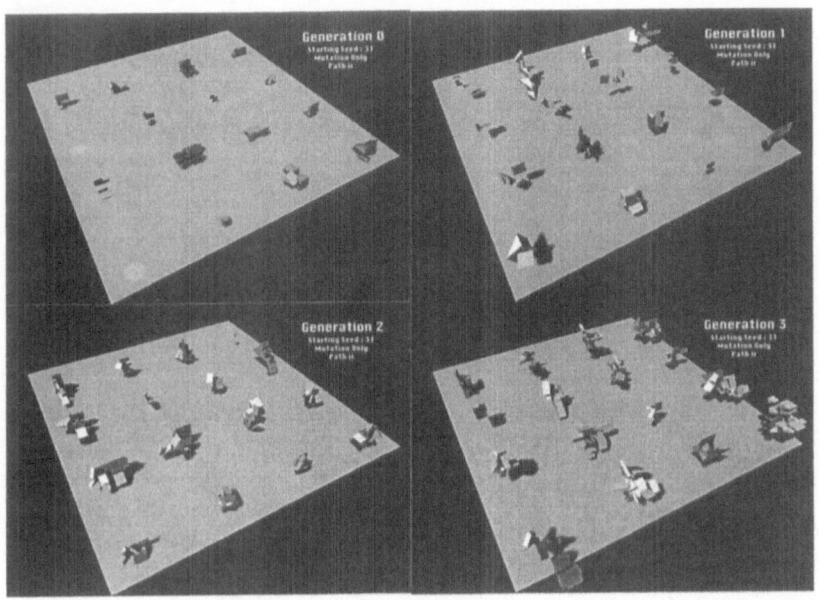

Fig 26 Mutation only

SPATIAL COMPUTER ABSTRACTION : FROM INTUITION TO GENETIC ALGORITHMS

ADAM JAKIMOWICZ
Technical University of Bialystok, Faculty of Architecture

JAVIER BARRALLO
The University of the Basque Country, Faculty of Architecture

ELIANA MARIA GUEDES
University of Taubate, Dept. of Architecture and Computing

Abstract

Many of the emblematic buildings constructed at present shows many formal and technological innovations that have not been satisfactorily resolved by the existing CAAD software. Frank O. Gerhy's Guggenheim Museum in Bilbao is a good example of architecture whose shapes and design are very advanced from the concepts and tools used by CAAD.

The search for new creative resources, from the educational and professional point of view, must be a priority. This will be the only way to get that CAAD contributes essentially in the process of architectural innovation, instead of merely being a reproduction tool.

From this viewpoint the computer exploration of the three dimensional form is presented in here. The concept of abstract art, that has been successfully applied to painting and sculpture in this century is used as a way to experiment, design and create architecture.

This paper juxtaposes three approaches, three different ways of understanding the abstract character, with the purpose to create new objects and environments, which are exclusively characteristic for computer space. This juxtaposition shows how creative and innovative activities in the field of CAAD can be developed using different intellectual bases : intuition, mathematical formulas and genetic algorithms.

917

R. Junge (ed.), CAAD Futures 1997, 917-926.
© 1997 *Kluwer Academic Publishers.*

1. Introduction

Nowadays, many emblematic buildings search for a singular character in their conception and execution that is usually obtained leaving the formal and aesthetic ordinary cannons. So, the well balanced forms, ordered and symmetric, are progressively being substituted by others of more sculptural character and abstract nature.

The Guggenheim Museum in Bilbao, by the architect Frank O. Gerhy, is an example of this type of architecture, whose shape and design is very advanced compared with the concepts and tools used by CAAD. It results paradoxical that a building symbol of modernity have not been conceived and represented by means of the current CAAD techniques, but by means of the use of models and scaled prototypes.

Although CAAD is fully integrated in all development phases of the contemporary architecture, specially when working in two dimensions, it seems to have less importance when used with hyper contemporary architecture. Maybe, this lack of interest is due to the great importance that many educators and final users give to architectural processes with a prevalence of merely representative tasks. Thus frequently, computers are understood just as an easy, fast and accurate way of representing architecture.

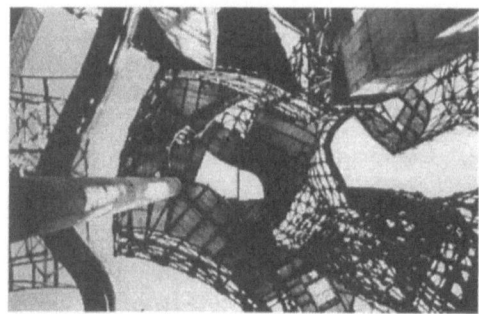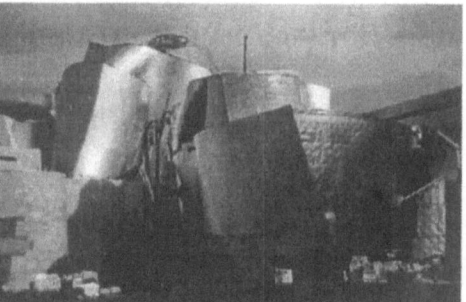

Figure 1. Guggenheim Museum (Bilbao, Spain) under construction.

Although we are living an incredible peak of high technology (Virtual Reality, Artificial Intelligence, Internet., ...), even professionals with a marked computer addiction use mainly manual methods in their first designs of a project. The reason can be found in the few facilities that computers support during the processes that need creativeness and imagination.

The search for new creative sources, technological or intuitive, and the exploration of the form in the three dimensions should be a priority, especially from the educational point of view. This article proposes the utilisation of the abstract art concept, successfully applied to painting and sculpture during this century, as a way to experiment, design and create architecture.

We present three different approaches to the understanding of the abstract character, with the purpose of creating new objects and atmospheres exclusive of the computational space. With this experience, it is pretended to show how the creative and innovative activities in the field of the CAAD can be developed using three different intellectual bases : intuition, mathematical formulas and genetic algorithms.

The work of Adam Jakimowicz studies the possibilities of abstract modelling as a mean for simulation and creation. Intuition and individual creativeness are employed in order to explore a new way of composing spatial forms, using a standard popular software. Javier Barrallo presents his Cybersculptures, a series of figures of mathematical nature generated by means of unusual geometrical objects and processes (Hyperrevolutions, parametric knots, Moebius bands, etc.).

The work of Eliana Maria Guedes is based on the ability of genetic algorithms for the creation of abstract models. The resulting figures have a marked organic character with many resemblance to living forms. The three collections described before constitute a representative sample of the possibilities of abstract models in CAAD.

2. Intuitive Abstract Modelling

No matter how much computers became popular and obvious devices in various spheres of the modern times, when they are already not only a domain of specific professions, but also private homes 'inhabitants', the sphere of interpretation, individualisation and improvisation, as conscious attitude to the possibilities they propose, is still hidden, concealed. This problem deeply concerns CAAD. The most obvious is the opinion, that there are only a few, proper and canonical uses of computers, which we could range as the one 'right' approach - i.e. rationalistic, effective, functional, and which should not be crossed or transgressed, which aims at a rigorous subordination to the rules imposed by software manuals. Such approach is an a 'priori' assumption.

The presented intuitive approach to computer based form making tries to face the mentioned problem. Computer, or rather 3D modelling software, as no other medium, enables us to visualise what is not yet rationalised or even mentally ordered, to produce certain forms, electronic realities, with very simple, basic, primitive input data (i.e. geometric primitives), as a material for geometric, formal exploration. This way of form making opens the user for the unexpected, for the new features of 3-dimensional space, visualising the idea before having it rationally formulated. All descriptions of modern arts (lets include architecture here too) are based on the subjective post-rationalisation of individual impression. Lets introduce it to the sphere of Computer Aided Architectural Design, not as an obligatory approach, but as an option, enriching architectural design as a whole. When Malevich introduced the idea of additional element to painting (...under Suprematism I understand the supremacy of pure feeling in creative art...), the affective approach to architectural computing (here - computer based architectural form making), can be an additional element in CAAD.

920

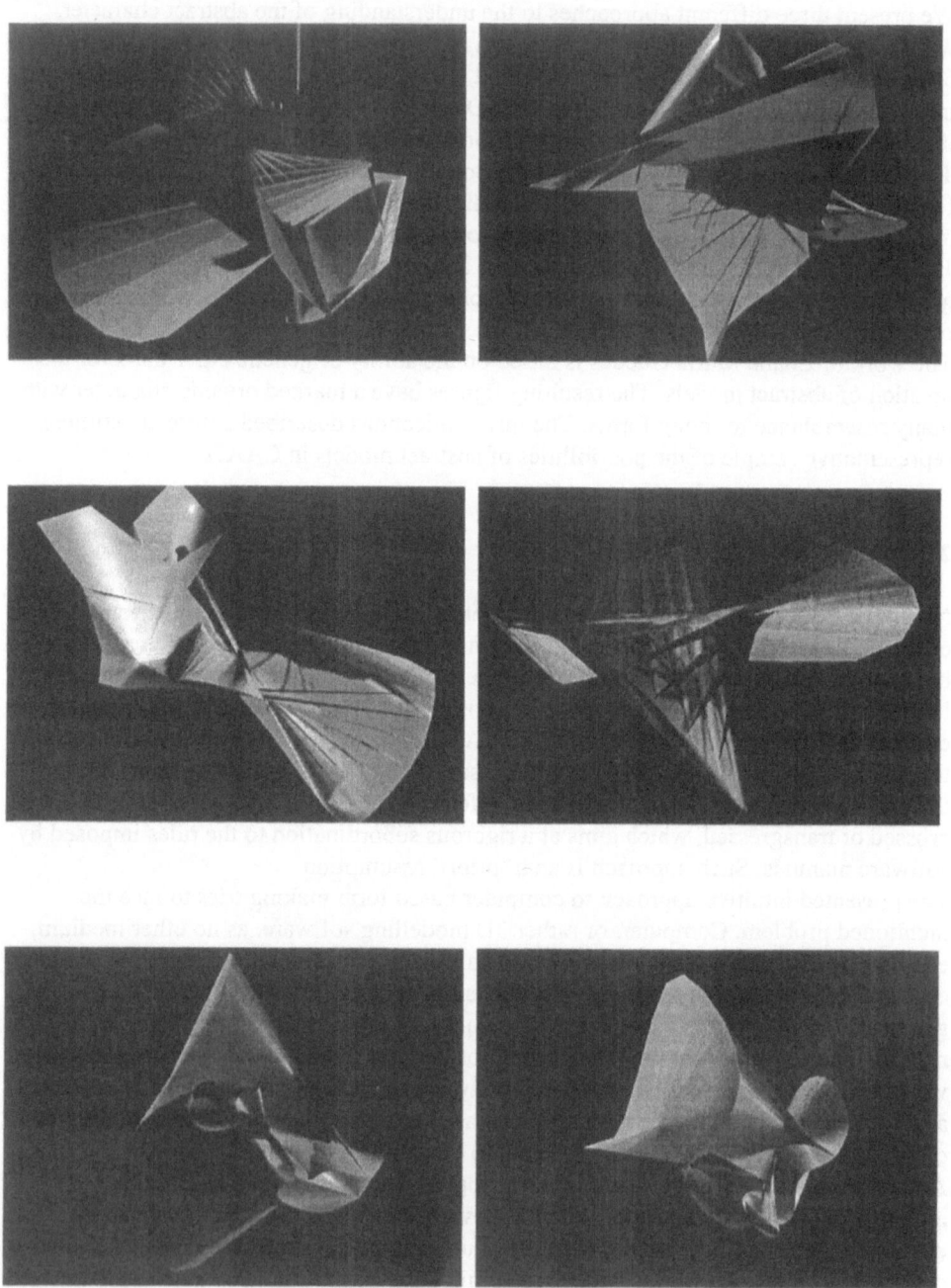

Figure 2. Intuitive abstract modelling.

Even when we have 2-dimensional images as the visible result, silent pictures of artificial reality, these simple experiments bring very deep message of the essential perceivable 'languages' of forms and relations with the value of discovering the new or even impossible.

All these are the games for the essential information. Rules are certain but liquid. Every formal system can be widened individually, even when our work is not received according to our intentions. The evaluation criterion is to be received 'hot', when receiver becomes emotionally involved. Receiver as well as the author is allowed to define his own perception of the given thing - and in such case the intention can be reversed.

In this context, we can here formulate two general approaches to computing in the context of architectural design:

- simulative approach
- creative approach

Simulative approach focuses on making a computer representation of the object, which can be regarded external to computer space. Computer space is used to simulate a reality of that object, which exists (or will exist) independently and externally. In most cases this approach concerns various sorts of analyses.

Creative (generative) approach, concerns making things, which are not just models, but original creatures of the environment they were done in. First of all this approach concerns synthesis : formal, geometrical or aesthetic.

The work presented here as an intuitive spatial abstraction intentionally represents one of the possible aspects of creative generative approach to architectural computing. Making this electronic, virtual domain a part of our reality, work, thoughts and life - let us not forget that it can be a source of the personal, and not necessarily selfish pleasure.

3. Mathematical Abstract Modelling

Since most ancient cultures, until the contemporaneous artist movements, mathematics have been specially relevant in many processes of artistic creation. There are many examples present in sculpture, painting or architecture, showing proportions or shapes based on the classic geometry (by example, the use of perspective in painting, the golden number in sculpture, or the geometric shapes in architecture).

Nevertheless, in the last years of the XIXth century and in the beginning of the XXth, century a group of mathematicians led by Koch, Peano, Cesàro, Hilbert, Julia, Pointcaré, etc. started the study of the possibilities of new geometries, clearly different from the shapes and basic principles used until this moment, and opposed in its conception to the euclidean geometry, preponderant until this moment.

Simultaneously to this precursors of modern geometry appears a new way of understanding and conceiving art, radically different from all the artistic tendencies developed until this moment and destined to be considered the most important artistic expression of the XXth Century : we are talking of abstract art.

922

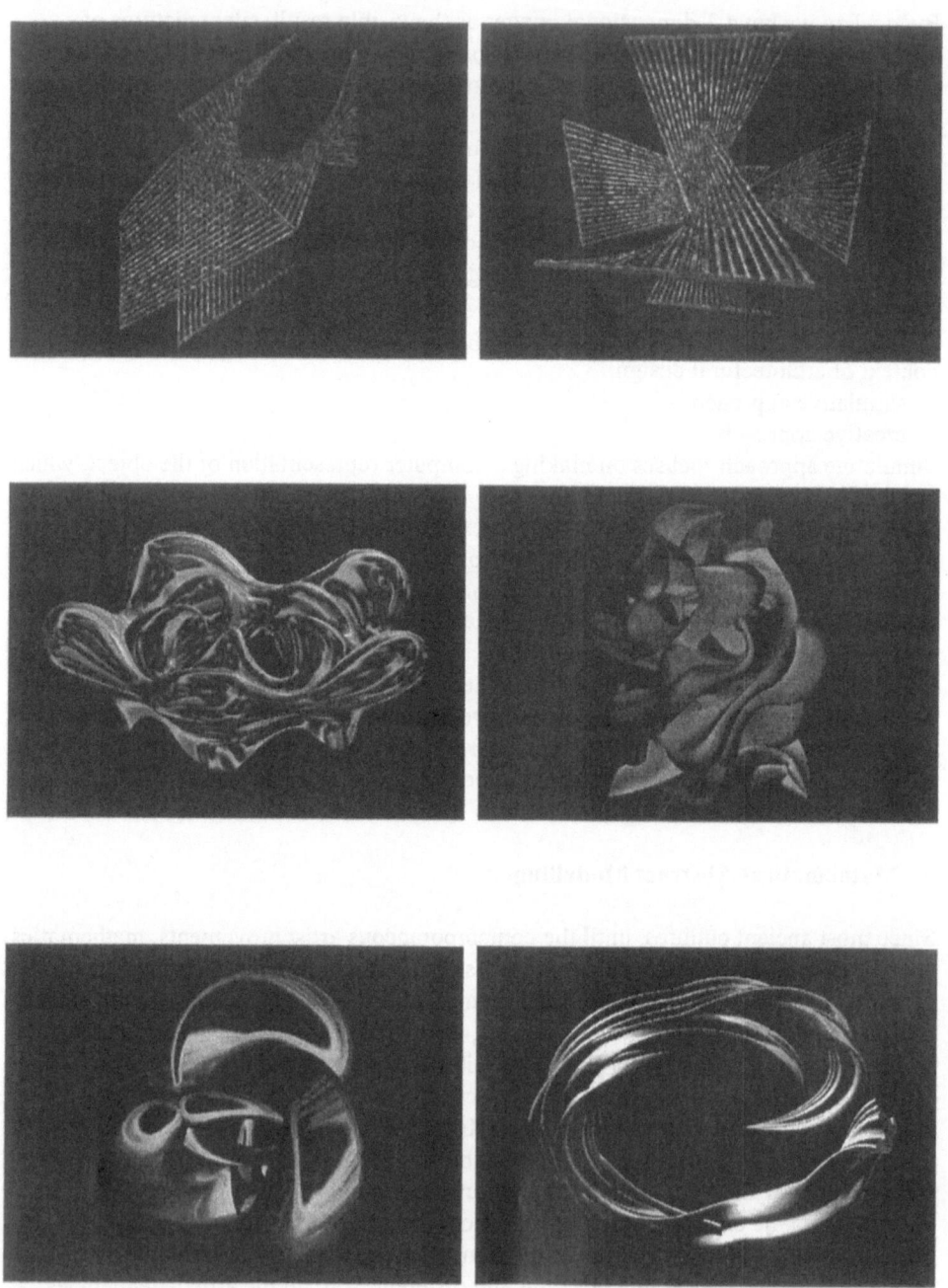

Figure 3. Mathematical abstract modelling.

The parallelism between the works made in the field of contemporaneous mathematics and abstract art is surprising, both in its revolutionary conception as in the use of an amazing conception of aesthetic, very different from the past. In this context, new branches of mathematics, like fractal geometry or chaos theory, cause a revolution in the scientific world generating an exciting family of images.

With the purpose of exploring and experimenting the three dimensional space, we have designed a new family of objects, whose aspect reminds the one of abstract sculpture and called 'Cybersculptures', due to its virtual character and the fact that their first exposition was made in the Cyberspace of Internet.

They are virtual three dimensional objects, so they can only exist in a computer, although can be visualised and animated from every scale and position.

Cybersculptures are created mathematically by mean of mathematical formulas programmed in C language. Their basic generation elements are not Euclidean objects manipulated under a CAD program, but formulas and mathematical processes belonging to contemporary Mathematics.

The mathematical character of these objects could seem that their creation is something automatic, cold and distant from the sensibility and intimacy that is supposed to creativeness, but this is not true. The use of mathematical formulas to model objects simply constitutes a tool to work, like the hammer and chisel of the sculptor or the brush of the painter.

It is true that a mathematical formula is a quite restricted tool according to its range of actuation, but modern Mathematics have many objects whose aesthetic is really amazing, and their artist possibilities almost infinite (multiple torus, hypercubes, Moebius bands, parametric knots, etc.)

The complex shape of these Cybersculptures, and the huge volume of calculations needed to represent them graphically convert the computer in an indispensable element in the creation process. But, far away from being a simply visualisation tool, the computer establishes a close relationship between the author and his work being converted in the umbilical cord joining both.

Factors like illumination, the point of view or the texture can change completely the final aspect of the Cybersculpture and allow the expression of sensations by means of the use of materials with unexpected shapes and colours, impossible lights and objects that float challenging the law of gravity. Is this lack of coherence with the real world what made easy the creation of virtual objects with no comparison with the shapes of our environment.

4. Genetic Algorithms Abstract Modelling

Genetic algorithms were originally developed and characterised by John Holland (1975) and are based on the principles of evolution by natural selection. With this programming technique, solutions to a particular problem are subjected to the action of genetic operators such as recombination and mutation.

924

Modelling with genetic algorithms is inspired by natural systems and how they often rely on the repetition of very simple steps such as crystal growth or the creation of stalagmites by water dripping in underground caverns. these natural systems have a huge potential for creating artistic forms. The first experimental system for art generation, 'FormSynth' was created by William Latham in 1989. It was a system for drawing on paper in which repeated applications of rules generate an evolutionary tree of unexpected forms, some of them with architectural quality.

The images presented here are based on 'Mutator', an algorithm created by William Latham and Stephen Todd. This algorithm displays geometric forms under the action of associated gene values.

The basis of the graphic process lies in a very simple gallery of euclidean objects (cubes, cylinders, pyramids, spheres, ...) and a collection of mutation processes based on natural systems (tree branch, spider web, DNA helix, ...). The combined action of one or more of these objects with a mutation process results in new forms with an imaginative and increasingly complex structure. The user must explore the resulting forms and select one of them, so only the most aesthetic form is allowed to survive and reproduce.

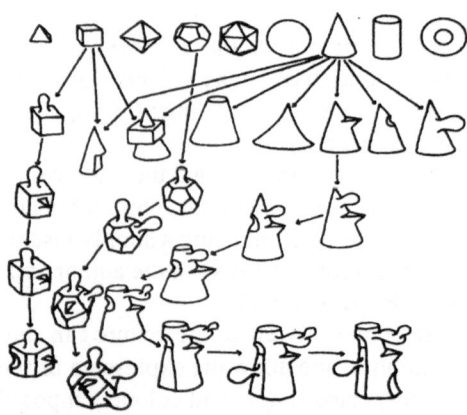

Figure 4. Small hand-drawn genetic tree.

These techniques provide a way to explore form space, making subjective decisions about the quality of forms. The computer does not allow changes based on structure definition, so the user does not need to have such analytic knowledge, only pure subjective exploration of form. In nature, new forms of life are created by mutation and marriage and the fittest survive in a process of natural selection.

Inside the computer space exists an artificial world by defining systems and structures for form generation and the user acts as a judge driving selection, using aesthetic judgements to breed artwork.

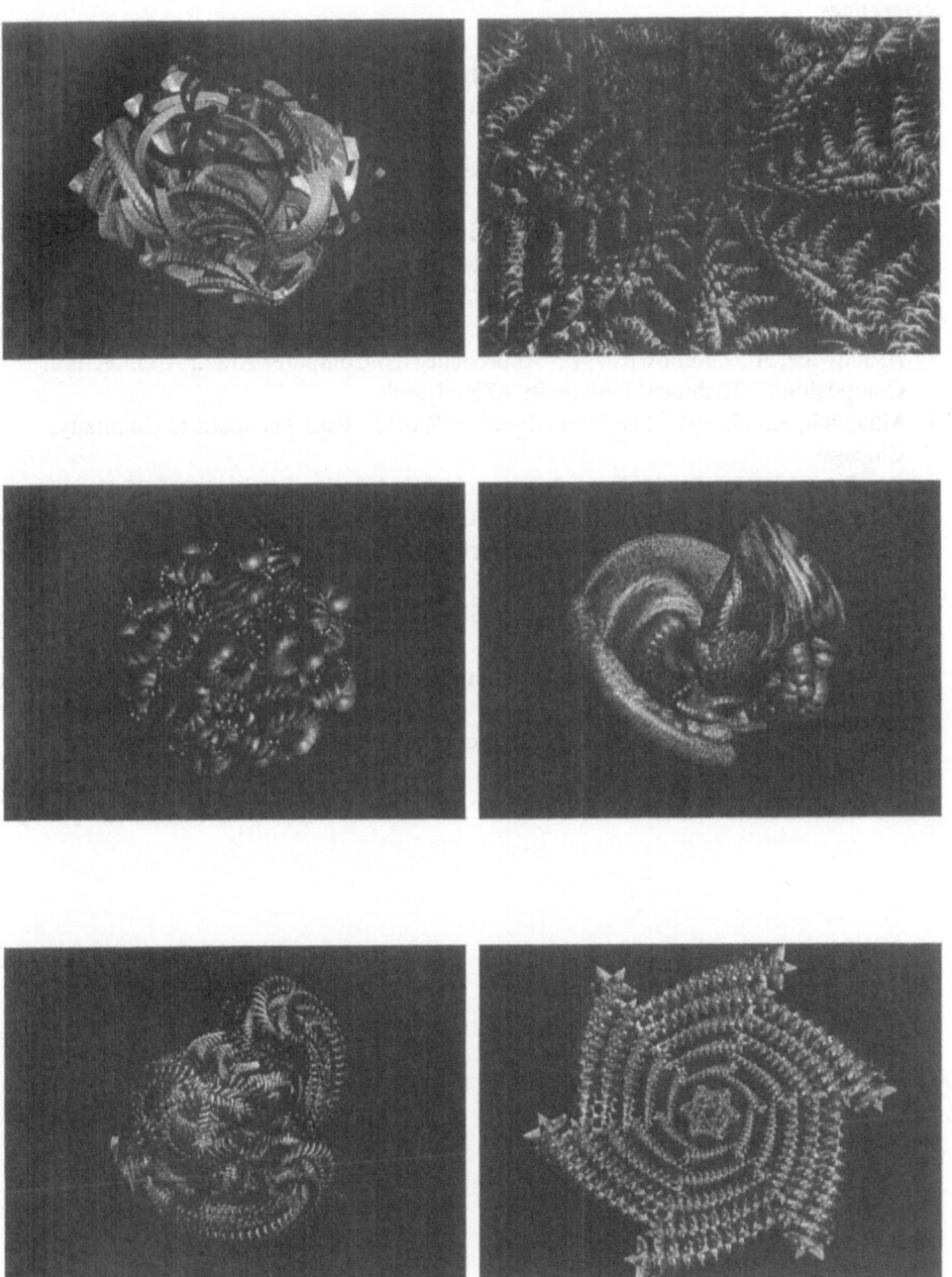

Figure 5. Genetic algorithms abstract modelling.

References

1. Jakimowicz, A., (1994) Abstract Modelling - Forming and Exploring, in Maver T., Petric, J., eds. The 'Virtual Studio', 12th ECAADE Conference proceedings, University of Strathclyde, Glasgow.
2. Carmagnola, F., (1992) Quality And The Aesthetic Nature Of Post - Industrial Technology, Domus 4/92.
3. Jakimowicz, A., (1995) Towards Affective Architectural Computing: An Additional Element in CAAD, in Asanowicz, A., Sawicki, B., eds., 'CAD Creativeness', Technical University of, Bialystok.
4. Jakimowicz, A., (1996), Creation Versus Simulation - Synthesis Versus Analysis, in Asanowicz, A., Jakimowicz, A., 'Approaches To Computer Aided Architectural Composition', Technical University of Bialystok.
5. Malevich, K., (1959), 'The Non Objective World', Paul Theobald & Company, Chicago.
6. Porebski, M., (1986), Sztuka A Informacja, Wydawnictwo Literackie, Krakow.
7. Barrallo, J. and Iglesias, A., (1995) Cybersculpture, in Asanowicz, A., Jakimowicz, A., 'Approaches To Computer Aided Architectural Composition', Technical University of Bialystok.
8. Ferguson, H., (1994), 'Mathematics in Stone and Bronze'. Meridian Creative Group, Pennsylvania.
9. Holland, J. H., (1975) 'Adaptation in natural and artificial systems'. The University of Michigan Press.
10. Todd, S. and Latham, W., (1992) 'Evolutionary Art and Computers'. Academic Press.

Author Index

Achten, Henry Eindhoven University of Technology, Faculty of Architechture, Building and Planning P.O.Box 513, 5600 MB Eindhoven, The Netherlands	Cao, Quinsan Dept. of Architecture, Virgina Tech., Blacksburg, VA 24061 USA
Barrallo, Javier The University of the Basque Country, Faculty of Architecture, Plaza de Oñati, 2. 20009 San Sebastiàn Spain email: mapbacaj@sa.ehu.es	Cicognani, Anna Key Centre of Design Computing, Dept. of Architectural & Design Science, University of Sydney NSW 2006 Australia email: anna,mary@arch.usyd.edu.au
Barth, Bertram Univ. Karlsruhe, Institut f. Industrielle Bauproduktion Englerstr. 7, 76131 Karlsruhe, Germany Tel.: ++49-721-608-2168, Fax.: ++49-721-661115 email: Bertram.Barth@ifib.uni-karlsruhe.de	Coyne, Richard Dept. of Architecture, University of Edinburgh 20 chambers St., EH1 1JZ Scotland email: Richard.Coyne@ed.ac.uk http://www.caad.ed.ac.uk/˜richard/ Tel.: ++44-131-6502332, Fax: ++44-131-6508019
Belblidia, S. CRAI (Research Center of Architecture & Engineering) School of Architecture, Nancy, France 2, rue Bastien Lepage, 54000 Nancy, France email: belblid@crai.archi.fr	Damski, J.C. Key Centre of Design Computing Dept. of Architectural & Design Science University of Sidney, NSW 2006 Australia email: jose@arch,usyd.edu.au
Bignon, J.C. CRAI (Research Center of Architecture & Engineering) School of Architecture, Nancy, France School of Architecture, Strasbourg, France 2, rue Bastien Lepage, 54000 Nancy, France email: bignon@crai.archi.fr	Dijkstra, J. Eindhoven University of Technology Faculty of Architechture, Building and Planning P.O.Box 513, 5600 MB Eindhoven, The Netherlands email: j.dijkstra@bwk.tue.nl
Bock, Thomas Universität Karlsruhe Seeweg 4, 74259 Widdern, Germany	Eastman, Chuck Georgia Institute of Technology, Architecture Bldg, Room 209, Atlanta GA. 30332-0155 email: chuck.eastman@arch.gatech.edu Tel.: ++1-404-894-911, Fax: ++1-40-8941629
Bourdakis, Vassilis Centre for Advanced Studies in Architecture (CASA) University of Bath, UK	Ehrhardt, Mark Sundance Laboratory for Computing in Design & Planning, College of Architecture &Palnning University of Colorado, Boulder, CO 80309-314, USA email: ehrhardt@ucsu.colorado.edu
Bridges, Alan Dept. of Architecture and Building Science 131 Rottenrow, Glasgow G4 0NG email: abacus@strath.ac.uk Tel.: ++44-141-552-4400 -3021, Fax: ++44-141-552 3997	Ekholm, Anders School of Architecture, Lund University, Sweden email: Anders.Ekholm@caad.lth.se
Bruton, Dean Dept. of Architecture, Landscape & Urban Design University of Adelaide, North Terrace SA 5000 Australia email: dbruton@drove.mtx.net.au Tel.: ++61-8-82983049, Fax: ++61-8-82962808	Elliott, Ame M. College of Environmental Design, University of Califonia, Berkeley CA, USA 94720 email: aelliott@ced.berkeley.edu
Caneparo, Luca Design Network Lab, Dept. Progettazione architettonica, Politecnico di Torino, v.el Mattioli 39, 10125 Torino, Italy email: media@centauro.polito.it	Elte, Nathanea Architecture and CAAD, Swiss Federal Institute of Technology- ETH Zurich Switzerland email: elte@arch.ethz.ch

Chang, Kuo-Chin Dept. of Architecture, Tunghai University, Taichung, Taiwan email: shuenn@s867.thu.edu.tw	Heitz, Dipl. Ing. Sandro Univ. Karlsruhe, Institut f. Industrielle Bauproduktion Englerstr. 7, D-76131 Karlsruhe, Germany Tel.: ++49-721-608-2168, Fax.: ++49-721-661115 email: Sandro.Heitz@ifib.uni-karlsruhe.de
Engeli, Maia Architecture and CAAD, Swiss Federal Institute of Technology- ETH Zurich Switzerland email: engeli@arch.ethz.ch	Hellgardt, Michael Prinsengracht 151, 1015 DR Amsterdam The Netherlands email: <michael@hellgar.iaf.nl>
Fridqvist, Sverker School of Architecture, Lund University, Sweden email: Sverker.Fridqvist@caad.lth.se	Hellgardt, Michael, Architect Prinsengracht 151, 1015 DR Amsterdam, The Netherlands email: michael@hellgar.iaf.nl
Gero, John S. Key Centre of Design Computing Dept. of Architectural & Design Science University of Sidney, NSW 2006 Australia email: john@arch,usyd.edu.au	Hermann, Dipl. Ing. Manfred Univ. Karlsruhe, Institut f. Industrielle Bauproduktion Englerstr. 7, D-76131 Karlsruhe, Germany Tel.: ++49-721-608-2168, Fax.: ++49-721-661115 email: Manfred.Hermann@ifib.uni-karlsruhe.de
Glaser, Daniel C. Dept. of Architecture, University of California, Berkeley 370 Wurster Hall, Berkeley, CA 94720, USA email: dcg@uclink4.berkeley.edu	Howe, A. Scott Kajima Corporation & The University of Michigan, School of Architecture & Urban Planning email: ash@ipc.kajima.co.jp
Gross, Mark D. Assistant Professor University of Colorado, Dept. of Planning and Design, College of Architecture & Planning Boulder, CO 80309-0314 email: mdg@cs.colorado.edu Tel.: ++1-303-492-6916, Fax: ++1-303-440-5277	Humbert, P. CRAI (Research Center of Architecture & Engineering) School of Architecture, Nancy, France 2, rue Bastien Lepage, 54000 Nancy, France email: humbert@crai.archi.fr
Guedes, Eliana Maria Univ. of Taubate, Dept. of Architecture & Computing Avenida Independencia 2405. Taubate, Sao Paulo 12032.000 Brazil email: emg@aquserv.aquarius.com.br	Jagielski, R. Key Centre of Design Computing Dept. of Architectural & Design Science University of Sidney, NSW 2006 Australia email: rom@arch,usyd.edu.au
Halin, G. CRAI (Research Center of Architecture & Engineering) School of Architecture, University of Metz, France 2, rue Bastien Lepage, 54000 Nancy, France email: halin@crai.archi.fr	Jakimowicz, Adam Technical University of Bialystok, Faculty of Architecture ul.Krakowska 9. 15-875 Bialystok, Poland email: jakima@cksr.ac.bialystok.pl
Hall, Theodore W. Dep. of Architecture, Chinese Univ. of Hong Kong Sha Tin, New Territories Hong Kong email: twhall@cuhk.edu.hk	Junge, Richard TU München, Fakultät f. Architektur Lehrstuhl f. Hochbaustatik & Tragwerksplanung Lehrgebiet für CAAD, Arcisstr. 21, 80290 München, Germany email: richard.junge@lrz.tu-muenchen.de
Haruyuki, Fujii Izumi Research Institute, Shimizu Corporation 2-2-2 Uschisaiwaicho, Chiyoda-ku, Tokyo 100 Japan email: haru@ori.shimz.co.jp Tel.: ++81-3-3508-8101, Fax: ++81-3-3508-2196	Kalay, Yehuda E. Dept. of Architecture, University of Califonia at Berkeley CA 94720, USA email: kalay@ced.berkeley.edu

Khemlani, Lachmi
Dept. of Architecture,
University of Califonia at Berkeley,
CA 94720,
USA
email: lachmi@ced.berkeley.edu

Kim, Sungah
Harvard University,
P.O.Box 1271, Atkinson, NH 03811 USA
email: skim@gsd.harvard.edu

Kohler, Niklaus
Univ. Karlsruhe, Institut f. Industrielle Bauproduktion
Englerstr. 7, D-76131 Karlsruhe, Germany
Tel.: ++49-721-608-2166, Fax.: ++49-721-661115
email: Niklaus.Kohler@ifib.uni-karlsruhe.de

Koile, Kimberle
MIT Artificial Intelligence Laboratory
545 Technology Square, NE43-826, Cambridge,
MA 02139 USA
email: kkoile@ai.mit.edu

Kolarevic, Branko
University of Hong Kong, Dept. of Architecture,
Knowles Building, Room 234, Pokfulam Road,
Hong Kong
email: branko@arch.hku.hk

Komatsu, Kiichiro
Institute of Policy & Planning Science
University of Tsukuba
1-1-1 Tennodai, Tsukuba-shi, Ibaraki,
Japan

Koutamanis, A.
Delft University of Technology, Faculty of Architechture
Berlageweg 1, 2628 CR Delft, The Netherlands
email: a.koutamanis@bk.tudelft.nl
http://caad.bk.tudelft.nl/koutamanis/index.htm
http://prive.bk.tudelft.nl/˜koutaman/index.html
http://www.bk.tudelft.nl/koutamanis/

Kundu, Sourav
Bionic Design Laboratory,
Dep. of Human and Mechanical Systems Engineering,
Faculty of Engineering, Kanazawa University
2-40-20 Kodatsuno, Kanazawa, Japan
email: sourav@kenroku.ipc.kanazawa-u.ac.jp

Kurmann, David
Architecture and CAAD,
Swiss Federal Institute of Technology- ETH Zurich
Switzerland
email: kurmann@arch.ethz.ch

Lam, Dr. Khee Poh
School of Architecture
Faculty of Architecture and Building
National University of Singapore
10 Kent Ridge Crescent
Singapore 119260, Republic of Singapore
email: akilamkp@nus.sg

Leonard, D.
CRAI (Research Center of Architecture & Engineering)
School of Architecture, Nancy, France
email: leonard@crai.archi.fr

Lewin, Jeniffer S.
Sundance Laboratory for Computing in Design &
Planning, College of Architecture &Palnning
University of Colorado, Boulder, CO 80309-314, USA
email: lewin@ucsub.Colorado.EDU

Li, Thomas Siu-Pan
Dept. of architecture, The University of Hong Kong
Pokfulam Road, Hong Kong
email: spli@hku.hk
Tel.: ++-852-2859-7961, Fax: ++-852-2559-6484

Liou, Shuenn-Ren
Dept. of Architecture,
Tunghai University,
Taichung, Taiwan
email: shuenn@s867.thu.edu.tw

MacCall, Raymond J.
Sundance Laboratory for Computing in Design &
Planning, College of Architechture & Planning
University of Colorado
Boulder Colorado, USA 80309-0314

Mahdavi, Ardeshir
Dept. of Architecture, Carnegie Mellon University
Pittsburgh, PA 15213-3890, USA
email: am4f@andrew.cmu.edu
Tel.: ++1-412-268-6389, Fax: ++1-412-268-6129

Maher, Mary Lou
Key Centre of Design Computing,
Dept. of Architectural & Design Science,
University of Sydney
NSW 2006 Australia
email: anna,mary@arch.usyd.edu.au

Mao-Lin Chiu
Dept. of Architecture
National Cheng Kung University, No. 1,
University Road, Tainan, Taiwan, R.O.C.
email: mc2p@mail.ncku.edu.tw

Müller, Christian Universität Karlsruhe (TH), ifib email: cmueller@ifib.uni-karlsruhe.de	Riegel, Jan Peter Universität Kaiserslautern, FB Informatik Postfach 3049, 67653 Kaiserslautern, Germany email: riegel@informatik.uni-kl.de Tel.: ++49-631-205-31 43
Naai-jung, Shih Dept. of Architecture, National Taiwan Institute of Technology 43, Section 4, Keelung Rd., Taipei, Taiwan, R.O.C. email: ashihnj@mail.ntit.edu.tw	Rutherford, Peter Univ. of Strathclyde, Dept. of Architecture & Buildg. Science 131 Rottenrow, Glasgow G4 ONG email: p.rutherford@strath.ac.uk Tel.: ++44-141-552-4400 -3017, Fax: ++44-141-552 3997
Nath. G. Key Centre of Design Computing Dept. of Architectural & Design Science University of Sidney NSW 2006 Australia email: nath_g@arch,usyd.edu.au	Sahnouni, Y. CRAI (Research Center of Architecture & Engineering) School of Architecture, Nancy, France 2, rue Bastien Lepage, 54000 Nancy, France email: sanouni@crai.archi.fr
Park, S.-H. Key Centre of Design Computing Dept. of Architectural & Design Science University of Sidney, NSW 2006 Australia email: soohoon@arch,usyd.edu.au	Salter, P. Faculty Design, Engineering & The Built Environment University of East London, Holbrook Centre, Holbrook Road, London E15 3EA, England email: CZES@hbmain.uel.ac.uk Tel.: ++44-181-590 7000/7722 (ext. 3222), Fax: ++44-181-849 3686
Parker, Laura Sundance Laboratory for Computing in Design & Planning, College of Architecture &Palnning University of Colorado, Boulder, CO 80309-314, USA email: parker@ucsu.colorado.edu	Schmitt, Gerhard Architekture and CAAD, ETH Zurch 8093 Zurich, Switzerland email: schmitt@arch.ethz.ch http://caad.arch.ethz.ch/~schmitt Tel.: ++41-1-633 2766, Fax: ++41-1-633 1050
Perrin, J.P. CRAI (Research Center of Architecture & Engineering) School of Architecture, Nancy, France 2, rue Bastien Lepage, 54000 Nancy, France email: perrin@crai.archi.fr	Schütze, Martin Universität Kaiserslautern, FB Informatik Postfach 3049, 67653 Kaiserslautern, Germany email: schuetze@informatik.uni-kl.de Tel.: ++49-631-205-31 43
Petzold, Frank Bauhaus Universität Weimar, Lehrgebiet Informatik Fakultät Architektur, Stadt- & Regionalplanung, 99421 Weimar, Germany email: caad@architektur.uni-weimar.de www.uni-weimar.de/iar Tel.: ++49-3643-584201, Fax: ++49-3643-584202	Shaviv, Edna Faculty of Architecture & Town Planning Technion - Israel Institute of Technology, Haifa, Israel email: arredna@techunix.technion.ac.il
Protzen, Jean-Pierre Dept. of Architecture, University of California at Berkeley, Berkeley, CA 94704, USA	Shen-Guan Shin Dept. of Architecture National Taiwan Institute of Technology, No. 43, Section 4, Keelung Road, Taipei, Taiwan, R.O.C. email: mc2p@mail.ncku.edu.tw
Radford, Antony D. University of Adelaide, School of Architechture, Landscape Architechture and Urban Design Australia 5005 email: archdept@arch.adelaide.edu.au Tel.: ++16-8-8303-5836, Fax: ++8-8303-4377	Silva, Neander F. Dept. de Projeto, Faculdade de Arquitetura e Urbanismo Universidade de Brasília, 70878-130, Brasília, DF, Brazil email: neander@guarany.cpd.unb.br Tel./Fax: ++55-61-340-1939

Siu-Pan LI, Thomas
Dep. of Architecture, Univ. of Hong Kong
Pokfulam Road, Hong Kong
email: spli@hku.hk
Tel.: (852) 2859 7961, Fax: (852)2559 6484

Wagter, Harry
Eindhoven University of Technology,
Faculty of Architechture, Building and Planning
Building Information Technology, Origin International
Technology, The Netherlands
email: harry.wagter@nlehvips.origin.nl

Sourav, Kundu
Bionic Design Laboratory, Faculty of Engineering,
Dept. of Human & Mechanical Systems Engineering,,
Kanazawa University, 2-40-20 Kodatsuno,
Kanazawa, Japan
email: <sourav@kenroku.ipc.kanazawa-u.ac.jp>

Watanabe, Shun
Institute of Policy & Planning Science
University of Tsukuba
1-1-1 Tennodai, Tsukuba-shi,
Ibaraki, Japan
email: shun@sk.tsukuba.ac.jp

Stouffs, Rudi
Architecture and CAAD,
Swiss Federal Institute of Technology Zurich
8093 Zurich-Hönggerberg, Switzerland
email: stouffs@arch.ethz.ch

Wenz, Florian
Architectural Space Lab., Chair for CAAD & Arch.
Dept. for Architecture, ETH Zurich
8093 Zuerich-Hoenggerberg, Switzerland
email: wenz@ethz.ch
http://caad.arch.ethz.ch/trace
Tel.: ++41-1-633 1050, Fax: ++41-1-633 2912

Timmermans, H.J.P.
Eindhoven University of Technology
Faculty of Architechture, Building and Planning
P.O.Box 513, 5600 MB Eindhoven, The Netherlands
email: h.j.p.timmermans@bwk.tue.nl

Woo, Sungho
Dept. of Environmental Engineering Osaka University
2-1 Yamada-Oka, Suita, Osaka, Japan 565
email: woo@env.eng.osaka-u.ac.jp
Tel.: ++81-6-877-5111 (ext. 3543)

Tomlinson, P.
Key Centre of Design Computing
Dept. of Architectural & Design Science
University of Sidney, NSW 2006 Australia
email: philt@arch,usyd.edu.au

Yan, Chie-Shan
Dept. of Architecture
National Taiwan Institute of Technology, No. 43,
Section 4, Keelung Road, Taipei, Taiwan, R.O.C.
email: shihnj@mail.ntit.edu.tw

Tsai, Daniel E.
Research Associate, Harvard University,
Visiting Lecturer, Massachusetts Institute of Technology
P.O.Box 1271, Atkinson, NH 03811 USA
email: dtsai@tiac.net

Yezioro, Abraham
Faculty of Architecture & Town Planning
Technion - Israel Institute of Technology,
Haifa,
Israel

Turner, James A.
College of Architecture & Urban Planning
The University of Michigan
email: turner@umich.edu

Yi-Luen Do, Ellen
College of Architecture, Georgia Institute of
Technology, Atlanta, GA 30332-0155, USA &
Sundance Lab. for Computing in Design & Planning,
University of Colorado, Boulder, CO 80309-0314, USA
email: ellendo@cc.gatech.edu
http://wallstreet.colorado.edu/Napkin
Tel.: ++1-303-492-2807, Fax: ++1-303-492-6163

Vakalo, Emmanuel-George
College of Architecture & Urban Planning
The University of Michigan, Ann Arbor,
MI 48109-2069 USA
email: egvakalo@umich.edu

Zimmermann, Gerhard
Universität Kaiserslautern, FB Informatik
Postfach 3049, 67653 Kaiserslautern,
Germany

Van Leeuwen, Jos P.
Eindhoven University of Technology,
Faculty of Architechture, Building and Planning
Building Information Technology, The Netherlands
email: j.p.v.leeuwen@bwk.tue.nl
http://www.calibre.bwk.tue.nl